GENERAL ZOOLOGY

Investigating the Animal World

DENNIS HOLLEY

Illustrated by George Barile Accurate Art Inc.
Production Manager Louis C. Bruno, Jr.
Interior Design by John Arbutante

First published by Dog Ear Publishing
4011 Vincennes Road
Indianapolis, IN 46268
www.dogearpublishing.net

ISBN: 978-145754-212-1

This book is printed on acid free paper.
Printed in the United States of America

CONTENTS

UNIT TWO ORIGINS AND PATTERNS OF ANIMAL LIFE

Chapter 4 Animal Architecture: Form and Function 81

UNIT THREE LOWER REALMS OF THE ANIMAL WORLD

Chapter 7 Phylum Porifera: The Simple Ones 189

UNIT FOUR MIDDLE REALMS OF THE ANIMAL WORLD

Chapter 11 Phylum Annelida: Masters of Coeloms and Segments 303

Chapter 13 Phylum Arthropoda: Sovereigns of the Terran Empire

Chapter 18 Reptiles: The Shattered Remains 551

Chapter 21 The Human Condition: Rise of the Cultural Ape...........683

PREFACE

Greetings zoology student and welcome to the always astonishing, sometimes strange, and occasionally even bizarre world of animals.

> *We patronize them for their incompleteness, for their tragic fate of having taken form so far below ourselves, And therein do we err. For the animal shall not be measured by man. In a world older and more complete than ours, they moved finished and complete, gifted with the extension of senses we have lost or never attained, living by voices we shall never hear. They are not our brethren. They are not our underlings. They are other nations, caught with ourselves in the net of life and time, fellow prisoners of the splendor and travail of the earth.*
> —Henry Beston

Zoology or any scientific endeavor should be thought of as consisting of two phases: the first being the *Investigation* and *Exploration* phase while the final is the *Accumulation* phase. Zoologists attempt to answer questions about the animal world by actively investigating animals through experimentation and by discovering new animal species through exploration. Investigation and exploration in turn lead to the accumulation of facts and information. These accumulated facts and information lead to even more questions that in turn lead to more investigation resulting in even more facts and information being accumulated. And the cycle continues.

In this course, you will confront the facts and concepts of zoology in your textbook (*Accumulation*). However, you will also be challenged to think, act, and work like a zoologist (*Investigation*) at certain points in your textbook, and especially in the laboratory segment of this course. As you investigate, you will use the same information, develop the same scientific skills, and employ the same scientific processes as do professional zoologists.

Science Process Skills

Organizing Information

- Classify
- Sequence

- Describe
- Summarize
- Explain
- Definition and proper use of terminology
- Accessing and using reference materials
- Reading comprehension

Critical Thinking

- Critical and creative thinking
- Observe
- Infer
- Compare and contrast
- Recognize cause and effect
- Formulate and use models

Experimentation

- Experimental design
- Formulate hypothesis/prediction
- Establish variables and controls
- Collect and organize data
- Accurate measurement
- Analyze data
- Draw reasonable conclusions

Graphics and Numbers

- Make and interpret graphs
- Construct and interpret tables
- Interpret scientific illustrations
- Calculate and compute

Communication

- Brainstorming
- Collaboration
- Communicating

Developing and using these skills effectively is very important if you are biology major, but even if you are not majoring in a scientific field, mastering these skills will help you function as a clear-thinking and scientifically literate citizen of a society that grows ever more science-based and technologically oriented.

Approach and Organization

Approach

Biology textbooks and related curricular materials at all levels have come under harsh but justified criticism by various scientific and educational groups in the past decade. From the inception of this text, it has been the goal to write a zoology program that acts on the criticisms and recommendations of those authorities and is based on current educational research. This textbook has been designed and written to be:

> **Readable and Interesting**. My goal has been to write a textbook in which the chapters read more like an interesting magazine or newspaper article and less like a dry and detailed technical entry from an encyclopedia. Increasing reader interest increases readability and to aid in that goal, I include out-of-the-ordinary things in each chapter that would not normally be found in zoology texts. I have also taken a different approach than other zoology books in that while I firmly believe that evolution is driving force and cornerstone of all things biological, I did not make the theoretical and often speculative aspects of origins and patterns of evolution the focal point of each chapter. Instead, I opted for a more concrete "here-and-now" approach in which our focus is mainly on animal systematics, phyla and class characteristics, and ecology. Hopefully, less emphasis on the theoretical translates into a work that is more relevant to you the student.

> **Understandable.** As I wrote this textbook I tried to avoid the "Huh? Factor" as much as possible. That is; students should not be obliged to reread a passage several times all the while armed with a biological dictionary to understand what they just read. The chapters of this textbook are centered on concepts and ideas. Specific facts, terms and terminology, and scientific names are used only when necessary and appropriate to illustrate and explain the concepts and ideas inherent in a particular chapter. This textbook is concept (idea) driven, not terminology (definitions) driven.

> **Connected.** Animals are all around us, on us and possibly in us, and they affect our daily lives directly and indirectly in ways we are continuing to uncover. In an attempt to connect you the reader directly to each animal group, each chapter concludes with a discussion on how the animals encountered in that chapter connect to humans economically, environmentally, medically, and even culturally.

> **Personable** Many textbooks are written by teams of writers, some of which are anonymous. As a result, the reader (student) lacks a personal connection with the author(s). Again, this text is different. First, this text was written in entirety only by the name you see stamped on the front of this book—Dennis Holley. Secondly, I have attempted to write each chapter in the tone of enthusiastic and passionate, but caring and concerned teacher speaking directly to you the student. Hopefully, I have succeeded. Lastly, personal notes will appear at the end of each chapter. In

these short conversations and dialogues, I may share an anecdote, look behind the headlines, or pose intriguing questions. You will encounter the first of these conversations and dialogues in this preface. Hopefully, you will find them interesting, thought-provoking and even amusing.

Organization

A quick glance at the table of contents reveals that what zoology is and how it works is detailed in Chapter 1. With this foundation in place, Unit One examines the place of animals in the bigger picture of the web of life. Unit Two delves into animal structure and behavior then investigates the possible origins of animals (phylogeny) as well as the system of organization (taxonomy) developed to bring scientific order to biologic chaos. In the remaining three units, we will voyage through the animal realm from the simplest life form—sponges to the most complex—humans. The progression of units is based on increasing morphological and anatomical complexity.

At the end of each chapter you will find both a set of *Review and Reflect* questions that will test your critical thinking skills while reviewing the main concepts of that particular chapter and a set of *Create and Connect* challenges that will help you develop and use important science process skills. Some or all of these questions and challenges may be assigned by the instructor as part of the assessment package for this course. In these assignments, you will be asked to write everything from formal scientific reports to essays to position papers to short stories. The exact format and details will be given with each assignment. Consult Appendix A—Scientific Writing for guidelines and suggestions for correct scientific writing.

I believe this textbook represents a major paradigm shift in the way college biology textbooks are written and presented because it was written by a teacher (not a research scientist) for students. I have labored to make this textbook accurate, understandable, and interesting so that you can and will read it. And if you do indeed bother to read it, I guarantee that you will gather not only a wealth of information, but also a never-ending respect for those amazing creatures we call animals.

A Personal Note from the Author

I am a biologist to the core, always have been, and always will be. My interest in all things living is broad and generic. If it's a living creature—plant, animal, or microbe—I find it fascinating. How did I get this way? Understanding parents and a nurturing habitat are to blame. My mother was constantly contending with tadpoles in jars, aquariums of fish, mice in cages, and occasionally rewashing the clothes she had just hung out to dry because my flock of pigeons flew too low overhead. She pretended to make a fuss but encouraged my every adventure. My father helped me build cages and traps and was quite adept at capturing and helping me rear the many kinds of small animals that constantly caught my attention and interest.

I was blessed with growing up in a very small rural village where my family's acreage was only several blocks from a meandering stream aptly known by the locals as "Muddy Creek." This brook was shaded by many huge overhanging trees and was full of snails, fish, frogs, turtles, and even beavers and muskrats. Many inquisitive hours were spent around and in that stream.

Two events sealed my fate and set me on my course. In my early high school years, my parents finally gave in to my pestering and bought me a small, simple microscope (which they couldn't afford even though it cost only around $30). This amazing black beauty came complete with a wooden box of slides and a few dissecting instruments. Once I dove into the microscopic world, I was hooked on all things biological. Later, I stumbled on Paul de Kruif's 1926 book, *Microbe Hunters* and was inspired to get the education that would allow me to become a professional biologist. At that point, I didn't know exactly what I wanted to do professionally, but I did know my future would have something to do with biology.

I eagerly devoured every biology course I could take in college, and while I flirted for a time with the idea of becoming a marine biologist, I eventually became an educator. For nearly forty years, high schools and universities have actually paid me for merely doing what I love—teaching biology and teaching others how to teach biology and science. I am a very inquiry-oriented, hands-on type of teacher whose philosophy as an educator is best and most simply articulated in the words of Louis Agassiz:

Study nature, not books.

My love of all things biological continues unabated to this day. As such, I would consider the day poorly spent were I not to stumble upon at least several biological "WOW! Moments" (*WoMos*) during the course of that day. Such moments are not hard to find for they are everywhere. You just have to be receptive to them. Stop, look, and appreciate the natural world around you.

It was my intent and it is my hope that through this zoology program, you will come to know and respect those amazing creatures we call animals and that you too will have many personal zoological *WoMos* as this course unfolds.

I would like to dedicate this book to my parents for their nurturing and understanding, my wife and family for their patience and support, and to my students—past and present—who have taught me more than they will ever know.

Dennis Holley

I want to address any complaints or problems you have with this textbook and seek input from you regarding any suggestions you might have for future editions. Also, please feel free to contact me if I might be of some assistance in your zoological endeavors either in the classroom or laboratory. Contact me for assistance or suggestions and join my blog at www.zoologytextbook.com.

Dennis Holley

ZOOLOGY: INVESTIGATING THE ANIMAL WORLD

Back to the sea again! Down there today I watched the goings-on of the sea snails, patellas (mussels with a single shell), and the pocket crabs, and it gave me a glow of pleasure observing them. No, really, how delightful and magnificent a living thing is! How exactly matched to its condition, how true, how intensely being! And how much I'm helped by the small amount of study I've done and how I look forward to taking it further.

—Johann Wolfgang von Goethe

Introduction

This planet pulsates with life. Interlaced within the larger physical world of the **atmosphere** (air), the **hydrosphere** (water), and the **lithosphere** (soil) are countless smaller biological worlds teeming with creatures of such variety of size, shape, and behavior as to boggle the mind. Of these many and varied biological realms, the one with which most people have the greatest familiarity, interest, and contact is the animal world.

Rise of Zoology

Zoology (Gr. *zoon*, animal + *logos*, to study) is the study of animals. In broadest biological and historical terms, zoology and its companion discipline botany were among the first investigations of the natural world made by humankind in the course of our evolution. Using trial-and-error methods to fulfill the basic need for food, humans quickly learned which animals and plants they could eat and which would eat or poison them. The result was the slow rise of the biological sciences we know as zoology and botany.

Humans have always attempted to understand animals, to enslave animals, and to capture the strength and power of animals. Mute testimony to this ancient and on-going fascination with animals lies deep within caves in France and Spain. Cast in yellow ochre, red hematite, and black charcoal, primal Figures of antelope, bison, horse, deer, lion, and bear seem to spring to life from the shadows. Cro-Magnon shamans may have entered these deep caves tens of thousands of years ago and in a trance-like state painted these animals in hopes of drawing the strength, speed, and courage of the animals depicted through the very walls of the cave itself (**Figure 1.1**).

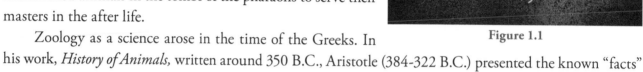

The allure of animals was so great to the ancient Egyptians of several thousand years B.C. that they elevated certain animals to the status of gods in their religious hierarchy and placed mummified animals in the tombs of the pharaohs to serve their masters in the after life.

Figure 1.1

Zoology as a science arose in the time of the Greeks. In his work, *History of Animals*, written around 350 B.C., Aristotle (384-322 B.C.) presented the known "facts"

for about 500 animals (**Figure 1.2**). Though riddled with errors, it was the grandest biological anthology of its time and served as the ultimate authority on all matters zoological for many centuries after his death. Aristotle's observations on the anatomy of marine invertebrates such as the octopus, cuttlefish and others were remarkably accurate and led to the conclusion that he must have had first-hand experience with dissection. He also delved into the social organization of bees and described the embryological development of a chick, and the chambered stomachs of ruminants. He distinguished whales and dolphins from fish and noted that some sharks give birth to live young as well. Aristotle also devised the first known classification system for animals in which he divided the animals into two groups: those with blood

Figure 1.2

and those without blood (or at least without red blood). This grouping corresponds closely to the distinction we make today between invertebrates and vertebrates.

Aristotle's Roman counterpart was Pliny the Elder (23-79 A.D.). In 77 A.D., this cavalry officer turned historian and scientist published the first ten books of his *Naturalis Historia*, a work that would eventually grow to 37 books after his death. In this treatise Pliny the Elder deliberates a wide range of topics: the mathematical and physical nature of the world, geography, anthropology and human physiology, zoology, botany and horticulture, pharmacology, and mineralogy. As with Aristotle's work, Pliny's dissertations served as the authority for many generations of scholars that followed.

With the fall of the Roman Empire, Christianity dominated the development of western civilization. It was a time of focus on the Bible and things religious rather than on science and things secular. Nevertheless, during this time a book called the *Physiologus* became widespread and second only to the Bible in popularity. Written in Greek by an unknown source around 200-300 A.D. in Egypt, this book depicted 49 separate animals, many of which were mythological. Because some of the source material was from the Bible and because the characteristics of animals in the book had a metaphorical significance in the Bible, it was difficult for anyone to doubt it. For example, unicorns are mentioned in the Bible and the *Physiologus* so their existence could not be questioned.

In the period of 1300-1500 A.D., the veil of darkness began to lift from Europe. The enlightenment of the Renaissance heralded a renewed interest in things natural and scientific as well as a curiosity of distant lands. A new age of science had begun.

The religious and social climate of the time promoted the spread of medieval bestiaries in which fictional creatures were storied and illustrated. Fueling the popularity and spread of these bestiaries were travel stories from the first European explorers such as Marco Polo, Sir John Mandeville, and others. Common people became increasingly enticed and enthralled with fictional beasts as spoken tales and stories of the strange lands of Africa, India, and Asia began to filter into Europe (**Figure 1.3**).

Figure 1.3

As the age of exploration continued apace in the 1500's and 1600's, individuals with at least some rudimentary scientific training often began to accompany voyages of exploration and conquest for the sole purpose of observing and recording the creatures of distant lands and returning specimens of those creatures, when possible, to their homeland for further study. The age of the naturalist collector had begun. Verification by collecting specimens instead of the accumulation of anecdotes, myths, and legends became more common, and scholars developed the faculty of careful observation. These early collectors and accumulators were the founders of zoology, and to this day, their modern naturalist, museum curator, and systematist counterparts play an important role in the continuing progress of zoology. Developing separately at that time was comparative anatomy in which anatomists and physiologists began to study the internal structure and functions of humans (the dawn of modern medicine) and other animals. Although arising somewhat separately in the beginning, it is now generally recognized that zoology and comparative anatomy are essentially synonymous with modern zoologists studying the anatomy (inside), morphology (outside), and ecology (habitat) of animals.

During the period of 1600-1800, the final piece of the foundation of modern zoology was placed when accurate observations and collected specimens began to be coupled with rigorous experimental methods into the structure and behavior of animals. This era also saw the evangelists of the new philosophy of experimental verification band together with the observers and collectors of nature into academic societies for mutual support and interaction. The first of these, the Academia Naturae Curisorium (1651), created in Schweinfurt, a Free Imperial City of the Holy Roman Empire, by four physicians; eleven years later (1662) the Royal Society of London was incorporated by royal charter, and a little later the Academy of Sciences was established by Louis XIV in Paris.

With the establishment of such societies and their enforcement of rigorous methods of inquiry, witchcraft, myth-as-fact, superstitions, mysticism, and other relics of the mediaeval thought began to disappear. (However, there is ample evidence to show such things have not been completely exorcised from the human psyche.)

Present day zoologists can be found both in the field observing and experimenting on animals and searching for new species of animals and in the laboratory examining the inner workings of animals and conducting controlled research activities. They may also be found in museums and natural history collections where they identify, catalog, and maintain important specimens. The main activity of zoologists working in museums, however, is studying the evolutionary relationships among groups of animals and in this regard these scientists are at the very heart of evolutionary biology. Whether they work with specimens of the past or animals of the present, all zoologists recognize the need for conservation and work tirelessly to maintain robust animal species and to protect those that have become endangered.

Doctrines of Modern Zoology

The biological principles of organic evolution, genetics, and ecology serve as the foundation upon which modern zoology rests and those same doctrines also guide and focus the ongoing investigations of zoologists. Combining the knowledge gathered by research in each of these areas gives zoologists a better understanding of animal species and populations and the animal kingdom as a whole.

Zoology and Organic Evolution

For millennia most of humankind faithfully clung to the religious dogma that living things, especially humans, were something apart from the rest of nature, specially created and specially cared for by a Divine Being. Surveys show that even to this present day a great many people continue to adhere to this belief as a plausible and acceptable explanation of the origin and existence of humans, animals, plants, and all other life forms.

From the time of the ancient Greeks on, however, a few dissenting voices to the Divine Creation dogma preached a different explanation, albeit mostly to deaf ears. These dissenters boldly suggested that living things developed or evolved (L. *evoltus*, unroll) by a slow process of trans-mutation in successive generations from simpler ancestors, and in the beginning from simple formless matter. The creationists attempted to beat this heresy down with two sticks of criticism. First, the early evolutionists could not demonstrate any mechanism by which evolution might necessarily occur, second, they could not offer any facts or observations that evolution had indeed ever occurred.

Figure 1.4 Louis Pasteur (1822-1895)

Nevertheless, dissension continued. Both Francesco Redi and his maggots in the late 1660s and Pasteur (**Figure 1.4**) with his microbes and swan-necked flasks in the mid-1860s demonstrated experimentally and conclusively that the widely-held tenets of spontaneous generation were wrong—neither complex nor simple life forms magically pop into existence.

Box 1.1
The Spontaneity of Life—A Classic Experiment

One of the great precepts or scientific laws of modern biology is that "life begets life." We accept without question that every living thing, regardless of size and complexity, attains life from the asexual or sexual reproduction of others of its kind and that life flows across the generations in an uninterrupted stream. However, for millennia, humankind believed just the opposite. Life was thought to spring from slime, putrefaction, or a host of other unlikely inorganic origins. This corruption of the truth was based solely on flawed observations and conjectures and began to unravel with the application of experimentation to the question. In 1688, the Italian, Francesco Redi, fired one of the first salvos in the war on the theory of spontaneous generation. Despite the clarity of Redi's results, the spontaneous generation theory survived. Its adherents were obliged to shift the battleground to the realm of microbiology where the decisive final battles were fought by Pasteur and Tyndall in the nineteenth century. Read, analyze, and above all, appreciate this classic experiment.

Experiments on the Generation of Insects by Francesco Redi

From *Experiments on the Generation of Insects* by F. Redi, translated from the 1688 edition by Mab Bigelow

"*Although content to be corrected by anyone wiser than myself, if I should make any erroneous statements, I shall express my belief that the Earth, after having brought forth the first plants and animals at the beginning by order of the Supreme and Omnipotent Creator, has never since produced any kinds of plants or animals, either perfect or imperfect; and everything which we know in past or present times that she has produced, came solely from the true seeds of the plants and animals themselves, which thus, through means of their own, preserve their species. And, although it be a matter of daily observation that infinite numbers of worms are produced in dead bodies and decayed plants, I feel, I say, inclined to believe that those worms are all generated by insemination and that the putrefied matter in which they are found has no other office than that of serving as a place, or suitable nest where animals deposit their eggs at the breeding season, and in which they also find nourishment; or otherwise, I assert that nothing is ever generated therein…*

At the beginning of June I ordered to be killed three snakes, the kind called eels of Aesculapius. As soon as they were dead, I placed them in an open box to decay. Not long afterward I saw that they were covered with worms of a conical shape and apparently without legs. These worms were intent on devouring the meat, increasing meanwhile in size, and from day to day I observed that they likewise increased in number; but, although of the same shape, they differed in size, having been born on different days. But all, little and big, after having consumed the meat, leaving only the bones intact, escaped from a small aperture in the closed box, and I was unable to discover their hiding place. Being curious, therefore, to know their fate, I again prepared three of the same snakes, which in three days were covered with small worms. These increased daily in number and size, remaining alike in form, though not in color. Of these, the largest were white outside, and the small ones, pink. When the meat was all consumed, the worms eagerly sought an exit, but I had closed every aperture. On the nineteenth day of the same month, some of the worms ceased all movements, as if they were asleep, and appeared to shrink and gradually assume a shape like an egg. On the twentieth day all the worms had assumed the egg shape, and had taken on a golden white color, turning to red, which in some darkened, becoming almost black. At this point the red, as well as the black ones, changed from soft to hard, resembling somewhat those chrysalides formed by caterpillars, silkworms, and other insects. My curiosity being thus aroused, I noticed that there was some difference in shape between the red and black eggs [pupae],[1] *though it was clear that all were formed alike of many rings joined together; nevertheless, these rings were more sharply outlined, and more apparent in the black than in the red, which last were almost smooth and without slight depression at one end, like that in a lemon picked from its stalk, which further distinguished the black egg-like balls.*

I placed these balls separately in glass vessels, well covered with paper, and at the end of eight days, every shell of the red balls was broken and from came forth a fly of gray color, torpid and dull, misshapen as if half finished, with closed wings; but after a few minutes they commenced to unfold and expand in exact proportion to the tiny body, which also in the meantime had acquired symmetry in its parts. Then the whole creature, as if made anew, having lost its gray color, took on a most brilliant and vivid green; and the whole body had expanded and grown so that it seemed incredible that it could ever have been contained in the small shell. Though the red eggs [pupae] brought forth green flies at the end of eight days, the black ones labored fourteen days to produce certain large black flies striped with white, having a hairy abdomen, of the kind that we see daily buzzing about butchers' stalls.…

I continued similar experiments with the raw and cooked flesh of the ox, the deer, the buffalo, the lion, the tiger, the dog, the lamb, the kid, the rabbit; and sometimes with the flesh of ducks, geese, hens, swallows, etc., and finally I experimented with different kinds of fish, such as swordfish, tuna, eel, sole, etc. In every case, one or other of the abovementioned kinds of flies were hatched, and sometimes all were found in a single animal. Besides these, there were to be seen many broods of small black flies, some of which were so minute as to be scarcely visible, and almost always I saw that the decaying flesh and the fissures in the boxes where it lay were covered with not alone with worms, but with the eggs from which, as I have said, the worms were hatched. These eggs made me think of those deposits dropped by flies on meats, that eventually become worms, a fact noted by the compilers of the dictionary of the Academy, and also well known to hunters and to butchers, who protect their meats in Summer from filth by covering them with white cloths. Hence great Homer in the nineteenth book of the Iliad, has good reason to say that Achilles feared lest the flies would breed worms in the wounds of dead Patroclus, whilst he was preparing to take vengeance on Hector.

Having considered these things, I began to believe that all worms found in meat were derived directly from the droppings of flies, and not from the putrefaction of the meat, and I was still more confirmed in this belief by having observed that, before the meat grew wormy, flies had hovered over it, of the same kind as those that later bred in it. Belief would be vain without the confirmation of the experiment, hence in the middle of July I put a snake, some fish, some eels of the Arno, and a slice of milk-fed veal in four large wide-mouthed flasks; having closed and sealed them, I then filled the same number of flasks in the same way, only leaving these open. It was not long before the meat and the fish, in the second vessels, became wormy and flies were seen entering and leaving at will; but in the closed flasks I did not see a worm though many days had passed since the dead flesh was put in them. Outside on the paper cover there was now and then a deposit, or a maggot that eagerly sought some crevice by which to enter and obtain nourishment. Meanwhile the different things placed in the flasks had become putrid and stinking; the fish, their bones excepted, had all been dissolved into a thick, turbid fluid, which on settling became clear, with a drop or so of liquid grease on the surface; but the snake kept its form intact, with the same color, as it had been put in but yesterday; the eels, on the contrary, produced little liquid, though they had become very much swollen, and losing all shape, looked like a viscous mass of glue; the veal after many weeks, became hard and dry.

Not content with these experiments, I tried many different others at different seasons, using different vessels. In order to leave nothing undone, I even had pieces of meat put under ground, but though remaining buried for weeks, they never bred worms, as was always the case when flies had been allowed to light on the meat. One day a large number of worms, which had been bred in some buffalo meat, were killed by my order; having placed part in a closed dish, and part in an open one, nothing appeared in the first dish, but in the second worms had hatched, which changing as usual into egg-shaped balls [pupae], finally became flies of the common kind. In the same experiment tried with dead flies, I never saw anything breed in the closed vessel."

[1] Throughout this work Redi uses the word "uova" where the context shows that pupa is meant. In this he followed Harvey, who called any embryonic mass an "egg."—*Translator*

In 1824, Rene Dutrochet discovered that "the cell is the fundamental element in the structure of living bodies, forming both animals and plants through juxtaposition." Work done in 1838 and 1839 by zoologist

Theodor Schwann and botanist Matthias Schleiden (**Figure 1.5**) would lead to the Cell Theory which initially stated that:

1. Cells are the basic unit of life.
2. All living things are composed of at least one cell.
3. Cells form by free-cell formation similar to the formation of crystals.

Figure 1.5

We know today that the first two tenets were correct but that the third was clearly wrong. The correct interpretation of cell formation as a result of cell division was enunciated by other scientists but finally enunciated by cytologist Rudolph Virchow (Figure 1.5) in a powerful dictum, *"Omni cellula e cellula"*....."All cells arise only from preexisting cells."

Theodor Schwann
(1810-1882)

Matthias Schleiden
(1804-1881)

Rudolph Virchow
(1821-1902)

The biological sciences were rocked to their foundations in 1859 when Charles Darwin (1808-1882) published his famous book *On the Origin of Species by Means of Natural Selection or the Preservation of Favored Races in the Struggle for Life* (**Figure 1.6**). In this profoundly influential and immensely controversial work, Darwin proposed the mechanism by which evolution might occur and offered many accurate and detailed observations to support his belief that not only had evolution occurred but that it was still occurring.

Darwin's theory of evolution and the modifications made to that theory over nearly 150 years are immensely important to zoologist attempting to fathom the source of animal diversity, understand differences in animal structure and adaptations, and explain family relationships within animal groups. For zoologists to truly and wholly comprehend any species of animal, they must understand the origin, evolution, and phylogenetic relationships of that species to other animals.

Figure 1.6
Charles Robert Darwin
(1845-1896)

Zoology and Genetics

The roots of present-day genetics can be traced to antiquity. The genetic power of selection and hybridization were undoubtedly employed by prehistoric plant growers and those who raised livestock, even though they did not understand the genetic principles involved. Babylonian stone tablets that are 6,000 years old have been interpreted as showing the pedigrees of several successive generations of horses, thus suggesting a conscious effort toward improvement.

In later eras Hippocrates, Aristotle, and other Greek philosopher scientists speculated on genetic precepts but what little truth their speculation held was overshadowed by what we now know were many glaring errors. Stories of unusual hybrids were invented by the Greeks and repeated with additional imaginative flourishes by Pliny and other writers of their times. The giraffe was supposedly a hybrid between the camel and the leopard whereas an ostrich was imagined to result from the mating of the camel and the sparrow, and so on.

In the seventeenth and eighteenth centuries as the fledgling fields of **cytology** (study of cells) and **embryology** (study of the development of embryos) began to emerge, scientists struggled to explain and understand how traits were passed from parent to offspring and the biological mechanisms involved in this transfer.

Around the time of the Civil War in the United States, the peaceful gardens of a monastery in Brno in the Czech Republic a revolution of another sort—a revolution in genetics—was brewing as an unassuming monk named Gregor Johann Mendel (1822-1884) (**Figure 1.7**) quietly labored to understand the inheritance of traits in the common garden pea (*Pisum sativum*). Between 1856 and 1863 Mendel cultivated and genetically tested some 29,000 pea plants. Based on this research, in 1865, Mendel presented the short monograph, *Experiments With Plant Hybrids,* to the Brünn Society for Natural History and published it in the *Proceedings of the Brünn Society for Natural History* in the following year.

Figure 1.7
Gregor Johann Mendel
(1822-1884)

Hampered by his lack of standing in the scientific hierarchy of the time and by the fact that other scientists of the day could not fully comprehend and understand the significance of his work, the man who would later be righteously dubbed "The Father of Modern Genetics" and his brilliantly intuitive investigations into the workings of heredity were quietly and quickly relegated to the dustbin of scientific oddities.

In 1900, Mendel's work was rediscovered and finally recognized for the monumental leap forward in genetic knowledge it represented. The first part of the 20th century saw a flurry of genetic revelations as discoveries piled one on the other. In the second half of the 20th century researchers came to realize that the nucleus of the cell is the site of genetic information storage and that this information is stored on long strands of a double helix molecule known as deoxyribose nucleic acid or DNA. Short segments of this nuclear DNA, now known as **genes**, are the "factors" that Mendel worked so diligently to understand.

Mendel's work opened a door that led to a quantum leap forward in our understanding of heredity. Building on Mendel's cornerstone, pivotal advances over the last fifty years have revolutionized genetics, including the discovery of the molecular structure of DNA in 1953 by James Watson and Francis Crick, deciphering the genetic code, the development of recombinant DNA technology, and the mapping of the human genome. Armed with this knowledge, humankind currently finds itself immersed in the throes of the next great biological revolution—the biotechnology revolution.

Although Mendel's ground-breaking work dealt with plants, we now know that all life forms—plant, animal, and microbe—are shaped and regulated by DNA thus making genetics a powerful unifying force for understanding all aspects of biology. Modern zoologists use the power of genetics to understand the form and function of individual animals, to detail evolutionary relationships between groups of animals, to discern the intricacies of the growth and workings of animal populations, and to fathom the diversity-producing mechanisms of evolution.

Zoology and Ecology

Although the number of types (species) of living things on this planet may stagger the imagination (Around 2.5 million species have been identified but there may 10-30 million more out there awaiting discovery.), the total number of individual living things (total biomass) defies comprehension.

Every single creature in this riotous multitude of life has certain requirements it must satisfy in order to survive. In the struggle for life, animals come in contact with other animals, other organisms such as plants and microbes, and with the physical environment. These contacts and dependencies form an unbelievably complex web of life in which all living things are intertwined with each other and with the physical components of this planet. Amazingly, the vast majority of all this life is crammed into about the first 100 feet above and below the surface of this planet in a layer known as the **biosphere**. The branch of biology that attempts to sort out and understand this thin, tangled veneer of life and physical environment comprising the biosphere is known as **ecology** (Gr. *okios,* house + *logos,* to study).

Zoologists use everything humans know about animal behavior, physiology, genetics, and the evolution of animals to understand how diverse ecological interactions determine the geographical distribution and abundance of animal populations. Such knowledge is vital for supporting strong, stable populations but critically important for ensuring the very survival of many populations weakened by changes to their natural environment wrought by the hand of humankind.

A Broad Discipline

Zoology is one of the broadest fields in all of science. The sheer enormity of the number of animal species and the complexity of those animals' bodies, internal processes, heredity, and evolutionary origins necessitates zoologists enlisting the aid of many other biological disciplines as they struggle to fully understand animals (**Table 1.1**).

Table 1.1	
Disciplines of Biology	
Discipline	**Description**
Ecology	Study of the interaction of animals with each other and with the physical environment
Anatomy	Study of the structure of entire animals and their parts
Physiology	Study of bodily processes
Cytology	Study of the structure of cells and the function of cell parts
Histology	Study of tissues
Embryology	Study of the development of animals from a fertilized egg to birth or hatching
Genetics	Study of the mechanisms of transmission of inherited traits
Molecular	Study of the interactions biology between RNA. DNA, and protein biosynthesis and how these interactions are regulated
Paleontology	Study of ancient animal life

Zoologists study all aspects of animal life from molecular chemical reactions to planet-sized ecological systems. Such a broad undertaking requires the input and perspective of many branches of biology.

Given the mind-boggling number of animal types, it is no wonder that zoologists usually specialize in one of the sub-disciplines of zoology centered on one particular animal phylum (**Table 1.2**). Due to the large number and diversity of animals within a single phylum, however, further specialization is usually the norm. For example, there are over 20,000 described bony fishes thus a zoologist who is an ichthyologist might further specialize in the study of eels only.

Table 1.2 *Subdisciplines of Biology*	
Specialization	**Area of Study**
Protozoology	Study of protozoa
Entomology	Study of insects
Herpetology	Study of amphibians and reptiles
Ichthyology	Study of fishes
Mammalogy	Study of mammals
Ornithology	Study of birds
Parasitiology	Study of parasites and their hosts

The animal world is so varied and diverse that zoologists usually specialize to maintain a practical and workable breadth and depth of focus and study.

The aim of zoology and other biological sciences is to understand life: its origin and evolution, its function and processes, and its organization and hierarchy. Zoology makes use of the concepts and methods of all other branches of biology plus those of chemistry, physics, and earth sciences. The animal world of present day is what it is, but the history of the inhabitants of this realm abounds with chance events and roads not taken. The deeper we dig, the more humbling it becomes to realize that things could have turned out very differently, especially for humankind.

Nature of Science

The task of biology is to understand the living world at all levels of the biological hierarchy—from the molecular level of the cell and the mechanisms of inheritance to the complex ecological interactions that shape the biosphere

Science cannot and should not address the realms of the mythical, the imaginary, the metaphysical, or the spiritual. For too long people accepted the musings of authority figures as truth. Often the more bizarre the speculation, the more eager people were (and, unfortunately, many still are) to believe it. The sole aim of science is to classify, understand, and unify the objects and phenomena of the material world. By using a combination of accurate observation and experimentation, logic and intuition, and the occasional fortunate happenstance of **serendipity**, biologists seek to divine and understand the rules that govern all levels of the natural universe

To understand and fully appreciate the art and craft of biology in general and zoology in particular, one must understand the nature of science, how scientists think, and how they develop those amazing tests known as experiments.

> *Discovery consists of seeing what everybody else has seen but then thinking what nobody else has thought.*
>
> —Albert Szent-Gyorgi

What is Science?

What is science? Simply put, *science is the search for natural truths by exacting individuals (scientists) using precise and reliable methods.* Although we find this a practical and workable definition for student use in our work-a-day educational world, there is more to it than that.

The answer to the opening question actually has two parts or components. The first component is *investigation*, and the second is *accumulation*.

Investigation. Science is an on-going and never-ending search (investigation) for the truth. However, this search must always be tempered by the realization that (1) we might not recognize the truth when we see it and that (2) what we regard as the truth is always subject to change. Investigative science must be grounded and guided by the understanding that there are not and never can be any absolute scientific "truths."

> *Our ignorance is sobering and boundless. Indeed, it is precisely the staggering progress of the natural sciences which constantly opens our eyes anew to our ignorance. With each step forward, with problem which we solve, we not only discover new and unsolved problems, but we also discover that where we believed that we were standing on firm and safe ground, are in truth, insecure and in a state of flux.*
>
> —Karl Popper

Accumulation. Science is also a body of knowledge obtained (accumulation) by exacting individuals (scientists) using precise and reliable methods. The knowledge (truths) accumulated through investigation are assimilated into our biological knowledge base. The sheer amount of biological knowledge is staggering, and human-kind continues to accumulate ever more of this knowledge at breath-taking speed. (Visualize the volume our knowledge of biology as an upside down pyramid growing upward and outward at an incredible pace.)

Unfortunately, science is often taught and learned solely as accumulation with prodigious amounts of information and terminology to be memorized and often quickly forgotten. Presenting such a myopic view as the entirety of science is a detriment to student understanding and appreciation of the true nature of science. As our society becomes ever more science-driven, a scientifically literate citizenry becomes ever more critical.

Science is constructed of facts as a house is of stones. But a collection of facts is no more science than a heap of stones is a house.

—Henri Poincare

Characteristics of Science

Investigative science rests on several cornerstones.

1. *Materialism.* Scientific explanations must be grounded in material causes and cannot violate natural law. Magic, myth, and mysticism have no place in science and only hinder the search for truths and obscure such truths as we might discover.

2. *Testability.* Science forms hypotheses that can and must be tested experimentally against the material world.

3. *Repeatability.* Results obtained through experimental testing must, for the most part, hold true time and again. If an experiment yields significantly different results each time it is conducted, none of the data sets collected from the experiment can be regarded as having any probability of being the truth. Regardless of who conducts the experiment or how many times it is repeated, the results must be substantially the same time and again. Nonreproducible results cannot be accorded any reliability.

4. *Self-Correcting.* A scientific theory makes a statement and draws some conclusion about the material world. If later observations or experiments contradict this conclusion, the theory must be revised or rejected. Ideally this should make science self-correcting thus validating the accuracy of any knowledge (truth) gained. However, such is not always the reality of the scientific endeavor as science historians, philosophers, and working scientists will attest.

How Do Scientists Think?

The single most important tool in a scientist's arsenal of discovery is clear and logical thought.

Just as nature has given to students, eyes with which to see her works, so has she also given them brains capable of understanding these works.

—Aristotle

Ancient Greeks, like Aristotle, attempted to explain the natural world through reason alone applying what is known as the **classical approach** to science. They felt that if the reasoning is sound, then the conclusion is trustworthy. This method works well in much of mathematics and logic, but in science it is not enough, because what appears true in the mind and what actually exists in nature are often not the same. **Empiricism**, the notion that reason alone is not sufficient and that ideas must be tested, appeared in Western culture at the time of the Renaissance. As the empirical approach—accurate observation, measurement, and experimentation—became the central aspect of science, modern science as we know it came to be.

Scientists apply empiricism to solve problems and answer questions in two ways:

1. *Induction or Inductive Reasoning* With induction, the researcher measures or carefully observes aspects of the phenomena being studied. Then those measurements and observations are analyzed, and generalizations (theories) are formed from the analysis. Using the induction approach results in reasoning progressing from the specific results to a general principle. For example, if finches, ostriches, and penguins possess feathers (specific results), one might logically reason that all birds have feathers (general principle).

   Inductive reasoning as an important tool of science blossomed in Europe during the 1500s and 1600s when scientists like Isaac Newton, Galileo Galilei, William Gilbert, Robert Boyle and others began to construct models of how the biological and physical world operates from the results of their experiments.

2. *Deduction or Deductive Reasoning* In deduction the reasoning progresses in a fashion opposite to that of induction, moving from general to specific. For example, livestock and pet animals need minerals, vitamins, and other trace elements to grow properly (general principle). With this general knowledge, we can formulate foods that contain the exact components needed by particular types of animals (specific results). Both deductive and inductive reasoning can lead to specific hypotheses that can then be tested.

How Does Science Work?

Although the working of science (or the *scientific method* as it is often known) is often taught and learned as a linear series of steps, it is, however, more accurate to visualize science as a circular enterprise—the science cycle (**Figure 1.8**).

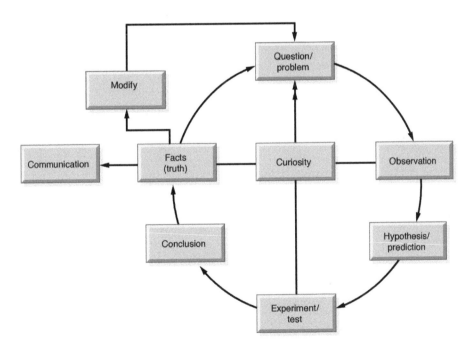

Figure 1.8 The Science Cycle. Science is a curiosity-driven cycle that usually generates more questions than answers.

Curiosity: Science begins and ends with curiosity. Curiosity is the fuel that drives the science machine and the central hub around which the science cycle spins. Without curiosity, humankind would not be driven to divine the truths of the universe. Albert Einstein once said, "I have no special gifts; I am only passionately curious."

Questions: Curiosity raises questions and questions present problems to be solved. Science is the enterprise humans have developed to discern the answers to our questions about nature and the universe around us.

> *The questions worth asking, in other words, come not from other people but from nature,*
> *and are, for the most part, delicate things easily drowned out by the noise of everyday life.*
> —Robert B. Laughlin

Observation: Humans are not born accurate observers. People have certain prejudices and preconceptions about how they perceive the world around them, especially as they grow older. Because of these prejudices, scientists and even the common person must strive to separate *reality* (the way things actually are) from *inference* (the way we think things are or the way we want or expect things to be) in any situation in which observation plays a role.

Hypothesis and Prediction: Curiosity raises questions. Accurate observations reveal information. We then use this information to suggest explanations and answers through hypotheses.

A prediction is a statement made in advance that states the results that will be obtained from testing a hypothesis. Scientists make predictions because predictions provide a way to test the validity of the hypothesis. If an experiment yields results inconsistent with the prediction, the hypothesis must be modified or rejected. If experimental results support the prediction, the hypothesis in turn is supported. The more experimentally supported predictions a hypothesis generates, the greater the validity of that hypothesis. What is the difference then between a fact and a hypothesis? Edward Teller (known as "the father of the hydrogen bomb") may have said it best when he penned, "A fact is a simple statement that everyone believes. It is innocent unless found guilty. A hypothesis is a novel suggestion that no one wants to believe. It is guilty until proven effective."

Experimentation: With the realization that speculation and prediction must be tested in controlled experiments, modern science was born. Scientists, however, must never lose sight of the fundamental fact that experiments are done to test hypotheses, not to confirm them.

Conclusion: What does the data mean? What conclusions (answers) may be drawn from the data? Things are not always clear cut. "Muddy data" can lead to "cloudy conclusions." The validity of any scientific conclusion rests squarely on the accuracy of the data and the interpretation of that data.

> *The great tragedy of science—the slaying of a beautiful hypothesis by an ugly fact.*
> —Richard P. Feynman

Communication: After performing a series of experiments that support the hypothesis, there comes a point at which the scientist must communicate and throw his or her work into the bright light of scientific scrutiny. This communication usually comes in the form of a written paper describing the experiment and its results. The paper is then submitted for publication in a recognized academic journal. However, before it is published, the paper is subjected to a process known as **peer review** in which the work is subjected to the scrutiny of others who are experts in the field. The process of peer review is the cornerstone of the self-correcting nature of the scientific endeavor. With the announcement of an important discovery published in a reputable scientific journal the peer review process continues as other scientists attempt to reproduce the work and gather the same results as the original. This multi-layered peer review approach provides the checks on accuracy and honesty that are such an important part of the repeatability characteristic of science.

Modify or Trash: If contradictions arise, a hypothesis should be modified to account for challenges to its validity. If, however, modification to the old hypothesis cannot bring it within the bounds of more recent findings and results, it must be trashed. (In reality, even experienced scientists are often reluctant to do this.) In science one must be comfortable with failure for failure, not success, is the norm. Actually, we learn more from being wrong than we do from being right because a wrong hypothesis opens up even more experimental pathways as we attempt to understand why our predictions were incorrect.

One cannot underestimate the power of insight, intuition, and imagination in the turning of the science cycle. It is because these qualities play such a large role in scientific progress that some scientists are so much better at science than others, just as some composers, artists, and athletes are better at their craft than are their contemporaries.

Mistakes are the portals of discovery.
—James Joyce

Terminology of the Truth

Not only scientists but the public in general must appreciate the idea that some knowledge (truth) is less likely to be (or has a lower *probability* of being) wrong than other knowledge. Unfortunately, many of the terms employed by investigative science to assign certain levels of truth at certain stages of the science cycle are often misused and misunderstood.

As curiosity leads to questions to be answered and problems to be solved, the science cycle begins to turn, and the search for the truth is on. A **hypothesis** is then offered as a tentative explanation or answer to the problem question. A hypothesis may be based on past observation or experiences of the scientist, the findings of other scientists, or simply a hunch. As such, a hypothesis is certainly more than mere opinion or a simple guess but lacking little evidential support, it must be considered as the lowest level or probability of truth.

If, over time, a hypothesis accumulates supporting evidence through observation and experimental results, it can be said to have achieved the level of a **theory**. Whereas a hypothesis has little factual support, a theory is supported by ever-increasing evidence. Therefore, a theory could be considered as having a median level of truth probability.

"Theory" can also be used to describe a collection of related concepts, supported by observation and experimental evidence that provides a framework for organizing a body of knowledge in some field of study. The quantum theory in physics and the cell theory or the theory of evolution in biology are examples. Thus scientist distinguish between **experimental theory** which is specific to a certain experiment and has a median level of truth probability from an **integrational theory** which is general to a large body of evidence in some area of study and has a high level of truth probability. The general public often confuses the former meaning with the latter which results, not surprisingly, in confusion and misunderstanding. Critics outside the fields of science are adept at utilizing whichever definition fits their agenda. For example, to suggest that evolution is "just a theory" is a misleading and even deceptive ploy to reduce evolution to experimental status when in reality the fact that evolution has occurred is an accepted scientific fact supported by overwhelming evidence.

If evidence continues to mount over a considerable period of time and it seems likely that a theory will not be proven incorrect or falsified, it moves to the category of the highest probability level of truth known as a **scientific law**. Exactly how and when a theory advances to become a law is somewhat ambiguous and arbitrary, and as a result, there are few true scientific laws. Although a scientific law has the highest degree of truth probability, it cannot and should never be regarded as the absolute truth. By its very nature science should preclude any and all absolutes for what we regard as the truth is, at best, a momentary illusion seen through a veil of ignorance.

Designing the Good Experiment

We defined science earlier as the search for the truth by exacting individuals (scientists) using precise and reliable methods. What are these so-called scientific methods? How do scientists design and conduct experiments to get at the truth? Experiments are a series of components usually carried out in sequence:

Component 1: The Problem. Curiosity leads to questions to be answered and problems to be solved. The first step must be to specify exactly what problem is to be solved. This is done through the *problem question*. The problem question should:

- Be in the form of a question.
- Be stated in clear and concise language.
- Present only one problem to be solved at a time.

It is no easy task to formulate a strong problem question and even professional scientists and researchers struggle with the process. It is, however, a crucial first step for as C.F. Kettering reminds us, "A problem well-stated is a problem half solved."

Component 2: Form a Hypothesis. A hypothesis may be formally defined as the possible explanation of some phenomenon. Realistically, a hypothesis is nothing more than an educated guess (but certainly more than a mere opinion) that the researcher makes as to the answer to the problem question. The hypothesis gives a prediction of the outcome of an experiment, and the analysis determines if the prediction is accurate.

Often the experiment is set up using a **null hypothesis** which is the opposite of the actual hypothesis. Better stated, the null hypothesis predicts the results that would be obtained if an actual hypothesis is not true. To use a simple example, say that the hypothesis is that variable A directly affects variable B. In this case the null hypothesis would be that B is not directly affected (is independent) by A. Analysis of the data then determines if the null hypothesis is or is not supported by the experimental results; if it is, then the actual (original) hypothesis can be rejected. If the null hypothesis is not supported, then the actual hypothesis can be accepted, at least from a statistical standpoint.

Don't let the fear of being wrong prevent you from freely expressing your ideas and hypotheses. In science, wrong doesn't matter, because right or wrong, a hypothesis can still get us to the truth. Any natural truth might be visualized as hidden in a thick concrete bunker of ignorance. Further visualize that the truth bunker has two doors. If over time and with scrutiny, a hypothesis proves correct, the researcher has entered the front door of the truth bunker. However, if a hypothesis proves false, the researcher has stilled entered the bunker, only through the back door. Either way, a truth has been learned for as John Dewey encourages us, "Failure is instructive. The person who really thinks learns quite as much from his failure as from his success."

Component 3: Methods and Materials. This component involves planning and then conducting the actual experiment and would include a step-by-step and highly detailed plan of exactly what would be done in the experiment. Within this plan would be found:

- Details on the specific kinds and amounts of materials and supplies that would be required to conduct the experiment.
- Details on manipulating variables and establishing control and test groups. Many factors or **variables** influence processes and outcomes. For example, to study the effect of a synthetic hormone on the growth rate of rats, rats would be put into identical cages, given identical amounts of water and food, and placed so they receive identical amounts of sunlight. The cages, water, food, and light are called the **independent (or control) variables**. The synthetic hormone would be called the **dependent (or test) variable** and might be dispensed to the test rats in their food or water. The rats receiving the hormone would be designated as the **test group** whereas the rats not receiving the hormone would be designated as the **control group**. If the test group grows larger faster than the control group, then it might be concluded that since the hormone was the only variable that differed between the two groups, the hormone did indeed increase the growth rate of the test rats. Without the control group, there is no way to guarantee that the hormone (dependent or test variable) produced an increased growth rate as there would be no group with which to compare results. Some types of experiments, such as field studies, do not lend themselves to controls. For example, suppose we hypothesize that a predator fish population determines the population of a prey species in a large lake and conduct an experiment to test this hypothesis. This may involve measuring the populations of predator and prey over a period of time and then trying to find a relationship between the two. In such an experiment, natural changes are merely monitored; establishing a control group is not practical or possible.

- Details on sample size and sampling methods. The rule of thumb on sample size is: *The larger the sample, the more accurate the results*. Would you be suspicious of a new medicine that was advertised to have miraculous powers if you discovered that the drug had only been tested on 20 people? 200 people? Certainly, as well you should. There is a practical limit to sample size that must be considered, however. Although it may be practical and possible to sample one million bacteria for a specific genetic trait, it would not be practical or possible to sample one million salamanders to determine the variations of a common gene.

- Details on what data will be collected and how the data will be organized. In our rat growth experiment example, one would carefully weigh the rats in both the test group and the control group over a given period of time organizing the weights obtained into a data table.

Scientists often distinguish between **qualitative data** (or soft data) and **quantitative data** (or hard data). We will consider hard data to be numerical data, such as the exact weights of the rats in the hormone experiment. Soft data will be considered to be observational or anecdotal data regarding both the control and test rats feeding in our rat growth experiment.

Component 4: Analyze the Data. Once you have the observed results (known at this point as "raw" data), what do you do with it? How can we make heads or tails out of what the numbers may be trying to tell us? Just looking at a jumble of numbers will probably reveal nothing of significance so scientists must analyze the data. One method of analysis is to graph the data. Graphs turn numbers into graphics which is useful because pictures are more easily understood than lists of numbers. Another analysis scientists perform is to run statistical tests on their raw numerical data. Statistical tests are mathematical processes that help determine if the data collected is significant and if so, how significant.

Component 5: Draw a Conclusion. Analysis of the data will hopefully reveal whether our original hypothesis was correct (supported by the data) or incorrect (not supported by the data). That in turn should answer the original problem question. In this component, scientists attempt to put the results in perspective, establish the significance of the results, and explain how the experiment fits into existing knowledge.

Statistically speaking, there are two major errors that can be made in the interpretation of experimental results. A *Type I error* is to reject a hypothesis that is correct, and a *Type II error* is to accept a hypothesis that is wrong. An important part of experimental design is to plan an experiment that minimizes the likelihood of both kinds of error. This is one reason the results of single studies should not be considered as conclusive evidence for some theory. The hard fact is that erroneous conclusions are always a possibility in scientific research and error is an integral part of the process.

It is also important to understand that negative results, where the data does not support the hypothesis, are valuable. Knowing what is not correct can be as useful as knowing what is correct. Whether the results are positive or negative should not determine the value of the research. However, the reality is that landing large grants, getting work published, and even entire careers often hinge on positive results.

Statistical Tests—Making Sense of the Data

People in general have the mistaken idea that "with statistics, you can prove anything." This is incorrect. With statistics one cannot *prove* anything, but statistics, if used properly, can indeed demonstrate certain degrees of relationship among factors or variables.

> *No discipline has been so misused as statistics.*
> —John Fennick

Statistics can be manipulated, however, and connections made that are artificial if not outright ridiculous at best. Did you know that living near the equator is dangerous to your health? Studies show that on average, people live about a half year longer for each 100 miles of resident distance from the equator. People who live 1,000 miles north of your present location have a life expectancy five years greater than yours. That is equal to what we have supposedly gained from the last thirty years of medical progress.

Can the microscopic parasite *Toxoplasma gondii* which infects the brains of rats, cats, and possibly 60 million symptom-free Americans, skew our gender ratios, increase the odds of having a traffic accident, and even "affect the cultures of nations?" You may scoff at the artificial contrivance and connection made in each of these cases but the evidence supporting these contentions is as good (and in some cases better) than that supporting many statistical "truths" that we sustain with billions of dollars of grant money each year. For example, it is better than the evidence that low-fat diets and exercise have reduced deadly heart attacks. Studies proclaiming doom or glory bombard us one after the other with the last often contradicting the previous. What choices are we to make? What lifestyle changes should we consider to improve and safeguard our health? At present, there is no clear-cut answer or easy choice.

What then is the best and proper use of statistics in biology? Statistical procedures are used to determine if the hypothesis is supported or rejected by the experimental data. The appropriate statistical test properly applied can verify whether our data shows any significant difference between our control group and our test group or if such difference could be attributed to mere chance.

What kinds of statistical tests do biologist conduct and in what experimental situations are each type of test appropriately used?

The **Chi square Test** (or goodness of fit test) is used to determine whether or not two or more samples (of seeds, bacteria, or humans) are significantly different enough from each other in some characteristic, trait, or behavior that we can generalize that the populations from which our samples were taken are also significantly different. Chi square tests are most often applied to data generated in genetics experiments or situations where there may be two possible outcomes. For example, what feather color(s) are dominant in a wild population of certain birds? The Chi square test could be used to answer this question.

The **t-test** assesses whether the means (simple average) of two groups are significantly different from each other in some characteristic or trait. For example, in our rat growth experiment, is the growth rate of the rats fed synthetic hormone greater than that of rats fed no hormone? The t-Test could be used to answer this question.

ANOVA (or Analysis of Variance) is a test applied when you wish to compare samples among more than two and up to any manageable number of groups. For example, a researcher wishes to determine the best concentration of synthetic hormone to be dispensed to rats to generate the fastest rate of growth. The rats would be divided into three equal groups—low hormone concentration, medium hormone concentration, and high hormone concentration. The total weight gain over a select period of time of the rats from each group is determined and recorded. The ANOVA test could verify which concentration of the hormone, if any, yielded a significantly faster growth rate. ANOVA essentially tests whether there is greater variance among than within groups than would be expected by chance, but it does not tell you which particular samples are different from each other.

A Closing Note
Embracing Failure

Science is not all fun and Nobel Prizes. It can be mind-numbingly boring and repetitive, mentally exhausting, and even physically dangerous on occasion. Because science is such a challenging prospect, doing science can be one of the most satisfying experiences a person can have when it is done correctly. Note I said, "done correctly", not "done successfully." Failure, not success, is the norm in science and those that truly understand science know that and live with it on a nearly daily basis. In fact, they embrace failure and turn it inside out gaining valuable knowledge and experience in the process. An educational case in point: I once had a young elementary teacher come to me for help on setting up an "experiment" detailed in her science textbook. The activity dealt with plants but was presented in such a cut-and-dried fashion that the students had no chance for input and no critical thinking on their part was required. I kept quiet and provided her with the necessary materials, interested to see where this might go.

Several weeks later she came to my classroom quite dejected. The whole project had been a failure. When I replied "Good!" her eyes widened, and her jaw dropped. I then counseled her to go back and do some real science by letting her students figure out why they didn't get the results the textbook anticipated. Over the next several weeks her students appeared ever so often at my door in need of additional glassware, lamps, plants, and other materials. I could tell something good was beginning to bubble.

About a month later, this young teacher who had doubted her credibility because an activity failed to produce the results expected again came to me, but this time with a huge smile on her face. After weeks of beakers full of plants under lamps everywhere in her classroom from the window sill to the storage closet and even inside a refrigerator, her students discovered the activity had failed initially due to low light conditions. Her students rode the science cycle (as discussed earlier in this chapter) and solved the mystery. Why? Because their teacher had the courage to cast aside her credibility concerns, embrace failure, and turn a negative situation into a positive learning experience. In the end, that is really what science is all about.

In Summary

- Zoology is the study of the animal world. Zoology and it companion discipline botany were among the first investigations of the natural world undertaken by humankind.

- Zoology as a science arose in the time of the ancient Greeks, but with the fall of Roman Empire, zoology became mythological and biblical based.

- With the rise of the Renaissance, zoology as an investigative science was reestablished.

- The biological principles of organic evolution, genetics, and ecology serve as the foundation of modern zoology. Those same doctrines guide and focus the investigations of zoologists.

- Zoology is one of the broadest fields in all of science. Zoologists enlist the aid of many other biological disciplines and zoologists usually specialize in one of the sub-divisions of zoology.

- To fully understand the art and craft of biology in general and zoology in particular, one must understand the nature of science.

- What is science? The uniquely human enterprise known as science has two components: investigation (the search for the truth) and accumulation (acquiring facts and knowledge).

- Science is (1) grounded in material causes, (2) testable, (3) repeatable, and (4) self-correcting.

- Scientists apply logical thought to solve problems and answer questions using two methods of reasoning: deductive reasoning and inductive reasoning.

- The working of science is most accurately viewed as a circular enterprise—the science cycle.

- Different scientific terms—hypothesis, theory, and scientific law—are used to describe the probability of truth achieved through scientific investigation.

- A well-designed experiment has a series of components carried out in sequence: problem question, hypothesis, methods and materials, data analyzation, and conclusion.

- Various types of statistical tests properly applied can verify whether data collected show any significant difference between control and test groups.

Review and Reflect

1. *Inspired*. Your younger sister has been looking through this zoology book and has been so inspired by it as to state, "I am going to become a zoologist." Knowing something of the depth and breadth of zoology you ask, "OK, but what kind of zoologist do you want to be?" Why would you ask that?

2. *Inspirational*. Your sister next asks, "In Chapter 1 of this book it says that organic evolution, genetics, and ecology are the foundation doctrines of zoology. What does that mean?" How would you reply?

3. *Define*.

   A. You have been commissioned to write a dictionary of scientific terms. When it comes time to write a definition for the word *science*, what would you say? Write your personal definition of science.

B. Some terms associated with the scientific process are often confusing to students and the general public and may be used in misleading ways. Develop a personal definition for the terms: *hypothesis, prediction, theory,* and *law?* Word your definitions in such a way as to show the scientific relationships between these terms and the truth probability of each term.

4. ***Characteristics of Science.***

   A. As part of the requirements for passing this course, your instructor has required you to attend an evening lecture by a visiting scientist. The lecture is entitled; *Materialism vs. Magic and Mysticism in Science.* In general terms, explain what this lecture will be about.
   B. Explain this statement—"Modern science was born with the realization of the need for testability."
   C. Explain this statement—"Science should be self-correcting."

5. ***Ride the Cycle.*** Why is it more realistic to think of science as a cyclic rather than a linear enterprise?
6. ***Figures Don't Lie.*** During a lecture on the concepts covered in this chapter, the person sitting next to you whispers, "I don't get this statistics stuff. What good is it anyway?" You quietly and politely promise to explain after the lecture. What will you say?

Create and Connect

1. ***A Touchy Subject.*** We expect and demand that scientists be honest and ethical but is that always the case? Imagine that you are the science writer for a large newspaper. Your editor has just assigned you the task of writing a newspaper column on scientific fraud. How would you proceed and what would you say?
2. ***Slimy Aardvarks.*** From a meteorite crash site, NASA scientists have collected a small quantity of a sticky green substance thought to have come from within the meteorite itself. For reasons they will not reveal; NASA has turned to you to design an experiment to test the problem question: *What are the effects of exoslime on aardvarks?*

   Guidelines:

   A. Following the tenets of a well-constructed experiment, your design should include the following components in order:

   ➢ The *Problem Question.* State exactly what problem you will be attempting to solve.
   ➢ Your *Hypothesis.* Although this is a fictitious experiment, word your hypothesis as realistically as possible.
   ➢ *Methods and Materials.* Explain exactly what you will do in your experiment including the materials necessary to accomplish the task. Be specific, take nothing for granted, and do not expect people to read your mind as they read your work.

> ➢ *Collecting and Analyzing Data.* Explain what type(s) of data will be collected and what statistical tests might be performed on that data. It is not necessary to concoct either fictitious data nor imaginary observations.

B. Your instructor may provide additional details or further instructions.

3. *A Classic by Redi.* For centuries generations of people believed that life could spring forth spontaneously from dead organisms or inorganic matter like mud. One of the major nails in the coffin of the spontaneous generation theory was the work of the Italian scientist Francesco Redi. Redi's experiments were simple yet had profound biological ramifications. Appreciate a classic experiment in biological history by carefully reading and thoughtfully analyzing Box 1.1 then answer the following questions.

A. Would you say Redi was a thorough experimenter? Explain.

B. What did Redi mean when he said, "Belief would be vain without the confirmation of the experiment."

C. Describe at least one of the several experiments that Redi discusses in this work. Your description should include the *Methods and Materials* Redi used and the *Results* he obtained from the experiment.

D. After conducting a number of experiments on many kinds of flesh, what was Redi's general conclusion about the origin of flies in relation to decaying meat?

2

THE ECOLOGY OF ANIMALS: POPULATIONS AND COMMUNITIES

When one tugs at a single thing in nature, he finds it attached to the rest of the world.

—John Muir

Introduction

Residing in the arm of one of countless spiral galaxies that adorn the universe is an unremarkable yellow star. In orbit approximately 93,000,000 miles off this star is a small planet that is anything but ordinary, a planet the many inhabitants call *Earth*.

From what we know of the rest of our own planetary system and the 1900 plus-and-counting extrasolar planets discovered so far by astronomers, Earth is extraordinarily unique in several respects. First, this world is mantled by an inordinate amount of liquid water, miles deep of it in fact. Furthermore, our atmosphere contains an unusual mixture and percentages of gases—N_2 78%, O_2 21%, and CO_2 0.03%. The other members of the inner circle of planets have no atmosphere (i.e. Mercury), a crushingly thick carbon dioxide dominated atmosphere (Venus), or a thin, wispy carbon dioxide atmosphere that barely still exists (Mars). Of the major gases that comprise earth's atmosphere, oxygen is the most difficult to explain. At 21% of the total atmosphere by composition (a bizarrely high amount), this gas is so reactive that were it found in the atmosphere at only slightly higher levels of around 25%, a single spark could ignite a firestorm that would sweep across the continent. However, were the oxygen level to fall to around 15%, you couldn't light a match no matter how hard you tried. Yet this highly reactive gas has remained stable at its current levels in the atmosphere for hundreds of millions of years.

As unusual and puzzling as the physical attributes of this planet are, the truly unique thing about our home world is the presence of life. And not just a few primitive life forms, like say, a homogenous green blanket of single-celled algae smeared over the planet with a few slimy slugs oozing their way through the green goo. Instead, we find here teeming legions of life forms numbering in the tens of millions of kinds, exhibiting every size, form, and lifestyle imaginable and inhabiting every conceivable nook and cranny of this place.

Matters of Ecology—An Awakening

On our way to the moon, we inadvertently stumbled upon the true nature of the Earth. Photos taken by the Apollo astronauts in the late 1960s first revealed our home planet to be a glorious blue water jewel inlaid with continental chunks of brown and green and flecked with the wispy whiteness of clouds. For those who understood what they were seeing, these powerful images were viewed not only with wonder and awe but also with a sense of concern and foreboding for the earth was finally exposed for what it truly was—a tiny, fragile isolated speck afloat in the cold, cruel velvety black of space (**Figure 2.1**).

Figure 2.1

Ironically, one of the main spin offs of the space program and the many off-world images it generated (and continues to generate) was that it provided both the stimulus scientifically to elevate the backwater field of ecology to a full-fledged and important branch of biology and also the momentum socially and politically to launch the modern conservation movement.

In 1866, the German biologist Ernst Haeckel conceived of studying not just organisms themselves, but also the environment surrounding those organisms. It was he who coined the term **ecology** (Gr. *oikos*, house or home + *logos*, to study) which means the "study of the household [of nature]". Modern ecologist expand the notion of "house" to include the whole populated outer layer of our home planet—the **biosphere**. Ecology can be defined then as the branch of biology that studies the interactions between living organisms (**biotic**) and the interactions between living organisms and the physical environment (**abiotic**). In modern parlance, the word "ecology" is also used as a synonym for the natural environment or environmentalism. Likewise, "ecologic" or "ecological" have come to symbolize anything that is environmentally friendly.

Ecology is a multi-disciplinary science. Because of its focus on the higher levels of organization and interplay of life and the physical environment, ecology draws heavily on biology, geology, geography, meteorology, chemistry, and physics. Ecology can also be subdivided according to the organisms of interest into fields such as animal ecology, plant ecology, and so on. Another means of subdivision and specialization is by the geographical area studied, e.g. polar ecology, tropical ecology, desert ecology and so on.

Given the broad scope of the field of ecology, one can find ecological investigators from mountain top to sea floor and everywhere in-between. The laboratory might be the setting for ecological experiments that can be conducted on a small scale and in the short term. However, to answer questions of larger scale and longer time periods, laborious and painstaking field work is required. Finally, the theoretical aspects of the interplay and interaction between and among populations and between populations and the physical environment can be studied by developing mathematical models. Such models, which might seem far removed from the real world, can frequently predict and reveal important ecological happenings and provide certain generalities about the living world. Of course, ecologists constantly test such models by comparing them to actual ecological systems and then refining or rejecting the models as necessary.

Ecologists face challenges that are quite formidable. First is the daunting task of understanding the structure and behavior of individual animals and how these individuals relate and interact as populations, both between members of the same species (**intraspecific**) and between different species (**interspecific**). Next comes the arduous assignment of understanding the physical components of the planet and how these components relate to and interact with the living components. The mission of understanding how two gigantic and highly dynamic components—abiotic and biotic—intertwine and relate is similar to the challenge that meteorologists face in trying to understand and predict the weather.

Biosphere—A Thin Veneer of Life

Life has a tenuous foothold on this planet at best. The total amount or volume of life (known as **biomass**) defies comprehension. The total biomass of the earth as measured in carbon is estimated (guesstimated?) to be around 1900 gigatonnes (418,874 trillion pounds) or about 3.7 kilograms (8.14 pounds) of carbon per square meter. The actual area and amount of the planet that supports this biomass, however, is shockingly small. The vast majority of life is compressed into occupying only about the first 100 feet above and below the surface, roughly 200 feet of inhabitable space (and that may be a generous estimate) on a planet some 7,900 miles in diameter. If the earth were shrunk to the size of a basketball and wrapped with cellophane, the cellophane would represent the amount of this planet suitable to support life. The biosphere truly coats the corrugations of the terrain and aquatic environs like a thin film. In ecological reality, life is clinging to this planet by its fingernails and this sobering and worrisome fact must serve as both the cornerstone and the driving force of any environmental conscience we may have as a species.

> *One touch of nature makes the whole world kin.*
>
> —William Shakespeare

Ecologists refer to this very thin life zone as the **biosphere** (Gr. *bios*, life + *sphaera*, round). Within this zone the living things contact and are supported by the physical components of the lithosphere, hydrosphere, and atmosphere.

The **lithosphere** (Gr. *lithos*, rocky + L. *sphaera*, round) consists of the rock, sand, and soil of the earth's outer crust to a depth of around 100 km (60 miles) and is the source of all mineral elements required by living organisms. All the water at or near the surface of the planet comprises the **hydrosphere** (Gr. *hydros*, water + L. *sphaera*, round) whereas the **atmosphere** (Gr. *atm*, air + L. *sphaera*, round) consists of the collection of wispy gases we commonly call air.

Box 2.1
Is Earth Alive?

Is it possible that the living (biotic) and nonliving (abiotic) components of the earth are a single complex interacting system which can self-regulate climate and atmospheric composition? Is the evolution of the species of organisms by natural selection and the evolution of the rocks and soil, atmosphere, and oceans a single tightly coupled on-going process? Is it possible that the Earth, in the grander scheme of things, is a single living system? Is the Earth alive?

This hypothesis, dubbed *Gaia* after the ancient Greek earth goddess, was first proposed nearly 40 years ago by atmospheric scientist James Lovelock. Lovelock discussed this novel idea with Nobel Prize-winning novelist William Golding during walks through the English countryside of Wiltshire. Golding suggested calling this system *Gaia* (pronounced GUY-ya) after the ancient Greek earth goddess. And that is just what Lovelock did.

Lovelock developed the notion of Gaia based on his work for NASA designing life-detecting instruments for Mars missions. After thinking the problem through, Lovelock decided that the best way to detect life on another planet was to look at the atmosphere. If a planet had life like Earth's, its atmospheric composition would reflect it. Lovelock turned his attention to looking at ways in which life might shape the development and stability of various physical components of the Earth.

Lovelock contends that Gaia, not conventional science, can best explain two profound mysteries regarding both the temperature and atmospheric composition of Earth. The Earth's mean temperature has remained constant and favorable for life for 3.6 billion years in spite of a rise in heat output from the Sun of 25-30%. Second, the Earth's atmosphere has a large percentage (21%) of the gas oxygen and the percentage of this very reactive gas found in appreciable amounts nowhere else in the solar system seems to have remained amazingly stable for at least 200 million years. Conversely, the percent of carbon dioxide in the atmosphere continues a gradual decline that has lasted hundreds of millions of years. Gaia explains these long-term anomalies by proposing that photosynthetic organisms such as green plants and phytoplankton draw down the greenhouse gas carbon dioxide during photosynthesis. As the sun gradually grows larger and hotter, the rate of photosynthesis gradually increases thus the percent of carbon dioxide in the atmosphere gradually decreases with the net result of a stable global temperature. The percent of oxygen gas in the atmosphere so vital to animal life can also be regulated by the rate of photosynthesis and other geochemical cycles.

When the public first heard about Gaia in the 1970s, it liked the message. Academia and scientists, on the other hand, were not impressed and rejected the idea as some sort of New Age science or ecological-based religion. Gaia has made much progress since then, and the creation of a Gaia Society has resulted in a growing number of scientists centering their work on the Gaian concepts. However, even among believers there is no real consensus as to what Gaia is and how it works.

Of those who object to the Gaia Hypothesis, no group has been more vehemently opposed than evolutionary biologists. The idea that some creatures waste effort making the world a better place for others doesn't make sense to them. Furthermore, complex systems come about by evolution, not by chance. Many different versions are tried out; only the best leave descendants. That's natural selection. Furthermore, evolutionary biologists are quick to point out that natural selection cannot apply to a whole planet, which has no competitors or ancestors.

Lovelock answers by saying, "All I ask is that you concede that there might be something in the Gaia theory, which sees Earth as a living system, to acknowledge Gaia at least for the purpose of argument. I do not ask you to suspend your common sense. All that I do ask is that you consider Gaia theory as an alternative to the conventional wisdom of a dead planet made of inanimate rocks, ocean and atmosphere merely inhabited by life. Consider it a real system, comprising all of life and all of its environment tightly coupled so as to form a self-regulating entity."

Ecological Hierarchy

The hierarchy of life reaches its zenith of size and complexity at the ecological level. If we work our way from the smallest and least complex ecological unit to the largest and most complex unit, we first encounter the **population** level consisting of a group or groups of individuals of the same species living in the same place. But just as individuals are not singular or solitary in the natural world, neither are populations. Populations of different species coexist in multifaceted associations at the **community** level. The higher the level of **species diversity** (total number of species that interact), the more complex the community is considered to be. Ecological communities in turn are part of a larger entity known as an **ecosystem**. At the ecosystem level, ecologists consider not only the interplay between and among communities, but also the physical environment in which these interactions unfold. Communities and ecosystems considered together over a large geographical area that is defined by a certain type of climate are known as **biomes**.

Ecology of Populations

When attempting to understand a population, ecologist investigate five major aspects that govern populations: the range of the population, access and barriers the population must contend with, the dispersal and spacing of the population, the density of the population, and the dynamic changes going on with the population.

Range

No species exists equally everywhere. Each population has its own requirements—space, moisture, humidity, temperature, food availability, and structural habitat—that determine where the species can and cannot survive. The geographical area(s) that provide the necessary requirements for a population is known as the species **range**.

Most populations have relatively limited geographical ranges, but the range of some species can be considered miniscule at best. The Laysan teal or Laysan duck (*Anas laysanensis*) is found only along the shores of a marshy lagoon of only 5 square kilometers (2 square miles) on the small island of Laysan. Even more restricted is the Devil's Hole pupfish (*Cyprinodon diabolis*) that can only be found in a *single* hot-water spring in southern Nevada. These and other small populations that occur in only one limited place such as the Northern white rhinoceros of Africa, the New Guinea tree kangaroo, and the Laysan teal to name only a few live on the razor's edge between survival and extinction.

To the other extreme are animals, such as the coyote and common dolphin, whose environmental requirements are so broad and general that they freely range widely across entire continents or oceans. Thus, the Devil's Hole pupfish is considered to have a very, very narrow **range of tolerance** whereas the coyote has a very, very wide range of tolerance. Determining range of tolerance for critical environmental factors is not an easy task for ecologists. A population may tolerate a wide range of conditions for one factor but much narrower ranges for other factors. Moreover, a population's range of tolerance for one factor may be widened or narrowed by other environmental factors. The many variables and complexity of measuring a range of tolerance in the field often necessitates that such determinations be made experimentally in a laboratory setting.

Population ranges are dynamic, not static and can often change dramatically over time. Such changes occur for two reasons. In some instances, the environment changes. Fire and volcanoes may open areas and present opportunities for some populations to expand their range into areas where they were not able to previously penetrate. With the retreat of the North American glaciers at the end of the last ice age about 10,000 year ago, many plant and animal populations expanded rapidly northward. This northward migration that flourished and then stabilized as the last glaciers disappeared seems to be on again, possibly as the result of global warming.

In addition, populations may find a way to breach and colonize previously unoccupied but suitable areas. Some time in the late 1880s, cattle egrets, a bird native to Africa, appeared in northern South America. It is believed that a few birds were able to make the nearly 2,000-mile transatlantic crossing aided by strong winds aloft. However they arrived, the egrets have steadily expanded their range northward as they marched first into Central America then Mexico, and onward so that they can now be found throughout a great deal of the United States. Another more familiar example is that of *Apis mellifera scutellata,* the African honey bee (or "killer bees" as they are commonly known in the popular press). In 1957, 26 African queen bees were accidentally released near Sao Paulo, Brazil. Not native to South or North America, the bees had been brought in from Africa in an attempt to hybridize various species of European honey bees to increase productivity in tropical regions. Since that time, the bees have spread northward at the astounding rate of almost two kilometers (about 1 mile) per day at their peak rate of expansion. Although confined mainly to the southern United States at this point, they have been seen as far north as Kansas City, Missouri.

The where and why of species distribution is the focus of a field of study in biology known as **biogeography**. As a branch of biogeography, **zoogeography** studies the patterns—latitudinal, longitudinal, and altitudinal/depth—of animal distribution both in time and space. Integrating information on historical and current ecology, genetics, and physiology of animals and their interaction with environmental processes, such as continental drift and climate change, biologists attempt to understand the geographic distribution of animal populations. Study of these patterns is key to our understanding of the processes that generate and maintain diversity over time. Furthermore, such information and understanding can be put to practical use in efforts to construct humane and ecologically sound animal habitats in zoos and in designing refuges and parks to aid animal conservation in the wild.

Access and Barriers

A species normally spreads gradually from its point of evolutionary origin. This spread, however, is not a uniform advance across the globe because insurmountable barriers eventually block access to certain acceptable but unreachable areas. Although the proverbial grass may be greener on the other side of the mountain, a species may not be able to cross the mountain. Thus, the mountain presents a **barrier** that denies a species access to an otherwise biologically and geographically acceptable area (**Figure 2.2**).

Access or denial is a constantly changing ecological variable that plays out biologically and geologically in both the short and long term. Rapid short-term changes such as fire, volcanoes, earthquakes, landslides and so on can either permit or deny access and colonization of an area but normally only on a local or regional scale.

Figure 2.2 Busy highways can serve as an insurmountable barrier to the movement of animals.

Playing out on a global scale over geologic periods of time, the slow ride and spread of crustal plates and continents has thrown up many physical barriers such as mountain ranges and oceans. Climatic change also plays a major role in whether an area is suitable for species habitation in the first place and for how long an area remains suitable. The grand and overwhelming effects of continental drift and climatic change are amply illustrated when we retrieve the fossils of tropical animals and plants from beneath a mile of Antarctic ice or realize that the dry, burning sands of northern Africa were once a blanket of lush plants and flowing streams populated by a large number of different animal types.

Dispersal and Spacing

The **means of dispersal** used by a population determines to a large extent what sort of environment will constitute a barrier. Large organisms often face fewer barriers and disperse (spread) more easily than smaller ones. With brute force of strength and endurance some mammals can travel great distances over land, and many birds fly amazingly long distances over land and water. Small creatures, however, face much stiffer dispersal challenges. To meet these challenges some rely on the wind. Thrusting their abdomen skyward, tiny spiders often release a loop of parachute silk that catches the breeze and sends them aloft in a process known as **ballooning**. Ballooning spiders have been known to strike airplane windshields thousands of feet in the air. Charles Darwin noted a mass landing of spiders on the decks of the H.M.S. Beagle when 200 miles off the coast of South America. There are so many tiny invertebrates (mainly insects and spiders) floating and drifting through the air that they have been collectively designated as **aeroplankton** or **aerial plankton**. Wind blown leaves may also serve as transport for a mixed crew of tiny creatures.

Other small creatures hitchhike on the feet, fur, or feathers of other animals, and some take a leisurely cruise on rafts of vegetation carried out to sea by storms or currents. Dig a large hole out on the prairie far removed from stream or lake, fill that hole with water, and in a surprisingly short time, you will have a pond that abounds with plant and animal life around and in the water. Blown by wind or carried in on other animals, seed and spore, egg and larvae will form a flourishing pond community populated by creatures that seem to magically appear out of the blue. The greatest dispersal mechanism on the planet, however, may well be the ocean currents that disperse both incredible volumes and diversity of marine invertebrates from their point of origin to worldwide destinations.

Normally, organisms crossing barriers fail to establish a viable population in the new territory. Occasionally, however, an active, reproducing population does take hold far beyond the normal range of that species—a process known as **jump dispersal**. Species colonization of far-flung oceanic islands was accomplished by jump dispersal.

Accidental or deliberate introduction of species by humans may also be a form of jump dispersal. At best, **introduced species** are benign new residents, but often they have a negative even devastating impact on their new surroundings and native populations. In a misguided attempt to establish in the United States every type of bird mentioned by Shakespeare, 100 starlings were introduced into New York City between 1890 and 1891. They spread like wildfire and by 1980, were found throughout the United States and into Alaska, Canada and Mexico, now numbering somewhere around 200 million. Starlings compete mightily with native birds for various food sources and hole-nesting natives for nest sites. They are regarded by many as the avian equivalent of the rat.

Think of populations not as a layer of butter spread more or less evenly across a piece of bread but rather as the noodles, carrots, chunks of chicken, and so on in chicken noodle soup. That is, populations are not situated uniformly (butter) in relation to each other but instead are clumped (soup) because the environment is not uniformly favorable and because biotic interactions with other organisms differ in different parts of the range. Much can be learned by ecologists about resource availability, variability of the environment, and interaction among individuals by studying the distribution of individuals and populations.

Were the environment uniformly favorable, and if the individuals of a population were neither attracted to nor repelled by others, individual spacing would be determined by chance and settle out into a random pattern. However, as is more usually the case antagonistic interactions between individuals or scarce food resources will lead to uniform spacing. Many oceanic birds form large nesting colonies on islands or along coast lines. These large colonies can be considered a vast clump offering nesting materials, group protection from predators, and access to mates. Within the colonial clump, each pair of nesting birds establishes their own territory around the nest. As a result, the nests tend to be uniformly spaced.

Density

To understand and assess changes that occur in populations over time, ecologists must also know how many individuals live in an area or the population's **density**. Population density is expressed as the estimated number of individuals in a given area at or over a given period of time.

The size of the designated area is determined by the species to be counted. For small tightly packed animals, such as barnacles on a wave-swept rock, square meters might an appropriate size unit, but for large mammals or birds, a more suitable unit would be square kilometers.

Although the size of the area to be estimated is relatively easy to discern, determining the exact number of individuals within that given area is difficult if not impossible to determine. Although counting the number of a certain species that is held in captive breeding but extinct in the wild might be easy, determining say the number of mountain lions that inhabit a certain area in and around the Grand Canyon presents the proverbial needle-in-a-haystack problem for population ecologists.

Because counting individuals in wild population is impractical if not impossible, ecologists must use density estimate strategies. One such practice is a technique known as **mark and recapture**. This involves ecologists capturing and counting a small portion of a population. The animals initially captured are marked in some way and then released. After a period of time, live traps are set again, and more animals are captured and counted. Ideally, the second trapping would contain some marked and unmarked animals. The ratio of marked to unmarked animals in the second trapping is proportional to the total number of animals in the entire population. The accuracy of such estimates derived from this technique depends on a number of factors, including the total number of animals trapped, the randomness of trap location, **mortality** (death) of marked individuals and **natality** (birth) of additional unmarked individuals, and behavioral factors, such as trap recognition. **Emigration** (individuals leaving the study area) and **immigration** (individuals coming into the study area) occurring during the investigation may also affect results.

Dynamic Changes

From generation to generation as well as over long periods of time, population distribution and density are constantly fluctuating and changing. Individuals are born, and others die. Some depart the population, and others enter. By looking at a population as a whole, ecologists can determine rates for each of these events. Those rates together with extrinsic factors such as environmental fluctuations, natural disasters, predation, and parasitism give ecologists an understanding of the dynamics of population change through time.

The dynamic of primary importance is birth rate (natality) and death rate (mortality). If natality increases, population sizes increases until and unless mortality rates increase as well.

Another major factor that must be considered is change due to individuals (usually of reproductive age) emigrating thus reducing the population size and density and/or individuals immigrating thus increasing the population size and density. In most situations, emigration and immigration roughly balance, creating a **population equilibrium**.

The most important dynamic seems to be the age distribution and gender distribution of individuals within the population. A population containing many reproductive females and at least some males of reproductive age has an immediate growth potential that is very high whereas a population with few females or with most individuals past reproductive age has a much lower growth potential.

A group of individuals within a population all of the same age is referred to as a **cohort**. The reproductive age cohort of each species has a characteristic birthrate or **fecundity**. In the wild, potential fecundity is never realized. A female mouse has a potential fecundity of 17 litters per year with each litter composed of 5 to 7 offspring. Due to predation, disease, and accident, the realized fecundity is far lower or fields and meadows would be carpeted with mice. In addition to fecundity, each species has a physiological longevity or **life span** of individuals that die of old age. The hard reality is that few individuals die of old age in the wild. A well-treated captive European robin lived for 11 years, for example, but robins in the wild seldom live longer than 2 years.

The relative number of individuals in each cohort determines a population's **age structure**. Because different cohorts have different fecundity and mortality rates, age structure has a critical impact on a population's growth rate. Analysis of age structure reveals whether a population is actively growing, stable, or declining.

The percentage of a population that survives to a given age is called its **survivorship**. Some individuals may die at birth or when very young, but others may live well beyond the average wild life span. At any time, the number of individuals at each age can be graphed to produce a **survivorship curve (Figure 2.3)**. Type I curves are typical of populations in which most mortality occurs among the elderly. Type II curves, in which rate of mortality is constant over all ages, is characteristic of some species of large birds and fish. As human beings, we are typically between Type I and Type II curves, with health playing a major role in deciding exactly where we fall.

Type III curves occur when juvenile mortality is extremely high. In other words, life expectancy increases dramatically for individuals who survive their risky juvenile period. Type III curves are typical of some invertebrates, such as clams and oysters, and of vertebrates like fish that produce vast numbers of eggs and offspring with only a very few surviving to reproductive age.

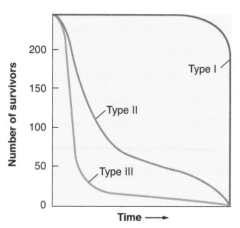

Figure 2.3 Three types of theoretical survivorship curves.

Population Growth

Populations often remain more or less at a relatively constant size, regardless of how many offspring are produced. This seeming contradiction illustrates that to truly comprehend populations we must understand how they grow and what factors regulate population growth.

The rate (*r*) of population growth may be calculated by determining the difference between the birth rate (*b*) (natality) and the death rate (*d*) (mortality) and correcting for any movement into (*i*) (immigration) or out of (*e*) (emigration) a given area. Therefore,

$$r = (b-d) + (i-e)$$

When *r* equals 0, a population maintains a stable size; when *r* is greater than 0, a population grows, and when *r* is less than 0, a population decreases in size.

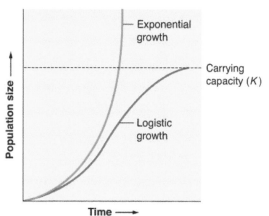

Figure 2.4 Population growth of a species in an unlimited environment (exponential) and a limited environment (logistic).

Populations grow exponentially or are at least inherently able to do so. This ability, shown by the symbol *r,* is known as the **intrinsic rate of increase**. The upward sloping line in Figure 2.4 shows **exponential growth (Figure 2.4)**. Although populations have the inherent capacity to grow exponentially and in some situations actually do so for a time in what are termed **population explosions**, such growth rates, fortunately, cannot be maintained for long. Otherwise, the earth's resources would quickly be exhausted, and mass extinctions would certainly follow. Consider this: a single bacterium asexually dividing three times per hour could potentially produce a layer of bacteria a foot deep over the surface of the planet in only 36 hours, and only one hour later, this layer would be up to our eyeballs and possibly over our heads.

Animals have much lower potential growth rates than bacteria, but they could achieve the same kind of growth over a longer period of time. Consider a single pair of houseflies, laying about 120 eggs per generation and producing about 7 generations in a year, could, over a year's time, potentially produce nearly 6 trillion descendants. In nature, such patterns of unrestrained growth prevail only for short periods, usually when an organism reaches a new area with abundant resources such as locust outbreaks or planktonic blooms in the ocean.

In actuality, populations grow logistically. That is, the population increases rapidly and then levels off at what is known as the **carrying capacity** of that area in which the population is found. Carrying capacity, symbolized as K, can be thought of as the best number of individuals that can be supported by the available resources—light, space, water, food—of an area for an indefinite period of time. In nature, populations tend to fluctuate slightly above or below the carrying capacity over time. Why do populations oscillate this way? First, some necessary factor in the environment becomes limited requiring a population to change its density. Second, there is always a time lag between the time when a necessary factor becomes limited and the time a population responds by lowering natality or increasing mortality. And lastly, **extrinsic** (outside) factors may work to stifle the growth of a population to a level below its carrying capacity. Extrinsic factors may be biotic—predation, parasitism, or disease—and/or abiotic—floods, fires, and storms.

Biotic factors act in a **density-dependent** manner. As the density of a population increases, predators and parasites respond to maintain that population at a fairly constant size. These sizes are usually below the actual carrying capacity because the population is being regulated by predation and parasitism and not by the limiting factors of their environment. Abiotic factors act in a **density-independent** manner. A tropical storm may kill most of the nesting young of shorebird populations, or a fire might eliminate an entire population of animals.

Populations that are adapted to thrive when the population is near carrying capacity are termed **K-selected** populations. In K-selected species such as pelicans, dolphins, and humans, resources are limited, and the cost of reproduction is high resulting in lower reproductive rates. By contrast, some populations prosper far below the carrying capacity. Such populations, termed **r-selected** populations have abundant resources and high reproductive rates. Examples of r-selected species include rats, mice, aphids, and cockroaches.

> *There are no passengers on spaceship earth. We are all crew.*
> —Marshall McLuhan

The Human Population

Studying a historical graph of human population growth and size (**Figure 2.5**) reveals the disturbing fact that human population growth has been explosive over the last three centuries. In fact, on a global scale, no other large creature has ever had a longer period of exponential growth that we know of than have humans, a dubious record at best.

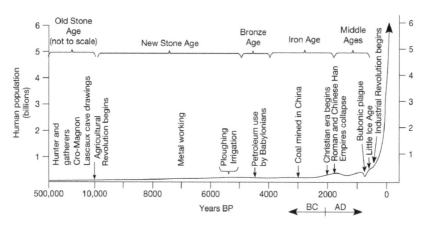

Figure 2.5 A visual history of human population growth.

Figure 2.5 shows that for millennia the global population of humankind slowly increased but remained more or less stable. The intrinsic and extrinsic regulators that control the population growth of other natural populations also held humankind in check. Wars, famines, and even the waves of "black death" (bubonic plague) that decimated much of Europe from the 14th to the 18th century were merely momentary setbacks. But late in the eighteenth century, humankind began to throw off the shackles of natural population control. With the Industrial Revolution in England and Europe and significant advances in our understanding of the nature and prevention of disease coupled with the discovery of new lands for colonization and better agricultural practices, the human population began to skyrocket (**Table 2.1**).

Table 2.1
How Long Has it Taken to Add an Additional 1 Billion People to the Human Population?
Word Population Reached
1 Billion in approximately 1800
2 Billion in 1930 (130 years later)
3 Billion in 1960 (30 year later)
4 Billion in 1974 (14 years later)
5 Billion in 1987 (13 years later)
6 Billion in 1999 (12 years later)
7 Billion in 2013 (14 years later)
8 Billion in 2027 (14 years later)
9 Billion in 2045 (18 Years later)

Source: United Nations Population Division. *Word Population Prospects: The 2004 Revision, February 2005,* and US Census Bureau, International Data base, 2005.

The dramatic increase in human population has been attributed to a combination of factors:

1. Settlement may have allowed women to bear and raise more children; freed from the nomadic lifestyle, they no longer had to carry children for long distances.

2. Children are more useful in agricultural societies than in hunter and gatherer societies and are, therefore, more highly valued.

3. Agriculture and domestication of livestock may have made softer foods available, which allowed mothers to wean their children earlier. Thus, women could bear more children over the course of their lifetime.

4. Agriculture and domestication of livestock led to food surpluses. With farming, one family or group could raise more food than they needed. This surplus led directly to the rise of civilization and cities and to manufacturing and technology.

What is the actual carrying capacity of the earth for humankind? To answer that question we need to consider not only the quantity and quality of resources, but the quality of life as well. This is an extremely complex issue that has yet to be satisfactorily answered. There may already be more people than the planet can sustain with current technologies. We produce enough food globally to feed our population of 6.6 billion, but an estimated 600 to 800 million of those people are undernourished (depending on the definition of "undernourished" you chose to accept).By some estimates, if all humans today lived at the standard of living in the industrialized world, two additional planet earths would be needed to sustain them!

On the other hand, some believe we have not even begun to approach the physical limits on human population size. With appropriate technology and distribution systems, some argue that Earth might be able to support 40-50 billion people, although for how long and at what standard of living would remain to be seen. Given current global food production, it has been suggested that our 6.6 billion could be fed adequately if everyone kept strictly to a vegetarian diet. To a certain extent, we can conceptually increase the world's carrying capacity by lowering the standards at which we are willing to exist.

In the last few decades, however, major environmental changes, such as global warming due to the buildup of greenhouses gases, the degradation of the ozone layer, accelerating habitat loss, and mounting pollution of land, water, and air have led many to conclude that we have finally reached, and perhaps begun to exceed Earth's carrying capacity for humans.

The present human population is distributed somewhat unevenly over the Earth even more so in terms of access to and use of resources (**Table 2.2**). Likewise, such basic statistics as infant mortality, crude birth and death rates, longevity, and so on vary widely from country to country.

Table 2.2
Population Distribution for World and Major Areas, 1750 to 2050 (in Millions)

Area	1750	1800	1850	1900	1950	2005	2014	2050
World	791	1,000	1,262	1,650	2,521	6,465	7,200	9,076
Africa	106	109	111	133	221	906	1,033	1,937
Asia	502	649	809	947	1,402	3,905	4,299	6,104
Europe	163	208	276	408	547	728	741	653
Latin America and Caribbean	16	25	38	74	167	561	575	783
North America	2	7	26	82	172	331	533	438
Oceania	2	2	2	6	13	33	38	48

Whereas the population growth curve for Europe is essentially flat, it is sharply rising for Africa and Asia. In fact, Europe has the lowest growth rate of any continent (0.1%), and Eastern Europe has a negative growth rate (-0.5%), which means that the population is declining. Early in the 21st century, only ten countries account for 60% of the world's population growth. The top two—China and India—accounted for about a third of the world's population growth, but the United States was in the top ten.

Rapid population growth and overpopulation have many far-reaching effects ecologically, economically, and societally. The increasing population is putting a greater and greater burden on the Earth's natural resource base and environment.

> *One can think of our species as having inherited from Earth a one-time bonanza of nonrenewable resources. These include fossil fuels, high-grade ores, deep agricultural soils, abundant groundwater, and the plethora of plants, animals, and microorganisms. These accumulate on time scales ranging from millennia (soils) to hundreds of millions of years (ores) but are being consumed and dispersed on time scales of centuries (fuels, oils) or even decades (water, soils, species).*
>
> —Paul Ehrlich

Ecological damage is not solely a function of more people. Given the discrepancies in affluence and consumption among the Earth's peoples, not everyone impacts equally on the environment. Persons in rich industrialized countries typically cause much more ecosystem damage per capita than persons in poor, non-industrialized countries. Based on such considerations, Paul Ehrlich has suggested that one new American baby and 250 new Bangladeshi babies pose an equal threat to the environment.

Between 1970 and 2005, the annual rate of growth of the world's population began to decrease, from about 2.0% per year to approximately 1.14%. According to some United Nations projections, the annual growth rate could continue to decline until it reaches about 0.47% in 2045 to 2050. In the last half of the 21st century, it may continue to decline until it reaches zero or possibly even a negative number (i.e. the population would shrink) in future centuries. This steady decline in the annual growth rate of the world population is significant—but any growth rate above zero means that the already huge global population is continuing to increase in size. If there are 9.3 billion people on Earth in 2050 and the growth rate at the time is 0.47%, that will mean that nearly 44 million people will be added to the world's population in 2050 (**Figure 2.6**). This, however, is less than the number of people currently added to the world on an annual basis.

Today, the world's population stands at 6.6 billion, and opinions differ as to where it will stand in the future. Is it possible the world's population may now finally be coming under control? Another billion people were added to Earth in the 12 years from 1987 to 1999; however, the growth rate continues to drop dramatically, and the end may be in sight. After the global population is stabilized at around 9 to 10 billion, it could even begin to fall during the twenty-second century as human populations age and fertility rates continue to decline.

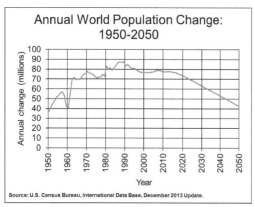

Figure 2.6 Annual increases in world population, 1950 to 2005, with projections to 2050.

Community Ecology

Ecologists define a community on many different levels. Generally speaking, the term community refers to the collective species that inhabit any particular locale. Obviously, the multitudes of species that inhabit a tropical rain forest constitute a rain forest community. But an ecologist might also define a smaller community living within the leaf mold on the forest floor or even within a single rotting log. In fact, there is a seldom-seen community in our own homes from the spiders and pill bugs in the basement to the flour beetles in the pantry to the dust mites in the carpet of the living room. Indeed, every habitable place on earth supports its own particular and sometimes peculiar array of communities.

Everything Changes Over Time

For the most part, species seem to respond independently to changing environmental conditions. As a result, the composition of communities change gradually as some species appear and flourish whereas others decline and eventually disappear. For example, if we examine the fossil record for the trees and small mammals that occurred in North America over the past 20,000 years, we find that prehistoric communities show little similarity to those of modern times. Many species that live together today were never found together in the past and species that used to live together in the distant past no longer do so today. These findings suggest that as the Ice Ages and other factors conspired to change the climate time and again, species responded to these changes independently (known as the **individualistic concept** of communities) rather than as an integrated unit (known as the **holistic concept** of communities).

Even when the climate and landscape of an area remain stable for a century or more, communities in that area will gradually change from simple to more complex in a process known as **ecological succession**. Ecological succession refers to the more-or-less orderly changes in the composition or structure of an ecological community. Such gradual change is familiar to anyone who has seen plowed fields left fallow overgrown with weeds or a small pond gradually grow smaller as vegetation steadily encroaches from all sides slowly filling the pond.

Nature does occasionally turn the succession clock completely back to zero in that bare rock may result after glacial melt, large landslides, lava flow, or certain human activities such as abandoned strip mines. But even barren rock can be made suitable for life. Soon after bare rock appears, nature goes to work in a process known as **primary succession**. Agents of weathering begin to break the rock down, first into cracks and then into pieces. The first primary pioneer species are mosses and lichens, r-selected species that are tolerant of the harsh conditions on barren rock. Arriving as hardy spores, these tiny plants (mosses) and algal-fungal associations (lichens) have very few requirements and begin to colonize the rock immediately. Lichens may degrade the rock surfaces chemically and mosses trap dust and organic matter. Both contribute to the formation of small pockets of soil. As soil begins to accumulate, seeds carried by the wind or birds find a suitable spot for germination and growth. These plants in turn continue the breakdown of the rock by mechanical means (root growth) and accumulate increasing amounts of organic material around them, which leads to the continued formation of even more soil. The organisms of early succession stages facilitate local changes in the habitat that favor the K-selected organisms that will follow. As the process continues, a threshold is gradually crossed, and secondary succession takes over resulting eventually in a climax situation of *K*-selected species (**Figure 2.7**).

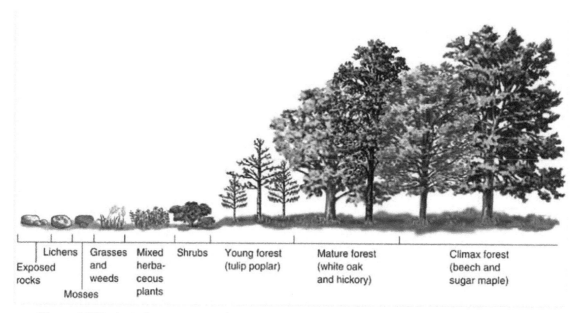

Figure 2.7 Ecological communities change over time. Primary communities of lichens on bare rock slowly change to intermediate communities that eventually terminate in climax communities.

Hawaii provides the perfect natural laboratory for ecological investigations of succession. All the undisturbed flows on Kilauea, Mauna Loa, and Hualalai volcanoes are recent enough to be in some degree of primary succession.

Ecologists have found that colonization of new flows begins almost immediately upon cooling. Certain native organisms are specially adapted to the harsh and barren conditions present on a newly cooled lava flow, and they soon begin to arrive from surrounding areas. Spiders and small insects are the first to take up residence. A steady deposition of organic material, spores, and seeds slowly accumulates in the cracks and pockets of the disintegrating lava. Eventually, some pockets of this protosoil retain enough moisture to support scattered `ohi`a seedlings and a few hardy ferns and shrubs. Over time, these colonizers and other species from nearby forests form a verdant cover of vegetation. A climax forest can develop in wet regions in less than 150 years but more slowly in dry areas. Except for the areas of newer lava flows, the windward (wet) surfaces of Kilauea are heavily forested, but the leeward (dry) slope is barren or only sparsely vegetated. Studies have also been done of coral colonization of newly formed underwater lava flows.

What is normally observed in most cases is called **secondary succession**. This kind of succession occurs in areas where existing communities have been disturbed or destroyed but where the soil remains intact. Fire, flood, erosion, catastrophic weather, or human activities, such as plowing and cultivating, can turn back the succession clock (but not back to zero) in an area forcing nature to start over. The first species of plants and animals that make their way back into newly disturbed land are known as **pioneer species**. Pioneer species tend to be small, *r*-selected types that reproduce rapidly. As the **pioneer stage** stabilizes and major changes gradually slow down, some larger, *K*-selected species with slower reproductive rates will appear and an intermediate stage or **seral stage** develops. The seral stage demonstrates a mix of *r*-selected species and *K*-selected species. Slowly the seral stage gives way to the **climax stage** or **climax community** in which the larger K-selected types dominate and eventually displace most if not all of the pioneers. As strange and contradictory as it seems, the pioneer types hasten their own demise by improving the abiotic and biotic conditions of an area such that it can support the larger, more complex types that will follow and eventually eliminate most of them (**Table 2.3**).

Table 2.3 *Trends in Ecological Succession*		
	Stage in ecosystem development	
Attribute	**Early**	**Late**
Biomass	Small	Large
Productivity	High	Low
Food chains	Short	Long, complex
Species diversity	Low	High
Niche specialization	Broad	Narrow
Feeding relationships	General	Specialized
Size of individuals	Smaller	Larger
Life cycles	Short, simple	Long, complex
Population control mechanisms	Physical	Biological
Fluctuations	More pronounced	Less pronounced
Mineral cycles	Open	More or less closed
Stability	Low	High
Potential yield to humans	High	Low

Source: Adapted from Smith R. *Elements of Ecology and Field Biology*. New York: Harper & Row, 1977. Table 8.9: *Trends in Ecological Succession from Elements of Ecology and Field Biology* by Robert Leo Smith. Copyright 1978 by Robert Leo Smith. Reprinted by permission of HarperCollins Publishers, Inc.

Since human disturbances are now widespread and natural disturbances are more common than once thought, the term "potential natural community" is replacing climax community as a more accurate and representative expression.

The potential natural community (or *PNC*) is what we construe would be established in a given area under present climatic conditions if succession were allowed to occur without human interference. However, pollution, soil erosion, or other conditions may present a threshold of damage so great that natural succession back to the historical climax community is no longer possible. In this case, an *altered* potential natural community replaces the natural climax. Thus, a vacant lot or field may not return to forest in Europe or the eastern United States nor will an empty wheat field return to mixed grass prairie in central Russia or Idaho.

The potential natural community perpetuates itself and is not followed by a different subsequent community. Unless there are local natural disturbances, we expect prairies to remain prairies and mixed hardwood forests to remain forests indefinitely. Climax communities were once considered stable or unchanging in their organization (the **equilibrium model**), but more recent research shows the climax plant and animal communities to be in a constant state of flux. As a result, many ecologists now implement **nonequilibrium**

models that are based on change, not stability. Most natural ecosystems experience disturbances at a rate and frequency sufficient to prevent arrival at a climax state. Disturbances may be widespread or local. Fires, floods, and drought may devastate large areas whereas a single tree overturned in a high wind may affect only the area where it falls and the ground around its dislodged root system.

Disturbances may act to increase the species diversity of an area. According to the *intermediate disturbance hypothesis*, communities experiencing moderate amounts of disturbance will have higher levels of species richness than communities that experience either little disturbance or great levels of disturbance. With low disturbance, competitive exclusion by the dominant species arises, but with high disturbance, only species tolerant of the stress can persist.

Tooth and Claw of Competition

Animal life at all levels struggles to compete for the food, water, space, light, air, mates and everything else that makes the life of the individual and thus, the existence of the species possible. In fact, many biologists consider competition the most common and important interaction in nature. For three decades, however, the relative importance of competition in the structure and regulation of animal communities has been intensely debated with consensus not likely to be achieved any time soon. Many ecologists believe that disturbance in various forms, including predation, is so pervasive that nonequilibrium in natural communities is the norm and employ nonequilibrial models of community organization based on this premise.

The place where an organism normally lives or where the individuals of a population live within their range is known as the **habitat** (L. *habitare*, to dwell). Habitats may be as small as a frog's intestine for certain protozoan parasites or as large as a forest for deer or an ocean for whales. Some physical characteristic and/or plant types typically characterize an animal's habitat, for example, grasses and sparse rainfall of a prairie habitat or the running water of a stream habitat. When speaking of an individual animal, the term **micro-habitat** is applied.

For a species to maintain its population, its individuals must survive and reproduce. Where a species can live and how abundant it can be at any one place within its range is determined by all the resources and physical conditions that species requires for its survival. These requirements are termed the **ecological niche**.

G.F. Hutchinson (1958) suggested that the niche could be modeled as an imaginary space with multiple dimensions, in which each dimension or axis represents the range of some condition or resource that is required by the species. Thus, the niche of a planktonic marine invertebrate might include the three dimensions of temperature, amount of light, and availability of food

The long course of evolution has endowed species with certain genetic characteristics and honed structural and behavioral adaptations to the point where any animal or population has the potential for success in fulfilling the ecological niche it has been dealt. A useful extension of the niche concept is the distinction between the fundamental niche and the realized niche of a species. The **fundamental** (potential) **niche** of a species includes the total range of environmental conditions that are suitable for existence discounting the influence of interspecies competition or predation from other species. However, niches and habitats of various species overlap, and because of competition, the actual niche that can be fulfilled and the actual habitat that be occupied by an animal or population is its **realized** (realistic) **niche.**

The competition faced by any animal or population is fierce, and the battle lines are drawn on two fronts. Animals of the same species within the same population compete against each other (**conspecies** or **intraspecies competition**), and they compete against individuals and populations of other species (**interspecies competition**). For example, before the prairies of North America became mainly farmland and pasture, grazing bison not only competed with each other (intraspecies) for grass to eat but also with a number of other herbivores (interspecies) ranging from pronghorns to prairie dogs to a multitude of insects. Such competition invariably results in a species occupying a smaller niche (realized niche) than it possibly could (fundamental niche) if competition were not a factor.

The ebb and flow of what a species' niche (fundamental) could be and what a species' niche (realized) actually is can be seen in the results of a classic study conducted by J. H. Connell of the University of California, Santa Barbara. Connell investigated the competitive battles between two species of barnacles that grow together on rocks along the coast of Scotland. Poli's stellate barnacle (*Chthamalus stellatus*) is usually found in shallow zones that are often exposed to the air when the tide retreats, the acorn barnacle (*Semibalanus balanoides*) (previously *Balanus balanoides*), on the other hand, is found only in deeper zones and is never exposed to air.

Space on the rocks where these barnacles live is at a premium and competition for a place is fierce (**Figure 2.8**). Connell found that in deeper zones *Semibalanus* always outcompetes *Chthamalus* by undercutting and crowding it out. Connell then removed first one species of barnacle and then the other. When *Semibalanus* was removed from an area, *Chthamalus* quickly and easily began to occupy the deeper zones previously denied to it. To Connell this indicated that it was failed competition with *Semibalanus* and not some physiological deficiency that kept *Chthamalus* out of the deeper zones. On the other hand,

Figure 2.8

when *Chthamalus* was removed from an area, *Semibalanus* did move to occupy the shallower now barren zones above it. This indicated to Connell that *Semibalanus* was not physiologically adapted for life anywhere but in the deeper zones. Thus, Connell concluded that the fundamental and realized niche of *Semibalanus* were essentially the same, but the fundamental niche of *Chthamalus* was restricted to a much narrower realized niche by the more competitive Semibalanus.

As contradictory as it sounds, the decline or absence of one species can lead to a smaller niche for another species. For example, the yucca plant of the prairies of North America can only be pollinated by the pollen-eating yucca moth (Proxidae). If yucca plants are grown outside the natural range of the moth or if the moth population in an area declines or disappears, the yuccas cannot form seeds. In this case, the decline or absence of one species—the yucca moth which is the only animal that pollinate the yucca—can result in the realized niche of another species—the yucca plant—being restricted or even eliminated.

When populations compete, something or someone must give. From a series of revealing experiments with the tiny protist, *Paramecium*, Russian ecologist G. F. Gause formulated what is now called the *principle of competitive exclusion*. Gause studied competition among several different species of paramecia, and he found that when certain species were mixed together in a culture, one species consistently dominated the other driving the less competitive species into extinction. Simply put, the principle of competitive exclusion states that no two species can occupy the same habitat and fulfill the same niche for long if they are directly competing for exactly the same resource(s). Head-to-head competition of this sort will result almost invariably in the local extinction of one of the competing populations.

Interspecies competition can represent a lose-lose situation (-/-) because both species are less able to convert resources into offspring due to competing species laying claim to the same resources. No species have their community or ecosystem to themselves. They share it with other species, and many species have fundamental niches that overlap. The competitive exclusion principle tells us such overlapping is not possible in the long-term. What happens then when two or more species whose fundamental niches overlap occupy the same community or ecosystem? Through a process called **resource partitioning**, species with overlapping fundamental niches subdivide the resources thus avoiding direct competition with each other. That is, they find a way to divide up the wealth so no one is consistently Figurehting over the same coin. The result is that, in real communities and ecosystems, a population is almost always utilizing only a part (realized niche) of the total niche (fundamental niche) they could occupy if they were the only species in that community or ecosystem.

One form of resource partitioning is known as **niche differentiation**. Niche differentiation is often seen in similar species that occupy the same geographical region. On the Serengeti Plain, the three major grazing populations are the wildebeests, the zebras, and the Thompson's gazelles. All three populations are migratory, but not all follow the same migratory patterns. Furthermore, structural differences of teeth, lips, and tongue allow each population to eat a different portion of the same grass plants.

Following the rains of March and April, the grasses, composed of stems, sheaths, and leaves, begin to flourish. First on the scene are the zebras. The zebras eat only the outer leaves. In May, June, and July, the zebras wander off and are gradually replaced by the wildebeests that eat only the grass sheaths. Finally, in July, August, and September, the gazelles come and eat the bare stems that are left. By fencing off different areas at different times, ecologists discovered that if an area was not grazed first by zebras, the wildebeests could not and would not graze there. Correspondingly, if an area were not first grazed by zebras and then by wildebeests, the gazelles could not and would not graze there.

Character displacement, another form of resource partitioning, permits natural selection to utilize the plasticity of animal forms to shape and change the very morphology (form and structure of an organism) of species from one habitat and niche to another.

In a landmark study of Galapagos finches, English ornithologist David Lack suspected that the bill sizes of two species, *Geospiza fuliginosa* and *Geospiza fortis*, depended on whether they occurred together on the same island. Recent work by American ornithologist Peter Grant confirms Lack's hypothesis. The two finches have bills of similar size on islands where they exist separately of each other (**allopatric**). However, on islands where they coexist (**sympatric**), the two species have evolved beaks of different sizes, one adapted to eating larger seeds and the other to eating smaller seeds. In the face of fierce competition, a slight alteration in morphology may be all that is needed to allow a peaceful coexistence.

Interspecies Interactions

As an animal or population struggles to survive within its niche and attempts to maintain or expand its habitat, it undergoes many interactions. Some of these interactions are with the physical environment, some with members of its own species (intraspecies), and many other encounters happen with other species or populations (interspecies) in what ecologist term **symbiosis**. Besides competition, other categories of symbiosis include the following:

Predation (+/-)

Commensalism (+/0)

Mutualism (+/+)

Parasitism (+/-)

(The plus and minus signs are a notation employed by ecologists to indicate the relative gain each participant obtains from the interaction.)

Predation is a state of affairs in which one animal, the **predator** (+) kills and consumes another animal (-), the **prey**. This is most ferocious type of interspecies competition, and it occurs at all levels from the microscopic realms up to the larger components of the animal world. Animals that prey on plants are referred to as **herbivores** whereas **carnivores** are animals that prey on other animals.

A predatory lifestyle dictates that predators must encounter prey (i.e. be in close proximity), then detect prey (i.e. sense they are in close proximity), capture prey, consume prey, and then successfully derive beneficial nutrients from the prey. Minimally, more energy and nutrients must be derived from the prey than the energy and nutrients spent by the predator to capture the prey. It therefore behooves the potential prey as an individual or as a population to minimize their numbers so as to be rarely found by predators, to be hidden and secretive in both coloration and behavior, to be capable of escaping if noticed, to be difficult to consume or to digest if captured, and to supply little nutritionally or even to be toxic to the potential predator.

The predator-prey relationship is an evolutionary arms race in which both sides are constantly adapting in subtle ways in order to gain some advantage over the other. Potential prey animals are adapted to avoid or repel predators in a number of ways. Some employ chemical defenses. Some insects, such as the monarch butterfly, concentrate and store distasteful or toxic chemical generated by the plants on which they feed. The caterpillar stage ingests the chemicals as it feeds on plants initially. These chemicals then pass with them through the chrysalis stage to the adult and even to the eggs of the next generation. A bird that eats a monarch butterfly quickly regurgitates it and learns to avoid the orange-and-black coloration that marks the adult monarch. However, some bird species, such as the black-backed oriole (*Icterus abeillei*) and the black-headed grosbeak (*Pheucticus melanocephalus*), have evolved the ability to tolerate the protective chemicals in the monarchs and they feed on them with near impunity. Orioles refrain from eating the cuticle of the butterfly that contains more toxins whereas grosbeaks eat the entire butterfly. They prefer male butterflies, which seem to contain less toxin.

Animals also manufacture a startling array of chemical substances within their own bodies and may use these chemicals both defensively and offensively. Some snakes, wasps, bees, scorpions, and spiders both kill prey and defend themselves with powerful toxins. The poison-dart frogs of Central and South America develop some of the most lethal biological toxins known. One type, the Kokoi poison-dart frog of Colombia, produces skin secretions so deadly that a mere 0.00001 grams (0.0000004 ounce) is enough to kill a full-grown man. The Choco Indians of the region poison the tips of as many as 50 darts with the secretions of one tiny frog and then use the darts to bring down large game such as monkeys and birds.

The poison-dart frogs and many potentially unpalatable insects advertize their toxic capabilities by adorning themselves in striking colors; a strategy known as **aposematic** or **warning coloration**. After all, it does no good for an animal to incur the biological expense of generating poisonous compounds if in the end it gets eaten, and both it and the predator die. Warning coloration is also employed by animals that may seriously bite or sting a potential predator or interloper. Animals displaying warning coloration are often decorated in blacks, reds, yellows, and oranges, the universal colors of warning in the animal world.

In a different approach, the forces of evolution have molded some animals into copycats. In a process known as **mimicry**, animals have coloration patterns, structural adaptations, and behaviors very similar to their truly poisonous or foul-tasting counterparts. Mimicry is employed in two different ways. **Batesian mimicry**, named for Henry Bates, the British naturalist who first brought it to attention in 1857, is a situation where an otherwise palatable prey species is adapted through coloration, structure, and behavior to closely resemble poisonous or unpalatable types. Many of the best examples of this type of mimicry occur among the moths and butterflies

Figure 2.9 In this example of Batesian mimicry, the color patterns of the palatable viceroy butterfly (a) mimic those of the foul-tasting monarch butterfly (b). If you were a hungry predator, could you tell the difference?

(**Figure 2.9**). **Mullerian mimicry**, first described in 1878 by the German biologist Fritz Muller, is a situation where several unrelated species come to resemble each other. For example, a number of harmless fly species closely resemble stinging wasps in color patterns, structure, and behavior.

Still other animals have adopted a defensive coloration strategy known as **cryptic coloration** or **camouflage** as it is more commonly known. Camouflage allows suitably colored animals to disappear in plain sight when viewed against appropriate backgrounds. Structural adaptations may also aid the deception allowing an animal to appear as a twig, a piece of bark, lichen on a rock, or even a brown leaf on the forest floor (**Figure 2.10**). Camouflage is the most widely employed use of defensive coloration.

Commensalism (L, *com* + *mensa* = sharing a table) is a relationship that benefits (+) one species, but has a neutral effect (0) on the other. The concept of commensalism is difficult to apply in a natural setting as it is impossible to determine whether the "unaffected" member in the relationship truly is completely unaffected. Some ecologists argue that relationships identified as commensal are likely mutualistic or parasitic in some subtle ways that have not yet been detected. In the absence of evidence for predation or mutualism, however, most ecologists assume a relationship to be commensalistic. Initially, the term was used to describe the use of waste

Figure 2.10 A lappet moth (*Gastropacha quercifolia*) employs cryptic camouflage to hide in plain sight on a tree branch.

food by second animals, such as carcass eaters like buzzards who follow large hunting animals but are forced to wait until the large carnivores are finished eating.

Other recognized forms of commensalism include:

- **Phoresy** or the use of a second organism for transportation. Examples would be the remora or pilot fish that attach to a shark or mites carried by dung beetles or damselflies.
- **Inquilinism** or the use of a second organism for housing. Examples are birds that live in holes in trees or the colorful clownfish that call a sea anemone home.
- **Metabiosis** or a mode of living in which one organism is indirectly dependent on another for preparation of the environment in which it can live. An example is the hermit crabs that move into empty gastropod shells to protect their bodies.

Mutualism describes a symbiotic relationship in which members of two different species benefit and neither suffers (+/+) from the relationship. In fact, the two species may be so tightly interwoven together that one cannot survive without the other. A classic example is the termite and the protozoans that inhabit its gut. The termite provides suitable shelter and food for the protozoans. The protozoans in turn digest the wood eaten by the termite thus providing food for the termite. When the protozoans are removed from the termite and exposed to normal environmental conditions, they perish. Lacking the enzymes necessary to digest wood cellulose, termites starve with a belly full of useless wood fibers when the protozoans are cleared from the termite using antibiotics. The mechanisms of coevolution have bound these two species so tightly together that they have, in essence, become one, a classic example of **obligate mutualism**.

Another spectacular example of obligate mutualism is the symbiosis between the siboglinid tube worms and bacteria that live at hydrothermal vents and cold seeps in the ocean. This relationship has resulted in the worm completely losing its digestive tract and becoming solely reliant on its internal bacterial companions for nutrition. The bacteria oxidize either hydrogen sulfide or methane that the worm supplies to them. These worms were discovered in the late 1980s at the hydrothermal vents near the Galapagos Islands and have since been found at deep-sea hydrothermal vents and cold seeps in all the world's oceans.

Most mutualisms, however, are facultative, meaning that both partners can live successfully apart. A terrestrial version of **facultative mutualism** is the relationship between the Egyptian Plover bird and the crocodile. In this relationship, the bird is well known for preying on parasites that feed on crocodiles. To that end, the crocodile openly invites the bird to hunt on its body, even going so far as to open its jaws to allow the parasite-picking bird access to its mouth. For the bird's part, this relationship not only is a ready source of food, but a safe one considering that few predators would dare attack a bird on or in the mouth a crocodile.

A fascinating example of facultative mutualism with a human twist is the relationship between the Boran people of Africa and a bird known appropriately as the Greater Honeyguide (*Indicator indicator*). Boran hunting parties are often joined by the honeyguide that leads them to bee colonies. According to the Boran, the honeyguide informs them of: direction, from the compass bearing of the bird's flight, distance, by the duration of the bird's disappearance and height of its perch, and arrival, by the "indicator call." On locating the bee colony, the Boran smoke out the bees, break open the honeycomb and remove the honey. The honeyguide gains easy access to food in the form of the bee larvae and the wax of the honeycomb when the humans leave.

Parasitism is a relationship in which the parasite is greatly aided (+), but the prey or **host** is harmed (-). In parasitic association, the parasite is usually much smaller than the host and usually remains closely associated with it.

Parasites feed either on the outside of a host (**ectoparasites**) or on the inside of a host (**endoparasites**). Animals with an ectoparasitic lifestyle include the usual list of subjects: leeches, ticks, mites, fleas, and lice. However, the most bizarre and grotesque ectoparasitic relationship may be that between the rose red snapper fish (*Lutjanus guttatus*) from the Gulf of California and a marine isopod crustacean (*Cymothoa exigua*), which is related to wood lice. When one of these crustaceans encounters a rose snapper, it enters the fish's mouth through its gills and steadily devours the fish's tongue. Once it has done this, the crustacean uses hooks on its underside to attach itself to the floor of fish's mouth, and thereafter serves as a replacement tongue for the fish (**Figure 2.11**).

Figure 2.11 The tongue-eating louse (*Cymothoa exigua*) enters a fish (here a Sand steenbras [*Lithognathus mormyrus*]) through the gills and then attaches itself to the fish's tongue which it slowly devours.

Endoparasites normally cause greater damage to their host than do their external counterparts, and endoparasites tend to be more simplistic structurally than other members of their phyla. A short list of endoparasites would include hookworms, tapeworms, liver flukes, lungworms and a host of protozoans such as those that cause amebic dysentery, sleeping sickness, and perhaps the deadliest of all—malaria.

Endoparasites usually do not kill their hosts outright as this would ultimately result in their demise as well. However, **parasitoids**, such as certain species of wasps and other insects that lay eggs on living hosts, feed on the host's tissues until the host dies. Shortly before or just after the death of the host, the next generation of adult parasites emerges from the shrunken husk of what was the host. If that sounds strange, consider these even more bizarre forms of parasitism: **Kleptoparasitism** involves the parasite stealing food that the host has caught or otherwise prepared. A specialized type of kleptoparasitism is **brood parasitism**, such as that engaged in by species of cuckoo birds. The female cuckoo never builds a nest nor does she care for her young. Instead, she lays her eggs in another bird's nest along side the eggs of the host. The host bird incubates and then feeds the young cuckoos until they are old enough to leave the nest, usually to the complete destruction of the host's eggs or chicks.

Epiparasitism is a strange relationship in which one parasite feeds on another parasite. For example, a wasp or fly larva may be an endoparasite of an Ichneumon wasp larva, which in turn is an endoparasite of a wood-boring beetle.

> *So, naturalist observe a flea has small fleas that on him prey; And these have smaller fleas still to bite 'em; And so proceed ad infititum.*
>
> —Jonathan Swift

Many endoparasites have complex life cycles that require several specific hosts in order for the parasite to reach adulthood and reproduction. Recent research has brought to light how certain parasites, using brain and behavior altering chemicals, may alter the behavior of their host in such away as to facilitate their transmission from one host to the next. Juvenile gordian worms parasitize grasshoppers, crickets, locust, and some beetles. After metamorphosing into adults, the worms drive the host insect to locate water and throw itself in. The worms then emerge and swim away to find mates. These gordian worms can even survive predation on its host. Grasshoppers and other host insects are often prey for fish and amphibians but even though their host may be ingested and killed, the worms escape by wiggling out of the nose, mouth, or gills of the predator.

The lancelet fluke, *Dicrocoelium dendriticum*, forces its host ant to clamp itself to the tip of grass blades, where a grazing mammal might inadvertently eat it. It is in the fluke's interest to get eaten, because only by getting into the gut of a sheep or deer can it complete its life cycle.

Indirect Interactions

In some cases, species interact indirectly with each other. The presence or actions of one species may influence another species in some fashion by way of an intermediary third, fourth, or even fifth species. Such effects are termed **trophic cascades**. For example, the number of coyotes in an area can influence the amount of grass in the same area. Coyotes are mainly carnivorous and do not eat plants so how can this be? There is an intermediary third species at work in this scenario—rodents. Rodents are the foundation of the coyote's diet and rodents most certainly eat plants and seeds. This grass-rodent-coyote example illustrates a top-down cascade but cascades can also occur from the bottom up. Were a prairie fire to sweep through a large area, most, if not all, the plants in that area would be reduced to ash. The rodents in the area would face the choice of starvation or immigration. With the rodent population reduced to next to nothing, the coyotes in turn would face the same dilemma—starve or leave. Although these examples may be somewhat simplistic, they do demonstrate how creatures can influence each other even though they never directly interact.

Some species have particularly strong effect directly or indirectly on the composition of communities. Such influential species are termed **keystone species**. Some keystone species influence their community by manipulating the environment as is the case with dams built by beavers or deep holes in lakes excavated by alligators. Others exert their power through the food web. If starfish are removed from tide pools containing a high diversity of life, fiercely competitive mussels explode in growth and effectively crowd out all other species. Without the starfish to control the numbers of mussels, the community collapses. Coyotes perform a similar function in their communities within the chaparral ecosystem.

A Closing Note
The Feeling is Mutual

Of all the interactions between species that ecologists have identified, I find that symbiosis we call mutualism (+/+) to be the most fascinating.

At the end of May, across the Midwest and Great Plains, yucca plants sprout a stalk on which white flowers open one-by-one into a striking inflorescence. Inside some of these flowers lurks the whitish yucca moth. Both the yucca plant and the yucca moth are powerful examples not only of mutualism but coevolution as well. The yucca plant has no ability to produce fertile seeds without the moth. The only pollinator of the plant is the yucca moth; bees and other insects are not attracted, and the wind cannot dislodge the sticky pollen.

Unlike most flowering plants, yucca pollen is not dispersed as individual grains. A pregnant female yucca moth is the only creature with the proper adaptations of mouthparts and behavior to collect and transfer yucca pollen. With specialized mouthparts, she collects and compacts yucca pollen into a golden ball. She then flies to a different yucca plant where she deposits her eggs into the flower's ovary. Then she climbs to the top of the ovary and presses pollen from the pollen ball she has been carrying all this time onto the receptive stigma of the flower. Thus, the yucca moth not only pollinates the yucca plant but also provide a means by which genetic variation may be achieved in the yucca plant.

Hundreds of seeds then develop in the ovary of the flower, some of which become the one and only food source for the hungry moth larva therein. There are no alternative host plants for the yucca moth. This system is so tightly coevolved that if one partner in the relationship becomes extinct for whatever reason, the other partner will go down as well. This amazing example is only one of many that support the realization that interactions of species over ecological time often translates into adaptations over evolutionary time.

In Summary

- Planet Earth, orbiting at just the right thermal distance from its star, has the proper combination of unique and somewhat puzzling physical characteristics to support an amazing array of life forms.
- Ecology is the branch of biology that studies the interactions between living organisms (biotic) and the physical environment (abiotic).
- The totality of life (biomass) is spread into a very thin life zone known as the biosphere around the outer surface of this planet.
- Ecological units—populations, communities, ecosystems, and biomes—form a nesting hierarchy based on size, physical conditions within each unit, and the type of living things within each unit.
- Population ecology studies the various factors of animal groups such as range, access and barriers, dispersal and spacing, density of populations, dynamic changes acting on populations, and population growth.
- Understanding the ecology of communities involves investigations into changes over time (succession), competition between animal groups, interspecies interactions, and indirect interactions.

Review and Reflect

1. *Tug Away*. Provide the ecological details that would explain the quote by John Muir that opens this chapter.
2. **Planet Goldilocks**. Scientists from a number of different disciplines have referred to Earth as the "Goldilocks Planet." Is there any biological and/or ecological validity to this description? Explain.

3. *Tribble Trouble*. Imagine that you are the commander of a starship. A pair of tribbles has somehow made their way aboard your ship. Tribbles, first encountered by Kirk, Spock, and the crew of the Enterprise, are warm, cuddly, peaceful little balls of fluff that reproduce at breakneck speed. In fact, their population doubles every 24 hours. Your ship quickly begins to fill up with tribbles. You and your officers decide to take action when the ship is half full of tribbles. Is this a wise decision? Explain.

4. *A Puzzling Relationship*. Imagine you are a field ecologist who has been observing the interactions between two different species of animals. You aren't sure whether the relationship between the two species is commensalistic or mutualistic. How could you determine the true nature of the relationship between these two species?

5. *Can They Coexist?* A small island in the Caribbean with beautiful beaches and lush tropical vegetation has been purchased by a major land development firm. They plan on putting a resort hotel on the island. The government has agreed to the sale with the provision that a small but stable population of rare birds on the island be protected. The developers have hired your ecology consulting firm to come up with a plan to protect the birds but provide for the tourists. What would be your plan? Explain

6. *Finding the Key*. As a freshwater ecologist, you suspect that a species of large snail is a keystone species in the shallow water communities of small lakes. Explain how you would experimentally test this hypothesis.

Create and Connect

1. *Home on the Range*. Animal territories are places within a larger area called a home range. A home range is the entire area an animal may cover to find food, shelter, water, and a mate. A territorial mammal doesn't usually defend its entire home range, only the area it marks out as its specific territory. What is the home range for a human? Answer that question by constructing your personal home range map for the dates given by your instructor.

Guidelines:

A. On a sheet of paper mark the center of the paper with the letter *H* to represent the location of your home (territory).

B. In the lower right-hand corner of the paper put the scale: ½ inch = 1 mile.

C. The top edge of the paper should be marked North, the bottom edge South, the right edge East and the left edge West.

D. Record your activities on your range map for the dates selected by the instructor. On your map, diagram and label on your map where you go for those days and how far and in what general direction each trip is. You may travel far enough that you will need to tape additional sheets of paper together. For example, if you go to the movie and you travel about ½ mile to there, draw a straight line about ½ inch long (see the scale) in the general direction you traveled to the movie. At the end of that line place a small dot and label the destination. Along the line, place the approximate distance traveled one way.

E. After you have mapped all your activities for the dates assigned, connect the outermost points on your map. These lines represent the perimeter of your home range.

F. Once your home range map is complete, analyze it and answer the following questions:

a. What is the difference between an animal's territory and its home range?

b. In what ways are range maps a useful tool for ecologists and zoologists?

2. *Earth Mother*. Thirty years ago James Lovelock proposed Gaia, a revolutionary and still controversial view of the planet and the living things that inhabit it (See Box 2.1). Research the Gaia Hypothesis and write a position paper on the topic.

Guidelines:

A. Compose a position paper, not an opinion paper. Defend your position with as many facts, Figures, quotes, and pertinent information as possible.

B. Your work will be evaluated not on the "correctness" of your position but on the quality of the defense of your position.

C. Your instructor may provide additional details or further instruction.

3. *Journey Back*. The time-traveling ecologist, Dr. Buckaroo Chronos, has taken you on a journey with him 500 years into the past. What would you see about you? Write a short report on your journey comparing and contrasting the appearance and structure of the ecosystem then with what it is today. Your report should include a description of the major plant and animal species of that time and a small food web or at least a food chain specific to that time.

THE ECOLOGY OF ANIMALS: ECOSYSTEMS AND BIOMES

I have killed the deer.
I have crushed the grasshopper
And the plants he feeds upon.
I have cut through the heart
Of trees growing old and straight.
I have taken fish from the water
And birds from the sky.
In my life, I have needed death
So that my life might be.
When I die, I must give life
To what has nourished me.
The earth receives my body
And gives it to the caterpillars
To the birds and to the coyotes
Each in its own turn so that
The circle of life is never broken.

—Native American tribe of Puebloan people

Introduction

The other planets of our solar system and, for the most part, those we have discovered orbiting distant stars, are hellish places where life as we know it could not possibly exist. By a series of fortuitous happenings that we struggle to unravel and fathom, our small world is perfectly equipped with a dynamic physical structure that supports and nourishes the flora, fauna, and microbe passengers that have come to populate it. Not only is our cosmic real estate perfectly constructed and landscaped, it is also perfectly located. Our planet is nestled just close enough to an immense ball of nuclear energy to benefit from that energy without being destroyed by it, aided, of course, by our planet-generated "sunscreen"—the earth's magnetic field. When viewed and understood from the proper perspective, our world is seen for what it truly is—a unique and irreplaceable gem that is to be cherished and protected at all costs.

Whereas the ecology of populations and communities deals with interactions between species and individuals, the study of ecosystems and biomes delves into the connectedness between living things and the physical properties of this tiny ball of rock, air, and water we call home.

> *We travel together, passengers on a little spaceship, totally dependent on its vulnerable resources of air, water, and soil, being preserved from annihilation only by the care, work, and love we give our fragile craft.*

> —Adlali Stevenson

Ecology of Ecosystems

At the community level of organization, ecologists are most concerned with the connections and interactions between living things. However, at the larger ecosystem level, the focus tends to be more on the interactions between biotic (living things) and the abiotic (physical) environment in an attempt to understand how **biogeochemicals** and energy circulate and move through the system.

Any ecosystem is composed of two entities: the living biota and the abiotic medium the life exists in. In the largest sense, *ecosystem* refers to continental ecosystems, such as forests, prairies, and deserts, freshwater ecosystems, such as lakes, ponds, streams, and rivers, and oceanic ecosystems, or what are called **macroecosystems**. However, the concept of an ecosystem can apply to areas of smaller size as well. A **microecosystem** might be only a large stone and all the life under it whereas a **mesoecosystem** could be a meadow in a small clearing within a forest.

Biogeochemical Cycles

The body of any animal, including you, is formed from the chemical elements found in the air, water, and soil of this planet. And when an animal dies, these chemical elements must be returned to the planet. Must be returned? Why? The Earth is sealed by the cold and airlessness of space and thus the supply of these elements is fixed and unchanging. That is, the earth is a **closed system.** The Earth has only the types and amounts of elements that were present when the planet first formed (Discounting the tons of space dust that rain down on the planet yearly). Because these chemicals operate on a closed system basis, they must be recycled throughout all of Earth's processes that use those chemicals or elements. These cycles include both the living (bio-), and the nonliving atmosphere, hydrosphere, and lithosphere (geo-).

Molecules of these chemical elements are sometimes held for long periods of time in one place or **reservoir**. Coal deposits are huge reservoirs that have been storing carbon for very long periods of time. Other times chemicals may be held for short periods of time in what are known as **exchange pools**. Generally, reservoirs are abiotic situations whereas the term "exchange pool" applies to living things. Examples of exchange pools include plants and animals, which temporarily hold and use carbon in their bodies for a relatively short time and then release the carbon back into the physical environment on their death. The amount of time an element or chemical remains in a reservoir or exchange pool is called its **residence time**.

Chemicals flow back and forth from the biotic to the abiotic and back in often complicated processes known as **biogeochemical cycles**.

Box 3.1
Worlds Within Worlds

On this planet there exist smaller worlds so alien to each other that should the inhabitants of one of those worlds chose to visit the other, they would have to wear spacesuits (humans in SCUBA equipment) or possess special evolutionary modifications (Mud skippers are able to spend time out on tidal flats because they possess the ability to breathe through their skin and lining of their mouth and they can move around out of water though somewhat clumsily because of their pectoral fins are modified into leg-like appendages). Consider TerraWorld and AquaWorld, places so totally different that they might be considered separate planets in their own right. TerraWorld is a sovereignty primarily of dry land and air and AquaWorld is the dominion of water.

Water and air are both considered to be fluids, but it is the different physical properties of the two fluids that make aquatic and terrestrial places quite different from each other. Compared to air, water has high density, high viscosity, low oxygen concentration, and high heat capacity and conductivity. Such differences in physical characteristics are reflected in the structure, physiology, and behavior of terrestrial and aquatic animals.

Density A liter (0.3 gallon) of water weighs 1 kg (2.2 pounds), whereas a liter of air weighs about 1.25 gram (0.044 ounce). Thus, water is more than 800 times denser than air. The high density of water is very supportive of an animal's bulk dramatically lessening the effect of gravity on that creature. This manifests itself in aquatic vertebrates that can grow to gigantic proportions (whales) or those—sharks, skates, and rays—that get by with softer cartilaginous skeleton rather than weight-bearing bone skeletons because they are close to neutral buoyancy in water and are thus supported by the water.

Viscosity Viscosity is a measure of how readily a fluid flows—the higher the viscosity, the slower the fluid flows. (Think of how fast cold water [low viscosity] flows in comparison to how fast cold syrup [high viscosity] flows.). Water is 18 times more viscous than air.

The combined effects of high density and viscosity are reflected in the hydrodynamic (streamlined) shape of aquatic animals versus terrestrial ones. Moving through water is much more physically and metabolically demanding than is moving through air.

Oxygen Concentration Water has a low oxygen content whereas air has a high oxygen content. There are 209 ml (0.4 pint) of oxygen in a liter (0.23 gallon)of air everywhere constantly resulting in oxygen composing 21 percent of the total volume of air. However, the oxygen content of water varies, but is never more than 50 ml (1.70 ounce) per liter of water and can be as low as 10 ml (0.34 ounce) or less.

The differences between water and air in regards to density, viscosity, and oxygen content have led to terrestrial and aquatic animals evolving different strategies for moving fluids in and out of their bodies and extracting oxygen from those fluids—lungs in terrestrial animals and gills in aquatic animals.

Heat Capacity and Heat Conductivity The specific heat of water (the energy required to increase the temperature of 1 gram of fluid by 1 degree) is nearly 3400 times that of air, and water conducts heat almost 24 times faster than air. As a result, water heats more slowly than air during the day but cools more slowly than air during the night. Thus, aquatic animals live in a more stable thermal realm than do animals on shore. However, because the heat conductivity of water is so high, the temperature at any given depth scarcely varies from one place to another, terrestrial animals have a wider variety of temperatures to choose from.

When viewed in this manner, one cannot help but come to the amazing realization that the entity we know as Earth is in actuality a set of smaller worlds within that larger entity.

Biogeochemical cycles of particular interest to ecologists include the water cycle, the carbon cycle, the nitrogen cycle, the oxygen cycle, and the phosphorous cycle.

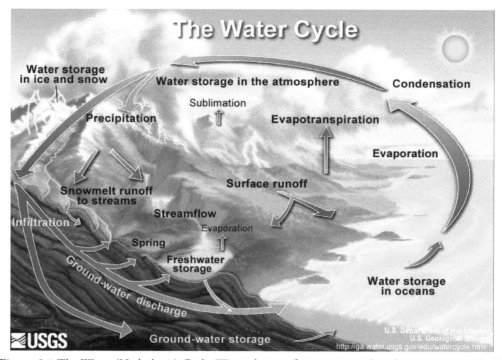

Figure 3.1 The Water (Hydrologic) Cycle. Water changes form as it circulates between air and land.

Hydrologic (Water) Cycle. Water is the most critical chemical on the planet as it is indispensable to the life and function of any organism. Not only do the bodies of living things consist mainly of this substance, but the chemical reactions that nourish and drive each cell of an organism must occur in water (**Figure 3.1**).

Driven by solar radiation, the water cycle is the continuous movement of water over, above, and beneath the Earth's surface. Only about 2% of all the water on earth is held in a reservoir of some sort—chemically bound into the bodies of organisms, frozen as ice, or residing in the soil. The remainder is free water circulating between the atmosphere, the ocean, and the land.

Movement of water through the cycle takes place by a variety of physical and biophysical processes. The main two transport processes responsible for moving the greatest quantities of water are precipitation and evaporation, thought to transport 505,000 km³ (121,156 miles³) of water each year.

As water moves around the hydrosphere, it changes state between liquid, vapor, and ice. The time taken for water to move from one place to another varies from seconds to thousands of years and the amount of water stored in different parts of the hydrosphere is estimated at up to 1.37 billion km³ (329 million miles³). Despite continual movement within the hydrosphere, the total amount of water at any one time remains essentially constant (discounting any that may be added from outer space on board comets or meteorites).

Carbon Cycle. The molecules that compose the physical structure (cells) of any organism have the carbon atom at their core. So important is this element to living things that it exists in living matter (bio) in amounts almost 100 times greater than its concentration in the earth (geo)—18% in living bodies vs 0.19% in the earth and 0.03% in the atmosphere.

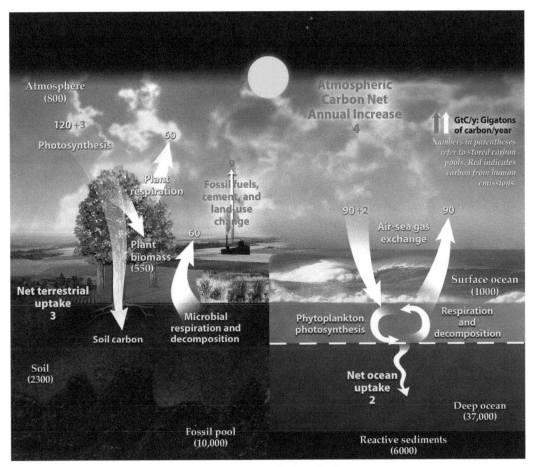

Figure 3.2 The Carbon Cycle.

The carbon cycle (**Figure 3.2**) can be thought of as six major reservoirs of carbon interconnected by pathways of exchange:

- *In the atmosphere* as carbon dioxide. At an estimated 700 billion metric tons, atmospheric carbon dioxide comprises only about 0.04% of the total atmosphere, although this number is rising possibly due to human-caused releases. Ice core studies indicate that the concentration of CO_2 in the atmosphere varies over geologic time and that the level of CO_2 in the past has been higher at times than present increasing readings.

- *In the ocean* as weak carbonic acid (H_2CO_3) and bicarbonate (HCO_3^-). Of the approximately 1 trillion metric tons dissolved in the ocean, more than half is found in the upper layers where photosynthesis occurs. Increases in atmospheric CO_2 translate into increasing acidity in the ocean, which has a negative impact on the formation and maintenance of the calcium carbonate shells of some marine invertebrates.
- *In the biosphere* where between 600 million and 1 trillion metric tons are locked up in the bodies of living things at any one time. Carbon enters the biotic world through the actions of photoautotrophs such as green plants and algae.
- *In carbonate rocks* such as limestone and coral ($CaCO_3$).
- *In fossil fuels* such as coal, oil, and natural gas.
- *In the soil* as organic matter (humus).

Nitrogen Cycle. Nitrogen is vital to living things as an essential component of organic compounds such as chlorophyll, nucleic acids, proteins, and amino acids. Weighing in at 4.11×10^{18} kg (9.05×10^{18} pounds) and comprising 78% of the total air, atmospheric nitrogen is the largest reservoir of that element. This reservoir is about one million times larger than the total nitrogen contained in the bodies of living organisms. Other stores of nitrogen include the organic matter in the soil and the ocean.

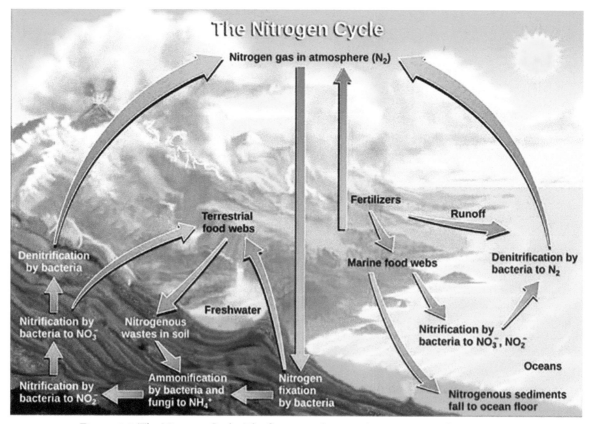

Figure 3.3 The Nitrogen Cycle. The first critical step is the conversion of atmospheric nitrogen by bacteria and fungi in the soil.

Atmospheric nitrogen exists as a very stable diatomic molecule in which two atoms of nitrogen are triple bonded together (N_2). However, most living things are not able to directly convert this abundant atmospheric nitrogen in the nitrogenous compounds they need. Although significant amounts enter the soil in rainfall or through the effects of lightning, the vast majority are **biochemically fixed** within the soil by specialized microorganisms such as some bacteria and fungi. Biologists estimate that biological fixation globally adds approximately 140 metric tons of nitrogen to ecosystems every year (**Figure 3.3**).

Large amounts of nitrogen are added by human activity to the environment in the form of agricultural fertilizers, the burning of fossil fuels, and other sources. For better or worse, humans now play a major role in the global nitrogen cycle

Oxygen Cycle. The fact that diatomic oxygen (O_2) makes up about 21% of the total atmosphere might lead one to conclude that oxygen's largest reservoir is the air as is the case with nitrogen. The silicate and oxide minerals in rocks of the crust (lithosphere), however, contain by far the largest reservoir of the Earth's total oxygen supply (99.5%). Only a small percent (0.01%) exists in the biosphere and only nominally more in the atmosphere (0.49%). Residence times for oxygen vary from the tightly locked oxygen in lithospheric rocks at around 50,000,000 years to diatomic atmospheric oxygen at around 4,500 years to biospheric oxygen which comparatively flies in and out of life cycles at roughly 50 year pulses.

Oxygen is removed from the atmosphere and water by the respiratory activities of animals, decay of organic matter, chemical weathering of rock, and shell building in some marine organisms. Chemical weathering, such as the formation of iron oxides (rust), consumes lithospheric oxygen. Oxygen is replaced into the atmosphere primarily by the action of green plants, algae, and cyanobacteria through the biochemical process of photosynthesis. Because of the vast amounts of oxygen in the atmosphere, however, even if all photosynthesis ceased this instant forever, it would still take about 2.5 million years to strip out all the oxygen from the air.

Phosphorous Cycle. The main reservoirs of phosphorous are as mineral salts in rocks bearing the element or in ocean sediments. Phosphorous is biologically important as a component of nucleic acids, phospholipids, and the energy storage molecule ATP. Furthermore, phosphorous compounds (phosphates) are very important to the nutrition of plants and algae.

As with nitrogen, millions of tons of phosphates are added to agricultural lands yearly in amounts often double or triple what is required and utilized by crops. The excess enters the groundwater or runs off into streams, lakes, and ponds. An excess load of phosphates in aquatic habitats can cause rapid growth of water plants and algae. The result is **eutrophication** in which rapid depletion of dissolved oxygen by the explosive growth of algae results in the suffocation of fish and other aquatic fauna.

Flow of Energy

Unlike the chemical elements that recycle through a closed system, the energy to power life and ecosystems occurs in an **open system**. The sun bombards the Earth with energy which in turn flows through the trophic levels of food chains and food webs. All the energy that comes into the system is eventually lost and is continually being replaced, not recycled.

Trophic levels (Gr., *trophos* = feeder) refer to how far removed from the original source of energy an organism is within a trophic structure whereas **trophic structures** are the feeding relationships within communities and therefore within ecosystems, a list of who's eating whom in other words. Organisms from each level, feeding on each other in turn, make up a sequence known as a **food chain**. Food chains illustrate specifically who does eat whom in a short sequence but sometimes it is beneficial to consider the bigger picture and investigate who could eat whom. By overlapping and interlocking the possible food chains in an ecosystem, we end up with a **food web** (**Figure 3.4**). A food web consists of all the food chains in an ecosystem.

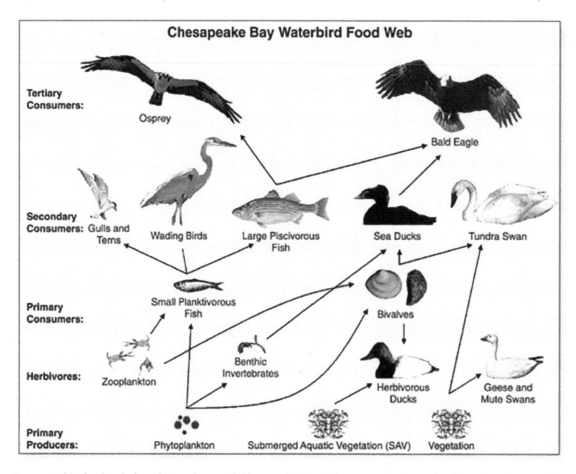

Figure 3.4 A food web found in and around Chesapeake Bay. The arrows indicate the direction of energy flow.

Energy enters food chains through the action of **primary producers** (or **autotrophs**) such as green plants, algae and some prokaryotes (bacteria). Once solar energy enters an ecosystem, almost exclusively as the result of photosynthesis, it is converted and passed from one trophic level to the next in the form of chemical energy.

The producers are fed upon in turn by the **consumers** (or **heterotrophs**). Several different levels of consumers exist. **Primary consumers** (or **herbivores**) feed directly on green plants, **secondary consumers** (or **carnivores**) in turn eat the herbivores, and **tertiary consumers** eat secondary consumers. **Omnivores** such as some fish, pigs, bears, rats, humans, raccoons, foxes, gulls, and crows can eat both plant material and flesh.

Close examination of the trophic levels of any ecosystem illustrate that it really is a "dog-eat-dog" world, but in the end, the Big Dogs eat everybody. Who or what are the Big Dogs? The answer lies with the **saprotrophs**. Saprotrophs include the **scavengers** (or **detritivores**) such as vultures, crabs, hyenas, burying beetles, and flies, that feed on the wastes given off by living organisms or the remains of dead organisms and the **decomposers**, such as bacteria and fungi, that complete the chemical breakdown of decaying organic matter. The eventual fate of any living creature is to be rendered back into air, water, and soil by the Big Dog saprotrophs.

Biological Productivity

Biological systems are not very energy efficient. Green plants, the primary producers of terrestrial ecosystems, capture only about 1% of the solar energy falling on their leaves converting it to food molecules. Phytoplankton, however, are substantially more energy efficient. Accountings for less than 1% of Earth's photosynthetic biomass, marine phytoplankton are responsible for more than 45% of our planet's annual net productivity.

The small amount of solar energy that is actually captured is further lost as it moves through the various trophic levels. First, most plants are not consumed by herbivores but die and become fodder for the saprotrophs. Second, much of the energy that an herbivore takes in is not assimilated into the herbivore's body but is utilized in growth and cellular respiration and lost through the feces. Finally, a great deal of the chemical bond energy is lost as heat produced by work at each trophic level, especially in the upper levels which are populated by endothermic (warm-blooded) animals.

When examining the productivity (efficiency) of ecosystems, ecologists speak of **biomass** (the total living matter in a given place), primary productivity, secondary productivity, and even tertiary productivity. **Primary productivity** is the productivity of the photosynthetic biomass in the form of terrestrial plants and photosynthetic phytoplankton. **Gross primary productivity** is the total amount of energy that is fixed by plants, whereas **net primary productivity** is smaller because it is adjusted for energy losses required to support plant respiration. If the net primary productivity of green plants and phytoplankton in an ecosystem is positive, then the photosynthetic biomass is increasing over time.

Gross and net **secondary productivities** refer to herbivorous animals whereas **tertiary productivities** refer to carnivores. Plants typically account for more than 90% of the total productivity of the food web, herbivores most of the rest, and carnivores less than 1%.

Because of differences in the availability of solar radiation, water, and nutrients, the world's ecosystems differ greatly in the amount of productivity that they sustain. Deserts, tundra, and the deep ocean are the least productive ecosystems, typically having an energy fixation of less than 0.5×10^3 kilocalories per square meter per year (kcal/m^2/yr). Grasslands, montane (moist, cool upland slopes below the timberline), and boreal forests, waters of the continental shelf, and rough agriculture typically have productivities of $0.5\text{-}3.0 \times 10^3$ kcal/m^2/yr. Moist forests, moist prairies, shallow lakes, and typical agricultural systems have productivities of $3\text{-}10 \times 10^3$ kcal/m^2/yr. The most productive ecosystems are fertile estuaries and marshes, coral reefs, terrestrial vegetation on moist alluvial deposits, and intensive agriculture, which can have productivities of $10\text{-}25 \times 10^3$ kcal/m^2/yr (**Table 3.1**).

Table 3.1
Ecosystem Productivity

Ecosystem Type	Area (106 km²)*	Net Primary Productivity per Unit Area per Year (g/m² or t/km² per Year)*		World Net Primary Production (10⁹ t per year)*
		Normal Range	Mean	
Tropical rain forest	17.0	1,000-3,500	2200	37.4
Tropical seasonal forest	7.5	1,000 -2,500	1600	12.0
Temperate evergreen forest	5.0	600-2500	1300	6.5
Temperate deciduous forest	7.0	600-2500	1200	8.4
Boreal northern forest	12.0	400-2000	800	9.6
Woodland and shrubland	8.5	250-1,200	700	6.0
Savanna	15.0	200-2,000	900	13.5
Temperate grassland	9.0	200-1,500	600	5.4
Tundra and alpine	8.0	10-400	140	1.1
Desert and semidesert shrub	18.0	10-250	90	1.6
Extreme desert, rock, sand, and ice	24.0	0-10	3	0.07
Cultivated land	14.0	100-3500	650	9.1
Swamp and marsh	2.0	800-3,500	2000	4.0
Lake and stream	2.0	100-1,500	250	0.5
Total continental	149	773	115	
Open ocean	332.0	2-400	125	41.5
Upwelling zones	0.4	400-1,000	500	0.2
Continental shelf	26.6	200-600	360	9.6
Reefs	0.6	500-4,000	2500	1.6
Estuaries	1.4	200-3,500	1500	2.1
Total marine	361		152	55.0
Full total	510		333	170

*t/km² = g/m² = metric tons/km² = approximately 2.85 tons per square mile.

ᵗ10⁹ t = 1 billion metric tons = approximately 1.102 billion tons.

ᵗ106 km² = approximately 386,000 square miles.

Source: Begon M. Harper J, Townsend C. *Ecology*, 2d ed. Cambridge, MA: Blackwell, 1990. Reprinted by permission of Blackwell Science, Inc.

Ecologists often employ a number of different **ecological pyramids** to diagrammatically represent the relationship between energy and the trophic levels of a given ecosystem. These pyramids include: a **pyramid of biomass** that shows the relationship between energy and trophic level by quantifying the amount of biomass (as dry mass) present at each trophic level, a **pyramid of numbers** that depicts the relative numbers of organisms at each trophic level, and what is probably the most useful of the various types of pyramids, and a **pyramid of energy** which measures the number of calories per trophic level.

In a famous study that would greatly influence the development of ecosystem ecology, H. T. Odum analyzed the flow of energy through a river and stream ecosystem in Silver Springs, Florida for four years

(**Figure 3.5**). Odum's objective was to develop a system for quantifying the energy flow between the trophic levels and to study the efficiency of the ecosystem's constant productivity output. Odum discovered that at each trophic level:

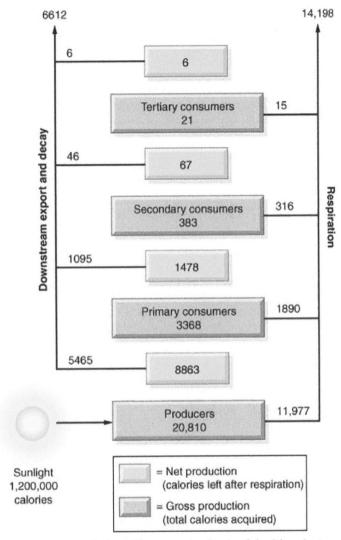

Figure 3.5 Energy flow through the different trophic levels of the Silver Springs ecosystem.

- Net production was only a fraction of gross production. Note that the difference between gross and net production is greater for consumers than for producers reflecting their greater activity.
- Much of the energy stored in net production was lost to the system by being carried downstream and through the process of decay.
- There were substantial losses in net production as energy passes from one trophic level to the next. The **conversion efficiency** (ratio of net production from one level to the next highest level) varied from 17% from producers to primary consumers to 4.5% from primary to secondary consumers. From similar studies in other ecosystems, ecologist now utilize 10% as the average conversion efficiency from producers to primary consumers.

- In this ecosystem, all the gross production of the producers ultimately disappeared in respiration and downstream export and decay. There was no storage of energy from one year to the next. We now know this is typical of mature ecosystems.

Odum's Silver Springs numbers may be used to construct visually revealing ecological pyramids of energy (**Figure 3.6**) and biomass (**Figure 3.7**). A pyramid of numbers for one acre of prairie is depicted in **Figure 3.8**.

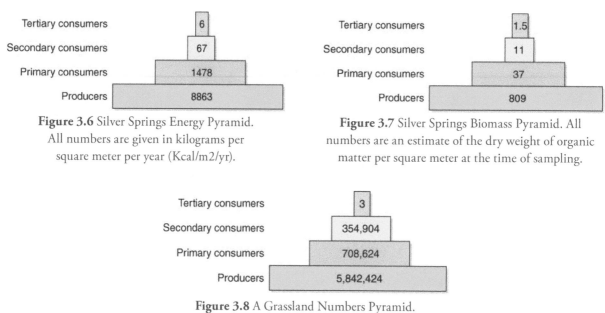

Figure 3.6 Silver Springs Energy Pyramid. All numbers are given in kilograms per square meter per year (Kcal/m2/yr).

Figure 3.7 Silver Springs Biomass Pyramid. All numbers are an estimate of the dry weight of organic matter per square meter at the time of sampling.

Figure 3.8 A Grassland Numbers Pyramid. All numbers are estimates of the population size of each trophic level on an acre of grassland.

Many ecosystems world-wide have been altered by humans through agriculture, mining, logging, and the construction of buildings and roads. In these disturbed ecosystems, the plants and animals of the original natural ecosystem may be found only in isolated pockets and patches. In their place humans have imposed economic and agricultural ecosystems.

> *A thing is right when it when it tends to preserve the integrity, stability and beauty of the natural community. It is wrong when it tends otherwise.*
>
> —Aldo Leopold

Ecology of Biomes

Biomes, the largest unit of the ecological hierarchy, are huge areas containing distinctive plant and animal groups which are adapted to the particular environmental conditions of a given area. Although terrestrial biomes are the most familiar and easily identified, ecologists also recognize aquatic biomes. Each biome consists of many ecosystems whose communities have adapted to the small differences in climate and other physical conditions within the biome.

Terrestrial Biomes

Terrestrial biomes are defined as large geographical areas defined by their climate, soil types, and characteristic vegetation. The North American continent holds examples of the major land biomes found around the world (**Figure 3.9**).

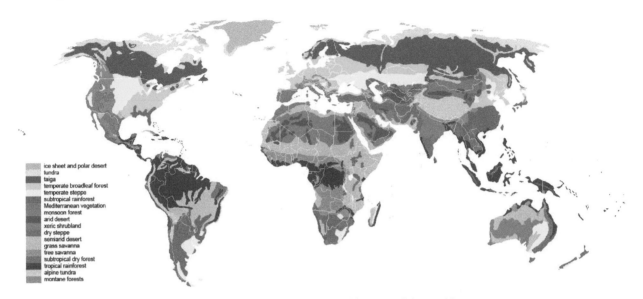

Figure 3.9 The major terrestrial biomes of the world.

Tundra. Arctic tundra encircles the earth south of the ice-covered polar seas in the Northern Hemisphere. Covering nearly 20% of the earth's land surface, it is a brutally cold realm that is characterized by long and even continuous darkness in the winter, but continuous light in the summer.

The tundra receives about 10-50 cm (4-20 inches) of moisture a year, mainly in the form of snow. In what passes for summer, the ground is covered with sedges and short-grasses with patches of lichens and mosses. Since the ground remains frozen (**permafrost**) except for the top several inches in the summer, large woody trees and shrubs cannot establish a foothold on the tundra. A few hardy animals, such as the lemming, ptarmigan, and musk-ox, live on the tundra year-round. The summer months may see the appearance of migratory creatures such as insects, waterfowl, caribou, reindeer, and wolves.

Taiga (Boreal Forest). The taiga, perhaps the largest of the biomes, is a wide band of coniferous forest that stretches across northern Eurasia and North America. The winters there are long and cold as evidenced by a yearly average temperature that ranges only from -5 to +3° C (23 to 37° F). The mean annual precipitation at less than 100 cm (40 inches) supports one of the largest plant formations on earth. The taiga is dominated by evergreen trees—pine, hemlock, cedar, fir, and spruce—which are adapted to withstand cold temperatures and heavy snowfall with their conical shape, needle-like leaves, and flexible limbs. Many herbivores, including hares, rodents, moose, elk, and deer, and such mammalian carnivores as wolves, foxes, wolverines, lynxes, weasels, martens, and omnivorous bears are adapted to year-round life in the boreal forests. Common birds include chickadees, nuthatches, warblers, jays, hawks, owls, and eagles.

In certain locations along the Pacific coast from Alaska down to California, the warm, moisture-laden winds off the Pacific Ocean drop large amounts of snow and rain providing enough moisture to morph what should geographically be taiga into a rare *temperate rainforest*.

Temperature Deciduous Forests. South of the taiga in eastern North America and most of Europe and eastern Asia are (or rather, *were*) great expanses of deciduous (hardwood) forests. Deciduous forests are associated with warmer continental and humid subtropical climates where the winters are mild and the summers warm. The 50-150 cm (20-60 inches) of precipitation is distributed throughout the year, and there is an approximately six month growing season.

Deciduous trees including oak, maple, beech, chestnut, hickory, ash, elm, and basswood, are those that drop their leaves with the onset of winter. Although the large hardwood trees dominate the forest in size and number, smaller plants form an **understory** (layers) beneath them. On sandy soils to the south and along the Gulf Coast, pines (softwoods) replace hardwood species. This is seen in the New Jersey pine barrens, the piney woods of the Deep South, and the tall, long-needled pines of Georgia and other areas of the Atlantic Coastal Plain.

Forest animals are herbivores such as deer, squirrels, and chipmunk, or omnivores including the raccoon, opossum, skunk, and black bear. Large carnivores have been mainly eliminated through the deliberate efforts of humans, but historically included wolves, mountain lions, and bobcats. Some birds—jays, woodpeckers, and chickadees—are year-round residents whereas insectivorous neotropical migrants such as warblers, wrens, thrushes, tanagers, and hummingbirds appear in the spring and depart in the fall. The milder winters also permit the year-round survival of salamanders, frogs, and other amphibians.

Temperate Grasslands. In the semiarid continental climates of the middle latitudes where the soil is rich but the summers hot and dry and only 30-90 cm (12-35 inches) of moisture is the norm, lie the temperate grasslands. Grasses are the dominant vegetation in these areas that are too wet to be a desert but too dry to support trees. Each continent has its own unique grassland system. In North America we have *prairies* (from the French meaning "a meadow grazed by cattle") whereas South America has its *pampas* and *savanna*. Central Eurasia is dominated by the *steppes* whereas Africa has the small *veld* and the larger *savanna*.

The North American prairie biome is composed of three ecosystems. These bands of grass stretch across the midsection of the continent from around western Indiana in the east to the Rocky Mountains in the west and from south-central Canada in the north to Texas and Mexico in the south.

Starting in the east and heading west, we first encounter the *tall grass prairie* ecosystem. Grasses, such as big bluestem, Indian grass and switch grass, receive enough moisture to grow up to six feet tall. These grasses grew so dense that historical accounts describe that when gentle winds rippled the top of the grass, it created the appearance of waves moving across the surface of an inland ocean of green, giving rise to the term *prairie schooner* to describe the wagons of early pioneers to the region. A profusion of perennial herbaceous forbs and wildflowers also dot the tall grass prairie. As we continue west, declining yearly moisture causes tall grass to give way to shorter grass in the *mixed grass* or *transitional prairie* ecosystem. As the designation implies, here we find a mixture of tall grass and short grass species with some forbs and wildflowers mingled in.

At the far western edge of the biome, we find the *short grass prairie* ecosystem nestled in the parching rain shadow of the Rocky Mountains. Semi-desert moisture conditions there permit the growth of only very short bunch grasses. Small herbivores including insects, rabbits, mice, ground squirrels, and prairie dogs, abound, but the large herds of bison and pronghorn antelope which historically dominated these ecosystems are all but gone. Prairie carnivores include snakes, hawks, owls, eagles, badgers, bobcats, coyotes, and foxes. As is so often the case, the large carnivores—bears, wolves, and mountain lions—that also once roamed the prairies have disappeared from their former range.

Wildfires were frequent visitors to the prairies in by-gone times. Ignited by lightning and driven by the relentless prairie winds, these fires would sweep majestically and rapidly across great swaths of the prairie. As contradictory as it sounds, prairie wildfires were not only beneficial, they were essential in maintaining the biological health of the prairie. Fire quickly recycled large amounts of stifling dead plant material into elemental compounds that could more easily be incorporated by prairie plants. Prairie plants and animals suffered little from such fires as they were adapted to them and depended upon them, especially the plants. However, the artificial economic and agricultural ecosystems humans have imposed on the former prairies are not fire friendly; as a result, the natural cycle of growth and fire on the prairies has been suppressed by humans.

Deserts. At around 30 degrees latitude north or south in the interior of the continents lie the world's great deserts. Deserts are dry places receiving less than 25 cm (10 inches) of moisture per year—an amount so low that vegetation is very sparse. In fact, the Sahara of Africa is almost totally devoid of any vegetation whatsoever. With few if any plants present and lacking cloud cover most of the time, temperature changes can be extreme with searing hot days followed by freezing cold nights.

The deserts of North America consist of the hot deserts of the southwest (which include the Mojave, the Sonoran, and the Chihuahuan) and rain shadow deserts created by various mountain ranges. Cacti, shrubs, and other desert plants have few or no leaves at all and other adaptations for dealing with little moisture. Animal life includes many insects, spiders and scorpions, lizards and snakes, rodents such as the kangaroo rat, birds including roadrunners, cactus wrens, vultures, and burrowing owls as well as coyotes, deer, peccaries, and rabbits.

Deserts are increasing in size. In the late 1800s, the amount of desert land on the earth was estimated to be around 9%. Today it is estimated to be 25% and climbing. The main culprits are thought to be extended droughts and overgrazing by livestock.

Chaparral. The chaparral (from the Spanish word meaning "small oak") is a shrubland ecosystem found primarily in California that is shaped and defined by a Mediterranean climate (mild, wet winters and hot, dry summers) and frequent wildfires. Scrub oaks and other low-growing drought resistant shrubs such as manzanita, ceanothus, chamise or greasewood, and mountain mahogany dominant the landscape, often growing so dense that they are all but impenetrable to large animals and humans.

Fire is a frequent visitor to the chaparral and the plants there and the seeds they produce are not only adapted to it but may actually require periodic burns every 10 to 15 years to maintain the ecosystem. As seems to be our need, humans have attempted to suppress wildfires in locations where settlement has encroached on the chaparral. In the face of low humidity, howling winds, and low plant (fuel) moisture, such efforts have failed time and again. In fact, the number of fires is increasing in step with human population growth pushing into the chaparral.

Tropical Forests. Aligned like a green belt along the equator where temperatures are constantly warm (20-25° C [68-77° F]) and rainfall plentiful (100 to 1000 cm [40 to 400 inches] yearly), are the tropical forests. This biome is a heterogeneous collection of several different types of forests.

- *Tropical rain forests* are found in South America, Africa, Indonesia, and New Guinea. Although rain forests now cover just 2 % of the globe, they are home to anywhere from half to two-thirds of all the living plant and animal species on the planet. It is not only the quantity of life, but diversity as well that makes the rain forests irreplaceable biological gems. Some of the strangest and most beautiful plants and animals are found only in the rain forest and this shady world may well hold hundreds of millions of new species of plants, insects, and microorganisms there still waiting discovery. According to a report by the National Academy of Sciences, a 1,000 hectare (4-square-miles) patch of rain forest can contain up to 1,500 species of flowering plants, 750 species of trees, 125 species of mammals, 400 species of birds, 150 species of butterflies, 100 species of reptiles, and 60 species of amphibians.

  Rain forests are organized into levels or layers (horizontal ecosystems) with an amazing number of plants and animals adapted for life in any particular layer. The canopy layer alone, by some estimates, may be home to 40% of all plant species and a quarter of all insect species on earth. A prospect that prompted American naturalist William Beebe to once declare that "another continent of life remains to be discovered, not upon the earth, but one to two hundred feet above it, extending over thousands of square miles."

- *Moist deciduous* and *semi-evergreen forests* are found in parts of South America, in Central America and around the Caribbean, in coastal West Africa, parts of the Indian subcontinent, and across much of Indochina. These forests experience a cooler winter dry season during which time many tree species drop some or all of their leaves.

- *Montane rain forests*, some of which are known as cloud forests, are found in mountainous areas with cooler climates.

- *Flooded forests* such as freshwater swamp and peat swamp forests, Brazil's Pantanal, the Amazon, and the Florida aquifer system.

Aquatic Biomes

Aquatic biomes may be classified as marine (saltwater) or freshwater. There are also some **estuary ecosystems** near the ocean where the freshwater from the land and the saltwater from the ocean mix producing brackish conditions. Some marine ecosystems, such as coral reefs and estuaries, are exceptionally productive and account for approximately 50% of the earth's primary biological productivity. Because the turn over rate is so high in aquatic systems, they produce about three times more animal production than do terrestrial systems.

Marine ecosystems. The marine biome consists of three major ecosystems: (1) the neritic zone, the shallow coastal waters, (2) the pelagic zone, the open waters above the ocean floor, and (3) the benthic zone, the actual seafloor (**Figure 3.10**).

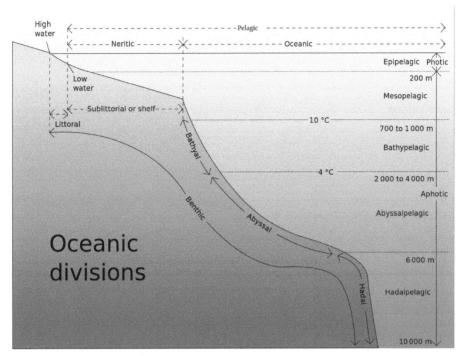

Figure 3.10 Oceanic ecosystems.

Where the ocean meets the land, we find the **intertidal** or **littoral region** of the neritic zone. Slammed by waves and exposed to air during low tides, this region is a very hostile environment where only the most highly adapted plants and animals can survive.

The **neritic zone** extends from the low-tide line outward to the edge of the continental shelf and reaches depths of around 200 meters (656 feet). Low water pressure, a fairly stable temperature, and an abundance of light for photosynthesis make this zone rich with life. In fact, the neritic zone together with estuaries account for essentially the total productivity of the ocean as evidenced by the fact that all the world's great commercial fishing areas are located in the neritic zone.

Beyond the continental shelf is the open ocean or **pelagic area**. This area contains a number of recognized subzones characterized mainly by their depth:

- *Epipelagic zone* reaches downward to around 200 meters (660 feet) below the surface. This sunlit layer contains practically all the life forms associated with the open ocean and is the home of the primary producers that support all the other living things in this region—microscopic organisms called **phytoplankton**. What phytoplankton lack in size they make up for in numbers, for collectively, these tiny green specks account for about 40% of all the photosynthetic productivity that takes place on the entire planet.

- *Mesopelagic zone*, extending downward from 200 m down to around 1000 m (3300 feet). This region of perpetual twilight is home only to a few animals such as luminescent shrimps, squids, and some fishes.

- *Bathypelagic zone*, from 1000 m down to around 4000 m (13,200 feet). a region of total and never-ending blackness characterized by near-freezing water temperatures, low oxygen, and crushing water pressure. The few animals that live here are delicate, colorless, and usually blind. Most survive by consuming the snow of **detritus** drifting down from the zones above, but some do prey on others. The legendary giant squid live at this depth where they are occasionally visited by deep-diving sperm whales that hunt them.

- *Abyssopelagic zone*, from 4,000m to just above the seafloor. From the Greek *abyss*, meaning bottomless, a holdover from the times when the deep ocean was thought to have no bottom. The few animals that survive here are delicate, colorless, and often blind.

- *Hadalpelagic zone* is the designation for ocean trenches, the deepest part of the marine environment. This zone extends from a depth of around 6,000 meters (20,000 ft) to the very bottom of a trench.

- *Benthic zone* is the actual seafloor and as with the land, it can be divided into smaller specific ecological zones. At ocean's edge we find the intertidal zone of rocky and soft bottom beaches, mudflats, and sand flats. Moving on out and deeper we encounter either the rocky subtidal zone that is home to kelp forests and coral reefs or the soft bottom subtidal zone where we find habitats such as estuaries, salt marshes, mangrove swamps, and sea grass beds. Starting at the base of the continental slope and rolling on for thousands of miles we find the flat **abyssal plains** of the deep sea floor. The deep sea floor itself is a thick blanket of mud and decaying matter (**ooze**) that have settled down from above and accumulated over millions of years. The crushing water pressure, near-freezing temperatures, velvet blackness, and seeming lack of food led to the belief that nothing could live on the deep seafloor. However, as technology has improved, and biologists have been able to better survey this region, it appears that the number of species living in this alien domain is quite high.

Freshwater Ecosystems. The lakes, ponds, rivers, streams, and wetland swamps and marshes that comprise the freshwater biome are scattered about and occupy a far smaller portion of the planet than either terrestrial biomes or the marine biome. Lakes and ponds cover about 2% of the earth's surface whereas streams and rivers cover about 0.3%.

Large standing bodies of freshwater are called lakes whereas smaller ones are referred to as ponds. Like the ocean, lakes and ponds have three zones (**Figure 3.11**). The **littoral zone** is the shallow edge around the shore where water meets land. The **photic** or **limnetic zone** is the open water from shore to shore and down as far as light penetrates. The **aphotic** or **profundal zone** is area below the limits of light penetration.

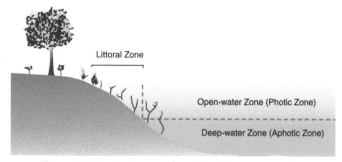

Figure 3.11 Biotic zones found in lakes and ponds.

Unlike the ocean, large temperate lakes undergo seasonal mixing of their zones in a process known as **overturn**. In the late fall, the upper layers become colder and thus denser than the warmer lower layers. This causes the surface waters to sink resulting in a displacement of the deeper layers—fall overturn. In spring, ice that formed on the lake over the winter melts and this cold water sinks below the warmer water in the lower zones again resulting in a rise of the deeper layers—spring overturn. Overturn occurs in temperate lakes but usually not in tropical ones.

In terms of productivity, fast-moving streams are the least productive of the various freshwater ecosystems. The moving water washes away the plankton and the only photosynthesis that supports the system is limited to attached algae and rooted plants along the stream margins.

Lakes can be divided into two categories based on productivity: **eutrophic lakes** and **oligotrophic lakes**. Eutrophic lakes contain an abundant supply of organic matter and minerals and support a rich variety of aquatic life. In oligotrophic lakes, organic matter and mineral nutrients are scarce. Such lakes are often deeper than eutrophic lakes and due to a lack of biological activity, they tend to have very clear, blue water.

Swamps, marshes, bogs, and other wetlands are among the most productive ecosystems on the planet and may be thought of as the freshwater equivalent to terrestrial rain forests and marine coral reefs in that regard. Most lakes are far less productive being limited by a lack of nutrients. Wetlands also play a key role ecologically by providing water storage basins that moderate flooding. Wetlands, however, suffer from the mistaken perception that these areas are useless and unproductive land and as such, they have been and continue to be seriously disrupted and eliminated by human "development."

Planet in Peril

Ecologists define an ecological crisis as a situation in which a species' adaptive capabilities can no longer keep pace with changes in its environment resulting in extinction of that species. Ecological crises may be local or global and last only a few months to millions of years. From the fossil record paleontologists know that world-wide ecological crises resulting in mass extinction has occurred at least five and as many as nine times in the history of the planet. The worst of these mass extinction events seems to have been the "Great Dying" that occurred around 251 million years ago as the Permian Period transitioned into the Triassic Period. During this time, an estimated 70% of all land species and 96% of all marine species became extinct.

A number of hypotheses have been advanced to explain these mass extinction events:

- Massive and sustained volcanism
- Sea-level rise and fall
- Asteroid and/or comet impacts
- Sustained global warming or cooling
- Oceanic overturn
- Nearby nova, supernova or gamma ray bursts
- Continental drift

A survey by the American Museum of Natural History found that 70% of biologists view the present geologic era as part of a human-induced mass extinction event, possibly one of the fastest ever. This present-day ecological crisis has been dubbed the *Holocene extinction event*. Harvard University biologist E.O. Wilson predicts that human destruction of the biosphere could cause the extinction of one-half of all species within the next 100 years.

Since the beginning, each generation has fought nature. Now, in the life-span of a single generation, we must turn around 180 degrees and become the protector of nature.

—Jacques-Yves Cousteau

What environmental challenges do this planet and its inhabitants face?

Pollution

Many chemicals released into the air, water, and soil by agriculture, industry, and domestic use are toxic to varying degrees and often persistent. These chemicals threaten not only the health of the biosphere but human health as well.

Unfortunately, as some of these toxic chemicals pass through the food chain they become increasingly concentrated in a process called **biological magnification**. The very symbol of our nation, the bald eagle, was nearly exterminated in the late 1960s when it was discovered that the insecticide DDT was magnified in many bird species, especially predator birds. The DDT caused the production of thin, fragile eggshells resulting in the eggs being easily broken. Although DDT was banned from production and sale (at least in this country) in time to save the birds from extinction, it and many other toxic chemicals no longer manufactured still persistently circulate in the ecosystem.

Ozone Hole

The "ozone hole" first appeared over Antarctica in 1975 and each year since, the layer of ozone is thinner and the hole larger. **Ozone** (O_3) is a form of atmospheric oxygen that blocks deadly ultraviolet radiation from the sun. The major factor in ozone depletion is the release of chlorofluorocarbons (CFCs), chemicals used in cooling systems, fire extinguishers, and Styrofoam containers. The CFCs percolate up through the atmosphere and reduce the O_3 form of oxygen to O_2 rendering it ineffective at blocking UV radiation.

Ultraviolet radiation is a serious human health concern. Every 1% drop in atmospheric ozone is estimated to lead to a 2% increase of UV light making it through to the surface which in turn could result in a 6% increase in the incidence of skin cancers. Furthermore, UV radiation has the potential to inflict radiation damage on many animals and could cause disruption of oceanic food chains given the sensitivity of plankton to such radiation.

Global Warming

Glaciologists tell us that by definition, the planet is still technically in an ice age due to the continued presence of large ice sheets both in the Arctic and Antarctic regions. Imbedded within this ice age have been dramatic shifts in the global climate system from warm **interglacial periods** to cold **glacial periods** and back again with the last cold glacial period ending about 12,500 years ago. Modern history and civilization as we know it have developed during the present warm interglacial period.

These interglacial-glacial-interglacial climate oscillations seem to have been occurring for the past 900,000 to 1,000,000 years as part of the current ice age. Clearly the earth has been warming and cooling as a part of a very long-term cycle. As debate rages and disagreement swirls around us regarding global warming, one must not lose sight of the fact that the global warming issue is about the answers to two basic questions: (1) Have human activities accelerated the natural warming of the present interglacial period? And if so, (2) what might be both the short-term and long-term consequences of such human-accelerated global warming?

Why are many scientists and governments so concerned? What is the compelling evidence that human activities are warming the planet?

- The temperature of the troposphere—the thick layer of atmosphere hugging the planet—and land surface are warming, whereas the stratosphere—the layer above the troposphere—is cooling. This has been interpreted as a sign that greenhouse gases are trapping energy and keeping that energy close to the surface of the earth. The Intergovernmental Panel on Climate Change predicts global temperatures are most likely to rise between 3.2 and 7.1 degrees Fahrenheit (1.8 and 4 degrees Celsius) by the year 2100. However, global temperatures during the **Paleocene-Eocene Thermal Maximum** (PETM) some 55-65 million years ago were much warmer than those presently and those predicted for the next century. At the height of the PETM, sea surface temperatures in the oceans rose 9 degrees F (5 degrees C) in the tropics and 11 degrees F (6 degrees C) in the Arctic. The oceans became more acidic, and 30 to 50 percent of the sea floor life went extinct.

- Although the temperature of the planet has fluctuated in the past, it got warmer or cooler over millions of years. What seems apparent is that we are seeing rates of increase in warming that exceed past natural rates by a factor of 100. Thus, humans may be doing in centuries what natural processes do over millions of years. But this begs the question—How do you accurately take the temperature of the entire planet both from the past and during the present?

- Both poles are getting warmer; in Greenland and Antarctica the surface of the ice is dropping, and there is less mass when the ice is measured from space. We know that the ice sheets have come and gone in the past. Why is this any different? The worrisome reason is that in the past, the ice sheets from the two poles didn't move together—one would lead and the other would follow. Presently, both the north and south icecaps are spewing ice into the global ocean and seem to be in an accelerating decline at the same time.

- Sea levels are on the rise. Have sea levels ever been as high as they are now? Yes, roughly 35,000 years ago. A few thousand years later, sea levels began to drop and by about 15,000 years ago, the oceans had declined by almost 500 feet. Then they started to rise and sea levels have slowly been going up ever since.

- Thanks to bubbles of ancient air held in deep ice cores, we know there has been a disturbing increase since the mid-20th century of the so-called **greenhouse gases**, predominantly carbon dioxide but also methane, nitrous oxide, and hydrofluorocarbons in the atmosphere. These gases are easily measured and the scientific community seems in agreement that not only are these gases increasing in concentration (being up around 35% since the beginning of the Industrial Revolution) but that they seem to be increasing at an alarming pace. Aside from the physiological processes of animals, greenhouse gases are released in the greatest amount by the burning of fossil fuels and clearing the land for agriculture. Increases in these gases are thought to lead to the warming of the surface of the planet as well as the lower atmosphere by increasing the **greenhouse effect**.

- Ocean chemistry is changing. When carbon dioxide dissolves in seawater, it forms carbonic acid, and in high enough concentrations carbonic acid is corrosive to the shells and skeletons of many marine organisms. If our current trends in burning fossil fuels and generating carbon dioxide continue, the ocean will become more acidic than at any time in the past 65 million years back to the PETM. At that time, high ocean acidity levels are thought to have been one of the contributing factors to a mass extinction event.

The probability and extent of **anthropogenic warming** (human-caused warming) and the future consequences of such planetary warming have become a very controversial and contentious issue worldwide. Multitudes of climatological investigations and observations both ground-based and from satellites have been and continue to be conducted. Data and observations from such studies are then fed into various computer climate models to search for trends and patterns and to predict future outcomes. A great many in the media and popular press have chosen to interpret these data, observations, and climate model results as supporting and verifying an impending global climate catastrophe. In what critics have dubbed the "Chicken Little Approach," these sources then bombard the public almost daily with visual and commentary doom-and-gloom linking possible anthropogenic global warming to melting glaciers, rising sea-levels, drought, wildfires, loss of habitat and diversity, seemingly wild swings in weather patterns, and a host of other perceived planetary ills.

Not only has this issue been popularized, but it has also been politicized. Government leaders of most industrialized countries and the United Nations have chosen to follow the lead of the popular press and conclude humankind faces an inevitable and catastrophic global climate shift. This interpretation of the many climatological studies and computer climate models serves as the driving force for proposed changes in governmental policies, procedures, and laws worldwide. But are these near-hysterical media prophecies and planned profound changes in governmental policies worldwide based on sound science? Critics contend that not only is the popular press behaving irresponsibly but that governments are headed down a path that will result in hundreds of billions to trillions of dollars being spent worldwide in planetary temperature abatement schemes with entire industries possibly laid waste, and even personal freedoms curtailed for no sound scientific reason.

The whole issue of anthropogenic warming and the possible catastrophic climate changes associated with it seem to have become what some have labeled as a "secular religion." And to disagree with its rigid interpretation that a global climate catastrophe is inevitable unless humanity changes its ways, is to be labeled a heretic. However, contrary to what you may have been led to believe, the scientific community worldwide is not unanimously in consensus on the subject of whether anthropogenic warming is actually occurring. In fact, a recent Senate Minority Report reveals that over 700 dissenting scientists from around the globe challenged human-caused global warming claims made by the United Nations Intergovernmental Panel on Climate Change (IPCC). In 2007, the number of dissenting scientists was listed at around 400 worldwide.

Can humankind afford the luxury of waiting for definitive answers to our global warming questions? Dare we wait? But even if through draconian measures we were to roll global emissions of greenhouse gases back to year-2000 levels, would it make any difference? System science such as is being applied to the global warming issue is very complicated because there are multiple possible outcomes. Scientific assessment of the situation requires understanding the relative likelihood of each of these outcomes. From that a value judgment about whether or not the risks are high enough to demand action. (Risk is what can happen multiplied by the probability of it actually happening.) Once science makes the risk levels clear, the risk management will fall to the will of the public and the politicians. As individuals and as a species we should carefully ponder and heed the words of Ken Caldeira when he states:

> *We are at a critical juncture in earth history. If we don't do the right thing and there are geologists around 50 million years from now, they'll be able to look at cores and see the remnants of a civilization that developed advanced technology but didn't develop the wisdom to use it wisely.*

Habitat Destruction

Habitat destruction results when one habitat type is removed and replaced with another habitat type. In the process, plants and animals which previously used the site are displaced or destroyed resulting in an alteration or reduction in the **biodiversity** (the number of different species of plants and animals) of the area involved. Fires, floods, and volcanoes are natural forces that may cause destruction of habitat but the most devastating and widespread demolition of ecological systems is perpetrated by human activities such as land clearing, development and agriculture. The most biologically productive areas are often the hardest hit. Consider that just a few thousand years ago rain forests flourished and formed a wide green belt around the equator. These magnificent verdant gems covered 14 percent of the planet's land surface, or around 2.1 billion hectares (8.2 million square miles). Humans have already destroyed more than half of this forest area, with most damage occurring in the last 100 years. With just 647 million hectares (2.5 million square miles) remaining, we continue, in our ignorance and apathy, to eradicate an estimated 150,000 square km (93,000 square miles) a year. As a result, experts predict that within only a few decades, there will be little undisturbed tropical forest left anywhere in the world.

The North America prairies once blanketed about 363 million hectares (1.4 million square miles) across the center of the continent. World-wide the rich fertile soil that characterizes grasslands has been their undoing, especially in North America. Now, instead of an ocean of grass stretching from horizon to horizon, cultivation or herds of grazing cattle reach as far as the eye can see and the wildflowers that once dotted the prairies have been replaced by towns, cities, and roads. The prairie that was exists for the most part in only small isolated patches accounting for only 1 to 2% of its former greatness. In its place humans have imposed an artificial ecosystem—the **agroecosystem**. Sadly, global wetlands and coral reefs are meeting the same fate and face the same dismal prospects for long-term sustainability.

A Closing Note
For the Frogs, the Icecaps, and the Rain Forests

With its bright red eyes and toes, green body, and bluish limbs, the red-eyed tree frog is my favorite animal. Truth be told, I have a soft spot in my heart for all things amphibian. That is why continuing global amphibian studies are increasingly distressing to me. At least 32 amphibian species have gone extinct in recent years, and at least another 26 are "missing," not having been seen for many years. The island nation of Sri Lanka has lost as many as 100 species in less than a decade, news that is not all that surprising given that 95% of that nation's rain forests have also disappeared in recent times.

Unfortunately, what is happening to amphibians is but one sad and distressing example of what is happening in general to our planet and the creatures that inhabit it. We all get a nearly daily dose of environmental doom-and-gloom from the media, and I don't mean to add to that,.but what clear-thinking and reasonable person can continue to be bombarded by environmental horror stories without becoming at least mildly concerned about the global environment? However, concern is one thing but acting on those concerns is quite another and many, if not most, people wring their hands in worry with good intentions in their heart but in the end continue to sit on the sidelines and do nothing. Why? Because people are frozen into inactivity in the face of the enormity and complexity of the global environmental problems we and all other life forms face. As individuals, these problems seem too much to bear, let alone solve, so we are content to blame industry or agriculture and wait for the government to do something.

This attitude of inactivity and apathy must change, and there are hopeful signs that more and more people world-wide are beginning to understand that. These are not top down problems, and the solutions to these problems will not shower down on us from above. Rather, meaningful solutions and positive environmental actions will come from people, from the bottom up. What's that you say? You think anything you could possibly do on your own would only amount to a drop in the bucket? It is understandable that you feel that way, but consider this—what is the ocean but a bunch of individual drops put together? There *are* things you can practically and realistically do to make an environmental difference. Research it. Find out what those things are and do at least one or some of them with confidence and pride. Plan your work and then work your plan and let this be your credo (with apologies to the original author who is unknown to me):

> *I am only one, but I am one.*
> *I cannot do everything, but I can do something.*
> *That which I can do, I will do.*

You have the power to be *One*, we all do. Will you accept the challenge and convince others to do the same? I hope so, not only for the sake of the red-eyed tree frogs, the polar bears, the whales, and the elephants but for all our sakes.

In Summary

- Planet Earth is perfectly constructed with a dynamic physical structure capable of supporting life. Furthermore, it is perfectly positioned from the sun to benefit from the sun's radiation without being destroyed by it.
- At the ecosystem level, ecologists study the interactions between living things and the physical environment in an attempt to understand how biogeochemicals and energy circulate through the system.

- Ecosystems come in several sizes: Macroecosystems are quite large (e.g. the North American deciduous forest), mesoecosystems are medium to small (e.g. a meadow clearing in a forest), and microecosystems are small to tiny (e.g. life under a rotting log in a forest).

- The most important biogeochemical cycles are: the water cycle, the carbon cycle, the nitrogen cycle, the oxygen cycle, and the phosphorous cycle.

- Earth is a closed system when it comes to biogeochemicals but an open system when it comes to energy and energy flow.

- Organisms from different trophic levels feed on each other in sequence forming a food chain (what does eat what). Overlapping and interlocking food chains form food webs (what could eat what).

- Photosynthetic organisms (autotrophs or producers) are the first trophic level. The producers are fed on by the primary consumers (herbivores) which in turn are fed upon by the secondary consumers (carnivores). Omnivores eat both plant material and flesh whereas saprotrophs (scavengers and decomposers) eat dead organisms or the waste given off by living organisms.

- Biological systems are not very efficient. Only about 1% of the solar energy that strikes photosynthetic creatures is converted into food molecules. Much energy is also lost as heat as it flows from one trophic level to the next.

- Ecologists employ a number of different ecological pyramids to graphically represent the relationship between energy and trophic levels of a given ecosystem—biomass pyramids, numbers pyramids, and energy pyramids.

- Biomes are the largest unit of the ecological hierarchy. Biomes are very large areas consisting of many interlocking and overlapping ecosystems. There are two main categories of biomes: terrestrial and aquatic.

- Many species on this planet are in peril of perishing in an ecological crisis that has been dubbed the Holocene extinction event. The factors driving this crisis are directly or indirectly tied to the global activities of humans.

Review and Reflect

1. *What's in a Quote?* Provide the ecological details that would explain the Taos Pueblo Indian poem that opens this chapter.

2. *Spaceship Earth.* Ecologically speaking, our planet has often been described as "Spaceship Earth." Is this indeed the case? Compare and contrast the Earth to the International Space Station.

3. *Scourge of the Ragu.* The evil Ragu from the planet Remlac wish to destroy all animal life on earth. They do not wish to attack in a direct and costly frontal assault militarily. Rather, they wish to destroy slowly, quietly, and undetected.

    Imagine you are Beldar from Remlac. You have been sent by the Ruling Council of High Ragu to destroy all animal life on earth by disrupting the biogeochemical cycles that regulate and replenish life on this planet. As you sit in your hidden base on the far side of the moon, you possess the technologies to do whatever you wish to any biogeochemical cycle. What cycles would you attack, why would you attack those particular cycles, and how would you attempt to disrupt those particular cycles?

4. ***The Dance of Death***. The graph shown in **Figure 3.12** illustrates a hypothetical relationship between a predator, such as a fox, and its prey, such as rodents. Analyze the graph and answer the questions that follow:

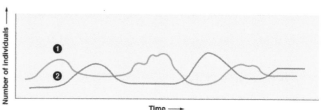

Figure 3.12 A hypothetical predator-prey growth graph.

A.   Which line represents the predator and which line represents the prey? Defend your answer.

B.   These hypothetical populations are not in sync but neither are such interacting populations in the real world. Why not?

C.   If the prey population were to decrease to zero, what would happen to the predator population and why?

D.   If the predator population were to decrease to zero, what would happen to the prey population and why?

E.   Which population regulates the other? Explain.

5. ***Webs and Pyramids***. Select any ecosystem—terrestrial, marine, or freshwater—of your choice. Prepare a small food web of organisms that inhabit the ecosystem you selected. From your food web then construct both an ecological pyramid of numbers and an ecological pyramid of energy.

Create and Connect

1. ***Alien Realms***. In regions of the deep sea, several strange almost alien ecosystems have been discovered—**hydrothermal vents** and **cold seeps** (sometimes called **cold vents**). Although the physical conditions in and around hydrothermal vents and cold seeps is very different, they are similar in that they are the only ecosystems known that do not depend on photosynthesis for food and energy production.

Write a short report in which you compare and contrast the physical characteristics of hydrothermal vents and cold seeps and the unique living communities that have developed around each.

Guidelines:

A.   Format your report in the following manner:

  ➢   *Title page* (including your name and lab section)
  ➢   *Body of the Report* (include pictures, charts, tables, etc. here as appropriate). The body of the report should be a minimum of two pages long—double-spaced, 1 inch margins all around with 12 pt font.

> *Literature Cited* A minimum of two references required. Only <u>one</u> reference may be from an online site. The *Literature Cited* page should be a separate page from the body of the report and it should be the last page of the report. Do NOT use your textbook as a reference.

B. The instructor may provide additional details and further instructions.

2. ***A Union Dispute.*** What if species formed unions? Imagine that both the union representing predatory species and the union representing prey species are demanding a very large increase in their respective populations. Other groups, such as parasites, are also demanding a seat at the negotiating table. You have been assigned to mediate this dispute. What would your decision be?

3. ***A Local Problem.*** Environmental problems and concerns are not far removed from your personal life as you might think or wish them to be. Regardless of how enormous global environmental problems might seem, communities and individuals can and are making a difference. The key is to "Think global but act local." In this investigation, you will evaluate environmental problems in and around your local community and determine the efforts being made by the community to deal with those problems.

Guidelines:

A. Select 1 or more *local* environmental problems you wish to investigate.

B. Prepare a chart listing specific examples of the problem(s) you have chosen to investigate.

C. Collect as much information as possible about each problem and the measures, if any, which have been or will be taken to address those specific problems. As part of this process, you should contact community government officials.

D. Use the Problems Points chart (**Table 3.2**) to rate each environmental problem you investigate.

Table 3.2 *Problem Points*	
Problem Points	**Guidelines**
0	No problem locally
1	Few mild problems
2	Some problems that could become significant
3	Moderate problems with the potential to become hazardous
4	Severe problems, becoming hazardous
5	Critical problems, hazards existing

E. Use the Solution Points chart (**Table 3.3**) to rate your community's past and present efforts to deal with each problem you investigate.

Table 3.3 Solution Points	
Solution Points	**Guidelines**
0	No pollution awareness or antipollution actions
1	Some community awareness; no community action or legislation
2	Definite awareness, citizens groups; some movement toward legislation
3	Strong awareness; much discussion and news coverage; legislation in process
4	Strong public and private community awareness; legislation in place
5	Community organized to deal with and prevent pollution; legislation enforced

F.   Once you have prepared your charts use them to answer the following questions:

a.   What environmental issues does your community deal with effectively and what issues are they neglecting or not dealing with effectively?

b.   Solution of community environmental problems depends on a triumvirate of community awareness, political action, and individual responsibility. Explain and give an example of each of these pieces to the solution of environmental problems.

c.   What suggestions might you offer to reduce or eliminate the environmental problem(s) you investigated?

4.   ***Small but Powerful***. It has been said, "We don't inherit the earth from our ancestors; we borrow it from our children." This solemn statement begs the question—what can each of us do personally to preserve the purity of this planet and the sanctity of the living things that inhabit it? (See *Closing Note* at the end of this chapter) Identify an environmental issue(s) around your home or on campus and then devise a practical and workable plan to help reduce or eliminate the problem(s). Plan your work then work your plan. HINT: Keep it simple. The simpler your plan, the more likely you are to actually implement it and the more effective your plan will be if and when you do implement it. Remember...You *can* make a difference!

4

ANIMAL ARCHITECTURE: FORM AND FUNCTION

The most wonderful mystery of life may well be the means by which it created so much diversity from so little physical matter.

—E. O. Wilson

Introduction

Armed with only a few basic types of living building blocks known as **cells**, nature has managed to fashion millions to perhaps tens of millions of different types of creatures—plants, animals, and microbes—that occupy every nook and cranny of this place. As a result, we are blessed to live on a planet bubbling and swarming with a riotous multitude of life forms. The mere contemplation of the beauty, mystery, and wonder of it all cannot help but invoke awe and reverence.

Properties of Life

Throughout time, humankind has raised and pondered questions of life and death. As our scientific wit has grown sharper, our life questions have become more focused and defined: Biologically speaking, exactly what is "life"? When, where, and how did this life force appear on this planet? When does an individual organism become living? How does life at the individual cell level translate into life at the organismal level? Conversely, what is "death" biologically speaking? When an organism dies what happens to its life force? All legitimate questions that impact each and every one of us in one way or another on any number of different levels—biologically, medically, socially, philosophically, and religiously. Unfortunately (or fortunately depending on your faith in humankind to properly handle such monumentally important knowledge), at the present time we have no definitive answers for any of these questions. Will biologists ever be able to actually create a living cell(s) from the requisite inorganic components? Is it possible to create an artificial life form? As unbelievably science-fictionish or even biblically profound as that prospect sounds, we may be inching ever closer to doing exactly that. The question at this point seems not to be "Will we ever create life from scratch?" but simply how much creating the life would cost. There is the belief among some that the number of steps that might be real potential roadblocks to such an undertaking has declined almost to zero. In fact, some researchers optimistically suggest that for less than a third of the approximately $500 million dollars spent to sequence (map) the human genome, it is conceivable that organic chemicals could be transformed into a single-celled organism that would grow, divide, and evolve and that such an amazing and profound feat could be accomplished in 3 to 5 years time.

The essence of life is a statistical improbability on a colossal scale.

—Richard Dawkins

Although most questions dealing with the biological nature of life still cannot yet be answered, we do know that all living organisms, regardless of their complexity (or the lack thereof), share a set of basic characteristics or properties. These commonly-held life properties should be regarded as the characteristics or signs of life rather than the definition of life.

1. *Biogenesis.* Any organism originates from an organism (asexual) or organisms (sexual) of the same species. Living things do not spontaneously spring into life.
2. *Organization.* All organisms are composed of one or more cells and these cells are constructed of the same basic types of atoms and molecules.
3. *Sensitivity.* All organisms respond to stimuli.
4. *Metabolism.* All organisms require energy (food) and produce waste products. Food energy is used to maintain internal order and grow.
5. *Homeostasis.* All organisms maintain stable internal conditions that are different than the surrounding environment.
6. *Movement.* All organisms are capable of self-generated movement at some scale and some stage of their life cycle.
7. *Reproduction.* All organisms are capable of reproduction.
8. *Life Stages.* All organisms grow and develop through definite stages—beginning, growth, maturity, decline, and death.

Hierarchy of Life

At first glance, one might not perceive much, if any, order to the chaos of life that inhabits this planet. However, through the work of many biologists in different disciplines over long periods of time, we now realize that the biological world is indeed organized and that this organization is hierarchal. That is, each level builds on the level below it. From the atoms and molecules that build cells to the planet-sized ecological systems that sustain whole species, there is an ordered regularity to this pandemonium we call life (**Figure 4.1**).

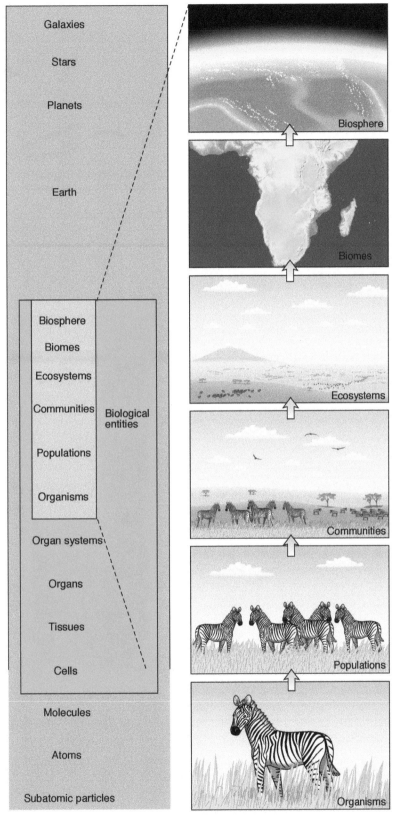

Figure 4.1 The Hierarchy of Life. A method by which order and organization may be imposed on chaos and confusion. Size and complexity increase from the bottom to the top of the figure—subatomic particles to galaxies.

Atomic and Molecular Levels

Ninety-two naturally-occurring **elements** (types of **atoms**) form the physical structure of this planet and its atmosphere. However, the main bulk of any living thing is composed primarily of carbon atoms, some hydrogen, oxygen, and nitrogen atoms, and trace amounts of a few other elements.

Atoms **bond** (join) together into clusters known as **molecules.** The molecules that constitute the cells of living things are large and complex assemblages called **macromolecules.** There are four basic types of biological macromolecules: carbohydrates, lipids, proteins, and nucleic acids.

Almost all aspects of life are engineered at the molecular level, and without understanding molecules we can only have a very sketchy understanding of life itself.

—Francis Crick

Cellular Level

Any animal or animal-like organism is vested with life at two levels—the cellular level and the organismal level. Each individual cell in an animal is alive but somehow the collective lives of each of those cells are transposed to a higher plane, and as a consequence, the entire organism lives. This amazing extrapolation of the life force results in any animal being far more than just the sum of its parts.

Organismal Level

Cells are grouped into tissues, such as muscle cells, which are grouped into organs, such as a heart, blood vessels, and blood, and organs are coordinated into organ systems to form a complex multicellular organism. A total of eleven different organ systems are found in **metazoans** (complex multicellular organisms): integumentary, muscular, skeletal, respiratory, digestive, circulatory, excretory, nervous, endocrine, immune, and reproductive. Only with the single-celled protozoans does the totality of life and death exist solely at the cellular level for animals or animal-like organisms.

Ecological Level

Individual organisms are intertwined into several hierarchical levels—*population, community, ecosystem,* and *biome*—within the larger physical world of this planet:

A **population** is a group of organisms of the same species living in the same area at the same time.

A **biological community** consists of all the populations of different species living together in the same place at the same time.

An **ecosystem** would be a complex of communities and their interactions with the physical environment—air, water, and soil—in a particular area.

A **biome** is the largest geographical biotic unit and consists of various ecosystems and their attendant communities with similar requirements of environmental conditions.

Examining a grasshopper gives us an example of the entire hierarchy in action. Atoms mainly of hydrogen, oxygen, carbon, and nitrogen are arranged into biological macromolecules of carbohydrates, proteins, fats, and nucleic acids which in turn are organized into living animal cells. These animal cells of various types are arranged into tissues, organs, and organ systems ultimately resulting in the complex multicellular organism we know as a grasshopper. A *population* of grasshoppers lives in a ravine *community* that is rich with many other species of animals and plants. This particular ravine is located within the mixed-grass prairie *ecosystem*. Finally, the three prairie ecosystems—short, mixed-grass, and tall—on this continent meld into the gigantic grassland *biome* of North America.

Zoological research is often very specific and quite precise. A working knowledge of and appreciation for this hierarchy allows zoologists to step back and better understand the implications and importance of their research within the bigger or smaller scheme of things.

Chemistry of Life

Cell theory tells us that all living things are composed of one or more cells and the products of those cells. If living things are composed of cells, then what is the chemical composition of cells? What is the chemistry of life?

The chemical compounds that compose cells are of two types: organic and inorganic. **Organic compounds** always contain carbon and hydrogen (as well as some other elements) bound (bonded) together whereas **inorganic compounds** never contain carbon and hydrogen bound together. The three major inorganic compounds found within living cells are water, oxygen, and carbon dioxide.

The major types of organic compounds associated with the composition of cells are carbohydrates, lipids, proteins, and nucleic acids.

Carbohydrates are compounds of carbon, hydrogen, and oxygen grouped as H—C—OH and occurring in the ratio of 1 C: 2 H: 1 O.

Through the wondrous process of **photosynthesis** plants and some photosynthetic bacteria and algae assemble the carbon, hydrogen, and oxygen molecules of water (H_2O) and carbon dioxide (CO_2) into simple sugars such as glucose, galactose, and fructose known as **monosaccharide**). Of these, **glucose** is the most biologically significant (**Figure 4.2**). Plants, animals, and most bacteria depend on glucose as their immediate source of biological energy. Because nearly all life forms on the planet depend on it to provide the chemical energy that drives

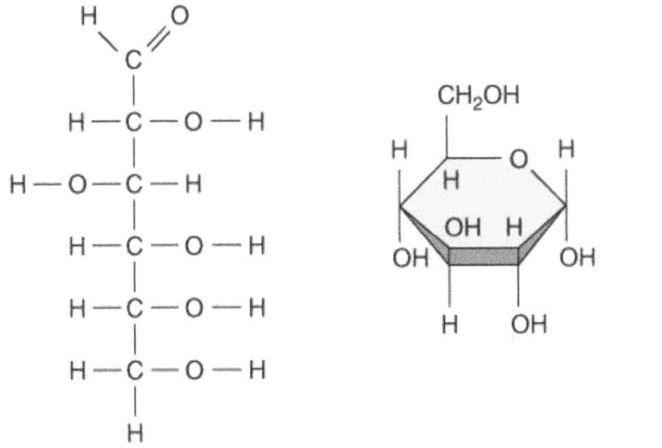

(a) Chain structure of glucose   **(b)** Ring structure of glucose

Figure 4.2 The Structure of Glucose (a) Chain structure (b) Ring Structure. The two forms are in dynamic equilibrium, but because the ring form is more stable in water, this form predominates.

their bodies, glucose is the single most important biological molecule on the planet. Plants produce the glucose initially through photosynthesis whereas animals satisfy their carbohydrate requirements by eating the plants or by eating the animals that ate the plants and digested out their glucose.

Glucose may also be chemically modified in such a way to make it suitable for long term storage and as a structural component for plants. To do this plants bond many glucose molecules together into large complex sugars known as **polysaccharides**. The most important of these polysaccharides are starch and cellulose. **Starch** molecules are more chemically stable (not soluble in water) than glucose and thus provide a long-term storage solution for excess glucose. With a different twist to their conFigureuration, starch molecules become **cellulose**, a polysaccharide that is so chemically tough (indigestible by most animals' enzyme systems) and stable (not soluble in water) that plants use it structurally to form their very bodies. The wood that forms the stem of a 300-foot tall redwood tree is nothing more than untold cellulose molecules bonded together. Due to the large biomass of photosynthetic organisms, cellulose exits in greater quantities than all other biological molecules combined. In animals, excess glucose is stored short-term in the liver and muscle cells as **glycogen** but stored long-term as **fat**.

Lipids are animal fats, plant oils such as olive oil and corn oil, and waxes such as beeswax and earwax produced by plants and animals. Fats are the major stored fuels of animals. These compounds can be oxidized back into glucose and released into the bloodstream as needed to meet cellular demand.

A fat molecule (or **triglyceride**) is a combination of three molecules of fatty acids with three molecules of glycerol (**Figure 4.3**). The three fatty acids of a triglyceride need not be identical, and often they differ markedly from one another. Animals store the energy of glucose mainly in the many C—H bonds of the triglyceride.

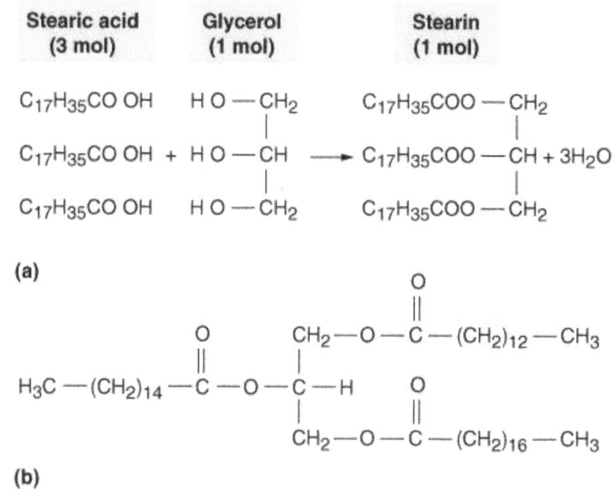

Triglycerides lack a polar end and thus are not soluble in water. Because of their insolubility, fats can be deposited at specific locations within an organism.

Stearic acid (3 mol)	Glycerol (1 mol)	Stearin (1 mol)
$C_{17}H_{35}CO\,OH$	$HO-CH_2$	$C_{17}H_{35}COO-CH_2$
$C_{17}H_{35}CO\,OH$ +	$HO-CH$ $\longrightarrow$	$C_{17}H_{35}COO-CH$ + $3H_2O$
$C_{17}H_{35}CO\,OH$	$HO-CH_2$	$C_{17}H_{35}COO-CH_2$

(a)

(b)

Figure 4.3 Neutral Fats (a) Formation of a neutral fat from three molecules of stearic acid (fatty acid) and a glycerol. (b) A fully formed neutral fat bearing three different kinds of fatty acids.

Fats are said to be saturated or unsaturated. **Saturated fats** have a structure in which all the carbon atoms in the fatty acid chain each bond to two hydrogen atoms whereas **unsaturated fats** have double bonds between one or more pairs of successive atoms on the fatty acid chain. If a given fatty acid has more than one double bond, it is said to be **polyunsaturated**. A polyunsaturated fat such as corn oil is usually liquid at room temperature and is called an oil, but saturated fats such as those in butter are solid at room temperature.

Phospholipids are fat derivatives in which one fatty acid has been replaced by a phosphate group. Because of their unique conFigureuration—one part of a phospholipid molecule is charged and polar but the remainder of the molecule is nonpolar—phospholipids can bridge two different chemical environments and bind water-soluble molecules, such as proteins, to water-insoluble materials making them important structural components of cell membranes.

Organisms contain many other kinds of lipids besides fats and phospholipids. **Terpenes** are the components of many biologically important pigments, such as chlorophyll and the visual pigment retinal. **Steroids**, such as cholesterol, are found in animal cell membranes whereas others, such as testosterone and estrogen, function in multicellular organisms as hormones. **Prostaglandins** are a group of about 20 lipids that act as local chemical messengers in many vertebrate tissues.

Proteins are very complex assemblages composed of smaller molecules known as **amino acids.** There are 20 kinds of amino acids that are joined together by **peptide bonds** gener-

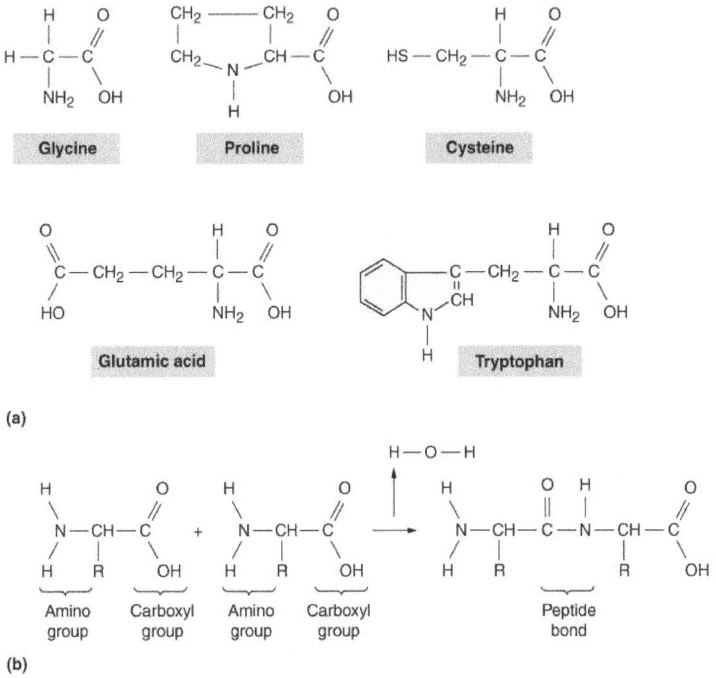

Figure 4.4 Amino Acids. (a) The structure of 5 of 20 amino acids. (b) The formation of a protein molecule through dehydration synthesis. R denotes the point of attachment for 1 of the 20 kinds of amino acids.

ated through **dehydration synthesis** (removal of a water molecule) to form proteins (**Figure 4.4**).

The 20 kinds of amino acids might be thought of as letters of a chemical alphabet. By varying the type and sequence of these "letters," a surprising number of different types of proteins can be constructed. Structurally, proteins are more than just a strand of amino acids. Biochemists recognized four levels of protein organization: primary, secondary, tertiary, and quaternary structures (**Figure 4.5**). It the type and sequence of amino acids in a protein that determines its shape and it is the shape and folding that determines the function of a protein. Proteins perform a wide variety of functions in animals (**Table 4.1**).

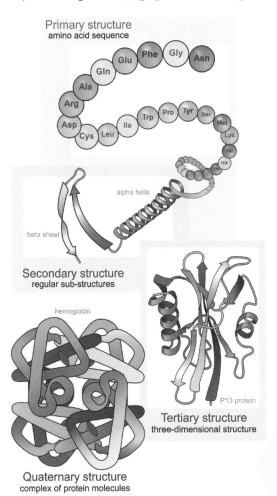

Figure 4.5 Different Levels of Protein Structure. Hydrogen bonding between amino acids in the primary structure results in folding and coiling that produce a secondary structure. Bends and helices in the secondary structure form the tertiary structure. The individual polypeptide chains of some proteins clump in a very convoluted and folded functional unit known as the quaternary structure.

Table 4.1		
Functions of Proteins in the Animal Body		
Type of Protein	**Example**	**Function**
Enzymes	Amylase	Promotes the breakdown of starch into the simple sugar glucose
Structural proteins	Keratin, collagen	Hair, wool, nails, horns, hoofs, tendon, cartilage
Hormones	Insulin, glucagon	Regulates the use of blood sugar
Contractile proteins	Actin, myosin	Contracts muscle fibers for movement
Storage proteins	Ferritin	Stores iron in spleen
Transport proteins	Hemoglobin	Carries oxygen in blood
	Serum albumin	Carries fatty acids in blood
Immunological proteins	Antibodies	Rid the body of foreign proteins
	MHC proteins	"Self" recognition
Toxins	Neurotoxin	Component of snake venom-attacks nerves
	Hemotoxin	Component of snake venom-attacks blood

Nucleic Acids Animal cells contain two types of nucleic acids: **deoxyribonucleic acid** (DNA) and **ribonucleic acid** (RNA). DNA is the genetic code for the formation of proteins and RNA serves as messengers and organizers in the construction of those proteins. DNA is found only in the nucleus of a eukaryotic cell and may be thought of as a chemical blueprint for the formation and maintenance of that cell. Both DNA and RNA are **polymers** (formed of repeating units) built of **nucleotides**. Each nucleotide contains three parts: a **sugar**, a **nitrogenous base**, and a **phosphate group** (**Figure 4.6**).

In simplest chemical parlance, a cell (and by extension) the entire body of an animal) is a watery bag of protein regulated by nucleic acid molecules and powered by the breakdown of carbohydrates and fats into CO_2 and H_2O.

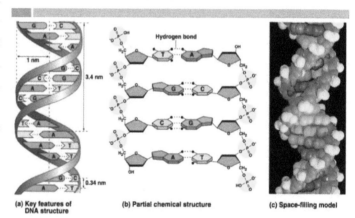

DNA helix structure

Figure 4.6 A Representative Segment of DNA. (a) & (b) depict the chemical nature of DNA in two dimensions whereas (c) depicts the true three dimensional structure of the molecule.

Cellular Nature of Life

Types of Cells

As we now accept, any living thing is composed of cells and cell products (such as bone or shell), be it a one-celled protozoan or the trillions of cells making up your body. If, as taxonomists suspect, there truly are tens of millions of different species of living things on this planet, the total number of cells that inhabit this planet could not even be imagined, let alone accurately determined. However, nature has smiled on biologists for all the untold multitude of cells on earth fall into only two general categories—*prokaryote* or *eukaryote*—based on inherent internal properties.

A **prokaryotic cell** (Gr. *pro*, before + *karyon*, kernel or nucleus) possesses DNA, but its DNA not isolated from the rest of the cell inside a membrane-bound nucleus. Instead, the DNA is a single loop floating free in the cytoplasm. In addition, the prokaryotic cell is less complex and much smaller than it eukaryotic counter-

part. Prokaryotic cells range in size from 1 to 10μm and lack internal membrane-bound **organelles** (L, *organum* = tiny organ), structures devoted primarily to the many metabolic tasks required to maintain the life of the cell. Most prokaryotic life forms are single cells, but some types are colonial.

The Prokaryotae, represented by the *Eubacteria* (or true bacteria) and the *Archaebacteria* (or ancient bacteria), are an extremely successful and diverse group that is thought to be the most ancient life form on earth (**Figure 4.7**). Bacteria occupy all terrestrial and aquatic habitats and even exist in great numbers on and inside the bodies of other living organisms, including humans. In our hubris, we declare humankind to be master of this place, but in biological actuality, bacteria rule this planet.

A **eukaryotic cell** (Gr. *eu*, true + *karyon*, kernel or nucleus) contains DNA that is complexed with DNA-binding proteins in complex linear chromosomes and located within a membrane-enclosed organelle known as the **nucleus** (**Figure 4.8**). Eukaryotic cells, ranging from 10 to 100μm, are larger than prokaryotic cells, and they contain numerous membranous organelles not found in prokaryotic cells (**Table 4.2**).

Once the two basic cell types originated, evolutionary pressures crafted variations of them resulting in a branching tree of life. The basic prokaryotic cell evolved into two bacterial lines—eubacteria and archaebacteria—which differ from each other in several ways:

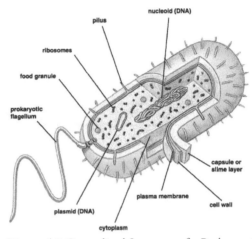

Figure 4.7 Generalized Structure of a Prokaryotic Cell. Beyond the basics of cytoplasm, plasma membrane and cell wall, some species also possess additional structures such as pili, flagella, photosynthetic pigments, and slime capsules.

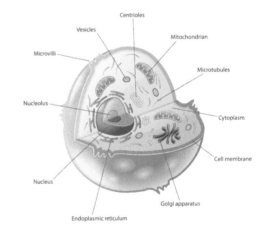

Figure 4.8 Generalized structure of an animal cell.

1. The structure of the cell wall, cell membrane, and ribosomal RNA is slightly different between the two.

2. The archaebacteria (aptly dubbed the extremophiles) live in very harsh environments such as acidic hot springs, high temperature hydrothermal vents, and water of extreme salinity similar to what conditions were thought to be on the Earth when life first appeared.

The basic eukaryotic cell has undergone even more modifications and adaptations than its prokaryotic kin. From it, four cell lines have evolved: protists (protozoa, unicellular and multicellular algae, and slime molds), fungi, plants, and animals. Furthermore, within large multicellular plants and animals, there are many variations of the basic eukaryotic cell. Animals are composed of tissues and organs consisting of muscle cells, nerve cells, blood cells, bone cells, connective tissues, reproductive cells, and epithelial tissues, all of which are revisions of the basic eukaryotic cell.

Cells let us walk, talk, think, make love and realize that the bath water is cold.

—Lorraine Lee Cudmore

Table 4.2
Comparison of Prokaryotic and Eukaryotic Cells

Characteristic	Prokaryotic Cell	Eukaryotic Cell
Cell size	Small (1-10 μm)	Large (10-100 μm)
DNA (genes)	Simple, circular DNA molecule free floating in cytoplasm	DNA complexed with proteins in complex linear chromosomes Located in membranous envelope
Nutrition	Mainly absorption; photosynthesis in some	Absorption then digestion; photosynthesis in some
Replication	Binary fission or budding; no mitosis	Some form of mitosis

Origin of Eukaryotic Cells

In the deep past some 3.5 to 3.8 billion years ago, the line from nonliving inorganic assemblages of molecules to living cells was apparently crossed. However, it is not biologically sensible to believe that even relatively simple prokaryotic cells such as bacteria sprang fully formed directly from inorganic molecules. A series of evolutionary intermediate steps would have most certainly have been required. However, the exact sequence and details of these steps has yet to be fully determined, it is known that three adaptations were required for true living cells to emerge:

1. *Self-replication* that allows a cell to replicate nearly identical copies of itself
2. *Metabolism* that allows a cell to feed and repair itself
3. *External membranes* that allow certain substances in (food, oxygen, and water) and others out (wastes and carbon dioxide)

Prokaryotic cells (Archaea and Eubacteria) seem to have been the only life form for several billion years until around 1.5 billion years ago when the first eukaryotic cells appeared. This has puzzled biologists for some time. Where did this new and substantially more complex type of cell come from? DNA evidence indicates that both Archaea and Eubacteria were major contributors to the origin of eukaryotic cells. This information has led to the rise of the **endosymbiotic theory** as an explanation for the rise of eukaryotic cells. According to this theory, certain cell organelles originated as free-living bacteria that were engulfed by another cell where they began to function as **endosymbionts** (an endosymbiont is an organism that can only live inside another organism, forming a relationship that benefits both organisms). Over millions of years, the capturing cells and the captured cells came to depend on one another for survival. This theory is supported by the fact that both mitochondria and chloroplasts possess their own DNA and form independently of other cell organelles through a process similar to binary fission.

Although the exact mechanisms that resulted in the origin of eukaryotic cells may never be known, the emergence of the eukaryotic cell and its adaptability and plasticity led to a dramatic increase in the complexity and diversity of life forms on this planet.

Discovering the Cellular Nature of Life

In an age when we can peer outward telescopically to nearly the edge of the known universe and microscopically inward to the atomic level, it is difficult for us to imagine a time when humankind did not even suspect the existence of a world smaller than what the unaided eye can discern. The actuality of this unseen world was first revealed nearly 350 years ago when Robert Hooke (1635-1703) used one of the first primitive compound microscopes to observe the fine structure of cork (**Figure 4.9**). In his major written work, *Micrographia,* Hooke described the honeycomb of cork cell walls as seen through his microscope by the term *cellulae* (L., small rooms) or *cells* as we term them today. Microscopists that followed Hooke came to realize that the cork cells that fired his imagination were actually dead remains with only resistant box-like cell walls remaining.

Figure 4.9

Hobbyist lens-grinder Anton von Leeuwenhoek, a contemporary of Hooke, would be the first to glimpse many types of living cells such as single-celled algae, bacteria, protozoans in pond water, sperm cells, and red blood cells. In a series of letters to the Royal Society of London from 1673 to 1723, Leeuwenhoek detailed his microscopic observations, often referring to the cells he had observed as "little animalcules" or "wee beasties." The realization quickly set in that cells were not the hollow boxes first seen by Hooke in cork, but rather, cells were filled will all manner of tiny exquisite structures; they were filled with life. It would take almost 200 years of technical improvements in the design of microscopes and lenses before Theodor Schwann, Matthias Schleiden, and Rudolf Virchow would state and refine one of the central pillars of biology now known as the *cell theory.*

In 1838 German botanist, Matthias Schleiden, proposed that all plant tissue was composed of cells. In 1834 Schleiden's countryman, Theodor Schwann, postulated that all animal tissue was composed of cells. In 1858, another German, Rudolf Virchow, advanced the proposition that all cells came from preexisting cells. Thus were laid the foundations for the cell theory—a proposition that all living organisms are composed of cells and cell products and that all cells come from preexisting cells. (It should be noted that biologist no longer consider this to be a theory in the experimental sense, rather the word "theory" in this case is used in a unifying and organizational sense.)

Microscopy has advanced greatly in the intervening years since the invention of the first microscopes and the discovery that living cells have contents. Improvements in light microscopes and the invention and perfection of the various types of electron microscopes have revealed the astonishing intricacies of cell structure.

Cell Structure

The internal architecture of the animal eukaryotic cell may be described according to two general features: *compartmentalization* and *macromolecular assemblages*. Most animal cells are divided into a number of compartments. Known as **organelles** (L., tiny organ), these compartments are usually surrounded by a membrane that serves to regulate the movement of molecules in and out of the organelle. Examples of organelles are the nucleus, mitochondria, Golgi apparatus, endoplasmic reticulum, and lysosomes (Figure 4.8).

Pervasive throughout the cell are macromolecular assemblages composed of specific macromolecules organized in a specific three-dimensional arrangement. Examples of macromolecular assemblages are the cytoskeleton, ribosomes, membranes, and the chromosomes.

In simplest terms, an animal cell is constructed of (1) a *system of membranes* that surround the cell and each of its organelles, (2) gelatinous *cytoplasm* that contains and supports the organelles as well as macromolecular assemblages and food storage products, and (3) the various *organelles* and *macromolecular assemblages* within the cytoplasmic matrix.

Membrane System Visualize membranes as closed bags. That is, membranes have an inner and outer surface but no edges. Membranes serve to separate an interior space from its surroundings and to separate the entire cell from its external environment. In a eukaryotic animal cell, the membrane system is two-fold and consists of the **plasma membrane** which surrounds and contains the entire cell, and internal membranes which surround and contain the organelles.

According to current understanding, known as the *fluid mosaic model*, the plasma membrane is constructed of a double layer (bilayer) of **phospholipids**. Phospholipids are lipids (fats) that have a phosphate group at the "head" end and a lipid "tail." Each layer of the plasma membrane consists of the hydrophilic ("water-loving") phosphate groups aligned so they all point out resulting in all the hydrophobic ("water-fearing") lipid tails pointing in (**Figure 4.10**). Associated with the plasma membrane are a number of diverse proteins. Some of these proteins simple adhere to the surface of the membrane whereas others are embedded in the bilayer. Many, such as the transmembrane proteins, act as carriers or channels that passively or

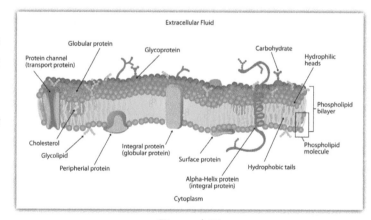

Figure 4.10

actively transport molecules through the bilayer. Membrane receptor proteins (glycoproteins) function as receptor sites first binding information molecules, such as hormones, and then transmitting signals from the bound molecules to the interior of the cell. Glycoproteins serve as a connection between the interior of the cell and the external environment around it. Attached to the microfilaments of the cytoskeleton, structural proteins insure the stability of the cell, and cell adhesion proteins allow cells to interact and identify each other. Membrane proteins may also exhibit enzymatic activity, catalyzing various reactions related to the plasma membrane.

Eukaryotic cells also have membranes enclosing the organelles. Similar in structure and function to the plasma membrane, the membranes surrounding the organelles allow the cell to segregate its chemical functions into discrete internal compartments.

Cytoplasm is the part of the cell enclosed by the plasma membrane. (Imagine a balloon filled with semi-solid gelatin. The balloon represents the plasma membrane, and the gelatin would represent the cytoplasm.) Cytoplasm is a complex matrix consisting of different components: cytosol, organelles, cytoskeleton, and inclusions.

- *Cytosol* is a translucent gel consisting of a complex mixture of water, salts, and organic molecules that compose around 70% of the total volume of a cell. The other cell components are suspended within the cytosol. The viscosity of cytosol is constantly changing.
- *Organelles* are membrane-bound compartments within the cytosol that have specific functions. Types and functions of the various organelles found in an animal cell follows.
- The *cytoskeleton* is a dynamic cellular scaffolding composed of protein filaments and tubules found within the cytosol. The cytoskeleton plays several important roles including maintaining the shape and stability of the cell, enabling intracellular transport, allowing some cellular movement, and assisting in cell replication (**Figure 4.11**).

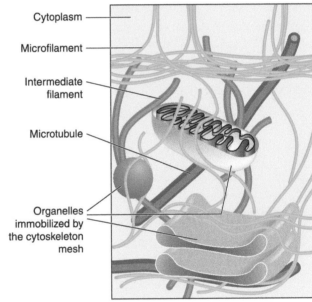

Figure 4.11

The cytoskeleton has three main structural components:

1. Microfilaments
2. Intermediate filaments
3. Microtubules

Microfilaments are fine thread-like protein fibers composed primarily of the contractile protein actin. In association with myosin, microfilaments help generate the forces used in basic cell movements such as contraction, cell crawling, amoeboid movement, and **cytokinesis**. Intermediate fibers are larger than microfilaments but smaller than microtubules. They provide tensile strength for the cell and anchor organelles to parts of the cell. **Microtubules** are tubular structures composed of a protein known as tubulin. Microtubules form the **mitotic spindle** which plays a vital role in the movement of chromosomes during cell replication (**Figure 4.12**). Microtubules are also involved in intracellular transport and are critical components of eukaryotic cilia and flagella.

- **Inclusions** are tiny particles or droplets of insoluble materials within the cytosol that are not bound by membranes. Although a wide range of different types of inclusions exists, the most common inclusion may be lipid droplets that are found in both prokaryotic and eukaryotic cells.

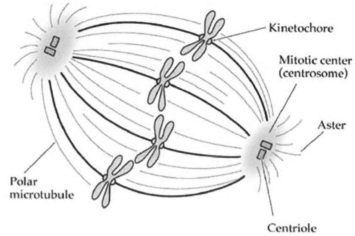

Figure 4.12 Microtubules radiate out from a microtubule organizing center, the **centrosome**, located near the nucleus. Within the centrosome is a pair of **centrioles**. Each centrioles is composed of nine triplets of microtubules.

Organelles The membranous organelles embedded/floating within the cytosol are diverse in structure and function:

- *Nucleus* Bound by the nuclear envelope, the nucleus is the largest of the organelles. Within the nucleus resides and the genetic material responsible for controlling and coordinating all the activities within the cell. Known as **chromatin**, the genetic material within the nucleus consists of long strands of DNA (**Deoxyribose Nucleic Acid**) complexed with proteins known as histones. Once a cell begins to replicate (prophase) the chromatin coils and thickens and is then referred to as **chromosomes (Figure 4.13)**. Pores in the nuclear envelope allow molecules to move from the nucleus out into the cytoplasm. Within the nucleus lies the **nucleolus**, a structure composed of proteins and nucleic acids not bound by membranes. The nucleolus functions to produce ribosomes (ribosomal RNA). Completed ribosomes detach from the nucleolus and move out of the nucleus through pores in the nuclear envelope.

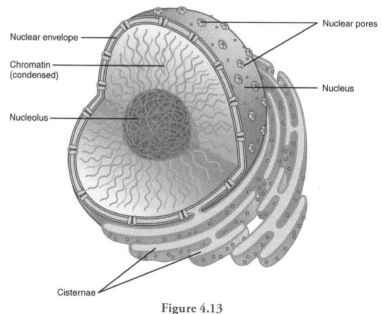

Figure 4.13

- *Endoplasmic Reticulum* Structurally, the ER is an extensive network of highly folded membranes supported and stabilized by the cytoskeleton that form tubules and vesicles. Their lumens (cavities inside the tubules) are interconnected and their membranes are continuous with the outer membrane of the nuclear envelope. The space between the membranes of the nuclear envelope "communicates" with the space between the membranes in the ER.

  The complex of membranes that is the ER is either covered on their outer surface with ribosomes (rough ER) or lacking ribosomes (smooth ER) (**Figure 4.14**). Rough endoplasmic reticulum (RER) is so named because its surface is studded with protein manufacturing ribosomes. However, the ribosomes bound to the RER at any one time are not an integral part of the RER membrane. Ribosomes attach to the RER membrane only during the process of protein formation and then release from the membrane once the protein has been

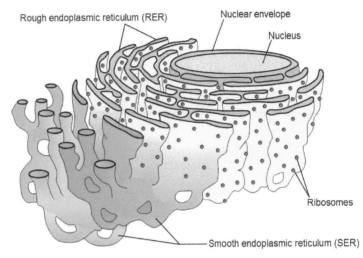

Figure 4.14

  formed. Smooth endoplasmic reticulum (SER) is so named because ribosomes do not attach to its membranes.

  ER performs a number of general functions in the cell:

  1. formation of transmembrane proteins and secreted proteins.
  2. transport system for proteins.
  3. synthesis and storage of steroids and lipids.
  4. metabolism of carbohydrates.
  5. regulation of calcium concentration.

- *Ribosomes*, complexes of RNAs and proteins that are produced in the nucleolus, are cellular components that serve as the site of protein synthesis in a cell. Ribosomes are divided into two subunits, one larger, one smaller. During protein synthesis the smaller subunit binds to messenger RNA (mRNA), but the larger subunit binds to transfer RNA (tRNA) and the amino acids. Once a ribosome attaches to the ER, mRNA from the nucleus attaches to the ribosome. Transfer RNA molecules holding specific amino acids then attach to the mRNA in a precise manner dictated by the code on the mRNA. The amino acids carried by the tRNA bond together into a protein chain. The newly synthesized proteins are sequestered in small transport vesicles and then move from the ER complex to the Golgi complex.

- *Mitochondria* (sing., *mitochondrion*) are organelles present in nearly all eukaryotic cells. They vary in shape, size, and number. Some are elongated whereas others are more or less spherical. The number of mitochondria in a cell varies widely by type of organism and tissue. Unicellular organisms may contain a single mitochondrion, but human liver cells may contain 1,000 to 2,000 mitochondria per cell, 1/5 of the cell volume (**Figure 4.15**). Mitochondria are the only animal cell organelles that contain their own DNA and thus are the only animal cell organelle that is self-replicating. Mitochondrial DNA, which hints at the possibility of the endosymbiotic origin of mitochondria, is in the form of a tiny, circular genome, much like the circular genome of prokaryotes only smaller. Mitochondria may be thought of as cellular power plants because it is here that a cell produces chemical energy in the form of ATP (adenosine triphosphate) from the oxidation of glucose molecules.

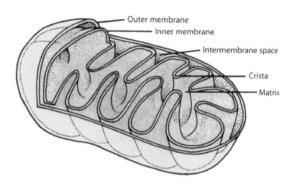

Figure 4.15

  Structurally, a mitochondrion contains both outer and inner membranes. The outer membrane is similar in structure to the plasma membrane and contains transmembrane proteins known as porins. Materials pass in and out of the mitochondria through the porins. The inner membrane is highly convoluted into complex folds known as cristae (sing., crista). The cristae greatly increase the inner membrane's surface area. It is on the cristae that glucose is combined with oxygen to produce molecules of adenosine triphosphate (ATP), the primary energy source for the cell.

- *Golgi complex* The Golgi complex is constructed of stacks of membrane-bound structures vesicles that serve to store, modify, and package protein products, especially secretory products.

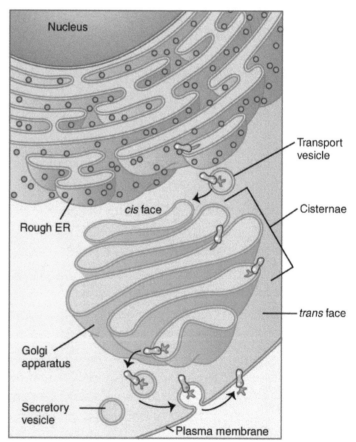

Figure 4.16

Small vesicles of the ER containing protein pinch off forming transport vesicles. The transport vesicles fuse with sacs on the *cis* face ("forming face") of a Golgi complex. After modification, the proteins are packaged into secretory vesicles on the *trans* face ("maturing face") of the complex (**Figure 4.16**). Secretory vesicles may contain contents from a glandular cell destined to be expelled to the outside of the cell. Some may contain proteins for incorporation into the plasma membrane whereas others may contain powerful enzymes and remain in the same cell that produces them. Known as **lysosomes**, these vesicles are involved in the breakdown of foreign protein or microbes, injured or diseased cells, and worn-out cell components. The interior of a lysosome is quite acidic (pH 4.8) compared to the slightly alkaline cytosol (pH 7.2) and contains enzymes for the digestion of starch, lipids, proteins, and nucleic acids. Lysosomal vesicles may also pump their enzymes into a larger membrane-body such as a **food vacuole** or **phagosome** in the process forming a **phagolysosome** (**Figure 4.17**).

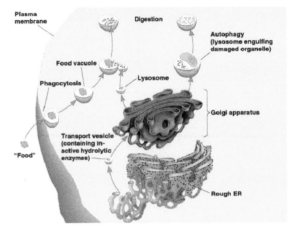

Figure 4.17

Table 4.3
Cellular Components—Structure and Function

Component	Description	Function
Plasma membrane	A double layered outer boundary composed of protein, and phospholipids.	Separate cell from other cells or the environment; protect cell; regulate passage of molecules into and out of the cell
Cytoplasm	Viscous gel-like substance filling the interior of a cell; consists of fluid cytosol and organelles	Houses all of a cell's organelles and dissolves certain substances
Cytoskeleton	Interconnected filaments and tubules forming a flexible cellular scaffolding	Provides support; assists in cell movement; aids in transport of vesicles; assists in chromosome movement during cell replication.
Centrosome (Micro-tubule organizing center)	An area near the nucleus where microtubules are produced. Within the centromere is a pair of centrioles with each centriole being composed of a ring of nine groups of microtubules.	Form microtubules that in turn function as conveyor belts within the cell moving vesicles, organelles, and chromosomes via special attachment/motor proteins

Nucleus	Large spherical structure contained by the nuclear membrane; contains nucleolus and DNA	DNA in the nucleus serves as template for the production of proteins that coordinate and control cell processes
Endoplasmic reticulum	A network of membranous tubules continuously extending throughout the cytoplasm from the nuclear membrane to the plasma membrane	Storage and internal transport; rough ER is a site for ribosome attachment; smooth ER makes lipids
Ribosomes	Composed of two subunits-one large and one small. Each subunit is composed of RNA and protein.	Link amino acids together into protein chains in the order specified by messenger RNA transcribed from the DNA in the nucleus.
Mitochondria	Spherical or rod-shaped; bound by a double membrane system with the inner membrane highly folded	Generate ATP, the source of chemical energy for the cell
Golgi complex (Golgi apparatus)	Stacks of disklike membranes	Sorts, packages, and routes cell's synthesized products
Vesicles	Small membranous sacs; contain secretory products or enzymes	Serve as vessels for storage, transport, or intracellular digestion

Cell Processes

For every instant of its existence, the internal components of a cell are involved in a frenzy of coordinated activities that occur with an accuracy and intricacy that boggles the mind and seriously challenges our best efforts to understand it all.

Cell processes can be described in terms of the transport and transduction (change) of energy, mass, and information required to maintain cell viability.

Cellular Transport A cell achieves its living individuality by being separated from its environment and other cells by a permeable barrier. To survive, however, a cell must communicate with its environment by the passage of mass, energy, and information through the permeable barrier. Cells employ several processes for moving materials in and out. (The same kinds of processes that allow entry can be used to achieve exit.) The different kinds of cellular transport can be divided into two categories: those that do not require energy (passive) and those that do require energy (active).

- *Passive transport by simple diffusion.* The plasma membrane is selectively permeable in that it allows only a few molecules in and even fewer out. This movement of molecules in or out is often accomplished passively through a process known as **diffusion** (**Figure 4.18**). Diffusion is considered a passive process because it will occur without the need of energy to power it forward. Diffusion occurs because molecules have the tendency to move from where there are many to where there are few until equilibrium is reached. Therefore, if there are more oxygen molecules outside a cell (higher concentration gradient) than within (lower concentration gradient), oxygen molecules will move through the plasma membrane into the cell until there are an equal number of oxygen molecules on both sides of the plasma membrane. Carbon dioxide, water (**Osmosis** is a term used to describe the diffusion of water only.), and a few other small, simple molecules also flow passively in and out of a cell through diffusion (or osmosis) due to differences in concentration gradient. Examples of diffusion are all around us such as

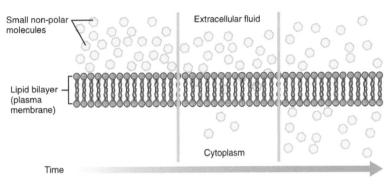

Figure 4.18 Diffusion. Over time molecules will move across permeable plasma membranes from a place of high concentration (extracellular) to a place of low concentration (cytoplasm) until equilibrium is reached.

when we add sugar to our tea (we stir the tea to speed up the diffusion process) or the smell of percolating coffee or cooking food waft throughout the whole house.

- *Passive transport by facilitated transport.* Another form of passive transport is facilitated transport (also known as passive-mediated transport). Due to the hydrophobic nature of the phospholipids that make up the plasma membrane, only small nonpolar molecules, such as oxygen can easily diffuse across the membrane. However, polar molecules (those with a positive charge [+] on one end and a negative charge [-] on the other) and charged ions cannot diffuse freely across the membrane. Small polar molecules are transported across the membranes through channels known as carrier proteins (**Figure 4.19**).

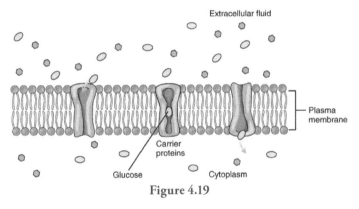

Figure 4.19

Ions move in or out through ion channels. Ion channels are protein pores in the membrane where ions flow according to their electrochemical gradient. Ion channels may allow ion diffusion at all times, or they may be gated channels, requiring a signal to open or close them. Larger molecules, such as glucose or amino acids are transported by carrier proteins, such as permeases that change their form as the molecules are carried through.

- *Bulk Transport.* Even larger molecules may be transported in, out or through the cell via vesicles. As we learned earlier in this chapter, vesicles are small membrane-bound sacs within the cell that store, transport, or digest cellular products. The membrane of a vesicle is structured in such a way that it can fuse not only with the plasma membrane but also the membrane of organelles within the cell. A cell may capture material from its external surroundings through a process known as **endocytosis**. On the other hand, unwanted material inside the cell may be expelled through a process known as **exocytosis**. Endocytosis occurs in three forms: phagocytosis, pinocytosis, and receptor-mediated endocytosis (**Figure 4.20**).

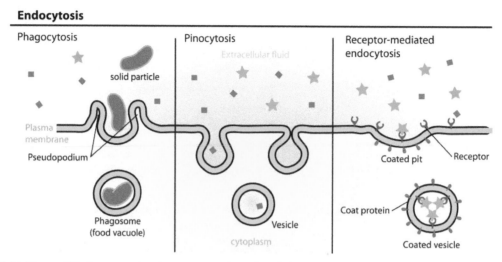

Figure 4.20 Types of Endocytosis. In phagocytosis, the cell membrane binds to a large particle and extends around it. In pinocytosis, small areas of the cell membrane . bearing specific receptors for small molecules or ions, invaginate to form caveolae (vesicles). Receptor-mediated endocytosis is a mechanism for selective uptake of large molecules in clathrin-coated pits. Binding of the ligand to the receptor on the surface membrane triggers invagination of pits.

- *Active Transport.* There are times when cells must transport materials against the concentration gradient (from low to high concentration), especially when the cell needs to accumulate high concentrations of vital materials, such as ions, glucose, and amino acids or remove unwanted ions that are diffusing freely into the cell. The process by which cells move materials in or out against a concentration gradient is known as **active transport** because the cell is required to expend energy to accomplish the task. The energy source that drives active transport is contained primarily in adenosine triphosphate (ATP) molecule. ATP powers specialized carrier proteins that "pump" materials against a concentration gradient. In fact, up to 40% of the ATP formed within a cell may be used up in the process of powering active transport (**Figure 4.21**).

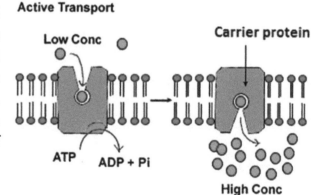

Figure 4.21 Active Transport. During active transport, a molecule combines with a carrier protein whose shape is changed as a result of the combination. This change in configuration , along with the energy of ATP, moves the molecule across the plasma membrane against the concentration gradient.

Metabolism. Cellular metabolism may be thought of as the sum total of all the biochemical processes occurring within a cell. Such biochemical processes are categorized as either anabolistic or catabolistic. Anabolic processes (anabolism) are those in which a cell constructs large macromolecular assemblages (polymers) from smaller units, such as building proteins from various amino acids. Anabolic processes are powered by catabolic processes (catabolism) in which large macromolecules are broken down into smaller units (monomers).

Proteins (polymer) ↔ amino acids (monomer)

Polysaccharides ↔ monosaccharides

Fats ↔ fatty acids

Cells use the monomers released from breaking down polymers to either construct new polymer molecules or degrade the monomers further forming the energy-carrying molecule adenosine triphosphate (ATP) and simple waste products (**Figure 4.22**).

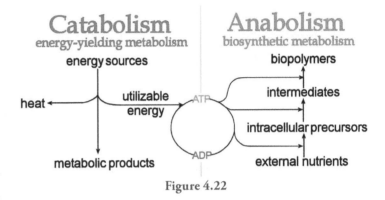

Figure 4.22

ATP is considered by biologists to be the "energy currency" of metabolism and thus life itself. Formed in the mitochondria, ATP is present in the cytoplasm of the cell and provides the energy required for every biochemical process within the cell. The importance of ATP can be seen in the fact that the human body produces 2×10^{26} molecules of ATP daily!

ATP may be thought of as a molecular "battery." It is discharged (energy released) when one of the phosphate groups on the "tail" of the molecule is lost. The ADP molecule that remains can be recharged through oxidative phosphorylation back into ATP.

Cellular metabolism encompasses:

- Degrading food molecules (catabolism)
- Synthesizing macromolecules needed by the cell (anabolism)
- Generating small precursor molecules needed by the cell such as some amino acids
- All reactions involving electron transfers

Metabolism occurs in a sequence of biochemical reactions known as pathways. Metabolic pathways may be simple linear sequences consisting of only a few steps or they may be highly branched and convoluted derivations from a central main pathway.

Some pathways can serve multiple functions. For example, the Kreb's cycle of aerobic respiration functions mainly to produce adenosine triphosphate (ATP), the molecule from which a cell derives most of its chemical energy, but it can also yield small precursor molecules necessary for many biochemical reactions within the cell. All biochemical reactions within metabolic pathways are catalyzed and controlled by proteins known as **enzymes**.

Cell Growth and Division

Humans (or any other complex animal that undergoes sexual reproduction) start life with a cellular bang. Amazingly it all begins with a single fertilized egg cell. Some 42 turns of the cell cycle later (a generational time of six to seven days per turn of the cycle) a human infant composed of approximately 2 trillion cells is born. Cell division is rapid during early embryonic development then slows with age eventually resulting in a human adult with approximately 60 trillion cells differentiated into over 200 different types of specialized cells working together in harmony. Approximately 2 million cells—about 25 million cells per second—are produced by an adult human body every 24 hours. How is such a near miraculous feat possible? The answer lies in the process of cell division (replication). Cell division is a method by which the genetic material (DNA) of a single cell (parent cell) is distributed equally and identically into two daughter cells.

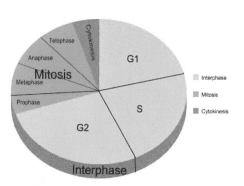

Figure 4.23

Cell division, however, is only one milestone in a cell's existence. The wholeness of a cell's existence is not a linear expression of its internal activities, but rather is cyclic—the cell cycle (**Figure 4.23**). The cell cycle is a progressive and repeating series of events in the life of a cell. The cell cycle is delineated into four distinct phases: G_1 phase, S phase, G_2 phase (collectively known as *interphase*), and M phase. The end result of the turning of the cell cycle is the production of two daughter cells genetically identical to the original parent cell.

Table 4.4		
Phases of Mitosis		
State	**Phase**	**Activity**
Interphase	*Gap 1* (G$_1$)	Also known as the growth phase because biosynthetic activity proceeds at an accelerated rate. This phase is also marked by the synthesis of certain enzymes and materials required during the S phase.
	Synthesis (S)	DNA replication occurs during this phase. During this phase, the amount of DNA in the cell has effectively doubled.
	Gap 2 (G$_2$)	During the gap between DNA synthesis and mitosis, increased biosynthetic activity again occurs, mainly involving the production of microtubules, which are required during the process of mitosis.
Cell division	*Mitosis* (M)	Cell growth stops at this stage, and cellular energy is focused on the orderly division into two identical daughter cells. M phase proceeds sequentially: • *Prophase* • *Metaphase* • *Anaphase* • *Telophase* • *Cytokinesis*

Quiescent/senescent	*Gap 0* (G_0)	A phase where the cell has left the cycle and is no longer dividing. Examples are neurons (nerve cells) and erythrocytes (red blood cells). G_0 cells usually do not reenter the cell cycle but instead will carry out their function in the organism until they die.

Command and Control of Cell Division

A cell spends most of its time in interphase, the time between mitotic activity when it performs its usual functions depending on its location within an animal's body. What prompts a cell to leave interphase and begin the mitotic process? In eukaryotic cells the passage of a cell through the cell cycle is controlled by genetically coded proteins in the cytoplasm, mainly cyclins and cyclin-dependent kinases.

Functioning much like chemical traffic signals, a series of checkpoints is built into the cell cycle:

G_1/S checkpoint This is the main checkpoint in the cell cycle. If cellular DNA is undamaged and healthy and necessary nutrients are present, proteins will initiate the S phase (DNA synthesis and replication). If condition for DNA synthesis and replication are not favorable, the cell cycle will stop at this point.

G_2 checkpoint If DNA has properly replicated, mitosis will proceed. If the DNA is damaged and cannot be repaired, apoptosis (cell death) will occur.

M checkpoint Mitosis will not continue if the chromosomes are not properly aligned on the mitotic spindle.

There are great differences in the generational time of cells that compose different types of tissues. In some, one turn of the cell cycle may be measured in hours whereas in others it is measured in days, months, or even years. In some cases cells divide only in the early stages of development. Muscle cells stop dividing during the third month of fetal development with further growth dependent on enlargement of fibers already present. Cells of the nervous system stop dividing early on in fetal development but persist throughout the life of the individual.

Collectively, cell replication in an animal is critical for growth, replacement of dying cells, and wound repair. However, each individual cell eventually leaves the cell cycle and enters into what cytologists call quiescent/senescent (Gap 0) leading to the death of the cell **(apoptosis)**. There are thought to be several ways in which a cell that is perfectly healthy and supplied with nutrients can die. One process is known as **programmed apoptosis**. As contradictory as it sounds, there are times when cell death is necessary for the continued well-being of the individual. This system works primarily to kill sick cells. For example, cells that have become infected with a virus behave differently biochemically. Nearby cells can "sense" this unusual activity and signal for that infected cell to undergo apoptosis. Apoptosis in select tissues is also critical in the formation of the fingers and toes of vertebrates.

Cells also die when their DNA accumulates enough mutations and damage that apoptosis is triggered. Finally, cell death can result because of the way DNA replicate. With every round of replication, the ends of the DNA molecules (**telomeres**) get shorter and shorter. As a result, critical genes will eventually be lost, and the cell will die.

Cell cycle checkpoints normally ensure that DNA replication and mitosis occur only when conditions are favorable, and all processes are proceeding correctly. However, unregulated cell division can lead to cancer. Mutations in genes that encode cell cycle proteins can lead to unregulated growth resulting in tumor formation and ultimately the invasion of surrounding tissue by cancerous cells.

Types of Cell Division—Mitosis and Meiosis

In eukaryotic organisms, cell division is a vital process at both the individual and species level. A type of cell division known as **mitosis** allows for the development (and in some cases regeneration), growth, and maintenance (cell replacement) of an individual animal whereas another type of cell division known as **meiosis** produces the gametes (sperm and egg) that allow for reproduction and the continuation of the species. Mitosis occurs in the somatic (body) cells of an animal, but meiosis occurs only in the gonads (testes and ovaries).

In the 1880s, chromosomes were discovered. A few years later chromosomes were found to segregate by an orderly process into the daughter cells formed by cell division as well as into the gametes formed by the division of reproductive cells (germ cells). Three important regularities were observed about the chromosome complement (complete set of chromosomes) of plants and animals.

1. During cell division, the chromosomes become visible and present as homologous pairs.
2. The number of chromosomes in somatic cells differs from the chromosome number of gametes.
3. Somatic cells, containing a full complement of chromosomes, are said to be diploid (2n) whereas gametes, containing a half set of chromosomes, are said to be haploid (n). A diploid individual carries two allelic copies of each gene present in each pair of chromosomes. The chromosomes occur in pairs because one chromosome of each pair derives from the maternal parent and the other from the paternal parent of the organism.

In multicellular organisms that develop from single cells, the presence of the diploid chromosome number in somatic cells and of the haploid chromosome number in germ cells indicates that there are *two* processes of nuclear division that differ in their outcome. One these (mitosis) maintains the chromosome number; the other (meiosis) reduces the number by half.

Table 4.5			
Diploid and Haploid Number for Select Animals			
Common Name	**Scientific Name**	**Haploid Number**	**Diploid Number**
Cat	*Felis domesticus*	19	38
Cow (domestic)	*Bos taurus*	30	60
Chimpanzee	*Pan troglodytes*	24	48
Dog (domestic)	*Canis familiaris*	39	78
Fruit fly	*Drosophila melanogaster*	4	8

Horse	*Equus caballus*	32	64
Human	*Homo sapiens*	23	46
Mosquito	*Culex pipiens*	3	6
Rhesus monkey	*Macaca mulatta*	21	42

Mitosis Through the process of mitosis, a single somatic parent cell duplicates its diploid genome exactly and precisely then partitions a complete and identical diploid genome into each of two new daughter cells (**Figure 4.24**). Mitosis is conventionally divided into four stages: **prophase**, **metaphase**, **anaphase**, and **telophase**.(If you have trouble remembering the order, you can jog your memory by using a mnemonic like *P*rince *M*ichael *A*te *T*oads.) Mitosis is preceded by *Synthesis* (S), a period when all the DNA in the nucleus is replicated resulting in two identical sets of genes.

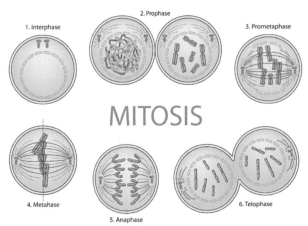

Figure 4.24

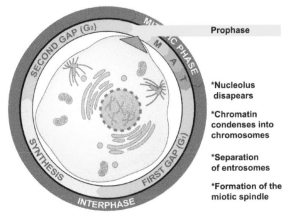

Figure 4.25

(**Figure 4.25**) *Prophase* In the early stages of prophase the chromatin is thread-like and not visible with a light microscope. As prophase proceeds, the chromatin begins to coil and then supercoil until it becomes visible as chromosomes (**Figure 4.26**). At this juncture, each chromosome is longitudinally double consisting of two closely associated subunits called chromatids. The chromatids in a pair are held together at a specific region of the chromosome called the centromere. At the end of prophase, the nucleoli disappear, and the nuclear membrane abruptly disintegrates and the chromatids move freely in the cytoplasm.

(**Figure 4.27**) *Metaphase* As the cell transitions into metaphase, the mitotic spindle forms as the centrioles move to opposite poles of the cell. The spindle is a football-shaped array of fibers consisting mainly of microtubules. A structure known as a kinetochore forms on the centromere of each chromatid pair. Spindle fibers then attach to each kinetochore and the chromatids align on the metaphase plate, an imaginary plane equidistant from each spindle pole. Proper chromatid alignment is an important cell cycle

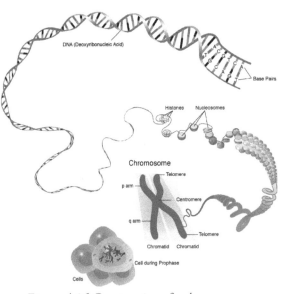

Figure 4.26 Construction of a chromosome.

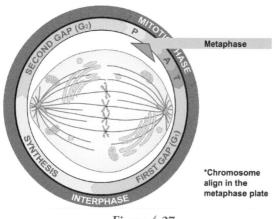

*Chromosome align in the metaphase plate

Figure 4.27

control checkpoint at metaphase in both mitosis and meiosis. In fact, proper chromatid alignment is so critical that mitosis will stop moving forward if the alignment is not perfect.

(**Figure 4.28**) *Anaphase* The centromeres of each chromatid pair separate and the two sister chromatids move toward opposite poles of the spindle as the fibers shorten. Once the centromeres separate, each sister chromatid is regarded as a separate chromosome in its own right. At the completion of anaphase, two identical sets of chromosomes lie near opposite poles of the spindle. Each chromosome set contains the same number of chromosomes and the same genes that were present in the pre*Synthesis* (S) interphase nucleus.

(**Figure 4.29**) *Telophase* As mitosis winds down, the nuclear membrane reforms, the chromosomes begin to uncoil, and the spindle disappears. Finally, in a process called **cytokinesis**, a plasma membrane grows inward between each cell separating them into individual entities.

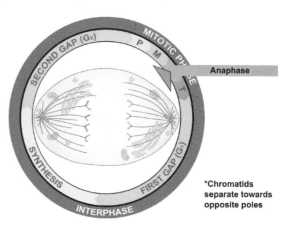

*Chromatids separate towards opposite poles

Figure 4.28

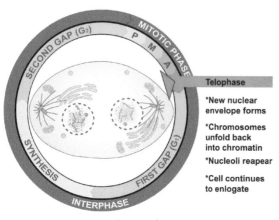

*New nuclear envelope forms

*Chromosomes unfold back into chromatin

*Nucleoli reapear

*Cell continues to enlogate

Figure 4.29

Meiosis Meiosis is a type of cell division that involves two successive divisions of a diploid (2n) eukaryotic cell of a sexually reproducing organism and results in four haploid (n) progeny cells (gametes), each with half of the genetic material of the parent cell (**Figure 4.30**). In animals other than sponges, meiosis takes place in gonads—male testes or female ovaries. In males, the meiotic process of spermatogenesis produces four haploid spermatids that each develops into a sperm cell. In females, the meiotic process of oogenesis produces a single ovum (egg) and three nonfunctioning polar bodies which eventually degenerate

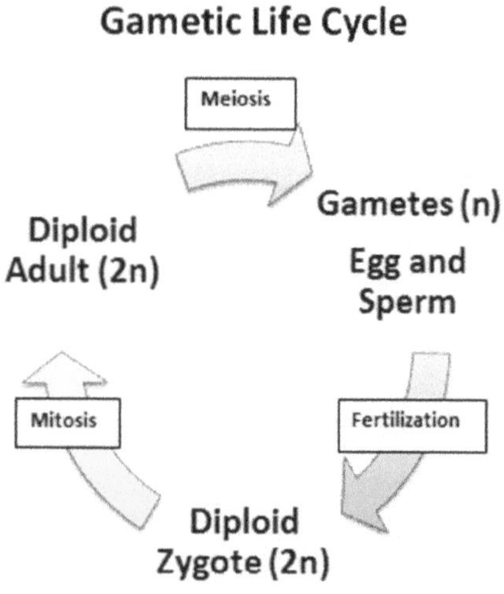

Gametic Life Cycle

Meiosis

Gametes (n)
Egg and Sperm

Diploid Adult (2n)

Fertilization

Mitosis

Diploid Zygote (2n)

Figure 4.30

Meiosis consists of two successive nuclear divisions: Meiosis I and Meiosis II (**Figure 4.31**).

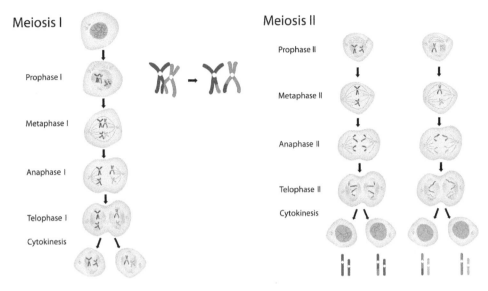

Figure 4.31

Meiosis I Meiosis I separates homologous chromosomes, producing two haploid (n) cell, thus meiosis I is referred to as a reductional division. A regular diploid human cell contains 46 chromosomes and is considered 2n because it contains 23 pairs of homologous chromosomes. However, after meiosis I, although the cell contains 46 chromatids, it is only considered as being n, with 23 chromosomes

Meiosis II Meiosis II is the second stage of the meiotic process, also known as equational division, Though the process is similar to mitosis, the genetic results are fundamentally different. The end result of Meiosis II is the production of four haploid cells (n) from the two haploid cells resulting from Meiosis I.

Table 4.6		
Meiosis vs. Mitosis		
	Meiosis	**Mitosis**
End Result	Normally four cells, each with half the number of chromosomes as the parent cell	Two cells, having the same number of chromosomes as the parent cell
Function	Sexual reproduction, production of gametes	Cellular replication, growth, and repair
Happens in	Gonads	Most somatic cells
Genetically same as parent cell?	No	Nearly always
Crossing over?	Yes in Prophase I	Sometimes
Pairing of homologous chromosomes	Yes	No
Cytokinesis	Occurs twice, once in Telophase I and again in Telophase II	Occurs once in Telophase
Centromeres split	Occurs in Anaphase II	Occurs in Anaphase
Time required	Days or even weeks	12 to 24 hours

Animal Architecture

The fact that our planet swarms with millions of different animal species results in a truly amazing diversity of animal forms. But is there any commonality to this multitude of types? Exactly what is an animal?

Animal Characteristics

Even though no one criterion fits all types, several characteristics are common to all animals:

Heterotrophy. All animals are heterotrophs. As such, they obtain energy and necessary organic molecules by ingesting other organisms.

Multicellularity. All animals are multicellular beings.

Lack Cell Walls. Unlike algae and vascular plants, the cells of animals lack rigid outer cell walls.

High Activity Level. Although plants are capable of self-movement, animals move more rapidly and in more complex ways than do plants. Plants move by growth and changes in water pressure, whereas animals employ muscles and nerve cells. One form of movement unique to animals is the ability to fly.

Embryonic Development. Most animals have similar patterns of embryonic development.

As the fertilized egg of an animal develops, different layers of cells form. Known as germ layers, these cells eventually develop into the external covering and internal structures of the animal (**Figure 4.32**).

Organization. The eukaryotic cells of all animals except sponges are organized into structural and functional units—cells are arranged into tissues; tissues are arranged into organs, and organs are arranged into organ systems (**Figure 4.33**).

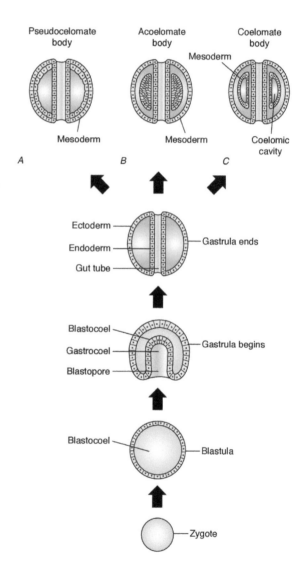

Figure 4.32 The formation of germ layers and the sequence of development from those layers.

Paramecium

Cellular Level. The "body of animals at this level consists of a single cell, the unicellular protozoans exhibit organization. Amazingly, single-celled organisms satisfy all the life processes necessary to sustain live in one cell, while organ system level animals require a complex body of tens of millions to billions or trillions of cell to accomplish the same thing. This shows there is no such thing as a "simple" or primitive" life form.

Volvox

Aggregate Level. Simple aggregates in which the cells are all identical in form with no division of labor are referred to as colonies. The volvox illustrated is an example of an advanced aggregate in which some cells are equipped with flagella and provide locomotion, while other cells asexually form daughter colonies within the parent aggregate. Sponges are considered to be the most complex aggregates, although some zoologists place them at the tissue level of organization.

Hydra

Tissue Level. Some organisms, such as the hydra and jellyfish (phylum Cnidaria), have layers of cells arranged into definite tissues; no organs are present, however. All animals at this level and above are referred to as metazoa.

Planaria

Organ Level. Definite organs are present at this level but they are not complex and not coordinated into organ systems. The planaria (phylum Platyhelminthes) is representative of such a level of organization. Planaria possess eyespots, a simple digestive tract, and even a reproductive system all formed from specific tissues.

Fish

Organ System Level. The animal phyla beyond Plathelminthes and through mammals exhibit this most complex of organization. Complex organs are arranged into systems — circulation, respiration, and so on all working in concert to support a very complex animal body.

Figure 4.33 Levels of Complexity Within the Animal World. Size and complexity decrease from the bottom of the diagram—organ system level—to the top—cellular level.

Symmetrical Arrangement Symmetry refers to the geometric design of the parts in animal bodies (**Figure 4.34**). A very few types are said to display **asymmetry** because the body lacks any definite form or geometry. Sponges and some protozoan display asymmetrical body plans.

Spherical symmetry describes a body that is basically round with its parts concentrically arranged around a central point. Spherical symmetry is usually found in animals that float such as some protozoa. Animal bodies in which the parts are arranged around and radiate outward from a central axis that is shaped like a pie, wheel, or column are said possess **radial symmetry.** Cnidarians, such as the jellyfish and sea anemone, and echinoderms such as the

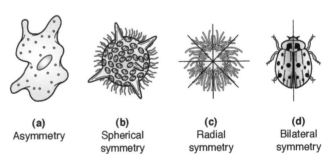

(a) Asymmetry   **(b)** Spherical symmetry   **(c)** Radial symmetry   **(d)** Bilateral symmetry

Figure 4.34 Planes of Symmetry. (a) Asymmetry as displayed by an amoeba(b) Spherical symmetry as displayed by a radiolarian. (c) Radial symmetry as displayed by a sea urchin. (d) Bilateral symmetry as displayed by an insect.

starfish demonstrate radial symmetry. Most animals exhibit **bilateral symmetry** (two-sided), a type of symmetry in which one side of the body is a mirror image of the other. Bilateral symmetry is correlated with motility (mobility) and cephalization (development of a head end). Bilateral animals are much better suited for directional movement and the concentration of brain and sense organs in a head bestows great advantages to an animal moving through its environment head first.

Box 4.1
A Matter of Size

In the kingdom of living animals, size is the supreme overseer of all matters biological and thus a driving force in animal evolution. The limitations of size govern not only the shape but also the scope of activities of all animals from the microscopic protozoa to the blue whale, a size range that represents a difference in mass of twenty-one orders of magnitude.

Size directly affects the physical proportions of any animal. An animal must have enough strength and rigidity to support its bulk and enough power to move that bulk. Even though the largest animal that has ever lived—the blue whale—is a vertebrate that does possess a bone endoskeleton, it collapses into a suffocating heap if removed from the water. The extreme density of water is absolutely necessary to this animal to support its record-holding mass. By contrast, the largest land animal is the male elephant with a mass roughly 1/15th that of the whale. Why such a discrepancy between the extreme size of aquatic versus land animals? The problem is that as the size and weight of an animal increases, the relative strength of its skeleton or other supporting structures decreases. Structural strength depends on the cross-sectional area of the support, say a leg bone. However, if you were to increase the size of that leg bone eightfold, the cross-section of that bone would increase only fourfold. Thus without the supporting force of water, a land animal is limited in weight by the strength of its bones and that translates at maximum into an animal the size of an elephant. (This holds true only for extant land animals. Some species of terrestrial dinosaurs were much larger than elephants. How did these creatures overcome the size vs. bone strength conundrum and not only reach gigantic size but also remain quite mobile? A number of different hypotheses have been advanced to explain this discrepancy, but it remains a mystery still.)

The rate of locomotion (speed) of an animal is also a function of size. Even though large animals move their body parts more slowly than small animals do, their overall rate of locomotion is faster. Bigger, in general, implies faster. Double the length of a running animal and it runs roughly twice as fast. The same hold true for swimmers (except the blue whale for reasons unknown) but the advantage is less pronounced for flyers.

Size also determines the metabolic rate of animals. Large animals have proportionally less surface area (outside) relative to their volume (inside) than small animals do. Surface area not only includes skin, but also all the membranes in the lungs and digestive system through which oxygen and digested food pass into the cells and waste products and heat pass out. Because increases in surface area do not keep pace with increases in volume, the metabolic rate of large animals is by necessity slower than those of small ones.

Size even influences longevity. Large-bodied animals and live longer than their small counterparts, but intriguingly, brain size may be an even better predictor of life span than body size. Biologists remain unsure about how size relates to longevity, but some experimental evidence links life span to metabolism.

Keeping an animal on a minimal diet (a nice way of saying near starvation) can substantially lengthen its life span. No other animal better illustrates the possible connection between size, metabolic rate, and longevity than does the shrew. With an off-the-charts metabolic rate and a heart rate of over 600 beats per minute, the tiny shrew's entire life span measures only a year or so making it an animal that truly "lives fast and dies young."

To describe the regions of animal bodies, zoologist use certain terms. These terms constitute what might be called the "compass points of animal morphology." (**Figure 4.35**). Anterior designates the head end and posterior the rear or tail end. Dorsal refers to the top or back side and ventral to the bottom or belly side. Distal parts are farther from the body; proximal parts are nearer. A frontal plane passes from anterior to posterior and from left to right sides. It is perpendicular to both the transverse and sagittal planes. The sagittal plane divides the animal into right and left halves whereas the transverse plane (also called a cross section) divides the animal into anterior and posterior portions.

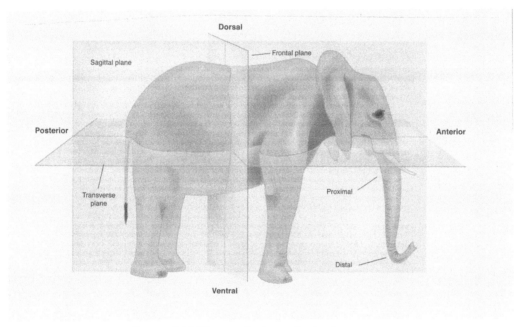

Figure 4.35 The terminology of animal symmetry.

A Closing Note
To Rise and Walk Again

In the course of a long career in public education, much of it in the same small rural community, I not only taught science and biology, but I also served as head football coach for a number of years. I have many fond memories of my coaching experiences, but there was one incident that was so devastating that it still haunts me to this day and probably will for the rest of my life.

Late in the game during a heated battle with one of our conference rivals, one of my football players, Shane, suffered some real or imagined indignation as our team punted. Letting his quick temper get the best of his judgment and training, Shane flew downfield like an angry blue missile and launched himself full speed and head first into our opponent's punt return man right in front of our bench, just a few feet from where I was standing helpless to prevent what I could see was coming. The collision that resulted launched the punt returner nearly to the track that surrounded our football field and dropped Shane, unconscious, nearly at my feet.

As the team and I visited Shane in the hospital that night after the game, we found him with a metal halo attached to his head, unable to move from the neck down. Sadly, Shane suffered a spinal cord injury that has and will leave him a quadriplegic for the rest of his life. The opposing player that Shane so violently collided with suffered a broken collar bone that, fortunately, healed quickly and normally.

Broken bones heal but severed spinal cords do not. That cruel fact is the reality that faces the nearly 400,000 people living with spinal cord injury (SPI) in the United States and the 7800 to 10,000 new cases of SPI that happen in this country alone every year. However, medical research into the healing powers of stem cells may be a candle of hope in the darkness that is SPI. In fact, as I write, a California bio-tech company has just begun testing an embryonic stem-cell drug treatment on a patient with spinal cord injuries; marking the first time a drug made with embryonic stem cells has ever been used on a human. The stem-cell drug, known as GRNOPC, contains cells that turn into **oligodendrocytes**, a type of cell that produces myelin, a coating that allows electrical impulses to move along nerves.

Stem cells, unspecialized cells found in all multi-cellular organisms, have the remarkable potential to develop into many different cell types. When a stem cell replicates, the resulting daughter cells can either remain unspecialized stem cells or develop into specialized cells with a specific function, such as muscle cells or spinal cord nerve cells. Stem cells may be thought of biologically as a lump of clay that holds the potential to be molded into whatever object (cell type) the body deems necessary.

Mammalian stem cells fall into one of two broad categories: **embryonic stem cells** which are found in the inner mass of a blastocyst and **adult stem cells** which reside in adult tissues. Human embryonic stem cells hold the greatest potential for use in cell-based regenerative therapies because unlike adult stem cells, they are genetically pliable enough to become any specific cell type plus they are relatively easy to grow in culture.

Human embryonic stem cells, directed to differentiate into specific cell types, could serve as a renewable source of cells and tissues to treat and possibly cure diseases such as Alzheimer's, heart disease, diabetes, osteoarthritis, and rheumatoid arthritis as well as injuries such as spinal cord injury, strokes, and burns.

The use of human embryonic stem cells is controversial, and the experimental use of human embryonic stem cells has been at the center of funding controversies. Because the research involves destroying human embryos, some have argued it is akin to abortion

The use of human embryonic stem cells to treat spinal cord injuries is in its infancy, too late for Shane to benefit. Such research, however, may eventually allow those rendered powerless and motionless by traumatic spinal cord injury to rise like the biblical Lazarus and walk again.

In Summary

- The riotous multitude of life forms that swarm over this planet are all fashioned from only a few basic types of living building blocks known as cells.
- There is an ordered hierarchy to this pandemonium we call life: Atoms and molecules are organized into cells. Cells are organized into organisms and organisms are organized into ecological zones.
- All organisms, regardless of their complexity, share a set of basic characteristics:

  1. Biogenesis
  2. Organization
  3. Sensitivity
  4. Energy and waste
  5. Movement
  6. Reproduction
  7. Life stages

- Cells are constructed of both organic and inorganic molecules.
- The major types of organic compounds associated with the composition of cells are carbohydrates, lipids, proteins, and nucleic acids.
- There are two main types of cells: prokaryotic cells that lack a defined nucleus and eukaryotic cells which possess a defined nucleus.
- The first eukaryotic cells are thought to have arisen some 3.5 to 3.8 billion years ago when membrane-bound assemblages of organic macromolecules developed the ability to undergo metabolism and self-replication.
- Since the invention of the first light microscope in the 1600s, humankind has slowly come to realize the true nature of life through the development of the Cell Theory.
- An animal cell is constructed of (1) a *system of membranes* that surround the cell and each of its organelles, (2) gelatinous *cytoplasm* that contains and supports the organelles as well as macromolecular assemblages and food storage products, and (3) the various *organelles* and *macromolecular assemblages* within the cytoplasmic matrix.
- Cell functions can be described in terms of the transport and transduction (change) of energy, mass, and information required to maintain cell viability.
- Materials are transported in, out, and through cells passively by diffusion, osmosis, facilitated transport, or bulk transport, and actively by active transport.
- Cellular metabolism may be thought of as the sum total of all the biochemical processes occurring within a cell. Biochemical processes may be categorized as either anabolic (building up) or catabolic (tearing down).
- Metabolism occurs in a sequence of biochemical reactions known as pathways. Biochemical pathways are controlled by enzymes.

- The life of a cell is not a linear expression of its internal activities, but rather a circular one—the cell cycle.
- The cell cycle is delineated into four distinct phases: *G_1 phase*, *S phase*, *G_2 phase* (collectively known as *interphase*), and *M phase*. The end result of the turning of the cell cycle is the production of two daughter cells genetically identical to the original parent cell.
- Mitosis allows for the development (and in some cases regeneration), growth, and maintenance (cell replacement) of an individual animal. Meiosis produces the gametes (sperm and egg) that allow for reproduction and the continuation of the species. Mitosis occurs in the somatic (body) cells of an animal, but meiosis occurs only in the gonads (testes and ovaries).
- Somatic cells, containing a full complement of chromosomes, are said to be diploid (2n) whereas gametes, containing a half set of chromosomes, are said to be haploid (n). A diploid individual carries two allelic copies of each gene present in each pair of chromosomes. The chromosomes occur in pairs because one chromosome of each pair derives from the maternal parent and the other from the paternal parent of the organism.
- Even though no one criterion fits all types, several characteristics are common to all animals:

1. Heterotrophy
2. Multicellularity
3. Lack of cell walls
4. High activity level
5. Embryonic development
6. Organization
7. Symmetrical arrangement

   A. Asymmetrical
   B. Spherical symmetry
   C. Radial symmetry
   D. Bilateral symmetry

Review and Reflect

1. *Signs of Life*. Trees move in the wind and boulders tumble downhill. Icicles and crystals grow. Are trees and boulders alive because they can move? Are icicles and crystals alive because they can grow? How many of the characteristics or signs of life must something demonstrate to be considered truly alive? Discuss.

2. *Prove It*. You take for granted that you are alive, but can you prove it? Use the characteristics of life discussed earlier in this chapter to make the case that you are, if fact, alive.

3. *What's in a Quote?* We opened this chapter with a quote from E. O. Wilson. React to Wilson's quote and explain in your own words. What do you believe Wilson is trying to say?

4. *Tiny Cells*. Your colleague believes she has just discovered a new species of microorganism. This previously unknown microbe is single-celled, and the cells range in size from 9-11 μm in size. What further information must she gather in order to determine if these newly-discovered cells are prokaryotic or eukaryotic?

5. *Cell Analogies*. "Analogies prove nothing that is true," wrote Sigmund Freud, "but they can make one feel more at home." Toward that end, we use analogies here to simplify the function of each of the organelles found within a cell. Answer the following:

   A. How is the **nucleus** of a cell analogous to *city hall*?
   B. How is a **mitochondrion** analogous to a *power plant*?
   C. How is the **endoplasmic reticulum** analogous to *highways and roads*?
   D. How are **ribosomes** analogous to *manufacturing plants*?
   E. How is the **Golgi apparatus** analogous to *packing and shipping facilities*?
   F. How are **lysosomes** analogous to *waste disposal and recycling plants*?

6. *Gateway to the Cell*. The plasma membrane is described as being "selectively permeable." What does this mean? Why is the permeability of its cell membrane critical to the life of any cell? What are the various mechanisms by which molecules enter and leave a cell through the plasma membrane (cellular transport)?

7. *Fear the Blob*? In the 1958 movie, *The Blob*, a small blob from outer space consumes living things (including humans) until it grows to gigantic size. Is it possible to take a single animal cell and nurture it until it grew as large as a car? Could you grow a super-size single animal cell in an orbiting laboratory?

8. *A Deadly Drink*. In his poem, *Rhyme of the Ancient Mariner*, Samuel Taylor Coleridge describes the plight of anyone cast adrift on the ocean.

   Water, water everywhere,
   and all the boards did shrink;
   water, water everywhere,
   nor any drop to drink.

   Explain why it is osmotic suicide for someone adrift on the ocean and dying of thirst to drink salt water.

9. *Lines or Circles*? Do cells live their lives in a circular mode or a linear one? Explain

10. *Coelom Confusion*. Your study buddy turns to you and says, "Zoologists make a big issue about between the number of germ cell layers and the presence or absence of a coelom in relation to the complexity of an animal's body. I don't get the connection." What would you say?

11. *Stymied by Symmetry*. Your study buddy seems to be struggling with this chapter. Now he wants to know, "What is bilateral symmetry and why do the most complex animals exhibit this type of symmetry?"

Create and Connect

1. ***Contemplating the End.*** Life in general is tenacious, however, the life of any individual organism is fleeting at best. Sooner or later, the Grim Reaper will visit your friends, your family, and you. Consider this as you write a short story in which the members of a family struggle with the decision of whether or not to maintain a terminally ill family member on life support.

   Guidelines:

   A. Set your story up in the following format:

      ➢ Appropriate Title
      ➢ Catchy Beginning. Catch and hold your reader's attention.
      ➢ Understandable Middle. Don't muddy up the middle.
      ➢ Believable Ending. Give believable (but not necessarily happy) closure.

   B. The instructor may provide additional details or further instructions.

2. ***The Wanderer Returns.*** The starship Wanderer has returned from a mission to a distant star system. While there, the crew collected several glowing gelatinous blobs. As head of the Xenobiology Division of the United Federation of Planets, the globs have been turned over to you and your team. Explain what tests you would run and what observations you would make in order to determine (1) whether the globs are alive based on earthly standards and if they are alive, (2) whether the globs are plant-like or animal-like in nature.

3. ***Stem Cell Controversy.*** The use of embryonic stems cells is controversial on several different levels—socially, medically, religiously, politically, and legally. Write a position paper on the use of embryonic stem cells in medical research.

   Guidelines:

   A. Compose a position paper, not an <u>opinion</u> paper. Defend your position with as many facts, Figures, quotes, and pertinent information as possible.
   B. Your work will be evaluated not on the "correctness" of your position but the quality of the defense of your position.
   C. Your instructor may provide additional details or further instructions.

ANIMAL BEHAVIOR— UNDERSTANDING ANIMALS

Any glimpse into the life of an animal quickens our own and makes it so much the larger and better in every way.

—John Muir

Introduction

Whatever the scale of size or contour of shape an animal's body displays that animal's actions are controlled by a scheme of developed and instinctive behaviors. If we define **behavior** to be the organized and integrated patterns of activity by which an organism responds to its environment, then even the single-celled protozoa, the simplest of animals, exhibit behavior.

An animal's suite of behavioral responses and patterns may be as simple as the feeding response of a hydra or as complex as the migratory patterns of snow geese. Whatever the behavioral pattern and level of complexity of the pattern, that pattern has been molded by evolution to precisely fit the needs and lifestyle of an individual animal and its species.

Biology of Behavior

The shadow that was the insightful genius of Charles Darwin was so long it cast across entire fields of knowledge that he brought into being, such as evolution, ecology and animal behavior. In 1872 Darwin published a pioneering work, *The Expression of the Emotions in Man and Animals* in which he attempted to explain how natural selection would favor specialized behavioral patterns in the struggle for survival. However, the scientific community in 1872 was struggling with the concept of natural selection itself let alone its role in the behavioral patterns of animals and humans. Thus, another 60 years would pass before evolutionary concepts would flourish within the framework of behavioral science.

With the awarding of the 1973 Nobel Prize in Physiology or Medicine to zoologists Karl von Frisch, Konrad Lorenz, and Niko Tinbergen (**Figure 5.1**) came the beginnings of ethology, the scientific study of animal behavior. Today a number of specific scientific disciplines fall under the general heading of animal behavior or **ethology** (Gr. *ethos*, custom + *logos*, study): behavioral ecology (the study of how adaptation shapes behavior and ensures reproductive success), behavioral genetics (the study of the role of genetics in the behavior of animals), and comparative psychology (The study of behavior as it relates to the mental existence of animals other than humans).

Figure 5.1 Early animal behaviorists. Nikolass Tinbergen (left) and Konrad Lorenz (right).

Behavioral biologists seek the answers to two basic questions: *how* do animals behave and *why* do animals behave the way they do? The *how* questions are concerned with **proximate causation** because they deal with immediate causes or with the way the behavioral patterns develop in the life of the animal. Proximate causation explains behavior in terms of immediate physiological or environmental factors. For example, male bowerbirds build exhaustively elaborate nests of sticks, stones, and bright objects (**Figure 5.2**). Females visit bower after bower meticulously inspecting each one. When after long last a choice is made, the female mates with the male architect of the particular bower she has chosen. A possible proximate explanation states that the appearance of the selected bower and the display of the male that constructed the bower leads to elevated hormone levels in the female resulting in copulatory behavior. The *why* questions are concerned with **ultimate causation** and seek to answers to the adaptive, reproductive, or ecological significance of the behavior and how it evolved in the first place. As we discovered, female bowerbirds demonstrate strong preferences for certain male display traits, such as elaborate nests. An ultimate explanation based on sexual selection states that females who display preferences tend to mate with the most vigorous and reproductively fit of the competing males.

Figure 5.2 The bower of a male Satin Bowerbird (*Ptilonorhynchus violaceus*). The male that built this nest clearly had a decorative affinity for shiny blue objects.

To answer the questions of how and why, behaviorists examine and study behavior from four perspectives: (1) *causation*, the causes of behavior, (2) *development* and *control,* the formation and regulation of the behavior in the development of the individual, (3) *function*, the significance of the behavior in assuring the survival and reproduction of the animal, and (4) *evolution* and *genetics*, the origin of the behavior in the evolution of the species as determined from comparative and phylogenetic studies (e.g. mapping behavior onto a phylogenetic tree to see how behavioral traits evolve within and among lineages), and the genetic component of the animal that results from that evolution.

Causation of Behavior

The core tenet of ethology is the belief that behavioral traits are identifiable and measurable entities like structural or physiological traits. To investigate behavioral traits, animal behaviorists gather data by field observations and experiments conducted both in a natural setting and in the more closely controlled confines of the laboratory.

> *We are recorders and reporters of the facts, not judges of the behaviors we describe.*
> —Alfred C. Kinsey

Such research has led to the understanding that the behavior of animals may be categorized as either innate (instinctive) or learned (associative). Lorenz and Tinbergen noted that some behavioral patterns were so stereotypical (appearing in the same way in different individuals of the same species) that those patterns must be instinctive or innate (based on preset neural pathways). In a landmark 1938 paper, they described the workings of innate behavior as it related to the egg-retrieval response of female greylag geese.

If an egg is disturbed from the nest or if an egg is presented a short distance from the nest, the female greylag will rise, extend her neck until the bill is just over the misplaced egg and then bend, pulling the egg carefully back into the nest with a side-to-side motion (**Figure 5.3**).

Although this behavior might seem intelligent and perhaps learned, Lorenz and Tinbergen noted that removing the egg once caused the goose to begin the

Figure 5.3 Egg rolling behavior of the greylag goose (*Anser anser*).

retrieval pattern. They also noted that if they removed the egg or the egg being retrieved slipped away and rolled down the outside of the nest, the goose would continue the retrieval motion pattern without the egg and then settle on the nest. Realizing that the egg was still missing, the goose would repeat the entire pattern anew. Once initiated, the egg-retrieval pattern apparently had to be completed in the same fixed series of steps. Lorenz and Tinbergen concluded that such behaviors must be set in the genetic blueprint (instinctive or innate) and cause animals to show essentially the same behavior throughout their entire lives from the first time it is triggered.

Further investigations demonstrated that the graylag goose was not particularly discriminating about what she attempted to retrieve. Geese will attempt to roll baseballs, small stuffed animals, balloons, and even beer cans back into their nests. Furthermore, once the foreign object is in the nest, the goose recognizes that they are not eggs and removes them, repeating the pattern over and over. Behavior performed in an orderly, predictable sequence is called **stereotypical behavior**.

Lorenz and Tinbergen hypothesized that any object similar to an egg outside the nest was a trigger or releaser that initiated the egg-retrieval pattern. As the animal responded to the sound, or to the shape, or to some other part of a releaser, that releaser was termed a **sign stimulus**. The sign stimulus is a "signal" of some sort in the environment that triggers (releases) a certain behavior pattern or response. Ethologists have found that sign stimuli always have highly predictable responses. For example, baby birds instinctively fear flying predators and will crouch and freeze (response) whenever a shadow passes over their nest (sign stimulus). Certain nocturnal moths will fall from the sky or take evasive maneuvers when they detect the ultrasonic cries of bats (**Figure 5.4**).

As he studied the stereotypical aggressive territorial response male three-spined stickleback fish make to other males, Tinbergen hypothesized that the red underbelly of the male acted as a releaser for other males to attack (response). Tinbergen utilized a series of models presented to the males. He found that territorial males vigorously attacked any model bearing a red underside, even a lump of wax. However, a carefully constructed model that closely resembled a male stickleback but lacking a red underside was attacked far less frequently

(**Figure 5.5**). In fact, a red postal truck passing by the window was enough to evoke attack behavior from males in their aquarium on a windowsill. In humans, a yawn (sign stimulus) by one person will invariably release yawning (response) in other nearby humans.

Why do most animals rely so heavily on preprogrammed instinctive (innate) behavior? Each animal species lives in a sensory world all its own where it responds to a complex of signals and stimuli of which other animals, including humans, are largely unaware. Because most animals are not raised and trained by parents, they must be equipped to respond to stimuli immediately and correctly as soon as they emerge as an individual. Whether or not they can learn certain behaviors, doing so might take time they can ill afford to spend on the learning process. Preprogrammed (innate) Instinctive behaviors triggered by sign stimuli enable animals to respond rapidly and appropriately when speed may be the key to survival.

The instinctive theory of behavior as developed by Lorena and Tinbergen has been modified since their time because it cannot be shown that a behavior develops without some environmental experiences. Critics of the instinctive theory have contended that genes code for proteins and not directly for behavior, so all forms of behavior depend on the interaction of the organism and environment. Given a different environment, even the most rigidly instinctive behavior patterns may be altered at least slightly.

Figure 5.4 Staying put on the underside of a leaf gives this moth the best chance for surviving against echolocating bats (*Myotis bechsteinii*).

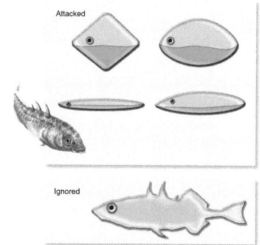

Figure 5.5 Models used to study territorial behavior in sticklebacks.

Instinctive behavior may be further categorized into *closed instinctive behavior* and *open instinctive behavior*. Closed instincts are preprogrammed; fixed motor patterns that are functional from the emergence of the individual and are usually not modified by the environment to any great extent. Some examples of closed instinctive behaviors are the sucking reflex in human infants, nest building and song in birds, and web building in orb-weaving spiders.

There is in all animals a sense of duty that man condescends to call instinct.

—Robert Brault

Open instincts are behaviors that are functional when first performed, but which can be modified as the result of interactions with a changing environment. Herring gull chicks peck at the beaks of the parents, which regurgitate partially digested food for the chick. This sign stimulus for the chick is a red spot on the beak of the parent. At first, much effort is wasted by the chicks as they clumsily miss the parents' beaks with their pecks. However, the chicks become much more efficient in this "begging behavior" as time goes by. Thus, the initial functional instinct is modified and refined by environmental interaction.

Learned or **associative behavior** develops from an interaction between the genotype of an animal and repeated inputs from a changing environment. In other words, learning is the modification of behavior through experience. Associative behaviors can take time to learn and are usually associated with those species in which at least a moderate amount of parental care is demonstrated.

The simplest form of learned behavior, known as **nonassociative learning**, is a behavioral change brought on by repeated presentation of one stimulus with no associated stimulus or event (such as reward or punishment). One form of nonassociative learning is **habituation**. As discussed earlier, young birds will instinctively crouch and freeze if a shadow or object moves overhead. However, if, over time no actual threat occurs, the young birds may habituate to such stimuli and stop responding. In the barrage of stimuli in a complex environment, habituation can be thought of as learning what signal stimuli can be safely ignored.

What has traditionally been called "learning" refers to what behavioral scientists know as **associative learning**. This term is derived from the fact that learning develops from the association of events. There are different ways of learning by association: *imprinting, conditioning, imitation,* and *instruction.*

Learning by Imprinting

As an animal matures, it must recognize and socially bond with others of its species. Such species recognition and bonding is called **imprinting**. The best-known form of imprinting is filial imprinting, in which a young animal learns the characteristics of its parent and bonds with them. Imprinting is what behaviorists categorize as *phase-sensitive social learning* because it occurs as a particular age or life stage. An extreme form of imprinting is exhibited mainly in birds such as goslings, ducklings and chicks. Upon coming out of their eggs, these young birds will follow and become socially bonded to the first moving object they encounter. The concept of imprinting was popularized by German ethologist Konrad Lorenz as a result of his work with greylag geese. Lorenz demonstrated how incubator-hatched geese would imprint on the first suitable moving object they encountered in what he called a

Figure 5.6

"critical period" of about 11-16 hours after hatching. Lorenz discovered that if greylag geese imprinted on him, they would follow him about and when they were adults, they would court him in preference to other greylag geese (**Figure 5.6**). In human terms, filial imprinting refers to the process by which a baby learns who its mother and father are and begins to bond with them.

Sexual imprinting is the process by which a young animal learns the characteristics of a desirable mate. Such imprinting is most prevalent in the avian realm and is demonstrated in over half the known orders of birds. For example, male zebra finches reared by female birds of another species, preferred mates with the appearance of the female who raised them rather than females of their own species.

Learning by Conditioning

Learning by conditioning can happen in several ways. One possibility is known as **classical conditioning** or **Pavlovian conditioning**, and the other is **operant conditioning** (or trial-and-error learning).

In the case of classical conditioning, what is achieved is that a response (the unconditioned response or UR) that was originally triggered by one stimulus (the unconditioned stimulus or US), can now be elicited by another stimulus that originally had no effect (the conditioned stimulus or CS). Originally attempting to understand the interaction between salivation and the digestive process in dogs, the Russian psychologist, Ivan Pavlov, came to realize that these processes were closely linked by reflexes in the autonomic nervous system (**Figure 5.7**). Pavlov observed that dogs begin to salivate (UR) at the sight or smell of food (US), and he wondered if external stimuli might affect this process. In one series of experiments, he repeatedly struck a bell or sounded a metronome at the same time the dogs viewed and then ate their food. By repeating this process over and over he eventually "conditioned" the dogs to begin salivating at only the sound of the bell or metronome without the necessity of presenting food. The sound acted as the conditioned stimuli (CS). Pavlov termed this form of associative learning a "conditioned reflex." Pavlov also discovered that there was a limitation as to how long the conditioned stimuli would provoke

Figure 5.7
Ivan Pavlov
(1894-1936)

the salivation response. If the bell or metronome sounds repeatedly and no food ever appears, eventually the dogs stop salivating at the sound, a situation known by behaviorists as extinction. Extinction happens in learned responses whereas habituation occurs in instinctive responses.

In the case of operant conditioning (trial-and-error learning), an animal learns to associate its behavioral response with reward or punishment. In a series of classic experiments, American psychologist B. F. Skinner studied operant behavior in rats. The rats were placed in an apparatus (which later became known as "Skinner boxes") with a lever sticking out

Figure 5.8 The Skinner box is a laboratory apparatus used to study both operant conditioning and classical conditioning in animals.

(**Figure 5.8**). During exploratory behavior, the rats would often press the lever by accident releasing a food pellet. At first the rats did not make the connection and would merely eat the food and continue exploring. However, the rats learned quickly and when hungry would spend all their time pressing the lever (behavioral response) and getting food (reward). Such trial-and-error learning is quite important in the development and survival of animals, especially vertebrates.

Learning by Imitation

Learning by imitation is a speedier and much more sophisticated means of building up behavior patterns than is learning by conditioning. Imitation depends on the ability of an animal to recognize and copy another member of its species, usually a parent or group member. In recent years, many behavioral scientists have come to see imitation as an important manifestation of intelligence in nonhuman species. Thus, it would seem to fit that there are a number of examples of learning by imitation in nonhuman primates. Rehabilitant orangutans in Tanjung Puting National Park in Central Kalimantan, Indonesia have provided some of the most complex examples of great ape behavior acquired, in part, by imitation. These orangutans copied many of the standardized human behaviors associated with life in the camp, including techniques for siphoning fuel from a drum, sweeping and weeding paths, mixing ingredients for pancakes, tying up hammocks and riding in them, and washing dishes or laundry. Perhaps the best-known examples of learning by imitation in nonhuman primates are those demonstrated by Japanese macaques. In 1963, a young female named Mukubili waded into a hot spring in the Nagano Mountains to retrieve soybeans that had been thrown into the water by the keepers. She liked the warmth, and soon other monkeys joined her. Over the years, the rest of the troop took up the behavior and they now regularly shelter in the hot springs to escape the winter cold (**Figure 5.9**). More recently, potato washing was observed in a young female named Imo. Researchers would put sweet potatoes and wheat along the beach to lure the macaques out in the open. Imo found that she could more quickly and completely remove sand from the sweet potatoes by dipping them into the river water rather than brushing the sand off with her hands as the other macaques were doing. This behavior was quickly imitated by her playmates and her own mother. Imo soon found that the potatoes tasted better if seasoned with salt water from the ocean. Soon most of the troop was once again following Imo's lead and seasoning their potatoes with salt by lightly biting into the potato then dipping into sea water to season it before consuming it. Seemingly the genius of her troop, Imo struck again when as an adult, she discovered wheat washing. Making a ball of wheat from grain thrown on the sand by researchers, she would throw the ball into the water. The wheat would float to the top where she could easily pluck it out and eat it sand-free. And yet again, the other members of her troop quickly imitated this novel food cleaning behavior. Even though Imo's peer macaques were quick to mimic her, the adult males, her superiors in the social hierarchy, refused to adopt any of her innovative behaviors.

Figure 5.9

Learning by Instruction

Learning by instruction (teaching) can be considered a higher level of learning than imitation for it involves conscious thought and intent and complex patterns of stimulus and response by both the "teacher" and by the "pupil." According to the accepted definition of teaching as applied to animal behavior, an individual is a teacher if it modifies its behavior in the presence of a naive observer, at some initial cost to itself, in order to set an example so that the other individual (pupil) can learn more quickly and completely. True teaching always involves feedback in both directions between the teacher and the pupil. That is, the teacher provides information or guidance for the pupil at a rate suited to the pupil's abilities, and the pupil signals to the teacher when parts of the "lesson" have been assimilated and that the lesson may continue.

Formal teaching was long thought the sole province of humans. Behaviorists then came to realize that learning by instruction was demonstrated by other types of vertebrates, especially our fellow primates and even other mammals. Researchers have now discovered the first example of formal teaching in an invertebrate species. Certain species of ants use a technique known as "tandem running" to lead another ant from the nest to a food source. Signals between the two ants control both the speed and course of the run and meets all criteria to qualify as teaching. Such research indicates that it could be the value of the information rather than the constraints of brain size that influences the evolution of teaching in species.

Development and Control of Behavior

Behaviorists have come to realize that even though behavior is strongly influenced by environmental influences and learning, it also has a genetic component. In the 1940s, Robert Tyson succeeded in breeding both a "maze-bright" colony and a "maze-dull" colony of rats that varied greatly as to their ability to learn the correct path through a maze. Furthermore, he discovered that the offspring of the maze-bright rats learned even faster than their parents did and that the offspring of the maze-dull rats were even poorer at negotiating mazes than their parents. The genes involved appeared to be very specific to this type of behavior as the two groups of rats did not demonstrate any other differences in performance behavior, including running a completely different type of maze. Continuing studies and experiments have revealed that behavior clearly has a heritable genetic component.

How can genes, which code only for proteins, result in behavioral actions and responses? Recent research into mutations associated with particular behavioral abnormalities has provided much greater detail as to the connection between genes and behavior. In the fruit fly *Drosophila*, individuals that possess alternate alleles for a single gene vary greatly in their feeding success as larvae and a variety of mutations that affect almost every aspect of courtship behavior in the fly have been identified.

In 1996 scientists discovered a new gene in mice, *fosB* that seems to determine whether or not female mice will properly nurture their young. Females with mutated *fosB* genes initially investigate their young but then ignore them. Females with normal *fosB* genes display caring and protective maternal behavior (**Figure 5.10**).

The genetic details of this behavior have been worked out, and it seems as if the cause of this maternal inattentiveness results from a chain of events:

1. When mothers of new babies initially inspect them, sensory information from auditory (sound), olfactory (smell), and tactile (touch) receptors is transmitted to the hypothalamus gland.

2. In the hypothalamus gland, the *fosB* genes are activated.

3. Gene activation in turn results in the activation of enzymes and other genes that mold the neural circuitry within the hypothalamus.

4. These modifications within the brain cause the female to behave in a nurturing maternal manner to her offspring.

Figure 5.10

In contrast, females with mutated *fosB* genes do not produce the proper activation proteins thus the brain's neural circuitry is not properly wired, and normal maternal behavior does not result.

The connection between environmental influences and genes in determining behavior quickly raise the old but ongoing and often controversial question of "nature-vs-nurture." That is, which plays the greater role in molding the existence of an individual, its *nature* (genes) or its *nurture* (environment)? Scientists attempt to answer that question in humans through the use of twin studies in which identical twins that were separated at birth and raised in different settings are investigated. Since identical twins have the same genome, any differences between them should be the result of being reared in different environments. However, the sample size of such investigations is relatively small and as such one must be cautious about drawing definitive conclusions, these studies have shown striking and even eerie similarities in personality, temperament, leisure-time activity preferences, and choices of everything from toothpaste to color and style of clothing.

More recently, investigators have added the search for pieces of DNA associated with particular behaviors, an approach that has been most productive to date in identifying potential locations for genes associated with major mental illnesses such as schizophrenia and bipolar disorder. Yet even here there have been no major breakthroughs, no clearly identified genes that geneticists can tie to disease. The search for genes associated with characteristics such as sexual preference and basic personality traits has been even more frustrating.

At present, the relative importance of genetics versus environment and the role each component play in behavior remains highly controversial and the subject of ongoing behavioral genetics research, especially in the relatively new field of human behavioral genetics. Understanding both the genetic and environmental contribution to variations in human behavior is a monumental task because:

• It is difficult to precisely define the behavior to be investigated. For example, consider the question. What is the role of intelligence in human behavior? Before one could even begin to answer the question, one would have to accurately define *intelligence*. Is intelligence a score on an IQ test? Is intelligence the ability to solve abstract problems? Are there different kinds of intelligence… social, financial, biological? In reality, there is no universal agreement on the definition of intelligence, even among those who study it for a living.

- Having established the specific parameters and definitions of the research, the investigator must still measure the behavior with acceptable degrees of scientific validity and reliability. That is especially difficult for basic personality traits such as shyness or assertiveness, which are the subject of much current research. Sometimes there is an interesting conflation of definition and measurement, as in the case of IQ tests, where the test score itself has come to define the trait it measures. This is a bit like using batting averages to define hitting prowess in baseball. A high average may indicate ability, but it does not define the essence of the trait.

- Behavior, like all other complex human traits, is the result of the action of multiple genes, a reality that complicate the search for genetic contributions exponentially.

- As with other studies in human genetics, studies of genes and behavior require the analysis of the pedigrees of families and populations for comparison of those that have the trait and those that do not. The explanatory power of heritability figures is limited, however, applying only to the population studied and only to the environment in place at the time the study was conducted. If the environment of the individual/population changes, the heritability may change as well. Thus, one would be attempting to conduct research on *two* variables at the same time, a definite "no-no" in rigorous and well-designed experiments. Suppose Uncle Bill and his nephew Mike both display certain aggressive behaviors. Unless both were brought up and live in exactly the same environment (making the environment a control factor and not a variable), it could not be determined what role genetics or the environment play in their shared aggressive tendencies.

Function of Behavior

Questions of how adaptation shapes behavior and how behavior may increase survival and reproduction are fertile grounds for investigation in the field of behavioral ecology. This branch of ecology, founded by Nobel laureate Niko Tinbergen and others, examines the adaptive significance of various kinds of behavior that animals have developed to increase their chances for reproductive success or fitness. To study the relation between behavior and fitness is to study the process of adaptation itself.

Behavioral ecologists understand that no animal or group of animals behaves in a vacuum thus all animal behavior is social behavior. As such, behavioral ecologists investigate a number of different animal social behaviors relating to fitness.

Agonistic (Competitive) Behavior

According to ethologists, agonistic (Gr. contest) behavior is any behavior associated with fighting, (i.e. aggression, defense, submission, etc.). Even though many animals possess specialized weapons of beak and claw or horn and tooth, inter-species aggression only rarely results in injury or death. This is because animals have, over time, invoked and utilized numerous ritualized threat displays. Mating fights, territorial fights, or fights for food are much more often symbolic jousts than they are battles to the death. Male gray catbirds will fluff their feathers, spread and lower their tails, and as a last resort, spread their wings during territorial boundary disputes with other males. Sparring for reproductive territory, fiddler crabs will display and wave their enormous claw but even if intense fighting breaks out, the crabs grasp each other in such a manner that injury does not occur. The banging and rattling of horns when bighorn sheep rams charge head long into each other can be heard for hundreds of meters yet their thick skulls and massive horns insure that only rarely and accidently does serious injury take place (**Figure 5.11**).

The silverback males of the mountain gorillas may demonstrate the most elaborate displays. With great hooting, throwing of vegetation, chest pounding, leg kicks, and side-

Figure 5.11

Figure 5.12 Darwin's principle of antithesis is often displayed by dogs. The antithesis of threat (a) is submission (b).

ways running, these massive animals attempt to intimidate each other without actually making contact. Nevertheless, their aggression can occasionally escalate into the death of a rival. Slashing with huge canines that can open gaping wounds, silverback mountain gorillas will sometimes battle to the death. Similarly, bull elephants sometimes turn to a type of combat that sees each warrior using tusks to attack weak points in the other's body. Most often the loser in a symbolic battle escapes with only minor injuries and either runs off or assumes a subordinate posture that signals defeat. In his book, *The Expression of the Emotions in Man and Animals* (1872), Charles Darwin described the opposite nature of threat and subordination displays as the "principle of antithesis," a principle still accepted by ethologists today (**Figure 5.12**).

Territorial Behavior

A **territory** is any space that an animal defends against intruders of the same species. Territorial defense is exhibited by nearly every type of animal, even humans. As with any competitive behavior, defending ground has benefits and costs. Possessing sufficient territory gives the holder access to forage areas, enhances the chances of attracting a mate, and reduces vulnerability to predators. The territory holder, however, usually maintains his ground only by expending much time and effort in its defense. Sunbirds, for example, can expend up to 3,000 calories per hour patrolling and defending territory. Such costs can only be paid and benefits realized if there is an abundance of food to support it. If food supplies are low, an animal may not gain enough energy to balance the energy used in defense. In such circumstances, it is not advantageous to be territorial. If food supplies are high, an animal can meet its daily energy needs without the added cost of being territorial.

From an energy standpoint, defending even abundant resources usually isn't worth the cost. Thus, territoriality usually occurs only at intermediate levels of food availability, where the benefits of defense offset the costs.

Territories vary greatly in size and are usually related to the mating system of a species. A male song sparrow may have a territory of approximately three-fourths of an acre whereas birds that congregate in large colonies, such as terns, gulls, gannets, boobies, and albatrosses, will precisely space their nests just beyond the pecking distance of the neighbors on all sides (**Figure 5.13**). Territorial behavior is more common in birds than it is in mammals. Birds have less difficulty patrolling a large area for enemies and competitors than do mammals. For this reason, many mammals establish a **home range**. The area an individual covers or patrols in its normal routine constitutes its home range. As the area covered is not exclusive to or solely defended by any one individual, a home range often overlaps the home ranges of the same species.

Figure 5.13 Nesting royal terns (*Thalasseus maximus*) nearly precisely aligned in relation to each other.

Sometimes the space defended moves with the individual. Known as *individual distance,* these traveling spaces can be observed in the spacing between birds on a wire or people waiting in line at the school cafeteria.

Mating Behavior and Reproductive Strategies

The patterns of mating and the types of mating systems found throughout the animal world are quite diverse. Behavioral ecologists differentiate mating systems by the patterns of interaction between males and females into four categories: (1) **monogamy**—one male and one female at a time, (2) **polygamy**—males or females have more than one mate, (3) **polygyny**—males mate with more than one female, and (4) **polyandry**—females mate with more than one male.

Mating systems represent reproductive adaptations to ecological conditions. The need for parental care and the ability of both parents to provide it are also important influences on the particular type of mating system any species develops within the ecological constraints of their situation.

Animals expend tremendous amounts of time and energy to find and attract mates and most have developed some type of sexual selection process to bring order to the haphazardness of reproduction. Sexual selection involves either *intrasexual selection* or interactions between members of the same gender and *intersexual selection* or mate choice. Sexual selection is particularly pronounced in birds. Male birds fighting with other males of their species to mate with a single female demonstrates intrasexual selection whereas a single female carefully selecting the male she will mate with demonstrates intersexual selection. Mate choice often involves elaborate and complex behaviors known as courtship rituals.

Animal courtship may involve complicated dances or touching; vocalizations and singing; displays of bright colored body parts and even fighting prowess (**Figure 5.14**). Most animal courtship tends to be secretive and out of human sight, so it is often the least documented of animal behaviors.

Foraging Behavior

Most animals are structurally and physiologically adapted to feed on a limited range of food and to gather this food in specific ways. For many animals, food comes in a variety of sizes. Larger foods may contain more energy but be harder to capture or obtain. Also, some food may be further away than other types. Hence, foraging (finding food) involves a trade-off between a food's energy content and the energy expended to obtain the food—the *net energy intake*.

Figure 5.14 As part of their mating ritual, red-crowned cranes (*Grus japonensis*) often "dance" with each other.

According to the optimal foraging theory, natural selection favors individual's whose foraging behavior is as energy efficient as possible. That is, the most successful animals tend to feed on prey that maximizes their net energy intake per amount of foraging time. That in turn should lead to increases in reproductive success. And that does seem to be the case. Studies on ground squirrels, captive zebra finches, and orb-weaving spiders have shown a direct relationship between net energy intake and the number of offspring raised.

Highly successful foraging strategies, however, can be complicated and compromised by matters of ecology, age, mate selection, and prey avoidance. An example of how differences in ecology can alter foraging strategies is found in the relatively unproductive marshes of northeastern United States. Great blue herons there must spend a great deal more time finding food than herons in Florida where marsh productivity is much higher, and food is easier to find.

Age plays a role in the foraging of the small North American bird known as the yellow-eyed juncos. Even though large prey contains more energy, young juncos lack the experience to handle large prey. They focus instead on smaller prey until they are older and experienced enough to easily dispatch larger prey.

Many studies have shown that a wide variety of animal species alter their foraging behavior by becoming less active and staying near cover when predators are present. Finding a mate, however, is such a driving force that males of many animal species throw caution to the wind by not only reducing their foraging time but also exposing themselves to predators in order to find a mate.

Communication

Animal communication is considered to be any behavior on the part of one animal that has an effect on the current behavior of another animal. The study of animal communication has played an important role in the development of behavioral ecology, ethology, and the study of animal cognition.

The sender and receiver of a communication may be of the same species (intraspecies) or different species (interspecies). Behavioral ecologists tend to focus on intraspecfic communication as it represents the majority of animal communication.

Animal communication takes a number of different forms. The most distinctive forms of communication tend to be visual displays of often brightly colored body parts or distinctive body movements. There may be no better example of coloration as communication than the amazing tail feathers of a peacock and no more complicated body movement communication than the dance language of the honey bee (**Figure 5.15**).

Vocalization is also an effective means of communication that has the added benefit of allowing the vocalizer to remain hidden from the prying eyes of potential predators. The best known vocalizers are birds and whales with their complicated, melodious, and haunting songs. However, many other animals also vocalize—frogs croak; king cobras will emit a growl-like hiss but the African puff adder produces a bell-like chiming sound; lions and tigers roar and bellow and many monkeys and apes screech and caterwaul. The strangest vocalization may be that of singing mice. Many anecdotal accounts of singing mice are on file, but the explanation for their vocal talents remains unclear. In his book, *Animal Curiosities* (1922), W. S. Berridge documents a quite amazing example of singing mice: "…a specimen kept by a lady as a pet was able to run up an octave when singing…it would often finish its vocal performance with a trill. When thus engaged it would vibrate and inflate its throat in the manner of a bird in song, and usually assume an upright posture upon its hind feet."

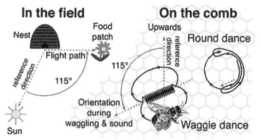

Figure 5.15 Dancing Bees. When a food source is very close to the hive (less than 50 meters), a forager performs a *round dance*. She does so by running around in narrow circles, suddenly reversing direction to her original course. If the food source is more than 150 meters from the hive, a *waggle dance* is performed. While running the straight course, the bee vigorously wags its body. The longer the waggle, the greater the distance to the food source. The orientation of the dancing bee during the straight line portion of the waggle dance indicates the position of the food source relative to the sun.

Less obvious is chemical communication. Chemical signals are likely the oldest and most widespread means of communication by which animal signal other members of their species. Any body secretion that functions as a mode of communication is called a **pheromone**. Pheromone communication has been discovered in nearly all organisms from unicellular protozoa to human beings.

Most pheromones act as releasers, eliciting a very specific but transitory type of behavior in the recipient. Others act as primers, evoking slower and longer lasting responses. In social insects releaser pheromones supplement visual and tactile clues in leading members of the colony back to a food source, as in the trail pheromones of ants. Some pheromones enable members of a colony to recognize one another, and some warn the colony (alarm pheromones) of foreign invaders. In many species, pheromones are important in attracting a mate and in gender recognition.

Some pheromones function over great relative distance and in often miniscule amounts. Sex pheromones released by some female lepidopterans (moths and butterflies) can be detected by a potential mate from as far as 10 km (6 miles) away. Special receptors on the antennae of male silk moths (*Bombyx mori*) are so exquisitely sensitive that they can detect a female sex pheromone known as *bombykol* in concentrations as low as one molecule of bombykol in 1017 molecules of air (**Figure 5.16**).

Altruism

Social groups of animals range in complexity from simple aggregations of unrelated individuals that may cooperate in chasing off a predator to huge nesting colonies of birds to the complex groups of ants, bees and other social insects in which there is a high degree of relatedness and a distinct division of labor. **Altruistic behavior**, in which one individual appears to act in such a way as to benefit others sometimes at a cost to itself, is frequently seen in more complex social groups. Such behavior is easily understood when the behavior is obviously cooperative or reciprocal, and all the cooperating members share in the benefits,

Figure 5.16

such as pack hunting in wolfs and lions. However, altruistic behavior that imposes a cost to an individual and how such behavior might be favored by natural selection presents a paradox that has long perplexed evolutionary biologists. If altruism imposes a cost or negative consequence to an individual, how can it be favored by natural selection?

True altruistic behavior may be explained at least in part by the inclusive fitness theory. This theory suggests that that an organism can improve its overall genetic success by altruistic social behavior. The population geneticist J. B. S. Haldane once remarked that, "I'd gladly lay down my life for two brothers or eight first cousins." Haldane's statement makes genetic sense in that two brothers, or eight first cousins (or four nephews) would pass on as many particular combinations of his genes to the next generation as he would himself. In essence, this statement explains the evolutionary core of the inclusive fitness theory. Natural selection will favor any behavior that increases the net flow of an individual's genes to the next generation

One specific form of inclusive fitness is kin selection. **Kin selection** may be explained as changes in gene frequency across generations that are driven at least in part by interactions between related individuals. In ant colonies, sterile workers help the queen raise their sisters, but they themselves never have offspring. Alarm calls in ground squirrels and male prairie dogs are another example of kin selection. Ground squirrels call most frequently when relatives are nearby, and male prairie dogs modify their rate of calling when closer to kin. These behaviors indicate that self-sacrifice is directed towards close relatives and that there is an indirect fitness gain. The larger members of the social shrimp, *Synalpheus regalis*, protect juveniles in the colony. By doing so, the large defender shrimp my increase their own inclusive fitness. Velvet monkeys (*Chlorocebus pygerythrus*) display a type of kin selection between siblings, mothers and offspring, and grandparent-grandchild known as allomothering, where typically an older female, sibling or grandmother will provide care to closely related infants.

With kin selection the "helper" individual(s) receive no direct benefits other than the advancement of their own genes. There are times, however, when the "helper' does directly benefit, and those situations are known as **reciprocal altruism** or **reciprocity**. Reciprocity is a hypothesis suggesting that individuals may form "partnerships" in which mutual exchanges occur because they benefit both participants. In such circumstances, "cheaters" (nonreciprocators) are discriminated against and are cut off from receiving future beneficial exchanges. The common vampire bat of Central and South America roost in hollow trees of groups of 8 to 12 individuals. These bats have a high rate of metabolism and may starve to death if they do not feed within at least 60 hours after their previous meal. About 30% of the immature bats and about 7% of the mature ones do not get a meal each night because they either do not leave the roost or do not find a mammalian host. The bats that do feed will regurgitate parts of their own blood meal to hungry roommates, especially with past reciprocators. Individuals that fail to give blood to a bat from which it received blood in the past will be excluded from future blood sharing.

Kin selection and reciprocity are demonstrated by vampire bats discussed above in that a significant percentage of the recipient bats are related to the donor. Another of many examples is seen in wild turkeys. Groups of males gather in a special mating territory where with ritualized tail spreading, wing dragging, and gobbling, they attempt to entice female turkeys to mate. Because of cooperation between males within a group, one group attains dominance over other groups, and its dominant member is the one that copulates most frequently with the females. Seemingly the males who helped establish the dominant group but who have low social status within the group gain nothing. Close analysis has shown, however that all members of the dominant group are brothers from the same brood. Since they share many genes with the successful male, they are indirectly perpetuating many of their own genes.

Humans may be the only species that perform acts of altruism that are on a higher plane than any other. Even though creatures as lowly as slime molds and everything in-between seem to exhibit some form of altruistic behavior, humans seem to be the only animal capable or willing to help others to which they are not genetically related and may not even know. Let natural disaster happen anywhere on the planet and all forms of relief—money, food, water, medicine, medical services, rescue teams and heavy equipment—will flow in from points around the globe to aid and assist the human survivors of that disaster.

Since it was first proposed by Oxford University biologist Bill Hamilton in the 1960s, inclusive fitness has been considered sacrosanct. For nearly 50 years hundreds of scientists have made careers observing degrees of kinship and reciprocity in termite mounds, anthills, ground squirrel colonies, and monkey troops. However, in an article published in the journal *Nature* in 2012, evolutionist and mathematician Martin Nowak, mathematician Corina Tarnita and the legendary sociobiologist E.O. Wilson, declared the mathematics of inclusive fitness so unwieldy as to be useless and that kinship and inclusive fitness are much less important than previously thought to explain the origin of complex insect societies such as those of bees and ants, or altruism towards kin in humans. In the new calculus, known as supercooperation, altruism emerged simply because it gave some individuals an edge in the struggle to survive. Over time, the descendants of these altruistic ancestors grouped together, and while the motive of any one individual may have been selfish, extreme cooperation was the happy result. In fact, Nowak contends that his mathematical models and elegant experiments clearly demonstrate that cooperation has been so important in human evolution that it drove not only the development of language but also the size of our brain. The notion of supercooperation was a firebomb thrown into the sacred heart of evolutionary biology, and the shock waves it generated are still reverberating.

Evolutionary biologists, ethologists, sociologists, and psychologists to present have not been able to provide definitive answers as the origin and development of the unique social behavior known as altruism nor can they precisely discern what advantages such an adaptation might impart on our species, although it may be a carryover from living in small related groups.

With the 1975 publication of E.O. Wilson's *Sociobiology: The New Synthesis*, a new discipline—**sociobiology**—was formalized within the field of behavioral ecology. Sociobiologists contend that social behavior has resulted from evolution, and they attempt to explain social behavior within that context. The result is an expansion of standard Darwinian evolutionary theory (which traditionally explains morphological adaptation) to a new domain: namely, animal sociality.

The foundation principle of sociobiology that animal social behavior is molded by natural selection and driven solely by genetics has always been controversial. The issue has become even more contentious with the application of sociobiological principles to human social behavior. Critics assert that sociobiological models are inadequate to account for human social behavior because they contributions of human intelligence and cultural influences. Furthermore, critics argue that the view that human behaviors (such as aggression) are genetic fixed and predetermined is scientifically implausible and socially dangerous.

The question, then, is this: Is the Darwinian theory of natural selection alone an appropriate framework for the study of human behavior or should human behavior best be studied apart from evolution and viewed as the product of human institutions, culture, and intelligence (social constructionism)? The debate continues.

Comparative Psychology

Comparative psychology is the study of the behavior and mental existence of animals other than humans. Since the 1990s, comparative psychology has undergone a reversal in its fundamental approach. Instead of seeking to explain human behavior from the principles of animal behavior, comparative psychologists started taking fundamentals that have been uncovered in the study of human cognition (conscious intellectual acts) and testing them in animals of different species testing them in animals of different species—animal cognition or cognitive ethology.

> *It is like man's vanity and impertinence to call an animal dumb because it is dumb to his dull perceptions*
>
> —Mark Twain

Studies on animal cognition develop out of comparative psychology but also draw on ethology, behavioral ecology, and evolutionary psychology. Working in settings ranging from the bush to the laboratory to the zoo, and using a range of methodologies, cognitive researchers investigate the animal analogs of human cognitive processes. These processes are:

Memory

The categories that have been developed to analyze human memory—short term, long term, and episodic memory—have also been applied to the study of animal memory. However, the study of **spatial memory** in animals such as Clark's Nutcracker (*Nucifraga columbiana*), certain squirrels, jays, and titmice whose scatter-hoarding behavior require them to remember the locations of thousands of hidden seed caches have been the most common type of animal memory studies performed.

Recent studies on scrub jays, gorillas, and chimpanzees seem to indicate that some animals can recall past event and use the information to plan for the future. For example, scrub jays stash food to recover later. Jays that had experienced avian robbery, however, were much more cautious about hiding their stashes in the future. A gorilla, communicating via sign cards, shows that he can remember who gave him certain foods even when his human caretakers cannot remember, and a chimp, after watching a variety of objects being hidden, can direct unknowing humans to the hiding places of the objects. Are these and other examples evidence that animals other than humans exhibit **episodic behavior** or the ability to recall the social content of an event and adjust future behavior accordingly? Cognitive researchers disagree as to whether the answer is yes, no or something in-between, but what they do agree on is that this fascinating idea is worthy of deeper exploration.

Tool Use

To decipher the existence of tool use in animals, one must first define what is meant by "tool use." A simple but workable definition is that a **tool** is anything not part of the body that is used to accomplish a given task. Given this definition, tools can vary from a stick poked into a termite mound by a chimp to a human using a hammer to drive nails to an elephant scratching itself against a tree.

A number of different species use tools as an essential part of their foraging behavior. To compensate for its short tongue, the woodpecker finch of the Galapagos Islands will use sticks and cactus spines to pry prey such as grubs from crevices in trees. The finches are very selective in picking the proper stick or spine and will even adjust the length of the tool to better fit their needs (**Figure 5.17**). However, these behaviors are often quite inflexible and cannot be applied effectively in new situations.

Figure 5.17 A Galapagos woodpecker finch (*Camarhynchus pallidus*) uses a stick to probe for grubs.

The primates seem to be the most capable and creative when it comes to using tools. Monkeys will throw branches to discourage pursuing individuals, use sticks for clubbing, throwing and to reach food, and wipe their wounds with lightly masticated leaves. The intellectually limited gorillas demonstrate very little tool use, leaving chimpanzees as the head of the nonhuman primate class, with orangutans not far behind. Chimps and orangutans not only to use tools but put a great deal of thought into accomplishing certain goals by using certain tools, some of which they make themselves. Chimps will use sticks to "reach and rake" and can dismantle objects by prying and levering. In captivity, chimps can dismantle entire "play sets" within their cages and will lever cage bars apart so as to stick their heads out. They will fish for termites, dip for ants, use sticks to touch things which they prefer to avoid such as unpleasant or dangerous objects or unfamiliar chimps, use poles for balancing and climbing, and clean themselves with leaves

In the wild, great apes and many monkeys are both willing and able to manufacture and use tools. These primates demonstrate a capacity to think and to act which is at least equal, if not superior, in intellectual level to that of human children.

Box 5.1
The Games Animals Play

Many animal behaviors seem to mimic human children's games. Puppies rough house with each other, kittens toy with dead prey; young monkeys play leapfrog, young hyenas engage in keep away with shreds of carcass and young otters love king-of-the-hill. On the face of it, humans find such behavior quite endearing, but is there something deeper going on? Ethologists are determined to learn the reasons why animals play.

Play may be regarded simply as actions resembling serious behavior in appearance, but not in purpose or outcome. Researchers recognize four basic forms of play behavior.

1. **Predatory play** involves mock chasing, pouncing, pawing and Figurehting and other such skills necessary for success as a predator.

2. **Locomotor play** consists of juvenile animals performing body movements that mimic the physical actions of adults. Locomotor play appears to function as motor skill training for animals that rely on agility and speed to avoid predation.

3. **Object play** is considered to overlap somewhat with predatory play. Non-predatory object play has been manifested primarily by the primates and to the greatest extent by those primates most closely related to humans.

4. **Social play** may serve two functions: safely teaching young the skills they will later use in aggressive social situations and strengthening the bonds between group members which will later serve to limit the amount of actual aggression between those group members.

Long regarded the provenance of mammals and to a lesser extent birds, particularly ravens, play has been recorded in reptiles. Pigface, a large African soft-shelled turtle was acquired by the National Zoo in Washington, D.C. in 1940. In the 1980s, Pigface began to claw at his own skin. In an attempt to distract the turtle from this destructive self-mutilation, zookeepers introduced balls, sticks and hoses into Pigface's enclosure. Surprisingly, zookeepers observed that PigFace began to spend a large amount of his time in what appeared to be object play behavior. Balls proved to be PigFace's favorite objects. He would approach a ball, push it away, approach again and push, over and over sometimes for hours on end. Once objects were introduced, PigFace opted for play over self-mutilation and his health improved. The power of play, even in a turtle, can be profound.

Although Shakespeare may have been correct when he stated, "Play needs no excuse," ethologists now know there are reasons *why* play is important in the life of young animals. It's not just all fun and games.

Reasoning and Problem Solving

Closely related to tool use is the study of reasoning and problem-solving in animals. It is clear that a considerable range of species, especially birds and mammals, are capable of solving a range of problems that can be argued to involve abstract reasoning.

In a series of classic experiments conducted in the 1920s, a chimp was placed in a room with an assortment of boxes on the floor and food hanging from the ceiling out of reach. After several unsuccessful attempts to reach the food by jumping, the chimp quickly proceeded to stack the boxes one on the other beneath the food until it could climb high enough to claim the food.

> *You can have the most beautifully designed experiment with the most carefully controlled variables, and the animal will do what it damn well pleases.*
>
> —The Harvard Rule of Animal Behaviour

Ravens have always been considered among the most intelligent birds, and recent experiments have revealed a grain of truth in this long-held belief. Captive-reared ravens in an outdoor aviary were presented meat hung from a branch on the end of a string. Ravens like to eat meat but cannot feed while hovering and thus were unable to get to the meat. After several hours of contemplation but no action, one raven flew to the branch, grabbed the string, pulled it up, and stepped on it. He repeated this action until he brought the meat close enough to grab. This raven, when faced with a complex and unfamiliar challenge, clearly demonstrated creativity and advanced problem-solving abilities (**Figure 5.18**).

Figure 5.18

Language

When investigating the existence of true animal language, the difficulty lies in separating actual language from mere communication. Some researchers argue that there are significant differences separating human language from animal communication and that the underlying principles between the two are clearly not related. Others argue that an evolutionary continuum exists between the communication methods of animals and human language. What properties separate human language from animal communication?

- *Discreteness.* Language is composed of discrete units that are used in combination to create meaning.
- *Displacement.* Language can communicate ideas about things that are not in the immediate vicinity, either spatially or temporally.

- *Productivity.* A finite number of units (alphabet) can be used to create an infinite number of words.
- *Arbitrariness.* There is no rational relationship between a sound or sign and its meaning.
- *Cultural transmission.* Language is passed from one user to the next.
- *Metalinguistics.* The ability to discuss language itself.

Even though some animal communication such as the waggle dance of bees, the vocal and object recognition abilities of some parrots, whale songs, prairie dog barking, the variety of color patterns and skin texture of some squid, and the interaction between humans and great apes through signs or symbols all demonstrate at least some of the properties of human language, none meet all the criteria to be considered a true language.

Consciousness

Do animals have a sense of consciousness or self-awareness? This question has been and continues to be hotly debated. Research involving the mirror test has been revealing but controversial. If an animal's skin is marked in some way when it is asleep or sedated, and it is then allowed to see its reflection in a mirror, any spontaneous grooming behavior directed to that mark could be taken as an indication that the animal is aware of itself (**Figure 5.19**). Self-awareness by this criterion has been reported for chimpanzees, orangutans, gorillas, and nonprimates such as dolphins, elephants, and magpies. The data and the interpretation of that data from such experiments, however, are contested. Critics contend that such experiments test the ability of an animal to understand their image more than the animal's actual awareness of their own self.

Figure 5.19

Another research approach based on a child's crib speech investigations is derived from passive speech research with talking birds such as macaws and African gray parrots. Further arguments revolve around the ability of animals to feel pain. Suffering implies consciousness.

Others contend that pain can be demonstrated by adverse reactions to negative stimuli that are non-purposeful or even maladaptive. A major hindrance is that experts don't necessarily agree on what "consciousness" means when applied to animals. In fact, about forty different meanings attributed to the term consciousness can be identified and categorized.

Deception and Empathy

Do animals deceive each other (lie) and can they show empathy (capacity to recognize emotions being experienced by another animal) or are these strictly human traits? Research to answer such questions has been carried out mainly on primates. Although many anecdotal accounts appear to support the idea that deception and empathy occur in some nonhuman primate species, such as baboons and chimpanzees, devising field-based experiments to test these ideas has proven difficult.

Emotion

Throughout human history animal lovers, scientists, and philosophers have considered the possibility that non-human animals feel and express emotions in the same sense that humans do. The whole matter is fraught with ambiguity and anthropomorphism (applying human characteristics to nonhuman animal behavior). On one hand society condemns animal cruelty and criminalizes such actions, and yet on the other hand it is far from clear scientifically whether animals "feel" in any meaningful sense. An animal may make certain distress movements and sounds, and show certain brain and chemical signals when injured in some way. But does this mean, as some persistent critics maintain, that the animal feels and is aware of pain in the same manner we do or does it mean that the animal is merely programmed to act a certain way in response to certain stimuli? To veterinary consultant Dr. Jean Swingle Greek there is no ambiguity. "It is amazing that time and resources still need to be wasted convincing some that what looks like distress in a rat, is, in fact, distress." Dr. Greek's contention is supported by mounting evidence that strongly suggests animals have emotions as people do, albeit lacking certain cognitive insights. Such investigations pose both a scientific dilemma—how can we tell?—and an ethical one—if true, what does it mean and what do we do about it?

A Closing Note
Jousting With Squirrels

Why do animals do what they do and how can we get them to do what <u>we </u>want them to do? People have always been more interested in the latter question than the former because we have always desired to manipulate animals and bend them to our will for many, many different purposes.

I am seriously into feeding birds and do so year round. As such, I have positioned a number of different feeders in and around a tree in my yard in such a manner that I can look down on the whole affair from the second story window of my home office. I find it very relaxing and quite fascinating to watch all the action around those feeders, especially when the snow falls. However, things do not always go as I would like in this little world I have created. You see, I find myself locked in a battle with three or four squirrels over the expensive bird seed and peanuts that I prefer they NOT eat.

Zoologists must constantly struggle to prevent the sin of anthropomorphism or attributing human traits, values, and actions to animals. I know this, but mea culpa. Even though I admire and respect their athletic persistence, I can't help seeing those squirrels as cocky and defiant. I also suspect they are mocking me and laughing at me behind my back.

I have tried so-called squirrel-proof feeders, squirrel baffles on feeder poles, covering my feeders with hail screen, sprinkling powered coyote urine around, and treating the bird seed with red hot cayenne pepper (which the squirrels can taste, but the birds cannot). I have bought a live animal trap in which to catch and relocate them. I quickly caught one squirrel in my trap but after that the other squirrels refuse to enter it regardless of how many tasty morsels I tempted them with. In hopes to keep them on their turf and not on mine, my latest tactic has been to feed the squirrels special food but place it across the street in my neighbor's stand of trees where the fuzzy-tailed little rats hang out. But alas, no strategy or defense I employ works very well if at all. In fact, when I think about it, the squirrels have been more successful in manipulating my behavior than I have been in manipulating theirs. In this battle of wills and jousting of wits, it pains me to admit it, but the squirrels seem to be winning.

In Summary

- If we define behavior to be the organized and integrated patterns of activity by which an organism responds to its environment, then even the single-celled protozoa, the simplest of animals, exhibit behavior
- Animal behavior may be examined from four perspectives:

  1. Causation
  2. Development and control
  3. Function
  4. Evolution and genetics

- The behavior of animals may be categorized as either innate (instinctive) or learned (associative). Instinctive behavior may be further categorized into closed instinctive behavior and open instinctive behavior
- Learned behavior develops from an interaction between the genotype of an animal and repeated inputs from a changing environment. Learning is the modification of behavior through experience.
- Learned behavior may be nonassociative or associative.
- Behaviorists have come to realize that even though behavior is strongly influenced by environmental influences and learning, it also has a genetic component. How can genes, which code only for proteins, result in behavioral actions and responses? Recent research into mutations associated with particular behavioral abnormalities has provided much greater detail as to the connection between genes and behavior.
- Questions of how adaptation shapes behavior and how behavior may increase survival and reproduction are questions investigated by behavioral ecologists.
- Behavioral ecologists investigate a number of different animal social behaviors relating to fitness:

  1. Agnostic (competitive) behavior
  2. Territorial behavior
  3. Mating and reproductive strategies
  4. Foraging behavior
  5. Communication
  6. Altruism

- Comparative psychology is the study of the behavior and mental existence of animals other than humans. Since the 1990s, comparative psychology has undergone a reversal in its fundamental approach. Instead of seeking to explain human behavior from the principles of animal behavior, comparative psychologists started taking fundamentals that have been uncovered in the study of human cognition (conscious intellectual acts) and testing them in animals of different species—animal cognition.

- Cognitive researchers investigate the animal analogs of human cognitive processes. These processes are:

1. Memory
2. Tool use
3. Reasoning and problem-solving
4. Language
5. Consciousness
6. Deception and empathy
7. Emotion

Review and Reflect

1. ***Instinct vs. Learning***. The behavior of animals may be categorized as either innate (instinctive) or learned (associative). Explain the differences between the two categories and give several examples in each category of behavior in action.

2. ***Filling in the Blanks***. Your friend fell asleep during lecture, and now he has some questions about what he missed. As you fill him in from your complete and well-kept lecture notes, he has some questions—"What are the differences between closed and open instinctive behavior? What are some examples of both closed and open instinctive behaviors in action?" How would you answer his questions?

3. ***Define It***. You have been asked to contribute to a biological dictionary. Among the word and phrases you have been assigned is the phrase *associative behavior* (learning). How would you define this phrase?

4. ***An Imprint***. You have been given the assignment of demonstrating imprinting in any animal species of your choice. Explain what animal species you would select as your test subject and the exact procedure you would follow to meet the challenge of demonstrating imprinting in that particular species of animal.

5. ***Pavlov's Dogs***. You did so well on your previous assignment that you have now been given the task of demonstrating both classical conditioning (Pavlovian conditioning) and operant conditioning (trial-and-error learning). Explain what type of animal(s) you would select as your test subject(s) to demonstrate each type of conditioning and the exact procedure you would follow to meet the challenge. Give an example of each type of conditioning in action.

6. ***Nature vs. Nurture***. Certain behaviors and behavior patterns seem to occur in families. When studying variations in human behavior, the exact contributions of genes (nature), and the environment (nurture) to these variations in behavior are exceeding difficult to determine. Why?

7. *The Puzzle of Altruism*. What is altruistic behavior and why has altruistic behavior long perplexed evolutionary biologists?

8. *Can Animals Reason*? Results of animal cognition (cognitive ethology) studies and the conclusions drawn from these results are often disputed and occasionally highly controversial. Research animal cognition (cognitive ethology) then write a position paper defending your answer to the question: Do animals display powers of higher cognition?

Guidelines:

A. Compose a position paper, not an opinion paper. Defend your position with as many facts, Figures, quotes, and pertinent information as possible.

B. Your work will be evaluated not on the "correctness" of your position but on the quality of the defense of your position.

C. Your instructor may provide additional details or further instruction.

9. *A New Language*? You are studying animal communication. As such, you have taught a chimpanzee a sign language system. Using that system, the chimp can sign an amazing number of words and form complete sentences. You go away for many months conducting research aboard. Before you go you leave your chimp under the care of a research primatologist who places your chimp in with hers. On your return you are surprised to discover that your chimp has not only taught your sign language system to the other chimps but that he has also made major modifications in the original system. Has your chimp developed its own language? Why or why not? Defend your answer.

Create and Connect

1. *Watch and Learn*. Observe a specific animal or a small group of animals of the same species and record detailed and accurate information derived from your observations. The instructor will provide details on possible animal subjects suitable for observation and on the number of observations to be made as well as the duration of the study.

2. *Marketing Widgets*. Imagine that you are the marketing director of GigantoCorp. Your company has just developed a new product known as a *widget*. Plan a campaign to market widgets to the public. Design your campaign around the behavioral concepts you learned in this chapter. Make sure all aspects of your advertising campaign are legal and ethical. The instructor will provide other details and requirements as needed.

3. *Time for School*. Learned behavior is believed to be a mental skill found almost exclusively in mammals and birds. (The invertebrate octopus, however, has a well-developed brain and sense organs and clearly demonstrates learned behavior and possibly logical thought.). If you have ever attempted to teach a pet some tricks, you have delved into learned behavior. We take for granted that humans and higher animals can learn but can you prove it? Design an experiment to test the

problem question: *Can* _____ *learn?* (Insert the specific mammal or bird you wish to investigate into the blank.)

Guidelines:

Following the tenets of a well-constructed experiment, your design should include the following components in order:

➢ The *Problem Question*. State exactly what problem you will be attempting to solve.

➢ Your *Hypothesis*. Even though this is a fictitious experiment, word your hypothesis as realistically as possible.

➢ *Methods and Materials*. Explain exactly what you will do in your experiment including the materials necessary to accomplish the task. Be specific, take nothing for granted, and do not expect people to read your mind as they read your work.

➢ *Collecting and Analyzing Data*. Explain what type(s) of data will be collected and what statistical tests might be performed on that data. It is not necessary to concoct either fictitious data or imaginary observations.

6

PHYLOGENY AND TAXONOMY: QUESTIONS OF ORIGINS AND ORGANIZATION

Nothing in biology makes sense except in the light of evolution.
—Theodosius Dobzhansky

Introduction

As this tiny and fragile craft that is our home planet pirouettes around the nurse star we call Sol, it carries aboard it a living crew of such amazing diversity that one can only marvel at the variety and complexity of it all. When evolutionary biologists contemplate our fellow passengers on this tiny mote, questions about this tooth and claw realm invariably spring to mind: Where did animals come from? (*origins*) How has animal life changed since it first originated? (*evolution*)

For all the questions about how and where animal life first appeared that we cannot answer, we do know with great certainty that once animals did appear, they flourished against the backdrop of ever-changing geologic conditions.

Questions of Evolution

Questions as to the origin of new species can be answered by examining the tenets of modern evolutionary theory. Please note that the word "theory" as it is applied to evolution in this chapter is used in an integrational sense to indicate a concept that unifies a set of ideas, not in the experimental sense to indicate an unproven hypothesis. Furthermore, the word "evolution" is defined here in its broadest sense to indicate change over time. Given that definition, no clear-thinking and rational person can deny that biological evolution (change) has occurred on this planet. To hold even a single fossil in one's hand is to possess overwhelming evidence that life on this planet has evolved over the course of geologic time.

History of Evolutionary Theory

Modern evolutionary theory is the core of biology from which springs all other aspects of life and living things. In the broadest sense, the task of biology is to understand the nature of the living world. To comprehend the whole of life, biologists must first understand the four pillars upon which the natural world rests:

1. *The unity of life.* All organisms on the planet rely on nucleic acids (DNA or RNA) for the transmission of genetic information across generations.
2. *The diversity of life.* Untold millions of species populate the earth with ever more slowly coming into existence.
3. *Biological adaptations.* All organisms fit their environment.
4. *The history of life.* Life on this planet extends back at least 3.4 billion years. New forms appear, diversify, and go extinct, leaving behind traces in both the fossil record and in their living relatives.

Only one set of ideas ties all these aspects of biology together and offers an overarching and complete explanation of the nature of life: modern evolutionary theory.

The evidence for biological change over long periods of time is indisputable. Even ancient people understood this and attempted to explain the mechanisms by which such changes occur. Rooted in antiquity, the possibility and processes of evolution were conceived as philosophical ideas during the times of the ancient Greeks and Romans.

From the Ancients to the Renaissance. The idea of evolution as biology and not just philosophy dates back to the sixth century B.C. to Anaximander, who advanced a theory on the aquatic descent of mankind. Later, Aristotle would articulate the *Great Chain of Being* (or *Scala Naturae*) that proposed all species be placed in order from the "lowliest" to "highest" in a ladder-like fashion, with worms on the bottom and God on the top. Because he assumed the universe to be ultimately perfect, Aristotle believed that the Great Chain could contain no empty links and that no link was represented by more than one species. Aristotle's perfect universe represented a very rigid and static view of the natural world where species could never change and must remain fixed. In a poem around 60 B.C. that sounds remarkably similar to the modern theories of the nebular origin of the solar system and the evolution of life, the Roman Lucretius described the development of the earth in stages from atoms colliding in the void as swirls of dust, then early plants and animals springing from the early earth's substance, to a succession of animals including a series of progressively less brutish humans.

Evolutionary thought gradually winked out in Europe after the fall of the Roman Empire. As Christianity arose, the Catholic papal state controlled and dictated most matters, including those of a scientific nature. Life and the planet were explained in religious terms, and evolutionary thought was considered rebellious and heretical. Although evolutionary ideas died out or were suppressed in European Christian society, they continued to be pronounced in the Eastern world as Islamic scholars embraced and advanced the works of those such as Aristotle and Galen, the great Roman physician.

In Europe, a spiritual view of the natural world developed that held species to be unconnected, unrelated, and unchanging since the moment of their creation. Humans were not considered part of the natural world in this view but were considered instead to be above and outside it. And Earth itself was thought to be unchanging and so young—perhaps only 6,000 years old—that species would not have had time to change.

The dawn of the age known as the Renaissance in Europe in the fourteenth century brought a refreshing questioning of the ideas and beliefs of antiquity in all aspects of human endeavor—culture, art, medicine, and especially science—as well as a healthy skepticism in the validity of those ancient values. From the English physician William Harvey who discovered the true nature of human blood circulation to the Dutch microscopist Anton van Leeuwenhoek, who first viewed the unimaginable complexity of life in a drop of pond water, naturalism was again on the rise. This new generation of naturalists envisioned life as machines and found that they could apply the same principles to life as they used in physics to invent machines.

The clergy of the time worried that this mechanistic approach to the study of life smacked of atheism and heresy. However, many of the naturalists themselves believed that they were actually on a religious mission. In fact, a number of them were both naturalists and theologians. By studying the intricate structures of an eye or a feather, these naturalists believed they could better appreciate God's benevolent design, an approach that came to be known as *natural theology.*

From the Renaissance to 1900 The word "evolution" (L. *evolutio,* unroll like a scroll) first appeared in the seventeenth century in reference to an orderly sequence of events, particularly one in which the outcome was somehow contained within it from the start implying the lack of need for divine intervention. Some evolutionary theories advanced from 1700 to 1850 proposed that the earth, life, and the entire universe developed without divine intervention or guidance. However, most contemporary theories of the time attempted to reconcile biology and spiritualism by postulating that evolution was fundamentally a spiritual process, with the entire course of natural and human evolution being "a self-disclosing revelation of the Absolute."

The work of French naturalist Georges-Louis Leclerc Comte de Buffon foreshadowed both evolutionary biology and modern geology. In his encyclopedia of the natural world, *Historie Naturelle,* Buffon toyed with the idea of common ancestry for the old world primates and humans. Although he believed in natural change, he could not come up with a mechanism for it. Charles Darwin would later state, "the first author who in modern times has treated it (natural selection) in a scientific spirit was Buffon."

In 1809, the French biologist Jean Baptiste de Lamarck attempted to explain the process of evolution in his book *Philosophie Zoologique.* Lamarck proposed that characteristics that an organism acquired during its lifetime could then be passed on to its offspring. Lamarck's theory of the inheritance of acquired characteristics would come to be known as "Lamarckism." One oft-stated example is how the giraffe acquired its long neck. Then (and now) it was believed that giraffes originally had relatively short necks that somehow grew longer over time. Lamarckism explained this phenomenon by contending that the giraffe's neck grew progressively longer as generations of giraffes stretched their necks reaching ever higher for food and then passed this stretched neck on to their offspring. Over the centuries, this stretching and passing along of longer and longer necks resulted in the modern giraffe's exceptionally long neck. Playing a central role in Lamarck's theory of the inheritability of acquired characteristics was **orthogenesis,** the hypothesis that life has an innate tendency to progress from simple forms to ever higher, ever more complex and perfect forms.

Charles Darwin not only praised Lamarck for supporting the concept of evolution and bringing it to popular attention, but he also accepted the idea of use and disuse and believed the inheritance of acquired characteristics to be not only plausible but likely. However, with the dawn of Mendelian genetics and a better understanding of the true nature of inheritance, genetic experimentation simply did not support the concept that purely "acquired traits" were inherited. Eventually, Lamarck and his ideas were ridiculed and discredited. An ironic turn of events, however, may have given Lamarck the last laugh. **Epigenetics**, an emerging field of genetics, has shown that Lamarck may have been at least partially correct all along. It seems that reversible and heritable changes can occur without a change in DNA sequence (genotype) and that such changes may be induced spontaneously or in response to environmental factors—Lamarck's "acquired traits". Determining which observed phenotypes are genetically inherited and which are environmentally induced remains an important and ongoing part of the study of genetics, developmental biology, and medicine.

The development of biological evolutionism was accompanied by an increasing understanding of geologic evolution and the true age of the earth. In his revolutionary work, *Principles of Geology* published in three volumes in 1830-33, the English geologist Charles Lyell established the principle of **geologic uniformitarianism.** The central argument in *Principles* was "the present is the key to the past." Highly controversial for its time, uniformitarianism proposed that the earth was and continued to be shaped by slow-acting natural forces over very long time periods. From his geological studies, Lyell concluded that the age of the earth must be measured in millions of years. These ideas were in direct conflict with the prevailing belief in **catastrophism**, which held that huge geologic changes occurred planet-wide over very short periods of time (thousands of years) as implied by Biblical chronology.

Box 6.1
A Voyage of Body and Mind

> *After having been driven back twice by heavy southwestern gales, Her Majesty's ship Beagle, a ten-gun brig, under the command of Captain Robert Fitzroy, R.N., sailed from Devenport on the 27th of December, 1831.*
>
> —From an account by Charles Darwin

So began a five-year round the world voyage of survey, exploration and collection that although modest in scope, produced profound repercussions whose importance reverberates to this day. Aboard the *Beagle* as the ship's naturalist was young Charles Darwin. When he sailed away, Darwin was a young university graduate filled with a life-long passion for plants, animals, and all things natural as well as an immense interest in all the sciences, especially geology. Darwin had been stunned and elated to receive an invitation to accompany the voyage because he had longed for some time to travel and explore the natural history of tropical lands. His father, Robert Darwin, however, had serious misgivings about this "wild scheme" and considered it reckless, dangerous, and unbefitting of the future clergyman he hoped Charles would become. Darwin sadly declined the offer but with only a month to go before sailing, Charles and his uncle, Josiah Wedgwood, convinced his father to allow Charles to make the voyage. In fact, Robert Darwin paid for all of Charles's expenses—no small consideration, since it was an unpaid position.

The original plan called for a two-year mission, but the voyage eventually stretched to five years (1831-1836). The British government backed the voyage and sent them off with the primary purpose of surveying the coastline and charting the harbors and islands of South America in order to produce better maps and protect British interests in the Americas. It was understood that secondary to the main mission, Darwin was to make scientific observation and collect specimens. Spending nearly two-thirds of his time ashore, Darwin explored the South America wilderness of Brazil, Argentina, and Chile as well as the remote Galapagos Islands. He filled dozens of notebooks with painstaking observations of plants, animals, and geologic formations as well as collecting and crating over 1,500 specimens, hundreds of which had never been seen before in Europe.

By the time he returned, Darwin was an established naturalist, known for the remarkable specimens he had sent ahead, and he had grown from a mere observer into a probing theorist. By any measure Darwin's labors were hugely successful, but more importantly, the trip gave him a life-time of experiences to ponder and planted the seeds of a theory he would work on for the rest of his life.

Darwin was presented Volume I of Lyell's *Principles* by Captain Fitzroy just before the *Beagle* departed. He received Volume II while in South America. These works had a major impact both on Darwin's interpretation and understanding of the geological world and fossils he encountered on his voyage and in the development of his own theory of organic evolution. On his return not only did Darwin and Lyell become friends, but Lyell would be the first scientist of standing to endorse Darwin's theory of evolution once it was published.

Darwin was barely off the ship before his next great journey—a journey of the mind—began. For the next six years, Darwin would ponder and question the biological wonders he had seen. He quietly gathered evidence from every possible source and sought out new ideas to support a notion regarding the transmutation of species that was gaining form and clarity in his thinking.

By the late summer of 1842, Darwin felt ready to commit an outline of his theory to paper. This rough sketch of his reasoning and arguments was a condensed version of his future masterwork, yet he kept his ideas secret for nearly two decades more. Why? Darwin feared the ridicule of other scientists, especially his friends and mentors as he knew his ideas would be very controversial and seen as an attack on religion and established society. He considered it prudent to wait and amass as much compelling evidence as possible to convince the world of the plausibility and possibility of such a radical idea.

By 1856, Darwin felt the time was right, and he began working to compile the findings from his vast research into a large thesis on the origin of species. Though his plans were to compile four volumes, it was not to be. Darwin had been corresponding with and receiving bird specimens from Alfred Russell Wallace, an English naturalist working in Malaya. In 1858, he received a manuscript from Wallace that summarized the main points of the very theory Darwin had struggled to piece together over the past two decades. Darwin was stunned. Darwin worked feverishly over the next twelve months in his preparation of an "abstract" to summarize his planned larger, multi-volume treatise. In November 1859, *On the Origin of Species by Means of Natural Selection, or the Preservation of Favoured Races in the Struggle for Life* was finally published. It took but one day for all 1,250 first-printing copies to sell out forcing the publisher to quickly rush 3,000 more copies to print. As Darwin suspected and feared, the book created instant controversy and misunderstandings that continue to this day

Even though frail in health, Darwin entered a very productive period producing book after book on evolutionary thought and theory for the next 23 years. He died on April 19, 1882 and is buried in Westminster Abbey. Darwin's ideas and theories stand among the greatest intellectual achievements of all time as they continue to influence scientific, religious, and social thought with ramifications that extend to the daily existence of each one of us (**Figure 6.1**).

I was a young man with uninformed ideas. I threw out queries and suggestions, wondering all the time over everything, and to my astonishment the ideas took like wildfire. People made a religion of them.
—Charles Darwin

Figure 6.1 Charles Darwin at age 51, one year after publishing one of the most revolutionary and controversial theories of all time.

From 1900 to 2000 As profound and revolutionary as Darwin's theory of evolution proved to be, it could not explain several critical components of the evolutionary process. Darwin's great frustration was his inability to explain the source of variation in traits within a species and the mechanism whereby traits were passed dependably from one generation to the next. The missing pieces began to fall into place in the early twentieth century with the emergence of the field of genetics from the landmark work on inheritance conducted the 1860s by Austrian monk Gregor Mendel.

In the years immediately following Darwin's death, evolutionary considerations splintered into a number of different interpretations. Although the scientific community generally accepted that evolution had occurred, many disagreed that it had happened as explained by Darwin. The ideas of the various disagreeing camps were brought together in the 1930s when Darwinian natural selection and Mendelian inheritance were combined to form **neo-Darwinism** or **modern evolutionary synthesis** as it later became known. Modern synthesis holds that the processes responsible for small-scale or micro-evolutionary changes can be extrapolated indefinitely to produce large-scale or macro-evolutionary changes leading to major changes and innovation in body form.

In the 1940s and 50s, the identification of DNA as the genetic material by Oswald Avery and colleagues, and the articulation of the structure of DNA by James Watson and Francis Crick revealed the true nature of genes and the genetic material. These discoveries launched the era of molecular biology and transformed our understanding of evolution into a molecular process. Since then, the role of genetics in evolutionary biology has become increasingly central.

In the 1960s Niles Eldredge and Stephen Jay Gould proposed the *theory of punctuated equilibrium*, which held that species remained persistently unchanged phenotypically for long periods of time with relatively sudden and brief periods of speciation resulting in phenotypic change. This theory has been called "evolution by jerks and creeps" as it stands in opposition to the more prevalent view that evolution progresses slowly and steadily.

Life is a copiously branching bush, continually pruned by the grim reaper of extinction, not a ladder of predictable progress.

—Stephen Jay Gould

With the emergence of modern evolutionary synthesis in the 1930s and 1940s, evolutionary biology as an academic discipline in its own right began to appear as well. However, it was not until the 1970s and 1980s that significant numbers of universities had departments specifically geared toward and termed evolutionary biology.

Some recent developments in evolutionary thought:

- Microbiology as an evolutionary discipline. Increasingly, evolutionary researchers are taking advantage of our extensive understanding of microbial physiology and microbial genomics to answer evolutionary questions.
- The combining of the disciplines of phylogenetics, paleontology, and comparative developmental biology into the field of evolutionary developmental biology (or "evo-devo" as it is sometimes dubbed). Developmental biologists compare the developmental processes of different organisms in an attempt to determine the ancestral relationship between organisms and how the developmental processes themselves evolved.

Today evolutionary investigators of all biological disciplines continue to refine and advance our understanding of the origin and evolution of the multitude of species that inhabit this planet.

Mechanisms of Evolution

Evolution is responsible for both the remarkable similarities we see across life and the amazing diversity of that life—but exactly how does evolution work? The modern theory of evolution may best be understood when viewed not as a single monolithic set of ideas but rather as the integration and melding of five different but mutually compatible subtheories or components:

In order for evolutionists to fathom the total process of evolution, they must first understand the workings of each Darwinian subtheory.

1. *Perpetual change.* This is the core on which the other theories rest. It states that the living world is constantly changing and, as it does so, organisms undergo transformation across generations throughout time.
2. *Common descent.* The second theory states that all forms of life descended from a common ancestor through a branching of lineages. Studies of organismal anatomy, cell structure, and macromolecular structures confirm the idea that life's history has all the components of a branching evolutionary tree.

149

3. *Multiplication of species.* The third theory states evolutionary processes produce new species by splitting and transforming older ones. Although evolutionists generally agree with this idea, there is still much controversy concerning the details of the process.

4. *Gradualism.* The fourth subtheory states that the large differences in anatomical traits that characterize different species originate through the accumulation of many small incremental changes over long periods of time. Evolutionists today concede that although gradual evolution is known to occur, it may not explain the origin and development of all the structural differences between species.

5. *Natural Selection.* The fifth theory, Darwin's most famous, holds that all members of a species vary slightly from each other and thus have different structural, behavioral, and/or physiological traits. Those organisms with variation in traits that permit them to best exploit their environment will preferentially survive and pass these beneficial traits on to future generations. Over long periods of time, the accumulation of such favorable variations produce new organismal characteristics and eventually new species. This proposition has been popularly termed "survival of the fittest" but might be more accurately described as "survival of the best adapted and most reproductively fit."

Evolution encompasses changes on two vastly different scales—from an increase in the frequency of a gene for colored spots on the feathers of a bird (**microevolution**) to something as grand in scale as the evolution of the entire bird lineage (**macroevolution**). Microevolution happens on a small scale and possibly over short time periods and is simply a change in the gene frequency of a single population. A relatable example would be the evolution of resistance to chemicals and antibiotics in certain organisms, such as insects becoming resistant to insecticides and bacteria becoming resistant to antibiotics. Taking long periods of time to occur, macroevolution encompasses the grandest trends, and transformations (speciation) in evolution and its patterns are what we see when we look at a tree of the history of life. Despite the scale on which it happens, evolution at both levels is driven by the primary mechanisms of natural selection mutation, genetic drift, gene flow, and speciation.

Natural Selection

Darwin's grand idea of evolution by natural selection is relatively simple, but often misunderstood. Let us use an imaginary population of small fish to illustrate the mechanism of natural selection:

1. *Variation in traits.* Some of the fish are a bright silver color on top (dorsal side) while others are a darker grayish color.

2. *Differential reproduction due to advantageous adaptation.* No environment can support unlimited population growth so not all individuals can reproduce to their full potential. If, in the fish population, the bright colored fish are more easily seen by predators and thus tend to get eaten more often, the dark colored fish will survive in greater numbers and therefore produce more offspring.

3. *Heredity.* The surviving dark colored fish pass this advantageous color trait (genes) on to their offspring. If you recall, this was the missing piece Darwin could not explain in his theory. He

knew that offspring tended to possess the traits of their parents; he just lacked a mechanism to explain it all.

4. *Long-term result.* The more advantageous trait, dark coloration, which allows the fish that possess the trait to have more offspring, becomes more common (higher gene frequency) in the population. If this process continues, eventually all (or at least most) of the fish in that population will be dark colored. The light-colored trait is likely to persist for a period of time. Fixation is the ultimate outcome of genetic drift, but there are other factors that may maintain the alternate trait in a population.

If you have variation, differential reproduction, and heredity occurring in a population, natural selection will result. A living example of what appears to be natural selection in action can be seen in the peppered moth (*Biston betularia*) of Britain. The moths come in two color variations or subspecies—light colored or *typica* and dark colored or *carbonaria* (**Figure 6.2**). Originally, the vast majority of peppered moths had the light coloration, which effectively camouflaged them against the light-colored lichen-covered trees on which they rested. However, with the onset of the Industrial Revolution and the burning of massive quantities of coal, the lichens died and the trees were blackened by soot. What had been an advantage now became a disadvantage as the typica moths became more visible against the trees. As predators found and ate more of the typica moths, the carbonaria moths flourished as their dark color now gave them an advantage.

With improved environmental standards, light-colored peppered moths have again become more common. These dramatic changes over relatively short periods of time has remained a subject of much interest and continued study and strongly implicates birds as the agent of selection in the case of the peppered moths.

Figure 6.2 Two color variations found in the peppered moth (*Biston betularia*). The dark melanic variant (a) is more visible to predators on unpolluted trees, whereas the lighter variant (b) is more visible to predators on bark darkened by pollution.

Mutation

A mutation is a change in the DNA of an organism, usually occurring because of errors in replication or repair. Evolutionists and geneticists have come to realize that mutation is the ultimate source of genetic variation.

Not all mutations serve evolution. **Somatic cell mutations** occur in the DNA of somatic (body) cells and are not passed on to offspring. Somatic cell mutations are at work in human cancer tumors but have proven useful in the development of navel oranges and red delicious apples. The only mutations that matter in evolution are **germ line mutations** because those mutations occur in reproductive cells—sperm or egg—and can be passed on to offspring.

Mutations may occur spontaneously at the molecular level in several forms, or they may be induced by **mutagens** such as certain chemicals and solar radiation. They may be small-scale (**point mutations)** affecting only a few nucleotides of a single gene or large-scale affecting the structure of entire chromosomes. Mutations may be beneficial, neutral, or harmful for the organism, but they are always totally random. Whether a mutation happens or not is unrelated to how useful or detrimental that mutation could be.

In our previous example of a population of different colored fish, the bright silver-colored fish may have arisen from a germ line mutation. This mutation may have proven detrimental in some situations but favorable in others. Bright colored fish in shallow water on a cloudy day would be easily seen by predators, but the same fish in deeper water near the surface on a bright sunny day would very hard for predators to detect. Because the mutation (adaptation) was beneficial, natural selection would then maintain the mutation in the fish population.

Genetic Drift

By the simple matter of chance, in each generation some individuals may leave behind a few more descendants that other individuals. These individuals passed on their genes due to a lucky accident, not because they were naturally selected. Unlike natural selection, genetic drift is a totally random process that doesn't work to produce new adaptations.

Returning to our imaginary fish population yet again, suppose fishermen looking for minnows (any small, shiny fish) to use as bait begin netting the mixed population of small fish in a single small lake (Lake A). They keep (remove) the bright colored ones but toss back the dark colored ones. Following this "catastrophe," genetic drift for at least one to possibly several generations could result in the removal of the unique genes carried by the bright colored fish.

Genetic drift acts more quickly to reduce variation in small populations than it does in large populations. This can create a **population bottleneck**, an evolutionary event in which a significant portion of a population is prevented from reproducing (**Figure 6.3**). Population bottlenecks result in increased sensitivity to genetic drift, increased inbreeding, and greatly reduced genetic variation in the remaining population. Suppose the fisherman seeking bait decided to keep all the small fish they netted from Lake A, regardless of color. If a substantial number of fish were removed from what was already a small population to begin with, a population bottleneck for that species in that lake would result, a catastro-

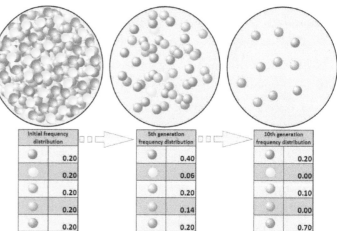

Figure 6.3 Representation of a population bottleneck. Colored balls represent the alleles present in the population. The population numbers 500 initially, but within five years the size of the population has dwindled to 50, and within ten years to just ten. As a consequence of the population bottleneck, there has been a random drift in the allele frequency distribution, and a loss of two of the original five alleles.

phe from which they would only slowly, if ever, recover. The incidences of real population bottlenecks are increasing in many species, especially large birds and mammals.

A special type of genetic drift known as the **founder effect** occurs when a new population is established by a very small number of individuals that splinter off from a larger population. Suppose the small lake (Lake A) containing our imaginary population of dark and bright colored fish of the same species is hit with a massive flood sweeping some of the fish from their ancestral lake. The founder effect would occur if a small group of the flood fish were to establish themselves downstream of their ancestral lake in a small pond(s) where their species had not existed before.

Apparently humankind has not been immune to such population bottlenecks in the past. Recent genetic research indicates that in our evolutionary path to modern *Homo sapiens*, our species may have experienced as many as four population bottlenecks and at one point there may have been as few as 15,000 individual humans on the planet. It seems probable that humanity has stared extinction in the face perhaps more than once but managed to avoid, at least for the time being, a fate that befell so many other animal species in the past. Incidences of population bottlenecks have increased in modern times, especially in large birds and mammals. These are the realities conservation biologists struggle with as they attempt to prevent the extinction of large animals such as the California condor, the whooping crane, the elephant, and the rhinoceros to name just a few.

Genetic variation within natural populations was a puzzle to Darwin and his contemporaries. As Mendel's landmark genetic work emerged, however, and became integrated into evolutionary thought, the picture became clearer. In the early 1900s, it was thought that dominant genes must, over time, inevitably swamp recessive genes out of existence. This theory was known as **genophagy** ("gene eating").

In 1908, an English mathematician, Godfrey Hardy, and a German physician, Wilhelm Weinberg, independently proposed a mathematical model that proved the theory of genophagy incorrect. The **Hardy-Weinberg Equilibrium model** as it has come to be known is based on probability and concludes that gene frequencies are inherently stable but that evolution should be expected in all populations virtually all of the time.

Evolutionary biologists and population geneticists came to understand that evolution will not occur in a population if seven conditions are met:

1. Mutation is not occurring.
2. Natural selection is not occurring.
3. The population is infinitely large.
4. All members of the population breed.
5. All mating is totally random.
6. All matings produce the same number of offspring.
7. There is no migration in or out of the population.

In other words, if no mechanisms of evolution are acting on a population, evolution will not occur, and the gene pool frequencies will remain unchanged. In the real world, however, few if any of the above conditions exist. Hence, evolution is the inevitable result.

The Hardy-Weinberg equilibrium would define evolution from the standpoint of population genetics as the sum total of the genetically inherited changes in the individuals who are the members of a population's gene pool. It is clear that the effects of evolution are felt by individuals, but it is the population as a whole that actually evolves. Genetically speaking, evolution is simply a change in the frequencies of genes in the gene pool of a population.

If a population is small, Hardy-Weinberg may be violated. Chance alone may eliminate certain members out of the proportion to their numbers in the population. In such cases, the frequency of a gene may begin to drift toward higher or lower values. Genetic drift may ultimately cause the frequency of the gene to represent 100% of the gene pool or, depending on the initial frequency of the alleles, completely disappear from the gene pool altogether. Genetic drift produces evolutionary change, but there is no guarantee that the new population will be more fit than the original one. Evolution by drift is aimless, not adaptive.

Gene Flow

Gene flow (also known as gene migration) results when genes are transferred from one population to another. If genes are carried to a population where those genes did not previously exist, gene flow can be an important source of genetic variation.

The most significant factor affecting the rate of gene flow between different populations is the mobility of the organisms comprising those populations. Animal populations are more mobile than plant populations and thus show a greater tendency to experience gene flow. **Migration** (new individuals coming in) can add new genes to an established **gene pool** while **emigration** (individuals leaving a population) may result in the removal of genes. If two populations maintain a gene flow between them, this can lead to a combination of both their gene pools, reducing the genetic variation between the two groups.

Barriers to gene flow are usually natural. They may include impassable mountains, open oceans, or vast tracts of deserts. Human development of the landscape has thrown up many artificial barriers. A multi-lane highway stretching for miles can be a major barrier for the movement of small animals. Human development and occupation can also fragment ecosystems into isolated islands and in the process seriously reduce the genetic variation of the creatures imprisoned on those islands. A great challenge facing conservation biologists is the maintenance or establishment of corridors or connections between these isolated fragments in order to maintain a healthy genetic variation through the facilitation of gene flow.

Although human constructions, occupation, and activity may pose barriers to other species, humans themselves are the most mobile and wide-ranging animal on the planet. Genetic analysis of humans of different geographical regions and races reveal tremendous rates of gene flow and a subsequent mixing of many human gene pools or the "dilution of the races" as it has been called.

Let us again consider our imaginary fish population. Suppose our mixed population of small fish—the dark-colored fish, representing the usual coloration of the species, and the bright-colored fish which arose through germ line mutation—being maintained by natural selection is found in a small lake (Lake A). A land barrier of a short distance separates Lake A from another small lake with a population of only the standard dark colored fish (Lake B). After torrential rains lasing for days, the lakes overflow and a few of the bright colored fish only from Lake A are washed into Lake B. This unintentional migration of bright colored fish has resulted in gene flow from Lake A to Lake B.

Speciation

> *Most species do their own evolving, making it up as they go along, which is the way Nature intended. And this is all very natural and organic and in tune with mysterious cycles of the cosmos, which believes that there's nothing like millions of years of really frustrating trial and error to give a species moral fiber, and in some cases, a backbone.*
>
> —Terry Pratchett

All signs indicate that a major extinction event is occurring at this moment with numbers of species disappearing at a frightening rate. But are any new species appearing through evolution to offset this great loss? To answer that, we must first define what constitutes a **species.** There are many different ways of defining a species with no definition being perfect. For our purposes we define a species simply as *the smallest cluster of organisms that possess at least one diagnostic character—morphological (structural), biochemical, or molecular— and that re reproductively isolated from its ancestral species.*

One of the criticisms leveled at the modern theory of evolution by the misguided or ill-informed is that if evolution is dynamic and ongoing, why have no new species appeared? In truth, molecular and genetic analysis reveals that a few new species of both plants and animals have come into being in modern times. These **speciation events** as they are dubbed are occurring, but at what rate? Unfortunately, there seems to be a lack of interest among biologists in pursuing this question. For one thing, the biological community considers this a settled question, so few researchers have bothered to look closely at the issue. Second, most biologists accept the idea that speciation takes a very long time (relative to human life spans). Thus, the number of speciation events that actually occur during the course of a researcher's life would be very small, if any.

What mechanism drives speciation? The first step in the process is that a population of interbreeding individuals of the same species must be splintered into at least two separate groups. These groups must then be isolated in such a way as to cause a reduction or elimination of the gene flow between them. Over a long period of time, as different mutations and adaptations accumulate in each group, the groups may become so reproductively isolated from each other that they could no longer interbreed even if they were brought back together again. Thus by our definition, a new species would have arisen.

A population may be separated into groups through geographic isolation as rivers changing course, the rise of mountain chains, the formation of canyons, continents drift, and the migration of organisms. However, even in the absence of a geographical barrier, reduced gene flow across a species' range can encourage speciation. An animal might occupy such a large range that eventually those individuals at the far edges of that range stop breeding with each other thereby reducing the gene flow between them.

Groups may be reproductively isolated from each other in any number of ways:

- They may evolve different mating times or mating rituals.
- A lack of "fit" between sexual organs may evolve. This seems to be especially true in insects where damselflies alone have nearly half a dozen different penis shapes. Is the lack of fit between sexual organs a cause of speciation, or the result? We have presented it here as a cause, but it could just as easily be the result.

- If physical mating is possible, inviable or sterile offspring may result. We do not mean to imply by this that sexual reproduction is necessary for speciation. It clearly is not.

Suppose the bright-colored fish taken from Lake A were carted hundreds of miles, and after some had been used as bait, the rest were dumped into an isolated lake that contained no other fish of that species. Thus geographically and physically isolated, all gene flow between these fish and others of their species would be eliminated. If the founder effect kicked in and they become established, mutations, adaptations, and behavioral changes will accumulate over long periods of time resulting in these isolated bright-colored fish becoming an entirely new species.

Evidence of Evolution

At the very core of evolutionary theory is the basic idea that life has existed in some form for billions of years and that life has changed drastically over time. The history of living things is documented through multiple lines of evidence that detail the story of life through time.

Fossil Record

Humans have attempted to understand and interpret fossils since the days of the ancient Greeks and Romans. More scientific views of fossils began to emerge during the Renaissance, but fossils were still largely misunderstood and misinterpreted. For example, in China the fossil bones of ancient mammals were thought to be "dragon bones" and useful in medicine when ground into powder. In the West, the presence of fossilized sea creatures high up on mountainsides was seen as proof of the Biblical flood.

In 1799, an English canal engineer named William Smith noted that in undisrupted layers of rock, fossils occurred in a definite sequential order, with more modern-appearing ones closer to the top. Because bottom layers of rock logically were laid down earlier and thus are older than top layers, the sequence of fossils could also be given a chronology from oldest to youngest.

When Darwin wrote the *Origin of Species*, the oldest animal fossils known were those from the Cambrian Period, about 540 million years old. The scarcity and relatively young age of fossils and especially the rarity of intermediate fossil forms held implications that worried Darwin concerning the validity of his theories. Since Darwin's time, the fossil record has been pushed back to 3.5 billion years before the present, and many of the gaps in the fossil record have been filled in by the research of paleontologists.

A **fossil** (L., *fossus*, having been dug up) is any remains, impressions, or traces of living organisms of past geological ages that have been preserved in the Earth's crust. Paleontological studies of fossils across geologic time and how they formed are vital to our understanding of the evolutionary relationships (phylogeny) between **taxa** (taxonomic groups). Large specimens (**macrofossils**) are more commonly found and displayed, although tiny remains (**microfossils**) are far more prevalent in the fossil record.

A number of different mechanisms can account for fossil formation:

- *Permineralization.* For permineralization to occur, the organism must become covered by sediment very soon after death, sealing it and preventing decay. (Sudden freezing, rapid desiccation (drying) or coming to rest in an anoxic (no oxygen) environment such as the bottom of a lake also suffices.) The degree of decay of the remains before being sealed determines the fineness of detail that will be found in the fossil. Mineral-rich groundwater slowly trickling down through the buried remains permeates the empty spaces of the dead organism and gradually, organic structures are replaced molecule by molecule by rock-like minerals. One example of fossilization by permineralization that most people are familiar with is petrified wood. Astonishingly, petrified wood can exhibit the original structure of the wood in all its detail, down to the microscopic level.

- *Molds, Replacement, and Compression.* When all that is left of the original organism is an organism-shaped hole in the rock, a *mold fossil* (or **typolite**) has formed. If this mold depression is later filled in with other minerals, a *cast fossil* has formed through the replacement of the original materials by new, unrelated ones. In some cases, replacement occurs so gradually and on such a minute scale that even microstructural features are preserved despite the total loss of the original organic material.

If the remains of an organism become pressed into soft mud or sand which then turns into rock, a *compression fossil* has formed. Usually, only plant stems and leaves form such depressions with only a general outline of the external structure of the plant part; internal anatomy is never preserved in compression fossils.

- *Traces.* Trace fossils are the indirect evidence of the presence and activity of some ancient organism including the remains of trackways, burrows, footprints, eggs and eggshells, and even fecal droppings (known as **coprolites**),

- *Resins.* In past geologic times as now, some species of trees excreted a sticky, syrupy material known as **resin**. Often, other organisms such as fungi, bacteria, seeds, leaves, insects, spiders, and rarely small invertebrates (collectively known as **inclusions**) became trapped in and covered by this sticky ooze. Over time, the resin hardened into amber, preserving in exquisite detail any inclusions entombed within.

To hold or even see a fossil is to transcend millions of years of time to a prehistoric world we can scarcely imagine.

> *When out fossil hunting, it is very easy to forget that rather than telling you how the creatures lived, the remains you find indicate only where they became fossilized.*
>
> —Richard Leakey

Homology and Analogy

In nature, animals look like one another for different reasons. Two beetles of the same species might look alike because they inherited their yellow and black spots from their parents. On the other hand, two flying animals, such as a bird and a bat, also appear similar to each other. The similarity between the beetles is inherited, but the similarity between the bird and the bat is not.

The skeletons of humans, whales, lizards, and birds are strikingly similar, despite the different lifestyles of these animals and the different environments they inhabit. Bone by bone, the similarity of these animals can be observed in every part of the body, including the limbs, yet a person throws a ball, a whale swims, an alligator shuffles, and a bird flies. Each of these animals is framed of bones that are different in detail but similar in general structure and relation to each other (**Figure 6.4**).

Evolutionary biologists regard such similar traits and structures as being **homologous** and have concluded that such similarities are best explained by descent from a common ancestor. Comparative anatomists investigate such homologies, working out relationships from degrees of similarity. Their conclusions provide valuable inferences into the details of evolutionary history, inferences that can be tested by comparison with the sequence of ancestral forms in the fossil record and through cladistic analysis. Different animals end up with the same sort of limb inherited from a common ancestor, just as cousins might inherit their hair color from their grandfather (**Figure 6.5**).

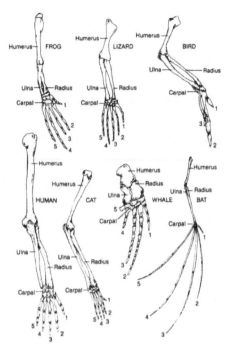

Figure 6.4 Homologous limb bones in tetrapods.

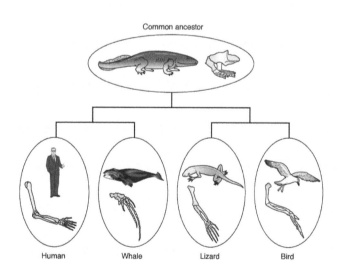

Figure 6.5 An evolutionary tree depicting the inheritance of homologous limb bones from a common ancestor.

Over time, a trait or structure inherited from a common ancestor may be changed and adapted taking on a different structure and function. Consider the teeth of the beaver and the elephant. The gnawing front teeth of the beaver and the tusks of the elephants are both basically incisor teeth. Although inherited in a basic form from an ancestor common to the beaver and the elephant, these teeth have been greatly modified by evolutionary mechanisms into the seemingly dissimilar teeth we see today in the beaver and the elephant.

At first glance, many traits and structures that seem dissimilar are homologous but some that seem homologous are not. The wings of a bird and those of a bat might appear to be homologous but are, in fact, **analogous**. Analogous traits or structures are those that evolve independently in organisms that are not closely related as a result of having adapted to similar environments or niches through a process known as **convergent evolution**.

Examples of convergent evolution abound in the animal kingdom. Several mammal groups have independently evolved prickly protrusions of the skin or spines, such as echidnas (monotremes), hedgehogs (insectivores), and porcupines (rodents). Claws and long, sticky tongues that allow them to open the nests of ants and termites are the analogous structures that mark mammals loosely grouped as "anteaters"—four species of true anteaters, eight species of pangolin, four species of echidna, the African aardvark, and the Australian numbat.

In birds, the Little Auk of the north Atlantic and the diving petrels of the southern oceans are remarkably similar in appearance and habits. Both Old World vultures and New World vultures have featherless necks and heads, search for food by soaring, and feed on carrion. However, Old World vultures are in the eagle and hawk family and use eyesight to find food, whereas the New World vultures are related to storks and use the sense of smell and sight to discover food.

Molecular Record

Evolutionary biologists believe that all living creatures have descended from simple cellular organism that arose about 2.5 billion years ago. Some of the characteristics of that earliest organism have been preserved in all living things alive today. By comparing the genomes (sequence of all the genes) of different groups of animals, we can specify the degree of relationship among the groups more precisely than by any other means.

All organisms store their biological history in their DNA. Therefore, evolutionary changes over time should involve a continued accumulation of genetic changes in the DNA. As a result, organisms that are more distantly related should have accumulated a greater number of these genetic differences than organisms that are more closely related. Thus, the genome of gorillas, which the fossil record indicates diverged from humans between 6 and 8 million years ago, differs from the genome of humans by 1.6% of its DNA. However, the genome of chimpanzees, which diverged about 5 million years ago, differs by only 1.2%. These echoes of the evolutionary past also show up protein sequences, such as hemoglobin, the oxygen-binding component of red blood cells. Humans and macaques, both primates, show less difference in their hemoglobin proteins than to more distantly related mammals, such as dogs. Nonmammalian vertebrates, such as birds, reptiles, and frogs, differ from primates even more.

Genes evolve at different rates because although mutation is a random event, some proteins are more tolerant of changes in their amino acid sequence that are other proteins. For this reason, the genes that encode these more tolerant, less constrained proteins evolve faster. The average rate at which a particular kind of gene or protein evolves has given rise to the concept of a **molecular clock**. Molecular clocks run rapidly for less constrained proteins and slowly for more constrained proteins, though they all time the same evolutionary events.

The concept of a molecular clock is useful for two purposes. It determines evolutionary relationships among organisms, and it indicates the time in the past when species started to diverge from one another. It should be noted, however, that the idea of a molecular clock ticking away in proteins is a controversial concept that is not totally embraced and supported by all biologists.

Another interesting line of evidence supporting evolution involves sequences of DNA known as pseudogenes. **Pseudogenes** (or "junk DNA") are remnants of genes that no longer function, but continue to be carried along in DNA as excess baggage. Pseudogenes also change over time as they are passed from ancestors to descendants, and in the process offer an especially useful way of reconstructing evolutionary relationships. Because they perform no function, the degree of similarity between pseudogenes must simply reflect their evolutionary relatedness. The more remote the last common ancestor of two organisms, the more dissimilar their pseudogenes will be.

The evidence for evolution from molecular biology is overwhelming and is growing steadily. In some case, this molecular evidence transcends paleontological evidence. Take whales for example. Anatomical and paleontological evidence indicates that the whale's closest living land relatives seem to be the even-toed hoofed mammals (cattle, sheep, camels, goats, etc.). Recent analysis of some milk protein genes have confirmed this relationship and have suggested that the closest land-bound living relative of the whales may be the hippopotamus.

Biogeography

Darwin in his time and biologists of today are fascinated by the inconsistency and variability in the distribution of plants and animals around the world. Why do the Galapagos Islands of South America and the Cape Verde Islands off Africa have strikingly different fauna and flora, despite having similar environmental conditions? Why does the Arctic have polar bears but no penguins yet the Antarctica has penguins, but no polar bears? In *Origin of Species*, Darwin states, "In considering the distribution of organic beings over the face of the globe, the first great fact that strikes us is, that neither the similarity nor the dissimilarity of the inhabitants of various regions can be wholly accounted for by climatic and other physical conditions."

Biogeography is the science that attempts to understand why (and sometimes more importantly, why not) plants and animals exist where they do. All creatures are adapted to the abiotic and biotic factors of their habitat. Thus, one might assume the same species would be found in a similar habitat in a similar geographic area, e.g. Africa and South America. That is not the case. Plant and animal species are discontinuously distributed around the planet. This discontinuity of distribution can be seen in African and South American fauna. Africa has short-tailed (Old World) monkeys, elephants, lions and giraffes, whereas South America has long-tailed monkeys, cougars, jaguars and llamas.

Before humans arrived 40,000 to 60,000 years ago, Australia had more than 100 species of kangaroos, koalas, and other marsupials but none of the more advanced terrestrial placental mammals such as wolves, lions, bears, horses. Land mammals were entirely absent from the even more isolated islands that make up Hawaii and New Zealand. However, each of these places had a great number of plant, insect, and bird species that were found nowhere else in the world. The most likely explanation for the existence of Australia's, New Zealand's, and Hawaii's mostly unique biotic environments is that the life forms in these areas have been evolving in isolation from the rest of the world for millions of years.

All creatures are adapted to the abiotic and biotic factors of their habitat. One might assume, therefore, that the same species would be found in a similar habitat in a similar geographic area, for instance, in Africa and South America. This is not the case. Plant and animal species are discontinuously distributed around the planet. This discontinuity of distribution can be seen in African and South American fauna. Africa has short-tailed (Old World) monkeys, elephants, lions and giraffes while South America has long-tailed monkeys, cougars, jaguars and llamas. The flora of North and South America show a similar discontinuity. Deserts in North and South America have native cacti, but deserts in Africa, Asia, and Australia have succulent native euphorbs that resemble cacti but are very different, even though in some cases cacti have done very well (for example in Australian deserts) when introduced by humans.

The discontinuity of species distribution is even more striking when it comes to islands. Biogeography divides islands into two categories. **Continental islands** are islands like Great Britain, and Japan that have at one time or another been part of a continent. **Oceanic islands**, like the Hawaiian Islands and the Galapagos islands, on the other hand, are islands that have formed in the ocean and have never been part of any continent.

Oceanic islands have distributions of native plants and animals that are unbalanced in ways that make them distinct from the flora and fauna found on continents or continental islands. Oceanic islands do not have native terrestrial mammals (other than bats and seals), amphibians, or fresh water fish. In some cases oceanic islands have terrestrial reptiles but most do not.

Starting with Charles Darwin, many scientists have conducted experiments and made observations that have shown that the types of animals and plants found, and not found, on such islands are consistent with the theory that these islands were colonized accidentally by plants and animals that were able to reach them. Such accidental colonization could occur by air, such as plant seeds carried by migratory birds, or bats and insects being blown out over the sea by the wind, or by floating from a continent or other island by sea. Many of the species found on oceanic islands are endemic to a particular island or group of islands, meaning they are found nowhere else on earth. However, they are clearly related to species found on other nearby islands or continents.

Geologic and Environmental Influences

It would be nearly impossible to understand the evolutionary patterns which shaped life on this planet without considering the canvas of geologic and environmental changes on which those patterns have been and continue to be painted.

The supercontinent Pangaea and the Carboniferous forests once found there no longer exist. Where did they go? What happened to them? Pangaea was eventually wrenched asunder by the slow but powerful process known as **continental drift**. Powerful movements in the hot putty-like upper mantle break the cool brittle crust into various sizes pieces known as **continental plates**. These tightly interlocked plates "float" on the mantle below them. As the mantle continues to churn, the plates are forced apart in some places, forced together in other places, and slip side by side in others. Over geologic periods of time, these movements change not only the shape but also the location of the continents (**Figure 6.6**). The North American plate broke from the remains of Pangaea, headed north, and became cooler and drier. As a result, the Carboniferous forests and the animals that inhabited the plate disappeared and in their place arose the plants and animals of present day North America. Continental drift also solves mysteries which had previously puzzled scientists such as the fact that beneath miles of ice on Antarctica we find the fossilized remains of tropical plants and animals and that Australia has many unique species of plants and animals found nowhere else on the planet.

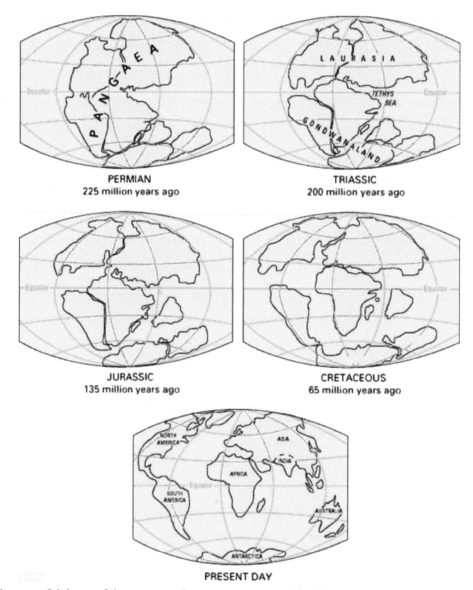

Figure 6.6 The powerful dance of the continents known as continental drift slowly makes and remakes the face of Earth.

As the continents changed, so did the flora and fauna occupying those continents. Some types of plants and animals became extinct yet others adapted and evolved into new species to fill the void. As a result, continental drift has caused the continents to become the burial ground for some species (extinction) but an evolutionary birth place (origins) for others.

The second major environmental influence on plant evolution has been ice, ice in the form of continental glaciers, polar ice sheets miles thick periodically forming and retreating in a series of cycles known as the **ice ages** (**glacial ages**). Starting around 2.4 billion to 2.1 billion years ago, ice ages have come and gone at least five times. Within each long-term ice age there are pulse of extra cold temperatures (**glacial periods**) mixed with intermittent warm periods (**interglacial periods**). The "ice age" colloquially refers to the most recent glacial period which peaked at the Last Glacial Maximum approximately 20,000 years ago, in which extensive ice sheets lay over large parts of the North American and Eurasian continents. The Earth is currently in a warm interglacial period known as the Holocene, that began 10,000 to 50,000 years ago.

As continental glaciers up to 2 miles thick slowly ground over the land, they obliterated all flora in their path, stripped away soil down to bare rock, and carved out many strange geologic features such as the Great Lakes. In fact, the tremendous weight of these glaciers deformed the crust in numerous places. When the glaciers retreated (melted) a barren, cold, dry landscape challenged plants and animals to adapt and evolve or perish.

All available evidence supports the central conclusions of evolutionary theory that life on Earth has evolved and that species share common ancestors. Biologists do not argue about these conclusions. What they are debating and attempting to discover are the details of the process:

1.   Does evolution proceed slowly and steadily or does it happen in fits and starts.
2.   Why are some clades (branches on the evolutionary tree) very diverse, whereas others are sparsely populated?
3.   How does evolution produce new and complex adaptations?
4.   Are there trends in evolution and if so, what processes generate these trends?

Questions of Phylogeny

Two main concepts comprise the core of modern evolutionary thought—living things have changed over time, and different species share common ancestors

Tree of Life

The process of evolution produces a pattern of relationships between species. As lineages with a common ancestry evolved and split and modifications were inherited, their evolutionary paths diverged (change over time). This produced a branching pattern of evolutionary relationships that we attempt to represent in the form of a tree—the Tree of Life (**Figure 6.7**).

This tree represents the **phylogeny** (history of organismal lineages over time) of all organisms. It implies that different species arise from previous forms via descent and that all organisms, from the tiniest microbe to the largest plants and vertebrates, are connected by the passage of genes along the branches of the tree that links all life. As such, biologists believe that the totality of life forms can be divided into three major domains: Archaea, Bacteria, and Eukaryota.

This tree and its many branches is supported by many lines of evidence but should be considered an on-going work-in-progress. As biologists gather even more information, they may revise particular hypotheses and rearrange some of the branches on the tree. For

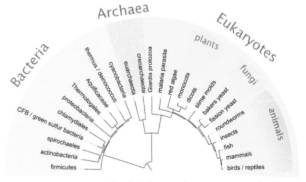

Figure 6.7 Tree of Life. This evolutionary tree shows the relationships among various biological groups. It includes data from DNA, RNA, and protein analysis.

example, evidence discovered in the last 50 years suggesting that birds are dinosaurs has required adjustments to several vertebrate "twigs."

From Aristotle through Darwin's time, biologists erroneously organized life as a ladder from "lower" to "higher" life forms. Similarly, it is easy to misinterpret phylogenies as implying that some organisms are more "advanced" than others. Phylogenies illustrate only patterns of relationships, nothing more.

Adding Time to the Tree

The stretch of geologic history is commonly referred to as "deep time" and is metered in hundreds of millions to billions of years. Can we measure deep time? Yes. Will we ever fully comprehend the immensity of it? No. However, to determine the origins of species and understand the flow of evolution, we must learn to cope with it.

The age of the Earth is calculated to be roughly 4.6 *billion* years. Although we cannot possibly grasp the enormity of that amount of deep time, we can cast it against time scales we can fathom. Let us first shrink all geologic time into 100 years (a century). On that scale, each year represents 46 *million* actual years. Each month corresponds to 3.8 million years, and each day signifies 127,000 years. An hour is 5,300 years; a minute is 88 years; and each second is a year and a half. By that time scale, the average human life span lasts only a minute or less.

Let us compress our time scale even more and squeeze all geologic time into one year (**Table 6.1**). Finally, if we condense just the 3.5 to 3.8 billion years of geologic time that has past since the first appearance of life on Earth into a single minute, we would have to wait about 50 seconds just for multicellular life to appear, another four seconds for vertebrates to invade the land, and then only in the last 0.002 seconds would *Homo sapiens* arise.

Table 6.1	
Geologic Time Compressed into 1 Year	
Time	**Event**
January 1- 12:00 AM	First forms from a planetary nebulae
February 25 -4:41 PM	First primitive living cells appear
July 17- 9:54 PM	First eukaryotic cells appear
September 3- :39 PM	First multicellular organism (algae) appear
November 8- 4:35 PM	Marine worms and jellyfishes appear
November 21- 7:40 PM	First fish appear
December 2- 3:54 AM	First land animals (amphibians) appear
December 5- 5:50 PM	First reptiles appear
December 12- 9:42 PM	First crocodiles appear
December 13- 8:37 PM	First dinosaurs appear
December 14- 9:59 AM	First mammals appear
December 28- 9:31 PM	First monkeys appear
December 31- 5:18 PM	First hominids appear
December 31- 10:30 PM (2.7 million ago)	First *Homo sapiens* appear

December 31- 11:46 PM (420,000 years ago)	Domestication of fire
December 31- 11:59:20 PM (20,000 years ago)	Invention of agriculture
December 31- 11:59:56 PM (20,000 years ago)	Roman Empire; birth of Christ
December 31- 11:59:58 PM (500 years ago)	Renaissance in Europe; emergence of the experimental method in science
December 31- 11:59:59 PM	Widespread development of science and technology; first steps in space exploration; advanced global communication

Instead of trying to view all of geologic time as a single span, scientists find it more meaningful and useful to break deep time down in a hierarchal system known as the *geologic timetable* (**Table 6.2**). The timetable has four major *eras* subdivided into *periods,* which are in turn subdivided into *epochs.*

Table 6.2
Geologic Ages and Associated Biological Events

Time Scale (eon)	Era	Period	Epoch	Millions of Years Before Present (approx.)	Duration in Millions of Years (approx.)	Some Major Organic events
P H A N E R Z O I C	Cenozoic	Quaternary	Recent last 5,000 years	0.01	1.8	Appearance of humans
			Pleistocene	1.8		
		Tertiary	Pliocene	5.3	3.5	Dominance of mammals and birds
			Miocene	23.8	18.5	Proliferation of bony fishes (teleosts)
			Oligocene	34	10.2	Rise of modern groups of mammals and invertebrates
			Eocene	55	21	Dominance of flowering plants
			Paleocene	65	10	Radiation of primitive mammals
	Mesozoic	Cretaceous		142	77	First flowering plants Extinction of dinosaurs
		Jurassic		206	64	Rise of giant dinosaurs Appearance of first birds
		Triassic		248	42	Development of conifer plants

Paleozoioc	Permian		290	42	Proliferation of reptiles Extinction of many early forms (invertebrates)
	Carboniferous	Pennsylvanian	320	30	Appearance of early reptiles
		Mississippian	354	34	Development of amphibians and insects
	Devonian		417	63	Rise of fishes First land vertebrates
	Silurian		443	26	First land plants and land invertebrates
	Ordovician		495	52	Dominance of invertebrates First vertebrates
	Cambrian		545	40	Sharp increase in fossils of invertebrate phyla
P R E C A M B R I A N	Proterozoic	Upper	900	355	Appearance of multi-cellular organisms
		Middle	1,600	700	Appearance of eukaryotic cells
		Lower	2,500	900	Appearance of plank-tonic prokaryotes
	Archean		4,000-4,400	1,400	Appearance of sedimentary rocks, stromatolites and benthic prokaryotes
	Hadean		4,560	160-560	From the formation of Earth until first appearance of sedimentary rocks; no observable fossil organisms

Dates derived mostly from Gradstein, F. M., et al., 2004. *A Geological Time Scale 2004.* Cambridge University Press, Cambridge. England, and from Geologic Time Scale, obtainable from http://www.stratigraphy.org, a Website maintained by the International Commission of Stratigraphy.

How do we know what happened when? What is the evidence to support scientific beliefs about the age of the earth and the sequence of events in the history of life on earth? Several methods are employed to determine past dates and events:

Radiometric dating is a technique based on the fact that certain elements in rocks, fossils and living things are naturally radioactive. Many elements have naturally occurring **isotopes**, varieties of the element that have different numbers of neutrons in the nucleus. Some isotopes are stable, but some are unstable or radioactive. Some radioactive isotopes **decay** (break down) almost instantaneously back into their stable atomic form; others take tens of thousands to millions or even billions of years to decay. Since the rate of decay of an isotope (known as its **half-life**) is predictable, we can use these radioactive isotopes to date rocks

Imagine you have been given charcoal briquettes like those used for outdoor grilling except yours have been very precisely formed and weighed. Let us also suppose that those briquettes (representing the initial amount of radioactive isotopes in a rock or fossil.) will reduce to ash at a known rate in a specific temperature of X° C You then begin to fire those briquettes at X° C. After allowing the briquettes to "decay" for a time, you stop the firing process. By carefully weighing the amount of ash (representing the now stable isotopes in a rock or fossil) and comparing it to the weight of the briquettes before firing began, you can use the rate of "decay" of the briquettes to determine how long they had been fired. By comparing the amount of radioactive isotopes in a present day rock or fossil to the amount of radioactive isotopes that was present when that rock or fossil formed, the rate of decay of that radioactive isotope can be used to determine the approximate age of that rock or fossil.

Radiocarbon or carbon-14 (C-14) is probably one of the most widely used and best known absolute dating methods. However, because C-14 has a half-life of only 5,730 years, it can only be used to accurately date carbonaceous materials up to 60,000 years old. Other isotopes, such as potassium-argon dating and uranium-lead dating, can be used to date fossils and rocks millions to even billions of years old.

Molecular clocks allow us to use the amount of genetic divergence between species to extrapolate backward to estimate dates. *Stratigraphy* uses the position of rock layers and the fossils within those layers to determine *relative* age. Reading from bottom to top, the **Law of Superposition** tells us that the lower a **strata** (layer) of rock or fossil is, the older it is.

Human Phylogeny

Unless humans fit within the evolutionary paradigm, the concept of evolution has no meaning for the study of human biology. Is humankind as much a part of the evolutionary process as any other species of living thing? Is there evidence that our own species is the product of descent from some ancestral type with modifications over time? Because the answer to both those questions is *yes*, the most controversial and contentious aspect of all modern evolutionary investigations has been the attempt to determine the origins and phylogeny of humankind. It is important to note that modern evolutionary theory does *not* profess that humans evolved from either monkeys or apes, but it does suggest that humans and other primates may have evolved from a common ancestor.

The same types of evidence gathered to understand the evolution of other species is applied to understanding human evolution. The fossil record has been of limited value in documenting human phylogeny. The main problem is the scarcity of fossilized human remains. Presumably, any common ancestor of our line and that leading to the great apes was most likely **arboreal** (tree-dwelling) and so lived in habitats unlikely to produce fossils. Clouding the issue was the tendency to classify every discovery as an entirely new species so that the sheer number of fossil species confused the picture of human ancestry as a whole. This problem was alleviated when paleontologists began to take a more populational view, realizing that different fossils were geographic variants of the same species rather than separate species or genera. Molecular evidence has helped fill in the blanks in the fossil record, leading evolutionary biologists to a better understanding of the relationships between extant primates.

A striking change occurred about 35,000 years ago. Before then, human artifacts were remarkably similar worldwide. Afterward, not only do previously existing tools begin to vary, but new tools also appear. Such regional differences in tool-kind suggest local cultural adaptations. This has been interpreted as a historical marker indicating when *Homo sapiens* became proficient in the use of language. The ability to communicate verbally resulted in an exchange of ideas and an explosion in creativity. That in turn led our species into a new sector of evolution, cultural or psychosocial evolution, and this new form of evolution has dominated our way of life ever since. As a species, we are still subject to organic evolution but on the whole, our evolution is in the realm of culture. Within this new realm, accelerated by the evolution of speech and the development of language, variability is in the form of ideas, inventions, traditions, laws, and customs, not genes; the vehicle of transmission is learning, not heredity; and the direction of cultural evolution is provided by imitation, cultural selection, foresight, and planning, not natural selection. Nevertheless, we cannot disregard our biological past for it tugs at us still and influences our behavior more than we realize or are willing to admit.

> *Man is the only animal for whom his own existence is a problem which he has yet to solve.*
> —Erich Fromm

From Few Came Many—Rise of Animal Forms

As we now understand it, the riotous multitudes of animal forms found on earth today are the evolutionary progeny of a few simpler forms. As the evolutionary tree grew and developed, critical branches and transitions of ever-increasing complexity sprang up (**Figure 6.8**).

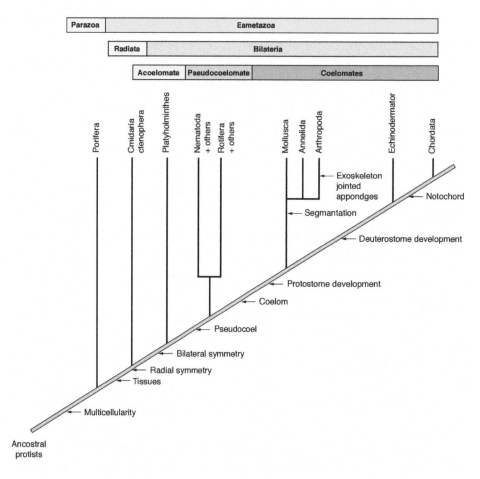

Figure 6.8 Origin of Animal Groups. This cladogram represents the possible phylogeny of the major phyla of the kingdom Animalia.

The animal branches of the tree may have sprung from ancestral single-celled protists. As these protists developed a more colonial existence, a threshold was crossed and multicellularity appeared. With the advent of multicellularity, the first branch appears as some new forms took multicellularity to the next level. Only one phylum—phylum Porifera—failed to take a new evolutionary direction. Known as *parazoa* ("beside the animals"), sponges never evolved a definite symmetry or achieved the tissue and organ state of complexity. All other animal forms sprang from the branch known as *eumetazoa*, resulting in bodies with a definite symmetry and tissues organized into organs.

The eumetazoan branch of the animal tree in turn split into two evolutionary paths—the *Radiata* and the *Bilateria*. Only two extant phyla—*Cnidaria* and *Ctenophora*—today represent the Radiata, creatures with only two layers of cells (diploblastic) and radial symmetry. All other living phyla come from the Bilateria branch and possess a triploblastic body with bilateral symmetry.

The bilateral branch in turn forked into animals with body cavities (coeloms) and those without. The hollow-bodied creatures were cleft by evolution into protostomes and deuterostomes that went their separate ways based on differences in embryological development. Evolution continued to refine and separate the deuterostomes into opposing groups based on such morphological traits as segmentation, endo or exoskeleton, and the presence of a notochord. Springing from simple roots billions of years ago, the animal portion of the tree of life has evolved and developed into a complex and interwoven tapestry of animal form and function.

Questions of Numbers

The answer to the question of *how many* depends on what is being counted, species or individuals. If we limit our counting only to species, we have a fair understanding of roughly only a third of all life forms. The complex species grouped as **Eukaryotes**—animals, plants, fungi and protists, those composed of cells with a nucleus—are much more familiar and far better documented than are the tiny, primitive bacteria that lack a true nucleus and comprise the other major group of life, the **Prokaryotes**.

Biologists believe we have cataloged and scientifically named about 1½ to 2 million species. However, because this estimate has remained seemingly static for at least the last two decades so, clearly no one has any idea as to precisely how many species we have identified let alone how many more may still be lurking out there. Although there are statistical ways of estimating the numbers of species in groups, the best biologists can do is guesstimate, and when they do, the number of yet-unidentified species that keeps popping up ranges anywhere from 10 to 100 million. Zoologists estimate that only 20% of all living animals have, to date, been accounted for and studied. Further, they believe that of all animal species ever to exist, less than 1% is known to us.

The guesstimate keeps rising because we are discovering new species at a frantic rate. From fish to monkeys and from dolphins to tree frogs, new species, especially insects seem to be popping up everywhere. A decade-long multi-billion dollar global census of marine life, the most ambitious ever attempted is underway. From this effort, an average of 1,700 new species is being cataloged in the ocean census database each year, and the scientists predict that the total—215,000+—could reach 2 million before the census is complete.

Terrestrial ecosystems are being scrutinized as well, especially those considered to be ecological hotspots. Within the last several years biologists have discovered at least 52 new species of animals and plants on the Asian island of Borneo alone, for a total of more than 400 new species discovered there since 1996.

Life on earth never looked more bountiful. This apparent abundance, however, is but an illusion. We are discovering new species as never before because we are looking for them with a heightened sense of urgency. Furthermore, new species become easier to find as their habitats shrink, and they have nowhere left to hide from us. Simply put, species are marching into extinction at a rate unprecedented in the short history of humankind. The noted paleontologist, Stephen Jay Gould, once estimated that 99% of all the animals and plants that have ever existed are now extinct with most leaving no traces or fossils.

Extrapolating from present day back into the distant past, it has been suggested that during the age of dinosaurs tens of millions of years ago, the earth lost one species to extinction every 1,000 years. Ten thousand to 12,000 years ago when humans just learned how to grow crops, the extinction rate was one species every 100 years. 1,000 years ago during the Middle Ages in Europe, the extinction rate was one species every ten years. By the time of the Industrial Revolution 300 years ago, the rate was up to one species lost every year. In modern times, the rate is thought to be as high as 100 species lost per day (around 40,000 per year)! At that rate biologists fear one-fourth of all current species may be extinct within the next 25-30 years. At this point, it seems it's not a race to save things; it's a race to know what's there before it's gone.

The human species alone numbers 6.6 billion individuals and rising. If you single out just one of those billions of humans, you will find they could provide a potential home to over 200 species of other organisms such as lice, tapeworms, fungi, and bacteria. There may be as many as 80 to 100 different species of bacteria in the human mouth alone. The total number of individuals we carry on and in our person boggles the mind. The average human body is covered with 2 square meters of skin. Microbiologists estimate that not only are there some 10 million individual bacteria inhabit each square centimeter of that skin but that the folds and cavities within the human body may harbor thousands of times more bacteria than skin does. In fact, the **genetic mass** (amount of DNA) of the organisms living on us combined has four times the genetic mass than we do ourselves. Each of us is truly a small "planet" with our own indigenous life forms, whereas we in turn are part of a larger scheme of species and individuals in numbers that defy comprehension.

Questions of Taxonomy

It grows increasingly evident that there may be tens of millions of different species on our little speck of a planet. Although it falls to the evolutionary biologist to sort out the origins and lineages of all those species, it is the taxonomist who is charged with bringing order to this riotous chaos of life. But how are they to accomplish this? How do you scientifically organize millions of species in the first place, yet at the same time design a system flexible enough to accommodate the tens of millions of more species that still await discovery? And as if those challenges weren't monumental enough, how do you scientifically name those species once you have designed a system for classifying them?

Imagine that you have won a free car through a drawing at a local convenience store. However, in your excitement you didn't read the fine print of the contract carefully, and when your vehicle is delivered, it is in several very large shipping crates, totally disassembled. The contents of all those crates have been emptied into one large pile. Fortunately, you were provided with a rather detailed instruction manual and schematics on how to assemble all those parts into your dream machine. What would you do first? How would you proceed? A logical first step would be to remove the parts one at a time from the one large pile and using the assembly manual as a guide, place each part in separate categorized piles—engine components, brake system components, electrical components, and so on. Within each of the smaller categorized piles, sub-piles would develop as you tried to understand each part and visualize how that part fits into the larger scheme. In other words, you would develop a taxonomic system for bringing order to chaos; a taxonomic system for car parts, but a taxonomic system nonetheless. That is essentially the same challenge facing biological taxonomists. Confronted with a huge pile of life (different species) that continues to grow larger, taxonomists labor to bring order to it all by imposing biological taxonomic systems upon it. Unfortunately, taxonomists have no assembly manual to reference. Developing such systems allows us to not only understand each species more thoroughly but also aids us in better comprehending and appreciating the intricate workings of all those species and their place within the blueprint of the biosphere.

Humankind has been practicing survival taxonomy (grouping) in one way, or another as long as we have walked the planet. Ancient humans quickly learned which animals and plants could provide food, power, fuel, transportation, clothing, and even medicines, and they grouped and named such creatures according to their specific culture and language. Biological **taxonomy** (Gr, *taxis*, order + *nomos*, rules) on the other hand is the scientific *classification* (grouping) and *nomenclature* (naming) of living things. As such, taxonomists face two challenges: develop not only practical but flexible classification systems, but also a uniform system of nomenclature to scientifically name each species. The science of taxonomy is the most encompassing field of biology. The modern taxonomist's training is broad, cutting across the fields of zoology and botany, genetics, evolution, paleontology, historical biology, biogeography, geology, ecology, and even ethology, chemistry, cellular biology and philosophy.

> *Taxonomy (the science of classification) is often undervalued as a glorified form of filing—with each species in its folder, like a stamp in its prescribed place in an album; but taxonomy is a fundamental and dynamic science, dedicated to exploring the causes of relationships and similarities among organisms. Classifications are theories about the basis of natural order, not dull catalogues compiled only to avoid chaos.*
>
> —Stephen Jay Gould

Modern taxonomists use a great variety of tools to classify species and study the relationships among taxa. These tools include the traditional but still useful techniques of comparative and functional anatomy as well as the more contemporary methods of embryology, immunology, serology, biochemistry, population and molecular genetics, and behavioral and physiological ecology. Although molecular and genetic analysis (cladistics) have proven to be a valuable tools for animal taxonomists, both traditional and cladistic methods in the form of integrative taxonomy are being used in our quest to understand the evolutionary patterns of animals and their proper taxonomic grouping.

Issues of Classification: Grouping

One of the first attempts at biological taxonomy was conducted by the Greek philosopher and scientist Aristotle. He classified all living organisms known at that time as either a plant or an animal. He further grouped animals according to their residence—air, land, or water. However, the roots of our modern system of biological taxonomy are to be found in the 18th century and can be traced mainly to the Swedish botanist and medical doctor Karl von Linne. (Latin was considered the language of science for that time and since most scholarly works were published in Latin, authors often Latinized their names. Hence, we more commonly know this man today as Carolus Linnaeus.).

Box 6.2
The Life and Times of the Father of Modern Taxonomy

Carl Linnaeus was born on May 23, 1707, at Stenbrohult, in the province of Smaland in southern Sweden. His father, Nils Ingemarsson Linnaeus, was a Lutheran pastor and an avid amateur horticulturalist in their small Swedish village. Nils' interest in plants and scholarly pursuits quickly rubbed off onto his son as Nils taught the boy not only about plants but also Latin, religion, and geography at an early age. In fact, family accounts state that Linnaeus could speak Latin before he could speak Swedish.

Figure 6.9 Carolus Linnaeus in Laponian attire dressed for exploration and field work. The plant he is holding was named after him—*Linnaea borealis*.

After two years of private tutoring, Linnaeus was sent to the Lower Grammar School at Växjö in 1717. Linnaeus rarely studied and often went out into the countryside in search of unfamiliar plants. In 1724, Linnaeus entered the Växjö Gymnasium where he studied a curriculum designed for boys preparing for the priesthood that focused on Greek, Hebrew, theology, and mathematics. There one of his professors, Johan Rothman, broadened Linnaeus's interest in both botany and medicine. Linnaeus would eventually come to live with Rothman and his family and with Rothman's help, Linnaeus enrolled in Lund University in Skåne at age 21. He was registered in the University as *Carolus Linnæus*, the Latin form of his full name, which he also used later for his Latin publications and the name by which we know him today.

Upon the advice of Rothman, Linnaeus entered Uppsala University in August 1728, a choice Rothman believed gave Linnaeus the best opportunity to study both medicine and botany. Linnaeus was selected to give the botany lectures at Uppsala, when only a second-year student.

In 1732 at age 25 Linnaeus received a grant from the Royal Society of Sciences in Uppsala to conduct the first scientific survey of Lapland, north of the Arctic Circle. On this trip, he collected plants and studied the animals and geology of the region. Linnaeus's passion for botanical exploration continued throughout his career, and he conducted numerous state-sponsored expeditions

In 1735, Linnaeus journeyed to the Netherlands where he promptly finished his medical degree at the University of Harderwijk. He then pursued further studies at the University of Leiden where in that same year he published the first edition of his classification of living things, *Systema Naturae*. During the next three years, he continued to develop his classification scheme all the while meeting and corresponding with Europe's greatest botanists. Returning to Sweden in 1738, he practiced medicine and lectured in Stockholm. In 1741, he was awarded a professorship at Uppsala where he quickly set about restoring the University's botanical garden arranging the plants according to his own classification system. At Uppsala, Linnaeus inspired an entire generation of students, and he was often instrumental in arranging for his students to accompany trade and exploration voyages to all parts of the world. His most famous student voyager was surely Daniel Solander, who served as naturalist on Captain James Cook's first round-the-world voyage. Cook and Solander brought back the first plants ever returned to Europe from Australia and the South Pacific.

Linnaeus was a popular Figure and who often led crowds of local people on field trips to the countryside. What a grand scene these trips must have been with Linnaeus leading the way followed by assistants in charge of note taking, collecting specimens, or carrying food and drink. Behind them came a happy throng of often a hundred or more citizens some beating drums, with others blowing horns, and lofting banners. Apparently a boisterously good time was had by all in the name of botany.

As more and more plant and animal specimens were sent to Linnaeus from every corner of the globe, he continued to revise and modify his concepts and *Systema Naturae* grew from one slim pamphlet to ten volumes. Linnaeus was also a botanical patriot who was deeply involved with ways in which to make the Swedish economy less dependent on foreign trade by importing and acclimatizing valuable nonnative plants, or by finding native substitutes. Unfortunately, Linnaeus's attempts to grow cacao, coffee, tea, bananas, rice, and mulberries proved unsuccessful in Sweden's cold climate.

He still found time to practice medicine, eventually becoming personal physician to the Swedish royal family. In 1758, he bought the manor estate of Hammarby, outside Uppsala, where he built a small museum for his extensive personal collections. In 1761, he was granted nobility, and thereafter was known as Carl von Linné to denote his enoblement. Unfortunately, his later years were marked by increasing depression and pessimism. Lingering on for several years after suffering what was probably a series of mild strokes in 1774, the man we now know as the Father of Taxonomy died in 1778.

Upon his death, Linnaeus's library, manuscripts, and natural history collections were sold to the English natural historian Sir James Edward Smith, who founded the Linnaean Society of London to house and care for the collection, a task they continue to faithfully perform.

Linnaeus was known for his collection and classification of flowers. He was also a prolific writer with over 180 books published in his lifetime, the most memorable being *Systema Naturae*. In this work published in 1735, Linnaeus separated all nature into three kingdoms: mineral, vegetable, and animal. He then divided the animal kingdom into five ranks or **taxonomic categories** or **taxons**—class, order, genus, species, and variety. Linnaeus used **morphology** (the form and structure of an organism) as the main criteria for placement in any particular rank. This scheme of arrangement is a nested hierarchal system in which each group grows more specific as you descend through the ranked groups. In the Linnaean system, the kingdom was the largest most inclusive group containing all known animal species. The kingdom in turn was composed of numerous classes each containing fewer but more specific groups of animals than the kingdom rank. Classes were composed of orders with even smaller and more specific groups of animals than the class rank and so on down to variety. The variety category would contain only one specific kind of animal.

From Aristotle in the 4th century through Linnaeus in the 18th century, a two-kingdom model was followed with all living things being classified into either *Kingdom Plantae*—plants or *Kingdom Animalia*—animals. This limited system was not adequate even for the time of Linnaeus as biologist struggled with how to classify fungi and microbes.

In 1894, Ernst Haeckel introduced the three-kingdom. This model incorporated the long-standing kingdoms Plantae and Animalia and added a third, *Kingdom Protista,* which included single-celled eukaryotes and bacteria (prokaryotes). Herbert Copeland proposed a four-kingdom system in 1956. This model included the three kingdoms Plantae, Animalia, and Protista and added a fourth. Reflecting the growing understanding that single-celled prokaryotes (bacteria) were fundamentally different than single-celled eukaryotes, Copeland suggested leaving the single-celled eukaryotes in Kingdom Protista but removing bacteria (prokaryotes) from that kingdom placing them in a kingdom of their own—*Kingdom Bacteria.*

Giving attention to concerns about those organisms classified broadly as plants, vegetation ecologist Robert Whittaker posed a five-kingdom model in 1959 that made two changes to Copeland's four kingdom model: (1) Copeland's wording of Kingdom Bacteria was changed to *Kingdom Monera* and (2) a fifth kingdom, *Kingdom Fungi*, was added.

In 1977, microbiologist Carl Woese used the emerging field of molecular biology to rework and expand Whittaker's system by replacing Kingdom Monera with two kingdoms, *Kingdom Eubacteria* and *Kingdom Archaebacteria*, resulting in a six-kingdom model. Woese proposed this split because genetic evidence indicated that archaebacteria (Gr., *archaio*, ancient) were as different from other bacteria as bacteria are from eukaryotes. In 1990, Woese again turned to evidence from molecular biology and proposed adding the rank of **domain** above the kingdom rank. His three-domain model places any organism into one of three domains: *Bacteria* (formerly Eubacteria), *Archaea* (formerly Archaebacteria), and *Eukarya* (all eukaryotic organisms). The Eukarya is in turn subdivided into four kingdoms: *Protista, Fungi, Plantae,* and *Animalia.* Many biologists accept and use the three-domain model, and it is currently favored in high school biology textbooks. However, like any other classification system of the past or present, the three-domain model is a work in progress. The exact origins, lineages, and classification of archaebacteria and bacteria have caused disagreement among taxonomists in the past and continue to do so today. Kingdom Protista, however, continues to be far more problematic than do the bacterial domains. The problem stems from the fact that this kingdom contains an assemblage of unicellular, colonial, and multicellular Eukaryotes that taxonomists shunt into Protista because they don't know what else to do with them. In effect, Kingdom Protista is a taxonomical junk drawer containing creatures as diverse as the animal-like protozoa, the plant-like algae, and the fungi-like slime molds and water molds.

Since the time of Linnaeus, the number of taxons has been expanded considerably. As more species are constantly discovered, taxonomists are forced to expand the system to accommodate them. The major taxons in turn can be subdivided into smaller ranks such as subkingdom, superclass, subspecies, and so on. To date, taxonomic ranks total more than 30. These additional categories are necessary when dealing with very large and complex groups of animals such as fish and insects.

Table 6.3 shows the traditional Linnaean hierarchal classification for several different animals utilizing the three domain model. Only the eight major taxons are used in the table. **Figure 6.10** demonstrates the nested hierarchal classification ranks for a specific animal.

Table 6.3
Traditional Hierarchal Classification of Select Animals

Categories	Human	Gorilla	Wolf	Dog	Frog	Fruit Fly
Domain	Eukaryota	Eukaryota	Eukaryota	Eukaryota	Eukaryota	Eukaryota
Kingdom	Animalia	Animalia	Animalia	Animalia	Animalia	Animalia
Phylum	Chordata	Chordata	Chordata	Chordata	Chordata	Arthropoda
Class	Mammalia	Mammalia	Mammalia	Mammalia	Amphibia	Insecta
Order	Primata	Primata	Carnivora	Carnivora	Anura	Diptera
Family	Hominidae	Hominidae	Canidae	Canidae	Ranidae	Drosophilidae
Genus	*Homo*	*Gorilla*	*Canis*	*Canis*	*Rana*	*Drosophila*
Species	*sapiens*	*gorilla*	*lupis*	*familiarus*	*pipiens*	*melanogaster*

The actual animal group that is placed at any particular level forms the taxon. Thus, the taxon Arthropoda is the appropriate phylum category of the fruit fly. The more morphologically similar two animals are, the more taxons they share in common. Accordingly, the dog and wolf are the two most closely related animals shown in the table.

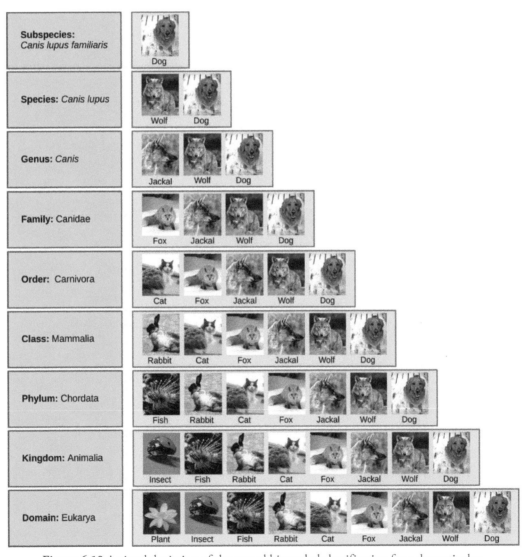

Figure 6.10 A visual depiction of the nested hierarchal classification for a domestic dog.

Phylogeny Determines Taxonomy

Most recently has been the emergence of an alternative method of taxonomy known as **phylogenetic systematics** or **cladistic taxonomy**. Traditional or Linnaean classification systems are based on morphology (similarity of traits) but do not take into account evolutionary relationships. **Phylogenetics** (Gr. *phyle,* tribe + *genetikos,* birth) is a philosophy of classification that arranges organisms only by their degree of evolutionary relatedness molecularly and genetically, not strictly by their morphological characteristics.

The goal of zoological systematics is to group animals in ways that reflect evolutionary relationships. One of the concepts most crucial to our understanding of evolutionary relationships is the determination of monophyletic taxa. A **monophyletic** taxon is a group of species whose members are related to one another through a unique history of common descent (with modification) from a common ancestor—a single evolutionary line resulting in a "single tribe" of animals.

To the consternation of taxonomists, what are known as **composite taxa** abound within various animal groups. There are two kinds of composite taxa—polyphyletic and paraphyletic. A **polyphyletic** taxon is a group comprising species that arose from two or more different immediate ancestors, whereas a **paraphyletic** group is an assemblage in which member species are all descendants of a common ancestor but which does not contain *all* the species descended from that ancestor. Composite taxa are the result of insufficient knowledge concerning a species or group of species. Through careful study taxonomists today struggle to eliminate polyphyletic and paraphyletic groups and taxons, reclassifying their members into appropriate monophyletic taxa (**Figure 6.11**).

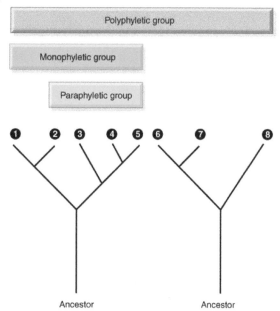

Figure 6.11 Evolutionary Relationships Between an Assemblage of Species. Species 1-8 represent a polyphyletic group because 1-5 and 6-8 have different ancestors. Species 1-5 are monophyletic because they share a common ancestor. Species 2-5 are paraphyletic because species 1 has been left out.

Three contemporary schools of systematic thought exist: evolutionary (traditional) systematics, numerical systematics, and phylogenetic (cladistic) systematics. Evolutionary (traditional) systematics, as the name indicates, is the oldest of the three approaches. The foundation principle of evolutionary systematics is the basic assumption that organisms closely related to an ancestor will morphologically resemble that ancestor more closely than they resemble more distantly related organisms. In other words, the more alike two organisms are morphologically, the more closely they are related evolutionarily.

Evolutionary systematists often depict the results of their works on what are called **phylogenetic trees**. Understanding a phylogenetic tree is a lot like reading your personal family tree. The root (base of the tree) represents the ancestor, and the tips of the branches represent the descendants of that ancestor. When speciation (lineage split) occurs, it is represented as a branching of the phylogeny on the tree. Phylogenies trace patterns of shared ancestry between lineages. Each lineage has a part of its history that is unique to it alone and parts that are shared with other lineages.

Although useful in visualizing evolutionary relationships and clarifying time scales, evolutionary-tree diagrams can be confusing and misleading in several ways. Trees depicting phyla or classes as ancestral groups are misleading because evolution occurs at the species (population) level, not at higher taxonomic levels (phyla or classes). Furthermore, evolutionary trees imply a slow progression of ever-increasing complexity and evolutionary success. This is misleading in that evolution has often resulted in the opposite—reduced complexity and body form and, ultimately, extinction. In spite of these problems, tree diagrams are useful enough to persist in scientific literature and texts.

Numerical taxonomy was developed in the 1950s and 1960s in an attempt to make taxonomy more objective. In this approach mathematical models and computer-assisted techniques are used to group select creatures according to their overall similarities. Although numerical taxonomy has fallen from favor, the computer programs developed by numerical taxonomists are used today by all taxonomists.

Phylogenetic (cladistic) systematics seeks to establish the genealogical (evolutionary) relationships between and among monophyletic groups (taxa) of organisms. The evolutionary relationship between select taxa may be analyzed phylogenetically (cladistically) using a number of sources of information including DNA or rRNA sequences (so-called molecular data), biochemical data, and morphological information. The result of such an analysis is a branching line diagram called a **cladogram**.

A grouping that includes a common ancestor and all the descendants (living and extinct) of that ancestor forms a **clade**. By definition, a clade is a monophyletic group. Determine what group of lineages form a clade by imagining clipping a single branch off the phylogeny—all the organisms on that pruned branch make up a clade and possess common **ancestral characters** (**Figure 6.12**). Clades are

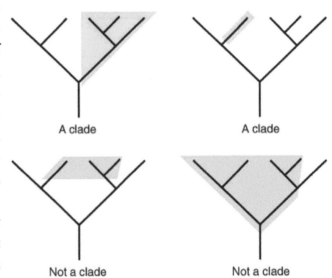

Figure 6.12 A clade is a grouping that includes a common ancestor and all the descendants (living and extinct) of that ancestor. It is easy to tell if a group of lineages form a clade. Imagine clipping a single branch off the phylogeny—all the organisms on that pruned branch make up a clade.

nested within one another, forming a nested hierarchy (**Figure 6.13**). When studying a group of seemingly related organisms, the trick is to find the **outgroup**. The outgroup can then be used to establish the common ancestral character(s), called **symplesiomorphies**.

Figure 6.13 Clades may be organized into nested hierarchies.

Depending on how many branches of the tree are included, the descendants at the tips might be different populations of a species, different species, or different clades, each composed of many species. The cladogram depicts only one kind of event—the origins or sequence of appearance of unique derived characters (**synapomorphies**) shared by the members of the study group. Identifying synapomorphies, or shared derived characters, constitutes the phylogenetic taxonomist's most powerful means of recognizing close evolutionary (genealogical) relationship.

Figure 6.14 depicts a hypothetical cladogram showing six taxa (1–6) and the derived characters (A–J) used to establish evolutionary relationships between the taxa. Note that taxon 6 is the outgroup for taxa 1 to

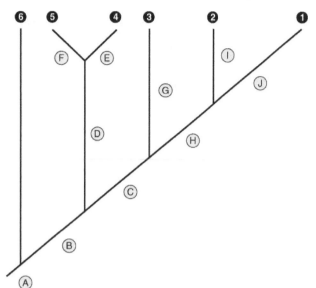

5. Character A is symplesiomorphic (ancestral) for both the outgroup and the study group. Characters B through J are synapomorphies having arisen after the divergence of the study group from a common ancestor. Taxa 1 through 5 in Figure B-4 share derived character B. This character arose after character A. Similarly, derived characters C and D arose more recently than B. Taxa 1 and 2 form a clade characterized by H.

Figure 6.14 Understanding Cladograms. Characteristics are listed as letters along the bottom of the cladogram, whereas species are listed as numbers along the top. Character A is the symplesiomorphic character for the assemblage of species 1-6. Species 6 is the outgroup because it shares only the symplesiomorphic character A with the rest of the assemblage 1-5. What characteristic is symplesiomorphic for species 1-3?

Construction of a cladogram can be a time-consuming process. The number of mathematically possible cladograms for more than a few species (taxa) is enormous—for 3 taxa there are only 4 possible cladograms, but for 10 taxa there are about 280 million possible cladograms. Needless to say, a thorough phylogenetic analysis of a group of several dozen species would be nearly impossible without serious computing power and the use of algorithm programs that generate cladograms by clustering taxa on the basis of nested sets of synapomorphies. Furthermore, more than one cladogram can be constructed from a particular set of data. In such cases, systematists employ the *rule of parsimony* that states that the simplest cladogram is probable the most accurate.

Evolutionary biologist Thomas Cavalier-Smith developed one of the first classification systems based on phylogeny. First proposed in 1981 and revised a number of times since, Cavalier-Smith's system is based on rRNA sequences as well as biochemical, ultrastructural, and paleontological data and groups all living things into two empires and eight kingdoms.

Empire Prokaryota

Kingdom Bacteria—Eubacteria and Archaebacteria

Empire Eukaryota

Kingdom Protozoa—amoebae, flagellates, ciliates
Kingdom Animalia—all animals from sponges to chordates
Kingdom Fungi—club fungi, sac fungi, molds
Kingdom Plantae—red algae, green algae, mosses, ferns, seed plants
Kingdom Chromista—diatoms, brown algae, oomycetes

This phylogenetic system more accurately reflects the true evolutionary relationships between the various groups of living things. The artificial contrivance of kingdom Protista has been eliminated, whereas kingdom Chromista housing the kelps, brown algae, diatoms, and water molds has been added.

Traditional and cladistic methods are being combined in what is called integrative taxonomy, an approach that combines data on specimens' distribution, morphology, genetics and molecular makeup, and ecology. Using integrative taxonomy, herpetologists have identified four new species of Peruvian glassfrogs. Based on a unique combination of traits and genetics, including finger webbing, liver shape, call pattern, and bone color (One species has green bones!), suspicions were confirmed that the frogs belong to four new species in the family Centrolenidae.

The perfect taxonomic system that is universally agreed upon has yet to be devised and probably never will be. Thus, biologists continue to struggle with the taxonomy of life as they have since the ancient times of Aristotle.

Issues of Nomenclature: Naming

> *What's in a name? That which we call a rose by any other name would smell as sweet.*
> —William Shakespeare

Linnaeus is best known for his introduction of the method still used to formulate the scientific name of every species. During the time of Linnaeus, scientific names were **polynomial** ("many names") and the descriptive words used in scientific names were not fixed to that specific creature. The European honeybee, for example, was named *Apis, pubescens, thorace subgriseo, abdomine fusco, pedibus posticis glabris utrinque margine ciliates*. The scientific name for humans from that time translated into English would read "hairy on top, bare on the bottom of the walking surfaces, bipedal, eyes forward with binocular vision, opposable thumbs, no wings, no feathers, capable of rational thought and the capacity to love." Linnaeus attempted to improve the composition and reduce the length of the many-worded names by abolishing unnecessary rhetorics, introducing new descriptive terms and defining their meaning with an unprecedented precision. In doing so, Linnaeus established a system of **binomial nomenclature**. In this Latin naming system, each species' name is made up of two italicized (underlined if handwritten or typed) words or epithets. The first epithet (capitalized) names the genus; and the second (lowercase) names the species of the genus. The only rules Linnaeus applied were that the epithets should be short and unique and that they should not be changed. Genera names were to refer only to single groups of organisms and could not be duplicated within the same genera of animals. Different genera could, however, share the same species epithet to differentiate species. According to traditional taxonomy, every organism technically has a Latin or Greek eight-word scientific name (Table 6.3). However, thanks to the binomial system of nomenclature developed by Linnaeus, the genus and species epithets for any particular animal are so specific that only those two words need be used to immediately identify precisely what animal is being discussed. *Sitta carolinensis* denotes the white-breasted nuthatch; *Drosophila melanogaster* is the fruit fly, and humans are *Homo sapiens*; no ambiguity or confusion, just preciseness and permanence.

Sometimes it is necessary for taxonomists to divide a species. In those cases, a **trinomial** nomenclature system is employed. A common example can be seen in our pets. The trinomy of the domestic dog would be *Canis lupus familiaris* so as to distinguish it from the closely related dingo of Australia, *Canis lupus dingo*, whereas the trinomy of the domestic cat would be *Felis silvestus catus* to distinguish it from the closely related African wildcat, *Felis silvestus lybica*.

But why use Latin and ancient Greek for scientific names? Why not use common names such as frog, cat, ladybug and so on? Common names are not suitable for use as scientific names for a number of reasons:

1. Common names are often not physically accurate. Imagine what the following animals would look like if their common names accurately reflected their physical structure or appearance: the house fly, the bald eagle, and the sea horse.

2. One common name may be applied to several different creatures. For example, the English robin is smaller with different feather coloration than the robin we know in North America.

3. A number of common names may be applied to the same creature. Puma, cougar, mountain lion and catamount are only several of the 40 different common names applied to the same large North America cat, *Puma concolor*.

4. Common names for the same animal differ not only between countries but also regionally within the same country. In our region, the fish we know as the bullhead and the crustacean we call a crayfish are known as the horn pout and the mudbug respectively in southern states.

5. Common names often do not translate accurately and meaningfully into other languages.

In the Middle Ages, Latin became the accepted language of the scholar scientist, and any scholarly or scientific work was published in Latin. As part of that tradition, the names of plants and animals were also given in Latin as they are to this day. However, Latin is not used solely for tradition's sake but because it is the perfect vocabulary for international communication in biology. Latin and ancient Greek solves all the problems presented by attempting to use common names for scientific purposes:

- Latin is a "dead" language. That is, Latin is no longer spoken as a national or even regional language anywhere in the world. Because of this, Latin words do not change. Although Greek is not technically a "dead" language, the ancient form of Greek is also no longer spoken. That is; no one is corrupting Latin or ancient Greek by developing slang words from it. Any spoken language develops slang words at an amazing rate. In fact, whole dictionaries are devoted just to explaining slang words. "Spider", for example, is a frying pan or skillet in the Deep South and you might be surprised to learn that "bruisecruise," "shingrater," and "stonesoup" are just several of the over 300 slang terms used to describe the same exact thing—crashing on a skateboard.

  Occupations develop their own unique slang. In the business world, a "404," from a computer error message meaning "file not found," denotes a mindless or distracted individual while "beepilepsy" describes the way people twitch when their pagers vibrate. In the world of science, a behavioral biologist might refer to the bluffing or posturing behavior of an animal as "gorilla dust," whereas an atmospheric chemist might measure "socks and knocks" (the oxides of sulfur and nitrogen commonly found in pollution). A "rat spill" (the accidental import of rats to a previously rat-free piece of land) or "voodoo stew" (water contaminated by sewage and other pollutants) couldn't help but concern any self-respecting ecologist, whereas an ornithologist might refer to a "jake" (male turkey) with especially long heel spurs as a "limb hanger." Trying to understand taxonomy may be "driving you bonkers" but "hang in there" and you will "get a handle on it."

- Latin and ancient Greek need not be translated when biologists from different countries communicate. For example, an American zoologist researching the mating behavior of the cat he commonly knows as the mountain lion might communicate with a colleague in Mexico about that animal. All parts of any communiqués between them would be translated from English to Spanish and back to English *except* the name of the cat (or any other animals) being discussed. The scientific name of the cat, *Puma concolor*, would be used instead of any common name and the scientific name would not be translated.

- Every species is assigned its own unique and permanent Latin or Greek name. This eliminates the problems of one animal having many names or many names being applied to the same animal.

- Linnaeus first established the system of binomial nomenclature that, with some refinement, we still employ today. In this system, each species has a Latinized name composed of two words or epithets (hence binomial). The first epithet names the genus; the second epithet names the species, which is specific to the species within the genus.

With the emergence of phylogenetic (cladistic) taxonomy, a new formal code of nomenclature, the PhyloCode, intended to deal with clades rather than ranks, is currently under development. The PhyloCode will be a formal set of rules governing phylogenetic nomenclature and is being designed to be used concurrently with the existing codes based on rank nomenclature.

A Closing Note
Travel Through Time Without Leaving the Farm

Some time ago, while digging a basement for an addition to his farm house, a former student of mine uncovered what appeared to be a large bone. Being a science teacher in a small rural community automatically makes you the "expert" on anything scientific so naturally I was the first person called to the scene.

As I scrambled down into the hole, I was thrilled and surprised to see a bit of what was obviously a very large bone sticking out of the dark soil. Always the teacher, I made plans to share this with my students and have them help me excavate the bone.

Student volunteers of all ages spent several afternoons after school carefully removing soil from around and off the bone using the same basic tools and techniques as actual paleontologists. On the last evening of our adventure, when we had almost totally exposed and cleaned what with my limited paleontological experience, I guessed to be a leg bone of a mastodon ("Real" paleontologists from the university would later confirm my suspicions.), we all sat down around the bone to rest. As the sun was setting, and shadows darkened the hole, I asked the students to all touch the bone and close their eyes. I then invited them to travel back in time and meet the creature whose bone we had uncovered. What was it appearance? What did its surrounding habitat look like? How did it die and how did this single leg bone (the only part of the skeleton we ever found) come to be in this place? In the cool quietness of the gathering evening, it was an almost surreal experience that I (and hopefully those former students) will never forget. And if you learn but a little of the life and earth of the past, you too can time travel, at least in your mind. By merely holding a fossil in your hand and closing your eyes, you can swim ancient seas, trudge prehistoric cycad forests, or slog through the muck of primordial swamps, all without leaving your chair, or in the case of our mastodon leg bone, the farm.

Playing the Name Game

I once had mudbug soup in Louisiana and I have eaten fried horn pout along the Savannah River between South Carolina and Georgia. However, these ramblings are not about exotic-sounding regional foods but rather about the huge challenge zoologists face as they attempt to group (classify) and name (nomenclature) animals.

When a claw appeared in a spoonful of my soup, I realized mudbugs to be what we call crayfish or crawdads in my area (However, in New Zealand, "crayfish" can mean salt-water rock lobster.) I helped catch and clean the bullhead catfish that my southern host knew so colorfully as horn pouts. The point is that great regional and cultural differences abound across North America as to the various common English names we assign animals. Names like "deer" and "sardine" can apply to dozens of different species world-wide. Compound that problem with having to translate these various common names into other languages and you have the potential for a chaos of confusion and misunderstanding.

Such uncertainty and impreciseness are intolerable to biologists who deal with the organism level of life and who must communicate with each other regarding these organisms with clarity and precision. Taxonomists have been struggling to concoct classification systems since Aristotle's first attempts hundreds of years B.C., and while our systems have grown more refined in the intervening centuries since, our modern taxonomic efforts and offerings are still a work in progress and probably always will be.

Although designing and perfecting classification systems may be an adventure in uncertainty, there are a couple of things of which I am very certain—Regardless of what you call them where you live, mud bug or crayfish or crawdad soup is really quite tasty and bullheads or horn pouts are fun to catch and tasty to eat, but not so much fun to clean and prepare.

In Summary

- Biologists have cataloged and named 1½ to 2 million species. As many as 10 to 100 million species could still be awaiting discovery.
- The apparent abundance of life is somewhat of an illusion as species are marching off into extinction at a rate unprecedented in human history.
- Evolution may be defined as change over time.
- The concept of biological evolution dates back to the sixth century B.C.
- During the Dark Ages, religious dogma concerning the origin and nature of life replaced scientific investigation of these ideas. However, with the dawn of the Renaissance, scientific interest in the origin and evolution of life was renewed.

- Modern evolutionary theory began in 1859 when Charles Darwin published *The Origin of Species.*

- Evolutionary biology as an academic discipline in its own right emerged in the 1930s and 1940s as a result of the rise of the modern evolutionary synthesis—the combining of Darwinian natural selection with Mendelian inheritance.

- Evolution happens on two levels: the genetic level (microevolution) and the taxonomic level (macroevolution).

- Evolution is driven by the mechanisms of mutation, genetic drift, gene flow, and natural selection and speciation.

- Evidence for evolution may be found in the fossil record, the homology and analogy of animal structures, and the molecular record.

- Since ancient times biologists have struggled with issues of taxonomy—grouping (classification) and naming (nomenclature)—in an attempt to bring order and organization to the riotous multitude of life forms on this planet.

- Traditional (Linnaean) classification systems are hierarchal and based on morphology (similarity of traits), but do not take into account evolutionary.

- Most recently has been the emergence of an alternative method of taxonomy known as phylogenetic systematics or cladistic taxonomy. Phylogenetics is a philosophy of classification that arranges organisms only by their degree of evolutionary relatedness and not strictly by their morphological characteristics.

- The evolutionary relationship between select taxa may be analyzed phylogenetically (cladistically) using a number of sources of information including DNA or rRNA sequences (so-called "molecular data"), biochemical data, and morphological information. The end result of such an analysis is a tree-like diagram called a cladogram.

Review and Reflect

1. *In Your Own Words*. Write you own personal definition of biological evolution.

2. *What Would You Say?* You are sitting in the front row as a world-renown evolutionary biologist serves as a guest lecturer to your zoology class on modern evolutionary theory. Suddenly the guest lecturer is taken ill. As she rushes toward the door, she grabs you and says, "You continue the presentation. Explain the mechanisms of evolution and how new species form (speciation) to them." What would you say?

3. *Hardy and Weinberg Speak*. Let us suppose that a trait in beetles, such as spots, is determined by the inheritance of a gene with two alleles—S and s. Fifty years ago, the frequency of S in a small population of beetles was determined to be 92% while the frequency for s was found to be 8%. Today we find S at 90% and s at 10%.

   A. Has evolution occurred in this population of beetles? Defend your answer using the Hardy-Weinberg equilibrium.

B. How might a population bottleneck and resulting genetic drift have played a role in the observed change of gene frequency in the beetles?

4. ***In the Blood of Cheetahs***. Clocked at 70 miles per hour while pursuing prey, the cheetah is by far the fastest and most agile member of the cat family. Superbly adapted for hunting, cheetahs are even more successful than lions in capturing prey. Yet the cheetah population is dwindling with estimates placing their numbers at anywhere from 1,000 to 20,000. What is happening? Why should the population of an animal with such extraordinary capabilities dwindle in this way?

It has been proposed that at some time in the cheetah's recent evolutionary history, an infectious disease or a natural catastrophe must have reduced its population, and a population bottleneck resulted. This in turn caused a reduction in the variety of alleles (gene pairs) in the population. The genetic drift resulting from the reduction of a population results in a surviving population that over time is no longer representative of the original population.

Biologists began to investigate the population bottleneck hypothesis by studying cheetahs at a breeding and research center in Africa. At this center, as with others, there was little success in breeding cheetahs in captivity; only a few reproduced and over 37% of the infant cheetahs died.

Biologists first examined semen from 18 male cheetahs and a similar number of male domestic cats. They found that the concentration of sperm in male cheetah semen was only one-tenth as great as the concentration of sperm in male domestic cats. Furthermore, 71% of the cheetah sperm had abnormal shapes with flagella that were either bent or coiled or sperm heads that were too large or too small. By comparison, only 29% of the sperm from domestic cats was abnormally shaped.

A. What conclusions can be drawn from this study?
B. Why were domestic cats included in this study?
C. The formation of sperm is directed by genes. What can you conclude about the genes that control sperm formation in the cheetahs studied?

When inbreeding occurs naturally or is carried out deliberately in a population, the number of heterozygous alleles (Aa) decline while the number of homozygous alleles (aa) increase. This results in a lack of genetic variability that can be detrimental to a population. For example, if an environmental change occurs, a lack of genetic diversity can reduce the populat9on's chance of adapting to the change.

A. Suppose a group of cheetahs all carried a gene that made them susceptible to the fatal disease *feline infectious peritonitis* (FIP). What might happen if cheetahs were exposed to this disease?

To test how genetically similar cheetahs were biologists took blood samples from both cheetahs and domestic cats then isolated various proteins. They determined whether each protein was the product of heterozygous or homozygous genes. Of 52 proteins tested, all of the cheetah proteins were found to be the products of homozygous genes. In domestic cats, a healthy mix of both heterozygous and homozygous proteins was found.

A. Does the protein evidence indicate that the cheetahs were seriously inbred?

B. Does the fact that the cheetahs are inbred support the population bottleneck hypothesis?

Next biologists studied the rejection rates of skin grafts on cheetahs. Usually, an animal rejects a skin graft unless it is from a close relative with a similar genotype. Domestic cats usually reject skin grafts from unrelated animals in 10 to 12 days. Biologists grafted three patches of skin onto a group of cheetahs: one patch from the cheetah itself, one patch from another cheetah, and one patch from a domestic cat.

A. Why were the cheetahs given grafts of their own skin?

B. From what you have learned so far, predict what you think the outcome of the skin graft experiment was.

C. Does the skin graft experiment support the population bottleneck hypothesis? Explain.

5. ***Defending the Theory***. You are in a heated discussion. One of the people in the discussion group does not accept the theory of evolution and believes instead that the origin of life and the diversity of species on this planet is best explained by divine creation (or what is often referred to as "intelligent design"). This person fires question after question at you. Explain how you would respond to each of the following statements or questions:

A. "How can evolution be true if biologists keep referring to it as evolutionary *theory*?"

B. "The only evidence to support evolution is a few fossils.

C. "The time frame for the age of the earth is all wrong. The earth is only about 10,000 years old, not billions of years old."

D. "How can you believe that humans evolved from apes?"

6. ***Is There Ever an End to It?*** In this chapter, we stated that our present taxonomic systems must be regarded as "works in progress." Why did we say that? Explain.

7. ***The Language of Taxonomy***. In a fit of frustration, your study buddy says to you, "What's with the Latin and ancient Greek in taxonomy? Are taxonomists the high priests of some secret society that uses these old languages as code that none of the rest of us can understand?" What would you say?

Create and Connect

1. ***How Old is Papa?*** A dispute has arisen over the age of a piece of fossil bone nicknamed the "Papa Femur." Some contend the bone is over 100,000 years old while others maintain it is less than

20,000 years old. Using radiocarbon-14 dating techniques, you have established that the Papa Femur contains 12.5 units of C-14. You know that such bone contained 100 units of C-14 when the organism was alive, and that C-14 has a half-life of 5,730 years. Which camp in this dispute is correct about the age of the Papa Femur? Defend your answer with the appropriate calculations.

Radiometric dating techniques are based on certain assumptions and have some limitations. Investigate the limitations of radiometric dating techniques and the assumptions on which these techniques are based.

2. *A Classroom Controversy.* How should the concept of biological evolution be taught in schools is a raging controversy that impacts society on a number of levels—educationally, legally, religiously, scientifically, and legislatively.

What are your feelings and beliefs on this controversy? Write a position paper in which you take a stand on this issue and defend the stand you take.

Guidelines:

A. Write a position paper, NOT an opinion paper. Defend your position with as many facts, Figures, quotes, and pertinent information as possible.

B. When completed your paper should detail: (1) your personal belief about the origin of life in general, (2) the origin of humans in particular (3) what should or should not be taught in public schools regarding these concepts, (4) who—persons or state or federal agencies—should decide through laws, rules, or policies what should or should not be taught in school regarding these concepts, (5) who—persons or state or federal agencies—should enforce any laws, rules, or policies established as to the teaching of these concepts, and (6) what punishment should there be for anyone—individual or school district—that violates established laws, rules, or policies.

C. Your work will be evaluated not on the "correctness" of your position but the quality of the defense of your position.

3. *An Alien Classification System.* A starship has just returned from a distant planetary system. While there, the ship's naturalists collected the species shown in **Figure 6.15**. As head of Star Command's Exobiology Taxonomy Division, you are charged with designing a classification system for these creatures.

Guidelines:

A. Start with only two or three large "kingdoms" (groups).
B. Use dichotomous branching.
C. Refer to the creatures by number. You may assign scientific names if you desire, but this is not necessary to complete your mission.
D. No creature can be in more than one group at the same time.

E. Use specific and precise wording to describe characteristics of each species (numbered animal).

F. Your goal is to end up with each creature in a group by itself.

Figure 6.15 Imaginary creatures from a fictional distant planet. Design a traditional classification scheme for these creatures.

PHYLUM PORIFERA: THE SIMPLE ONES

If sponges could express themselves, as is the desire of so many of us today, I think a well-speaking sponge might address biologists somewhat in the following fashion: 'I realize that we are not as widely known as some others and yet I feel that our family memoirs show that we are not an uninteresting race.'

—H. V. Wilson

Introduction

The phylum Porifera (L., *porus*, pore + *fera*, bearing) contains those animals commonly referred to as **sponges**. When contemplating a sponge, the casual observer of today might easily make the same mistake as early biologists and misjudge the sponge (or its remains) to be a creature best suited for the plant kingdom. Closer examination, however, would reveal that the sponge actually falls on the animal side of the sometimes fine evolutionary lines that separate the two kingdoms. In fact, sponges are the simplest multicellular beasts (**metazoa**) in the animal realm.

History of Porifera

Poriferans are an ancient, but successful animal group whose lineage can be traced back to the beginnings of animal life. The 9,000 or so living species of sponges represent the only parazoa branch on the Tree of Life (**Figure 7.1**). Fossil evidence and molecular data places poriferan origins at somewhere around 650 million years ago, nearly back to the dawn of animal life itself. DNA analysis indicates that the sponge's immediate evolutionary predecessors are the protistan choanoflagellates, a group biologists long suspected could have been the nearest thing to animals (metazoa) without actually being animals. These single-celled creatures with whip-like flagella bear an amazing resemblance to the choanocyte cells of present day sponges. In one species, a particular molecule previously found in only multicellular animals was discovered leading scientists to conclude that the choanoflagellates appear to contain the "genetic tool kit" (i.e. genes for inter-cell adhesion, signaling, and differentiation.) from which the first animals were made.

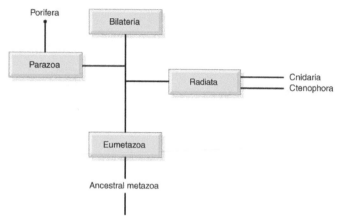

Figure 7.1 The probable position of phylum Porifera on the phylogenetic tree of life.

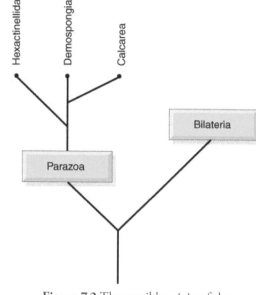

Figure 7.2 The possible origin of the various taxa comprising phylum Porifera.

Molecular evidence supports the hypothesis that choanoflagellates and metazoans are sister groups as are poriferans and Eumetazoa. **Figure 7.2** depicts the possible origin of the various classes comprising phylum Porifera. Classes of sponges are distinguished on the basis of the form and chemical composition of their spicules and the complexity of their body form. However, to date, evolutionary relationships between poriferan classes have not been clearly distinguished nor fully understood.

Diversity and Classification of Porifera

The variety of size, shape, and color exhibited by living sponge species is truly amazing. Some are round, whereas others are flat and crusty or vase-like in shape. Many species are gray or drab brown, but scores of sponge species display brilliant colors of red, orange, yellow, blue, violet, and even black. Many species also harbor symbiotic bacteria or unicellular photosynthetic protists that can add color and hue to their construction.

Sponges range in size from a few millimeters (tenths of an inch) to giants more than 2 meters (7 feet) in height or diameter. Such gigantic sponges may be hundreds of years old.

The approximately 9,000 total species of **extant** (living) sponges are mostly marine and can be found living from the intertidal zone at the ocean's edge to abyssal depths miles down. Freshwater sponges number about 150 species, with approximately 27 species found in the lakes and streams of North America. Terrestrial sponges do not and cannot exist.

Recent molecular analysis of sponge phylogeny suggests that the traditional phylum designation of Porifera should be eliminated. In its place the class Calcarea should be delegated as a separate phylum with the other sponges placed in phylum Silicarea. Such taxonomic changes, however, have yet to become widely accepted and prevalent in the scientific literature (and textbooks), and thus we present here the more traditional classification of sponges.

Domain: Eukarya
    Kingdom Animalia
        Sub-Kingdom Parazoa
            Phylum Porifera
                Class Calcarea
                Class Hexactinellida
                Class Demospongia

Taxonomists carve two subkingdoms from the larger animal kingdom: Parazoa and Eumetazoa. The parazoa (Gr. *para,* alongside + *zoa,* animals) exhibit no symmetry and lack tissues and organs, whereas the eumetazoa (Gr. *eu,* true +*meta,* later + *zoa,* animals) are both radically and bilaterally symmetrical and do posses tissues and organs. Sponges are the only living parazoans on the planet; all other animal types are eumetaozoans.

Phylum Characteristics

Despite the fact that sponges are large-bodied multicellular animals, they function largely like organisms at the unicellular grade of complexity. Their strategies of nutrition, cellular organization, gas exchange, reproduction, and response to environmental stimuli are all very protist-like.

The members of phylum Porifera exhibit the following general characteristics:

- Their multicellular body is a loose aggregate of cells without true tissues; adults are asymmetrical or superficially radially symmetrical.
- They are all aquatic, and most are marine.
- All adults are **sessile** (do not move about freely) suspension feeders; larval stages are **motile** (move about freely).
- Their "body" is perforated by **ostia** (pores), canals, and chambers that serve for the passage of water.
- Water currents move through the chambers and canals due to the action of unique flagellated cells known as **choanocytes.**
- Skeletal elements are composed of calcareous or siliceous crystalline **spicules,** often combined with variously modified collagen fibers (**spongin**).
- Asexual reproduction is by regeneration, buds, or gemmules and sexual reproduction by eggs and sperm; larvae are ciliated and free-swimming.
- Their cells tend to be **totipotent** in that they retain a high degree of mobility and are capable of changing form and function. This potential for plasticity has helped sponges survive for over 500 million years. Sponges can respond to unfavorable environmental changes by simply crawling slowly away, rebuilding their canal system as they go. If one is damaged by an angelfish (one of the few fishes that eat sponges), instead of building new tissue to repair the damage like a more advanced animal might, a sponge merely mobilizes existing cells to move over and cover the

wound, often leaving a scar. Sponges use this plasticity potential to change shape and spread out, filling irregular open spaces on reefs and rocks crowding out more structurally complex animals. The prize is the rarest commodity on the shallow seafloor, space in which to settle down and live.

Class Characteristics

Historically, the classes of Porifera have been defined by the presence and nature of their spicules. Certain fossil groups whose organization was consistent with that of living sponges, however, were not placed in the phylum Porifera due to having a solid calcareous skeleton that lacked spicules. The discovery of more than 15 extant species also having a solid calcareous skeleton radically changes our view of poriferan phylogeny. It is now widely accepted among poriferan biologists that the Calcarea and the Demospongia are more closely related to each other than either is to the Hexactinellida. Furthermore, poriferan biologists realized that a traditional fourth class of Porifera, the Sclerospongia or coralline sponges, was not a monophyletic grouping and has thus been abandoned.

Class Calcarea (L. calis, lime)

Figure 7.3 A Calcareous Sponge. The yellow sponge *Clathrina darwinii* from the Mediterranean Sea.

The calcareous sponges are so named for their calcium carbonate spicules. The spicules, laid down as calcite, may have two to four points and lack a hollow axial canal.

All the sponges in this class are marine dwellers most commonly found in shallow tropical waters. They are found from the intertidal zone down to around 200 meters (660 feet), although one species is known from a depth of 4,000 meters (13,200 feet).

Calcareous sponges are typically small, measuring about 8 to 10 cm (3-4 inches) and they demonstrate relatively simple shapes being either purse-, vase-, pear-, or cylinder-shaped (**Figure 7.3**). This class is considered to be the ancestral group of sponges, and it is the only class demonstrating all three sponge body plans—asconoid, synconoid, and leuconoid. Calcareous sponges lack the hollow canals that mark the morphology of other sponge types. Instead, their skeleton has either a mesh or honeycomb structure making them structurally stronger than other sponges. The fossil record indicates that Calcarean sponges first appeared during the Cambrian Period 540 to 480 million years ago and that their diversity reached its zenith during the Cretaceous Period 146 to 66 million years ago.

Class Hexactinellida (Gr. hex, six + aktis, ray)

Upon close internal examination, Hexactinellida sponges can be easily distinguished from other sponge types. The skeleton of a hexactinellid is composed entirely of silica in the form of six-pointed siliceous spicules. The spicules are composed of three perpendicular rays giving them six points. Furthermore, the spicules of hexactinellid sponges are fused together imparting a rigidity of structure not found in other sponge types. When the living tissue is removed, the cylindrical skeletons often have the appearance of spun glass and resemble ornate jewel-like objects (**Figure 7.4**). Hence their common name—glass sponges.

Hexactinellid sponges occur in all oceans worldwide, mostly at depths between 200 and 1000 meters (650-3,300 feet), although they are more abundant and diverse at shallower depths of polar regions. They are especially abundant and diverse in Antarctic waters. Frequently, they are the most conspicuous form of benthic life in these frigid waters, and it appears that they may even be important in structuring biodiversity on the continental slopes, as well as the continental shelf of Antarctica. Large mats of their spicules provide a suitable substratum that may allow a greater number of species to exist in a given area.

Recently, it has been discovered that some hexactinellid sponges form impressive reefs off the coast of British Columbia, Canada. Found in waters 180-250 meters (600-820 feet) deep, these reefs with nearly vertical sides tower as high as 18 meters (60 feet) above the surrounding sea floor. Biologists have learned that glass sponges fall into two main categories based on their spicule structure: lyssacine glass sponges with a loose spicule skeleton and the dictyonine glass sponges with a fused spicule skeleton that forms a rigid exterior. Glass sponge reefs are formed by the dictyonine type. As with the coral reefs will we study in a later chapter, glass sponge reefs form as new generations of sponges grow atop the remains of the previous ones pushing the crystal palace ever upward. This results in the underlying structure of the reef being a solid interlocked mass of silica.

All glass sponges are upright, and most appear to be radially symmetrical. They are pale in color and typically cylindrical, but they may also be cup-shaped, urn-shaped, or branching. The average height of a hexactinellid is 10-30 cm (4-12 inches), but some can grow to be much larger. They possess a cavernous central cavity (atrium) through which water passes and although they superficially resemble the syconoid body plan, they differ too much internally to be considered truly syconoid.

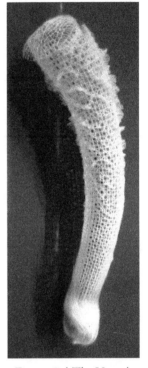

Figure 7.4 The Venus's flower basket sponge (*Euplectella aspergillum*) is an ornate hexactinellid sponge. The glass fibers at the bottom of the sponge are used for attachment. The skeleton and fibers of this sponge may hold the secret to more efficient fiber optics and solar cells and cheaper semiconductor devices.

Hexactinellids are completely sessile; even the larvae seem to display no movement outside of their ability to disperse small distances within currents. Furthermore, unlike other sponges, hexactinellids do not contract when stimulated. Because of these and other major distinctions, some researchers consider them sufficiently unique and deserving of their own phylum—Symplasma.

The oldest known sponge fossil of any type is a hexactinellid. By the Cambrian, relatively simple forms are known worldwide. They probably reached their maximum diversity during the Cretaceous.

Class Demospongia (Gr. demos, people + spongos, sponge)

Figure 7.5 Some demosponges are brightly colored as brilliantly demonstrated by this azure vase sponge (*Callyspongia vaginalis*)

The Demospongia is not only the most diverse group of sponges; it is also the largest class of sponges containing over 90% of all the known living sponge species. Demospongia can range in size from a few millimeters (tenths of an inch) to giants over 2 meters (6-7 feet) in their largest dimension. The Demosponges come in a bright pallet of colors, including bright yellow, orange, red, purple, and green (**Figure 7.5**). Produced by pigment granules in amoebocyte cells, the significance of such bright hues is uncertain but protection from solar radiation and warning coloration have been suggested for some species.

Marine Demospongia are found from the warm, shallow intertidal zone to dark, cold abyssal depths where they grow as lumps, finger-like projections, or urn shapes. Only this class of sponges also includes freshwater species. Freshwater demosponges are fragile and often colored green from the algae living within them. In form and appearance they live as branching masses or as thin encrustations on plants, submerged logs and twigs, and rocks. Only one species, *Spongilla lacustris*, has been found growing out of soft bottom sediments in standing water.

Freshwater demosponges (genus *Spongilla*) are found in clean lake waters and slow-moving streams throughout North America. They are more common in northern and eastern lakes and streams, however, than in southern and western ones. In general, freshwater sponges are found in areas with frequent physical disturbances, high pollution levels, or turbid waters containing high levels of silt.

The skeletal elements of Demospongia consist of siliceous spicules having one to four rays and/or coarse spongin fibers. Demospongia predominantly exhibit the leuconoid body structure.

Demospongia do not possess skeletons that easily fossilize so the record of their past is not as easily determined as is that of other sponge types. From what we do know, it seems likely that several of the major lines of Demospongia were already established in the lower Paleozoic. By the beginning of the Cretaceous, all orders of Demospongia were represented.

Poriferan Body Plan

Does a sponge possess a true "body?" Early investigators thought not and viewed sponges as essentially colonial animals. A later refinement of this outlook held that each excurrent opening and its attendant flagellated surfaces constituted an "individual." By this reasoning, an encrusting sponge growing on a rock and possessing ten oscula would be a colony of 10 individuals. Another view, and the one we prefer holds that a single individual consists of any and all sponge material bounded by a continuous outer cellular covering. Facts would seem to indicate that a whole sponge grows a whole body, determined largely by environmental factors and pressures. As an entity, a sponge grows by continually adding new cells that differentiate as needed, and there is the existence of some coordinated cell functions and responses in sponges. Neither of these characteristics is demonstrated by colonial organisms and supports our view that each sponge is in its entirety an "individual."

Although traditionally described as sessile creatures, sponges are not the permanent fixtures they were once thought to be because we now realize that adult sponges can move. Biologist Calhoun Bond, then at the University of North Carolina at Chapel Hill, found in 1986 that sponges don't just sit still—many actually move. Using time-lapse photography to study ten sponge species, he documented sponges not only slowly moving over the bottom of their containers but up the sides as well. The speedster of the group was a lavender beauty, *Haliclona lossanoffi*, which clocked in at more than 4 millimeters (0.16 inch) a day. Movement consisted of extending flat foot-like appendages and dragging the rest of the sponge behind, often discarding bulky pieces of skeleton on the way.

Sponges have managed to successfully overcome the limitations imposed on them by their primitive parazoan level of organization through three unique organizational features: the water current system, or the **aquiferous system**; the highly totipotent nature of sponge cells; and the general plasticity of their body form. These peculiarities have driven the evolution of sponges and accounts for the tremendous diversity and variation among sponges in size and form. Cellular totipotency allows a plasticity of form that enables sponges to radically alter their form and function as dictated by environmental demands. The aquiferous system compensates for the lack of tissues and organs by bringing water through the sponge and close to each cell. The water carries in food particles and oxygen for the cells and carries away the excretory and digestive wastes and, on certain occasions, reproductive cells. In a sense, the water performs much the same function as do the organs of a complex eumetozoan animal. Your organs and the blood supply that connects them live inside you, but as a wholly aquatic animal, a sponge lives inside its "blood" and "organs." The morphological/anatomical trick the sponges have managed to master quite successfully is to bring life-giving water into close contact with each cell.

Body Structure and the Aquiferous System

Most sponges are vase- or tube-shaped with a large excurrent opening, the **osculum**, at the top (**Figure 7.6**). The interior cavity, called the **atrium** or **spongocoel**, opens to the outside through the osculum. A sponge is composed of a number of cell types (**Figure 7.7**). The outer surface cells of a sponge make up the **pinacoderm**, which consists of a layer of flattened cells or **pinacocytes** (Figure 7.7a). Perforating the pinacoderm and forming openings known as **incurrent pores** are the ring-shaped **porocyte** cells (Figure 7.7b). The atrium (spongocoel) is lined with flagellated choanocytes that together comprise the **choanoderm** (Figure 7.7c). Both the outer pinacoderm and the inner choanoderm are only one cell thick. Sandwiched between these two thin cellular sheets is a layer of mesenchyme, called the **mesohyl**, containing ameboid cells of different types—**collencytes**, which secrete the dispersed fibrous collagen found

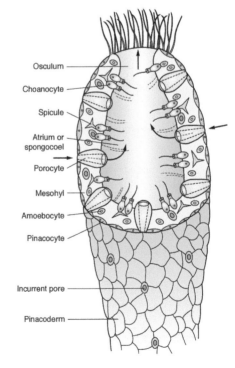

Osculum
Choanocyte
Spicule
Atrium or spongocoel
Porocyte
Mesohyl
Amoebocyte
Pinacocyte
Incurrent pore
Pinacoderm

Figure 7.6 A partially sectioned asconoid sponge. Arrows indicate the direction of water flow.

intracellularly in virtually all sponges; **spongocytes** (Figure 7.7d), which produce the supportive collagen referred to as spongin; **sclerocytes** (Figure 7.7e), responsible for the production of calcareous and siliceous spicules; and **archaeocytes** large, highly motile amoeboid cells that are the sponge's wondrous workhorses. They can, as needed, change into any and all of the other types of sponge cells. They help build the skeleton, produce or nourish developing eggs, and, by wandering through the sponge, distribute food and mediate the chemical and physiological responses that enable the sponge to react to environmental stimuli in a very low-level manner.

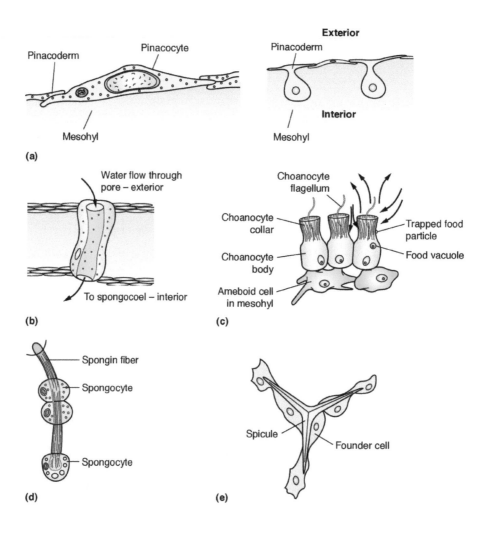

Figure 7.7 Sponge Cell Types. (a) Pinacocytes and their relationship to mesohyl. (b) Porocytes demonstrating their tubular nature. (c) Choanocytes and their current-generating flagella. (d) Spongocytes working in series to secrete a spongin fiber. (e) Sclerocytes secreting a spicule. The founder cell is one of three that join to begin secreting the spicule in the first place.

These cell layers are draped over and wrapped around a supporting skeletal structure. The skeleton of sponges consists of spicules (mineral crystals) of calcium carbonate or silicon dioxide secreted by sclerocytes. Spicules may be a single needle-like ray, or several rays joined together at certain angles to each other. Spicules are generally microscopic and unconnected, and each species of sponge possesses a characteristic combination of several different spicule types (**Figure 7.8**). Traditionally, spicule structure has been an important characteristic in the classification of sponges.

Another skeletal element found in sponges is coarse fibers of protein known as **spongin** (**Figure 7.9**), a substance similar to collagen in other animals. Some sponges have skeletons consisting entirely of either spongin fibers or spicules, whereas the skeleton of many sponges is formed by a combination of spicules and spongin fibers. Whatever its composition, the skeleton is highly organized with regards to other features of sponge structure. It plays a major role in supporting the internal water canals and maintaining the integrity of the sponge in the face of strong currents and surges and the prodding of pesky animals.

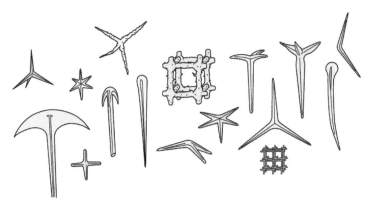

Figure 7.8 Spicules come in a wide variety of forms and complexity of structure.

Sponges vary in the complexity of their skeleton with three basic grades of structure being recognized—asconoid, syconoid, and leuconoid (**Figure 7.10**). It should be noted that these groups are meant to be morphological, not taxonomic in nature.

Asconoid sponges are usually a simple vase-like shape and are small, not more than a few centimeters tall. Size is severely limited in asconoid sponges because, as the volume of the spongocoel increases, the flagellated surface area does not increase pro-

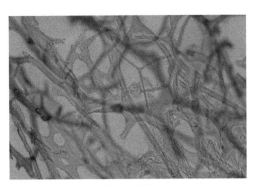

Figure 7.9

portionately. Consequently, a large asconoid sponge would contain more water volume than its choanocytes could effectively move, thus depriving the sponge cells of food and suffocating them in their own wastes.

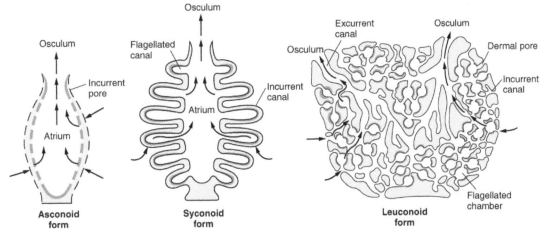

Figure 7.10 The Three Body Forms of Sponges. Dark layers indicate Choanocytes and arrows indicate the direction of water flow.

This size versus volume problem was overcome with the development of folded sponges. The first level of folding is exhibited by **syconoid sponges**, in which the flagellated layer is evaginated outward into finger-like projections. These evaginations are called **flagellated canals**, and the corresponding invaginations of the surface between the "fingers" are called **incurrent canals**. Pores are located between the incurrent and flagellated canals. The flagellated canals open into a central atrium devoid of choanocytes. Folding of this sort increases the surface area of flagellated cells, allowing the sponge to process a greater amount of water at a faster rate than is possible with the asconoid plan. Thus, syconoid sponges can grow considerably larger than their asconoid counterparts.

Taking the benefits of folding to the extreme are the **leuconoid sponges**. In these sponges, the surface area of the flagellated layer has been increased by the formation of many small chambers in which the choanocytes are located. In leuconoid sponges, water enters through dermal pores on the body surface (formed by porocytes) and passes through a system of incurrent canals, eventually reaching the flagellated chambers. Many leuconoid sponges have no atrium; water leaves the body through converging excurrent canals opening to the exterior through a number of **oscula**. Within the system of canals and chambers water always flows in the direction of the oscula.

Leuconoid sponges are very efficient at moving large volumes of water. One leuconoid sponge 10 cm (4 inches) high by 1 cm (0.4 inch) in diameter was found to possess some 2,225,000 flagellated chambers and could pass 22.5 liters (6 gallons) of water through its body in 24 hours. As a result, giant leuconoid sponges are found measuring several meters in height and width.

Feeding and Digestion

Working in unison, the flagella of the choanocytes drive water away from these cells, creating a current that pulls water in through the incurrent pores and drives it out the spongocoel through the oscula. Sponges are filter feeders that remove bacteria, protists of various sorts, and very fine organic matter from the water. In some tropical sponges, 80% of the organic matter taken in as food was of a particle size visible only through an electron microscope. The other 20% was composed primarily of bacteria and protists. Sponges are also size-selective feeders, with the size of pores and passageways permitting only certain size particles to enter.

Large food particles (1–50 micrometers or 0.00004–0.002 inches) are **phagocytized** (engulfed) mainly by the archaeocytes along the sponge's system of canals carrying water from the ostia to the spongocoels. Approximately 80% of a sponge's food is trapped by the choanocytes and consists of smaller particles (0.1–1.0 micrometers or 0.000004–0.00004 inches). In the case of archaeocyte phagocytosis, digestion takes place in the food vacuole formed at the time of capture. In the case of choanocyte capture, food particles are partially digested in the choanocytes and then quickly passed on to a mesohyl archaeocyte for final digestion. In both cases, mobility of the mesohyl cells ensures transport of nutrients throughout the sponge body. Thus, digestion is entirely **intracellular** as opposed to the extracellular digestion that occurs in the digestive organs of eumetozoa animals. The exception (and there is always an exception) are the carnivorous sponges of the family Cladorhizidae. These small, white sponges bearing 30 to 60 sticky barbed filaments were first discovered in the Mediterranean on rocks in shallow caves where water is trapped all year long and thus has a constant low temperature of 13 to 15°C (55–58°F). These sponges lack an aquiferous system and choanocyte chambers. Their diet usually consists of small crustaceans that blunder into them and become ensnarled in their sticky filaments. New filaments grow over the prey until it is completely covered with sponge cells. Using **extracellular** digestion, these encrusting cells then slowly dissolve the prey over the course of several days.

Recently, a new species of Cladorhizidae was found off the Aleutian Islands that does not inhabit shallow, stagnant caves. These sponges capture prey floating in the water column by impaling organisms with tiny spines. Encrusting and extracellular digestion then follow as with the Mediterranean types. This presents the question of whether these different types of carnivorous sponges result from convergent evolution due to similar selection pressure or whether these two groups share a common ancestor.

Many sponges harbor photosynthetic symbionts such as cyanobacteria and certain species of protists. In fact, one-third of the body mass of some sponges has been found to be symbiotic cyanobacteria. Through photosynthesis, these green partners of the sponge produce an excess of food they share with their host.

Gas Exchange and Waste Removal

Excretion (primarily ammonia) and gas exchange (oxygen in and carbon dioxide out) are accomplished by simple diffusion into and out of the water currents passing through a sponge. The efficiency and structure of the poriferan body plan are such that no cell is farther from water than about 1.0 mm, the distance beyond which gas exchange by diffusion becomes efficient. Freshwater sponges also possess water expulsion vesicles that serve to rid the body of excess water.

Response and Defense

Although there is no evidence that sponges possess anything resembling nerve cells or discrete sense organs, they are capable of responding to a variety of environmental stimuli. The usual effect of these responses is to reduce or stop the flow of water through the aquiferous system. For example, when suspended particles around them become too large or too concentrated, sponges typically respond by closing the oscula and immobilizing the choanocyte flagella. With a hand lens or low-power microscope one can see that the direct physical stimulation of running one's finger across the sponge's surface will cause contractions of the dermal pores or oscula.

Sponges respond to certain internal factors as well. For example, periods of reproductive activity or a major growth phase result in a decline in activity levels and a drop in pumping rates. Even under "normal" conditions, variations in pumping rate occur with some sponges ceasing activity for a few minutes or hours at time; others cease activity for several days at a time.

The switch from full pumping to complete shutdown requires at least several minutes. This is a fairly short response time considering the simplistic level of organization of sponges. The spread of stimulation and response in sponges seems to be by simple mechanical stimulation from one cell to adjacent cells with the diffusion of certain chemical messengers possibly playing a role as well.

Even a casual examination of the shallow water off practically any beach in the world would reveal that sponges are just about everywhere. Being immobile and seemingly defenseless, sponges should be easy targets for predators, but they aren't. Why not? One factor must certainly be the cost-effectiveness of having to deal with trying to bite or nibble off pieces of the spiky, coarse skeleton of the sponge to get very little flesh (cells) per bite. Most poriferan biologists suggest, however, that the primary defense mechanism in sponges is biochemical in nature.

Research over the last several decades has revealed that sponges produce an amazing spectrum of biotoxins, some of which are quite potent. For example, *Neofibularia nolitangere,* or the touch-me-not sponge as it is commonly known (*nolitangere* is Latin for "do not touch"), can cause a painful rash on human skin similar to that of poison ivy. Touching or walking on these sponges with bare skin can result in an agonizingly painful laboratory session in marine toxicology.

Sponges engage in "chemical warfare" as they compete for space in the crowded underwater world where solid space is at a premium and planktonic larvae of sedentary invertebrates settle on every available uninhabited surface niche. Sponges, corals, colonies of bryozoans, tubeworms, and algae continuously engage in battles to the death, pushing and crowding each other for space. In diverse habitats such as coral reefs, evidence suggests that complex competitive hierarchies of various sponges may exist, each sponge competing for space with its own arsenal of biochemicals.

Regeneration and Reproduction

Sponges have several options when it comes to reproduction as they have the capacity for both asexual and sexual reproduction. Asexual methods include regeneration, budding, gemmules, and possibly the formation of asexual larvae. Probably all sponges are capable of regenerating viable adults from small fragments, as are some other invertebrates. Sponges, however, have taken **regeneration** to an even higher level than fragmentation. H.V. Wilson (1907) disassociated sponges into individual cells and cell clusters. He noted that the disassociated cells and clumps began to fuse (**reaggregation**) and eventually formed reconstituted sponges with all the appropriate accoutrements. In fact, a single sponge cell has the potential to become a fully formed and functional adult sponge.

Box 7.1
Investigating Sponge Regeneration

In the case of germ cells, many protist forms, amoeboid lymphocytes, and metastasizing tumor cells, a high degree of cell independence is apparent. A most convincing series of demonstrations illustrating the high degree of cellular individuality and independence is provided by experiments with sponges, hydroids, and polyps. In these experiments, an entire highly differentiated, multi-cellular organism may be reassembled from cells that have been wholly detached one form another by certain isolation techniques. Such studies have had an important influence on a number of fundamental problems such as metabolic gradients, wound healing, organ replacement, and tissue differentiation.

Coalescence and Regeneration in Sponges by H. V. Wilson
From the *Journal of Experimental Zoology* Vol. 5, No. 2, pp. 245-253 (1907)

In a recent communication, I described some degenerative and regenerative phenomena in sponges and pointed out that a knowledge of these powers made it possible for us to grow sponges in a new way. The gist of the matter is that silicious sponges when kept in confinement under proper conditions degenerate in such a manner that even though the bulk of the sponge dies, the cells in the certain regions become aggregated to form lumps of undifferentiated

---

tissue. Such lumps or plasmodial masses, which may be exceedingly abundant, are often of a rounded shape resembling gemmules, more especially the simpler gemmules of marine sponges (Chalina), and were shown to possess in at least one form (Stylotella) full regenerative power. When isolated they grow and differentiate producing perfect sponges. I described moreover a simple method by which plasmodial masses of the same appearance could be directly produced (in Microciona). The sponge was kept in an aquarium until the degenerative process had begun. It was then teased with needles so as to liberate cells and cell agglomerates. These were brought together with the result that they fused and formed masses similar in appearance to those produced in this species when the sponge remain quietly in aquarium. At the time, I was forced to leave it an open question whether the mass of teased tissue were able to regenerate the sponge body.

During the past summer's works at the Beaufort Laboratory I again took up this question and am now in a position to state that the disassociated cells of silicious sponges after removal from the body will combine to form syncytial masses that have the power to differentiate into new sponges. In Microciona, the form especially worked on, nothing is easier than to obtain by this method hundreds of young sponges with well-developed canal system and flagellated chambers. How hardy sponges produced in this artificial way are and how perfectly they will differentiate the characteristic skeleton are questions that must be left for more prolonged experimentation.

Taking up the matter where it had been left at the end of the preceding summer, I soon found that it was not necessary to allow the sponge to pass into a degenerative state, but that the fresh and normal sponge could be used from which to obtained the teased out cells. Again in order to get the cells in quantity and yet as free as possible from bits of the parent skeleton, I devised a substitute for the teasing method. The method is rough but effective.

Let me briefly describe the facts for Microciona. This species (M. prolifera Verr.) in the younger state is incrusting. As it grows older it throws up lobes and this may go so far that the habitus becomes bushy. The skeletal framework consists of strong horny fibers with embedded spicules. Lobes of the sponge are cut into small pieces with scissors and then strained through fine bolting cloth such as is used for tow nets. A square piece of cloth is folded like a bag around the bits of sponge and is immersed in a saucer of filtered sea-water. While the bag is kept closed with fingers of one hand it is squeezed between the arms of a small pair of forceps. The pressure and the elastic recoil of the skeleton break up the living tissue of the sponge into its constituent cells, and these pass out through the pores in the bolting cloth into the surrounding water. The cells, which pass out in such quantity as to present the appearance of red clouds, quickly settle down over the bottom of the saucer like a fine sediment. Enough tissue is squeezed out to cover the bottom well. The cells display amoeboid activities and attach to the substratum. Moreover they begin at once to fuse with one another. After allowing time for the cells to settle and attach, the water is poured off and fresh sea-water is added. The tissue is freed by currents of the pipette from the bottom and is collected in the center of the saucer. Fusion between the individual cells has by this time gone on to such an extent that the tissue now exists in the shape of minute balls or cell conglomerates of more or less rounded shape looking to the eye much like small invertebrate eggs. Microscopic examination shows that between these little masses free cells also exist, but the masses are constantly incorporating such cells. The tissue is this shape is easily handled. It make be sucked up to fill a pipette and then strewn over cover glasses, slide, bolting cloth, watch glasses, etc. The cell conglomerates which are true synctial masses throw out pseudopodia all over the surface and neighboring conglomerates fuse together to form larger masses, some rounded, some irregular. The details of later behavior vary, being largely dependent on the amount of tissue which is deposited in a spot, and on the strength of attachment between the mass of tissue and the substratum.

Decidedly the best results are obtained when the tissue has been strewn rather sparsely on slides and covers. The syncytial masses at first compact and more or less rounded, flatten out becoming incrusting. They continue to fuse with one another and thus the whole cover glass may come to be occupied by a single incrustation, or there may be in the end several such. If the cover glass is examined at intervals, it will be found that differentiation is gradually taking place. The dense homogeneous syncytial mass first develops at the surface a thin membrane with underlying connective tissue (collenchyma). Flagellated chambers make their appearance in great abundance. Canals appear as isolated spaces which come to connect with one another. Short oscular tubes with terminal oscula develop as vertical projections from the flat incrustation. If the incrustation be of any size it produces several such tubes. The currents from the oscula are easily observed, and if the cover glass be mounted in an inverted position on a slide the movements of the flagella of the collar cells may be watched with a high power (Zeiss 2 mm). This degree of differentiation is attained in the course of six or seven days when the preparations are kept in laboratory aquaria (dishes in which the water is changed answer about as well as running aquaria). Differentiation goes on more rapidly when the preparation is hung in the open harbor in a live-box (a slide preparation inclosed in a coarse wire cage is convenient). Sponges reared in this way have been kept for a couple of weeks. The currents of water passing through them are certainly active and the sponges appear to be healthy. In such a sponge spicules are present, but some of these have unquestionably been carried over from the parent body along with the squeezed out cells.

The old question of individuality may receive a word here. Microciona is one of that large class of monaxonid sponges which lack definite shape and in which the number of oscula is correlated simple with the size of the mass. While we may look on such a mass from the phylogenetic standpoint as a corm, we speak of it as an individual. Yet it is an individual of which with the stroke of a knife we can make two. Or conversely it is an individual which may be made to fuse with another, the two forming one. To such a mass the ordinary idea of the individual is not applicable. It is only a mass large or small having the characteristic organs and tissues of the whole and the number of organs are indefinite. As with the adult so with the lumps of regenerative tissue. They have no definiteness of shape or size, and their structure is only definite in so far as the histological character of the syncytial mass is fixed for the species. A tiny lump may metamorphose into a sponge, or may first fuse with many such lumps, the aggregate also producing but a single sponge, although a larger one. In a word we are not dealing with embryonic bodies of complicated organization but with a reproductive or regenerative tissue which we may start on its upward path of differentiation in almost any desired quantity. A striking illustration of this nature of the material is afforded by the following experiment. The tissue in the shape of tiny lumps was poured out in such wise that it formed a continuous sheet about one millimeter thick. Such sheets were then cut into pieces, each about one cubic millimeter. These were hung in bolting cloth bags in an outside live-box. Some of the pieces in spite of such rough handling metamorphosed into functional sponges.

Immunocompetence (the ability to distinguish between self and nonself) has been demonstrated in some sponges. Such sponges will accept a graft from themselves but will reject a graft from another individual of the same species.

Many marine sponges produce **buds** of various types. They appear as bumps or club-shaped protrusions arising on the sponge surface. The buds break from the parent sponge and may be carried along by water currents for a brief time before attaching to the substratum and slowly forming a new individual.

In freshwater sponges of the family Spongillidae, small spherical structures called **gemmules** (L. *gemmula*, little bud) are produced with the onset of winter or during times of drought. Archaeocytes collect in the mesohyl and become surrounded by a tough spongin coat reinforced with siliceous spicules. (**Figure 7.11**) Gemmules are highly resistant to both freezing and drying. When the parent sponge dies, the gemmules survive and remain dormant, preserving potential new individuals during periods of freezing or severe drought. When environmental conditions are again favorable, the micropyle opens, and the archaeocytes flow out in waves. The first wave flows onto the substratum where they begin to construct a framework of new pinacoderm and choanoderm. The second wave of archaeocytes colonizes this framework and gives rise to every cell type of the adult sponge in an amazing testimony to the totipotency of sponge cells. Many marine species form asexual reproductive bodies similar to gemmules though not as complex in structure.

Figure 7.11 Gemmule formation in the freshwater sponge *Spongilla lacustris*. The gemmules appear as golden orbs within the sponge.

Some demosponges seem capable of producing asexual larvae. It has been theorized that such a process might assure production of a free dispersal stage when fertilization has failed.

Sexually, most sponges are **hermaphroditic** (equipped with both male and female sexual capabilities), but they produce sperm and eggs at different times. This insures that cross-fertilization takes place.

Sperm appear to arise primarily from choanocytes whereas eggs arise from choanocytes or archaeocytes. Both develop within the mesohyl but are not located within distinct ovaries or testes. Surprisingly, external fertilization does not occur in sponges; it always takes place within the body of the receiving sponge. Mature sperm are released into the environment through the aquiferous system. Sperm release,

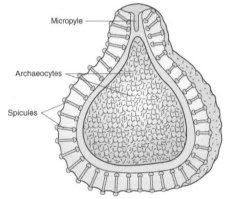

Section through a freshwater sponge gemmule.

which may be synchronized in a local population or restricted to select individuals, can produce a visually striking phenomenon known as "smoking sponges." Once in the water, the sperm are drawn into neighboring sponges containing mature eggs. There the sperm are trapped by the choanocytes of the receiving sponge and transferred to an adjacent ripe egg. Fertilized eggs may be released into the excurrent water stream and left to develop outside the parent sponge, but in most sponges the fertilized eggs are brooded within the parental mesohyl until they develop into free-swimming larvae, a stage critical to the dispersal of sessile animals. The flagellated larvae exits through the aquiferous system, and after a brief free-swimming existence, settles to suitable substratum and begins development into a mature sponge (**Figure 7.12**).

Sponges are pretty successful. They have existed for millions of years, and the fact that they still exist in very simple forms indicates that they are able to do what they do without increased complexity. Bacteria are simple, but they rule the world. Innovation does not ren-

der the old way of doing things obsolete or unusable; the complex (new) way just lets you do different things like grow larger, become mobile, move to land, and chase prey.

—Matthew Nelson

Poriferan Connection

Economically

For millennia, humans have used the dried fibrous but flexible and porous skeletons of certain Demospongia for bathing, mopping, scrubbing, painting, even as canteens. In the latter half of the nineteenth century, a thriving sponge harvesting industry developed off the warm coastal waters around Florida. In its heyday, the sponge fleet of Key West, Florida numbered over 350 ships and over 1400 people were employed in the industry. This harvesting came to a grinding halt in 1946 and 1947 as over collection and blight thought caused by toxins produced by a **red tide** (a bloom of toxic red algae) or perhaps fungal infection wiped out the sponge beds. The first synthetic sponges were introduced at about the time the sponge beds were collapsing, additionally contributing to the decline of the sponge-harvesting industry. Natural sponges harvested in very limited numbers are still available and are preferred over synthetic varieties by certain artisans and craftsmen such as industrial cleaners, leatherworkers, potters, silversmiths, lithographers, and cosmeticians. Currently,

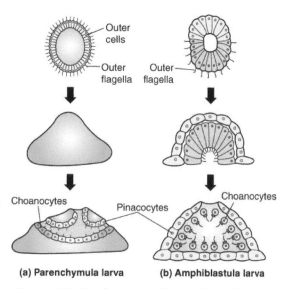

(a) Parenchymula larva **(b) Amphiblastula larva**

Figure 7.12 Development of Sponge Larval Stages. (a) Parenchyma lava. Flagellated cells cover most of the larva's outer surface. After the larva settles and attaches, the outer cells lose their flagella, move to the interior, and form choanocytes. Interior cells move to the outside and form Pinacocytes. (b) Amphiblastula larva are hollow with half of the larva bearing flagellated cells. On settling, the flagellated cells invaginate into the interior of the embryo and form choanocytes. Nonflagellated cells overgrow the choanocytes and form the pinacocytes.

sponge harvesting is around a $40 million dollar a year industry. The worldwide demand, however, far exceeds the supply. Most products sold today as sponges are synthetic, and even a natural sponge is nothing more than the trimmed, cleaned, and dried remains of the once-living animal. Consequently, attempts using aquaculture techniques to increase the supply of natural sponges in a sustainable manner are underway in the Torres Straits between Australia and New Guinea.

Certain glass sponges such as the Venus' Flower Basket sponge (Figure 7.4) have a tuft of fibers that extends outward like an inverted crown at the base of their skeleton. These fibers are between 50 and 175 mm (2-7 inches) long and about the thickness of a human hair. These fibers have been found to work as optical fibers that are surprisingly similar to the artificial fibers used in modern telecommunications networks. Unlike commercial fiber, it is possible to tie them in knots without cracking or breaking and these natural fibers have the additional advantage of being produced by chemical deposition at low temperatures whereas commercial optical fibers must be produced in high-temperature ovens with expensive equipment.

Ecologically

Even though sponges may have a bristly texture and often produce strong irritating chemicals that are toxic to some small animals and can cause skin irritation and rash in humans, they are fed upon by certain fishes, sea stars, and slugs. Imagine your surprise if you were to see a sponge crawl away when you reached for it! Closer examination would reveal that sponge was attached to the top of a crab. Such crabs will carefully cut pieces off larger sponges and attach them to their top side relying on small hooks in their exoskeleton to hold the sponge in place. This strange behavior on the part of the crab benefits both parties. The crab with a sponge "hat" acquires instant camouflage and protection whereas the movements of the crab impart an indirect mobility to the sponge allowing it greater access to possible food sources.

Many different kinds of marine animals have evolved the ability to mechanically or chemically drill into hard substrates—shell, coral, limestone, compact clay, and wood—and they do so in search of food or for their own protection. Sponges usually do not come to mind when one thinks of boring marine animals, but they do exist. Members of a common and widespread family of sponges (Clionidae), often called boring sponges, attach to the shells of mollusks and corals, tunneling down into them and eventually killing them (**bioerosion**). The mollusks or corals are not eaten; the sponge seeks not food, but protection for itself by sinking into the hard structures it erodes. The active boring process involves specialized archaeocytes called **etching cells** that chemically remove fragments or chips of the calcareous material.

Sponge bioerosion has had a significant impact on coral reefs in some areas. More important than the actual erosion is the weakening of the coral resulting in much coral loss during heavy tropical storms. Any shell-strewn beach will contain many specimens perforated by boring sponges. Even this seemingly destructive process, however, has some beneficial effects in that it is an important part of the method by which calcium is recycled.

The greatest ecological role sponges play in their natural habitat is to provide a home for a wide variety of creatures both microscopic and macroscopic. The porous nature of sponges makes them ideally suited as biological boardinghouses for habitation by a host of smaller invertebrates and often by fishes such as gobies and blennies. Ecologist A. S. Pearse once carefully picked apart a loggerhead sponge (*Spheciospongia vesparium*) the size of a kitchen sink with a total volume roughly equivalent to that of a 55-gallon drum and counted and classified all the creatures inside it. He recorded 17,128 total animals. Around 16,000 were alphaeid shrimp, but many other kinds of animals were found, including several fish up to 13 cm (5 inches) long. Lobsters, crabs, shrimps, fish, and brittle stars retreat to into large sponges using their many openings and folds as a refuge when danger or daylight threaten. Zoanthids—small anemone-like animals resembling yellow daisies—often crowd together on the outsides of sponges as do some tubeworms in their thin and brittle parchment-like tubes.

Most residents use their host only for space and protection, but some rely on the sponge's water current for a supply of suspended food particles. A classic example is the bond between a male-female pair of shrimp (*Spongicola*) and the hexactinellid sponge known as Venus' flower basket (*Euplectella*). The shrimp enter the sponge when young and small, only to become trapped in the glasslike case of their host as they grow too large to leave (prisoners of love?). Appropriately, this sponge with its mummified guests is a traditional wedding present in Japan—a symbol of the lifetime bond between two partners.

Sponges also form mutualistic associations with bacteria and protists. In some demosponges, the meso-hyl bacteria population can account for some 38 percent of the body volume of the sponge, far exceeding the actual sponge-cell volume of only 21 percent. Presumably, the sponge matrix provides a nourishing medium for bacterial growth and the host benefits by having food within it that can be easily phagocytized. Commensalistic relationships have also been reported between sponges and filamentous green algae, red algae, and diatoms.

Medically

Unable to flee predation, sponges resort to a bristly texture and a concoction of toxic chemicals to defend themselves. These chemicals may also play a role in competition between sponges as they jostle each other for precious space. Some of these chemicals have been found to possess beneficial pharmaceutical effects on humans. In fact, pharmaceuticals prepared from sponges have been used since ancient times as referenced in the writings of Aristotle and Hippocrates. In modern times, one of the most effective treatments for leuke-mia for over 30 years has been based on derivatives of sponges and AZT, the first drug to turn HIV from a death sentence into a controllable disease, was derived from the marine demosponge *Tectitethya crypta*.

Researchers continue to **bioprospect** sponges as sources for potential new antibiotics to kill bacteria and fungi, new drugs to combat cancer and certain viruses, and as possible aids in controlling asthma, inflammation, and certain forms of arthritis even chemical blocks against ultraviolet radiation. In the con-tinuing assay of chemical compounds for possible therapeutic use in humans, sponges are running neck and neck with sea squirts, and gorgonian sea fans, two other chemically rich groups of marine animals.

> *If we go out onto a reef and see something that looks like a large chunk of food—poorly protected, soft-bodied and easy to grab—and nothing is eating it, we assume it has chemical protection. Organisms that appear to defend themselves chemically rather than by shells and spines, or the mobility to run away and hide, are of great interest to us.*
>
> —John Faulkner

A Closing Note
The Immortal Sponge

As soon as one acquires even a rudimentary understanding of the workings of cells, tissues, and organs, it becomes clear that there is no such thing as a "simple" or "primitive" living creature (the subtitle of this chapter notwithstanding). Although such terms might have their place when speaking comparatively, they have no biological validity when it comes to individual organisms. And in no group of animals does this ring truer than it does with sponges. The seemingly lowly poriferans certainly are at the bottom of the complexity heap compared to other animals, but there is far more to them than first meets the eye for a sponge may potentially be immortal.

I have a culture of freshwater sponges that is decades old. Since its establishment in the late 1970s, all I have ever done to maintain the culture is to replace the water that evaporates from the dish and to occasionally throw in a few flakes of fish food. There it sits day by day, year by year changing form with the seasons. In warm weather it exists as a thick, organized sheet, but in cooler temperatures it disassociates onto the bottom of the dish as a mass of cottony fluff dotted with dark gemmules. Ever so often I find myself staring at it and wondering, "Are you a true organism and is it possible that you are immortal (or at least the closest thing to it in the animal kingdom)?"

Sponges may technically be immortal because in undisturbed natural surroundings they never die. The only natural death threat they face is total desiccation or being frozen solid. Macerate a sponge into tiny pieces and most of those pieces will regenerate a whole new sponge. In fact, if you force a sponge through a fine silk mesh and break it up into individual single cells, each of those individual cells has the potential to regenerate an entire sponge. (Imagine the mess you would have if you tried that with a chunk of human flesh, let alone an entire human body.) Unlike any other animal, sponges just keep on keeping on.

If you ever have the opportunity to confront a living sponge, do not dismiss it as a primitive being lacking even a single organ. Rather, respect this strange entity that would laugh (if it had the body structures to laugh with) at the fragility and short life span of our so-called complex body. Knowing this, I can't help wondering who will care for my old friend after I am long gone, for it will surely not only outlive me, but potentially any other human alive today as well.

In Summary

- Sponges are ancient animals whose origins date back 650 million years.
- The approximately 9,000 species of extant sponges come in a variety of shapes and sizes and a multitude of colors.
- Phylum Porifera is composed of three classes:

  1. Class Calcarea
  2. Class Hexactinellida
  3. Class Demospongia

- The characteristics of phylum Porifera are as follows:

  1. Their multicellular body is a loose aggregate of cells without true tissues; adults are asymmetrical or radially symmetrical.
  2. They are all aquatic; most are marine.
  3. All adults are sessile (not moving about freely) filter feeders; larval stages are motile.
  4. Their body is perforated by ostia (pores), canals, and chambers that serve for the passage of water.
  5. Water currents move through the chambers and canals due to the action of unique flagellated cells known as choanocytes.
  6. Skeletal elements composed of calcareous or siliceous crystalline spicules, often combined with variously modified collagen fibers (spongin).

7. Asexual reproduction by buds or gemmules and sexual reproduction by eggs and sperm; larvae are ciliated and free-swimming.

8. Their cells tend to be totipotent in that they retain a high degree of mobility and are capable of changing form and function.

- Class Calcarea or calcareous sponges possess calcium carbonate spicules. All species in the class are marine, and they are most commonly found in shallow tropical waters.

- Class Hexactinellida or glass sponges possess siliceous spicules. All species in this class are marine, and they occur at depth world-wide, they are most abundant and diverse at shallower depths in polar regions.

- Class Demospongia possesses both siliceous spicules and coarse spongin fibers. Demospongia is the largest class of sponges. They are found in all ocean zones, and a few species inhabit fresh water.

- Poriferans:

  1. Come in three basic designs—asconoid, synconoid, and leuconoid

  2. Possess a skeleton that is a loose aggregation of spicules in marine types and spicules and spongin fibers in freshwater types.

  3. Lack tissues and have no internal organs. A single layer of cells of several different types lives on and in the skeletal structure.

  4. Are filter feeders that draw water in through their porous structure.

  5. Extract oxygen from the water flowing into them and dump carbon dioxide and metabolic wastes into the water as it flows out of them.

  6. Asexually reproduce by several methods—regeneration, budding, and gemmules.

  7. Sexual reproduction also occurs.

- Poriferans connect to us in important ways economically, ecologically, and medically.

Review and Reflect

1. *Prove It*. Imagine you have been assigned to a small aquarium containing a living sponge about the size of a tennis ball. As part of a lab practical exam, your instructor has challenged you to prove that your sponge is an animal and not a plant. How would you proceed? (Assume any equipment or supplies you might need are available.)

2. *Terrestrial Sponges?* An explorer has returned from a remote area of the world proclaiming the discovery of the world's first known land sponge. Is this possible? If not, why not?

3. *Muddy Waters*. If you were to attempt to collect living freshwater sponges, would you go to a clear slow-moving stream or a slow-moving stream that was turbid with silt eroded from nearby farmland? Explain.

4. ***Feed Me Seymour.*** If you were to keep a small living marine sponge at home in an aquarium as a pet, what would you feed it?

5. ***Eat Without a Mouth?*** Your little sister is also studying sponges in her middle school science class. She is confused and asks you, "We learned that a sponge has no organs, so that means it has no mouth, no stomach nor intestines. Then how do sponges eat and digest their food?" What would you say?

6. ***A Skeleton Without Bones*** Your little sister has more questions about sponges. Now she asks, "Our science textbook keeps talking about the skeleton of a sponge but how can a sponge have a skeleton when sponges don't have organs?" What would you say?

7. ***A Gem of an Adaptation*** Why is it vital for freshwater sponges to form gemmules, but not as critical for marine sponges to do so?

8. ***Degenerate Then Regenerate*** Carefully read H.V. Wilson's investigations into sponge regeneration in Box 7.1 and then answer the questions that follow:

   A. Wilson opens by reviewing in general terms previous experiments he had conducted. Briefly describe what Wilson had learned from these earlier experiments.

   B. What method did Wilson use to disassociate the sponges in his previous experiments?

   C. What question had Wilson not answered with his previous experiments?

   D. The following summer Wilson took up the question again. What did he discover?

   E. Wilson developed an alternative method for disassociating sponges from that he had used previously. Describe this "rough but effective" method.

   F. Describe step-by-step what happened to the sponge cells and cell aggregates resulting from the disassociation of a parent sponge.

   G. Wilson found that sponges regenerated faster when placed in the open harbor in live-boxes than they did in laboratory aquaria. Hypothesize as to the reasons why this would be so.

9. ***No Body? No Problem*** Sponges are the masters of regeneration and reaggregation. What can humans learn and apply to the human condition from studying regeneration and reaggregation in sponges?

Create and Connect

1. ***The Lowly Sponge?*** Imagine you are the science editor of a large newspaper. As such you write a weekly column on various physical science and biology topics. You are planning to write a column on sponges but when you propose this idea to your managing editor she replies, "No, a column on sponges will be boring. Sponges are primitive blobs that are barely alive, just barely qualify as animals, and just sit there and do nothing." How would you respond?

2. ***Pump and Filter.*** Many sponges can filter their entire body volume in less than one minute! In fact, 1 cubic centimeter of sponge can propel anywhere from 0.002 to 0.84 milliliters of water per second or roughly 20 liters (5.3 gallons) of water per day. With 7,000 to 18,000 chambers per

cubic centimeter of sponge, and each chamber can filter up to 1,200 times its own volume in water daily. These seemingly inanimate creatures are fantastic pumps, filtering tons of water to harvest a few ounces of food.

If you were to visit an aquarium and observe a single large marine sponge in a 100 gallon (13.4 cubic feet) tank of water, what details would you have to know in order to calculate how long it would take that sponge to pass all the water in that tank through its body? Using the hypothetical figures for sponge size and tank volume, solve the problem of how long it would take a sponge to pass all the water in its container through its body. Use the pumping rates mentioned in this chapter and assume a constant pumping rate that never varies.

3. ***Shining a Light***. As you discovered in this chapter, sponges possess nothing resembling the nervous system or brain of higher animals yet they are said to be able to respond to environmental stimuli. Can sponges really respond to the world around them? Design an experiment to test the problem question: *Do sponges respond positively or negatively to light?*

Guidelines:

A. Following the tenets of a well-constructed experiment, your design should include the following components in order:

> - The *Problem Question*. State exactly what problem you will be attempting to solve.
> - Your *Hypothesis*. Although this is a fictitious experiment, word your hypothesis as realistically as possible.
> - *Methods and Materials*. Explain exactly what you will do in your experiment including the materials necessary to accomplish the task. Be specific, take nothing for granted, and do not expect people to read your mind as they read your work.
> - *Collecting and Analyzing Data*. Explain what type(s) of data will be collected and what statistical tests might be performed on that data. It is not necessary to concoct either fictitious data nor imaginary observations.

B. Your instructor may provide additional details or further instructions.

PHYLUM CNIDARIA AND PHYLUM CTENOPHORA: A RADIAL EXISTENCE

Corals have done more to change the face of the earth than any other creature. The mighty pyramids of Egypt look like a child's sandbox project if you compare them to the dramatic changes that have been wrought on the landscape by coral reefs.

—George Fichter

Introduction

Over 700 million years ago in ancient seas at the dawn of animal time, small balls of metazoan animal cells provided the stem that branched into three different evolutionary paths. One pathway became the Parazoa represented now only by the Porifera (sponges), whereas the second pathway became the Radiata represented by the Cnidaria and Ctenophora. Although both the parazoa and radiata pathways are populated with animals that are highly successful, these pathways apparently led to no further branching and became evolutionary dead ends. Those metazoans that traveled the third evolutionary path would become the Bilateria, the stem on the tree of life from which would evolve all the other animal types known today.

PHYLUM CNIDARIA

History of Cnidaria

Molecular analysis indicates that the cnidarians are the most ancient of animal phyla, seemingly eclipsing even the Porifera in their antiquity. From the famous Ediacara Hills of South Australia come the fossils of several kinds of medusae and sea pens reaching back to around 700 million years ago. These fossil cnidarians predate even the sponges. Indeed, the origin of the cnidarians seems intimately tied to the origin of the metazoa themselves (**Figure 8.1**).

The classic interpretation of cnidaria phylogeny holds that the Hydrozoa were the ancestral radiates and that the Anthozoa represent the most advanced form, with the Scyphozoa and Cubozoa falling in-between. Recent molecular data, however, turn the classical hypothesis upside down in that they indicate primitive anthozoans were most likely the ancestral cnidarian stock (**Figure 8.2**). Radial symmetry appears to be the ancestral cnidarian characteristic (symplesiomorphy), whereas the absence of synapomorphies (specifically the presence of cnidocils) in the Anthozoa strongly indicates they are the outgroup for phylum Cnidaria.

Although the Cnidaria and Ctenophora share certain morphological characteristics, such as radial and biradial symmetry, diploblastic cellular organization, and gastrovascular cavities, major differences in embryological development and appearance of the adult body form make it impossible to derive Ctenophores from any cnidarian group. Ctenophora and Cnidaria seem at best to be only distantly related.

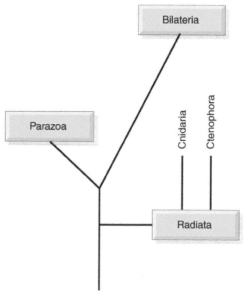

Figure 8.1 The probable position of phylum Cnidaria and phylum Ctenophora on the phylogenetic tree of life.

Diversity and Classification of Cnidaria

The present day phylum Cnidaria (Gr. *cnidos*, stinging nettle) is a highly diverse assemblage of soft, gelatinous animals armed with stinging tentacles whorled around a single opening that leads into a hollow baglike body. Floating or slowly pulsing below the surface of the ocean are the hydroid and scyphozoan jellyfish and the cubozoans ("box jellyfish"). Dotting the seafloor substratum are the sessile sea anemones and corals. In fact, the accumulated mineral accretions of corals *are* the sea floor in certain locations. Some, such as the *Hydra*, have successfully invaded fresh waters

The approximately 9,000 species that comprise this phylum range in size from nearly microscopic polyps and medusae to giant individual jellyfish 2.5 meters (8 feet) across trailing tentacles up

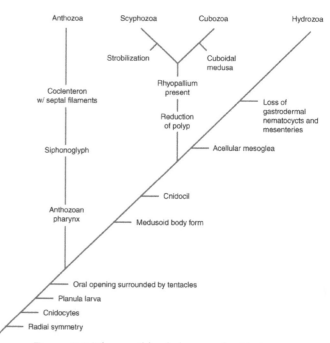

Figure 8.2 The possible phylogeny of cnidarians.

to 30 meters (100 feet) long. Corals form reefs in the tropical waters of every sea covering a surface area of approximately 280,000 square miles, biological constructs so large their shimmering outlines can easily be seen from space.

As was the case with sponges, the nature of cnidarians was long debated. Renaissance scholars considered them plants, and it was not until the eighteenth century that the animal nature of the cnidarians was truly recognized. Nineteenth-century naturalists classified them along with the sponges and a few other groups under Linnaeus's Zoophytes, a taxonomic category for organisms deemed intermediate between plants and animals. Jean-BaptisteLamarck instituted the group Radiata for medusoid cnidarians, ctenophores, and echinoderms in reference to their common radial or biradial symmetry. Rudolf Leukart eventually recognized the fundamental differences between the "radiate" groups, and in 1847 established the group *Colenterata* (Gr. *kiolos*, hollow + *enderon*, gut) to include Porifera, Cnidaria, and Ctenophora. In 1888 Berthold Hatschek split Leukart's Coelenterata into the three separate phyla we recognize today: Porifera, Cnidaria, and Ctenophora. Although, there are some that are inclined to lump the cnidarians and ctenophores together in Coelenterata (or even Radiata), these two groups are almost universally recognized as distinct phyla.

    Domain: Eukarya
        Kingdom Animalia
            Sub-Kingdom Eumetazoa
                Phylum Cnidaria
                    Class Hydrozoa—Hydroids
                    Class Scyphozoa—True jellyfish
                    Class Cubozoa—Box jellyfish
                    Class Anthozoa—Sea anemones and Corals

Phylum Characteristics

Whereas sponges function at the cellular level, cnidarians are more complexly organized into a tissue grade of construction. Although it may be debated whether sponges are little more than glorified colonies of cells, cnidarians have a clearly defined body with true tissues. However, their body is hollow (though not a true coelom) and lacks any semblance of true organs.

The members of phylum Cnidaria exhibit the following general characteristics:

- Their bodies exhibit radial, biradial, or quadriradial symmetry (Radiata) along a longitudinal axis with two opposite ends but no definite head. Cnidarians are dimorphic and thus demonstrate two basic body forms: the polypoid form and the medusoid form (**Figure 8.3**).

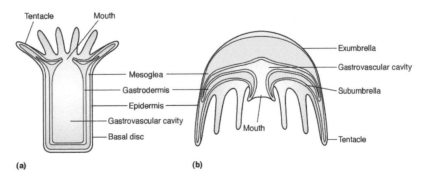

Figure 8.3 The Dimorphic Body Forms of Cnidarians. (a) Polypoid or polyp form. (b) Medusoid form.

213

- Their bodies are **diploblastic** (two cell layers) consisting of an outer layer of cells known as the **epidermis** and an inner layer of cells known as the **gastrodermis**. The cell layers are separated by a layer of ectodermally-derived acellular **mesoglea** or partially cellular **mesenchyme**. In hydrozoans, the mesoglea is thin and contains few or no living cells; in anthozoans it is a thick and rich with many cells. Scyphozoans and cubozoans have a few living cells in their thick mesoglea layer.

- The **gastrovascular cavity** or **coelenteron** is saclike or branched cavity with a single opening that serves as both mouth and anus. Extensible tentacles usually encircle the mouth or oral region. The gastrovascular cavity is the only body cavity (no true coelom) and serves as the site of digestion.

- They possess unique stinging or adhesive structures called **cnidae**. The most common cnidae are the stinging **nematocysts**.

- They lack a central nervous system (no brain) but are equipped with a simple nerve net with some sensory structures.

- They possess a simple epitheliomuscular system consisting of an outer layer of longitudinal fibers at the base of the epidermis and an inner layer of circular fibers at the base of the gastrodermis.

- They lack a circulatory, excretory or respiratory system. As a result, they are entirely aquatic with most being marine, but a few species are found in fresh water.

- They exhibit alternation of asexual polypoid and sexual medusoid generations, but there are a number of variations on this basic theme.

Phylum Cnidaria is populated by some of the strangest and loveliest creatures on the planet: branching, plantlike hydroids, colorful and flowerlike sea anemones, free swimming and nearly transparent jellyfish, and the ocean's greatest construction crew—the horny corals (sea fans, sea whips, and others) and the stony corals whose diligent calcareous house-building has produced majestic reefs and coral islands.

Class Characteristics

Class Hydrozoa *(Gr. hydra, water serpent + zoon, animal)*

The class Hydrozoa includes the hydras and many colonial species called hydroids. Although hydrozoans are abundant, they usually are small and not as conspicuous size-wise as are the larger jellyfish and sea anemones. Most species are found in the ocean where they exist mainly as fixed colonies such as *Obelia*, (**Figure 8.4**) although the Portuguese man-o-war (*Physalia*) is a floating colony composed of several different types of both polypoid and medusoid individuals. (**Figure 8.5**) Some, such as the hydras, are found in freshwater and are mainly solitary individuals (**Figure 8.6**).

Hydrozoans may exhibit a medusoid body form or a polypoid body form, or both during their life history. Hydrozoan mesoglea contains few or no cells; cnidocytes form only in the epidermis, and the gametes usually develop within the epidermis. Hydrozoans are distinguished by four recurring features: colony formation, the presence of skeletons, polymorphism, and medusa reduction.

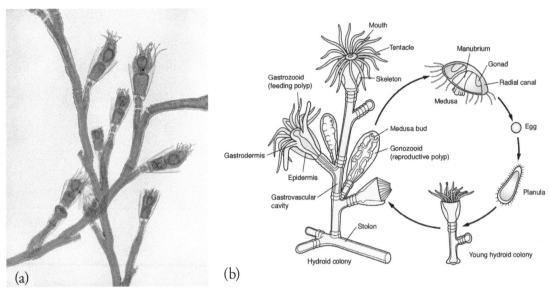

Figure 8.4 The Colonial Hydroid *Obelia*. (a) Photomicrograph of a stained portion of an *Obelia* colony. (b) Basic structure and life cycle.

Colony Formation Most marine species of hydrozoans are colonial. Colonies form when asexual buds remain attached to the parent when they mature. The individuals, or polyps, of a hydroid colony are usually attached to a main stalk, which in turn is anchored to the substratum (algae, rocks, shells, or human constructs such as wharf pilings) by a rootlike stolon. (Figure 8.4) The arrangement of polyps on the stalk and the branching of the stalk vary with the species. The tissue layers of the stalks, stolons, and polyps are continuous forming a common gastrovascular cavity for the entire colony.

Skeletal Formation The solitary polypoids, such as *Hydra*, have no solid skeletons, and the hydroid colonies in which the polyps arise directly from the substratum have only anchoring skeletons. Most hydroid colonies, however, are 3 to 10 cm (1.2 to 4 inches) in height and supported by an external chitinous skeleton secreted by the epidermis. This skeletal envelope may be limited to the stalk in some whereas in others it extends around the body of the polyps but is open at the oral end allowing for the emergence of the tentacles and mouth.

Polymorphism Many hydroid colonies exhibit morphological polymorphism. That is, they have two or more structurally and functionally different kinds of individuals within the same species and often within the same colony. Among polymorphic hydrozoids, the most common

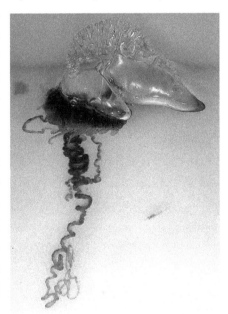

Figure 8.5 Named for a 16th century armed sailing ship, the floating Portuguese Man-o'-War (*Physalia Physalis*), a colony of several specialized zooids, wanders at the mercy of wind and currents.

type of individual is the **gastrozooid** or feeding polyp, which resembles a hydra. Such polyps capture food and carry out initial extracellular digestion. This partially digested food then moves throughout the colony via the continuous gastrovascular cavity until it is absorbed and digested intracellularly by individual cells.

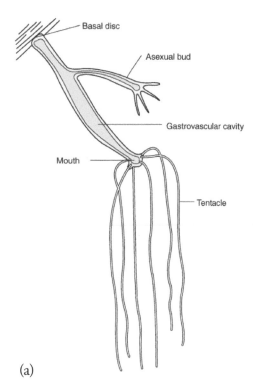

(a)

Figure 8.6 The Freshwater *Hydra*. (a) General structure and typical posture. (b) The solitary freshwater green hydra (*Hydra viridissima*) patiently awaits some small, unsuspecting prey animal to blunder into its delicate but deadly tentacles.

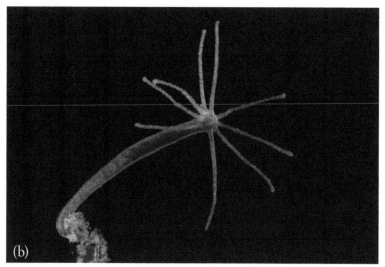

(b)

Most colonial hydrozoans consist of groups of attached polyps of various types. However, members of order Siphonophora form large pelagic (floating) colonies composed of both medusoid and polypoid individuals with long tentacles. Some individuals form the float bag or a pulsating swimming bell from which hang the feeding polyp individuals with tentacles. The most common siphonophoran is the Portuguese man-of-war or Bluebottle (*Physalia*). These hydroids have a single gas-filled float from which dangle feeding polyps (**dactylozooids**) with tentacles up to 13 meters (42 feet) in length. (Figure 8.5) Although most siphonophorans float beneath the surface, the Portuguese man-of-war floats on the surface. A gas-filled sac above the surface is the man-of-war's sail, and it allows the animal to move off at an angle of up to 45 degrees from the direction of the wind. The man-of-war can control the curvature and the height of the sail; it can also affect the sailing angle and the speed by using its tentacles as a sea anchor. The fishing tentacles of a man-o-war are armed with powerful nematocysts (stinging cells) capable of paralyzing sizable fish. A man-of-war was once observed catching over 20 small fish at one time. Captured prey is transferred to the gastrozooid polyps for digestion.

Human encounters with these dangerous hydroids occur mainly in two scenarios: swimmers can become entangled in their tentacles in deep water; a potentially deadly situation or strong winds may wash them ashore where unsuspecting bare-footed beach strollers may step on them.

Medusae The medusoid body form represents the sexual reproducing structure in nearly all cnidarians. Although most hydroids exhibit mainly or exclusively the polypoid form as adults, some types, known as hydromedusae, exist mainly or exclusively in the medusoid form. The hydrozoan medusas are small, only a few centimeters in diameter and should not be confused with the larger "true jellyfish" of the class Scyphozoa.

The hydromedusae display a varying number of tentacles hanging from the margin of the bell (umbrella). (**Figure 8.7**) A fold of the body wall (the **manubrium**) hangs from the center of the undersurface of the bell (the **subumbrella**) and surrounds the mouth. A shelf-like inward projecting fold of the bell margin, the **velum**, increases the force of the water jet. As they weakly pulse along, the hydromedusae rely on statocysts (gravitational detectors) and ocelli (photoreceptor cells) located between or at the base of the tentacles to orient themselves in respect to up (light from above) and down (pull of gravity below).

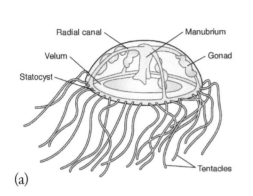

Figure 8.7 Hydromedusae. (a) General structure. Note the velum. This thin circular flap of tissue is a useful feature in distinguishing between hydromedusae (velum present) and scyphomedusae (velum absent). (b) *Aequorea victoria*, also known as the crystal jelly, is a bioluminescent hydromedusa found off the west coast of North America. Most hydromedusae are negatively buoyant and must swim to maintain themselves in proper feeding posture in the water column.

Hydromedusae feed on all manner of small planktonic organisms. The gastrovascular cavity is not a single sac, as it is in the polyp body form. The mouth opens into a central stomach from which extend four **radial canals**. These in turn connect with a **ring canal** that extends around the bell margin.

Life Cycles Hydrozoans exhibit great variation in their life cycles. Although the polyp is the more conspicuous and persistent stage in most groups, some types lack the medusa phase, whereas others lack the polyp phase.

Hydrozoan polyps reproduce asexually by budding. Certain hydromedusae also undergo asexual reproduction by budding or by direct longitudinal fission (splitting). All hydrozoan cnidarians have a sexual phase in their life cycle and hydrozoans may be either hermaphroditic or dioecious. (**Figure 8.8**) The gametes develop from interstitial cells that form aggregates or gonads in specific locations in the epidermis or gastrodermis.

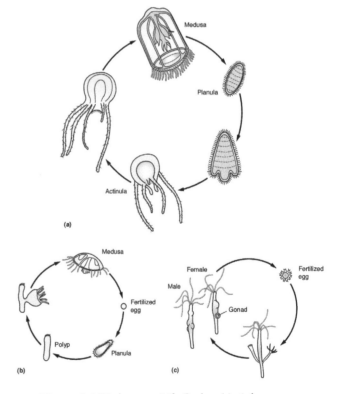

Figure 8.8 Hydrozoan Life Cycles. (a) *Aglaura*, a hydrozoan that has no polypoid stage. Planula larva develops into an actinula, which develops directly into a medusa. (b) *Gonionemus* has a solitary polyp. (c) *Hydra* is a freshwater hydrozoan in which the medusoid stage has disappeared and the planula larva is suppressed.

Fertilization is commonly external as the eggs are shed into the seawater. Floating and drifting as plankton, fertilized eggs develop first into a hollow blastula and then usually into a **stereogastrula**. The stereogastrula rapidly elongates to become a solid or hollow nonfeeding, ciliated, free-swimming **planula larva** (Figure 8.8). Although radially symmetrical, the planula larva possesses distinct "anterior" and "posterior" ends. After free swimming for several hours to several days, the planula larva attaches to an object by the anterior end. The mouth and tentacles develop on the unattached oral end as the larva metamorphoses into a young polyp. In some species of hydromedusae, fertilization and early development occur on the manubrium or even in the gonads. In other hydromedusae that have no polypoid form, the planula larva develops into **actinula larvae**, bypassing the sessile polypoid phase (Figure 8.8).

The mode of reproduction—asexual or sexual—may be seasonal and, in temperate regions at least, is closely correlated with water temperature. Freshwater hydras usually asexually bud during the warm spring and summer months but the cooler temperatures of fall triggers a flurry of sexual reproduction. The fertilized eggs formed during this time have a thick covering that assures they will survive the winter even if their soft-bodied parents do not. Quietly resting through the cold, these eggs will spring to life as the water around them warms in the spring, and a new generation of hydra will appear.

When the time is right, hydrozoans that produce a medusae stage may liberate enormous numbers of them in a short time. A 7-cm (3 inch) colony of *Bougainvillia* released 4450 medusae over a three-day period.

Class Scyphozoa (Gr. *skyphos*, cup + *zoon*, animal) and Class Cubozoa (Gr. *kybos*, a cube + *zoon*, animal)

Members of these classes are those cnidarians commonly referred to as jellyfish. Classically, the Cubozoa were viewed as a subgroup of Scyphozoa. When the cubozoan polyp and life cycle were found to be distinctly different than those of the scyphozoans, however, they were assigned their own class rank. Class Scyphozoa contains the majority of species in this class (**Figure 8.9**). The relatively small number of cubozoans, or box jellies are all tropical and semitropical in habitat (**Figure 8.10**). In contrast to their scyphozoan kin, these cnidarians are cube-shaped (as their class name suggests), the bell margin is not scalloped, and they bear four tentacles or tentacle clusters. Both the scyphozoans and the cubozoans are strictly marine. The so-called freshwater jellyfish that one hears about occasionally are in reality medusoid hydrozoans, and not true jellies.

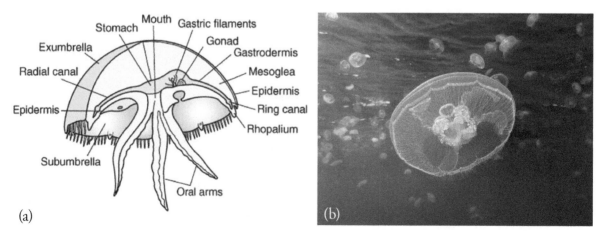

(a)

(b)

Figure 8.9 *Aurelia*, a Scyphozoan Medusa. (a) Side view of structure in section. (b) *Aurelia aurita* or moon jelly as they are commonly known. This genus is found throughout most oceans and is the world's most common jellyfish.

In both classes the medusa is the dominant and conspicuous form in the life cycle, the polypoid form being strictly a larval stage. Scyphomedusae and cubomedusae are generally larger than hydromedusae, most having a bell diameter ranging from 2 to 40 cm (0.7—16 inches); some species considerably larger. The bell of the Lion's mane jellyfish (*Cyanea capillata*), for example, may reach 2.5 meters (8 feet) in diameter and trail tentacles up to 30 meters (100 feet) long. Also in contrast to hydromedusae, the mesoglea of scyphomedusae may be cellular; some cnidocytes are present in the gastroderm, and in both scyphomedusae and cubomedusae the gametes develop within the gastrodermis rather than within the epidermis. In most species, tentacles of varying numbers and lengths hang down from the margin of the transparent bell which is often tinted orange, pink, purple, and other colors. The convex upper (aboral) surface of the bell is called the **exumbrella** whereas the concave lower (oral) surface is the **subumbrella** (Figure 8.9a). The mouth is located in the center of the subumbrella, often suspended on a tubular extension called the **manubrium**. The manubrium of many common species is drawn out into four or eight often frilly **oral arms**, which bear cnidocytes and aid in the capture and ingestion of prey.

Figure 8.10 The aptly-named cubozoan (box) jellies look like boxes with tentacles and often appear square when viewed from above.

Although they drift with currents or waves, most can swim vertically and horizontally, and some can do so with amazing speed. Swimming movements are accomplished by bell contractions like those of the hydromedusae.

Scyphozoans and cubozoans feed on animals of various sorts, including fish that come in contact with the tentacles or oral arms. Some species, such as *Aurelia*, feed on plankton that adheres to the surface of the subumbrella. Ciliated surface cells sweep the plankton to the bell margin where it is wiped by the oral arms and then transferred from there to the mouth. Unlike the hydromedusae, the mouth opens through the manubrium into a central **stomach**, from which extend four **gastric pouches** (Figure 8.9b). Between the pouches are vertical **septa**, each of which contains an opening to help circulate water. The margins of the septa bear a large number of **gastric filaments** containing cnidocytes (nematocysts) and gland cells. Radiating canals typically run from the stomach to the bell margin. Digestion is essentially the same as described in hydrozoa. The gastric filaments are the source of extracellular digestive enzymes, and the gastrodermal nematocysts are probably used to quell prey that is still active.

As in hydromedusae, statocysts and ocelli are present but they are housed in specialized projections called **rhopalia** (Gr. *rhopalon*, club) located between the scallops of the bell margin (Figure 8.9a). Cubozoans have complex eyes containing a lens and a retina-like arrangement of sensory cells. Many scyphozoans display a distinct **phototaxis** (response to light) often coming close to the surface during cloudy weather and at twilight but moving downward in bright sunlight and at night.

With few exceptions, scyphozoans and cubozoans are dioecious. Gametes develop in the gastrodermis of the stomach floor, and the gonads are often conspicuous within the transparent body. The eggs and sperm are shed through the mouth. Fertilization and development may occur in the seawater, or the fertilized eggs may be brooded on the oral arms. In either case, a planula larva develops. The planula larva transforms into a polypoid larval stage, called a **scyphistoma**, which is about the size of a hydra and lives attached to hard bottom surfaces (**Figure 8.11**). Existing for about one or two years, the scyphistoma feeds and produces more scyphistomae by budding. At certain times, it ceases feeding and undergoes a special form of budding to produce young medusae. These buds are produced at the oral end of the body and stack up like plates (**strobila**). In the cubozoans, the juvenile medusae are not produced by budding; rather the entire scyphistoma larva transforms into a young medusae and swims away.

Class Anthozoa (Gr. *anthos*, flower + *zoon*, animal)

The anthozoans constitute the largest class of cnidarians and include the familiar sea anemones and the various types of coral as well as sea fans, and sea pansies (**Figure 8.12**). Anthozoans are either solitary or colonial polypoid cnidarians in which the medusoid stage is completely absent. Another distinctive feature of anthozoans is the presence of a pharynx and mesenteries. The **pharynx** is a tube that hangs from the mouth into the gastrovascular cavity like a sleeve. (**Figure 8.13**) The **mesenteries** are sheet-like partitions that extend from the outer body wall into the gastrovascular cavity. Each is composed of two layers of gastrodermis separated by a layer of cellular mesenchyme. At least eight of the mesenteries, called complete mesenteries, connect with the mesoglea and gastrodermis of the pharynx. Below the pharynx, all of the mesenteries are incomplete so that there is a common central region into which all

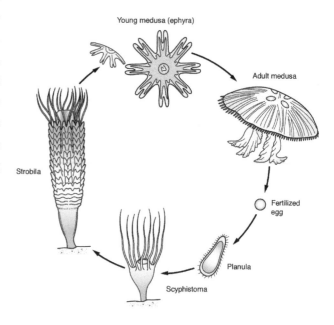

Figure 8.11 The Life Cycle of *Aurelia*. The polypoid stage is a larva and produces medusae by transverse budding.

Figure 8.12 Although anemones are often brightly colored and resemble fleshy flowers, their "petals" are actually stinging tentacles.

of the sections of the gastrovascular cavity open. The mesenteries are arranged in couples, and the free edge of each mesentery bears a glandular ciliated band called the **septal filament**. On reaching the aboral end of the mesentery, the septal filament terminates in a free thread called an **acontium** or **acontial filament**. These filaments are the location of the gastrodermal cnidocytes, of enzyme production for extracellular digestion, and of intracellular digestion and absorption. When prey is swallowed and the sea anemone contracts downward, the many free edges of the mesenteries with their filaments would be pressed against the prey's body.

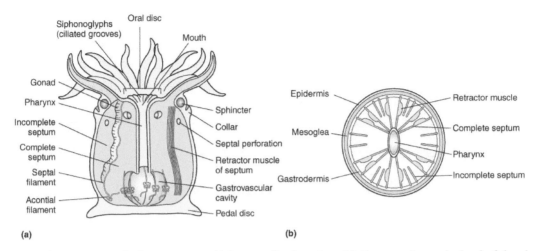

Figure 8.13 The Structure of a Sea Anemone. (a) Longitudinal section. (b) Cross-section at the level of the pharynx.

The function of the mesenteries is not fully understood. It might seem that they would provide greater surface area for digestion and absorption, but studies show that this is not the case. They most likely contribute to the support of the animal. Under slight muscle tension and the shape limitations imposed by the mesenteries, the water in the gastrovascular cavity acts as a hydrostatic skeleton.

Sea Anemones The large solitary and often brilliantly colored polyps that are sea anemones live attached to hard substrates. Most are several centimeters in diameter, but there are giants among the group. A species on the Great Barrier Reef of Australia and one on the north Pacific coast of the United States attain a diameter of over 1 meter (3+ feet) at the oral end. Sea anemones inhabit deep or coastal waters throughout the world but are particularly diverse in tropical oceans. A few species are commensal on other animals, such as the shells of hermit crabs.

The body of a sea anemone is mostly a thick column. (Figure 8.13) At the oral end, the column flares slightly to form the **oral disc**, which bears eight to several hundred hollow tentacles. In the center of the oral disc is the slit-shaped mouth, bearing a ciliated groove called a **siphonoglyph**, which drives water into the gastrovascular cavity even when the mouth is closed. This current of water functions to maintain a hydrostatic skeleton against which the muscular system can act, and also provides for the exchange of gases through the gastrodermal surface. At the aboral end of the column, there is a flattened **pedal disc** for attachment.

The muscle fibers are entirely gastrodermal. Circular fibers are located in the gastrodermis of the body wall, and longitudinal retractor fibers are located in the mesenteries. When sea anemones contract, the upper collar-like rim of the body is pulled over the oral disc giving the impression that the sea anemone has swallowed its own tentacles.

A primitive nervous system, without centralization, coordinates the processes involved in maintaining homeostasis as well as biochemical and physical responses to various stimuli. There are no specialized sense organs.

Sea anemones feed upon other invertebrates and small fish, and they display a considerable range of feeding adaptations. Some live just at the edge of the subtidal zone on rocky shorelines where they feed on crabs or bivalve mollusks dislodged by waves or predators higher above them. Some live lodged within the rock crevices from which long fishing tentacles emerge. The prey is caught by the tentacles, paralyzed by the nematocysts, and carried to the mouth where it is swallowed.

There are species with very short tentacle that feed on fine particulate matter trapped on the oral surface and tentacles. Cilia on the surface beat toward the oral disc, and cilia on the tentacles beat toward the tentacle tips. The tentacles then fold over and deposit the food in the mouth.

Sea anemones may appear to be passive animalistic "flowers," but that presentation is deceptive. Although they are essentially sessile animals, many species can change location by slow gliding on the pedal disc, by crawling on the side of the column, or by walking on the tentacles. A few species can detach the pedal disc and swim briefly with lashing motions of the body column or the tentacles. Most sea anemones produce a noticeable sting on contacting human skin, and a few anemones occur in European and American waters whose nematocysts are quite virulent producing a severe toxic reaction in humans. The West Australian *Dofleinia armata* is believed to be the most toxic sea anemone.

Box 8.1
Anemone Clone Wars

Battles rage throughout the natural world. Some battles such as the competition between species for space and resources, are waged on the ecological scale whereas others, such as those swirling within our own bodies as our immune system battles foreign invaders, occur on the microscopic level. Nowhere are such confrontations less obvious than in the realm of creatures that do not appear to move, such as the sea anemone.

With the appearance and bright coloration of fleshy chrysanthemums (although some look like undulating patches of grass and others like stalks of cauliflower) and an apparent lack of mobility, sea anemones resemble oceanic flowers. Being named for the buttercup family of flowers and sporting a generally peaceful and sessile countenance belies the reality of an animal that is not only an ferocious predator but also an aggressive combatant in the on-going war for limited space in its habitat.

Sea anemones use mucous secretions and the muscular action of the pedal disc to attach themselves to rocks, pilings, shells, plants, and even other animals. But at one time or another, almost all of them move. Their pace is usually around a centimeter (0.04 inch) per hour, a rate revealed only to time-lapse photography or the most dedicated of sea anemone researchers. Anemone watchers have known for some time that these animals move, but it was believed that they did so only in response to unfavorable conditions or the threat of a predator. It has come as a surprise then to learn that not only do individuals move but that whole genetically identical anemone colonies behave as sophisticated armies surging against each other.

The sea anemone *Anthopleura elegantissima* lives in large colonies or patches of genetically identical clones on boulders in the intertidal zone. Where two colonies meet, they form a distinct boundary zone. When one colony or the other tries to penetrate this neutral zone, a battle swiftly ensues.

David J. Ayre from the University of Wollongong, Australia and Rick Grosberg from the University of California Davis have discovered that patches of *A. elegantissima* are organized into distinct castes of scouts, warriors, and reproductives. The smaller scouts occupy the fringes of the patch whereas the warriors, armed with inflatable tentacles ripe with nematocysts called **acrorhagi**, surround the reproductives in the middle of the patch.

When the tide is out, the polyps are contracted and quiet. As the tide returns and covers the colonies, the army begins to mobilize. Small "scout" polyps move out into the border looking for space to occupy. (These scouts typically get stung a few times and then retreat to the safety of their own lines.) Larger, well-armed "warrior" polyps inflate their acrorhagi and begin to swing them around. Towards the center of the colony, poorly armed "reproductive" polyps stay out of the fray and conduct the colony's business of breeding. And it's not just those polyps along the border between two clone armies that clash. Warrior polyps three or four rows away from the front will reach over their comrades to engage in battle.

The nematocysts left behind when the acrorhagi strike an opponent remain embedded in the enemy's body where they cause **necrosis** (tissue death) around the area penetrated by the nematocysts. Severely battered individuals can die as a result of widespread necrosis.

When anemones from opposing colonies come in contact, they usually Figureht but after about 20 or 30 minutes of battle the sides settle into an uneasy truce until the next high tide. In some situations, however, borders between colonies can remain stable for years.

Lacking a brain and possessing only a rudimentary nervous system, sea anemones appear to respond and organize based on chemical messages that flow between them. Being genetically and thus chemically unique, each colony reacts differently to similar stimuli resulting in many different variations of organization and levels of aggression towards neighbors.

Studying anemone clone wars could shed light on the interplay of environment and genetics and clearly demonstrates that a very complex, sophisticated, and coordinated organization and behaviors can emerge at the level of even morphologically simple organisms. Thus the ancient anemone, an animal built on contradictions, continues to amaze and delight us.

Asexual reproduction is common in sea anemones. One such method is **pedal laceration**, in which parts of the pedal disk are left behind as the animal moves with the remnants forming new anemones. Many reproduce asexually by longitudinal (lengthwise) fission, and a few by transverse (crosswise) fission. Most sea anemones are hermaphroditic, but usually only one type of gamete is reproduced at any one time. The gametes develop in gastrodermal bands just behind the free edge of the mesenteries. Fertilization and early devel-

opment may occur externally in the sea water or within the gastrovascular cavity. The planula larva develops into a ciliated planktonic polypoid larva in which mesenteries and pharynx appear. This polyp soon settles and becomes attached as a proper young sea anemone.

Scleractinian (Stony) Corals The stony corals are the group with which the name coral is generally associated. Their common name derives from the fact that the polyps of stony corals produce a rock-like calcareous skeleton in which they live (**Figure 8.14**). The polyps of stony corals are similar to those of sea anemones but lack siphonoglyphs and are usually smaller averaging 1 to 3mm (0.04—0.12

Figure 8.14 A close-up of stony coral reveals tiny polyps each inhabiting a rock-like calcareous cup. Colonies consisting of many individual polyps can become massive in size. Note that one polyp on the bottom is ingesting a piece of food.

inch) in diameter. Although there are some solitary scleractinian corals, most are connected together in colonies.

The skeleton is composed of calcium carbonate and is secreted by the epidermis of the lower half of the column as well as by the pedal disc. This secreting process produces a skeletal cup in which the polyp resides. The floor of the cup contains thin, radiating calcareous **sclerosepta** (**Figure 8.15**). Each scleroseptum projects upward into folds in the base of the polyp. The living colony thus has the form of sheet resting on top of or wrapped around a skeleton and as long as a colony lives, calcium carbonate is deposited beneath the living tissues.

The skeletal conFigureurations of various species of corals are due in part to the growth pattern of the colony and in part to the arrangement of polyps in the colony. Some are low and encrusting; others are upright and branching. Some are large and heavy whereas others are small and delicate. When the polyps are well separated, the coral appears to be pitted or pockmarked. If the colonies are arranged in rows, troughs or valleys separated by skeletal ridges are the result. Skeletal growth rate varies depending on the species and water temperature. Flat dome and plate corals grow only 0.3 to 2 cm (0.12—0.8 inch) a year but the branched corals grow more rapidly, increasing in the linear direction as much as 10 cm (4 inches) per year.

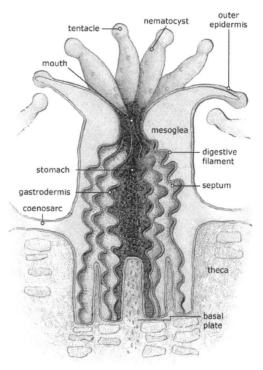

Figure 8.15 The Structure of a Stony Coral Polyp. The theca is the rocky calcium carbonate cup in which the polyp lives.

The density of the secreted calcium carbonate is not the same throughout the year, the change being governed by seasonal shifts of temperature and light. As a result, many coral skeletons exhibit seasonal growth rings similar to trees, which can be used to determine the age and growth rate of the coral.

Corals feed like sea anemones and their prey ranges from zooplankton to small fish. When expanded, the outstretched and often waving tentacles of adjacent polyps present a broad, continuous mesh that prey might touch. In addition to capturing prey, many corals also collect fine organic particles in mucous films or strands, which are then driven by cilia to the mouth. Corals also resemble anemones in their aggressive battle for limited space, although some types are only slightly combative. If a coral touches another species it may extrude filaments and digest the intruder's tissues. Or, like sea anemones, it may deploy specialized defensive tentacles.

Virtually all reef-dwelling corals possess symbiotic zooxanthellae algae within their gastrodermal cells. Some cnidarian biologists have speculated that there is more algal than cnidarian tissue in many corals, and no doubt the algae account for high productivity of coral reefs. The nutritive needs of the coral are supplied in part by the prey on which it feeds and in part by the organic compounds passed to it from the photosynthetic processes of the algae. In fact, as much as 50% of the algal production may be transferred to the coral host.

The algae symbionts give most of their coral host a yellow-brown to dark brown color. **Coral bleaching**, in which corals expel their zooxanthellae, has been observed in some areas in recent years and seems to be increasing. Coral bleaching has been shown to be correlated with seasonal elevated water temperatures and may be evidence of a global rise in temperature. Although, the coral is stressed by the loss of its symbionts, it can recover if the condition is not prolonged.

Sexual reproduction is similar to that in the sea anemones in that they are both dioecious and hermaphroditic species. The planula larva produced by sexual reproduction attaches and the subsequent first polyp becomes the parent of all other members of the colony through asexual budding.

Octocorallia or *Octocorals* Most of the remaining anthozoans, including sea pens, sea pansies, sea fans, whip corals and pipe corals are known as the octocorals or horny corals. Each polyp contains eight mesenteries (*octo*) and eight tentacles with side branches as does a feather. The octocorals are colonial cnidarians with polyps similar in size to those of stony corals. However, the interconnections that bind the colony together are quite different from that of stony coral.

Amebocytes in the mesoglea secrete separate or fused calcareous spicules and a central, horny rod. Thus, the skeleton of the Octocorallia is internal, somewhat flexible, and an integral part of the tissue. Such a skeleton provides not only support, but the flexibility necessary to bend in a wave surge.

The two most common groups of octocorallians are the gorgonian, or horny, corals which include the whip corals, sea feathers and sea fans (**Figure 8.16**) and the soft corals. The erect, rodlike branching growth of the gorgonian corals which can exceed 2 meters (7 feet) in some species, allows for the exploitation of vertical water column with only a small space needed for substratum attachment.

Figure 8.16 Sea fans such as these are found primarily in tropical or subtropical waters are one of over 1,200 species of Gorgonians (horny coral).

Figure 8.17 Leather or soft coral are often highly branched in their growth pattern.

Soft or leathery corals are most abundant on reefs of the Indo-Pacific. These octocorallians possess soft fleshy or leathery colonies that may reach up to 1 m (3.3 feet) in diameter. The colonies are irregular in shape, some encrusting and some massive, often with lobes or finger-like projections. (**Figure 8.17**)

Sea pansies and sea pens are inhabitants of soft bottoms. (**Figure 8.18**) They are quite different than other coral types in that they have a large primary polyp with a stemlike base which is anchored in the sand. The upper part of the primary polyp gives rise to secondary polyps, the most conspicuous of which are termed **autozooids**. Highly modified polyps, called **siphonozooids**, pump water into the interconnected gastrovascular cavities, thereby keeping the fleshy colony turgid and erect.

Cnidarian Body Plan

The radially symmetrical cnidarians show marked morphological advances over the sponges. They possess two embryonic germ layers (diploblastic)—the ectoderm and the endoderm—which become the adult epidermis and gastrodermis, respectively. The middle mesoglea or mesenchyme in adults is derived largely from ectoderm but never produces the complex organs seen in **triploblastic** (three germ layers) eumetozoa.

In spite of their simple symmetry and lack of organ-generating germ layers, the cnidarians are a very successful and diverse group. Their dimorphic lifestyles contribute greatly to their ecological success as this plasticity of form allows them to exploit different resources and environments by leading a double life.

Body Form

Practically all cnidarians fit into one of two morphological types (dimorphism): a *polypoid form*, which is adapted to a sedentary or sessile existence, and a *medusoid form*, which is adapted for a floating or free-swimming lifestyle. Attached or unattached matters not as both forms have the same basic body plan rendering any differences between the two of them variations on a theme. (Figure 8.3)

Figure 8.18 With a featherlike appearance reminiscent of antique quill pens, sea pens lead a benthic existence. Sea pens (*Ptilosarcus gurneyi*) are colonial animals with multiple polyps specialized to specific functions. A single polyp develops into a rigid erect stalk (the rachis), forming a bulbous "root" (the peduncle) at its base. The other polyps branch out from this central stalk, forming water intake structures (siphonozooids), feeding structures (autozooids) with nematocysts, and reproductive structures.

Polypoid Form Most polyps, such as hydras, sea anemones, and corals, have tubular or cylindrical bodies with the top or **oral end** bearing the mouth and tentacles and the bottom or **arboral** end attached to substratum by means of a basal disc in hydras or a pedal disc in anemones.

Medusoid Form Free medusae occur in all cnidarian classes except Anthozoa. The medusoid form is far less diverse than the polypoid form and demonstrates uniformity consistent with their similar lifestyles in open water. Medusae and polyps are morphologically similar, but each is adapted to very different lifestyles. Medusae are bell- or umbrella-shaped, and generally possess a thick, jelly-like mesogleal layer—hence the common name "jellyfish," which of course is a misnomer as they are neither composed of jelly nor are they fish or fish-like in any manner.

Cell Layers The epidermis is composed of five principal types of cells (**Figure 8.19**).

1. **Epitheliomuscular cells** These cells rest on the mesoglea and consist of a columnar cell body and two, three, or more flattened, contractile myofibril basal extensions called **myoneme (Figure 8.20**). The myoneme are oriented parallel to the axis of the body stalk and tentacles and interwo-

ven with those of other epitheliomuscular cells. Collectively, they form a cylindrical, longitudinal, contractile layer.

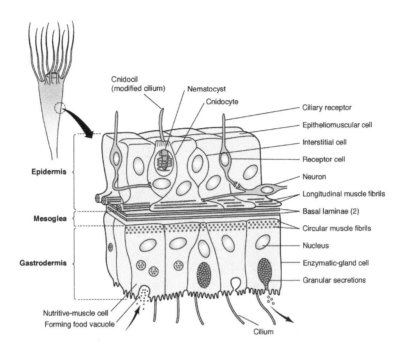

Figure 8.19 The cell layers and tissues of a cnidarian as demonstrated by a longitudinal section through the body wall of a hydra.

2. **Interstitial cells** Located beneath the epidermal surface and wedged between the epitheliomuscular cells are the small rounded **interstitial cells**. Possessing an extremely large nucleus, these cells are totipotent and function as the germinal, or formative, cells of the animal giving rise to the sperm and eggs as well as to other types of cells.

3. **Cnidocytes** These specialized cells, unique to and characteristic of all cnidarians, are located throughout the epidermis but are especially

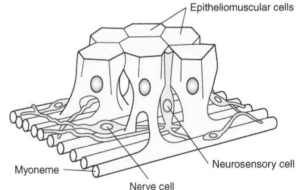

Figure 8.20 The epitheliomuscular cells and a portion of the nerve net of a cnidarian.

abundant on the tentacles and around the oral region. They serve a variety of functions, including prey capture, defense, locomotion, and attachment. Cnidocytes are produced by cells called **cnidoblasts,** which develop from interstitial cells in the epidermis. A nematocyst is not an organelle. It is manufactured by the Golgi apparatus within a cnidoblast cell and as such, qualifies as a secretory product, not a cells structure. In fact, a nematocyst is the most complex secretory product known to exist. Once fully formed, the product is properly called a **cnidocyte** in which resides the **cnidae**. Over 30 kinds of cnidae have been identified, but they can all be ascribed to three basic types:

- **Nematocysts** which are harpoon-like tubes connected to toxins sacs that are used to impale and paralyze prey (**Figure 8.21**).

- **Spirocysts** which are sticky tubes or loops that wrap around and stick to the victim rather than penetrating it. Nematocysts occur in members of all four cnidarian classes, but spirocysts are found only in certain anthozoans.

- **Ptychocysts** differ both morphologically and functionally from nematocysts and spirocysts. These specialized cnidocytes occur only in ceriantharians (burrowing anemones) where they help form the tube the animal lives in. *Ceriantheopsis ameri-*

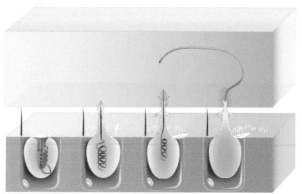

Figure 8.21 The discharge process of a cnidarian nematocyst.

canus or the American tube anemone is a burrowing anemone-like anthozoan found in soft mud and sandy bottoms where it resides in a tube. Within minutes after removal from its tube, one can notice a fine white sheath beginning to form over the column of the animal, representing the beginnings of a new tube. After five minutes, it is possible to see long, white filaments radiating out from the epidermis of the column. Examination of these filaments, as well as the newly formed tube material under a light microscope reveals a dense mat of intertwined threads and capsules of thousands of discharged cnidae. The ptychocysts are primarily, if not entirely, responsible for the strength and formation of the tube. The "stickiness" and interweaving capabilities of the long ptychocyst threads form a tube of extreme toughness and resiliency as well as allowing for the incorporation of sand and other sedimentary material into the outer matrix of the tube.

To regulate their use, cnidocytes are connected in "batteries" containing several types of cnidae connected to supporting cells and neurons. The supporting cells contain chemoreceptors that detect chemicals found in prey. Together with the trigger-hair mechanoreceptor **cnidocil** (Figure 8.21) on the cnidocyte, these chemoreceptors allow only the right combination of chemical and mechanical stimuli to cause the cnidocyte to discharge. Evidence suggests, however, that the animal does have at least some control over the action of its cnidae. Starved anemones seem to have a lower threshold than satiated animals, and the discharge of cnidae is inhibited in some cnidarians during certain activities (e.g., during locomotion in some anemones).

Once discharged, the coiled capsular tube is forcibly everted and thrown out of the bursting cnidocyte at great speed to penetrate or wrap around an unwary victim. The cnidae take only milliseconds to fire, and the everting tube may reach a velocity of 2m/sec (7 feet/sec)—an acceleration of about 40,000 g—the fastest physical reaction known in the animal realm.

Despite their small size, nematocysts are extremely efficient weapon systems. Although nematocyst toxins vary in strength, as a class they are potent biological poisons capable of subduing large active prey. Most appear to be neurotoxins, and some are powerful enough to seriously affect humans with the most venomous being the cubomedusan box-jellies such as the Sea Wasp (*Chironex fleckeri*). According to the Australian Institute of Marine Science, "the Sea Wasp is the most venomous marine animal known. It causes excruciating pain to humans, often followed by death. The chance of survival if stung while swimming alone is virtually zero." Certain scyphozoan jellyfish and the hydroid Portuguese man-of-war (*Physalia*) can also cause extremely painful and sometimes fatal stings. Nematocysts and other cnidae are used but once; new cnidocytes are formed from nearby interstitial cells usually within 48 hours.

4. **Mucous-Secreting Cells** Glands that secrete mucus are found in the epidermis. They are particularly abundant in the adhesive basal disk of *Hydra*.
5. **Receptor and Nerve Cells** Receptor cells are elongated cells oriented at right angles to the epidermal surface (Figure 8.19) Receptor cells are particularly abundant on the tentacles. The nerve cells, which superficially resemble the neurons of higher animals, are located at the base of the epidermis next to the mesoglea.

I've never been hurt by a sea creature, except jellyfish and sea urchins.
—Peter Benchley, author of *Jaws*

The gastrodermis consists of **nutritive-muscle cells** and **enzymatic-gland cells** (Figure 8.19). Nutritive-muscle cells are similar in structure to the epitheliomuscular cells of the epidermis and form both longitudinal and circular contractile layers. These large elongated vacuolated cells are usually flagellated. These cells engulf small food particles by phagocytosis, and digestion takes place intracellularly in food vacuoles. Interspersed among the nutritive-muscle cells are the enzymatic cells. These wedge-shaped ciliated cells produce the enzymes necessary for extracellular digestion to occur.

In many cnidarians the gastrodermal cells contain photosynthetic yellow-brown unicellular algae known as **zooanthellae**. These algae impart a similar color to their host cnidarian, but more importantly excess products from their photosynthetic processes provide food for their host.

The polypoid stage occurs in all four classes of cnidarians although it is greatly reduced in the Scyphozoa and Cubozoa. Given their capacities for asexual reproduction and colony formation polyps tend to be much more diverse than medusae.

Support and Movement

Cnidarians use a number of different support mechanisms. Polypoid forms often rely upon a **hydrostatic skeleton**, which consists of a water-filled interior constrained by the muscular walls of the body. In addition, the colonial anthozoans may incorporate bits of shell fragments into the column wall for further support. In medusae, the middle layer functions as the main support mechanism. It ranges from a thin and flexible mesoglea to a thick and fibrous mesenchyme. The mesenchyme of some medusae is so stiff with thick fibers that it is almost cartilaginous in composition.

Hard skeletal structures are also found among the cnidarians. Horny **axial skeletons** occur in several groups of colonial anthozoans such as gorgonians, sea pens, and the black corals. Such skeletons consist of a stiff internal axial rod embedded in the body mass. In the black corals, the axial skeleton is so hard and dense that it is ground and polished to make jewelry.

Rocky and often massive **calcareous skeletons** are formed by two hydrozoan orders, but the best known are formed by the true, or stony, anthozoan corals. In these corals, epidermal cells in the lower half of the polyp secrete a calcium carbonate skeleton. Such skeletons can assume a wide variety of sizes and shapes, from simple cup-shaped structures in solitary corals to large branching forms in colonial species.

The muscle cells of cnidarians are the most primitive in the animal kingdom. Unlike higher eumetazoa forms, the muscle system of cnidarians consists not of discrete cells and tissues but of modified epidermis (epitheliomuscular cells) and gastrodermis (nutritive muscle cells), whose basal regions are capable of contraction (Figure 8.19).

Most polyps are sessile (permanently attached) and their movements consist primarily of extending, contracting, and bending. In hydras, the gastrodermal muscle fibers are so poorly developed that movement is due almost entirely to the contractions of the longitudinal, epidermal fibers working in conjunction with the fluid-filled hydrostatic skeleton. By taking in water though the mouth as a result of the beating of gastrodermal flagella, a relaxed hydra may stretch to a length of 20 mm (0.8 inch), whereas contractions of the epidermal fibers can reduce it to a mere 0.5 mm (0.02 inch).

Body musculature is most highly developed in the anthozoans, particularly the sea anemones. Most anemones can creep about slowly by use of pedal disc musculature, and many can bend far enough to allow the tentacles to touch the substratum. Both anemones and hydras can bend over, touch, and then release their hold, somersaulting to avoid a predator or move to a new location (**Figure 8.22**). Adhesive cnidae may aid in anchoring the tentacles during these gymnastic-like maneuvers. A few anemones can detach from the substratum and actually swim away by flexing of bending of the body or by thrashing the tentacles. Hydra are also known to detach and float upside down by means of a mucus-coated gas bubble on the bottom of their basal disk. Hungry hydra in laboratory culture jars will often detach and float at the surface behaving more like medusae than polyps.

Probably the oddest form of polyp locomotion is that of the Pom-pom anemone (*Liponema brevicornis*), which is capable of drawing itself up into a ball and rolling freely along the sea floor pushed by the bottom currents.

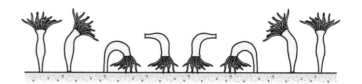

Figure 8.22 The peculiar somersaulting movement occasionally demonstrated by hydra and anemones.

In medusae, the epidermal muscles predominate, whereas the gastrodermal muscles so important in polyps are reduced or lacking altogether. Circular sheets of epidermal muscle fibers—called **coronal muscles**—around the bell margin and over the subumbrellar surface contract rhythmically. These pulsations of the bell provide hydropropulsion by driving water out from beneath the subumbrella resulting in a thrust upward.

Most medusae spend their time pulsing upward in the water column, then sinking slowly down to capture prey by chance encounters with their tentacles. Some can change direction as they swim, and those containing zooxanthellae are strongly attracted to light.

Feeding and Digestion

All cnidarians are blunder feeding carnivores. Small forms, such as hydras and corals, feed mainly upon tiny planktonic crustaceans, but the larger jellyfish and sea anemones can capture and consume small fish and clams. Cnidarians do not pursue their prey but depend on the prey to blunder into them. Once contact is made, the nematocysts fire, stinging and ultimately paralyzing the prey. The tentacles then carry the food to the mouth where often with great stretching, it is ingested whole. The first stage of digestion is extracellular and occurs within the gastrovascular cavity. Aided by the mixing action of the gastrodermal flagella, secretions from the mucous gland cells and the enzymatic gland cells of the gastrodermis reduce the prey to a soupy broth, particles of which are taken into the nutritive cells (Figure 8.19). In the nutritive cells, the final stage of digestion is completed intracellularly within food vacuoles. Completely digested food molecules from the food vacuoles of the nutritive cells are distributed to all other cells by the process of diffusion.

Gas Exchange and Waste Removal

As with sponges, cnidarians lack circulatory, respiratory and excretory organs. Also like sponges, cnidarians depend on water as their "blood" and thus all cnidarians, by necessity, are aquatic. The exchange of O_2-CO_2 occurs from the water into and out of the body surface.

Nitrogenous wastes in the form of ammonia diffuse out into the water through the body surface. Indigestible food material, such as the exoskeleton of prey animals is expelled back out through the mouth opening by contractions of the body. In freshwater species, there is a continual influx of water into the body. This excess water creates an osmotic stress that is relieved by periodic expulsion of fluids from the gastrovascular cavity. Thus, the gastrovascular cavity acts like a giant version of the contractile vacuoles found in protozoa or a nephridium found in the kidneys of more advanced animals.

Nervous System and Sense Organs

Cnidarians lack a centralized brain but they do possess neurosensory cells. Unlike the nerve cells of higher animals, the cnidarian neurons are arranged as **nerve nets**, one between the epidermis and the mesoglea and another between the gastrodermis and the mesoglea, rather than as nerve bundles. (**Figure 8.23**) Most of the neurons are nonpolar. That is, impulses can travel in either direction along the cell and a sufficient stimulus can send an impulse that spreads out in every direction.

Polyps have very few sensory structures. Minute hair-like projections found across the general body surface and most abundantly on the tentacles serve as mechanoreceptors, and perhaps as chemoreceptors. They are involved in tentacle movement toward prey and in general body movement. Oddly, although these structures are clearly sensory in nature, they do not appear to be directly connected to the nerve nets.

As might be expected from their more mobile lifestyles, medusae have more sophisticated nervous systems and sense organs than do sedentary polyps. In many groups, especially the hydromedusae, the epidermal nerve net is concentrated into two **nerve rings** near the bell margin. One nerve ring connects with nerve fibers in the tentacles, muscles, and sense organs whereas the other ring stimulates rhythmic pulsations of the bell. Distributed around the bell margin are ocelli, statocysts, and patches of chemoreceptors.

231

Ocelli are discs or pits of photoreceptor cells that allow the animal to respond to light. Many medusae will use directed swimming motions to maintain a particular light regime. As one might expect, such behaviors are seen especially in those medusae that harbor populations of photosynthetic protists. **Statocysts** are pits or closed vesicles containing a tiny calcareous bit (**statolith**) adjacent to a sensory cilium. Movement of the statolith against the sensory cilium helps the animal orient itself to the direction of the pull of gravity. When the statocysts are stimulated, they inhibit muscular contraction on the stimulated side of the bell. By contracting muscles on the opposite side, the medusa rights itself.

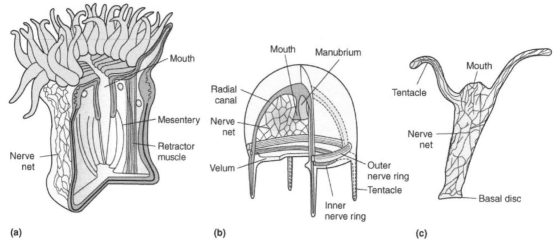

Figure 8.23 Cnidarian Nerve Nets. (a) Nerve net of an anemone (Anthozoa). (b) Nerve net of a hydromedusae (Hydrozoa). (c) Nerve net of a hydra (Hydrozoa).

Cubomedusae posses as many as 24 well-developed eyes located near the bell margin, and they can orient accurately to the light of a match as far away as 1.5 m (5 feet).

Bioluminescence is common in cnidarians and has been observed in all classes except Cubozoa. In some forms, luminescence consists of single flashes whereas in others bursts of flashes propagate as waves across the body or colony surface.

Regeneration and Reproduction

The basic cnidarian life cycle involves an asexually reproducing polyp stage alternating with a sexually reproducing medusoid stage. (**Figure 8.24**) There are variations on the theme, however. The medusa dominates the life cycle of members of the classes Cubozoa and Scyphozoa but is absent altogether in class Anthozoa.

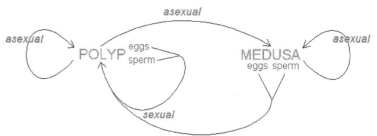

Figure 8.24 The generalized life cycle of cnidarians.

Polyps perform asexual reproduction by **budding** in which an outgrowth from the body wall separates to form a new polyp or medusa. The budded polyps may remain attached to the parent (colonial types) or detach and move away from the parent (solitary types). Medusae primarily reproduce sexually, but the medusae of some cnidarians may also form polyps by budding.

Most cnidarians are **dioecious** (separate sexes) with the gametes (sperm and egg) being formed by interstitial cells and then shed into the water to accomplish external fertilization. The fertilized egg goes through a period known as **gastrulation** resulting in the formation of a free-swimming planula larva. The larva in turn develops into a polyp or medusa.

Relative simplicity of structure and the potential of totipotency infuse the cnidarians with considerable powers of regeneration. These powers are especially well developed in the hydras where a polarity exists from oral to aboral end. If the body stalk of a hydra is cut into several sections, each regenerates into a new individual. Furthermore, the original polarity is retained, so the tentacles always form on the end that was closest to the oral end of the original intact animal, and a basal disc forms at the other end. Even more amazing is the hydra's capacity to totally reorganize itself. In 1744, Swiss naturalist Abraham Trembley inserted a knotted thread through the basal disc of a hydra and pulled it out through the mouth, turning the animal inside out. Surely no animal could survive such a drastic blow to its bodily integrity. However, to Trembley's delight, the reversed form survived quite well with gastrodermal and epidermal cells easily reversing functions. Trembley also performed the first permanent graft of animal tissues when he successfully fused two hydras by placing one inside the other; he pushed it in tail first through the mouth of the second individual.

Cells removed from the body of a *Hydra* also have a modest degree of reaggregative ability, but not to the same degree as sponges. In some cases, entire animals can be reconstituted from cells taken only from the gastrodermis or epidermis. Such capacity for cellular reorganization is a reflection of the primitive state of tissue development in the members of this phylum.

Cnidarian Connection

> *The last word in ignorance is the man who says of an animal or plant: 'What good is it?'*
> —Aldo Leopold

Economically

There is no doubt that corals and the reefs they build have been and continue to be the most useful types of cnidarians. Humans have utilized this resource in many ways—for food animals living around the reef, as building materials, filtering agents, jewelry, decorations, and for aquarium pets. The physical structure of the coral reef also provides a buttress against damaging wave action along coastlines.

Ecologically

The real value of coral reefs lies in their importance to ocean ecology. On and in the rocky outcrops of the reefs lives a dazzling array of strange and wonderful creatures. Some reefs contain as much solid structure and as many living organisms as the largest human cities. These watery wonderlands, often called the "rainforests of the ocean" because of their diversity of life, are to be prized and protected not only for the practical and mundane products they yield but more importantly for the preciousness of their beauty and uniqueness. Many reefs in numerous locations are under assault from human activities, such as deforestation, runoff of agricultural chemicals and sewage, overfishing of food fish, over collection of living coral, invertebrates and certain fish species for the aquarium trade, and tourism as well as possible climatic and oceanic changes due to global warming. It has been estimated that over 75% of reefs worldwide are now suffering some sort of human-inflicted damage and some are feared to be completely dead. In just the Caribbean, biologists have determined that more than 80% of the live coral cover in the Caribbean reef system has vanished since the 1970s.

Many species of jellyfish are capable of congregating into large swarms or "blooms" consisting of hundreds or thousands of individuals. The formation of such blooms seems to be a result of a complex interplay between ocean currents, nutrients, temperature, and oxygen content of the water. In recent times the frequency of such blooms worldwide has increased dramatically. Some biologists believe that increasing jellyfish blooms are an indication that in certain areas the ecology of the ocean is askew as the result of human activity. Speculation is that the jellyfish are replacing fish that have been overharvested or perhaps agricultural runoff provides nourishment for the small organisms on which jellyfish feed. Such nutrient-rich waters often suffer **eutrophication** (low oxygen levels) that favors jellyfish as they thrive in oxygen-poor waters that fish cannot tolerate. Whatever the reasons, the increasing numbers of jellyfish blooms is a symptom that something is ecologically awry in many oceanic ecosystems.

Cnidarians play a role in aquatic food chains, where they are aggressive predators (if, of course, the prey gets close enough) and in turn they are prey to sea stars, sea slugs, and some types of fish, turtles, and sea birds that are seemingly immune to their venom. The stinger of a Portuguese man-of-war can penetrate a surgical glove, but it apparently does not bother the alimentary tracts of either the loggerhead turtles or sunfish, both of which feed upon it. Frigate birds also scoop jellyfish from the sea.

The oddest enemy of the jellyfish and hydromedusae are a number of nudibranchs or sea slugs. In a phenomenon known as **kleptocnidae** (literally "stealing stingers"), these slugs consume their cnidarian prey, ingesting their unfired nematocysts and storing them in bumps on their dorsal surfaces. Once in place, the nematocysts of the now-digested cnidarian serve as a workable defense for the nudibranch.

Given the prickly virulence of most cnidarians, one would expect other marine creatures to stay at a safe distance. Some creatures keep such close company with cnidarians, however, that they have developed commensal relationships with them. Fish are among them. Some swim among the tentacles of scyphomedusae and hydromedusae whereas the anemone fish actually make a home not only among the tentacles but also within the hollow body of large anemones. A single lion's mane jellyfish in the Caribbean was found to have over 200 fish within its tentacles, and 305 young fish were counted swimming inside a single ten-pound medusa. The fish obviously use the cnidarian's tentacles as protection from predators as they quickly flee back into their tentacled fortress when danger threatens. The fish may also share in part of the prey captured by the cnidarian whereas they in turn keep their host free of debris and possible parasites. Why don't these fish themselves fall victim to the stinging tentacles? The available evidence suggests that the cnidarian does not voluntarily fail to sting its fish partners; rather, the fish alter the chemical nature of their mucous coating, perhaps by accumulating mucus from the cnidarian, thereby masking the normal chemical stimulus to which the anemones cnidae would respond. Even when accidentally stung, cnidarian-dwelling fish show a much higher survival rate than other fishes of the same size.

Several species of hermit crabs and true crabs carry anemones on their shells, exoskeletons, or claws and use them as living weapons to deter would-be predators.

Medically

Although not so terrifying as the jaws of a huge shark, nor so dreadful as the entangling arms of a cartoon octopus, getting stung by a large cnidarian can be excruciatingly painful at best and possibly lethal at worst. Being wrapped in the tentacles of a large cnidarian has been likened to being embraced by a swarm of angry bees. Such human-cnidarian encounters result in 10 to 30 or more fatalities a year worldwide. However, medical researchers have been bioprospecting the potential beneficial uses for the potent toxins of cnidarians. These powerful chemicals are being used to study nerve cell function and to track calcium in chemical reactions. They may also prove useful in treating inflammation, cancer and arthritis and as anti-venom against other types of biotoxins.

Researchers studying bioluminescence in jellyfish discovered green fluorescent protein (GFP), a protein that fluoresces green when exposed to blue light. As such, GFP has proven to be very useful in the development of biosensors and biological markers and tags. Jellyfish have even flown into space aboard the space shuttle to help humans study the effects of weightlessness on living things. In May 1991, 2500 moon jellies flew aboard Space Shuttle Columbia where scientists studied how their balance organs (statocysts) develop in weightless conditions.

PHYLUM CTENOPHORA

Diversity and Classification

Numbering around 100 species, ctenophores are exclusively marine and occur in all the world's seas, at all latitudes. Most swim but one group creeps along the bottom. They display a variety of shapes and range in size from a pea to that of a golf ball. Some deep-water species the size of watermelons, however, have been observed from submersibles and some bizarre ribbon-like forms over 1 meter (7 feet) have also been discovered.

Although the brilliantly luminescent ctenophores have been seen from ships since ancient times, the first recognizable Figures of ctenophores were drawn by a ship's doctor and naturalist in 1671. Linnaeus classified them in his group Zoophytea with other "primitive" invertebrates such as poriferans and cnidarians. Only in the 19th century were ctenophorans recognized as a standalone taxon. Two classes of ctenophores are currently recognized: Class Tentaculata whose members possess tentacles and Class Nuda whose members do not possess tentacles. However, molecular genetic analysis and cladistic studies have indicated that the overall classification of the group needs to be revised. For the time being the phylogeny and taxonomy of ctenophora must be regarded as unsettled though it remains an active area of research.

Domain: Eukarya
   Kingdom Animalia
      Sub-Kingdom Eumetazoa
         Phylum Ctenophora
            Class Tentaculata
            Class Nuda

Phylum Characteristics

Slowly moving and drifting as part of the open-ocean plankton are transparent, gelatinous animals trailing long comb-like tentacles—the ctenophores (Gr. *cten*, comb + *phero*, to bear). Commonly known as comb jellies, sea gooseberries, or sea walnuts, these exceptionally delicate animals form a major portion of the planktonic biomass in many areas of the ocean.

Ctenophores somewhat resemble medusoid cnidarians and so were thought to be an offshoot from them. However, there are fundamental differences, and any similarities are now thought to represent more convergence features resulting from adaptations to an open-ocean lifestyle than common ancestry.

The members of phylum Ctenophora exhibit the following general characteristics:

- Their arrangement of internal canals and the position of the paired tentacles result in **biradial symmetry**, a combination of radial and bilateral symmetry. They are generally ellipsoidal or spherical in shape with radially arranged rows of **comb plates** for locomotion (**Figure 8.25**).

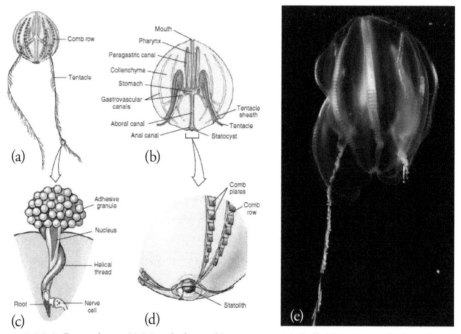

Figure 8.25 A Cetnophore. (a) Morphology. (b) Anatomy. (c) Colloblasts are cells found on the tentacles of ctenophores. The adhesive granules rupture and release an adhesive substance that sticks to prey. (d) The statocyst is a balance sensor consisting of a statolith, a solid particle supported on four bundles of cilia, called "balancers", that sense its orientation. (e) As colorfully lit as a tiny cruise ship, the flashing iridescent comb row displays of the ctenophore *Mertensia ovum* are beautiful and mesmerizing. One of the two feeding tentacles is deployed while the other is retracted.

- All species are marine.
- The **gastrovascular system** consists of a mouth, pharynx, stomach, a series of anal canals, and anal pores.
- They lack stinging nematocysts but do possess **colloblasts** (adhesive cells).

- They lack a brain but are equipped with a subepidermal plexus (bundle) of nerves concentrated under each comb plate. They also possess a statocyst at the aboral end.
- They are monomorphic, without alternation of generations and without any kind of attached life stage

Ctenophoran Body Plan

Ctenophores are diploblastic possessing an ectoderm and endoderm separated by a thick cellular mesenchyme composed of jelly-like material strewn with fibers and amebocytes. The ctenophoran mesenchyme differs from that of the cnidarian mesoglea in that true muscle cells develop within the mesenchyme itself, a condition characteristic of triploblastic metazoa. However, there is disagreement over the presence of true mesoderm in ctenophora.

Body Form

Ctenophores are mainly globular-shaped transparent creatures so delicate and fragile that they must be collected in jars individually by divers. Though they are usually colorless, there are red, orange, or even black colors in certain species. Their transparent bodies glisten like fine glass, brilliantly iridescent during the day and luminescent at night. One of the most delightful characteristics of ctenophores is the light-scattering produced by the beating of the eight rows of cilia, which appears as a changing rainbow of colors running down the comb rows. (Figure 8.25) One might assume this rainbow effect is bioluminescence, but in reality it is simple light diffraction or scattering of light by the moving cilia, iridescence, not luminescence. Most ctenophores are also bioluminescent, but that light (usually blue or green) can only be seen in darkness. Some ctenophores can also release great clouds of luminescent particles to blind or confuse a potential predator.

Their radially symmetrical bodies are distinguished by eight rows of ciliary combs (plates) that propel the body in swimming, oral end forward. Many species have two opposing retractable tentacles emerging somewhere near the midpoint of the body. (Figure 8.25) From these central tentacles branch additional filaments. Unlike cnidarians, the tentacles of ctenophores contain no nematocysts. Instead, they possess **colloblasts** Gr. *kollo*, glue + *blastos*, bud) or "lasso cells" that burst open when prey comes in contact with the tentacle.

Support and Movement

Ctenophores rely primarily on their elastic mesenchyme for structural support. Most ctenophores simply drift with the currents. They can, however, swim with modest locomotor power provided by the beating of the cilia on the comb rows. They are the largest animals to use cilia for propulsion and can reach speeds of about five centimeters (2 inches) a second.

The mesenchymal musculature that sets ctenophores apart from cnidarians is usually used to maintain body shape and assist in feeding, pharyngeal contractions, and tentacle movement.

Feeding and Digestion

Ctenophores are entirely carnivorous, feeding on plankton animals. The tentacles trail passively or are "fished" by various swirling motions of the body. When prey contacts the tentacles, the adhesive colloblasts fire a sticky spiral filament binding the prey to the tentacles. As the tentacles accumulate prey, they are periodically wiped across the mouth by muscular retraction of the tentacles. Colloblasts are only used once and usually the entire tentacular branch is lost and regenerated.

The mouth opens into an elongated, highly folded and flattened pharynx that is richly endowed with gland cells that produce digestive juices. Extracellular digestion begins as large food particles are tumbled within the pharynx by ciliary action. The pharynx leads to the stomach and a system of gastrovascular canals that branch through the ctenophore extending to the comb plates, tentacles, and elsewhere. Within this complicated gastrovascular canal system, digestion—extracellular and intracellular—is completed. Indigestible wastes collect in **anal canals** and are expelled to the outside via small **anal pores**. (Figure 8.25) Functioning as a primitive anus, the anal pores assist the mouth in the elimination of indigestible and metabolic wastes.

Gas Exchange and Waste Removal

As with the cnidarians, ctenophores lack organized circulatory, respiratory, and excretory systems. The surrounding water serves as "blood" with gas exchange occurring across the general body surface and walls of the gastrovascular system. The gastrovascular system probably also picks up metabolic wastes from the mesenchyme for expulsion out of the mouth and anal pores.

Nervous System and Sense Organs

The nervous system of ctenophores is a noncentralized (no brain) nerve network that is particularly well developed beneath the comb rows. The only sense organ is a statolith on the end of the body opposite the mouth. (Figure 8.25). Resting on four tufts of balancer cilia in a deep pit, the statolith exerts selective pressure on the balancer cilia which can then change the rate and direction of the beat of the comb row cilia. Removal of or injury to the statolith can result in a ctenophore that is unable to maintain a vertical posture. The comb rows are very sensitive to contact; when a comb row is touched, many species retract it into a groove formed in the gelatinous body.

Regeneration and Reproduction

Ctenophores possess extraordinary powers of regeneration; even if half the creature including the statocyst is destroyed, the remaining half can reconstruct itself. Bottom-dwelling ctenophores can reproduce asexually by a process that resembles pedal laceration in anemones.

Almost all members of the phylum are hermaphroditic. The gonads develop in the form of bands in the gut pouches. The gametes are shed via the mouth into the surrounding sea water where fertilization takes place. The embryos quickly grow into planktonic **cydippid larvae**. Whereas most medusae grow to adult size before beginning to produce sperm and eggs, Ctenophores seem to produce small numbers of gametes before they reach adult size, and thus have very rapid generation times, resulting in rapid population growth. After attaining their adult size, ctenophores will spawn eggs and sperm daily for weeks, as long as there is sufficient food available.

Human-Ctenophoran Connection

Ecologically

Ctenophores are predatory feeding on small crustaceans and the larvae of other marine invertebrates such as clams and snails. In turn, ctenophores are eaten by many species of medusae as well as by sea turtles and some fish.

Ctenophores are hardly noticeable, and their influence on an ecosystem is seemingly very low. However, they can become marine invaders capable of wrecking entire ecosystems. To everyone's amazement, the accidental introduction via a ship's water ballast of an American ctenophore, *Mnemiopsis leidyi*, into the Black Sea in the early 1980's, caused a full ecosystem fisheries collapse within less than 10 years. How could this happen? It seems this rapidly-reproducing ctenophore managed to outcompete the native planktonic fishes for food, mostly by eating nearly all of the zooplankton before the fish eggs hatched, leaving little food for the native fish larvae. The adult fish remaining in the Black Sea system were also in poor condition as a result of having to compete with the *Mnemiopsis* ctenophores for food. Millions of tons of fish were replaced by millions of tons of (inedible) ctenophores. *Mnemiopsis* populations in the Black Sea have finally come under control in the last few years with the "spontaneous" appearance of a predatory ctenophore, *Beroe ovata*, probably also an import of American waters. However, this ecosystem is still dominated by exotic ctenophores and the jellyfish *Aurelia*. A similar scenario is playing out in the nearby Caspian Sea after *Mnemiopsis* traveled between the two bodies of water, probably via the canal system that connects them. The *Mnemiopsis*-eating *Beroe* also arrived within short order meaning that things may play out somewhat differently in the Caspian Sea, but undoubtedly the arrival of these two invasive ctenophores will bring great changes to the Caspian Sea greatly affecting the four countries that surround the Caspian Sea as they did the six countries that surround the Black Sea.

A Closing Note
A Stinging Ball of Snot

I once had the opportunity to hunt for fossil marine animals along the Chesapeake Bay in an area known as Calvert Cliffs. The cliffs, which now stand up to 110 feet tall, were once the bottom of a shallow, warm sea that covered all of what is now southern Maryland. Weathering out of these cliffs are Miocene fossil deposits regarded as the most extensive such assemblages in the eastern United States. We were there to see what we could find but the main prize everyone sought was fossilized sharks' teeth.

The fossils from the cliffs or the bottom of Chesapeake Bay end up on the beach, especially after strong storms. Since we would be wading in shallow water along the beach, we had been advised the previous evening to dress in long pants and sturdy footwear to protect against possible contact with jellyfish. The shoes were no problem but as it was late summer, I hadn't packed any long pants. "No problem," I thought, "I'll wear shorts but just be extra cautious."

When we arrived bright and early that morning we could see a few jellyfish several yards off shore. They were green and about the size of a softball and looked, as one of our group described them "like giant balls of snot" In short order I found my first shark tooth and all caution flew to the wind. I was bent over facing the shore in about knee-deep water intently sorting through the shell debris that littered the beach when it felt like someone had touched the back of my leg with a red-hot poker. I let out a screech of pain and jumped about two feet up the beach out of the water. Whirling around I saw the culprit—a jellyfish that wave action had washed against my leg.

Our group leader (A zoologist from the University of Maryland) quickly scraped my calf with the edge of a credit card to remove any nematocysts from my skin. He then applied a mud poultice made from soil out of the nearby cliffs. That and aspirin helped dull the pain and I was able to continue. I came back from that expedition with several excellent fossils and a painful red welt that took weeks to fully heal. Remember, if you ever find yourself on the beach in shallow water on a warm, bright sunny day, don't turn your back. You never know when a stinging ball of snot may sneak up on you.

In Summary

Phylum Cnidaria

- Cnidarians are the most ancient of animal phyla.
- There are approximtely 9,000 species of cnidarians organized into four classes:

  Class Hydrozoa
  Class Scyphozoa
  Class Cubozoa
  Class Anthozoa

- The characteristics of phylum Cnidaria:

  1. Their bodies exhibit radial, biradial, or quadriradial symmetry (Radiata) along a longitudinal axis with two opposite ends but no definite head. Cnidarians are dimorphic demonstrating two basic body forms: the polypoid form and the medusoid form.
  2. They are entirely aquatic with most being marine, but a few species are found in fresh water.
  3. Their bodies are diploblastic (two cell layers) consisting of an outer layer of cells known as the epidermis and an inner layer of cells known as the gastrodermis. The cell layers are separated by a layer of jelly-like mesoglea.
  4. The gastrovascular cavity or coelenteron is a saclike or branched cavity with a single opening that serves as both mouth and anus. Extensible tentacles usually encircle the mouth or oral

region. The gastrovascular cavity is the only body cavity (no true coelom) and serves as the site of digestion.

5. They possess unique stinging or adhesive structures called cnidae.

6. They lack a central nervous system (no brain) but are equipped with a simple nerve net with some sensory structures.

7. They possess a simple epitheliomuscular system consisting of an outer layer of longitudinal fibers at the base of the epidermis and an inner layer of circular fibers at the base of the gastrodermis.

8. They lack a circulatory, excretory or respiratory system.

9. They exhibit alternation of asexual polypoid and sexual medusoid generations, but there are a number of variations on this basic theme.

- Class Hydrozoa—hydras and hydroids—are small and not as conspicuous size-wise as are the larger jellyfish and sea anemones. The majority of species are found in the ocean where they exist mainly as fixed colonies such as *Obelia*, although the Portuguese man-o-war (*Physalia*) is a floating colony composed of several different types of both polypoid and medusoid individuals. Some, such as the hydras, are found in freshwater and are mainly solitary individuals.

- Class Scyphozoa and Class Cubozoa are the jellyfish. Class Scyphozoa contains the majority of species in this class. The relatively small number of cubozoans, or box jellies are all tropical and semitropical in habitat. Both the scyphozoans and the cubozoans are strictly marine. In both classes the medusa is the dominant and conspicuous form in the life cycle, the polypoid form being strictly a larval stage.

- Class Anthozoa constitute the largest class of cnidarians and include the familiar sea anemones and the various types of coral as well as sea fans, and sea pansies. All species are marine. Anthozoans are either solitary or colonial polypoid cnidarians in which the medusoid stage is completely absent

- Cnidarians:

1. Have a hollow bag- or tube-like body composed of two cell layers with tentacles surrounding an oral opening,

2. Are soft-bodied and rely on a hydrostatic skeleton for support. Movement is provided by contractile cells of the epidermis and gastrodermis.

3. Are blunder feeding carnivores that use stinging cells in their tentacles to subdue prey.

4. Lack a circulatory, respiratory, or excretory system.

5. Lack a centralized brain but do possess neurosensory cells.

6. Have a life cycle in which an asexually reproducing polyp stage alternates with a sexually reproducing medusoid stage.

- Cnidarians connect to us in important ways economically, ecologically, and medically

Phylum Ctenophora

- Numbering around 100 species, ctenophores are exclusively marine and occur in all the world's seas, at all latitudes. Most swim but one group creeps along the bottom. They display a variety of shapes and range in size from a pea to that of a golf ball. Some deep-water species the size of watermelons, however, have been observed from submersibles and some bizarre ribbon-like forms have also been discovered.

- Phylum Ctenophora is composed of two classes:

  Class Tentaculata
  Class Nuda

- Ctenophores, commonly known as comb jellies, sea gooseberries, or sea walnuts, display unique characteristics:

  1. Their arrangement of internal canals and the position of the paired tentacles result in biradial symmetry, a combination of radial and bilateral symmetry. They are generally ellipsoidal or spherical in shape with radially arranged rows of comb plates for locomotion.
  2. The gastrovascular system consists of a mouth, pharynx, stomach, a series of anal canals, and anal pores.
  3. They lack stinging nematocysts but do possess colloblasts (adhesive cells).
  4. They lack a brain but are equipped with a subepidermal plexus (bundle) of nerves concentrated under each comb plate. They also possess a statocyst at the aboral end.
  5. They exhibit neither polymorphism or dimorphism.

Review and Reflect

1. *A Radial Existence.* Radial symmetry is usually found only in animals that live in water whereas those that live on land tend to be bilaterally symmetrical. Why is radial symmetry better suited to an aquatic existence than to a terrestrial lifestyle?
2. *Cnidarian Monsters?* Hydra, medusa, and gorgon corals are terms that are applied to types and forms of cnidarians, but are also the names of monstrous creatures from Greek mythology. Investigate this connection and propose an explanation for this connection.
3. *Cnidarian Reincarnation.* Reincarnation is an important part of some religious beliefs. Suppose you had to be reincarnated as a hydra. Which species would you want to be—brown (*Hydra vulgaris*) or green (*Chlorohydra viridissima*)? Defend your selection.
4. *Gargling Hydra?* Freshwater hydra can be observed periodically discharging water out of their mouth. Why would they do this?
5. *Model It.* Your zoology class has been invited to a sixth-grade classroom to make presentations regarding a number of invertebrates. You have been assigned the task of making models of cnidae.

Explain what common materials you would use to construct models of the nematocysts, spirocysts, and ptychocysts of cnidaria and the colloblasts of ctenophores.

6. *Biodiversity Run Amok*. Coral reefs are often described as the marine equivalent of tropical rain forests. Do you agree with this statement? Explain.

7. *Teach It*. Your zoology class did such a great job teaching invertebrate zoology to those six graders that you have been invited to do the same for a high school biology class. You have been assigned the task of teaching those students the fundamentals of cnidarian reproduction. What would you say?

8. *End the Confusion*. Your study buddy is having difficulty understanding why cnidarians and ctenophores are classified as separate phyla. Teach them by preparing a chart in which you compare and contrast the two groups.

Create and Connect

1. *Coral Reefs—A Call to Action*. A United Nations committee on the environment has been convened to examine the possibility of banning all commercial exploitation of coral reefs worldwide. You have been called to testify before this committee. Write a position paper in which you explain what you will say in your testimony.

   Carry this further. Is there anything you can do personally to help protect our precious reefs? Investigate the problem and develop a plan of action to do what you can. We are all part of the problem so we must all become part of the solution.

2. *The Prey's Perspective*. In this book we examine what different kinds of animals eat and how they eat it but have you ever considered what it would be like to be a prey animal? Write a short story in which you detail a deadly encounter between a small prey animal and a cnidarian from the perspective of you being the prey. Your story should include: (1) specifically what kind of prey animal you are, (2) specifically what kind of cnidarian you have blundered into, and (3) the whole process of contact, stinging, and ingestion.

   Guidelines:

   A. Set your story up in the following format:

   - Appropriate Title
   - Catchy Beginning. Catch and hold your reader's attention.
   - Understandable Middle. Don't muddy up the middle.
   - Believable Ending. Give believable (but not necessarily happy) closure.

   B. The instructor may provide additional details or further instructions.

3. ***Does Temperature Matter?*** Analyze the graph below and answer the questions that follow (**Figure 8.26**).

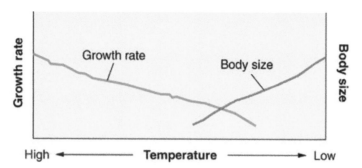

Figure 8.26 A generalized graph of growth rate versus body size in hydra.

A. What specific information is shown on this graph?

B. What is the relationship between reduction of temperature and growth rate?

C. What is the relationship between reduction of temperature and body size?

D. If you had a culture of hydra and you wanted to increase their growth rate, what would you do? What would you do to increase the body size of your hydra?

E. What would you have to do to develop the largest hydra possible in the shortest amount of time?

PHYLUM PLATYHELMINTHES AND PHYLUM NEMERTEA: LIVING THE ACOELOMATE BODY PLAN

Flatworms have both male and female sex organs in the one animal. If you think about it, that's a great way to be. If you're rare and you are looking for a mate and you're traveling alone you don't have to meet another female or another male. You just have to meet another one.

—Leslie Newman

Introduction

The primeval ocean was the womb of life in which the first living cells formed some 4 billion years ago. It is believed that within those ancient seas complex assemblages of chemicals engaged in ever-changing reactions, possibly powered by the heat from volcanic vents. Inorganic eventually became organic, and those organic aggregates eventually developed the structures and processes required to make the leap from inanimate to animate. Life appeared.

Evolution has led to many "experimental models" some of which failed and some of which gave rise to abundant and dominant species that inhabit the world today. Still others produced a small number of species, some of which persist, whereas others were formerly more abundant, but are now in decline.

An Explosion of Complexity

The ocean has been the wellspring of animal life pouring forth creatures of ever increasing complexity in a series of evolutionary leaps and bounds. First came single-celled animals then about 1 billion years ago a great leap forward resulted in the appearance of the choanoflagellates (metazoa). Around 700 million years ago the next evolutionary lurch produced the Radiata—cnidarians and the ctenophores—radial metazoa arranged into two cell layers (diploblastic) with a central cavity. For reasons not fully understood, the Cambrian period 535 to 530 million years ago proved to be an unusually productive and even explosive time in evolutionary history. For over 3 billion years before this time, evolution had forged little more than microscopic prokaryotes and unicellular eukaryotes. Then, within the space of a few million years, all other major phyla of macroscopic invertebrates had formed and become established, including phylum Platyhelminthes. This was the Cambrian Explosion, the greatest evolutionary surge the planet has ever known.

Some researchers believe that a planuloid ancestor gave rise to Radiata, free-floating or sessile radial animals (cnidarians), and Bilateria, creeping animals with a bilateral form (flatworms). The first bilateral animals (Bilateria) evolved into two separate branches on the tree of life—Protostomia and Deuterostomia. Protostome animals develop from an embryo whose first opening (**blastopore**) becomes the animal's mouth (**schizocoely**), whereas in deuterostomes the blastopore becomes the anus (**enterocoely**). From the Protostomia branch sprang two more branches—Lophotrochozoa and Ecdysozoa (**Figure 9.1**).

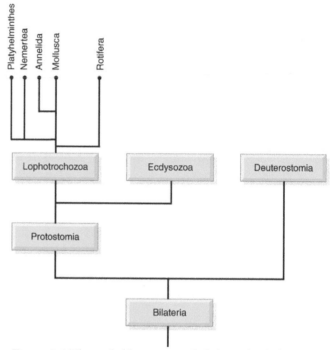

Figure 9.1 The probable position of phylum Platyhelminthes and phylum Nemertea on the phylogenetic tree of life.

The lophotrochozoa ("crest-bearing animals") are a diverse phylogenetic group that molecular data analysis indicates was only recently formed. The name "lophotrochozoa" combines two previous taxon names—Lophophorata (which feed via a fringe of hollow tentacles, a **lophophore**) and Trochozoa (which have a unique larval stage, the **trochophore larva**). The other branch of protostomes are the Ecdysozoa ("molting animals"). Phyla developed from this branch are characterized by a cuticle that is molted (**ecdysis**).

PHYLUM PLATYHELMINTHES

History of Platyhelminthes

Biologists believe that a flatworm-like animal was the first bilateral **triploblastic** (3 cell layers) creature to develop a definite head (**cephalization**), rudimentary organs including a brain, paired senses, and a tail. These worms were the first to move forward and thus the first to hunt for food and mates. This more advanced body design was an enormous evolutionary leap forward.

Creatures possessing these adaptations had huge advantages over animals that could only sit and wait for food to float to them (sponges and cnidarian polyps) or simply pulse aimlessly through the water in hopes of accidentally bumping into a meal (medusoid cnidarians). This ancient pioneering lineage is represented in modern times by the flatworms of phylum Platyhelminthes (Gr. *platy*, flat + *helmins*, worm).

Being small and soft-bodied, ancient platyhelminths left little fossil record. A few traces in fossilized mud that were probably made by platyhelminths have been found, and trematode (fluke) eggs have been found in Egyptian mummies and the dried dung of the Pleistocene ground sloth. Trematode larvae that parasitized mollusks may leave pits or thin spots on the inside of the mollusk's shell, and these pits may be detected even after the shell fossilizes.

Diversity and Classification

The phylum Platyhelminthes includes about 25,000 living species of both free-living and parasitic worms. Platyhelminths (or flatworms as they are commonly known) display a variety of body forms and successfully inhabit a wide range of environments ranging from freshwater and marine to land with some being symbiotic (parasitic) in or on other invertebrates. In their habitat, the free-living flatworms are carnivores and scavengers.

As their common name suggests, platyhelminths are highly flattened dorsoventrally (top to bottom) and many are not much thicker than a piece of paper, although the body shape varies from broadly oval to elongate and ribbon-like. They range in size from less than 1 mm (0.04 inch) in some free-living forms to about 30 cm (12 inches) in some parasitic forms. The largest flatworms are certain tapeworms that attain lengths of several meters (7-10 feet). The longest flatworm ever recorded was a tapeworm over 27 meters (90 feet) long!

In his first edition of *Systema Naturae* (1735), Linnaeus established two phyla encompassing all the invertebrates. He assigned the insects to one phylum with the rest of the invertebrates in the other. This latter group Linnaeus called Vermes (Gr. worms). By the writing of the thirteenth edition of *Systema Naturae* (1788), the various groups of flatworms were placed together in the order Intestina. Though rejected by many leading biologists of the day, this taxon persisted through the early 1800s and came to serve as a dumping ground for almost any creature with a wormlike body.

During the nineteenth century, the unique features of flatworms were gradually realized, and they were eventually separated from most other groups of worms and wormlike creatures. In 1851 Vogt established a single taxon for the flatworms and nemerteans, which he called the Platyelmia, a name changed to Platyhelminthes by Gegenbaur in 1859.

Because the phylum lacks a single unique synapomorphy (defining characteristic), some authorities contend that the phylum Platyhelminthes is not a valid monophyletic grouping (all classes derived from one common ancestor). Recent molecular studies suggest that the phylum as a whole may even be polyphyletic (having arisen as two independent groups from different ancestral stock). Considerable controversy surrounds the taxonomic validity of this phylum and the monophyly of some of the traditional classes (especially the Turbellaria) within the phylum. In fact, molecular studies leave little doubt that (1) Turbellaria is clearly paraphyletic (includes some but not all of the descendants of a common ancestor), and (2) Cestodes and monogenetic flukes should be classified together as a single taxonomic unit outside the phylum Platyhelminthes (**Figure 9.2**).

We classify Platyhelminthes here in a more traditional fashion because the taxon remains prevalent in the zoological literature and because there is no current consensus of how to revise the classification of the phylum. (We do stray somewhat from tradition by ranking Monogenea as a separate class and not as an order in the class Trematoda.)

Domain Eukarya
    Kingdom Animalia
        Phylum Platyhelminthes
            Class Turbellaria—Planarians
            Class Trematoda—Digenetic flukes
            Class Monogenea—Monogenetic flukes
            Class Cestoda—Tapeworms

Phylum Characteristics

The flatworms display some of the most important advances found in the body plan of animals. In fact, some hypotheses hold that these acoelomate triploblastic bilateral animals represent the plan from which many of the remaining triploblastic groups of animals may have ultimately been derived.

The members of phylum Platyhelminthes exhibit the following general characteristics:

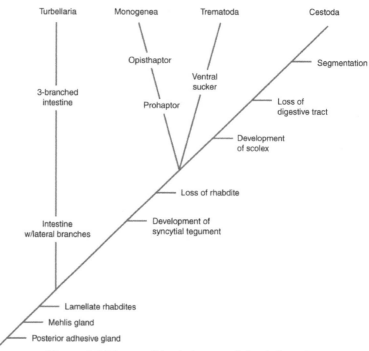

Figure 9.2 The possible phylogeny of platyhelminths

- They are unsegmented worms that are either parasitic or free-living in their mode of existence. (Class Cestoda or tapeworms have many individual sections called **proglottids** that appear to be segments, but structurally are not.)
- They are triploblastic, possessing three germ layers—ectoderm, mesoderm, and endoderm. The middle layer of mesoderm is significant in that it is the germ layer from which organs and organ systems arise.
- They are **acoelomate** (lacking a true coelom). Their bodies are filled with mesodermal **parenchyma** cells ("packing tissue"), and their mesenchyme contains more cells and fibers that do the mesoglea of cnidarians. Thus, there are no internal spaces inside them other than the digestive tube.
- They are bilaterally symmetrical (Bilateria) and flattened dorsoventrally. In general, the longer the worm, the more pronounced the flattening.
- They display cephalization (possess a definite head end) and possess a central nervous system consisting of a pair of anterior ganglia ("brain") with longitudinal nerve cords connected by transverse nerve cords ("nerve "ladder") running the length of the body. Some forms possess simple sense organs such as eyespots.

- They are the simplest animals with an excretory system for removing liquid metabolic wastes.
- They possess a single opening or one-way digestive system consisting of a mouth, pharynx, and a highly invaginated (pouched) intestine. Cestodes have no digestive system.
- Most forms are monoecious (hermaphroditic) with a complex reproductive system consisting of well-developed gonads, ducts, and accessory organs. Many of those that are internal parasites have complicated life cycles involving several hosts.

Although it is likely that the trematodes and cestodes both evolved from the turbellarian form, over the course of time they lost much of the original turbellarian complexity, especially the cestodes.

Class Characteristics

Class Turbellaria (L. *turbellae* [pl], stir + *aria*, like or connected with)

Turbellarians, for the most part, are free-living flatworms that are found both in the ocean and freshwater. A few species have even managed to break the bonds of a totally aquatic existence and are found on land albeit in humid densely vegetated areas (**Figure 9.3**). They are typically small creeping worms 5mm to 50 cm (0.2 inch to 20 inches) that combine muscular movement with ciliary action to achieve locomotion. Turbellarians are taxonomically aligned into over 3,000 species comprising 12 orders.

Figure 9.3 The Turbellarians. (a) The planaria *Dugesia sp.* is common in freshwater around the world. (b) Planaria have a definite head equipped with a number of sensory organs. (c)Ruffled marine worm *Pseudoceros* sp.). These worms often resemble the colorful nudibranch molluscs. (d) The shovel-headed worm (*Bipalium adventitum*) is a large terrestrial turbellarian that preys on earthworms. (e) Generalized anatomy of a turbellarian.

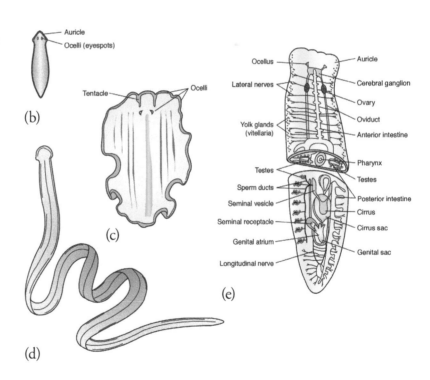
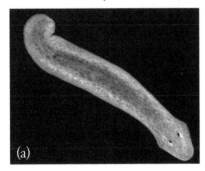

The marine types, composing the majority of species in this class, are bottom dwellers that live in sand or mud, under stones and shells, or on seaweed. Many marine turbellarians are common members of the **interstitial fauna**, a hidden world of tiny creatures that inhabit the water-filled spaces between grains of sand. The animals of this dynamic Lilliputian world represent nearly every invertebrate phylum. However, this complex, shifting labyrinthal realm is especially well suited to small worms or those creatures with a worm-shaped body.

Freshwater turbellarians live in lakes, ponds, streams, and springs where they are carnivores and scavengers feeding upon small invertebrate prey and decaying organic matter.

Another indicator of the evolutionary advance and environmental adaptability of the flatworm body plan is the fact that a few species of the genus *Bipalium* manage to live on land. Poriferans, cnidarians and most platyhelminths are totally dependent on water—salty or fresh—to bring them food and oxygen, to carry away their wastes, and to spread their gametes and fertilized eggs. The land planarians have managed to take an aquatically-oriented body plan and adapt it to terrestrial conditions. Although turbellarians are the epitome of bodily complexity among the members of phylum Platyhelminthes, they have a much simpler reproductive cycle than do the flukes and tapeworms.

Class Trematoda—Digenetic Flukes (Gr. *trematodes*, with holes + *eidos*, form)

The trematodes are endoparasites (internal) of all classes of vertebrates from fish to mammals including humans, our livestock, and our pets. Taxonomists recognize two subclasses of trematodes—Aspidogastrea and Digenea. Consisting of about 80-100 species, the Aspidogastreans with their simple life styles usually involving only one host, are primarily endoparasites on molluscs. The approximately 6,000 species of Digeneans are endoparasites on all types of vertebrates and have complicated reproductive cycles often involving at least two hosts.

Flukes are leaf-shaped or oval in form with adults ranging in length from approximately 0.2 mm to 6.0 cm (0.008 inch to 2.4 inches). (**Figure 9.4**) Flukes are structurally adapted to a parasitic lifestyle, being equipped with specialized glands for penetration or cyst production, suckers and hooks for attachment, and an increased reproductive capacity. In fact, egg production in trematodes is 10,000 to 100,000 times greater that it is for the turbellarians. Otherwise, trematodes share a number of structural characteristics and similarities with the turbellarians in their digestive system, excretory system, and nervous system as well as musculature and parenchyma that are only slightly modified from those of the turbellarians. Trematodes are said to be digenetic flukes because their complicated life cycle involves two or more host organisms, and their development is indirect as the larvae pass through several stages of development before becoming adults.

Class Monogenea—Monogenetic Flukes (Gr. *mono*, single + *gene*, origin or birth)

Traditionally, monogenetic flukes were placed as an order of Trematoda. Morphological and molecular data, however, indicate differences significant enough to warrant a separate class.

Monogeneans are all parasites, primarily of fish where they are found as endoparasites in the gut or as ectoparasites (outside) on the external surfaces or gills. A few types, however, turn up in the urinary bladder of frogs and turtles, and one species parasitizes the eye of the hippopotamus. Curiously, the monogeneans do not seem to cause serious damage to their hosts compared to the digenetic flukes.

Because they are often attached to the skin of a fast-moving host, monogeneans have a large posterior attachment organ, the **opisthaptor**, which bears hooks and suckers allowing the parasite to cling tenaciously to its host (Figure 9.4c).

(a)

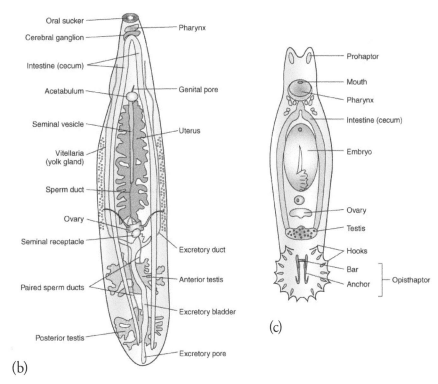

(b)

(c)

Figure 9.4 Flukes. (a) Photomicrograph of a stained liver fluke. (b) Generalized anatomy of a liver fluke. (c) *Gyrodactylus* sp., a monogenean ectoparasite on fish.

Unlike their digenetic (Gr. *di*, two + *gene*, origin or birth) kin, the monogenetic flukes (Gr. *mono*, one + *gene*, origin or birth) usually require but a single host (mono) in their life cycle and the development of their egg is direct lacking the developmental stages that mark digenetic flukes.

There are several interesting variations of the general reproductive style of most monogeneans. *Gyrodactylus elegans* produces four succeeding generations from one ova (egg). An adult *G. elegans* gives birth to a single live young. In this young is another embryo and this embryo contains another even smaller embryo. As they grow and develop, each one of these animals gives birth to the embryo within it. This, however, is not, in fact, four generations but four sisters who all develop from the same egg, the biological equivalent of a wooden puzzle box.

Another unique monogenean is *Diplozoon paradoxum*. Juvenile members of this species do not become sexually mature until they encounter another member of the same species of the opposite gender. They then achieve sexual maturity and mate. Unlike most animals, however, they remain mated for life—literally and physically. After their first mating, the two animals form a permanent union attaching themselves together at their midsections. And, thus they remain for the remainder of their lives resembling a cross with movable arms.

Class Cestoda (Gr. *kestros*, girdle + *eidos*, form)

The approximately 8,000 species of cestodes (tapeworms) are all endoparasites with the majority adapted for living in the guts of various kinds of vertebrates. As with trematodes, there are two subclasses of cestodes—Cestodaria and Eucestoda. Consisting of only a few species, the Cestodaria with their unsegmented and oval-shaped bodes are endoparasites in the body cavities and intestine of primitive fish. The Eucestoda conform more closely to the standard tapeworm body plan and are regarded as the true tapeworms.

> *To parasites, the body of another living thing is the buffet bus of life.*
> —Rob Dunn

On the flatworm branch of the traditional tree of life tapeworms were the last to appear. One might expect the terminus group of a phylum branch to be the most highly developed and complex group of that branch. That is certainly not the case with the cestodes, however, for these worms lack many of the accruements and structures that distinguish the turbellarians and flukes. As a result, humans often view the tapeworm as a "degenerate" or "primitive" class of worms. The tapeworm would undoubtedly scoff (if tapeworms could scoff) at such a notion. They might view themselves as the more advanced creature for they live a life of ease totally encased, warmed, fed and protected by the bag of flesh that is their host. As such, they require none of the excess organ baggage—digestive systems, brains and nervous systems and so on—required by other animals. Possibly living for decades, they face none of the problems and challenges presented by a free-living mode of existence. Do tapeworms deserve the last evolutionary laugh? Did they, in fact, take the wiser evolutionary path?

In reality, the ribbon-like tapeworms are essentially nothing more than a reproductive machine (**Figure 9.5**). Their body consists of three parts: a head region or **scolex** equipped with attachment structures like suckers and hooks, the **neck** immediately behind the scolex followed by the **strobila.** The strobila consists of many repeating individual sections known as **proglottids** with each proglottid containing both male and female reproductive organs and ducts. In fact, it has been proposed by some that a tapeworm is actually a colonial creature whose individual sections make up the colony. The first anterior section, the scolex, is modified for attachment whereas all the other individuals are modified for reproductive purposes. Even if this nontraditional view of tapeworms is not supported by the authors, it is presented here as another perspective for your consideration. The animal world is such a complex place that the more angles we view it from, the more we can learn about it.

Platyhelminth Body Plan

The evolution of bilateral symmetry and triploblastic cell layering almost certainly occurred in concert with the evolution of sophisticated internal organization of organs and organ systems. The addition of these features allowed for a far more active lifestyle than that of the cnidarians. These evolutionary advances allowed these newly adapted creatures to move around freely and thus exploit survival strategies that were previously impossible.

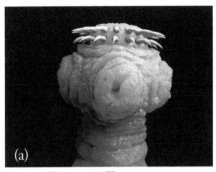

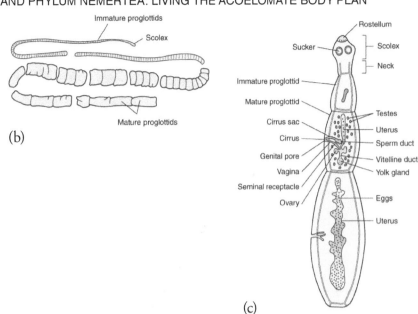

Figure 9.5 Tapeworms.
(a) Photomicrograph of a tapeworm scolex. The scolex burrows in to the intestine of the host where the hooks and suckers aid in the attachment.
(b) Generalized morphology of a beef tapeworm (*Taenia saginata*).
(c) Generalized anatomy of the dog tapeworm (*Echinococcus granulosus*).

This revolutionary body plan is not without constraints, however. An active lifestyle demands high energy (more food). That in turn requires a digestive system (possessed by platyhelminths except for cestodes), an efficient circulatory system (which none of the platyhelminths possess) to move digested food through the body, and a respiratory system (also not found in any platyhelminths) for O_2-CO_2 exchange. How did the platyhelminths overcome the lack of these crucial organ systems to not only survive and but to flourish? The answer lies in their flatness, for regardless of length, all platyhelminths are nearly paper-thin. The physical process of diffusion can move gases and materials around, but it is effective for doing so only over short distances. Thinness maintains the short distances required for diffusion to be effective. Flattening the body places the branched digestive system and digested food materials close to all body cells allows incoming O_2 and outgoing CO_2 to easily diffuse the short distances required to enter or leave through the body wall. Although some tapeworms can reach amazing lengths, one must never confuse length with complexity. The measure of a truly complex animal body plan that allows for a large volume and bulk is one that is marked with a circulatory system and a respiratory system.

The basic platythelminth body plan was an evolutionary revolution in animal structure when it first appeared and it has proven to be amazingly resilient since then.

Body Covering

Turbellarians are covered by a simple ciliated epidermis in which each cell bears many cilia. The swirling motion of particles close to the surface of free-living flatworms is responsible for their class name Turbellaria, which means "whirlpool." Gland cells and sensory nerve ending are distributed through the epidermis. Beneath the epidermis is a fibrous **basement membrane**, which is often thick enough to lend some structural support to the body (**Figure 9.6**).

Typical of almost all turbellarians are numerous membrane-bound, rod-shaped secretions known as **rhabdoids**. The most common kind of rhabdoid is the **rhabdite**. Reaching the surface through intercellular spaces, rhabdites produce copious amounts of mucus that may help protect the animal from **desiccation** (drying) and possible predators. Rhabdites may also be responsible for the release of noxious chemical defenses employed by some turbellarians.

The flukes and tapeworms possess an external covering known as a **tegument**, formed of nonciliated extensions of large cells within the mesenchyme (**Figure 9.7**). The tegument not only provides protection, but it is the site of exchange between the inner body and the outside environment; everything moves into or out of these worms across the tegument. In fact, in tapeworms, the ingestion of nutrients occurs solely across the tegument. The absorptive surface area of the tegument is greatly increased by many tiny folds or **microtriches** comparable to the villi of our own intestines. The tegument of flukes and tapeworms is a living outer layer opposed to the non-living cuticle of some invertebrates you will be introduced to later in this book.

How can tapeworms or indeed any internal parasite thrive in an environment of strong hydrolytic enzymes (powerful digestive chemicals)? This is an intriguing question that has been little investigated. One possibility is that gut parasites produce enzyme inhibitors ("antienzymes"). *Hymenolepis diminuta*, a tapeworm common in rats and mice, has been shown to release proteins that inhibit trypsin activity. This tapeworm can also regulate the pH of its immediate

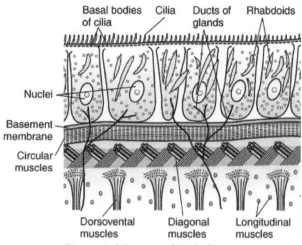

Figure 9.6 Section of Turbellarian epidermis and body wall.

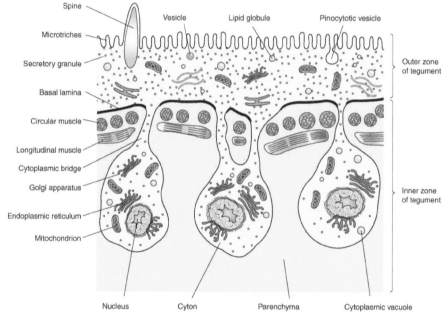

Figure 9.7 Section through the body wall and tegument of a digenean fluke.

environment to about 5.0 by excreting organic acids that may also inhibit trypsin activity. The tegument also provides some mechanical protection against powerful digestive enzymes and acids.

Support and Movement

Only a very few flatworms possess any sort of skeletal rudiments. In a few turbellarian species, tiny calcareous (calcium carbonate) plates or spicules are embedded in the body wall. Body support in all other flatworms is provided by the hydrostatic qualities of the mesenchyme and parenchyma, the general body musculature, and the elasticity of the body wall.

Locomotion is powered by a series of muscle fibers located beneath the basement membrane. These muscle fibers run circularly, longitudinally, and diagonally throughout the worm. (Figure 9.6). Most turbellarians glide along on cilia on their ventral surfaces. Mucus provides lubrication as the worm moves, and serves as a viscous medium against which the cilia can act. Turbellarians can twist and turn their bodies through muscular action allowing them to move and steer through interstitial spaces and debris.

Adult flukes lack external cilia, and movement depends on the fluke's body wall muscles or the movement of the body fluids of the host. Some larval stages do possess cilia, however, and are highly motile. Adult tapeworms have little need to move at all, but they are capable of muscular undulations of the body.

Once established within a host, it is advantageous for parasitic flukes and tapeworms to remain more-or-less in one place. To achieve a stability of position, flukes and tapeworms are equipped with a variety of attachment devices. The monogenetic flukes typically have both an anterior and posterior adhesive organ called the **prohaptor** and **opisthaptor** respectively. The prohaptor consists of a pair of adhesive pads, one on each side of the mouth. The opisthaptor on the posterior end is usually larger than the prohaptor and is equipped with one or more well-developed suckers armed with hooks or claws. The digenetic flukes possess two hookless suckers. The **oral sucker** surrounds the mouth whereas the other, the **acetabulum**, is located on the ventral surface. Tapeworms remain attached to the host's intestinal walls by the hooks, spines, and suckers of the scolex (**Figure 9.8**).

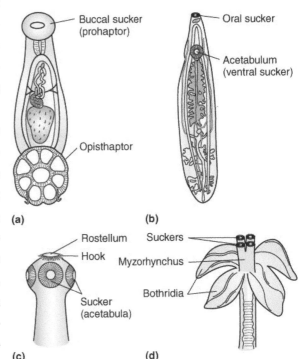

Figure 9.8 The Attachment Organs of Flukes and Tapeworms. (a) A monogenetic fluke. (b) A digenetic fluke. (c-d) Different types of cestode scolices. The foliose bothridia often have suckers at their anterior ends.

Feeding and Digestion

The term "free-living" denotes creatures that are not parasitic. In the phylum Platyhelminthes that term can be applied to the majority of turbellarians. Most turbellarians are carnivores or scavengers feeding on nearly all available animal matter, living or dead. A few types are herbivores on algae and some species switch from herbivory to carnivory as they mature.

It should be noted that there are a handful of turbellarian species known to be symbiotic with other invertebrates. Some of these are simply commensals with structural modifications for temporary attachment whereas others feed upon their hosts, causing various degrees of damage.

The prey of free-living turbellarians includes almost any invertebrate small enough to be captured and ingested (e.g. protists, rotifers, insect larvae, small crustaceans, nematodes, and annelid worms). In some turbellarians such as planaria, feeding behavior is triggered by the detection of substances emitted from the potential food sources (**chemoreception**). Prey is subdued by producing sticky mucal secretions from mucous glands and the rhabdites. In addition to entangling the prey, the mucous may contain poisonous or narcotizing chemicals. A few flatworms harpoon their prey with a sharp **stylet** located on the copulatory organ.

The general plan of the turbellarian digestive system includes a mouth and a pharynx leading to a blind intestine. The location of the mouth, the type of pharynx, and the amount of branching of the intestine are highly variable among the turbellarians. In some, the mouth is on the anterior end but in others it resides on the posterior end. Some, like the common planaria, have their mouth midventral (**Figure 9.9**). Most turbellarians have a blind intestine with but a single opening (mouth) and no anus. Thus, indigestible food materials must be excreted back out through the pharynx.

Once prey is captured and subdued, the pharynx goes into action. In some turbellarians, the pharynx is everted over or into the food source where it secretes digestive enzymes that partially digest the prey to a soupy consistency prior to swallowing. Many other types swallow their prey whole by the action of powerful pharyngeal muscles. Digestion is finalized in the intestine with the products of digestion being dispensed throughout the body by diffusion.

Adult flukes actively feed on the host's tissues and bodily fluids and possess a digestive system and digestive processes similar to those of the turbellarians.

Tapeworms lack any vestige of a mouth or digestive tract, and all nutrients must be absorbed directly across the tegument. Uptake of nutrients most likely occurs by pinocytosis and diffusion across the increased surface area of the many microtriches.

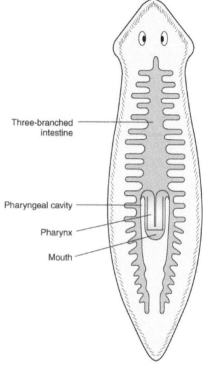

Figure 9.9 Generalized Digestive System of a Turbellarian. Turbellarians and flukes have a single opening digestive tract, whereas tapeworms lack a digestive system completely.

Circulation and Gas Exchange

The platyhelminths lack any specialized circulatory or respiratory structures. This forces these worms to rely solely on diffusion to move digested materials through the body and to power the O_2-CO_2 exchange process. Diffusion, however, is effective only over short distances. This lack of systems and the shortcomings of diffusion as a circulatory mechanism have necessitated two major evolutionary adaptations in turbellarians: a highly branched intestine and a very flat body. A highly branched intestine ensures that no cell is ever further than effective diffusion distance from digested food, and a very flat body ensures that the insides are never far from the outside and the life-giving oxygen out there.

Excretion and Osmoregulation

One major structural advance of flatworms over the diploblastic cnidarians can be seen in the development of an excretory system. The system consists of single or paired **protonephridia** connected to a network of closed collecting tubules that lead to one or more excretory pores called **nephridiopores** (**Figure 9.10**).

Within the protonephridia is the **flame cell**. The flame cell has a nucleated cell body with a tuft of cilia projecting downward into a cup-shaped projection. The beating cilia resemble tongues of flame, and their actions bring water and wastes into the canals for eventual elimination.

Turbellarian protonephridia function primarily as **osmoregulators** (excess water removal). The protonephridia remove the excess water which collects in the tubules and is eventually eliminated out through the nephridiopores. Nitrogenous metabolic wastes, mainly in the form of ammonia, are excreted across the body wall into the surrounding water by diffusion.

Flukes also possess protonephridia that connect to two **nephridioducts**. In contrast to the many nephridiopores of turbellarians, flukes collect excess water in a **bladder** and that water is eliminated through a single posterior nephridiopore (monogenetic flukes) or a pair of anterior pores (digenetic flukes).

Tapeworms have numerous protonephridia that drain to nephridioducts running along the dorsal and ventral sides. In adult tapeworms, the nephridioducts open separately to the outside on the margin (edge) of the hindmost proglottid. The exact functioning of the protonephridia in cestodes has yet to be determined, but the evidence seems to indicate that they function in the excretion of both metabolic wastes (excretion) and excess water (osmoregulation).

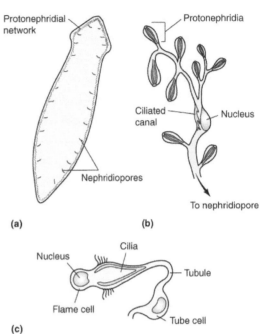

Figure 9.10 Generalized Excretory System of a Turbellarian. (a) The network of protonephridia and nephridiopores in a freshwater turbellarian. (b) A number of protonephridia and their common collecting tube highly magnified. (c) Structure of a flame cell.

Nervous System and Sense Organs

The active free-living lifestyle of the turbellarians demands a more advanced and centralized nervous system and sense organs than is required by the sedentary or slow-moving existence of the cnidarians. As one would expect from a bilateral animal with unidirectional movement, the turbellarians have a nervous system and sense organs that are centralized in the anterior (head) end of the worm. In the turbellarian head, we find a well-developed **cerebral ganglion** (rudimentary brain). From the cerebral ganglion, two longitudinal nerve cords run the length of the body. These longitudinal nerve cords are joined at regular intervals by transverse (crosswise) connections which together give the system a ladder-like configuration (**Figure 9.11**).

Whereas tactile (touch) receptors are abundant over much of the body surface as sensory bristles projecting from the epidermis, most turbellarian sense organs are concentrated in the

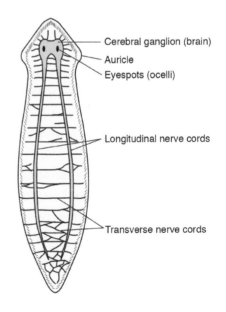

Figure 9.11 The generalized nervous system and sense organs of a turbellarian.

head end of the worm. Most turbellarians are equipped with **chemoreceptors** that aid in the location of food. Some species, such as the familiar planarians, have chemoreceptors in spike-like protrusions called **auricles** on either side of the head. Other types have these chemical detectors in ciliated pits, on tentacles, or distributed more-or-less over the whole anterior end.

Statocysts (balance organ) are common in swimming or interstitial species in which orientation to gravity could not be accomplished by sense of touch. Some turbellarians orient to water movements by employing **rheoreceptors** (current direction) located on the sides of the head.

Most turbellarians also possess **photoreceptors** in the form of **eyespots** or **ocelli** which detect the presence of light, but do not form images (Figure 9.11). Most free-living turbellarians are creatures of shade and shadow that avoid light (negatively phototaxic). The placement and orientation of the eyespots facilitates not only the detection of light intensity but the direction of the light as well. When viewed through a microscope, the location of the eyespots gives the common planaria a comical "cross-eyed" appearance due to the placement of the pigment within each eyespot.

The nervous system of flukes is also ladder-like and very similar to that found in many turbellarians. As might be expected with an attached parasitic animal, the sense organs of flukes and tapeworms consist nearly entirely of only tactile receptors associated with the organs of attachment.

Regeneration and Reproduction

Asexual reproduction is common among freshwater and terrestrial turbellarians, and is accomplished by either **transverse fission** (splitting across) or budding. The common freshwater planarians split in half behind the pharynx, (a process known as "dropping tail"); each half goes its own way eventually regenerating the lost parts (**Figure 9.12**). Budding occurs in several turbellarian genera whose species are quite small. The buds, known as **zooids**, differentiate along the length of the parent's body and form a chain before separating into new individuals.

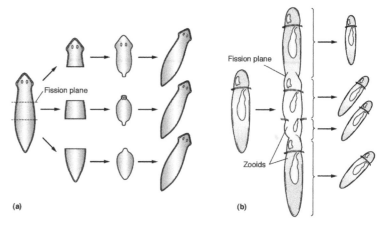

Figure 9.12 Asexual Reproduction in Turbellarians. (a) Transverse fission in a planarian. (b) Budding in *Catenula lemnae*.

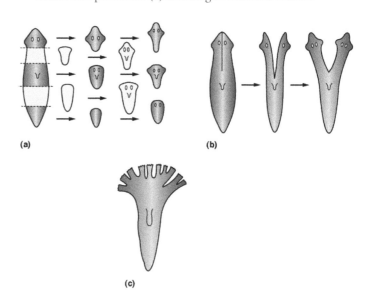

Figure 9.13 Regeneration in Planaria (*Dugesia*). (a) Each piece regenerates, but the rapidity with which the head develops depends on the location and size of the piece. (b) Polarity of regeneration shown by splitting the anterior end. (c) Repeated splitting of each anterior end can result in a worm with many heads.

The remarkable regenerative powers of the turbellarians have been intensely studied for some time, especially the freshwater planarian *Dugesia*. What has become apparent is the fact that the cells of *Dugesia* are not **totipotent** (the ability of a single cell to divide and produce all the differentiated cells in an organism). Instead, there exists an anterior to posterior polarity in terms of the regenerative capabilities of the cells. A piece taken from the middle of the body always regenerates a new anterior and posterior end (**Figure 9.13**). If a small piece of the tail is separated from the rest of the body, however, the larger piece will grow a new posterior end, but the small tail piece lacks the capability of producing an entire new anterior end. There is also a lateral polarity. This is demonstrated by the fact that a two-headed (or more) planaria can be produced by cutting the anterior end along the midline. This gradient of cell potency has intrigued cytologists and medical researchers for some time because of its relevance to wound healing and the potential for regeneration in higher animals.

Box 9.1

Lost a Piece? Grow it Back: The Wonders of Regeneration

Humans have long been intrigued by the possibilities of accelerated healing of wounds and even the complete regeneration of damaged or diseased organs. In John Milton's classic work, *Paradise Lost* (Book VI), the character of Satan possess the power of accelerated healing. In more modern times, Wolverine of X-Men comic and film fame also has the power of accelerated healing and in the television series Heroes, the characters of Claire Bennett and Adam Monroe can both "regenerate." According to the story line, this ability has allowed Monroe to live for about 400 years without apparent aging.

In biological parlance, an organism is said to have regenerated a lost or damaged part if the part regrows so that the original function is restored. The capacity to regenerate seems to be inversely related to complexity; in general, the more complex an animal is the less regeneration it is capable of. As discussed in this chapter, planaria have an amazingly enhanced capacity for regeneration. This power seems to be the result of clusters of stem cells retained within the bodies of adults. When damaged occurs, these stem cells migrate to the parts of the body that need repair then divide and differentiate to replace the required missing or damaged tissue.

Newts can regenerate severed limbs, but the regenerative process takes a different pathway than in planaria. In these amphibians, limb regeneration occurs in two major steps, first some adult cells redifferentiate back into a stem cell state similar to embryonic cells, and then these cells differentiate to replace the required missing or damaged tissue.

Mammals, on the other hand, have very limited abilities when it comes to regeneration. Each of us has minor and perhaps major examples of how cuts, abrasion, and even surgical incisions will heal, but if you sever a finger, prepare to live the rest of your life with an odd number of digits. The human liver demonstrates the greatest vigor for regeneration of any of our internal organs being able to regenerate fully from as little as 25% of its original tissue. This ability is due to hepatocyte cells in the liver. Loss of liver tissue induces the remaining hepatocytes to proliferate until the lost mass is restored.

Biomedical researchers are diligently investigating any and all avenues that might someday allow humans greater powers of regeneration. Imagine a world where those with spinal cord injuries are not doomed to entrapment in a body they cannot control, where our military combatants no longer face a life of impairment due to traumatic brain injury, or where a child with a failing heart is given the miracle of growing old. These possibilities and more await if only researchers can uncover our inner planaria.

Turbellarians are hermaphroditic (monoecious) but undergo cross-fertilization. There is typically simultaneous and mutual copulation with the penis of each worm being received by the female system of the other worm. Some turbellarians have two penises and one or more genital pores for receiving a unique two-tailed sperm delivered during copulation. Others have a penis modified into a sharp, hollow stylet. Sperm transfer in such species is by **hypodermic impregnation**, whereby the copulating partners stab each other with their penis, injecting sperm through the body wall of the other. The marine flatworm *Pseudobiceros hancockanus* engages in an odd reproductive behavior that has been dubbed **penis fencing**. During penis fencing, each flatworm tries to pierce the skin of the other using one of its penises. The first to succeed becomes the de facto male, delivering its sperm to the other, the de facto female. This is serious business to these worms and mating is truly a battle because the worm that assumes the female role must then expend considerable energy caring for the developing eggs.

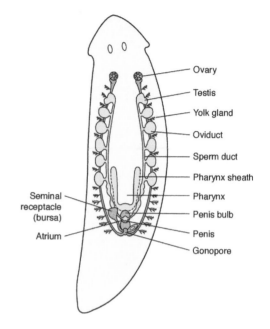

Figure 9.14 Generalized turbellarian reproductive system.

A typical male system (**Figure 9.14**) usually contains many pairs of testes. A small duct connects each testis with a **main sperm duct** that extends along each side of the body. The sperm ducts join to form an ejaculation duct, which exits through the **penis**. The penis is commonly located in a chamber, the **atrium**, which may also contain the terminal portion of the female system.

The female system usually contains many pair of **ovaries**, but only a single pair of **oviducts** is present. (Figure 9.14) A **seminal receptacle** (or copulatory bursa) that stores sperm, a **vagina** or **atrium**, and glands for the production of egg envelopes are typically present. In many turbellarians **yolk glands** are located along the length of the oviduct. Yolk cells are released as the egg travels down the oviduct, and the eggs, when deposited, are surrounded by yolk.

Because of their small size, turbellarians produce relatively few eggs, and these precious eggs are never recklessly spawned. Some freshwater turbellarians produce two kinds of eggs: **summer eggs** which have a thin shell and hatch rapidly, allowing for a rapid population increase and **winter** or **dormant** eggs which have thick, resistant shells and are capable of withstanding cold and desiccation. Development of the egg is direct and the young hatch as juveniles.

Flukes demonstrate a great deal of variation in the details of their reproductive system, but most are constructed around a common plan similar to that of certain turbellarians. Like the turbellarians, flukes are hermaphroditic and typically engage in mutual cross-fertilization to transfer sperm. Self-fertilization can occur but only in rare cases. Unlike their turbellarian kin, egg development in flukes is indirect, involving one or more hosts and independent larval stages and flukes produce 10,000 to 100,000 times more eggs than do turbellarians. The fertilized eggs of flukes pass out of the host in its feces.

It has been said "the business of animals is reproduction." If this be true, no other group of animals demonstrates it more clearly than do the tapeworms. Each proglottid of a tapeworm contains a complete reproductive system packed with as many as 100,000 eggs with seven to nine of the terminal fertile proglottids released *each day*! If you do the math, one cannot help but view such a staggering reproductive potential with awe.

Cross-fertilization in tapeworms is the rule where there is more than one worm in the host's gut, but self-fertilization between two proglottids in the same strobila or even within the same proglottid is known to occur. The terminal proglottids, packed with fertilized egg capsules, break away from the strobila and exit in the host's feces. All the while new proglottids are constantly being formed just behind the scolex in the **germinative zone**.

Platyhelminth Connection

Economically

According to some health authorities an estimated 250-300 million people worldwide are infected with some type of parasitic flatworm. These estimated Figures, however, are regarded by some experts as being far too low. These insidious worms ply their biological trade primarily in the underdeveloped countries and regions of the tropics and subtropics where they seriously impact the health, social structure, and economic development of areas where they are especially prevalent. The slow downward spiral of the health of individuals infected with these worms translates into untold billions of dollars in increased health costs and lost productivity worldwide yearly. These worms can also have a very negative effect on the health and productivity of those animals that we raise as livestock and those that we cherish as pets. It is estimated that parasitic worms decrease feed efficiency in cattle, turkeys, and chickens by up to 10% with cattle losing as much as a pound of gain a day to such worms.

Ecologically

As a result of their scavenging behavior, the freshwater planarians play a positive ecological role in recycling dead animal matter. In addition, planarians serve both as predator to small invertebrates and prey to larger animals in aquatic food chains. Research shows, however, that these small and apparently defenseless worms are not the easy mark for predators they appear to be. Some chemical defense is thought to help planarian hold most predators at bay.

A terrestrial flatworm introduced to Ireland from New Zealand is eating through the annelid (segmented worms) fauna of the Emerald Isle at an alarming rate. The ecological damage to Ireland, its soil, its crops, and its natural landscape has yet to be fully calculated and understood, but the situation is worrisome. The wandering broadhead planarian, *Bipalium adventitium*, is now widespread and abundant in North America, but data are sketchy at best as to the exact impact they are actually having here.

Medically

The greatest impact parasitic worms have is on the lives and health of those unfortunate individuals infected with them. The vast majority of those infected are found in countries or geographical areas where clean water and proper sewage treatment may not exist nor may there by adequate fuel to fully cook food, especially meat.

Host humans plagued with worms suffer effects that range from nausea, anemia, diarrhea and weight loss, to enlargement of the liver and spleen, lung damage, bleeding of the bladder, and even slow death. Unfortunately, these complications can manifest themselves for a long time as flukes can live as long as five years and tapeworms even longer (**Table 9.1**).

Table 9. 1
Pathogenic Flatworm Parasites of Humans

Parasite	Adult Location in Host	Pathology	Transmission
Trematodes			
Schistosoma japonicum	Veins of small intestine	Causes schistosomiasis enlargement of liver and spleen, destruction of intestinal tissue, general debilitation, toxic reactions	Cercariae (larvae) released from snails penetrate the skin
Schistosoma mansoni	Veins of large intestine	Similar to S. *Japonicum* except large intestine is more affected than small intestine	Cercariae (larvae) released from snails penetrate the skin
Schistosoma haematobium	Veins of urinary bladder	Painful urination; bleeding and ulceration of bladder wall	Cercariae (larvae) released from snails penetrate the skin
Opisthorchis sinensis	Bile ducts	Gastrointestinal disorders; damage to bile ducts	Eating raw fish containing encysted metacercariae
Paragonimus westermani	Lungs	Some respiratory tissue damage with coughing, chest pain, excessive mucus production	Eating freshwater crabs and crayfish containing encysted metacercariae
Cestodes			
Diphyllobothrium latum	Small intestine	Anemia and various gastrointestinal disorders	Eating freshwater fish containing intermediate stage
Hymenolepis nana	Small intestine	Mild toxic reactions	Ingestion of eggs passed in rat, mouse, or human feces
Taenia solium	Small intestine	Cysts in brain, skeletal muscle, heart; reactions to adult worm may include diarrhea and weight loss	Ingestion of cysticerci in raw or undercooked pork or ingestion of eggs passed in feces
Taeniarhynchus saginatus	Small intestine	Nausea, abdominal pain, weight loss, diarrhea	Ingestion of cysticerci in raw or undercooked beef
Echinococcus granulosus	Cysts in liver or other organs or tissues	Various reactions to increasing size and toxicity of cyst	Ingestion of 6995 passed in feces

Flukes tend to inflict greater harm to their host than do tapeworms. Adult *Taenia* tapeworms are found passively attached only in the small intestine whereas flukes, depending on the species, are found in the veins of the small and large intestine, veins of the urinary bladder, bile ducts, and lungs. Flukes feeding aggressively on bodily tissues and fluids can clog ducts with their numbers and can trigger the formation of bile stones resulting in **hypertrophy** (excessive enlargement) of the liver. Flukes have complicated life cycles as illustrated by the life histories of the Chinese liver fluke (*Clonorchis sinensis*) and the blood fluke (*Schistosoma mansoni*).

The lungs, bile ducts, pancreatic ducts, and intestines are common sites of trematode infection. The Chinese liver fluke takes up residence in the bile ducts of humans, dogs, cats, and pigs. The intermediate hosts penetrated first by miracidium and then by cercaria are a snail and a fish, respectively. Human infection has been common in the Orient because human feces are often used to fertilize fish ponds and because the fish in these ponds are often eaten raw, in part

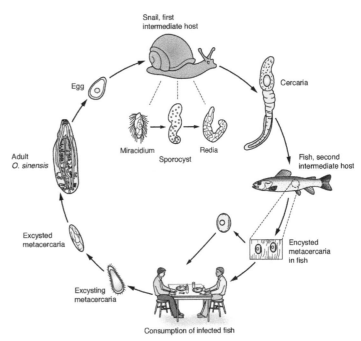

Figure 9.15 The life cycle of the Chinese liver fluke (*Clonorchis sinensis*)

by preference and in part because of limited cooking fuel. A few worms cause no problems or symptoms, but several hundred can cause destruction of liver tissue, clogging of ducts, formation of bile stones, and hypertrophy of the liver. Millions of people in Asia are believed to be infected with this species of trematode (**Figure 9.15**).

The members of three families of flukes live in the vertebrate circulatory system. Of these blood flukes several species of the genus *Schistosoma* infect humans and are responsible for the terribly debilitating disease **schistosomiasis** or **bilharzia** after the German pathologist Theodor Bilharz who first realized the cause of the disease was parasitic worms.

The adults of *Schistosoma* are about 2 cm (0.7 inch) long and about 1 mm (0.04 inch) long. The sexes are separate, and the narrower female fits into a longitudinal groove on the male. Depending upon the species, they live in the veins of the urinary bladder, small intestine, or large intestine. Fertile eggs break out of the veins into the lumen of the intestine or bladder. Here the eggs are passed out of the host with feces or urine, and if deposited in water, hatch as a miracidium (**Figure 9.16**). The miracidium enters certain species of freshwater snails and passes through several generations of sporocysts that eventually give rise to **cercariae**. The cercariae leave the snail and on contacting a human wading or bathing in a stream or pond, penetrate the skin, using enzymes and muscular movement. They are then carried by the circulatory system to the lungs, liver, and eventually the intestinal veins. Serious tissue damage can be caused by the passage of eggs into the lumen of the intestine or bladder, the lodging of eggs in uncharacteristic sites, and larval development in the lung and liver. Schistosomiasis ranks with malaria and hookworm infections as one of the three great scourges of humans. Some 300 million people in tropical Africa, the Caribbean, eastern South America, eastern Asia, and the Middle East are thought to be hosts to these menacing parasites.

Flukes of one type or another are found in parts of South America and the Caribbean, the Middle East, Africa, and Southeast Asia. In North America (excluding parts of Mexico), proper water sanitation and sewage treatment as well as the overall cleanliness of the food supply, insure that fluke infestation in humans is a rare occurrence. Thus in North America, these sinister worms are confined primarily to large herbivores, both domestic such as sheep, goats, pigs, cattle and horses and wild such as deer, moose, and elk but can also

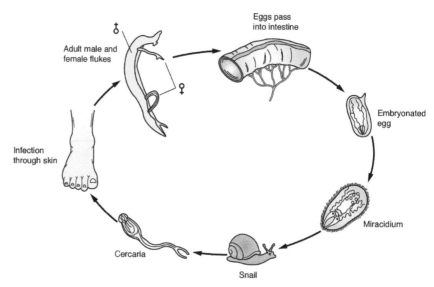

Figure 9.16 The life cycle of *Schistosoma mansoni*, a trematode fluke that causes schistosomiasis in humans.

be found in amphibians, reptiles, birds, fish and snails. Fluke infestation in dogs and cats are rare compared with infection rates for nematodes and tapeworms. In general, flukes do not cause serious effects in cats and dogs unless they are present in high numbers.

No parasite, however, has burrowed into our collective psyche quite like the tapeworm. They are the stuff of horror stories and the mere mention of one is enough to make most people squirm. Yet tapeworms are much more host-friendly than are flukes. Attaching to our small intestines, they do not feed on us but rather absorb the nutrients we have digested for them. Problems arise when the tapeworm(s) becomes too large and starts blocking the bowel or robbing us of vital nutrients, especially vitamins such as B 12. One strange symptom of tapeworm infection in humans is an enlarged abdomen. A large lump of tapeworm protein in the gut generates an immune response from the host human and the **ascites** (accumulation of excess fluid in the abdomen) that results gives the infected person a pot belly even though they may actually be malnourished.

Like flukes, tapeworms have complicated life cycles requiring an intermediate host(s) as illustrated by the life history of the beef tapeworm, *Taenia saginata* (**Figure 9.17**). If an infected person defecates in a pasture or if human feces are used as fertilizer, the eggs may be eaten by grazing cattle, sheep, or goats. On hatching within the intermediate host, an **oncosphere larva** bores into the intestinal wall, where it is picked up by the circulatory system and transported to striated (skeletal) muscle. In the muscle, the larva develop into a **cysticercus** sometimes known as a bladder worm. If raw or undercooked beef is ingested by humans, the cysts breaks open, the scolex evaginates, and the larva begins to develop into an adult worm within the gut. A severe infection of adult tapeworms may cause diarrhea, weight loss, and reactions to the toxic wastes of the worm.

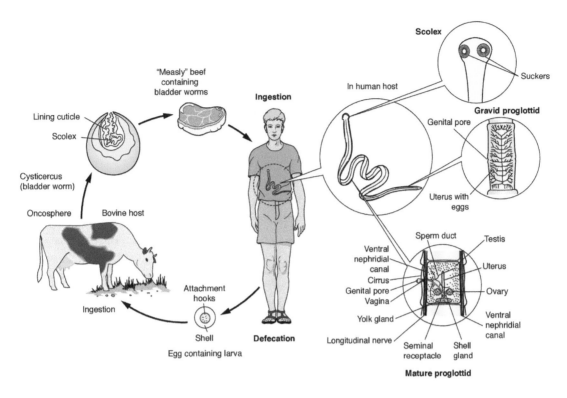

Figure 9.17 The life cycle and reproductive structure of the beef tapeworm (*Taenia saginata*).

From the human perspective, the nastiest of the tapeworm bunch are the tissue tapeworms also known as hydatid tapeworms. These tapeworms do not reach an adult stage in the human host. Instead, they infect humans as their intermediate or cyst stage. In North America, humans usually ingest tapeworm eggs from dog feces. These eggs then begin to form cysts typically found in the liver and lungs, but they may also be found in the brain (**Figure 9.18**). As they grow larger, they gradually replace the tissue in which they grow. Worse still, should a cyst rupture, the release foreign parasite proteins into the body can trigger a massive and sometimes fatal allergic reaction. The *Journal of Forensic Sciences* (Kok, A.N. 1993 A Sudden Death Due to a Hydatid Cyst JFSCA 38, 978-980) features a report on a young Turkish teenager who collapsed and died suddenly after being tackled in a football game. An autopsy revealed that he had a

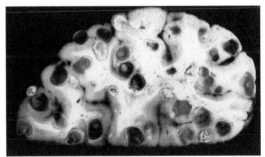

Figure 9.18 A human brain riddled with tapeworm cysts.

single golf ball-sized hydatid cyst in his liver which was ruptured during the tackle. The ruptured cyst triggered a massive allergic reaction which killed him in minutes. The only means of treating hydatid cysts is through careful surgical removal.

Adult tapeworms parasitize primarily vertebrates, and there are species adapted to each class of vertebrate—fish, amphibians, reptiles, birds, and mammals, including humans, our livestock, and our pets. There are nine major species of *Taenia* tapeworm in North America, seven of which have cats or dogs as their **definitive host** (animals that harbor the adult form of the parasite). The intermediary hosts range from lice to fleas and snails to other classes of vertebrates, including humans.

Although certainly worse in some areas than others, these worms are cosmopolitan across the globe and are stealthy in nature with their eggs and larva hidden in fish, veiled in pork, and buried in beef. What then is one to do in order to prevent becoming infected by flukes and tapeworms and what of our dear pets? Our stoutest defense against these worms remains our supply of clean water and the proper treatment of sewage along with the sanitary handling and inspection of our food supply. As well as those defenses have worked in the past and continue to work today; they may not be enough. Consequently, health authorities strongly advocate the following to ensure you (and your pets) remain worm-free:

- Thoroughly wash all fresh fruits and vegetables if you do not know their origin. In the global economy of today, much of the fresh produce we purchase may come from beyond our shores where animal manure and human sewage may be used as fertilizer. Feces of any kind can carry worm eggs so err on the side of caution and scrub your produce.
- Provide adequate and appropriate preventative endoparasitic (worms) and ectoparasitic (fleas, lice, and ticks) medication for your pet cats or dogs. Many parasitic worms are **zoonotic** in nature in that they are not species specific and can infect both animals (your pets) and humans (you).
- Never let your pet cat or dog lick you on the face. Although your pets may receive adequate worm control medication, other pets may not and any wild animals in your area—foxes, coyotes, raccoons, etc.—certainly do not. If your pet comes in contact with the powdered feces of an infected animal, they may pick up worm eggs on their fur. Animals often groom themselves by licking their fur. Get the picture?
- Wash your hands after handling *any* animal.
- Cook all meat until it is well done. Think twice before eating raw fish—sushi, sashimi, or cerviche—as you face an increased risk of acquiring not only parasitic worms but also pathogens like Hepatitis A and food poisoning when you do so, especially if you do so in a foreign country. Although the World Health Organization estimates that 40 million people worldwide are infected by foodborne parasitic worms and that raw or undercooked fish is most frequently the carrier, the risk of acquiring these worms in this country or in Europe is significantly lower because the fish will have been frozen and freezing kills the worms. Trained sushi chefs have also learned how to identify and remove worms or worm eggs in fish before preparation. In addition, reputable packing houses use a process called **candling** (holding up fish filets up to bright light) to check for worms. The bottom line is that when you eat raw or undercooked meat of any kind, you are rolling the parasitic dice.

PHYLUM NEMERTEA

History of Nemertea

The origin of the nemerteans is puzzling, and unfortunately the fossil record is of no use in establishing the foundation of these worms. One must also take into account the unsettled controversial position of platyhelminths on the Tree of Life as well as the exact origin of the acoelomate condition. One view is that nemerteans arose from early turbellarians stock and that the platyhelminths and the nemerteans constitute a true monophyletic clade, both exhibiting the acoelomate condition (**Figure 9.19**).

Diversity and Classification of Nemertea

The phylum Nemertea (Gr. *nemerta*, sea nymph) comprises about 900 species of predominantly benthic marine worms. The many common shallow-water nemerteans live beneath shells and stones, in algae, or burrow into mud and sand. Some species form semipermanent burrows lined with mucous or distinct cellophane-like tubes. A few types, however, are planktonic, and some are symbiotic in mollusks and other marine invertebrates. There are a few freshwater and terrestrial genera

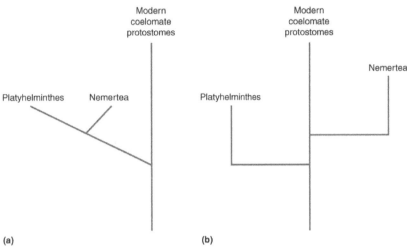

Figure 9.19 Several Different Hypotheses as to the Position of Nemertea on phylogenetic tree of life. Either (a) they are a monophyletic group or (b) they are not monophyletic.

known. The land varieties are confined to the tropics and subtropics. In contrast to the platyhelminths, there are only a few parasitic species of nemerteans.

Commonly known as ribbon worms or proboscis worms, nemerteans range in length from less than 1 cm (0.4 inch) to several meters (6-7 feet). Many can stretch easily to several times their contracted length. One appropriately named specimen, *Lineus longissimus*, was found to be nearly 60 meters (197 feet) long! Many ribbon worms are drab in appearance, but some are adorned with bright colors of red, green, purple, and orange and some bear distinctive stripes or markings (**Figure 9.20**).

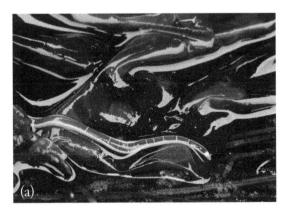

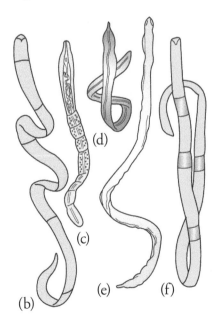

Figure 9.20 The Nemerteans. (a) Ribbon worm *Micrura verrilli* (b) The tube-dwelling *Tubulanus rhabdotus* sans tube. (c) An annulated sand-dwelling paleonemertean. (d) The burrowing *Caninoma tremaphoros*. (e) The burrowing *Cerebratulus lacteus*. (f) The crevice-dwelling *Lineus socialis*.

267

The earliest report of a nemertean, described as the sea long-worm, comes from the mid-eighteenth century. For nearly a century, most zoologists placed the ribbon worms with the turbellarian flatworms. It was not until 1851 that substantial evidence for the distinctive nature of the ribbon worms was published by Max Schultze, who described the unique features of these worms: an extensible proboscis, the presence of nephridia, and a true anus. It was not until the mid-twentieth century, however, that the ribbon worms were fully accepted as a valid and separate phylum.

The nemerteans may be classified as follows:

> Domain Eukarya
>> Kingdom Animalia
>>> Phylum Nemertea (Rhynchocoela)
>>>> Class Enopla
>>>> Class Anopla

Phylum Characteristics

Nemerteans and the platyhelminths are similar enough in body plan that these two groups are often viewed together as the triploblastic acoelomate bilateria. There are structural similarities in the over-all architecture of the nervous systems and types of sense organs as well as the structure of the excretory system. In other respects, however, the ribbon worms differ greatly from the platyhelminths: nemerteans possess an anus (a complete digestive system), a closed circulatory system, and an eversible (extendable) proboscis that sits within a hollow cavity called the **rhynchocoel** ("nose cavity"). The proboscis and rhynchocoel are unique to this phylum and represent its most distinguishing characteristic. Although we have chosen to go with the more traditional name for this phylum, there are some who have dubbed it phylum Rhynchocoela after the uniqueness of the rhynchocoel and proboscis.

The members of phylum Nemertea exhibit the following general characteristics:

- Their bodies are triploblastic, acoelomate, and bilaterally symmetrical. However, some zoologists contend that the rhynchocoel should be regarded as a true coelomic cavity.
- They possess a complete digestive tract with a mouth on one end and an anus on the other.
- Protonephridia function as organs of excretion.
- They possess a closed circulatory system.
- Their body spaces are filled with partially gelatinous parenchyma.
- They possess an eversible proboscis which lies free within the rhynchocoel.
- They possess a four-lobed brain connected to a pair of longitudinal nerve trunks.

Although variations exist, the "typical" nemertean may be viewed as an active benthic dweller that vigorously hunts invertebrate prey as it aggressively moves through the nooks and crannies of its habitat.

Class Characteristics

Class Enopla (Gr. *enoplos*, armed)

As their class name suggests, these ribbon worms have a proboscis armed with sharp stylets for spearing prey. Their mouth opens in front of the brain.

Class Anopla (Gr. *anoplos*, unarmed)

The proboscis of these ribbon worms lack stylets, and their mouth opens below or posterior to the brain. (It should be noted however, that some authorities question the validity of this class as a monophyletic group.)

Nemertean Body Plan

The basic body plan of nemerteans is presented in **Figure 9.21**.

Body Covering and Body Wall

The body wall of nemerteans consists of an epidermis, a dermis, relatively thick muscle layers, and a mesenchyme of varying thickness packed with partially gelatinous parenchyma. The epidermis is composed of ciliated columnar epithelium cells. Mixed in among the epidermal cells are sensory cells (tactile most likely) and mucous gland cells. Below the epidermis is the dermis and beneath that are well-developed circular and longitudinal muscles. Inside the musculature is a dense more-or-less solid mesenchyme filled with parenchyma.

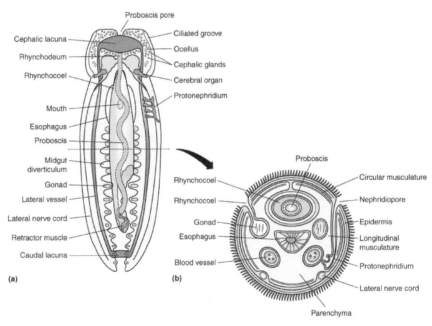

Figure 9.21 Basic Body Plan of Nemerteans. (a) dorsal view. (b) Cross-section.

Support and Movement

As with platyhelminths, the nemerteans lack any rigid skeletal elements. Instead, support is provided by the muscles and the hydrostatic properties of the parenchyma. The smaller forms are propelled by the action of epidermal cilia whereas larger types employ peristaltic waves of the body to propel them over moist surfaces or through soft substrata or undulatory swimming as a means of locomotion.

Feeding and Digestion

Most ribbon worms are active predators on small invertebrates (polychaete worms and amphipod crustaceans are especially favored), but some are scavengers. Well-endowed with chemoreceptors, some types are capable of tracking prey over considerable distances and may fire their proboscis along the trail ahead of them in hopes of snagging prey.

The proboscis apparatus is a complex arrangement of tubes, muscles, and hydraulic systems (**Figure 9.22**). This organ serves double duty in that it is used both for food capture and for manipulating the food to the mouth for ingestion. When prey is encountered, the proboscis shoots out and wraps around the victim. The prey is not only subdued by the proboscis but may also be killed by toxic secretions from the proboscis. Nemerteans possessing a probos-

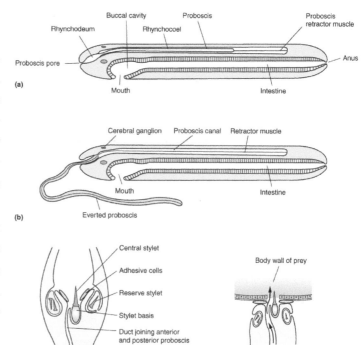

Figure 9.22 Nemertean Proboscis. (a) Retracted unarmed. (b) Extended unarmed (c) Retracted tip of proboscis. (d) Armed proboscis penetrating prey.

cis armed with stylets use the stylets to pierce the body of the prey and introduce the toxin. Once captured, the prey is drawn to the mouth by retraction of the proboscis where it is generally swallowed whole.

Unlike flatworms, nemerteans possess a one-way digestive system with a mouth on one end and an anus on the other. (**Figure 9.23**) The digestive system consists of a mouth that leads to a bulbous **buccal cavity**, a short **esophagus** and a **stomach**. The stomach empties into an elongated **intestine** bearing many **lateral diverticula** (side pouches). Digestion occurs solely within the system (extracellular) with digested nutrients then being carried throughout the body by the circulatory system. Indigestible materials move through the gut and eventually are eliminated out the anus.

Circulation and Gas Exchange

The ribbon worms are the first branch on the Tree of Life that we have so far encountered that possesses a closed circulatory system. This closed

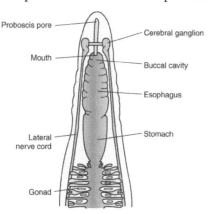

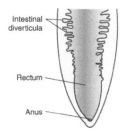

Figure 9.23 Nemertean digestive system showing anterior and posterior regions of the gut.

system consists of vessels and thin-walled spaces called **lacunae**. The blood is a colorless fluid that is forced around the body primarily by general body movements (no true heart).

Nemerteans lack a respiratory system, so gas exchange occurs solely across the moist body surfaces. As with any creature that lacks a respiratory system and is forced to "skin breathe," desiccation is a constant threat to both the platyhelminths and nemerteans, especially terrestrial forms. If they dry, they die.

Excretion and Osmoregulation

The excretory system of nemerteans consists of anywhere from several to thousands of protonephridia similar to those found in turbellarian flatworms. Given the close association of the protonephridia with the circulatory system (in some types the protonephridia are bathed directly in blood), it is likely that nitrogenous wastes (ammonia), excess salts, and other metabolic break-down products are removed from the blood and surrounding mesenchyme by the protonephridia. The protonephridia may also play a role in osmoregulation, especially in freshwater and terrestrial species.

Nervous System and Sense Organs

Although the basic organization of the nemertean nervous system resembles that of the turbellarian flatworms, it tends to be a bit more elaborate and complex. Nemerteans are also more highly cephalized than are the flatworms, especially in the placement of the mouth and feeding structures as well as sensory and other nervous system elements.

Reflecting on their active lifestyle, ribbon worms possess a variety of sensory receptors concentrated at the anterior (head) end. Ribbon worms are sensitive to touch and possess eyespots, but most of the sensory input important to them is in chemical form. These worms are very sensitive to dissolved chemicals in their environment and utilize this sensitivity in food location, mate selection, and substratum testing.

Regeneration and Reproduction

Many nemerteans demonstrate remarkable powers of regeneration. Larger species display such a tendency to fragment when irritated that it makes large, intact specimens difficult to collect. Species of *Lineus* engage in asexual reproduction on a regular basis by breaking along preformed lines (**annulations**). Smaller pieces may form a cyst in which they then regenerate whereas larger fragments may regenerate directly. (**Figure 9.24**)

The majority of nemerteans are dioecious with rows of gonads alternating with the intestinal diverticula to form a regular row on each side of the body.

Figure 9.24 Fragmentation and Regeneration in a Nemertean (*Lineus vegetus*). Each fragment regenerates into a complete worm.

After maturation of the gametes, a short duct grows from the gonad to the outside, allowing the gametes to escape. Each ovary produces 1 to 50 eggs depending on the species. The shedding of eggs or sperm does not necessarily require contact between two worms, although some species aggregate at the time of spawning or a pair of worms may occupy a common burrow or secreted cocoon. Fertilization is external in most nemerteans, and the eggs are either shed into the water or deposited within the burrow or tube, or in gelatinous strings.

A Closing Note
Eaten Alive

In the realm of weird and wonderful creatures that inhabit Animal World, parasites are king of the bizarre. No other beasts strike us with such fear and loathing as do parasites. Such feelings are well-placed as parasites have inflicted casualties and consequences on humanity that we may never be able to fully count or even fathom. Whereas the flukes and tapeworms detailed in this chapter certainly extract a heavy toll on their various hosts (globally, about one in every four people hosts parasitic worms), the most deadly and bizarre of parasites may be those beyond our sight. One single protistan type, the sporozoal *Plasmodium*, inflicts on humanity the scourge of malaria, a disease that in the whole of human history has killed as many as half the people who have ever lived. In fact, an estimated 250,000,000 malaria cases occur each year according to the World Health Organization, resulting in nearly a million deaths, mostly children under age 5.

We need to fear the parasite, for *Plasmodium* is not alone. Joined by a host of others, ranging from microscopic bacteria to 30-foot long tapeworms that attack our insides and those such as ticks and leeches that burrow in from the outside, they eat us alive and often disfigure and eventually kill us in the long run. Depending on how loosely you define the term parasite, every human that now or ever has walked the earth has had or currently has some sort of parasite on them or in them, including you.

The parasitic realm is inhabited by bizarre creatures but *Toxoplasma gondii*, a single-celled protistan parasite that lives primarily in the guts of cats but that may also be found in other animals such as rats and even humans, may be the strangest. *Toxoplasma* forms cysts throughout the intermediate host's body, including the brain. And yet the Toxoplasma-ridden host animal appears perfectly healthy and symptom-free. Scientists at Oxford, however, discovered that the parasites can change infected rats in one subtle but critical way—infected rats lose their fear of cat odor. The scientists speculate that that *Toxoplasma* secretes some substance(s) that alter the patterns of brain activity in the rats making them more susceptible to predation by cats, the preferred host of Toxoplasma.

When humans handle egg-laden soil or kitty litter, they too may play host to *Toxoplasma*. It has been estimated by some that perhaps half the people on the planet are infected but only those with compromised immune systems, such as in AIDS, or unborn fetuses show serious symptoms. This begs the question—could a microscopic parasite be subtly altering the behavior of millions or even billions of people? Various studies have recently linked *Toxoplasma* in humans to changes in personality, altered gender ratios and an increased risk of schizophrenia, prompting one researcher to proclaim that, "*Toxoplasma* has altered the cultural diversity of entire nations." Such hypotheses are highly controversial and strongly disputed by many scientists.

Fear them we should, and guard against them we must. However, one cannot help but respect and even admire the elegance of their design and the sophistication of their adaptive capabilities. As long as animals continue to walk, swim, and fly the planet, they will continue to be eaten alive by the magnificently adapted parasites.

In Summary

- The first bilateral animals (Bilateria) evolved into two separate branches on the Tree of Life—Protostomia and Deuterostomia. From the Protostomia branch evolution crafted two more branches—Lophotrochozoa and Ecdysozoa.

- Biologists believe that a flatworm-like animal was the first bilateral triploblastic (3 cell layers) creature to develop a definite head (cephalization), rudimentary organs including a brain, paired senses, and a tail.

- The phylum Platyhelminthes includes about 25,000 extant species of both free-living and parasitic worms.

- Platyhelminths (or flatworms as they are commonly known) display a variety of body forms and successfully inhabit a wide range of environments ranging from freshwater and marine to land with some being symbiotic in or on other invertebrates. In their habitat, the free-living flatworms are carnivores and scavengers.

- Platyhelminths are highly flattened dorsoventrally (top to bottom) and many are not much thicker than a piece of paper, although the body shape varies from broadly oval to elongate and ribbon-like.

- Platyhelminths are classified into four classes:

  Class Turbellaria
  Class Trematoda
  Class Monogenea
  Class Cestoda

- The characteristics of phylum Platyhelminthes:

  1. They are unsegmented worms that are either parasitic or free-living in their mode of existence.

  2. They are triploblastic possessing three germ layers—ectoderm, mesoderm, and endoderm. The middle layer of mesoderm is significant in that it is the germ layer that gives rise to organs and organ systems.

  3. They are acoelomate (lacking a true coelom). Their bodies are filled with mesodermal parenchyma cells ("packing tissue") and their mesenchyme contain more cells and fibers that do the mesoglea of cnidarians. Thus, there are no internal spaces inside them other than the digestive tube.

  4. They are bilaterally symmetrical (Bilateria) and flattened dorsoventrally (top to bottom). In general, the longer the worm, the more pronounced the flattening.

  5. They are cephalized (possess a definite head end) with a central nervous system consisting of a pair of anterior ganglia ("brain") with longitudinal nerve cords connected by transverse nerve cords (a nerve "ladder") running the length of the body. Some forms possess simple sense organs such as eyespots.

  6. They are the simplest animals with an excretory system for removing liquid metabolic wastes.

7. They possess a single opening or one-way digestive system consisting of a mouth, pharynx, and a highly invaginated (pouched) intestine. Cestodes have no digestive system.

8. Most forms are monoecious (hermaphroditic) with a complex reproductive system consisting of well-developed gonads, ducts, and accessory organs. Many of those that are internal parasites have complicated life cycles involving several hosts

- Class Turbellaria—planarians—are mostly free-living flatworms that are found both in the ocean and freshwater. They are typically small creeping worms 5mm to 50 cm (0.2inch to 20 inches) that combine muscular movement with ciliary action to achieve locomotion. Freshwater turbellarians live in lakes, ponds, streams, and springs where they are carnivores and scavengers feeding upon small invertebrate prey and dead animal matter.

- Class Trematoda—digenetic flukes—are all endoparasites of all classes of vertebrates from fish to mammals including humans, our livestock, and our pets. Flukes are leaf-shaped or oval in form and adults range in length from approximately 0.2 mm to 6.0 cm (0.008 inch to 2.4 inches). Flukes are structurally adapted to a parasitic lifestyle being equipped with specialized glands for penetration or cyst production, suckers and hooks for attachment, and an increased reproductive capacity.

- Class Monogenea—monogenetic flukes—are all parasites, primarily of the gills, gut, and external surfaces of fish. A few species are found in the urinary bladder of frogs and turtles, and one type parasitizes the eye of the hippopotamus.

- Class Cestoda—tapeworms—are all endoparasitic with the majority adapted for living in the guts of various kinds of vertebrates.

- Platyhelminths:

1. Are covered by a simple ciliated epidermis is which each cell bears many cilia (planaria) or a cuticle-like tegument (flukes and tapeworms).

2. Are powered by a series of muscle fibers beneath the basement membrane that run circularly, longitudinally, and diagonally in turbellarians. Adult flukes lack external cilia, and movement depends on the fluke's body wall muscles or the movement of the body fluids of the host. Adult tapeworms do not move around much at all, but they are capable of muscular undulations of the body.

3. Are free-living or parasitic. The prey of free-living turbellarians includes almost any invertebrate small enough to be captured and ingested (e.g. protists, rotifers, insect larvae, small crustaceans, nematodes, annelid worms). Adult flukes actively feed on their host's tissues and bodily fluids whereas tapeworms lack any vestige of a mouth or digestive tract, and absorb nutrients directly across the tegument.

4. Lack any specialized circulatory or respiratory structures. This forces these worms to rely solely on diffusion to move digested materials through the body and to power the O_2-CO_2 exchange process.

5. Possess a rudimentary excretory system in the form of protonephridia tubules. Turbellarians and tapeworms excrete metabolic wastes through series of excretory pores (nephridiopores) whereas flukes collect metabolic wastes in a bladder before release.

6. Posses a centralized ganglion (rudimentary brain) and two longitudinal nerve cords. Turbellarians have tactile, chemo-, and photoreceptors whereas flukes and tapeworms possess only tactile nerve receptors for the most part.

7. Are hermaphroditic and usually cross-fertilize, although self-fertilization sometimes occurs in tapeworms. Turbellarians also have great powers of regeneration.

- Platyhelminths connect to us in important ways economically, ecologically, and medically.

- The phylum Nemertea comprises about 900 species of predominantly benthic marine worms. The many common shallow-water nemerteans live beneath shells and stones, in algae, or burrow into mud and sand. Some species form semipermanent burrows lined with mucous or distinct cellophane-like tubes.

- Nemerteans are classified into two classes:

  Class Enopla
  Class Anopla

- The characteristics of phylum Nemertea:

1. Their bodies are triploblastic, acoelomate, and bilaterally symmetrical.
2. They possess a complete digestive tract with a mouth on one end and an anus on the other.
3. Protonephridia function as organs of excretion.
4. They possess a closed circulatory system.
5. Their body spaces are filled with partially gelatinous parenchyma.
6. They possess an eversible proboscis which lies free within the rhynchocoel.
7. They possess a four-lobed brain connected to a pair of longitudinal nerve trunks.

- Class Enopla have a proboscis armed with sharp stylets for spearing prey. Their mouth opens in front of the brain.
- Class Anopla lack stylets on their proboscis and their mouth opens below or posterior to the brain.
- Nemerteans:

1. Possess a body wall consisting of an epidermis, a dermis, relatively thick muscle layers, and a mesenchyme of varying thickness packed with partially gelatinous parenchyma.
2. Support is provided by the muscles and the hydrostatic properties of the parenchyma.
3. Are active predators on small invertebrates (polychaete worms and amphipod crustaceans are especially favored), but some are scavengers. The proboscis is used to impale or ensnare prey which is swallowed whole.
4. Possess an excretory system consisting of from two to thousands of protonephridia similar to those found in turbellarian flatworms.

5. Possess a nervous system resembles that of the turbellarian flatworms, but nemerteans are more highly cephalized than are the flatworms, especially in the placement of the mouth and feeding structures as well as sensory and other nervous system elements.

6. Are mainly dioecious with rows of gonads alternating with the intestinal diverticula to form a regular row on each side of the body. Fertilization is external in most nemerteans and the eggs are either shed into the water or deposited within the burrow or tube, or in gelatinous strings.

Review and Reflect

1. *On Safari*. Your instructor has assigned you the task of collecting live flatworm specimens for use in the laboratory. Explain where you would go and what you would have to do to collect live turbellarians (planaria), flukes, and tapeworms.

2. *A Switch Hitter*. Explain why it is beneficial to the planaria to be both a carnivore and a scavenger.

3. *Platyhelminth Reincarnation*. If you had to be reincarnated as a flatworm, which type—turbellarian, fluke, or tapeworm—would you chose to become? Defend your answer.

4. *Explain It*. Why are flatworms flat?

5. *Egg Machines*. Tapeworms can produce about 100,000 slowly-maturing eggs per proglottid. Each day an average of 8 fertile proglottids detach and pass out of the host in the host's feces. Assuming this rate to be constant, calculate the total number of tapeworm eggs produced per day, per week, per month and per year.

6. *Darn That Dam*. In 1970, the Aswan High Dam was completed across the Nile River in Egypt. This monumental achievement brought the promise of increased irrigation, flood control and electric power. However, no one suspected another more sinister effect for with the dam came an increase in the rate of schistosomiasis in people living around the dam. What is the connection?

7. *Break the Cycle*. Schistosomiasis, or bilharzia as it is also known, is a serious illness caused by blood flukes. This health problem is thought to affect 250-300 million people worldwide, second only to malaria in the number of people infected. Carefully examine the life cycle of these flukes and suggest as many strategies as possible to break the blood fluke life cycle in as many places as possible.

8. *Are All Worms the Same*? In lab you and your partner have been observing preserved specimens of both flatworms and ribbon worms. Your lab partner says to you, "I don't get it. These worms seem pretty much alike. Why are they placed in separate phyla?" Prepare a table in which you compare and contrast the similarities and differences between phylum Platyhelminthes and phylum Nemertea.

Create and Connect

1. *Animal Rights*. Do animals have "rights?" Should vertebrate animals be accorded more rights than lowly invertebrate animals such as sponges, jellyfish, and worms? Who decides? Should ani-

mal rights be in the form of legislation? Who would design and enforce such legislation? What would be the penalties for failure to comply with such legislation? Prepare a position paper on animal rights in which you discuss these questions and others that may apply to your position.

Guidelines:

A.   Compose a position paper, not an <u>opinion</u> paper. Defend your position with as many facts, Figures, quotes, and pertinent information as possible.

B.   Your work will be evaluated not on the "correctness" of your position but the quality of the defense of your position.

C.   Your instructor may provide additional details or further instructions.

2.   *Drawing Conclusions*. Numbers (data) speak to us as researchers but they whisper faintly when they do so. The trick is to find ways of amplifying and understanding those whispers. The data shown in **Table 9.2** was gathered from an experiment on the optimum temperature conditions for best survival rate following regeneration surgery on planaria. Although the data is accurate (and based on actual research), our research assistant Igor entered the data in a jumbled fashion. Organize this data and then use it to prepare graphs. (Your instructor may also ask you to run appropriate statistical tests on the data.) Once you have analyzed the data, you should have an answer (conclusion) to the initial problem question of this experiment—For best survival rate of planaria after regeneration surgery, should the pieces be kept at room temperature or in the refrigerator? Defend your conclusion.

Table 9.2		
Experimental Data on the Effect of Temperature on the Survival Rate of Planaria Pieces After Surgery		
Survival Rate (number of pieces alive) At Room Temperature (21°C)		
Day 12:15	Day 14:14	Day 13:14
Day 2:22	Day 5:19	Day 16:13
Day 3:20	Day 15:14	Day 18:12
Day 10:16	Day 6:19	Day 19:10
Day 1:30	Day 8:17	Day 11:15
Day 7:19	Day 20:9	Day 17:13
Day 4:20	Day 9:17	
Survival Rate (number of pieces still alive) in Refrigerator (10°C)		
Day 11:28	Day 5:29	Day 4:30
Day 14:26	Day 12:27	Day 2:30
Day 10:28	Day 1:30	Day 20:23
Day 9:28	Day 18:25	Day 13:26
Day 16:26	Day 3:30	Day 7:29
Day 8:28	Day 17:25	

Tasting the Water. Planarians are equipped with chemoreceptors located in each of two lateral projections called *auricles* on their head. How important are the auricles in finding food? Devise an experiment to test this problem question.

Guidelines:

A. Your design should include the following components in order:

➤ The *Problem Question*. State exactly what problem you will be attempting to solve.
➤ Your *Hypothesis*. Although this is a fictitious experiment, word your hypothesis as realistically as possible.
➤ *Methods and Materials*. Explain exactly what you will do in your experiment including the materials necessary to accomplish the task. Be specific, take nothing for granted, and do not expect people to read your mind as they read your work.
➤ *Collecting and Analyzing Data*. Explain what type(s) of data will be collected and what statistical tests might be performed on that data. It is not necessary to concoct either fictitious data or imaginary observations.

B. Your instructor may provide additional details or further instructions.

PSEUDOCOELOMATES: THE RISE OF HOLLOWNESS

Evolution is a tinkerer.
　　　　—Francois Jacob

Introduction

The first appearance of body cavities (coeloms) is hidden in the mists of deep time, but the evolution of coeloms was a critical step because the presence of body cavities was crucial for the rise of increasingly large and ever-more complex animals.

Terminology of Hollow Spaces

When it comes to body cavities, evolution resulted in three morphological types among the triploblastic animals that comprise Bilateria: (**Figure 10.1**)

- *Acoelomate* animals, such as the flatworms and ribbon worms have no body cavity other than the gastrovascular system. The mesoderm is solid, and the organs have

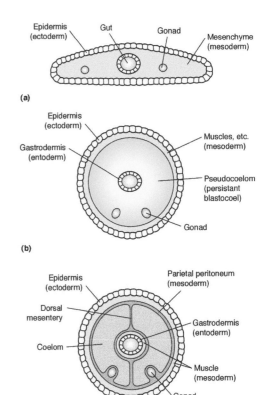

Figure 10.1 Generalized Body Plans of Eumetazoa as Seen in Diagrammatic Cross-Section. (a) The acoelomate plan. (b) The pseudocoelomate plan. (c) The eucoelomate plan.

direct contact with the epithelium. Semi-solid mesodermal tissues between the gut and body wall hold the organs in place.

- *Pseudocoelomate* (Gr. *pseudo*, false + *coel*, hollow cavity) animals have a **pseudocoel** ("false cavity") which is a fully functional body cavity that develops between the mesoderm and the endoderm. However, tissue derived from mesoderm only partly lines the fluid-filled body cavity of these animals. Thus, whereas the organs are held in place loosely, they are not as well organized as in eucoelomate animals.

- *Eucoelomate* animals have a fluid-filled body cavity called a **coelom** lined with a slick epithelial layer known as the **peritoneum** that is derived from the mesoderm. The peritoneum allows organs to be attached to each other so that they can be suspended in a particular order yet still being able to move freely within the coelom. The internal organs are suspended from the walls of the coelom by membranous sheets called **mesenteries**.

    Eucoelomates are said to have a "tube-within-a-tube" body arrangement in which the secondary body cavity lies between the two tubes. The outer tube (body wall or **somatic tube**) contains the sense organs and muscles whereas the inner tube (gut or **visceral tube**) typically contain structures that control an organism's internal environment—digestion, circulation, excretion, and reproduction. The nerve functioning of the somatic tube components are typically voluntary in nerve function whereas those of the visceral tube are involuntary.

Importance of Hollow Spaces

The acoelomate body plan marked by a single body cavity imposes severe limitations on the bearer. The muscle contractions that power locomotion squeeze and distort the gastrovascular cavity thus restricting the flow of nutrients. As a consequence, acoelomate animals must rely on diffusion or muscle contractions for the transport of nutrients, waste products, and respiratory gases. Both methods of transport are highly inefficient compared to the circulatory systems that have evolved in the more spacious eucoelomate plan.

As a result, the acoelomate phyla all have comparatively simple body arrangements. Most are small in size and characterized by a highly flattened and elongated shape suited to diffusion as a transport mechanism. Even the pseudocoelomates are small, flat and lack circulatory systems.

The evolution of coeloms led to the development of complex but separate digestive and circulatory systems that are very efficient. Increased efficiency in both digestion and circulation allows for the evolution of larger body sizes and increased metabolic rates. Not only does a coelom provide space, but it also accords protection in that fluid-filled spaces cushion and shield the internal organs from injury.

Coeloms can serve other functions as well. In some types, the fluid-filled, pressurized coelom provides hydrostatic support. Earthworms (Annelida) and slugs (Mollusca) rely on the hydrostatic "skeleton" provided by the coelom for support and locomotion. In starfish (Echinodermata), the water vascular system that provides locomotion via tube feet is derived from the coelom.

Within the eucoelomates, two different groups may be distinguished based on the way the coelom forms during development—the protostomes and the deuterostomes.

History of Pseudocoelomates

The various phyla known collectively known as the pseudocoelomates are a diverse group of animals. The absence of synapomorphies strongly suggests that each phylum evolved independently of the others and that any similarities between them are the result of convergent evolution. Not only do the various groups of pseudocoelomate animals not constitute a monophyletic taxonomic assemblage, but some of the phyla lumped in under the pseudocoelomate label are functionally acoelomate. Being soft-bodied, their fossil record is meager, and their phylogenetic relationships and affinities are murky at best.

Molecular evidence suggests that sometime after the ancestral deuterostomes and protostomes diverged from each other in the Precambrian, the protostome lineage splintered evolutionarily again into two branches: Lophotrochozoa, animals that do not molt and Ecdysozoa, animals that do molt during development (**Figure 10.2**).

Both the Lophotrochozoa and Ecdysozoa branches of Protostomia contain member phyla most (but not all) of which are pseudocoelomate in nature. Although the members of each phylum represent a success story in their own right, some are far more abundant than others. Some of these phyla comprise only a few known species, whereas others (e.g. Nematoda) include thousands of described species.

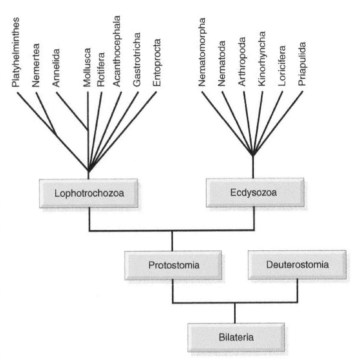

Figure 10.2 The position of Protostomia on the phylogenetic tree of life.

Diversity and Classification of Pseudocoelomates

Pseudocoelomate is an anatomical and descriptive term, not a taxonomic one. Not only do the various groups of pseudocoelomate animals not constitute a monophyletic taxonomic assemblage, but as we will discover later in this chapter, some of the phyla lumped in under the pseudocoelomate label are functionally acoelomate. Contradictions and exceptions abound when it comes to understanding and classify pseudocoelomate animals.

Many of the creatures now described as pseudocoelomates were discovered centuries ago. By the late nineteenth century, much of the morphological infrastructure of the classification of these groups had been established. With the rise of molecular analysis as a tool of classification in the mid1990s, systematists began to attempt to piece together the phylogeny and evolutionary relationships of various animal groups. Through molecular analysis, it soon became clear that many of the morphological taxa in traditional classification schemes were artificial contrivances. Melding phylogeny with morphology into a single classification system that everyone (or even the majority) agrees on has proven to be a monumental challenge that is ongoing and that probably will not be resolved for some time to come. Whole taxonomic groups have disappeared; some groups have been merged, and entirely new groups have been created.

Hypotheses fly, debate rages, and reorganization occurs on a regular basis but probably in no portion of the Tree of Life is revision more hotly contested than it is for those phyla of animals we loosely designate as the pseudocoelomates.

As many as 25 phyla branch off from the protostome segment of the Tree of Life with most of those being pseudocoelomate in form. Space considerations and time constraints prevent us from investigating all the pseudocoelomate phyla thus limiting us to only a brief consideration of select phyla:

> Kingdom Animalia
>> Superphylum Lophotrochozoa
>>> Phlyum Rotifera—rotifers
>>> Phylum Acanthocephala—spiny-headed worms
>>> Phylum Gastrotricha—bristle backs or hairy backs
>>> Phylum Entoprocta—goblet worms
>>> Phylum Platyhelminthes—flatworms
>>> Phylum Nemertea—ribbon worms
>>> Phylum Annelida—segmented worms
>>> Phylum Mollusca—molluscs
>> Superphylum Ecdysozoa
>>> Phylum Tardigrada—water bears
>>> PhylumNematoda—roundworms
>>> Phylum Nematomorpha—horsehair worms
>>> Phylum Kinorhyncha—mud dragons
>>> Phylum Loricifera—brush heads
>>> Phylum Priapulida—penis worms
>>> Phylum Arthropoda—arthropods

The whole shifting and ever-changing enterprise of how zoologists interpret the phylogeny of the many phyla of protostomes and then devise functional classification schemes from that information is a potential quagmire for both authors and students. And though we must tread that quagmire, we will do so as lightly and as quickly as possible.

Phylum Characteristics of the Lophotrochozoan Pseudocoelomates

Phylum Rotifera (L. *rota*, wheel + *fera*, to bear).

The rotifers are a widespread group of about 1,800 described species, some of which are found worldwide. A few rotifer species reach lengths of 2-3 mm (0.0800-0.12 inch), but most are less than 1 mm (0.04 inch) long. Although most are transparent, some have beautiful colors, and bizarre shapes seem the norm (**Figure**

10.3). Their shapes are often correlated with their biological station in life. Floaters are usually globular and sac-like; creepers and swimmers are elongated and even wormlike whereas sessile types are commonly vase-like. Most exist as solitary individuals, but there are some colonial types.

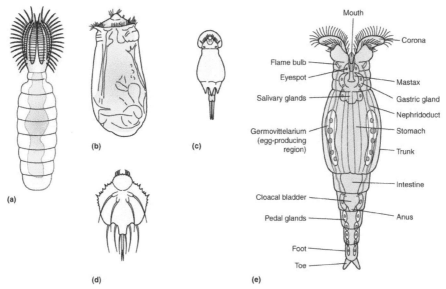

Figure 10.3 The Variety of Form in Rotifers and Generalized Rotifer Structure. (a) *Stephano ceros* has five long coronal lobes with short bristles that it uses to catch prey. (b) *Asplanchna* is a pelagic predatory species with no foot. (c) *Squatinella* has a semicircular nonretractable transparent extension covering the head. (d) *Machrochaetus* is highly flattened dorsoventrally. (e) Generalized structure of *Philodina* sp.

Most species are found in freshwater, a few are marine, some are semiterrestrial living in damp soil or the water film on mosses, and some live as symbionts on other animals, including a few that are parasitic. Most species are benthic, living on the bottom or in vegetation of ponds or along the shores of freshwater lakes. Pelagic forms are common in the surface waters of freshwater lakes and ponds where they may exhibit **cyclomorphosis**, variations in body form resulting from seasonal or nutritional changes.

Many species of rotifers can withstand long periods of desiccation. In a dried state, rotifers can endure amazing environmental extremes from years of desiccation to temperatures hundreds of degrees below zero.

Rotifers are triploblastic, bilateral, unsegmented pseudocoelomates that exhibit the following characteristics:

- Their body is covered with a well-developed cuticle.
- Their anterior end often bears variable ciliated arrangements as a **corona**. (Figure 10.3e) When viewed through a microscope, these whirling cilia give the impression of rapidly rotating propellers. Coronal cilia are used both for locomotion and feeding. Historically, rotifers were first called "wheel animalcules" by Leeuwenhoek in 1703.
- Their posterior end often bears a **foot** with **toes** and adhesive glands (Figure 10.3e). The foot usually bears one to four toes, and the cuticle of the foot may be ringed so that it is telescopically retractile. The foot is an attachment organ and contains **pedal glands** that secrete an adhesive material used by both sessile and creeping types. In swimming pelagic forms, the foot is usually

reduced. Rotifers can move by creeping with leechlike movements aided by the foot, or by swimming with the coronal cilia or both.

- Their digestive system is complete. Some rotifers feed by sweeping minute organic particles or algae toward the mouth with the coronal cilia. Their pharynx (**mastax**) is equipped with a muscular portion containing hard jaws (**trophi**) for grinding up food particles. The mastax of suspension feeders takes a crushing and grinding form whereas that of predatory species that feed on protozoa and other small animals, is grasping and piercing in form. Some types trap their prey with a funnel-shaped area around the mouth that folds in on the hapless victim.

- Protonephridia provide an excretory system, but no circulatory or respiratory system is present.

- The nervous system consists of a bilobed brain, dorsal to the mastax, with paired nerve cords branching out to the sense organs, mastax, muscles, and organs. Depending on the species, sensory organs include paired eyespots, sensory bristles, and papillae, ciliated pits and dorsal antennae.

- Rotifers are technically dioecious, and males are usually smaller than females. **Parthenogenesis** (a mode of reproduction involving the development of an unfertilized egg into a functional individual), however, is the rule among some rotifers but it is also a common and usually seasonal occurrence in others where it tends to alternate with sexual reproduction. (**Figure 10.4**) This cycle is an adaptation to freshwater habitats that are subject to severe seasonal changes. During

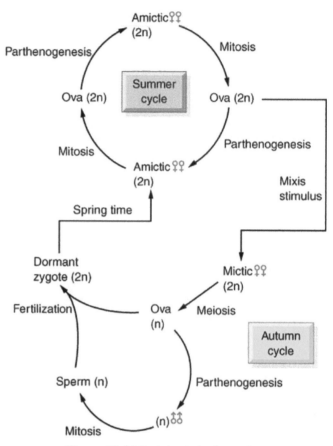

Figure 10.4 Mictic/amictic alternation in the life cycle of rotifers.

favorable conditions, females reproduce only **amictic** females through parthenogenesis. When an environmental trigger of sort—changes in day length, temperature, food sources, or increase in population density—occurs, however, the amictic females begin to produce ova that develop into **mictic** females that produce haploid **mictic ova** by meiosis. The mictic ova also develop by parthenogenesis into haploid males that produce sperm through mitosis. These sperm fertilize the other mictic ova, producing diploid, thick-walled resting zygotes. These resting zygotes are extremely resistant to low temperatures, desiccation, and other adverse environmental conditions. Upon the return of favorable conditions, the resting zygotes develop and hatch as amictic females, completing the cycle.

Phylum Acanthocephala (Gr. *akantha*, spine or thorn + *cephalo*, head)

Commonly known as "spiny-headed worms," the members of this phyla derive their name from a most distinctive feature, a cylindrical, eversible **proboscis** bearing rows of recurved spines. (**Figure 10.5**) The rest of the body forms a cylindrical or flattened trunk, often bearing small rings of small spines. As adults, the 1,000 described species of spiny-headed worms are **obligate intestinal parasites** in birds, mammals, and especially freshwater teleosts (bony fish). No species routinely parasitizes humans, although occasionally infections are reported.

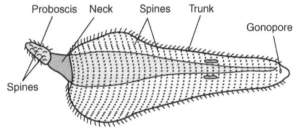

Figure 10.5 *Corynosoma*, a parasite in birds and seals, with its proboscis partially everted.

Most spiny headed worms are less than 20 cm (11-12 inches) long, but a few species can reach nearly a meter (3.3 feet) in length. The digestive tract has been completely lost. And, except for the reproductive organs, there is significant structural and functional reduction of most other systems, as related to their parasitic lifestyle.

Acanthocephalans are triploblastic, bilateral, unsegmented, **vermiform** (resembling a worm in form or movement) pseudocoelomates that exhibit the following characteristics:

- Their anterior end bears a spiny or hooked proboscis. The proboscis is used by adult acanthocephalans to attach to their host's intestinal wall.
- The digestive system has been completely lost; all nutrients are absorbed through the body wall, which comprises a thin outer cuticle underlain by a **syncytial** (multinucleate mass of fused cells) epidermis.
- The epidermis houses a complex system of circulatory channels called the **lacunar system**.
- Reflective of their parasitic life style, the lack of digestive, circulatory, respiratory, and excretory systems results in exchanges of nutrients, gases, and waste products occurring by diffusion across the body wall. Internal transport is by diffusion within the body cavity and probably by the epidermal lacunar system.
- The nervous system and sense organs are also greatly reduced. The proboscis bears several sensory receptors presumed to be tactile.
- Internally they possess a unique system of ligaments and ligament sacs that partially partition the body cavity.
- Acanthocephalans are dioecious, and females are usually somewhat larger than males. In both sexes, the reproductive systems are associated with ligament sacs. Direct copulation results in fertilized eggs that complete much of their early development within the body cavity of the female. Eventually a shelled larva (**acanthor larva**) is formed and leaves the mother's body eventually being released with the host's feces. The larval acanthocephalan must then be ingested by an arthropod intermediate host—usually an insect or crustacean—to continue its life cycle. In the intermediate hosts, the acanthor eventually develops into a **cystacanth**. When the intermediate host is eaten by an appropriate definitive host, the cystacanth attaches to the intestinal wall of the

host and matures into an adult. Multiple infections can do serious damage to an animal's intestines, and perforations can occur. (**Figure 10.6**)

Phylum Gastrotricha (Gr. *gasteros*, stomach + *trichos*, hair)

This phylum consists of about 450 species of small, ventrally flattened animals found in marine, brackish, and freshwater environs. They range in size from 1 mm (0.04 inch) to a few at 3 mm (0.12 inch) and have a superficial resemblance to rotifers or even large ciliate protists (for which they are often mistaken) (**Figure 10.7**).

The gastrotrich body is typically divisible into head and trunk and a few possess an elongated "tail." Externally, these animals bear various assortments of spines, bristles, and scales or plates all of which are derived from the cuticle. They are also equipped with two or more **adhesive tubes** (Figure 9-8) containing cement glands that secrete attachment substances used for temporary adherence to objects in the environment.

Characteristics of phylum Gastrotricha are as follows:

- The area between their gut and body wall is filled with loose organs and mesenchyme, effectively creating an "acoelomate" condition.
- The digestive system is complete and is made up of a mouth, a muscular pharynx, a stomach-intestine, and an anus. Food is mainly algae, protozoa, bacteria, and detritus, which are directed to their mouth by their head bristles.
- They lack circulatory and respiratory systems. The excretory system is protonephridic.
- The nervous system consists of a relatively large bilobed cerebral ganglion (**neutropil**) with a lateral longitudinal nerve cord

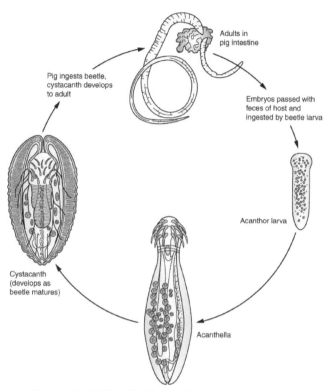

Figure 10.6 The Life Cycle of *Macracanthorhynchus hirudinaceus*. The adults reside in the intestine of the definitive host, and embryos are released with the feces. The encapsulated embryos are ingested by the secondary host, in this case, beetle larvae. Within the secondary host, the embryo passes through the ancanthor and acanthella stages, eventually becoming a cystacanth, while the beetle grows. When the beetle is ingested by a pig, the juvenile matures into an adult, thereby completing the life cycle.

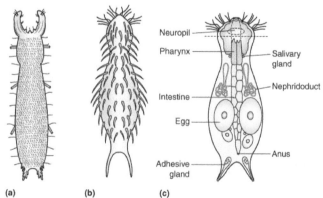

Figure 10.7 Marine Gastrotrichs. (a) *Pseudoturbanella* (b) *Halichaetonotus*. (c) Generalized gastrorich anatomy.

arising from each lobe of the brain and extending to the posterior of the body. Sensory receptors are predominantly tactile in nature.

- Gastrotrichs are typically hermaphroditic. After mutual cross-fertilization, the zygotes are released singly or a few at a time by rupture of the female's body wall. Both thick-shelled dormant eggs and thin-shelled rapidly developing eggs (sexually maturity within a few days of hatching) are produced in response to seasonal environmental conditions.

Phylum Entoprocta (Gr. *entos*, inside + *proktos*, anus)

The 150 species of entoprocts are mostly microscopic (none is more than 5 mm [0.2 inch] long), sessile, solitary or colonial stalked creatures that superficially resemble hydroids and ectoprocts. (**Figure 10.8**) With the exception of a single group that occurs in freshwater, all entoprocts are marine forms. They are widely distributed from polar regions to the tropics where they live attached to various substrata, including algae, shells, and rock surfaces. Most call the intertidal zone home, but some forms are known from depths as great as 500 meters (1650 feet). Freshwater entoprocts occur on the underside of rocks in running water.

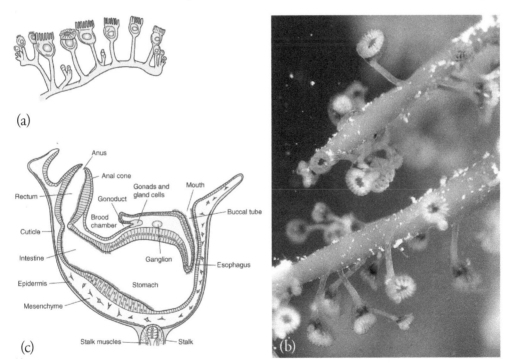

Figure 10.8 Representative Entoprocts. (a) Portion of a *Pedicellina* colony. (b) The feeding locophores of *Loxosoma*. (c) Entoproct structure as displayed by a zoid of *Pedicellina* in sagittal section.

The body consists of a cuplike **calyx** from which arises a whorl of ciliated tentacles. Both the mouth and anal openings are within this whorl of tentacles. The 8-30 tentacles making up the crown are ciliated on their lateral and inner surfaces, and each can move individually. The tentacles may be rolled inward to cover and protect the mouth and anus, but they cannot be retracted into the calyx.

The following are characteristics of phylum Entoprocta:

- Although technically bilateral, the individual zooids of colonial types have in many respects assumed a functionally radial form.
- The area between the gut and body wall is filled with gelatinous mesenchyme rendering them functionally acoelomate.
- The digestive system is complete, and the gut is U-shaped. (Figure 10.8c) Entoprocts are ciliary suspension feeders. These animals are oriented with their ventral side away from the substratum (up); the dorsal surface is attached to the anchoring stalk (down). Long cilia on the sides of the tentacles create a current of water that draws in protozoa, diatoms, and particles of detritus from dorsal to ventral, flowing between the tentacles. Short cilia on the inner surfaces of the tentacles capture the food and move it downward to the mouth.
- The excretory system consists of one pair of protonephridia, but circulatory and respiratory organs are absent.
- As is so often the case in small sessile invertebrates, the nervous system is greatly reduced. A single small ganglionic mass lies ventral to the stomach. Unicellular tactile receptors are concentrated in the tentacles and scattered over much of the body surface.
- Colony growth occurs by budding from the stalk. Most, perhaps all entoprocts are hermaphroditic; some are **protandric** (the gonad at first producing sperm and then later eggs). Sperm are released into the water and then enter the female reproductive tract where fertilization occurs. The fertilized eggs develop within a brood chamber. In time, ciliated and free swimming larva are released from the brood chamber. Eventually, the larva settles on a suitable substrate and metamorphoses into an adult zooid. In some species, the developing embryos are retained within the female tract where special cells of the adult provide nutrition in a pseudoplacental arrangement.

Phylum Characteristics of the Ecdysozoan Pseudocoelomates

Phylum Tardigrada (L. *tardigrada*, slow walker).

Tardigrades, commonly known as water bears or moss piglets because of their appearance and slow-moving gait, are an ancient group with fossils dating back to the Cambrian period some 530 million years ago. Ranging in size from 0.3 to 0.5 millimeters (0.012 to 0.020 inches) in length, tardigrades are cylindrical, segmented animals with eight extensions of the body wall that act as clawed legs (**Figure 10.9**).

The 1,150 species of tardigrades are found in marine, freshwater, and semiaquatic terrestrial habitats worldwide ranging from the tops of mountains to the deep sea floor and from polar regions to the equator even in hot springs. Tardigrades require a film of water around them to thrive thus on land they are found in damp soil and on damp leaf litter, moss, and lichens. Some types, however, live on the legs of isopods, on the gills of mussels or as parasites on the epidermis of barnacles.

The characteristics of phylum Tardigrada are as follows:

- The body is covered with a chitinous cuticle that is molted periodically.
- The body consists of a head segment, three body segments with a pair of legs each, and a caudal segment with a fourth pair of legs. The lobopodial limbs (poorly articulated) are not jointed, and the feet have four to eight claws each.
- A tubular mouth is armed with stylets, which are used to pierce plant cells, algae, protozoa, or small invertebrates releasing body fluids or cell contents upon which the animal feeds by sucking them up with a bulbous bucco-pharyngeal apparatus. Some types are entirely carnivorous.
- They lack both a respiratory system and a circulatory system. They breathe across their cuticle and possess a hemocoel that provides very limited circulation.
- They do possess a well-developed muscular, excretory, and digestive system.
- The nervous system consists of a multilobed brain attached to a double ventral nerve cord that runs the length of the body. Two inverted pigment-cup eyes are present as are numerous sensory bristles over the head and body.
- Tardigrades are dioecious (males and female individuals in the population). Each has a single gonad which lies dorsal (above) the gut. Mating usually occurs during molt with females depositing 1 to 30 eggs in the shed cuticle which is then covered with sperm by the male. In some types, the eggs are laid directly on the substratum. Development is direct (no larval stages) with juveniles hatching from eggs.

Figure 10.9 Short and plump with four pairs of legs, tardigrades are commonly known as water bears or moss piglets. They are water animals that live on moist moss and lichens where they feed on plant cells, algae, and small invertebrates.

Box 10.1
One Tough Critter

If there was a Guinness Book of World Records category for the biologically toughest animal, what animal would be the record-holder? Enter the water bear (or moss piglet if you prefer), a tiny, plump, bear-like creature that waddles around on stumpy legs. Don't be fooled by their small size, however, because these miniscule creatures are much tougher than your average bear.

Water bears can survive incredibly hostile environmental conditions:

- *Pressure*. They can withstand unbelievable pressure extremes from the near vacuum of space to 6,000 atmospheres, nearly six times the pressure of water at the bottom of the deepest ocean trench.

- *Temperature.* They can survive a few minutes either being heated to 151° C (304° F) or flash frozen to -273° C (-523° F), but days at a frosty -200° C (-328° F). Some have even survived being boiled in alcohol.
- *Dehydration.* Water bears have been shown to routinely survive drying for up to ten years and there are reports of live water bears being revived from dried moss kept in a museum for over 100 years although it is unclear whether the revived water bears actually survived or whether a few merely showed a momentary leg movements.
- *Radiation.* They can withstand up to 1,000 more times radiation than other animals. Eggs and developing young, however, are far less tolerant of radiation than are adults.

Water bears can even survive the cold vacuum and solar radiation of outer space. On September, 2007, a robotic spacecraft exposed groups of dried water bears to only the cold vacuum of space and to both the cold vacuum and solar radiation of space. Upon return to Earth, attempts to revive the water bears revealed a survival rate of 68% for those shielded from radiation, but far lower survival rates for that group exposed to both. The wonder is that any survived at all.

How do water bears defy death by environment? Tardigrades have evolved an adaptation known as **cryptobiosis** in response to various environmental stresses. Cryptobiosis is a situation in which metabolic processes come to a standstill. During cryptobiosis the metabolic rate can drop to 0.01% of normal, and the water content of the body can fall to 1% of normal, a true death-like state. Different types of cryptobiosis exist:

- cryobiosis (low temperature)
- anhydrobiosis (drying) Water bears survive desiccation by curling up into a little ball called a **tun**. In this state, the metabolic rate drops and a protective sugar known as trehalose is synthesized and moves into the cells replacing lost water.
- osmobiosis (increased solute concentration (salt)
- anoxybiosis (lack of oxygen)

Although tardigrades can *survive* extreme environmental conditions, they are not considered extremophiles because they are not adapted to *live* in those conditions.

Research into cryptobiosis has led to improved methods of preserving sperm, seeds, blood, and food, as well as the possibilities of cryosurgery, and suspended animation that might make long space voyages possible.

Phylum Nematoda (Gr. *nematos*, thread)

Nematodes may be the most abundant group of animals on Earth. Approximately 25,000 species have been named but potentially but undoubtedly many more await discovery. In fact, it has been estimated that there may be 500,000 species or more of nematodes. In one study 90,000 nematodes were found in a single rotting apple, and another study turned up 236 different species in just 6.7 cc of marine beach mud. Their amazing abundance inspired N. A. Cobb to write of them in the 1914 edition of the Yearbook of the United States Department of Agriculture:

If all the matter in the universe except the nematodes were swept away, our world would still be dimly recognizable, and if, as disembodied spirits, we could investigate it, we would find its mountains, hills, vales, rivers, lakes, and oceans represented by a thin film of nematodes. The location of the towns would be decipherable, since for every massing of human beings there would be a corresponding massing of certain nematodes. Trees would still stand in ghostly rows representing our streets and highways. The location of the various plants and animals would still be decipherable, and had we sufficient knowledge, in many cases even their species could be determined by an examination of their erstwhile nematode parasites.

There is no corner of the biosphere where nematodes are not found. They are plentiful in the ocean, freshwater, and on land. Furthermore, they parasitize nearly all groups of animals and plants, and as a result, some cause serious damage to crops and livestock and humans and their pets. In terms of number of species and individuals, marine nematodes are the most abundant and widespread group of animals, occurring from the mud of beaches to the bottom of the deep abyss. Some marine environments yield as many as 3 million nematodes per square meter. Fertile topsoil may contain many billions of nematodes per acre. Some species are generalists, but many have very specific habitats. Case in point, one nematode species is found only on and in the felt coasters used under beer mugs in only a few towns in eastern Europe, specialization carried to the extreme.

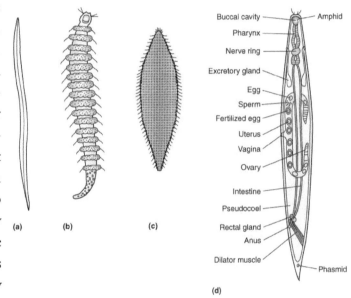

Figure 10.10 Representative Nematodes. (a) the marine *Echinotheristus*. (b) The marine interstitial *Demoscolex*. (c) The marine *Greeffiela*. (d) Generalized nematode anatomy as displayed by a female nematode.

Nematodes are vermiform with thin, unsegmented bodies that are usually distinctly round. Many are microscopic, and most are less than 5 cm (2 inches) long, but some parasitic types are often more than 1 meter (3 feet) in length (**Figure 10.10**). The largest nematode ever recorded at 8.5 meters (28 feet) in length was discovered parasitizing the placenta of a sperm whale.

Characteristics of phylum Nematoda are as follows:

- The body is covered with a layered cuticle; growth in juveniles is usually accomplished by molting.
- Their body wall muscles contract longitudinally only. This produces a somewhat undulatory pattern of movement. There is no circular muscle layer.

- The digestive system is complete. The mouth is surrounded by lips bearing sense organs. Free-living nematodes feed on bacteria, yeasts, fungal hyphae, and algae. Some types scavenge detritus and fecal matter. Predatory species may eat rotifers, small annelid worms, and other nematodes. As previously mentioned, a great many also parasitize virtually every type of animal and most plants. The effects of nematode infestation on crops, domestic animals (livestock and pets), and humans make this the most economically and medically important of all parasitic animal groups. Infection occurs by eating uncooked or undercooked meat with larva in it, by entrance into unprotected cuts or direct burrowing through the skin, and via transmission by blood-sucking insects. Nematodes commonly parasitic on humans include whipworms, hookworms, pinworms, ascaris worms, trichina worms, and filarial worms (**Table 10.1**) and (**Figure 10.11**).

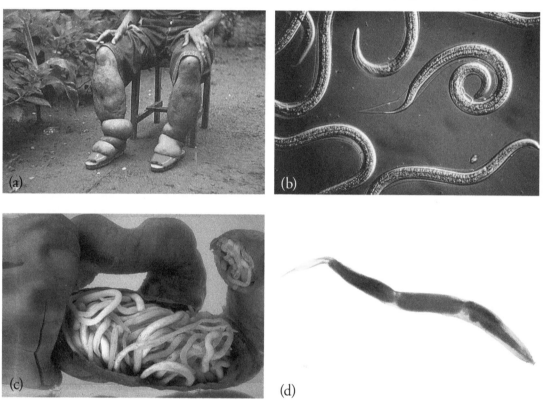

Figure 10.11 Parasitic Nematodes. (a) The disfiguring condition known medically as lymphatic filariasis and commonly as elephantiasis, results from the accumulation of lymphatic fluid due to massive numbers of filarial nematodes in the lymphatic system. (b) Entomopathic nematodes inhabit the soil where they parasitize insects in both the larval and adult forms. (c) A piece of intestine, totally blocked by *Ascaris lumbricoides*, removed from a 3-year old child in South Africa. The child survived. (d) *Enterobius vermicularis*, or pinworm as it is commonly known, is a rather innocuous parasitic worm that plagues many families at one time or another (See *A Closing Note* at the end of this chapter).

- Correlated with their complete lack of cilia (even their sperm cells lack cilia), nematodes do not have protonephridia; their excretory system consists of one or more large gland cells opening into an excretory pore or a canal system.
- There are no circulatory or gas exchange structures in nematodes.

- Their nervous system consists of a ring of nerve tissue and ganglia that surround the pharynx. Nerve cords running both anteriorly and posteriorly come off the nerve ring. The most abundant sense organs of nematodes are various papillae and setae that serve as tactile receptors. Chemoreceptors are found both on the head (**amphids**) and near the posterior end (**phasmids**).

- Most nematodes are dioecious. The males are usually smaller than the females, and their posterior end bears a pair of **copulatory spicules**. Fertilization is internal, and eggs are usually stored in the uterus until deposition. Juveniles grow through a number of stages each marked by a molt, or shedding, of the cuticle. As with most endoparasites, the parasitic nematodes often have quite complicated life cycles, alternating between several different hosts or locations on the host's body. Many parasitic types have free-living juvenile stages whereas others require an intermediate host to complete their life cycles.

Table 10.1	
Common North American Nematode Parasites of Humans	
Parasite	**Transmission and Prevalence**
Hookworm (Ancylostoma duodenale and Necator americanus)	Juveniles burrow into skin upon contact with infected soil; common in southern states
Pinworm (Enterobius vermicularis)	Inhalation of dust containing eggs and by finger contamination; common in entire United States
Intestinal roundworm (Ascaris lumbn'coides)	Ingestion of eggs in contaminated food; common in rural areas of Appalachia and southern states
Trichina worm (Trichinella sp.)	Ingestion of rare or undercooked meat containing cysts; occasional throughout North America
Whipworm (Trichuris trichiura)	Ingestion of food contaminated by eggs or by unhygienic habits; common in southern states

Adapted from Hickman, C.P., et al. Integrated Principles of Zoology, 13th edition. 2006. McGraw-Hill Higher Education.

One species of nematode, *Caenorhabditis elegans*, has proven to be the most important living experimental model in biology. In 1963 Sydney Brenner first started studying this small free-living nematode. In the years since, biologists have traced the origin and lineage of all 959 cells in its body from zygote to adult (adults always have the same number of cells, a phenomenon known as **eutely**), deciphered its neural network down to each neuron and the connections between each neuron, and mapped and sequenced its entire genome. This lowly worm has provided a wealth of biological knowledge and understanding. Who knows what secrets *C. elegans* may share with us next and how this knowledge may be practically applied.

Phylum Nematomorpha (Gr. *nema*, thread + *morphe*, shape)

Commonly called "horsehair worms" from the superstition that they formed when horse hairs happen to fall in water, the 320 species of nematomorphs are long elegant vermiforms that tend to twist and turn upon themselves in such a way as to give the appearance of complicated knots. Because hundreds of individuals can be found in seemingly undecipherable tangles during mating, they are also sometimes dubbed "Gordian worms" from the Greek myth of King Gordius and his legendary knot.

Worldwide in distribution, they are free living as adults but as juveniles they are parasitic on beetles, cockroaches, grasshoppers, and crustaceans, especially crabs. Most adult nematomorphs live in fresh water among the litter and algal mats near the edges of ponds and streams. A few species are terrestrial in damp soil (**Figure 10.12**).

The following are characteristics of phylum Nematomorpha:

- The body is covered with a well-developed cuticle.

- The body wall has only longitudinal muscles.

- The gut is reduced to various degrees, but is always nonfunctional in adults. The larval forms absorb food from their arthropod hosts across their body wall. Until recently, adults were thought to live entirely on stored nutrients. Recent research, however, has shown that adults absorb organic molecules in much the same way as juveniles.

- They possess no specialized excretory, circulatory, or respiratory structures. Very little is known about the physiology of these worms.

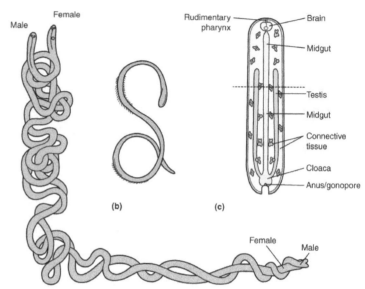

Figure 10.12 Representative Nematomorphs. (a) Copulating *Gordius robustus*. (b) The marine *Nectonema agile* displaying lateral swimming seta. (c) Generalized male nematomorph anatomy.

- The nervous system consists of a nerve ring around the pharynx and a midventral nerve cord. Nematomorphs are touch-sensitive, and some are apparently chemosensitive for adult males are able to detect and track mature females for some distance. The structures associated with these sensory functions, however, are only a matter of speculation.

- Adults are dioecious and mate shortly after emerging from their host. Within a month, females lay long strings of up to 6 million eggs in water, after which they soon die. Larvae hatch from the eggs within 20 days, and penetrate and encyst in the first aquatic animal they encounter, however, they will develop normally only in an appropriate arthropod host. After several months in the **hemocoel** (a body cavity that contains blood or hemolymph) of the host, the mature worm wriggles out into water to mate. On maturity, the nematomorph *Spinochordodes tellinii*, which parasitizes grasshoppers, somehow stimulates the grasshopper to seek and drown itself, thus returning the worm to water. These worms are driven to find water at all costs, and they have even been seen wriggling out of the predator that just ate their host arthropod.

Phylum Kinorhyncha (Gr. *kinein*, to move + *rhynchos*, beak)

About 150 species of these small (usually less than 1 mm [0.04 inch]) marine worms have been described to date. They are found from pole to pole and from intertidal zones to 8,000 meters (26,250 feet) deep. Most live in mud or sandy mud, but some have been found in algal mats, sponges, or other marine invertebrates. The body of kinorhynchs, flattened ventrally and arched dorsally, is divided into 13 segments (**zonites**), which bear spines but have no cilia (**Figure 10.13**).

Characteristics of phylum Kinorhyncha are the following:

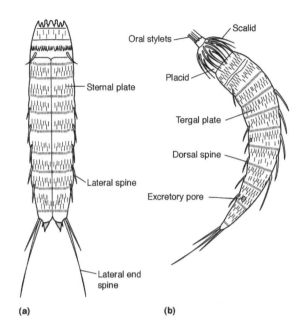

Figure 10.13 Morphology of the Kinorhynch *Echinoderes*. (a) Dorsal view with head retracted. (b) Ventral view with head extended.

- The retractile head bears a circlet of spines and is equipped with a small retractile proboscis. A kinorhynch cannot swim. In the silt and mud where it commonly lives, it burrows by extending the head into the mud, anchoring it with spines, and then drawing its body forward until its head is retracted into its body. The whorl of spines around the mouth and their propensity for burrowing in mud combine to give them the common name "mud dragons."

- Their body wall is composed of a cuticle, an epidermis, and longitudinal muscle cords, much like those of nematodes. Circular and diagonal muscle bands are also present. The kinorhynchs are clearly pseudocoelomates. Their similarities with the arthropods, however, are striking and include metamerism (segment body), nature of the cuticle, and **ecdysis** (molting or shedding of the outer cuticle).

- The digestive system is complete. Kinorhynchs feed on diatoms or by ingesting organic material from the surface of the mud particles through which they burrow.

- Protonephridia on each side of the tenth and eleventh segments serve as their excretory system.

- There are no circulatory or respiratory structures present.

- Their nervous system is composed of a nerve ring encircling the pharynx and a double nerve cord running down the ventral side of the body. Sensory receptors include tactile bristles and spines on the body and photoreceptors are present on the pharyngeal nerve ring in at least some species.

- Sexes are separate, but males and females are usually indistinguishable from one another. Mating has never been seen; thus egg laying and early development have not been adequately studied. The juveniles emerge from the egg case with 11 of the 13 zonites already formed. Juveniles molt periodically passing through six juvenile stages to adulthood.

Phylum Loricifera (L. *lorica*, corselet + *ferre*, to bear)

Danish zoologist R. M. Kristensen first described this phylum in 1983 from marine sediments collected off the coast of France. Since then, the species count of qualifying members of this phylum has grown to only 22 described species. There are approximately 100 more that have been collected, and not yet named. They are strictly marine and are found at all depths in different coarse sediment types, and at all latitudes.

These tiny animals (0.25 mm [0.01 inch]) live in spaces between grains of gravel to which they cling so tenaciously with adhesive glands that they are difficult to extract from sediment samples. This tenacity of attachment may account for them only being recently discovered. Details of this phylum are scarce because most specimens die before they can be examined and thoroughly studied.

The following are characteristics of phylum Loricifera:

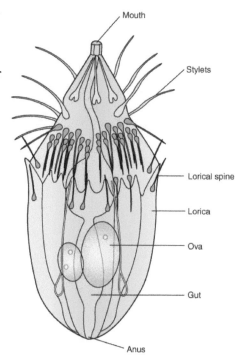

Figure 10.14 Generalized morphology and anatomy of an adult female loriciferan *Nanaloricus mysticus.*

- The major part of their body is encased within a circular **lorica** composed of plates. These plates give the phylum its name. The cone-shaped head end or **introvert** bears many recurved oral stylets (**scalids**). The entire forepart of their body can be retracted into the lorica (**Figure 10.14**).
- The digestive system is complete. Their diet probably consists of suspended organic particles, algae, and bacteria. Current thinking seems to be that they pierce the bacteria or algae with their oral stylets and then suck out the contents.
- A pair of protonephridia, uniquely situated inside the gonads, provides an excretory system but there appear to be no specialized structures for circulation or respiration.
- A complex muscle system is present. Adults can crawl by using their many scalids and their mouth cones.
- A large brain with a ventral nerve cord fills most of the introvert.
- Sexes are separate in loriciferans. They have complicated life cycles with both sexual and asexual modes of reproduction demonstrated.

These tiny seemingly insignificant creatures may prove to be quite economically useful to humans. For reasons yet to be discerned, they are found in large numbers where methane seeps up through the crust and bottom sediments. In the future, we might use loriciferans as indicators for the presence of commercially significant pockets of methane gas, a cleaner alternative fuel to oil.

Phylum Priapulida (from *Priapos*, the Greek god of reproduction)

The priapulids are a small group consisting of only 16 species of marine worms found in colder oceanic waters. Most are burrowing predators that live in bottom sediments at depths from the intertidal zones to several thousand meters down.

They range in size from 0.55 mm (0.02 inch) to20 cm (8.0 inches) in length. Their body consists of a proboscis (introvert), a necklike **collar**, trunk or abdomen, and sometimes **caudal appendages (Figure 10.15)**. Their eversible proboscis is ornamented with papillae and ends with rows of curved spines that surround the mouth.

Characteristics of phylum Priapula are as follows:

- They are covered with a thin cuticle that is periodically molted throughout their entire life.
- The digestive system is complete. Most are predators that orient themselves upright in mud with their mouth at the surface where they use their extensible introvert and toothed pharynx to snare various soft-bodied invertebrates such as polychaete worms that wander by. Smaller species may feed on the organic detritus found in bottom sediments.
- The many protonephridia associated with the gonads form a urogenital system.
- The caudal appendages may serve for gas exchange, but they have no circulatory system.
- The nervous system consists of a nerve ring around the pharynx and a midventral nerve cord. Little is known of the sense organs in priapulids. The caudal appendage contains tactile receptors, and so may many of the bumps and spines on the body surface.
- The sexes are separate and

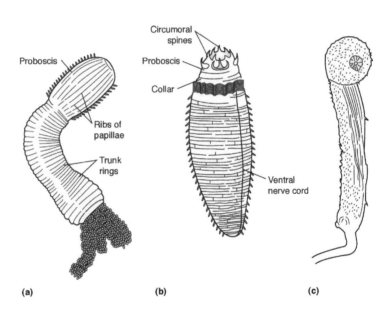

Figure 10.15 Representative Priapulids. (a) *Priapulus caudatus*. (b) *Halicryptus*. (c) *Tubiluchus corallicola*.

fertilization is external. A larval form known as a **loricate lava** for its resemblance to loriciferans eventually forms. The loricate larva lives in the mud where it goes through several molts as it grows into a juvenile priapulid.

A Closing Note
Passing the Pinworms

In this chapter, you were presented the opportunity for an academic encounter with pseudocoelomates. Chances are good, however, that you may have also had a more personal biological encounter with some of them.

I once had a junior high student bring me a large *Ascaris* worm in small jar of rubbing alcohol. Supposedly this student's younger sister had passed the worm while on the toilet and his mother wanted to know what it was and what she should do about it. As the only biology teacher in that small rural school, people viewed me as the resident expert on all things living. So regarded, I was constantly presented with all manner of specimens for identification and explanation. Although the young man who brought in the specimen found the whole affair quite funny (until he learned that he was probably infected as well), the mother and little sister were mortified when they learned the identity of the creature.

Even more widespread than *Ascaris* is the pinworm. Health authorities estimate (guesstimate?) that as many as 50 million people in the United States alone, mainly children aged 5-14, contract pinworms in one year. Pinworms can live in human intestines and survive by ingesting some of the nutrients found in predigested form in the human intestine. The most common symptom of pinworms is an itchy rectal area. Symptoms are worse at night (How do the worms know it is night?) when the worms are most active and crawl out of the anus to deposit their eggs. When an infected person scratches this area, the eggs can become lodged under the fingernails and spread to anything or anyone the infected person touches.

We discovered our family had been victimized by pinworms when my very young daughter started becoming very restless at night and was constantly scratching her behind. Our family doctor suggested we wait until she was asleep then pull her butt cheeks apart and examine her rectal area with a flashlight. And when we did, there they were—a number of tiny, white wriggly worms around her anus. (My wife nearly fainted on the spot.) Treatment consisted of the entire family (because we were all probably infected) taking a course of red liquid medicine that turned our stools some amazing colors.

After the first course of treatment, we waited several weeks and then repeated the treatment to make sure we had gotten all worms out of our systems. (Not to worry—although pinworm infections can be annoying, they do not usually cause any serious long-term health problems.) To this day my wife and our now grown daughter regard the whole affair as a dark family secret so embarrassing as to never be spoken of again. In truth, most people would feel that way about an encounter with parasitic worms, and understandably so. Whether you like it or not or even realize it or not, many of us have had personal guests from that parade of phyla we call the pseudocoelomates.

In Summary

- When it comes to body cavities, evolution resulted in three morphological types from the triploblastic animals that comprise Bilateria:

1. Acoelomate animals have no body cavity at all. Their organs have direct contact with the epithelium and semi-solid mesodermal tissues between the gut and body wall hold the organs in place.

2. Pseudocoelomate animals have a pseudocoel that is a fully functional body cavity. However, tissue derived from mesoderm only partly lines the fluid-filled body cavity of these animals.

3. Eucoelomate animals have a fluid-filled body cavity called a **coelom** with a complete lining called **peritoneum** derived from the mesoderm. The peritoneum allows organs to be attached to each other so that they can be suspended in a particular order yet still free to move within the coelom.

- "Pseudocoelomate" is a morphological term, not a taxonomic one, in that the various groups of pseudocoelomate animals do not constitute a monophyletic taxonomic group.
- Both the Lophotrochozoa and Ecdysozoa branches of Protostomia contain member phyla most (but not all) of which are pseudocoelomate in nature. As many as 25 phyla branch off from the protostome segment of the Tree of Life with most of those being pseudocoelomate in form. In this chapter we encountered:

    Superphylum Lophotrochozoa
        Phylum Rotifera—rotifers.
        Phylum Acanthocephala—spiny-headed worms
        Phylum Gastrotricha—bristle backs or hairy backs
        Phylum Entoprocta—goblet worms
    Superphylum Ecdysozoa
        Phylum Tardigrada—water bears
        Phylum Nematoda—roundworms
        Phylum Nematamorpha—horsehair worms
        Phylum Kinorhyncha—mud dragons
        Phylum Loricifera—brush heads
        Phylum Priapulida—penis worms

Review and Reflect

1. *Establishing Identity*. You are given three different pseudocoels—an acanthocephaln, a rotifer, and a priapulid—and are asked to identify them correctly to phylum. What features would you use to distinguish between these three pseudocoels?

2. *Rotifers Reborn*. Your zoology professor is away on a sabbatical and has left you in charge of feeding her rotifer culture daily, maintaining proper temperature in the lab, making sure the lights are turned on in the morning. Three weeks in, however, the snowstorm of the century, happens, and you are unable to get into the lab for a week. When you finally return, you dread what awaits with the rotifer culture because the lights have been off the whole time, the room temperature has

been very low, and the rotifers have had no food for the past week. To your surprise there are some rotifers that survived this disturbance, and you even find some reproductive females. Explain the difference in reproduction that likely occurred in the rotifers after the snowstorm.

3. *Lacking Systems.* Many of the pseudocoels lack a digestive system. Determine which ones lack a digestive system and explain how those types obtain food.

4. *Feed Me Seymour.* You have been given live samples of four different pseudocoels—lorciferans, kinorhynchs, entoprocts, and rotifers—and charged with caring for these cultures for several weeks. Explain what and how you would feed each type.

5. *What are the Advantages?* Compare and contrast the advantages and disadvantages of being an acoelomate, pseudocoelomate, and eucoelomate. Provide specific examples of animals in each category.

Create and Connect

1. *The Nematode Threat.* You are an advisor to the president of the World Health Assembly. As such, you have been asked to assess the potential threat of nematodes to the human population (both direct and indirect risks). Research one current health situation caused by a nematode species and write a press release that describes this animal and it threat to human beings. Next, develop an assessment plan (paragraph description) describing how to manage this nematode and reduce its negative effects on the human population. In your plan be sure to include at least two ways of dealing with the nematode and the potential impact of your plan on other species, humans, and the environment.

2. *What is it?* During a collecting trip to the tropics, you discover a small animal in the sediments along the coast. You believe it to be a pseudocoelomate, but you are unable to conclusively identify this organism, so you take it back to the lab for closer examination. Develop a plan for how you will identify the phylum of this organism. Make sure you eliminate all other possible pseudocoel phyla before you come to a conclusion.

3. *Rotifers With a Sprinkle of Salt.* Design an experiment to test whether changes in salinity will alter the reproductive cycle of a rotifer.

Guidelines:

Your design should include the following components in order:

➤ The *Problem Question.* State exactly what problem you will be attempting to solve.
➤ Your *Hypothesis.*
➤ *Methods and Materials.* Explain exactly what you will do in your experiment, including the materials necessary to accomplish the task. Be specific, take nothing for granted, and do not expect people to read your mind as they read your plan.

> ➤ *Collecting and Analyzing Data.* Explain what type(s) of data will be collected and what statistical tests might be performed on that data. It is not necessary to concoct either fictitious data or imaginary observations.

4. ***What's in a Quote?*** Using the N. A. Cobb quotation about nematodes under the chapter heading *Phylum Nematoda* as inspiration, do the following:

    A. Use your knowledge of the life history, reproduction, and feeding patterns of nematodes to explain why nematodes are so abundant and found in nearly every imaginable habitat on this planet.

    B. Research and describe the life history (feeding mechanism, reproduction, and movement) of a species of nematode that lives in each of the following habitats: fertile soil, lakes, and ocean. Explain how this species has adapted to its respective habitat.

PHYLUM ANNELIDA: MASTERS OF COELOMS AND SEGMENTS

It is a marvelous reflection that the whole expanse has passed, and will again pass, every few years through the bodies of worms. The plough is one of the most ancient and most valuable of man's inventions; but long before he existed the land was in fact regularly ploughed, and still continues to be thus ploughed, by earthworms. It may be doubted whether there are many other animals which have played so important a role in the history of the world as have these lowly organized creatures.

—Charles Darwin

History of Annelids

Fossil worm tracks from what was the muddy sea floor 540-560 million years ago hint at the next great evolutionary lurch forward in animal complexity. Somewhere around this time the ancestors of the annelid worms are thought to have first appeared. Probably resembling the modern polychaetes (sea worms), these worms bore for the first time two great evolutionary advances necessary for the development of complex animal bodies—a true coelom and **metamerism** (segmentation).

The ancient annelids have undergone extensive adaptive radiation over the course of their evolution. The basic body structure of the polychaetes has proven to be adaptively plastic and has lent itself to a bewildering array of modifications and habitats (**Figure 11.1**).

Diversity and Classification

The approximately 22,000 species of worms known taxonomically as annelids, or more commonly as segmented worms, are round, cylindrical animals with segmented bodies composed of muscular rings (L. *annelus*, little ring + *ida*, pl. suffix). This group includes the familiar earthworms and leeches as well as a bewildering variety of marine "sandworms," "tubeworms," "plumed worms," and others.

With a great affinity for water, the annelids have successfully occupied virtually all habitats where suitable moisture is available. They are particularly bountiful in the ocean but also proliferate in freshwater; many live in damp soil on land. They also run the full gamut of ecological relationships with other animals as there are predatory, parasitic, mutualistic, and commensal species of annelid worms.

The terrestrial annelids are found in damp soil practically everywhere, although the dryness of deserts and polar cold seriously limits their occurrence in those climes. The vast majority of annelids, however, are at home in aquatic habitats. From freshwater pools and streams to ocean beaches and bottoms, the water teems with aquatic annelids in numbers and diversity of form and appearance that spark the imagination and boggle the mind. One study found the average density of marine worms on the ocean floor off Tampa Bay, Florida to be an amazing 13,425 per square meter. They are so prevalent that such worms may compose 40-80% of the total fauna on the upper continental slope and the deep ocean floor.

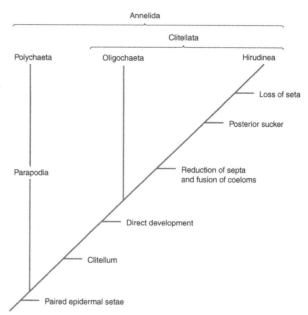

Figure 11.1 In this interpretation of the evolultionary relationships between traditional annelid taxa, paired epidermal setae is the symplesiomorphic characteristic and polychaetes are the outgroup. The lack of synapomorphic characteristics indicates that the oligochaetes were most likely ancestral to the leeches.

In size annelids range from tiny aquatic worms not even a millimeter (0.04 inch) long as adults to the giant earthworms between 3 to 4 meters (10-12 feet) long found in Australia, New Zealand, South Africa, and South America. *Lamellibrachia luymesi* is a cold seep marine tube worm that reaches lengths of over 3 meters (10 feet) and possesses such a slow growth rate that some individuals are a record-setting 250 years old or more.

Leeches are larger on average than other annelids with most species being 2 to 5 cm (0.8-13 inches) long, but the monster of the class is the Giant Amazonian leech, *Haementeria ghilianii*, which can reach 50 cm (20 inches).

Annelids were customarily classified based primarily on the presence or absence of fleshy lobes called **parapodia** and bristles known as **chaetae** or **setae** as well as other select morphological features. Traditionally, the phylum Annelida has been classified:

> Domain Eukarya
>> Kingdom Animalia
>>> Phylum Annelida
>>>> Class Polychaeta—Sea worms
>>>> Class Oligochaeta—Earthworms
>>>> Class Hirudinea—Leeches

However, recent cladistic analysis has presented other possible interpretations of the classification of annelids. Primarily because earthworms and leeches both possess a clitellum that functions in cocoon formation and because both groups are hermaphroditic (monoecious) with direct development, some authorities believe a more accurate classification of phylum Annelida to be:

Domain Eukarya
    Kingdom Animalia
        Phylum Annelida
            Class Polychaeta—Sea worms
            Class Clitellata
                Subclass Oligochaeta—Earthworms
                Subclass Hirudinea—Leeches

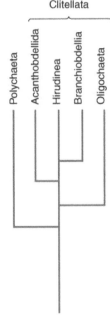

(**Figure 11.2**) Lamarck first established the taxon in 1809 and since that time studies attempting to sort out the phylogenetic relationships within the phylum and to other phyla have often resulted in interpretations that are often radically different and sometimes conflicting. Questions abound:

- Should the phylum Annelida continue to be recognized as a monophyletic (single) group at all? Recent cladistic studies suggest the polychaetes arose independently of the oligochaetes and leeches. This has prompted some authorities to propose that the taxon "Annelida" be abandoned altogether.

Figure 11.2 The possible phylogeny of the various taxa comprising phylum Annelida.

- Because they probably all originated from an ancestral metameric species, should Annelida and Arthropoda be united under a single taxon called Articulata?
- Should the oligochaetes and leeches be combined into Clitellata and placed as a clade within Polychaeta?
- What of the Branchiobdellida? The branchiobdellians are small worms that are parasitic or commensal on freshwater crayfish. They have 15 segments and a head sucker, and although they show similarities to both oligochaetes and leeches and have been included with both groups, they are considered a separate class by some authorities.
- What of the Acanthobdellida? Once considered only a separate genus of leech, these annelids have some characteristics of both leeches and oligochaetes and thus may warrant separation from other leeches into a special class.
- What of the Pogonophorans (beard worms), Echiurans (spoon worms), and Vestimeniferans? These three groups were formerly placed in a separate phylum, but recent work suggests that they are most likely polychaetes.

Clearly issues of taxonomic and phylogenetic relationships within the phylum Annelida are numerous and the subject of intense research and debate. With uncertainty ruling the taxonomy of annelids, we believe it educationally prudent to focus only on those groups traditionally investigated as a part of introductory zoology courses—sea worms, earthworms, and leeches.

Phylum Characteristics

The members of phylum Annelida exhibit the following general characteristics:

- Their bilaterally symmetrical body is metameric (segmented) both internally and externally.
- Except for leeches, they possess a true coelom that is well developed and divided by **septa**. Coelomic fluid supplies turgidity and functions as a hydrostatic skeleton.
- The body is covered with a thin transparent moist cuticle secreted by the epidermis.
- They possess chitinous setae only (oligochaetes), setae on fleshy appendages called parapodia (polychaetes), or neither (leeches).
- The digestive system is complete with some specialization.
- They possess a closed circulatory system but no true heart. Respiratory pigments such as hemoglobin, hemerythrin, or chlorocruorin are often present in the blood.
- The excretory system typically consists of a pair of nephridia in each segment (metamere).
- Gas exchange occurs across moist skin, parapodial flaps, or through gills.
- The nervous system typically consists of a mass of cerebral ganglia (primitive brain) located on the dorsal end in the head region, connected by a ring of nerves to a ventral nerve cord that runs the length of the body. The nerve cord branches to lateral nerves and ganglia in each segment. Sense organs include tactile organs, statocysts (organs of equilibrium), photoreceptor cells and eyes with lenses in some species, as well as chemoreceptors.
- They are hermaphroditic or separate sexes with asexual budding occurring in some species.

Class Characteristics

Class Polychaeta (Gr. *polys,* many + *chaite,* long hairs)

Polychaetes, also known as sea worms or sandworms, are the largest class of annelids. Sea worms have distinct heads with eyes and tentacles, and most segments exhibit parapodia (fleshy or feathery paired lateral appendages) bearing tufts of many chaetae or setae (bristles).Polychaetes lack a clitellum, usually are a single sex—male or female (**dioecious**), and usually develop from a **trochophore larvae**. Many polychaetes are also strikingly beautiful, colored of red, pink, green or a brilliant combination of colors.

The polychaetes are the dominant group of annelids both in number of species with their 10,000 or so species comprising two-thirds of the species in the phylum and in the more advanced construction of their bodies, especially the head end. This class of annelids lives exclusively in or near the ocean. Some burrow in the sand and mud, some form tubes in bottom sediments or attached to submerged objects, whereas others swim freely about as part of the pelagic plankton population. Many are filter feeders that sift detritus and plankton from the water with feathery gills or tentacles. Others eat their way through organic rich bottom sediments, whereas some types, such as *Nereis* the clam worm, are predatory. With astonishing speed and dexterity, these worms thrust out their proboscis equipped with chitinous hooks to impale small invertebrate prey such as other worms. Clam worms (also known as ragworms or sandworms) also scavenge dead fish and will even eat algae.

Based on their level of activity, morphology, and ecology, polychaetes can be divided into two morphologic groups: the errantia (free-moving) sea worms and the sedentaria (sedentary) sea worms.

Figure 11.3 Representative Errant Polychaetes. (a) *Hermodice carunculata* is commonly known as the bearded fireworm because the whitish tufts of bristles (beard) can penetrate human skin and inject a powerful neurotoxin causing an intense burning sensation (fire). (b) Head region of *Nereis* from the left side. (b1) Proboscis extended. (b2) Proboscis retracted.

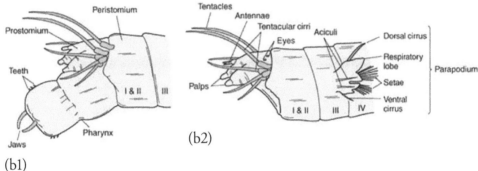

Errant polychaetes. Errant polychaetes tend to be active sea worms that are found crawling or swimming much of the time. They may, however, spend some of their time in burrows or crevices and under rocks. Most errant polychaetes have a well-developed head bearing eyes and sensory tentacles. Many errant polychaetes are predatory whereas some are scavengers whereas others are herbivores, using their jaws for tearing plants. Some jawed polychaetes even have poison glands at the base of the jaws (**Figure 11.3** and **Figure 11.4**).

Sedentary polychaetes. Most sedentary polychaetes are adapted to a life of permanent confinement in a tube or burrow. Some, such as the lugworm (*Arenicola*), burrow through soft sand or mud sediments. The majority of sedentary polychaetes, however, are tube builders. (**Figure 11.5**) The tubes of different species vary greatly in their composition and structure. Some tubes, known as shaggy tubes, may be constructed of particles of sand, shell, or other materials all held together with mucus. In others, the tube may be made entirely of substances secreted by the worm itself that harden on contact with water. Life in a tube dictates the necessity of developing a greatly modified head for specialized feeding. Some have tentacles they throw out from the tube across the mud or sand and then drag them about to pick up organic deposits. Others are filter feeders. The striking feather-duster worms extend a whorl of feathery, ciliated tentacles from the tube opening to trap small planktonic organisms drifting in the water. This ornate crown can be pulled back down into the tube with amazing speed if the worm is threatened or disturbed, sometimes by nothing more than a shadow.

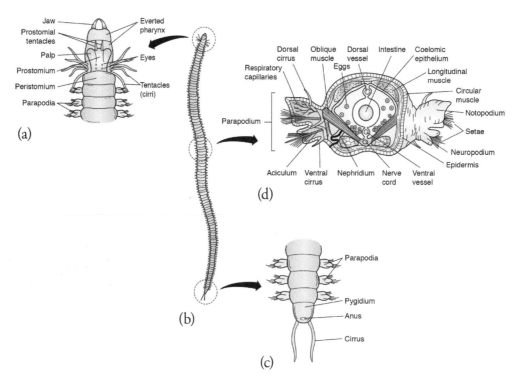

Figure 11.4 The Body Plan of an Errant Polychaete: *Nereis* (clamworm). (a) Anterior (head) end. (b) General morphology. (c) Posterior (rear) end. (d) Anatomy as displayed by a cross-section slice.

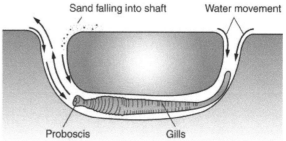

Figure 11.5 Representative Sedentary Polychaetes. (a) The extended feathery feeding branchiae (gills) of the magnificent feather duster worm (*Sabellastarte magnifica*) (Sabellidae). (b) The lugworm (*Arenicola*) lives in a U-shaped burrow on intertidal mud flats. It burrows with its proboscis and uses peristaltic movements of its body to keep water coming in. This worm ingests the food-rich sand falling into the burrow. (c) The spaghetti worm (*Amphitrite*) at the top of its burrow. Outstretched tentacles collect food.

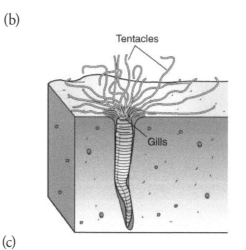

Class Oligochaeta (Gr. *oligos*, few + *chaite*, long hairs)

Members of this class do not possess a distinct head, lack parapodia, and have only a few small chaetae per segment. (**Figure 11.6**) They do possess an external mucus-producing band known as a **clitellum**. Oligochaetes are usually **monoecious** (possessing both male and female gonads in the same body), and develop directly with no larval stage (**Figure 11.7**).

People are the most familiar with the terrestrial oligochaete earthworms that are seen in soil when it has been overturned or when the worms come out on the sidewalk after a heavy rain. Such worms eat their way through the soil digesting out the organic matter therein.

> *Earthworms though in appearance a small and despicable link in the chain of nature, yet, if lost, would make a lamentable chasm… worms seem to be the great promoters of vegetation, which would proceed lamely without them.*
>
> —Gilbert White

In North America, much of Canada was stripped of its native earthworms during the last Ice Age, but relatively little of the United States and none of Mexico were affected. There are currently 17 native species and 13 European species in the eastern United States. The spread of the introduced European species has been far less successful in western areas as the European worms do not adapt well to dry conditions. Based on habitat and feeding behavior, the over 3,000 earthworm species can be divided into three general groups. Again, these groups are morphologic rather than taxonomic in nature.

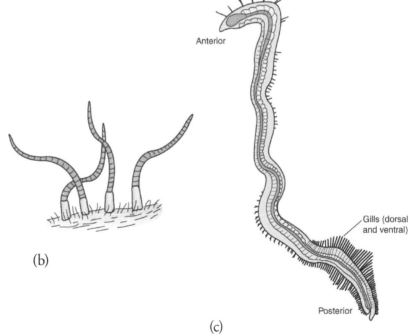

Figure 11.6 Representative Oligochaetes. (a) *Lumbricus terrestris* or night crawler is widely distributed in damp soil worldwide. (b) The freshwater tube dwelling *Tubifex*. The posterior end of the worm extends from the tube. (c) The freshwater bottom-dwelling *Branchiura sowerbyi*.

- **Epigeic species**. Living near or on the surface, these worms ingest large amounts of undecomposed litter. However, they are exposed to climatic fluctuations and extreme predator pressure resulting in worms with small bodes and rapid reproduction rates. A common example is *Eisenia foetida* (redworm or manure worm) which is used in **vermiculture** or **vermicomposting** (earthworm "farming").

- **Endogenic species**. Foraging below the surface, these worms ingest large quantities of organic rich soil and build burrows that are mainly horizontal in nature.

- **Anecic species**. Although they build permanent, vertical burrows that penetrate the soil deeply, these worms come to the surface to feed on partially decomposed litter, manure, and other organic matter. Their burrows create a microclimatic gradient, and the worms can be found shallow or deep in their burrows depending on the prevailing conditions. Anecics have profound effects on organic matter decomposition, nutrient cycling, and soil formation. The most common example is the nightcrawler, *Lumbricus terrestris*, sold by fish-bait dealers.

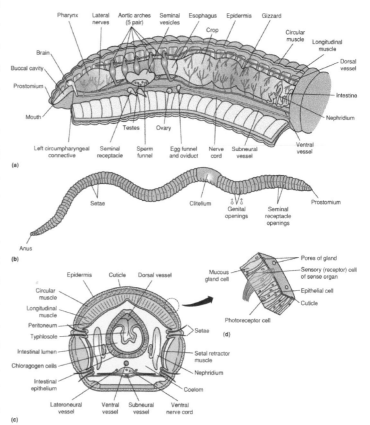

Figure 11.7 The Body Plan of an Oligochaete: Earthworm (*Lumbricus*). (a) Lateral view of the anterior end anatomy. (b) General morphology. (c) Anatomy as displayed by a cross-section slice. (d) A piece of the epidermis.

Found creeping along the bottom or burrowing into the mud of stagnant pools and ponds, are the freshwater annelids. Although usually smaller than the earthworms, they tend to be more active and mobile than their terrestrial counterparts. Their main food sources are algae and detritus, which may be harvested by mucus-coated mouthparts or swept in on currents generated by cilia. The burrowers of the group swallow mud and digest out the organic content. A few species are **ectoparasites** on other pond animals.

Class Hirudinea (L. *hirudo*, leech + *ea*, characterized by)

Leeches have a fixed number of segments (normally 34) with many **annuli** (rings). Parapodia and setae are absent, but like the oligochaetes, leeches are hermaphroditic (monoecious), and possess a clitellum that appears only during the breeding season. Unique to the leeches are suckers and leeches possess two of them—a small anterior (oral) sucker surrounding the mouth and a large posterior sucker. Black, olive green, brown and red are common leech colors but striped and spotted patterns are not unusual (**Figure 11.8**).

Leeches are found predominantly in and around fresh water. Few species can tolerate rapid currents and most prefer the shallow, vegetated margins of ponds, lakes, and sluggish streams. A few types, however, are marine, and some have even adapted to a strictly terrestrial life, although they are limited in this existence to warm, moist tropical areas. Although leeches are found throughout the world, they are most abundant in northern temperate lakes and ponds.

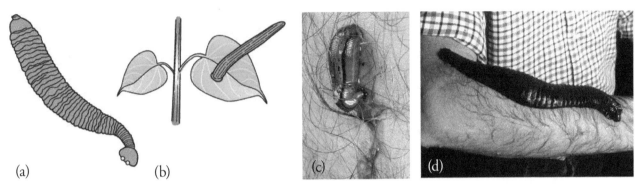

Figure 11.8 Representative Hirudineans. (a) The fish leech (*Pontobdella muricata*). (b) A tropical blood-sucking leech (*Haemadipsa*) poised on a leaf awaiting prey to brush by. These leeches will converge and drop down onto a person or other mammal remaining still for a time. (c) A feeding medicinal leech (*Hirudo medicinalis*). (d) *Haementeria ghilianii*, the giant Amazon leech, is the world's largest species of leech.

If an aquatic environment is high in organic pollutants, the number of leeches can reach epic proportions. More than 10,000 individuals per square meter have been reported in such conditions. During times of drought, some species survive by burrowing into the mud at the bottom of a pond or stream where they **estivate** (state of summer torpor). In such situations they can survive the loss of as much as 90% of their body weight and return from near death when moisture levels increase.

Leeches are the smallest (around 500 species) but most specialized group of annelids. Most of their distinguishing characteristics are not found in the other two annelid groups but they do share some common features with the oligochaetes (**Figure 11.9**).

Annelid Body Plan

Of all the protostomic animals loosely referred to as "worms," the annelids are the most structurally complex. As such, annelids stand roughly midpoint on a scale that would place the simplest animals on one end and the most complex on the other end.

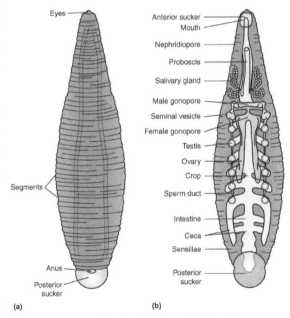

Figure 11.9 The Body Plan of a Hirudinean: Leech (*Hirudo medicinalis*) (a) Morphology. (b) Anatomy. Adult leeches feed on mammal blood whereas young leeches feed on frog blood.

The coelom, segmentation, closed circulatory system, specialized gut, and the nature of the excretory structures act in concert to free the annelids from many of the constraints imposed by the acoelomate and pseudocoelomate body plans. The annelids are bound neither to the small size nor the extremely flattened shapes dictated by the necessity of short diffusion distances. The efficiency and specialization of musculature and metabolic organ systems made possible through segmentation also allow them the most active lifestyle of any animal type so far discussed in this book.

Segmentation

The bilaterally symmetrical annelid body is **metameric** (segmented) both internally and externally. Internally, each segment has a fluid-filled cavity (**coelom**) separated and divided from neighboring segments by thin, fleshy walls known as **septa**. Structures such as the numerous respiratory, locomotory, and excretory organs are repeated with each segment. The gut and longitudinal blood vessels perforate the septa of each segment and run the entire length of the worm (Figure 11.7a).

During embryonic development, the solid mass of mesoderm that occupies the region between the ectoderm and endoderm on either side of the embryonic gut (**schizocoel**) is split segmentally giving rise to body cavities. The body cavity of annelids arises by the segmented splitting of a solid mass of mesoderm that occupies a region between ectoderm and endoderm on either side of the schizocoel. Each segment is marked externally by one or more rings (annuli) and is covered with a thin, moist cellophane-like cuticle overlying a layer of epidermal tissue (**epidermis**). In oligochaetes and leeches, a series of swollen segments in the anterior half of the worm form a girdle-like band around the worm called the **clitellum** (Figure 11.7b). The clitellum secretes mucus for copulation and for forming cocoons around the eggs. Secondarily, this mucus also keeps the skin moist allowing for more efficient respiration, and it also helps lubricate the worm for easier movement through soil tunnels.

One advantage to the segmental arrangement of fluid-filled coelomic spaces is the formation of hydrostatic sections. The constant volume and pressure of fluid within these cavities serves as a **hydrostatic skeleton** against which the muscles can operate. In leeches, however, the coelom is filled with tissue and reduced to a series of narrow canals.

A second advantage of metamerism (segmentation) is that it reduces the impact of injury. Although a few segments might be damaged, the fact that they are walled off from the other segments increases the likelihood that the worm will survive the trauma.

Thirdly, metamerism permits the modification of segments for specialized functions, such as locomotion, reproductions and feeding, through a process known as **tagmatization**. Tagmatization in polychaetes has resulted in a true head on which the **prostomium** is equipped with antennae, **sensory palps**, and ciliated pits or grooves known as **nuchal organs**. The first anterior segment, the **peristomium**, surrounds the mouth and bears sensory tentacles (**cirri**). The peristomium in predatory types may also bear chitinous hooks whereas ciliary forms may bear a crown of feathery tentacles that open like a fan.

Box 11.1
The Significance of Segmentation

Charting the path of animal evolution from the earliest beginnings reveals a pattern of milestones of ever-increasing complexities that form the foundations that later groups build upon. One of those milestones was the appearance of segmentation (metamerism) in the ancestors of that group of animals we call annelids. Segmentation is widespread among animals, occurring in annelids, arthropods, and most chordates.

A fully segmented animal is like a classic train with an engine in front, a string of cars, and a caboose on the end. That is, segmented animals typically have a specialized anterior segment(s) usually in the form of a head and a posterior segment(s) with a varying number of intermediate segments in-between. In examples of near perfect segmentation, appendages, muscles, nerves, blood vessels, coeloms, and excretory and reproductive systems are faithfully replicated in each segment.

Segmentation is an efficient method of constructing an animal sequentially from embryo to fully-formed individual, but once the pattern is set it is apparently unnecessary for animals to remain extremely segmented. Some segmented animals, like most annelids and centipedes, remain extremely segmented as adults; others like flies or vertebrates lay out their developing embryos using a segmented pattern and then modify this pattern as development continues until it almost disappears in the adult. In vertebrates, segmentation is readily apparent only in the axial skeleton (vertebrae), nerves, and muscles (like the "six-pack" abdominal muscles so highly prized by body builders).

Zoologists believe that segmentation evolved independently several times over the course of animal evolution. Why? What is the significance of segmentation? The evolution of segmentation, like the evolution of multicellularity or colonies, provided a framework for specialization. In colonies, such specialization is apparent in the polymorphism of zooids, whereas, in segmented animals, it results from regional specialization of segments. In annelids, regional specialization has resulted in a body that may be weakly divided into a head, thorax, and abdomen, although these same three divisions are much more prominent in many arthropods. In the case of higher vertebrates such as humans, regional specialization is carried to its most extreme. Humans have a well-developed head "segment" equipped with a large centralized brain and several major sensory systems. Our thorax and abdominal segments are fused and hollow containing many large and complex organ systems. The human body is what it is morphologically because of the extreme regional specialization that occurs within the first primitive segments that appear in our early embryonic stages.

Segmentation is also an important event in the orderly development of form, a process known as **somitogenesis** (the formation of somites or body divisions). Models on the mechanisms controlling this process date back some three or four decades. Understanding the genetic and molecular control of somitogenesis has been gained only recently by the discovery of molecular oscillators or segmentation "clocks" (Notch signaling pathway and Wnt signaling pathway) and gradients of signaling molecules (the FGF8 gradient and the Wnt gradient), as predicted by early models. However, some important pieces of this puzzle are still missing, though the overall picture is taking shape.

Thus, we realize that be we earthworm, fruit fly or human, we owe the development and form of our body to the evolution of segmentation.

* Adapted from *Invertebrate Biology*. Sixth Edition. Edward Ruppert and Robert Barnes. Saunders College Publishing. 1994.

Body Covering

The exterior of an annelid is covered with a thin transparent **cuticle**, marked with fine cross striations that produce a slight iridescence. The nonliving cuticle, which is never shed or sloughed off, is secreted by the epidermis which lies just beneath it. The epidermis consists of a single layer of columnar epithelium in which are embedded numerous **unicellular glands** producing mucus that passes out through pores in the cuticle. The cuticle is also perforated by pores over many **sensory cells** in the epidermis. (Figure 11.7c) Both types of pores can be easily seen in pieces of cuticle that have been stripped off and spread to dry on a glass microscope slide. The epidermis rests on the dermis and beneath dermis are the muscles.

Setae

Distinct to annelids are structures called chaetae. **Setae** are bundles of chitinous bristles. They are produced within follicles or sacs that arise from the invagination of certain epidermal cells. Setae show a tremendous amount of variation, from long, thin filaments to stout pronged hooks (**Figure 11.10**).

Polychaetes (*poly* = many) carry many more chaetae (setae) than the other types of annelids. The setae of polychaetes are found at the distal ends (outer margins) of parapodia where they function for digging into the substrate, holding a worm in its tube or burrow, or increasing the surface areas of appendages for swimming. The tropical fireworm (*Hermodice*) has setae that are composed of calcium carbonate rather than chitin (Figure

Figure 11.10 Magnified view of the setae of an earthworm (*Lumbricus terrestris*). These stiff bristles provide traction for movement and help prevent predators from pulling the worm straight out of the ground.

11.3a). These brittle spikes are hollow, and contain a toxin that can cause a painful, burning reaction in the skin of any human unlucky enough to brush against them. The polychaetes known as scaleworms are named for the overlapping scales that cover the animal's upper surface. In one type of scaleworm, known as the sea mouse (*Aphrodita*), these scales are completely covered with long slender, felt-like setae projecting from the parapodia giving the worm a hairy or fuzzy mouse-like appearance.

Oligochaetes possess four pair of setae per segment. In a few leech species, setae appear only on several anterior segments but for the most part, leeches lack setae entirely.

Support and Movement

Muscles develop from the mesodermal layers associated with each segment. A layer of circular muscles lies below the epidermis, and a layer of longitudinal muscles, just below the circular muscles, runs between the septa that separate each segment. In addition, some polychaetes have oblique muscles, and leeches have dorsoventral muscles. Both polychaetes and oligochaetes rely heavily on their well-developed hydrostatic skeleton for support and movement. Body support in leeches, however, is provided by the mostly solid construction of the body, bands of muscles, and the hydrostatic qualities of the coelomic cavities.

Contraction of the longitudinal muscles causes the body to shorten, whereas contraction of the circular muscles causes it to lengthen. The resultant changes in the shapes of groups of segments provide the basis for swimming, burrowing, and crawling in polychaetes and oligochaetes.

Most leeches slink along with looping movements of the body, attaching first one sucker and then the other and pulling the body along the surface. Aquatic leeches swim with a slow, graceful undulatory movement.

Feeding and Digestion

The digestive system of annelids consists of a single unsegmented straight tube (**gut**) running the length of the worm from the mouth on the underside of the head on the anterior end to the anus on the terminal **pygidium** segment at the posterior end. The gut perforates the septa of each segment and is held in place through the middle of the worm by the membranes of septa and mesentery tissue. In polychaetes, the gut is differentiated into a **pharynx** (or **buccal cavity** if the pharynx is absent), short esophagus, stomach (in sedentary species), intestine, and anus. The stomach and upper intestine produce enzymes for extracellular digestion. The intestine, often folded to increase its surface area, is the site of food absorption.

Errant polychaetes are typically predators and scavengers. The tube-bound sedentary polychaetes filter suspended organic particles from the water, whereas the burrowers consume organic material on or in the sediments. Lug worms (*Arenicola*) live by burrowing through marine sediments and consuming the organic matter in the sediments (Figure 11.5b). In the shallow water littoral zones where they are especially abundant, lug worms may ingest as much as 1,900 tons of sand per acre per year. Over several years time, these worms can redistribute virtually all of the sand in their vicinity of activity to a depth of about 60 cm (24 inches).

Tube-dwelling polychaetes face a problem—how are feces to be removed from the tube after defecation? In those that live in tubes with two openings, the wastes are readily removed by water circulating through the tube. Those that live in tubes with only one opening may turn around in the tube and thrust the pygidium out of the opening during defecation. Still others have ciliated grooves in which the feces are slowly moved by the cilia from the anus up and out the tube opening.

In oligochaetes, (Figure 11.7a) the mouth, situated beneath the prostomium, opens into a small buccal cavity, which in turn opens into the bulb-like pharynx. In earthworms, the pharynx acts as a pump whereas pharyngeal glands produce salivary secretions containing mucus and enzymes.

The pharynx opens into a narrow, tubular esophagus which in turn leads to a large thin-walled storage chamber known as the **crop**. The crop leads to the muscular **gizzard** where food particles are ground up.

A characteristic feature of the terrestrial oligochaete gut is the presence of **calciferous glands** in the esophagus. These glands secrete calcium carbonate into the esophagus in the form of calcite crystals. These glands and the calcite crystals they form may function to remove excess calcium taken in with food, remove CO_2 via the gut when surrounding soil CO_2 are very high, and regulate the acid-base balance of body fluids.

The intestine forms the remainder of the digestive tract. The anterior (front) half of the intestine is the site of secretion and digestion, and the posterior (back) half is the primary site of food absorption. The internal surface area of the intestine is increased in many types of earthworms by a ridge or fold called a **typhlosole**.

Most oligochaete species are scavengers and feed on dead organic matter, particularly vegetation. Earthworms feed on decaying matter at the surface and may pull plants into the burrow. They also ingest organic matter from mud or soil as they burrow.

Although the blood-sucking behavior of some leeches gives the entire group a disgusting reputation, many leeches survive as scavengers or as active predators on small invertebrates such as worms, snails, and insect larvae. Some leeches are equipped with a **proboscis** that can extend out of the mouth and forced into the soft tissue of the host or prey. Other types lack a proboscis and possess a sucking pharynx and bladelike jaws instead (**Figure 11.11**).

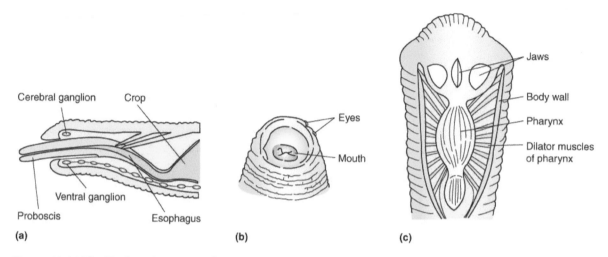

Figure 11.11 The Feeding Apparatus of a Leech. (a) Sagittal section through the anterior end of a proboscis-feeding leech with proboscis extended. (b) Ventral view of the oral region of a terrestrial blood-sucking leech. The jaws are not exposed. (c) Ventral dissection of the anterior end of a medicinal leech (*Hirudo medicinalis*) showing the jaws.

A skillful leech is better far, than half a hundred men of war.

—Samuel Butler

Over half of the known species of leeches are blood-sucking ectoparasites that attack a variety of hosts. Some feed only on invertebrates, such as snails, oligochaete worms, crustaceans, and insects, but most species prefer vertebrates. Parasitic leeches are rarely confined to one host, but they are usually confined to one class of vertebrates. Fish leeches will not attack reptiles and leeches that feed on reptiles rarely attack amphibians or mammals. Furthermore, some species of leeches that are exclusively bloodsuckers as adults are predacious during juvenile stages of their development.

Bloodsuckers feed infrequently, but when they do feed, they can consume enormous amounts of blood, ranging anywhere from two to ten times their own body weight. Digestion of the blood takes place very slowly and so these leeches can tolerate long periods without a meal. Medicinal leeches may require 200 days to digest a meal and thus need not feed more than twice a year in order to grow.

Leech digestion is peculiar in a number of respects. The gut secretes few digestive enzymes but is equipped with a huge flora of symbiotic bacteria. These bacteria are thought to be responsible for a considerable part of the digestive process in all leeches. The bacteria may also produce vitamins and other chemical compounds useful to the leech.

Circulation and Gas Exchange

Annelids usually possess a well-developed closed blood-vascular system (Figure 11.7a). The blood in most annelids never leaves the blood vessels as it moves in a loop anteriorly (forward) in a **dorsal vessel** situated over the digestive tract, laterally (down) in smaller vessels and capillaries in each segment, around and through the internal organs, and into a **ventral vessel** that carries it posteriorly beneath the digestive tract. Some of the lateral vessels are thick-walled and muscular. Their contractions serve as "hearts" propelling the blood through the system. In some annelid types, **coelomic sinuses** (openings) act as a supplemental circulatory system, particularly in leeches. In fact, some leeches have no circulatory system at all, and the coelomic sinuses and coelomic fluid serve as the sole internal transport system. The blood of most annelids contains the respiratory pigment hemoglobin or hemoglobin-like chlorocruorin and ranges in color from clear (no pigments) to red (hemoglobin) to green (chlorocruorin).

Respiration (gas exchange) occurs primarily across moist skin in oligochaetes and leeches, but across parapodia and gills in various species of polychaetes but mainly through gills in the majority of polychaetes. Again, life in a tube presents respiratory restrictions that demand survival adaptations. The tube-dwelling polychaetes often drive water through their burrows by undulating or peristaltic contractions of the body in a process known as **ventilation.** Research has shown that ventilation increases the worm's oxygen requirements as much as 15-fold, but there is approximately a 20-fold increase in oxygen uptake.

Low oxygen levels and even complete lack of oxygen also presents survival challenges to aquatic worms. The oligochaete sewage worm (*Tubifex*) ventilates in stagnant water by sticking its posterior end out of the mud and waving it about. If oxygen levels continue to drop, *Tubifex* can switch to anaerobic respiration and live up to 20 days in an environment completely devoid of oxygen.

Although earthworms are considered terrestrial animals, they are not adapted to life on the surface in contact with the drying effects of air. Respiratory exchange must occur by diffusion through the general body surface, which is underlain by a network of capillaries. The surface of the worm must be thin and must be kept moist for this exchange to occur for they dry, they die. Moistening of the surface is accomplished by mucus glands that occur in the epidermis and by the release of coelomic fluid that issues from pores located in grooves between segments.

Excretion and Osmoregulation

Annelids filter wastes from their bodies through a system of tubes known as **nephridia** found in each segment of the annelid body. (**Figure 11.12**) There are two types of nephridia. One type, the **protonephridium** (a tubule with a closed bulb at one end and an opening to the outside at the other end), is found only in some polychaetes. Most annelids possess **metanephridia** that consist of an open ciliated funnel called **nephrostome** on one end and opening to the outside on the other end. In both types, filtered wastes are discharged to the outside through an

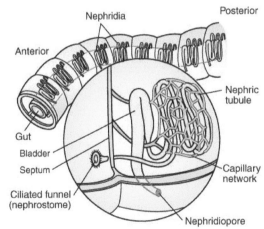

Figure 11.12 The Nephridium of an Earthworm. Wastes are drawn into the ciliated nephrostome in one segment, passed though the loops of the nephridium, and then expelled though the nephridiopore of the adjoining segment.

317

opening known as a **nephridiopore**. These tubes perform the function of kidneys in that they take in coelomic fluid as well as salts and organic wastes from the blood capillaries around them, filter these fluids, and then discharge wastes to the outside.

The nephridia are not the only means of excretion in annelids. The coelomic lining surrounding the intestine and the main blood vessels is modified into special **chlorogogen cells**. Nitrogenous wastes such as ammonia and urea extracted from the blood accumulate in the chlorogogen cells. These cells eventually detach and float in the coelomic fluid. Some of the chlorogogen detritus is removed by the metanephridia directly whereas the remainder may be carried by the blood or eliminated through the body surface.

In terrestrial earthworms, some of the chlorogogen cells are engulfed by amoeboid cells of the coelomic fluid, which wander into the tissues and disintegrate, leaving their wastes as deposits of pigments in the body wall. These pigments help shield the underlying tissues from the light, particularly ultra-violet, which is very harmful to earthworms. One hour's exposure to strong sunlight causes complete paralysis in some worms, and several hours' exposure is usually fatal.

Nervous System and Sense Organs

The more active forms of annelids—errant polychaetes and oligochaetes—have the most complex and differentiated brain whereas sedentary burrowing, or tube-dwelling worms have simple brains with little differentiation. (**Figure 11.13**) The cerebral ganglia are connected by a looping pair of nerves to a double ventral nerve cord. This nerve cord lies beneath the digestive tract and runs the entire length of the worm. Lateral nerves branch off the ventral nerve cord up into the body wall, muscles and gut of each segment.

Polychaetes have the most highly developed sense organs of all the annelids. The principal sense organs, found mainly on the head, are the eyes, nuchal organs, and statocysts. Eyes range from simple eyespots to well-developed visual systems. The wandering errant polychaetes tend to have the most developed eyes.

Nuchal organs are sensory pits or slits located in the head region. These areas function as chemoreceptors for detecting food. **Statocysts** are sensory chambers that aid some burrowing and tube-building polychaetes in proper body orientation. Spicules, diatom shells, and quartz grains may be found in these chambers where they stimulate nerve endings that signal the orientation of the head of the worm. For example, if an aquarium containing a worm that always burrows head down is tilted 90 degrees, the worm makes a compensating 90-degree turn in its burrowing. The body wall of polychaetes also contains numerous ridges and bands of tactile (touch) receptors.

Seemingly defenseless and blessed with none of the well-developed sense organs of the polychaetes, earthworms have managed not only to survive but to thrive. Although they have no specialized sensory organs, their bodies are covered with various sensory receptors that

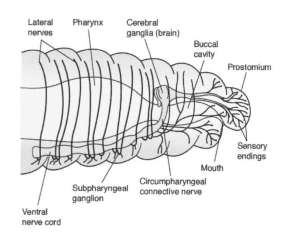

Figure 11.13 The Anterior Portion of the Nervous System of an Earthworm. Note the high concentration of sensory nerve endings in this region.

allow them to react to a variety of stimuli. Photoreceptors help them avoid strong light whereas chemoreceptors help them not only find food but also sense its texture, acidity, and calcium content. Free nerve endings in the epidermis give the worm a tactile sense of the world around it, and as anyone knows who has ever tried to pluck night crawlers from their burrow on a warm, wet night or to see them rapidly disappear back into their burrows, they are sensitive to even the slightest vibration, such as a nearby footfall.

As with earthworms, leeches lack highly organized, concentrated sense organs. They too can respond to low levels of many types of stimuli, however. The sensitivity of response in leeches is usually a function of finding prey or a host. Attracted by body heat, tropical blood-sucking leeches will crawl over vegetation and converge on a stationary warm-blooded animal whereas fish leeches respond to moving shadows and water pressure vibrations and predatory or blood-sucking leeches will attempt to attach to an inanimate object smeared with the appropriate host substances, such as fish scales, oil secretions, or sweat.

Regeneration, Reproduction and Development

Polychaetes have great powers of regeneration. Palps, tentacles, and even entire heads ripped off by predators are soon replaced. Some species have break points that allow the worms to sever off body sections when a predator grabs them. Such replacement is commonplace in burrowers and tube dwellers. Earthworms have a very limited capacity to regenerate some segments, especially at the posterior end, whereas leeches completely lack the ability to regenerate any lost parts or segments. Contrary to popular belief, there is no known species of annelid that when cut in half, will form two, viable complete worms. There are many more species of annelids out there awaiting discovery, however, and many more worm mysteries await resolution.

Some polychaetes and freshwater oligochaetes are known to reproduce asexually by budding or division of the body into two parts, or into a number of fragments. Earthworms and leeches do not undergo asexual reproduction.

The vast majority of polychaetes reproduce only sexually, and the majority of species are diecious. They have no permanent sex organs, however, as gonads develop only during the breeding season. Gametes develop from masses of tissues in the connective tissue associated with such structures as septa, blood vessels, and the lining of the coelom. The gametes are often shed in an immature form into the coelom. When the worm matures, the coelom is packed with egg or sperm. The mature gametes leave the worm through the nephridiopores, specialized gonoducts, or a rupture in the body wall of the worm resulting in the death of the worm. Because the sex cells are capable of being fertilized for only a short time after they are released into sea water, some polychaetes undergo changes that make them much more free-swimming during breeding season allowing them to swarm together in massive numbers. The fireworms of Bermuda come to the surface to swarm and spawn each month a few days after the full moon

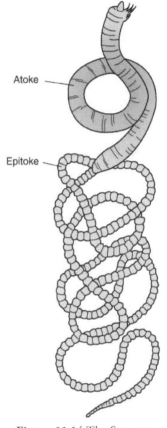

Figure 11.14 The Samoan Palolo Worm (*Eunice viridis*). The epitoke region is packed with gametes, and each segment is equipped with an eyespot on the ventral side. During their once-a-year swarmings, the epitokes detach and rise upward, attracted to the moonlight. Near or at the surface the epitokes explode releasing so many gametes the surrounding water turns milky.

319

at about an hour after sunset. The females appear first and circle about, emitting a greenish phosphorescent glow at intervals. The smaller males then dart rapidly toward the female, emitting flashes of light as they go. When the two sexes come in close proximity, they burst, shedding eggs and sperm into the sea. The spent adults, reduced to shards of tissue, quickly perish.

In some tube-dwelling or burrowing worms a remarkable transformation takes place as breeding season approaches (**Figure 11.14**). The posterior portion of the worm's body, known as the **epitoke,** develops gonads and begins to fill with gametes. The anterior portion of the worm, known as the **atoke,** does not differentiate and carries on normal functions. When the time is right, the epitokes break free from the atokes and swarm to the surface where they burst releasing their sperm and eggs into the sea. The atokes remain behind and slowly begin to asexually develop a new epitoke for the next breeding season.

The palolo worm of the South Pacific occupies rock and coral crevices below the low-tide mark but just at dawn one week after the November full moon the sexual epitokes of this worm rise to the surface in countless millions. The ocean takes on the color of milk as the epitokes discharge spilling out their precious cargo of eggs and sperm. Below the writhing, exploding swarm the atokes begin again to regenerate another generation of epitokes. The natives of Samoa and other islands are familiar with the habits of the palolos. They consider them a great delicacy and eagerly await their breeding season. When the day arrives, they scoop them up in buckets and prepare a great feast, gorging themselves as we do on Thanksgiving day, knowing there will not be another treat like it until exactly the same day of the next year.

Although most polychaetes shed their eggs into sea water, some retain the eggs within their tubes or burrows or lay them in mucous masses that are attached to tubes or to other objects. Once fertilized, the eggs of polychaetes develop into swimming larvae that then shortly morph into a more recognizable worm form.

The reproductive system and patterns of development of oligochaetes are markedly different from that of polychaetes. First, oligochaetes are monoecious (hermaphrodites) and they possess distinct and permanent gonads found only in a few specialized genital segments. Secondly, the entire reproductive process in terrestrial oligochaetes is an adaptation of the challenges of a life on land and in air. Naked sex cells cannot simply be discharged to the exterior as they are in polychaetes as they would dry and die very quickly in the open air. Terrestrial oligochaetes have had to adapt every aspect of their reproduction to meet the threat of drying that hangs over the entire process.

The complexities and adaptations of the terrestrial oligochaete reproductive structures and the development of the young are best illustrated by the common earthworm. Earthworms reproduce during much of the year but most actively on warm, moist nights. Earthworms, unlike polychaetes, are hermaphroditic, each individual having a complete male and female sexual apparatus. This arrangement makes possible two exchanges of sperms, instead of only one, for each meeting of two individuals that are not and cannot be sexually gregarious by nature (**Figure 11.15**).

An earthworm's sex organs are located in the anterior end of the worm, each organ in a particular segment (Figure 11.7a). The male reproductive system consists of two pairs of **testes,** a ciliated **sperm funnel** connecting to a **vas deferens** that leads to the **male pore.** The testes and funnels are contained in (6) two pairs of large sac-like **seminal vesicles.** Sperm are formed in two pairs of minute testes, located in segments 10 and 11. Immature sperm cells separate from the testes and complete their differentiation in the seminal vesicles. During copulation, mature sperm move into the sperm funnels, travel through the vas deferens, and exit through male pores on segment 15. The female reproductive system includes a pair of **ovaries** in segment

13 that discharge mature ova (eggs) into the coelom, where they are collected by two **oviduct funnels** with egg sacs connecting to the **oviducts** opening on segment 14.

Mating occurs at night and requires 2 or 3 hours to complete. Two worms stretch out from their burrows and bring their ventral (bottom) anterior (front end) surfaces together with their heads pointing in opposite directions. The worms contact each other so that the clitellum of one worm is opposite segments 9-11 of the other. Special ventral setae on the clitellum of each penetrate the body of the other to aid in holding the worms together. Mucus is then secreted until each worm becomes enclosed in a slime tube. When the sperm issues forth from the male pores in segment 15, they are carried backwards in longitudinal grooves and through the mucus to the openings of the sperm receptacles on segments 9 and 10 of the mating partner. The worms then separate having accomplished reciprocal fertilization and cross-fertilization.

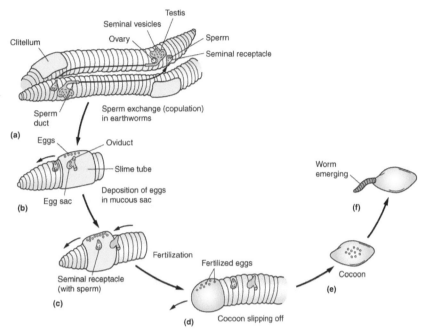

Figure 11.15 Earthworm Reproduction and Egg Cocoon Formation. (a) Sperm is exchanged during copulation. (b) Eggs are deposited in a mucous sac. (c) Eggs are fertilized. (d) Egg cocoon slips off. (e) Egg cocoon is deposited (d) Young worms emerge from the cocoon in 2 to 3 weeks.

Egg-laying starts when the gland cells of the clitellum secrete a mucus ring which glides forward over the body of the worm. As it passes the opening of the oviducts on segment 14, it receives several ripe eggs and then as it passes the opening of the sperm receptacles in segments 9 and 10; it receives sperm which were deposited there previously by another worm in the mating process. Fertilization of the eggs takes place within the mucus ring, which finally slips off the anterior tip of the worm and closes at both ends to form a sealed capsule known as a **cocoon**. Within the capsule, which lies in the soil, the fertilized eggs (**zygotes**) develop directly into young worms and then escape. As in other land animals, there is no free-swimming larval stage comparable with that of polychaetes.

Leech reproductive structures and processes and development of the young are very similar to those of the earthworm in practically all aspects. The one major difference being that most leeches have a penis with which sperm is transferred from one worm to the mating partner.

Annelid Connection

Economically

Earthworms play a variety of important roles in the worldwide yearly agricultural output:

- Their feeding and burrowing mix organic residues (crop stubble, etc) and added amendments into the soil enhancing decomposition, humus formation, soil structural development, and nutrient cycling.
- Their burrows persist as **macropores** that provide channels for root growth, water penetration, and gas exchange.

Earthworm "farms," (vermiculture) which raise earthworms on a commercial scale, market live worms to anglers as bait and to gardeners and farmers as a low-cost, low-tech way to improve the quality of marginal soils. A lucrative market also exists with farmers and gardeners for **worm casts**, the excrement that passes out of the worm. Worm casts contain five times more nitrogen, seven times more phosphorus, and 11 times more potassium than ordinary soil, are rich in humic acids which condition the soil, have a perfect pH balance, and contain plant growth factors similar to those found in seaweed. It's no wonder then that farmers and gardeners are willing to pay a premium price (nearly $7.00 per pound) for what they regard as the perfect soil amendment.

Ecologically

The ecological importance of earthworms has been known for a long time. The ancient Greek philosopher and scientist Aristotle called earthworms "the intestines of the soil." Centuries later Charles Darwin published his classic work, *"The Formation of Vegetable Mould Through the Action of Worms"* in which he explained how worms enrich and loosen the soil by bringing subsoil to the surface and mixing it with topsoil.

Functioning as detritivores, the material worms eat in turn helps recycle detritus and organic matter. Earthworms and sea worms are especially useful in this regard. Some sandy beaches may harbor over 30,000 burrowing marine worms per square meter collectively ingesting and excreting 3 metric tons (3.4 tons or 6,800 pounds) of sand per year. Earthworms, on the other hand, can ingest their own weight in soil every 24 hours and the 50,000 or so worms per acre of fertile soil pass 9 to 16 metric tons (10 to 18 tons or 20,000 to 36,000 pounds) of soil, minerals such as potassium and phosphorus, and organic matter through their intestines each year. Such churning loosens, aerates, and enriches the soil to a high degree. Biologists estimate that if all the material that has ever passed through the guts of earthworms was piled up on the surface of the planet, the heap would rise thirty miles, more than five times the height of Mt. Everest.

All annelids serve as important links in food chains both on land and in water. They are eaten by nearly every type of animal imaginable, including humans. As offensive as it sounds to our North American palettes, in parts of India, Australia, and South America, people regularly cook and eat worms.

Medically

The use of leeches to treat certain medical conditions is an ancient practice first appearing in the tomb paintings of Egyptian pharaohs more than 3,000 years ago. Later the classical Greek theory of bodily functions being regulated by good or bad "humors" required bleeding to drain bad humors. Applying leeches for such bloodletting was regarded simply as a medical device to remove such bad humors. Although leeching and other forms of blood-letting were discredited long ago as beneficial medical practices, they were used regularly right up until the 1960's as treatment/cure for cure numerous medical conditions including headaches, brain congestion, obesity, hemorrhoids, eye disorders, and mental illness.

In the 1980's interest in the use of leeches for therapeutic purposes resurfaced with the development of cosmetic and reconstructive surgery. One of the consequences of such surgery is the formation of blood clots that stop the flow of fresh blood between old and new tissue, especially in very tiny blood vessels. In a practice known as **hirudotherapy**, leeches are applied to restore blood circulation to grafted or severely damaged tissue. Leeches received FDA approval as a medical device in 2004, and they are now commonly used in hospitals around the world. The International Medical Leech Centre in Russia produces over 3 million medicinal leeches yearly. In the United States, the medicinal leech, *Hirudo medicinalis*, is the most commonly used species.

Leeches attach themselves to the patient by means of their two suckers. One of these suckers surrounds the leech's mouth. In the mouth are three sets of jaws that bite through the patient's skin. The saliva of the leeches contains a number of chemicals that result in the benefits sought through hirudotherapy.

- Local anesthetic—this numbs the area so that the patient feels little if any pain.
- Anticoagulant enzymes—various chemicals in leech saliva block different states in the process of blood coagulation. One of these chemicals—**hirudin**—names both this form of medical therapy and the taxonomic class leeches occupy.
- Vasodilator and prostaglandin—help reduce swelling.

Without any clotting, blood flows more freely within damaged tissue and, naturally, from the host to the leech. This anticoagulant no doubt prevents the blood from forming solid lumps in the digestive system of the leech. The feeding process of the leech usually lasts about 20-40 minutes where 10-15 mL of blood is ingested, and their body size may increase 8-11 times their original size.

Hirudotherapy is used mainly in trauma and cosmetic and reconstructive surgery to salvage severed fingers, ears, lips, noses or tissue flaps and skin grafts. But with up to 100 different and untested proteins in leech saliva, physicians are discovering other conditions where the practice may have a role to play. These include:

- Boils and abscesses
- Venous congestion
- Ear infections
- Sinusitis
- Tonsillary abscess
- Thrombosis (blood clot)

A Closing Note
A Fine Line

There seem to be some animals that humans are hardwired to either hate, fear, loathe, or despise (or all of the aforementioned). In Chapter 9 it was tapeworms and flukes, but in this chapter that "honor" belongs to the leeches. I think people are much more cognizant of their loathing of the ectoparasitic leeches than they are of their disgust for the endoparasitic tapeworms and flukes because as the saying goes, "out of sight, out of mind."

I once had a junior high student who referred to all the animals she feared or found disgusting as the "Icky Poos." And leeches were high on her quite long list. Personally, my father provided me with a positive perspective toward leeches on my very first encounter with them. I had come home late one afternoon after having spent the entire day in and around my favorite haunt—Muddy Creek. Unbeknownst to me, I had a small leech attached to my back. My father discovered it and calmly told me it was there. Not being able to see it, I started twisting my head around and reaching back with my hands frantically trying to dislodge it. In short, I started to freak out. My father grabbed my shoulders and looked me straight in the eye and said, "Calm down. It's just a little leech, and it can't and won't hurt you." With that he gently pulled it off my back and handed it to me. As a combat infantry man in World War II, my father had seen and survived unspeakable horrors. To him, and now to me, a little leech was a thing to be appreciated and admired, not feared.

It has been said, and it is true that, "people fear what they don't understand." The line between fear and respect is thin, but it can be crossed. It is my hope that as we continue our journey through the dominion of animals, our encounters with those animals on your personal "Icky Poo" list will result in a change in your perspective about them. Hopefully, what you learn in these pages will allow you to cross the fine line and replace your fear and loathing with understanding and respect.

In Summary

- The ancestors of modern annelids first appeared 540-560 million years ago.
- The approximately 15,000 species of worms we know taxonomically as annelids or more commonly as segmented worms, are round, cylindrical animals with segmented bodies composed of muscular rings.
- Traditional classification schemes feature three classes of annelids:

  Class Polychaeta
  Class Oligochaeta
  Class Hirudinea

- Cladistic classification systems propose only two classes of annelids:

  Class Polychaeta
  Class Clitellata
  Subclass Oligochaeta
  Subclass Hirudinea

- The following are characteristics of phylum Annelida:

1. Their bilaterally symmetrical body is metameric (segmented) both internally and externally.
2. Except for leeches, they possess a true coelom that is well developed and divided by septa. Coelomic fluid supplies turgidity and functions as a hydrostatic skeleton.
3. The body is covered with a thin transparent moist cuticle secreted by the epidermis.
4. They possess chitinous setae only (oligochaetes). setae on fleshy appendages called parapodia (polychaetes), or neither (leeches).
5. The digestive system is complete with some specialization.
6. They possess a closed circulatory system but no true heart. Respiratory pigments such as hemoglobin, hemerythrin, or chlorocruorin are often present in the blood.
7. The excretory system typically consists of a pair of nephridia in each segment (metamere).
8. Gas exchange occurs across moist skin, parapodial flaps, or through gills.
9. The nervous system typically consists of cerebral ganglia (primitive brain) located on the dorsal end in the head region, connected by a ring of nerves to a ventral nerve cord that runs the length of the body. The cord branches to lateral nerves and ganglia in each segment. Sense organs include tactile organs, statocysts (organs of equilibrium), photoreceptor cells and eyes with lenses in some species and chemoreceptors.
10. Hermaphroditic or separate sexes with asexual budding in some species.

- Class Polychaeta—sea worms—is the largest class of annelids. These worms with bristly parapodia and well-developed heads are found in all oceanic zones from intertidal to deep sea floor.
- Class Oligochaeta—earthworms—do not possess a distinct head, lack parapodia, and have only a few small chaetae per segment. They do possess an external mucus-producing band known as a **clitellum**. Earthworms are found primarily in damp soil worldwide, but a few species inhabit stagnant freshwater pools and ponds.
- Class Hirudinea—leeches—lack parapodia and setae but they do possess a clitellum and two suckers—one on the anterior end and one on the posterior end. Leeches are found predominantly in and around fresh water with most prefer the shallow, vegetated margins of ponds, lakes, and sluggish streams. However, a few types are marine and some have even adapted to a strictly terrestrial life, although they are limited in this existence to warm, moist tropical
- Annelids:

1. Are segmented both internally and externally and exhibit bilaterally symmetry. Internally, each segment has a fluid-filled cavity (coelom) separated and divided from neighboring segments by thin, fleshy walls known as septa.
2. Are covered with a thin transparent cuticle. The nonliving cuticle, which is never shed or sloughed off, is secreted by the epidermis that lies just beneath it. Beneath the epidermis lies the dermis and muscles.
3. Except for flukes, annelids are equipped with bundles of chitinous bristles known as setae (chaetae)

4. Possess a digestive system consisting of a single unsegmented straight tube (gut) running the length of the worm from the mouth on the underside of the head on the anterior end to the anus on the terminal pygidium segment at the posterior end.

5. Have a layer of circular muscles below the epidermis and a layer of longitudinal muscles just below the circular muscles that run between the septa that separate each segment. In addition, some polychaetes have oblique muscles, and leeches have dorsoventral muscles

6. In the case of earthworms and leeches, have a digestive system of consisting of a single unsegmented straight tube (gut) running the length of the worm from the mouth on the underside of the head on the anterior end to the anus on at the posterior end. Tapeworms lack a digestive system.

7. Possess a well-developed closed blood-vascular system in which muscular quiverings in certain vessels move the blood around.

8. Exchange gases primarily across moist skin in oligochaetes, leeches, and a few types of polychaetes, but mainly through gills in the majority of polychaetes.

9. Filter metabolic wastes from their bodies through a system of tubes known as nephridia found in each segment of the body.

10. Possess a single mass of cerebral ganglionic tissue (brain) residing in the head or anterior end. Various sensory capabilities—tactile, chemical, light, orientation—exist among members of the phylum.

11. Have powers of regeneration that vary from great (polychaetes) to very limited (oligochaetes).

12. Reproduce almost exclusively by sexual means and most are dioecious.

- Annelids connect to humans in ways that are important economically, ecologically, and medically.

Review and Reflect

1. *Comparing Complexity.* Are annelids more biologically complex than the platyhelminths and nematodes we studied in the previous chapters? Defend your answer with specific reasons and examples.

2. *An Annelid Safari.* If you wanted to collect the most annelids of a single species in the smallest space, what types of habitats or environments would you seek out? Which types of habitats or environments would you avoid?

3. *Sort it Out.* Your instructor has just presented you with three unlabeled jars; each containing a preserved annelid worm specimen and challenged you to place each specimen in its proper class. One jar contains an earthworm; one jar contains a large sea worm, and one jar contains a leech. What external morphological features (or lack of) on each worm would you use to meet this challenge? What are the visual (morphological) differences between classes of annelid worms?

4. *Significance of Segmentation.* Your friend is also taking this zoology course, and she is struggling with several of the concepts that the instructor has presented regarding annelid worms. She

asks, "Why did the instructor make such a big deal about segmentation in annelids?" How would you reply?

5. **Worms Have Hair?** Your friend seeks further annelid enlightenment and asks, "And for that matter, what are those little hairy things called chaetae or setae and what purposes do they serve in annelid worms anyway?" How would you reply?

6. **Giddy Up Lil' Worm.** Compare and contrast the locomotion (movement) of a pelagic (free-swimming) sea worm, an earthworm, and a leech.

7. **Feed Me Seymour.** Your instructor is away attending an important conference on annelid worms. Before leaving, she put you in charge of the biology department's collection of living annelid worms. The collection includes several predatory errant polychaetes, several sedentary polychaetes, many earthworms, some predatory leeches, and some blood-sucking leeches. Explain what type of food you will need to give to each category of living worm.

8. **Can Earthworms Drown?** You have probably seen earthworms come to the surface after a heavy rain and perhaps observed dead worms in puddles after such deluges. Most people believe that worms come to the surface when rain is heavy to avoid drowning in their burrows and that they do drown when trapped in rain puddles. But is that the truth of the matter? Look into it.

9. **Wormy Love.** Compare and contrast the reproductive patterns of sea worms, earthworms, and leeches.

10. **Leeches to the Rescue.** Imagine that you are a vascular surgeon and that you have just reattached the severed finger of a small child. You are on your way to the waiting room to inform the parents that the surgery went well and that you plan to use Hirudotherapy on their child. What would you say to them to explain your choice of post-surgical treatment for their child?

Create and Connect

1. **Darwin Defends.** Charles Darwin wrote a classic treatise entitled *The Formation of Vegetable Mould Through the Action of Worms With Observations on Their Habits.* Published six months before his death in 1882, this 139-page work detailed elaborate experiments Darwin performed to study the behavior and "intelligence" of earthworms. Imagine that you are Charles Darwin. You have studied earthworms for many years which has prompted you to write, "It may be doubted whether there are many other animals which have played so important a part in the history of the world as have these lowly organized creatures." Few of your colleagues agree with your statement. Write a short essay in which you explain and defend this statement.

2. **School of Worms.** Though they lack visible external sense organs, earthworms are sensitive to their external environments. You may accept that because that is what experts tell you, but can you prove it? Pick one sensory reaction from the following list: phototaxis (reaction to light), chemotaxis (reaction to chemicals), or thigmotaxis (reaction to contact) and design an experiment to prove whether earthworms do or do not exhibit the specific taxis selected.

Guidelines:

A. Your design should include the following components in order:

> ➢ The *Problem Question*. State exactly what problem you will be attempting to solve.
> ➢ Your *Hypothesis*. Although this is a fictitious experiment, word your hypothesis as realistically as possible.
> ➢ *Methods and Materials*. Explain exactly what you will do in your experiment including the materials necessary to accomplish the task. Be specific, take nothing for granted, and do not expect people to read your mind as they read your work.
> ➢ *Collecting and Analyzing Data*. Explain what type(s) of data will be collected and what statistical tests might be performed on that data. It is not necessary to concoct either fictitious data or imaginary observations.

B. Your instructor may provide additional details or further instructions.

3. ***Old MacDonald Had Some Worms.*** Imagine that you are a corn farmer. You know pesticides will kill harmful insects and increase your yield, but you also know such chemicals can seriously harm beneficial creatures, such as earthworms, that also live in the soil. In **Table 11.1**, you see the results of an earthworm survivability study done on three different pesticides.

Table 11.1 *Earthworm Survivability Versus Pesticide Applied*		
	Number of Worms per Square Meter (Experimental)	**Number of Worms per square Meter (Control)**
Pesticide X		
Before Pesticide Application	75	75
After Pesticide Application		
Week 1	21	92
Week 2	12	118
Week 3	11	127
Week 4	18	136
Week 5	39	151
Pesticide Y		
Before Pesticide Application	75	75
After Pesticide Application 90		
Week 1	76	90
Week 2	74	112
Week 3	75	130
Week 4	76	141
Week 5	75	147

Pesticide Z		
Before Pesticide Application	75	75
After Pesticide Application		
Week 1	87	91
Week 2	117	117
Week 3	131	133
Week 4	142	141
Week 5	153	155

Source: Holley; Dennis. *Animals Alike-An Ecological Guide to Animal Activities*. Lanham, MD; Roberts Rinehart Publishers. 1997.

Construct a line graph for the data on each pesticide. Use the graphs you generate to answer the following questions:

A. Exactly what did you just graph?

B. What was the difference between the control group of worms and the experimental group of worms?

C. Which pesticide

   i. Had no effect on the number of worms?

   ii. Reduced the number of worms?

   iii. Did not reduce the number of worms but apparently kept the worms from reproducing?

D. As the farmer, which pesticide would you use?

E. Suppose the pesticide that harmed the worms the most was the cheapest or the most effective at killing soil insects. Would this change your opinion about which pesticide to use?

PHYLUM MOLLUSCA: A SCHEME OF SHELLS AND TENTACLES

In essentials molluscs are one of the most compact groups of animals; but there can be few phyla that show such wide diversity imposed on such a uniform plan.

—J. E. Morton

Introduction

Of all the evolutionary tinkering that led to the many branches on the animal tree of life, one of the grandest experiments must surely be the lineages that led to the extremely diverse group we know as phylum Mollusca. Molluscs most likely arose during the Precambrian times, but we have no good record of molluscs before they evolved shells. And what of the ancient fossilized shells we do find? Can we be certain that ancient shells are really the remains of molluscs, particularly if the mollusc lineage they appear to represent is now extinct?

For some time most molluscan specialists (**malacologists**) proposed the idea of a "hypothetical ancestral mollusc" (HAM) as the single ancestor for all lineages of modern molluscs. HAM was viewed as a small (about 1 mm), bilaterally symmetrical, benthic marine wormlike creature inhabiting Precambrian seas. The lack of evidence for HAM and the analysis of the phylogenetic history of the molluscs by cladistic methods, however, cast serious doubt on the idea of a single ancestral type for all lineages of molluscs. It should be noted that phylogenetic trees and cladograms relating to molluscs in textbooks (including this one) are still structured around a HAM.

History of Molluscs

Molluscs (L. *molluscus*, soft) are extremely diverse and because so many taxa below the class level are artificial (polyphyletic or paraphyletic), efforts to trace their evolutionary history have proven frustrating at best. However, since many molluscs form a hard shell, the fossil record for some mollusc groups is quite good, and this record seems to indicate that molluscs first appeared in ancient seas about 600 million years ago.

About 400 million years ago evolutionary pioneers of the bivalve affiliation began to inhabit the world's freshwater streams and lakes. It took another 300 million years for certain gastropods to follow bivalves not only into freshwater habitats but also to make the leap to land. Since then, the development of new habitats and new lines of molluscs to fill those habitats seems to have gone hand-in-hand. The fossil record shows that over the course of their evolutionary history molluscs have undergone great changes. Some taxa (groups) grew larger in number, some smaller, and others became extinct. Some types developed shells; some reduced their shells whereas others lost their shells altogether. The shells of some types coiled tighter, some looser, and some lost their coil entirely. Some, like the ammonids, evolved into a huge number of species and then mysteriously disappeared forever. Today evolutionary biologists speculate that the modern gastropod molluscs seem to be at the pinnacle of their evolution whereas the cephalopods seem to be a declining group (**Figure 12.1**). The exact relationship between the member classes of phylum Mollusca continues to be debated, however.

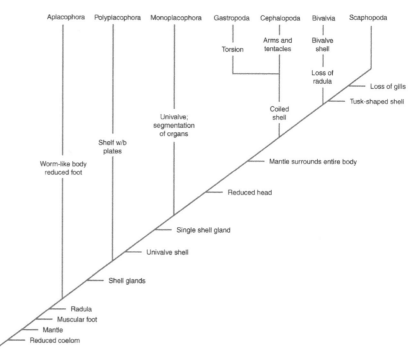

Figure 12.1 One Possible Interpretation of the Evolutionary Relationships Between the Taxa of Phylum Mollusca. In some cases synapomorphies that identify the various clades have been modified or even lost in some lines of descent.

Diversity and Classification

The number of extant (living) species of molluscs is estimated at anywhere from 50,000 to 110,000 depending upon the authority consulted, whereas 70,000 extinct species are known only by their fossilized remains. Many malacologists believe it that up to 500,000 species will eventually be formally identified and that most newly discovered species will probably come from the deep ocean, tropical forests, and freshwater habitats. Eclipsed today only by the arthropods, the molluscs may yet one day wear the crown of the largest and most diverse phylum of animals.

It is beyond the scope of this book to delve into the taxonomic history of the molluscs other than to say that that history has been a long and convoluted enterprise in which hundreds of names for various taxa have come and gone. Molluscan classification at the genera and species level remains incredibly jumbled with many species bearing anywhere from a few to hundreds of different names (synonyms).

Considering only the taxa with extant members, the molluscs may be classified:

Domain Eukarya
    Kingdom Animalia
        Phylum Mollusca
            Class Gastropoda—Snails, conchs, limpets, slugs
            Class Bivalvia—Clams, oysters, mussels, scallops
            Class Cephalopoda—Octopus, squid, nautilus
            Class Polyplacophora—Chitons
            Class Scaphopoda—Tusk shells
            Class Aplacophora
            Class Monoplacophora

Phylum Characteristics

Molluscs can be found in nearly every ecosystem on earth from high, barren mountains to rolling, grassy plains and from calm, clear lakes to roaring, rushing rivers. They may be plucked from the pounding surf and recovered from the dark, abyssal depths of every ocean. Some types creep about on land; some burrow into the bottom sediments of fresh and salt water whereas others swim freely about the ocean actively pursuing prey. This phylum includes herbivorous grazers, predaceous carnivores, filter feeders, detritus feeders, and parasites.

Molluscs range in size from minute clams and snails less than a millimeter (0.04 inch) long as adults, to the massive giant squid *Architeuthis* that can reach 30 meters (100 feet) in length, weigh up to 900 kg (1,980 pounds), and is exceeded only in size by some of the larger whales. The vast majority of mollusks (perhaps as many as 80%), however, are shy of 10 centimeters (4.0 inches) in length.

The members of phylum Mollusca exhibit the following general characteristics:

- They are bilaterally symmetrical coelomate protostomes with the majority possessing a definite head.
- Their coelom is reduced to small spaces around the heart, kidney, and gonads.
- Their ventral body wall is specialized into a large, well-defined muscular **foot** used for locomotion or burrowing.
- Their dorsal body wall forms a pair of fleshy folds or lobes called the **mantle** that secretes calcareous spicules, shell plates, or shells.
- Most possess a rolled extension of the mantle known as a **siphon** or **funnel**. Usually tube-like in form, the siphon allows for the movement of water into and out of the aquatic molluscs and serves as inhalant tube in terrestrial gastropods.
- The viscera (internal organs) is concentrated into a **visceral mass**.

- A **mantle cavity** houses the visceral mass. The mantle cavity opens to the outside and functions in gas exchange, excretion, elimination of digestive wastes, and the release of reproductive cells.
- Except for the filter feeding bivalves, molluscs possess a file-like set of hooked teeth-called the **radula**.
- Except for the Cephalopoda, molluscs possess an open circulatory system consisting of a heart with distinct ventricle and atria, blood vessels, and blood with respiratory pigments.
- Molluscs possess an excretory system consisting of one or two kidneys (**metanephridia**) that usually empty into the mantle cavity.
- Gaseous exchange is accomplished through the gills, lungs, mantle, or body surface.
- Molluscs possess a complete and complex digestive system with the anus usually emptying into the mantle cavity.

Class Characteristics

Class Gastropoda (L. *gaster*, stomach + *pous*, foot)

Gastropoda is the largest and perhaps best-known class of all the molluscs with anywhere from 60,000 to 75,000 living species and an additional 15,000 fossil forms. This class contains a vast number of marine and freshwater species as well as being the only class of molluscs with members that are permanently terrestrial. Species include the familiar land and water snails, land and sea slugs, abalone, limpets, cowries, whelks, periwinkles, sea hares, and conchs (**Figure 12.2**). Without a doubt, the most flamboyant gastropods are the brightly colored sea slugs. By virtue of the fact that their gills are often in the form of a feathery plume on their backs, sea slugs are also known as nudibranchs (L. *nudus*, bare + Gr. *brankhia*, gills).

Although breathtaking to humans, the vibrant colors of nudibranchs camouflage individuals against the coral reefs where they live or serve as a clear-cut and unmistakable warning to predators that they are distasteful or poisonous. The champions of colorful warning signals must surely be the Chromodorids (**Figure 12.3**). These nudibranchs feed on hydroids and incorporate hydroid nematocysts (stinging cells) into their own body wall. Remarkably, the hydroid nematocysts remain functional providing the nudibranch with a potent and painful defense against any who would disturb it, a fact that it advertises quite vividly.

> *Orange and speckled and fluted nudibranchs slide gracefully over the rocks, their skirts waving like the dresses of Spanish dancers.*
>
> —John Steinbeck *Cannery Row*, 1945

Gastropods are also the most environmentally successful group of molluscs occupying almost every conceivable habitat on the planet. As a result, gastropods are found from scorching deserts to high, cool mountains and from verdant fields and forests to lakes, ponds, swamps, streams, and rivers. They exist, however, in the greatest numbers in the ocean where they inhabit every niche from the intertidal zone to the deepest ocean trenches. This cosmopolitan group exhibits virtually every type of feeding mode possible. Members of this class can be herbivores, carnivores, scavengers or detritivores, ciliary feeders, or parasites.

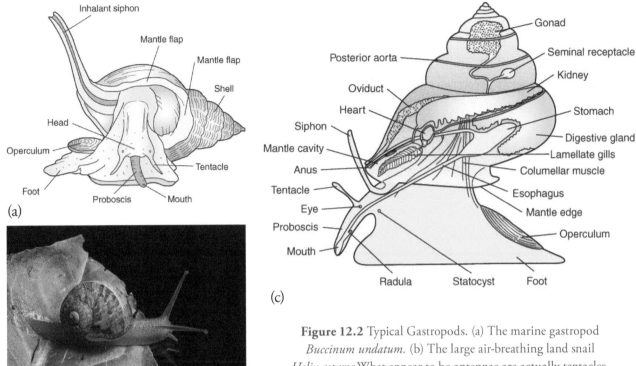

(a)

(b)

(c)

Figure 12.2 Typical Gastropods. (a) The marine gastropod *Buccinum undatum*. (b) The large air-breathing land snail *Helix aspersa*. What appear to be antennae are actually tentacles. (c) The anatomy of a female coiled-shell gastropod.

Most members of this class have a one piece shell that is typically coiled or spiraled and that usually opens on the right-hand side (as viewed with the shell apex pointing straight upward). This was not the case early in the evolutionary history of the gastropods. The ancestral shell form of the gastropods was spiral-shaped but flat (**planospiral**) resembling a coiled garden hose lying flat on the ground. Such a shell was not very compact nor would it have been easy to carry around as the diameter could become quite great (Some fossil gastropod shells have been found measuring 2.5 m (8 feet) in diameter!). As the evolutionary path of the gastropods unfolded, a significant change occurred in the form of **torsion** (twisting) of both shell and body. Most of the body located behind the head, including the visceral mass, mantle, and mantle cavity, was twisted 180 degrees counterclockwise (in a right-handed direction). Given there are exceptions to nearly everything in biology, it should be no surprise to learn that some molluscs were twisted the other direction with the result that many land and freshwater species and even some entire genera are "left-handed." Furthermore,

Figure 12.3 The brilliant colors of the chromodroid nudibranchs are clear warning of the deadly appropriated weaponry they carry.

some types have an **operculum** that functions as a trapdoor to close and seal the shell against predators and water loss. This operculum is usually made of a horn-like material, but in some species it is calcareous.

About 500 million years ago, during the Cambrian period, three different strains with different shell plans arose. Although most living gastropods today bear a single asymmetrical coiled (**conispiral**) shell,

335

some, such as the limpets and abalone have a flat saucer-shaped shell. In the land and sea slugs, the shell is absent entirely or greatly reduced, and the body is streamlined in such a manner that its torsion is relatively inconspicuous.

The terrestrial gastropods may be the most highly adapted members of this class. The challenges of living on land have forced the development of a head equipped with numerous sense organs, a large, flat foot adapted for locomotion over a variety of surfaces, a pulmonary (lung) system for gaseous exchange in air, and a robust radula with an aggressive feeding behavior to accompany it.

An intriguing puzzle of gastropod evolution is the appearance of the shelless "slug" form. Despite the fact that evolution of the coiled shell led to great success for the gastropods—80 percent of all living molluscs are snails—secondary loss of the shell occurred many times in this class. In forms such as the land and sea slugs, the shell may persist as a small vestige (e.g. some sea hares) or it may be lost altogether (e.g. the true nudibranchs).

Class Bivalvia (L. *bi*, two + *valva*, valves or plates)

Practically all the 30,000 species of bivalves are marine with the majority live in the intertidal (littoral) zone. Some types, however, are found at all depths down to the deepest abyssal plains. Most of the rest have adapted to life in brackish and freshwater environments whereas a few live commensally with other marine creatures, and a few have evolved to become parasites.

The clams, oysters, mussels, scallops, and others that comprise this class are equipped with typical molluscan features and structures. The molluscan traits in bivalves, however, have evolved along slightly different patterns than those of their gastropod kin. For one thing, bivalves lack the definite and well-developed head found in gastropods. Contemplating the shell of a bivalve reveals that the soft body of the bivalve is encased not in a single spiral shell as are the gastropods but rather in two (bi) rock-like calcareous shells (valves). The shape of bivalve shells vary greatly—some are rounded and globular, others flattened and plate-like, whereas still others, such as the razor clam have become elongated as an aid to burrowing (**Figure 12.4**).

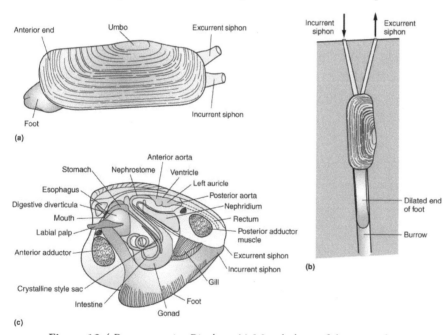

Figure 12.4 Representative Bivalves. (a) Morphology of the razor clam (*Tagelus plebeius*). (b) The razor clam in it normal habitat burrowed into soft substratum. These clams are common off the southeastern coast of the United States, where they may occur in enormous numbers. (c) Anatomy of a typical bivalve. Most of the gills have been removed in this view.

And burrow they do into the expansive mud, silt, and sand bottoms of ocean and freshwaters. Some such as the infamous shipworms (Teredinidae) can burrow into wood and even rock and cement. In fact, most species of bivalves have evolved to become burrowers out of necessity for they have little power of locomotion. Not only are their shells heavy and poorly shaped for movement, the muscular foot that is so large and propulsive in gastropods has been reduced and flattened in bivalves into a blade-like form that allows it to be extended out between the shells as an aid to burrowing and anchoring but little else. Because of the shape of bivalve foot, this class has also been known as *Pelecypoda* (L. *pele*, hatchet + *pous*, foot).

Although many bivalves, such as clams, are **infaunal** (burrowers), some, such as mussels and oysters, are **epifaunal** (bottom-dwelling but not burrowing) and attach themselves to objects in the water by means of organic cementation (**Figure 12.5**). Others, such as scallops, are unattached bottom dwellers that are capable of short bursts of "jet-propelled" swimming, that is accomplished by rapidly clapping the shells together.

Bivalves also differ markedly from gastropods in their mode of feeding. Not only are bivalves confined within two heavy shells with limited mobility, but they also lack the scraping file-like radula that gastropods wield so effectively. Given these restrictions, how is food to be obtained? The answer lies with the extra-large gills of the bivalves and the cilia that cover these gills. The beating of the many cilia on the gills creates a current that brings water carrying food into the shells. Thus, bivalves are said to be ciliary or filter feeders. Settled in sediments, the animal draws water in through the incurrent siphon, extracts from the water, and sends the water out the excurrent siphon (**Figure 12.6**).

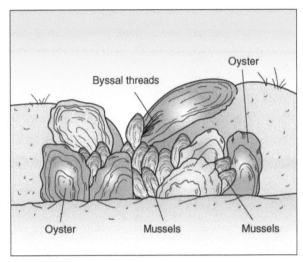

Figure 12.5 A Gathering of Bivalves. Mussels (*Mytilus*) anchored to oysters (*Crassostrea*). The oysters are permanently attached to the rocks whereas the smaller mussels are attached to the oysters by byssal threads.

Class Cephalopoda (L. *cephalus*, head + *pous*, foot)

The cephalopods are an ancient and very successful group, albeit at 650-700 species not a very large one. The squids, octopuses, cuttlefish, and nautilids that comprise this class are considered not only to be the most morphologically and behaviorally complex members of phylum Mollusca, but in terms of their intelligence and the complexity of their sense organs, the most advanced of all the invertebrates. Cephalopods have differentiated from the ancestral molluscs and from the living gastropods and bivalves mainly through the loss of the shell (except in nautilids) and in the modification of the foot into long, slender arms and tentacles.

Figure 12.6 Pea clam (*Sphaerium*) with both its nearly transparent siphons extended.

337

Few types of marine animals have generated more myths or been the source of more fearful speculation and imagination than have the cephalopods. For centuries the cephalopods have been portrayed as fearsome sea-monsters such as the Kraken of ancient Norse legend or hideous and even devilish creatures. In fact, for many years the octopus was referred to (and in some places and situations is still referred to) as the "devil-fish." The nineteenth-century writer Victor Hugo graphically described the octopus as "a disease embodied in monstrosity," adding: "The octopus, O Horror!, inhales a man."

We now know that such fears and prejudices for so long portrayed in myth, print, and in more modern times on film have no biological basis in fact. The small shallow-water squids and octopuses that humans do encounter usually flee, or they may move a safe distance all the while viewing the human interloper into their realm with a detached curiosity. Like their shallow-water kin, the large deep-water octopuses show no aggressive tendencies to humans whatsoever and will passively and serenely sail on when such encounters do occur. Large deep-water squid have been known to attempt to drag divers down to greater depths, and just a few of these large squid could probably kill and devour a human. In spite of myth, legend or the chills and what authors and filmmakers would like us to believe, cephalopods do not drag ships and their unfortunate crews into the briny deep nor do they pluck scantily-clad beauties off surfboards and gruesomely devour them, or any other such nonsense. Fortunately, attitudes toward cephalopods seem to be changing for the better.

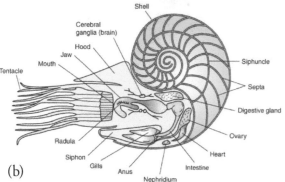

(a)

(b)

Figure 12.7 The Chambered Nautilus (*Nautilus pompilius*) (a) The nautilus is the only cephalopod with an external shell. Its eyes are simpler than those of other cephalopods and function like a pinhole camera. (b) Anatomy of body and gas-filled chambers shown in longitudinal section.

All cephalopods are strictly marine creatures although a few types can tolerate brackish water. They live in all oceans of the world and are most abundant in shallow, coastal areas. Cephalopods are mobile, agile, and hostile (but only to prey animals, not to humans) being active and aggressive predators, although occasional scavenging is not out of the question if the opportunity presents itself. Cephalopods snare prey such as fish, crustacean arthropods, and other molluscs with their muscular tentacles and crunch it to pieces with a hard, chitinous beak located in the center of the arms or tentacles. Too small to overpower prey in the customary manner, some young octopuses turn to the poison of other creatures instead. First, they tear off long pieces of toxic jellyfish tentacles, that they secure to the underside of their own arms with tiny suction cups. Then the small octopuses attack and paralyze prey with these commandeered "weapons."

The living genera of cephalopods can be placed into three informal groups based on their ecology and morphology. First, the *nautilids* are found primarily in the Southwest Pacific in tropical waters from 5 to 550 m (16 to 1804 feet) deep. Bearing up to 94 tentacles, *Nautilus* is a nocturnal predator that stays deep during the day then rises at night to seek prey (**Figure 12.7**). The nautilids are the only cephalopod group to possess an external shell. Although a useful adaptation for defense, a large, heavy external shell does present challenges when it comes to locomotion. To move freely about, the *Nautilus* employs a different strategy for each axis of movement. To move back and forth (horizontal axis), the *Nautilus* employs a hydro-propulsion system in which water is taken into the mantle cavity and then forced out through the tube-like siphon. To steer they simply adjust the direction of the siphon. Up and down (vertical) movement is accomplished by changing the buoyancy of the shell. To rise, the animal adds air to the many chambers in its shell through a small tube called the **siphuncle**. This increases the buoyancy of the shell which in turn allows the animal to rise upward. To sink, the animal replaces some of the air in the chambers with liquid called the **cameral fluid** decreasing the buoyancy of the shell. The shell then becomes heavier and the animal sinks. Movement in any axis can be made in a surprisingly precise manner given the cumbersome nature of the shell.

The second group, the *squids* and *cuttlefish* occur in all oceans of the world at depths from near the surface to around 3,000 m (9,843 feet), although the giant squid lives at depths around 5,000 m (16,500 feet) (**Figure 12.8**). Squids and cuttlefish possess 8 arms and 2 longer tentacles furnished with **denticulated** (toothed) **suckers**, and although they lack an external shell, they do have a small, thin internal shell called a **gladius** (also **pen** or **cuttlebone**). Squids and cuttlefish are very mobile and extremely agile animals capable of making hovering and darting movements with great precision and speed. Such agility is accomplished by employing muscular flaps (fins) and a water siphon system for hydro-propulsion. Some squid can reach speeds of up to 25 miles per hour by this means of water propulsion and zip along so fast that they can literally "fly" out of the water to avoid predators.

> *You'd see these pulsating white blobs coming out of the blackness. They were definitely curious. A squid would come up at you, feel you, touch your hand. Even when four or five at a time were in feeding mode, they were delicate and retiring. It was pretty moving. It was as close as you could imagine to meeting alien intelligence.*
>
> —William Gilly

Recent studies indicate that cuttlefish may be the most intelligent of all the many invertebrate species (although the octopuses might contest this supposition), and certainly they must be ranked as one of the most colorful groups as well. In fact, cuttlefish are sometimes called the "chameleon of the sea" because of their remarkable ability to alter their skin color at will. Flashing and pulsating with colors and patterns that rival any neon light display ever invented by humans, cuttlefish communicate with other cuttlefish and camouflage themselves from predators. This color-changing capability is produced by groups of red, yellow, brown, and black pigmented **chromatophores** above a layer of reflective **iridophores** and **leucophores** (white), with up to 200 of these specialized pigment cells per square millimeter. Cuttlefish can not only influence the color of the light that reflects off their skin, but also the polarization of the light as well, that can be used to signal other marine animals, many of which can also sense polarization. All cephalopods have chromatophores and are capable of camouflage; however, the cuttlefish seems to display the greatest array of colors and patterns.

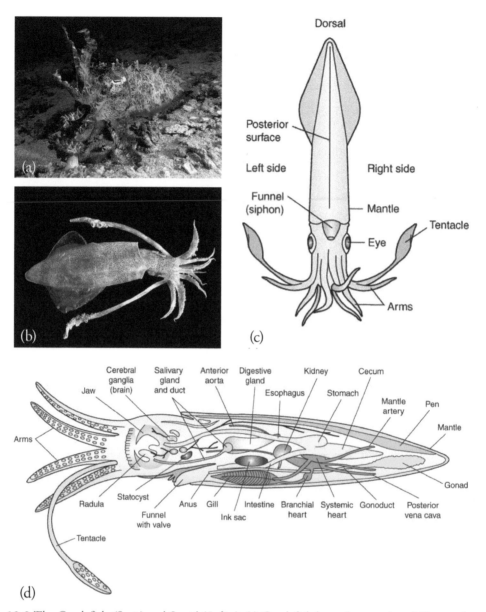

Figure 12.8 The Cuttlefish *(Sepia)* and Squid *(Loligo)*. (a) Cuttlefish have the amazing ability to change not only the color of their skin but its very texture as well. (b) With their long extensible grasping tentacles, squids are the most agile and aggressive of the cephalopods. (c) Morphology of a squid. Squids have eight arms and two tentacles, which are used for grasping prey, (d) General anatomy of a squid in longitudinal section.

The octopus has an amazing skin, because there are up to 20 million of these chromatophore pigment cells and to control 20 million of anything is going to take a lot of processing power... These animals have extraordinarily large, complicated brains to make all this work.

—Roger T. Hanlon

The majority of squids and cuttlefish are no more than 60 cm (24 inches) long. The giant squid *Architeuthis,* however, may reach 13 m (43 feet) in length), although the possibility certainly exists that far larger individuals lurk in the inky black depths. On September 30, 2004, researchers from the National

Science Museum of Japan and the Ogasawara Whale Watching Association took the first still photos of a live giant squid in its natural habitat. The same team successfully filmed a live giant squid for the first time on December 4, 2006.

In 2003, a large specimen of an apparently abundant but poorly understood species known as the Colossal Squid (*Mesonychoteuthis hamiltoni*) was discovered. This species may grow to 14 m (46 feet), making it the largest invertebrate.

Finally, the third group, the *octopuses* is found primarily in shallow water in nearly all oceans of the world, with some types ranging as deep as 5,000 m (16,500 feet). To avoid predators, the shallow water species live in dens and lairs ranging from holes in rocks to shells, bottles, and discarded cans on the ocean floor. Jacques Cousteau, the legendary French ocean explorer, told of trying to salvage the sunken cargo of an ancient Greek ship consisting mainly of jars of wine. The jars had lost their stoppers, and when Cousteau and his comrades hauled them up, they found an octopus in almost every one. Sporting eight arms of equal length with suction cups, octopuses range in size from the California octopus only 1 cm (0.38 inch) long to the giant octopus at 7 m (23 feet) from arm tip to arm tip (**Figure 12.9**).

Unlike most other cephalopods, the majority of octopuses have entirely soft bodies with no internal or external shell. A hard chitinous beak, similar in shape to a parrot's beak, is the only hard part of their body. This allows them to enter incredibly narrow slits and crawl through amazingly tiny openings between underwater rocks in search of prey or to hide from predators. As an escape artist, the octopus has few rivals. One octopus in a research lab was able to wiggle out of a box through a hole only the size of one of its own arms in less than two minutes.

Figure 12.9 The octopus skulks over reefs and rocks using its arms to pluck prey from cracks and crevices.

Octopuses are highly intelligent (possibly about the same as a cat) but the exact extent of their intelligence and learning capability is much debated among biologists. Maze and problem-solving experiments have shown that octopuses do have both short- and long-term memory. Because octopuses have very little if any contact with their parents and thus learn almost no behaviors from their parents, there has been much speculation to the effect that almost all octopus behaviors are independently learned rather than instinct-based, although this remains largely unproven.

In laboratory experiments, octopuses can be readily trained to distinguish between different shapes and patterns and they have been observed in what some have described as play: repeatedly releasing bottles or toys into a circular current in their aquariums and then catching them over and over. Octopuses break out of their aquariums and sometimes into others in search of food. They have even been known to board fishing boats opening holding tanks to get at the tasty crabs held therein. In fact, their intelligence is so highly regarded that in some countries, octopuses are on the list of experimental animals on which surgery may not be performed without anesthesia. In the United Kingdom, cephalopods such as octopuses are regarded as *honorary vertebrates* in the Animals Scientific Procedures Act of 1986, according them protections not normally afforded to invertebrates.

Octopuses lack the muscular flaps and fins of the squids and cuttlefish, but they do possess and efficiently use a similar water siphon system for hydro-propulsion. The primary mode of locomotion for an octopus, however, is to crawl using the arms in a sculling motion to push and pull itself along a surface or the ocean floor. The ability to crawl in octopuses is not fully understood by biologist since they do not seem to have full cerebral control over their arms. They seem to simply send out a high-level command for movement and somehow it sorts itself out. There are several species that "walk" in the classic sense, using two alternating arms in a rolling gait whereas the other arms are utilized for camouflage. *Amphioctopus marginatus* resembles a walking coconut with legs, and *Octopus aculeatus* appears to be a moving clump of floating algae.

Lacking shells, octopuses are soft, tasty morsels for many kinds of predators. To protect against this onslaught, the octopuses employ a number of different defense mechanisms: release of ink, camouflage, autotomising of limbs, and mimicry.

When threatened or alarmed, most octopuses and squids can eject a thick black or brown cloud of ink. This fluid contains a high concentration of melanin pigment (the same pigment that gives humans their hair and skin color) and mucus. This ink cloud not only confuses predators by forming a "dummy" image to cover the retreat of the animal but may also dull chemoreception (sense of smell), that is particularly useful for evading predators that depend on smell for hunting.

As with squids, an octopus can change the apparent color, opacity, and reflectiveness of their epidermis (skin). Such color and shape-shifting serves not only to camouflage the octopus but as communication with other octopuses as well.

When under attack, some species can detach their own limbs (**autotomise**), in a manner similar to the way skinks and other lizards detach their tails. The detached crawling arm serves as a distraction to would-be predators allowing the octopus to make its escape.

A few species, such as the Mimic Octopus, have a fourth defense mechanism—**mimicry**. These cephalopods can combine their color changing talent with the ability to change the shape of their highly flexible body in order to accurately mimic the appearance of dangerous sea animals such as lionfish, sea snakes, and eels. And two species of Indonesian octopus mimic flatfish. Not only can they change shape but they can also change the very texture of their skin to further heighten their camouflage. In an astonishing instant, these shape-shifters can take on the appearance and color of ruffled seaweed, or the scraggly, bumpy texture of rock, among other amazing disguises. The evolution of camouflage and ink production coupled with high mobility and complex behavior has allowed the cephalopods to radically modify the basic molluscan body plan with great success.

Class Polyplacophora (L. *poly*, many + *plac*, plate + *phor*, carry)

The polyplacophorans, commonly known as *chitons*, are considered to the most morphologically primitive of all existing molluscs. Chitons have an ovoid, flattened body covered by eight overlapping calcareous shell plates (**Figure 12.10**). The posterior margin of each plate projects backwards, and the anterior lateral margins of each bears a large wing that projects forward. These projections then fit beneath the plate immediately in front, and each plate overlaps the plate behind; a tight arrangement perfect for defense. Individual plates are sometimes called "butterfly shells." The largest chiton is the brick-red gumboot chiton (*Cryptochiton stelleri*) of the Pacific Northwest at around 13 cm (5.2 inches).

Chitons are strictly marine with the majority of species inhabiting the rocky intertidal (littoral) zone throughout most of the world where their low dome-shaped bodies are well suited to withstanding the violent surge of ocean waves. Some species, however, have been found as deep as 6,000 meters (20,000 feet). Wherever they are found they cling tenaciously to the hard substratum, and if it is dislodged from its rock, a chiton will roll up into a ball to protect its fleshy underside.

Slowly creeping along on their large foot, chitons eat algae, bryozoans, diatoms, and sometimes bacteria by scraping the rocky substrate with their well-developed radula. Some species have a large anterior **girdle** used for trapping other small invertebrates such as shrimp and possibly even small fish. The girdle is held up off the surface and then clamped down on any unsuspecting prey that may have taken shelter beneath it.

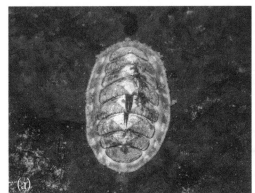

Figure 12.10 Chiton Morphology. (a) Dorsal view of the Blue-lined chiton (*Tonicella undocaerulea*) showing the characteristic eight overlapping calcareous shell plates. (b) Ventral view of the leather chiton (*Katharina tunicata*) showing the mouth (top) and massive muscular foot.

Class Scaphopoda (L. *scaphe*, boat + *pous*, foot))

This class of only around 200 living species, commonly known as the tusk or tooth shells, have the simplest shell structure and anatomy of all the molluscs. Scaphopods are relatively inactive creatures that burrow in sand at depths ranging from 6 to 200 meters (18 to 600 feet). A few do inhabit shallower waters, however, and some have even been found in the deepest of ocean trenches. Their food consists primarily of one-celled foraminiferans.

Class Aplacophora (L. *a*, without + *plac*, plate + *phor*, carry)

This class of approximately 320 living species is populated by exclusively benthic molluscs that have deviated from the normal molluscan form. These small creatures around 2.5 cm (1 inch) in length are worm-like without a shell, and there is no fossil evidence to suggest that any members of the class ever had one. Rather, the mantle is embedded with calcareous spicules. Furthermore, the foot is absent or vestigial at best leaving the Aplacophora with no visible means of locomotion. These out-of-the-ordinary molluscs live mostly in deep water where some feed as carnivores on cnidarians or annelids whereas others are mostly detritivores.

Class Monoplacophora (L. *mono*, one + *plac*, plate + *phor*, carry)

This class of molluscs was thought extinct until April, 1952 when ten living specimens were discovered by the Danish Galathea expedition. In what has been called "one of the most dramatic cases in the history of malacology," Dannish biologist Dr. Henning Lemche recognized these animals for what they truly were and named them *Neopilina*.

Little is known about the monoplacophorans as only a dozen living species have been discovered so far. They have a single, flat rounded bilateral shell that is often thin and fragile and range in size from 0.5 to 3.0 cm (0.2 to 1.2 inch) in length. All the presently known living species live deep down in ocean trenches where they seem to feed on microscopic organisms in mud or bottom detritus.

Molluscan Body Plan

The zoological novice may be confused as to how snails and clams—animals with external shells—can be classified in the same phylum with the squids and octopuses—animals with no external shells and grasping tentacles. How can animals that look so morphologically different be classified together? Actually, these creatures have many similarities that are not readily apparent. Structurally, their bodies consist of the same basic parts—foot mantle, and visceral mass—with these parts being slightly to greatly modified in each group (**Figure 12.11**). The muscular foot in gastropods is flat for crawling but blade-like in bivalves for burrowing, whereas in cephalopods it is lengthened into arms and tentacles. The mantle is a thin layer of delicate tissue that secretes the shell of gastropods and bivalves or forms the muscular body of the cephalopods. The tube-like siphon in bivalves (double siphon) and cephalopods (single siphon) moves water into and out of the animal whereas in gastropods the siphon serves merely as an opening into the respiratory organs. Within all the internal organs are concentrated into the visceral mass.

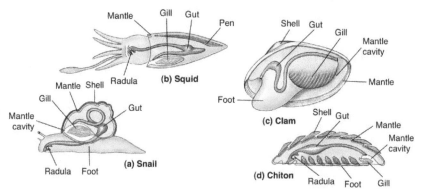

Figure 12.11 Unique molluscan characteristics as illustrated by (a) the snail, (b) the squid, (c) the clam, and (d) the chiton.

Body Wall

The body wall of molluscs consists of three recognizable layers: the cuticle, the epidermis, and muscles. The cuticle is not chitinous as was initially thought to be the case, but rather it is composed of a protein called **conchin**. The epidermis is usually a single layer of cuboidal to columnar cells; over much of the body these epidermal cells bear cilia. A great deal of the epidermis is secretory in function. Some epidermal cells secrete the cuticle; some are especially abundant of the ventral surface where they secrete mucus, whereas others constitute the **shell glands** that secrete the calcareous spicules or shells characteristic of this phylum. The body wall usually includes three distinct layers of smooth muscle fibers: an outer circular layer, a middle diagonal layer, and an inner longitudinal layer.

Mantle and Shell Formation

The mantle is the dorsal body wall covering the visceral mass. As development occurs in most molluscs, the mantle forms one or two folds. The outward growth of these folds creates a space (mantle cavity) between the mantle fold and the body proper. The mantle cavity may be in the form of shallow grooves, or one (gastropods) or two (bivalves) large chambers through which water is passed by ciliary or muscular action. In most cases the mantle cavity houses the gills (**ctenidia**) and receives the fecal matter from the anus as well as products of the excretory and reproductive systems. The exposed, fleshy body surface of squids and octopuses is the mantle itself. Water passes through the mantle cavity of gastropods and bivalves by ciliary action, but the mantle of cephalopods, unconstrained by an external shell, expands and contracts to draw water into the mantle cavity and then be forcefully expel it out through a narrow muscular funnel or siphon. This jet of water provides a means of rapid locomotion for most cephalopods.

Molluscan shells are noted for their wonderfully intricate sculpturing and often flamboyant color patterns, but very little is known about the evolutionary origins and functions of this signature molluscan feature. Molluscs may have one shell, two shells, eight shell plates, or no shell. Although molluscan shells vary greatly in number, shape and size, they all adhere to the basic construction plan of calcium carbonate laid down in layers and often covered by a thin organic surface coating called a **periostracum**.

The blood of molluscs is rich in calcium and during shell formation the calcium is concentrated out of the blood and crystallized as calcium carbonate ($CaCO_3$) in the form of calcite or aragonite crystals by the epidermal shell glands of the mantle. In a process known as **biomineralization**, these crystals are arranged into two layers: an outer, chalk-like **prismatic** layer and a smooth inner pearly **lamellar** or **nacreous** layer. Conchin protein is incorporated into these layers to help the calcium carbonate crystals bond together. Some molluscs also incorporate protein-rich pigment compounds into the construct, which accounts for the fabulous colors of some seashells. The shell grows gradually over time as the animal adds calcium carbonate to the leading edge of the shell. Thus, the shell gradually becomes larger and wider to better accommodate the growing animal within.

A mollusc shell is formed, repaired, and maintained by the mantle. Any injuries to or abnormal conditions of the mantle are usually reflected in the shape and form and even color of the shell. When the animal encounters harsh conditions, or the food supply becomes severely limited, the mantle often ceases to produce the shell secretions. When conditions improve again, and the mantle resumes production, a "growth line" that extends around the entire shell is produced. The pattern and even the colors on the shell after these dormant periods

Figure 12.12 The smooth nacreous layer lining an abalone shell.

are sometimes quite different from previous colors and patterns. **Conchologists** (a branch of malacology devoted to the study of the shells of molluscs) are intrigued and puzzled by the surprising degree of variation in the exact shape, pattern, ornamentation, and color of the shell within some species of molluscs.

Nacre, also known as **mother of pearl**, is an iridescent aragonite-based secretion that is continually deposited onto the inner surface of shells of some bivalve molluscs. Deposition of this material serves to thicken, strengthen, and smooth the inner surface of the shell itself and as a defense against parasitic organisms and potentially damaging detritus (**Figure 12.12**).

Box 12.1
From the Sea to Thee—Artificial Bone

The field of **biomimetrics**, defined as "the science and art of transposing biological designs into human technologies such as engineering, materials design, and computing", has been a desire of humankind for a long time as evidenced by Chinese writings detailing the desire to make artificial silk 3,000 years ago. Only in modern times, however, have humans had the technologies and advanced techniques to fully exploit biomimetrics. One of the first practical examples is that of George de Mestral who, in the 1950s, perfected the hook and loop fastener we know as Velcro® after microscopically investigating why pesky seed pods gathered from a walk in the fields with his dog clung so tenaciously to his socks. Today designers are biomimetrically crafting everything from highly maneuverable robotic fish with fins and commercial fiber optics from the Venus flower basket sponge to sophisticated lenses from the eyes of the starfish and sticky plasters based on the Gekko lizard's ability to walk up glass to name a few.

One of the Holy Grails of biomimetrics is the creation of artificial bone. As our population grows older, the demand for replacement bone continues to increase (as this aging baby boomer author who has had total hip replacement in both hips will attest to).

The natural product that has piqued the attention of biomimetrics researchers as a possible candidate for artificial bone is the nacre of mollusc shells. Nacre is, in essence, a natural ceramic, in this case, a hard nonmetallic type of the mineral calcium carbonate known as aragonite. Unlike grandmother's prized teacup or your run-of-the-mill coffee mug, nacre is not at all brittle but is incredibly tough. Tiny cracks inevitably appear in human-created ceramic objects and spread throughout the brittle structure eventually causing full-blown failure. (Think of how many ceramic plates or mugs you have seen with tiny cracks in them.) Nacre, however, does not crack.

Nacre is secreted by the mantle tissue of some species of mollusc, pearl oysters primarily, onto the inner surface of the shell. Microscopic examination reveals nacre to be hexagonal platelets of aragonite arranged in continuous parallel sheets. The thin ceramic aragonite sheets are separated and held together by an even thinner layer of elastic biopolymers such as chitin, lustrin, and silk-like proteins. Nacre is so tough because whereas one ceramic sheet is easy to fracture, the crack stops when it hits the gluey layer. Human bone is also made up largely of a ceramic known as hydroxyapatite (calcium apatite). Like nacre, hydroxyapatite is strong without being brittle, but its structure is different than nacre in that an organized network of collagen fibers strengthens bone, and a lattice of tiny struts forms a spongy, energy-absorbing framework for most bones. Obviously, nacre sculpted to replace shattered kneecaps, or arthritic pelvises would certainly be a better natural match than the variety of metals, plastics, and ceramics presently used.

Although the structure and remarkable properties of nacre have been known for thirty years and appear simple, no one has been able to make synthetic nacre. Most attempts have focused on alternating a layer of ceramic with a thin wash of glue and repeating over and over. The thickness of the ceramic layers, however, is hard to control, and it takes thousands of cycles to produce even a modest quantity of material. More important is the fact that although ceramic bone implants currently on the market wear well, they are more brittle than healthy bone.

A promising new approach takes advantage of the properties of freezing water to make a finely layered composite that's amazingly tough. The technique involves adding granules of hydroxyapatite to water, then freezing the mixture at a very low temperature. The result is a finely layered composite of ice and ceramic. After the ice is sublimated away by freeze-drying, an epoxy glue is added to the dried material in a vacuum; the epoxy infiltrates the spaces between the plates of hydroxyapatite where the ice used to be mimicking the organic glue layer of nacre. This process is still evolving as researchers struggle to achieve even thinner layers of their faux nacre and to develop practical methods of making the material in bulk and molding it to exact specifications.

As bioengineers continue to struggle with the challenge of artificial bone, the hope is that someday soon a product can be mimicked that goes from the deep blue sea to your shattered knee and beyond.

When a mollusc is invaded by a parasite or irritated by a foreign object that the animal cannot eject, a process known as **encystation** entombs the offending entity in successive, concentric layers of nacre. This process eventually forms what we call a **pearl** and continues for as long as the mollusc lives. Almost any species of bivalve or gastropod is capable of producing pearls, even molluscs that do not produce nacre. However, only a few species, such as the famous *pearl oysters*, can create the highly prized pearl (**Figure 12.13**).

Foot and Locomotion

The foot in gastropods takes the form of a flat, ventral creeping **sole**. A large **pedal gland** supplies substantial amounts of mucus, especially in terrestrial species that must move over relatively dry surfaces. The foot of terrestrial gastropods is so tough, and the mucus it produces is so cushioning and protective that these snails can crawl along a razor blade or over broken glass with ease and impunity.

Figure 12.13 Pearls. At the center of every lump of nacre that is a pearl lies an irritant that triggered the formation of the pearl in the first place.

Muscular ripples in the foot of gastropods propel them slowly but smoothly along in a gliding motion. In some gastropods the muscles of the foot are separated by a midventral line, so the two sides of the sole operate somewhat independently of each other. The right and left sides of the foot alternate in their forward motion, almost in a stepping fashion.

In bivalves, the foot is usually bladelike and flattened, as is the body in general. Most bivalves live in soft benthic sediments and through a combination of muscle action and hydraulic pressure, their foot is used for burrowing and anchoring into benthic habitats (**Figure 12.14**). Some bivalves such as the scallops are capable of short bursts of jet-propelled swimming accomplished by rapidly clapping the valves together. Other bivalves permanently attach to the substratum by either fusing one valve to a hard surface (e.g. rock oysters) or by using special anchoring lines called **byssal threads** (e.g. mussels) (Figure 12.5). The foot is greatly reduced in attached bivalves and plays little role in the life of the animal.

The cephalopods have abandoned the generally sedentary habits of other molluscs and have become swift and agile high-speed predators. All cephalopods can move rapidly by forcefully expelling water from the mantle cavity out through the siphon. The siphon is highly maneuverable and can be pointed in nearly every direction allowing the animal to steer precisely. The streamlined squids are also equipped with fins and the cuttlefish with a muscular flap, which allows them to swim effectively and swiftly. Most octopuses are benthic and lack the fins and streamlined bodies of the squids and cuttlefish. Although they still use water-powered propulsion on occasion, octopuses more commonly rely on their long suckered arms for crawling about the sea floor.

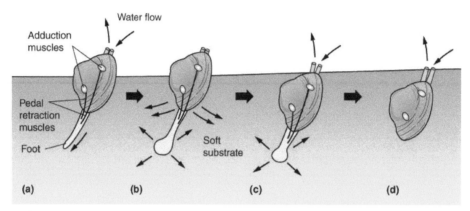

Figure 12.14 Digging In. The burrowing sequence of a bottom-dwelling bivalve. (a) Pedal retractor muscles and adductor muscles relax. Foot muscles contract, causing the foot to expand into soft substrate. (b) Hemolymph is pumped into the tip of the foot causing it to swell and form an anchorage. (c) Pedal retractor muscles contract forcing (pulling) the clam deeper. (d) The clam settles in and extends it siphon tubes.

Feeding

Three main feeding modes occur among the species of Mollusca: rasping herbivory, suspension feeding, and predation. Gastropods use a ribbon of recurved chitinous teeth known as a radula to rasp (scrape) food particles such as algae from surfaces for ingestion (**Figure 12.15**). Terrestrial gastropods use their radula to grind off pieces of soft, green vegetation. The gastropods known as the cone snails (*Conus*) have a highly modified radula that has been reduced to a few isolated poison-injecting teeth. The harpoon-like teeth are set at the end of a long proboscis that can be thrown out rapidly to impale and poison prey such as fish, worms, or other gastropods that is then pulled into the gut as the proboscis is withdrawn.

Suspension filter feeding is employed by rather sedentary gastropods or burrowing bivalves and involves modification of the gills that enable the animal to trap particulate matter carried in the mantle cavity current. The gills are greatly enlarged and coated with a thick layer of mucus. Cilia on the gills create currents drawing water in through the incurrent siphon, over the gills and out the excurrent siphon. Food particles in the water current are trapped in this mucus and carried to the mouth where they are ingested. The suspension feeding lifestyle

Figure 12.15 The radula of *Abyssochrysos melanioides*, a species of sea snail. The radula is a toothed chitinous ribbon that is used for scrapping or cutting food before it enters the esophagus. The radula is a structure unique to molluscs and is found in every class of molluscs except bivalves.

has reached its zenith in the vermetids or "worm gastropods." Vermetids are permanently attached to the substratum. A special pedal gland produces copious amounts of mucus that is released outside the shell as a sticky web or net to trap plankton. Periodically, the net is hauled in, and a new net is quickly secreted.

Cephalopods are active and aggressive predators. In squids and cuttlefish, prey is grabbed by the arms or tentacles and then moved to the powerful beak-like jaws where it is bitten and torn into smaller pieces. The radula then grinds these larger chunks into a smaller size suitable for ingestion. In contrast, the feeding pattern of octopuses is rather like that of spiders. The prey is injected with poison from modified salivary glands and then flooded with digestive enzymes. The partially digested tissue that results is then ingested. The vampire squid (*Vampyroteuthis infernalis*) is an exception to the notion that cephalopods are predators. Lacking the squid's standard feeding tentacles and dwelling where typical cephalopod prey is scarce, the vampire squid uses feeding filaments to detect and collect the decomposing remains and feces of minute crustaceans and plankton, as well as nutritious gelatinous debris from filter feeders. As bits of food stick to a filament, the squid wipes them across its webbed arms, where they become encased in mucous en route to the mouth. Some octopuses have the radula modified as a drill to bore into heavy shell of bivalves. A chemical softening agent is secreted in the drilling process, and once the shell has been breached, poison is injected directly into the shell. The *Nautilus* will also scavenge along the bottom if the opportunity presents itself.

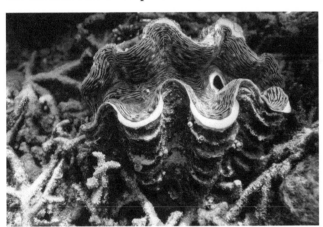

Figure 12.16 The giant blue clam (*Tridacna gigas*) is native to the shallow coral reefs of the South Pacific and Indian oceans. These clams can weigh more than 200 kg (440 lbs), measure as much as 120 cm (47 in) across, and live more than 100 years in the wild.

Some molluscs live symbiotically with other creatures. The giant clams (Tridacnidae) contain symbiotic zooxanthellae (the dinoflagellate *Symbiodinium*). The mantle tissues harbor the zooxanthellae that the clams expose to sunlight through a large gape in their shell (**Figure 12.16**). Many giant clam species have special lenslike structures that focus light on zooxanthellae living in deeper tissues. An even stranger symbiotic arrangement occurs in certain sea slugs (e.g. *Placobranchus*). These slugs obtain functional chloroplasts from the algae upon which they feed and incorporate them into their tissues, where they remain active for a period of time producing products of photosynthesis (sugar) utilized by the host slug.

Circulation and Gas Exchange

Most molluscs have an open circulatory system that functions more-or-less like that of the water circulation pattern in a reef aquarium. Water from the tank collects in a sump and is then run through an aerator that oxygenates the water, before a pump pushes the water back up into the tank. The open circulatory system of most molluscs in that blood flow is only partially contained in vessels is similar to that pattern. The blood, known as **hemolymph,** (water) is pumped into an open space known as the **hemocoel** (tank). Oxygenated hemolymph is collected from the **ctenidia** or gills (aerator) and pumped into a series of **sinuses** (open spaces) where the tissues and organs are bathed in oxygen-rich blood (like the organisms in a reef tank would be bathed in oxygen-rich water). As the hemolymph passes over the tissues in the sinuses, it pools in the **ctenidal vessels** (sump) where gas exchange occurs and it is once again drawn into the three-chambered heart (pump) to be pumped back to the hemocoel.

With a circulatory system that is largely closed and equipped with an extensive system of vessels, the cephalopods possess the most complex and effective circulatory system of any mollusc. This advanced system allows cephalopods to move more rapidly and for longer periods of time than can other types of molluscs. The circulatory system of cephalopods consists of one **systemic heart**, two **branchial hearts**, and the blood vessels (Figure 12.8d). The two brachial hearts, which sit at the gill base, collect unoxygenated blood from all the body parts. They then contract and send this unoxygenated blood coursing throughout the capillaries of the gills and into the auricles of the systemic heart. The two auricles of the systemic heart then drain the now oxygen rich blood into the ventricle of the systemic heart where it is pumped out into the body through progressively smaller vessels and finally into tissue capillaries. Wholly terrestrial gastropods lack gills and exchange gases directly across a vascularized region within the mantle cavity known as a **lung**.

The blood (hemolymph) of molluscs utilizes **hemocyanin**, a copper-containing protein, rather than hemoglobin to transport oxygen. As a result, their blood is colorless when deoxygenated and turns blue when exposed to oxygen (air). Because hemocyanin is substantially less capable of carrying oxygen than is hemoglobin, the molluscan heart must pump a higher blood flow than animals that utilize hemoglobin for oxygen transport.

Excretion and Osmoregulation

The basic excretory structures of molluscs are tubular metanephridia (kidneys). Most molluscs have a single pair of metanephridia. The nautilids have two pairs, however, whereas the gastropods have lost the right nephridia due to torsion and now possess only a single nephridium (Figure 12.2c). Mollusc nephridia are rather large and saclike, U-shaped, and their walls are greatly folded to increase surface area. In bivalves, one arm is glandular and functions to remove wastes, whereas the other arm has been modified into a simple bladder.

Excretion involves the kidneys draining the **pericardial coelom** (around the heart) by way of **renopericardial canals** (blood vessels from the heart to the kidneys) and emptying the collected wastes via **nephridiopores** into the mantle cavity.

Terrestrial gastropods have evolved several water-saving adaptations in order to survive the drying conditions of an existence in air. Aquatic gastropods, like most aquatic invertebrates, excrete wastes in the form of ammonia or ammonia compounds. Terrestrial gastropods, however, convert this ammonia into relatively insoluble uric acid and water. This helps them conserve their valuable body moisture. Behaviorally, some have become nocturnal to avoid the heat of the day whereas others live beneath moist, decaying vegetation. During hot periods or in the colder months in the temperate regions, they become inactive and their metabolic rate approach zero. In this condition, known as **estivation**, they can survive many years awaiting the return of favorable conditions.

Nervous System and Sense Organs

The molluscan nervous system is patterned along the lines of the basic protostome plan consisting of an anterior arrangement of ganglia (brain) with paired ventral nerve cords coming out of the brain. In molluscs, there are typically two pairs of ventral nerve cords; one pair called the **pedal cords** innervate the muscles of the foot, whereas the other pair, known as the **visceral cords,** serve the mantle and viscera (internal organs). As a result of torsion, the posterior portion of the gastropod nervous system is twisted into a figure-eight, a condition known as **streptoneury.**

The nervous system of the cephalopods is developed to a degree unequaled among invertebrates. In cephalopods, we see a larger brain and more pronounced cephalization (development of a true and distinct head) than in other molluscs. As a result, the intelligence of the cephalopods, especially the cuttlefish and octopuses, is readily apparent as discussed earlier in this chapter. Cephalopods have definite brain centers that control specific activities. The subesophageal region of the brain has several ganglionic subdivisions with nerves to the arms, funnel, mantle and viscera all arising from this region. Centers for arm movements, respiratory movements, control of eye muscles, the iris, pupil and chromatophore activity are also found in this area.

Molluscs possess various combinations of chemosensory tentacles, photoreceptors and eyes, statocysts (equilibrium organs), and **osphradia** (chemoreceptor patches on the gill or mantle wall). Most gastropods have one or two pairs of **cephalic** (head) **tentacles** that may bear eyes as well as tactile and chemoreceptor cells (Figure 12.2a). Motile gastropods also usually possess a pair of closed statocysts in the anterior region of the foot.

In bivalves, most sense organs are located in the margin of the mantle. Many species possess **pallial tentacles** that contain tactile and chemoreception cells, and a statocyst is often found near or embedded in the pedal ganglia. The bivalve statocyst consists of a fluid-filled sac containing **statoliths** (little stones) that help to convey relative position. In some bivalves, **ocelli** (small simple eyes) are present along the edge of the mantle or on the siphon.

Tactile senses in all cephalopods are quite well developed with the rim of each sucker being the most sensitive area on the body. Studies have shown that even when blindfolded, an octopus can differentiate between objects of various sizes and shapes, often with remarkable accuracy: Some can even detect size differences in spheres, for example, which most humans cannot do. The most highly developed sense organs in cephalopods, however, must surely be their eyes. Cephalopod eyes are superficially similar to those of vertebrates, and the two are often cited as classic examples of convergent evolution. Cephalopod vision is acute, and experiments have shown that the octopus can distinguish the brightness, size, shape, and horizontal versus vertical orientation of objects. Cephalopod eyes are also sensitive to the plane of polarization of light. Surprising given their ability to change color, most are probably color blind.

Nautilids have statocysts that provide information on static body position and on body movement and they are the only cephalopod with olfactory organs known as **osphradia** that most likely monitors the amount of sediment in the incoming currents of water. Nautilids have a relatively simple eye compared to the other cephalopods. *Nautilus* eyes have a large pigment cup and no cornea, lens or other refractive parts.

Reproduction and Development

The great species diversity of Mollusca is further reflected in the many different patterns of reproduction present among the members of the phylum. Depending on the species, gastropods are either dioecious or monoecious (hermaphroditic). In either conFigureuration, one of the gonads has been lost, and the remaining one is usually coiled within the visceral mass. Furthermore, the right nephridium serves the reproductive system as a genital duct and not the excretory system as a urinary duct.

In nearly all gastropods, there is copulation and internal fertilization. (Hermaphroditic types will simultaneously inseminate each other.) Sperm from the male's inserted penis is transferred to the female where it is then stored in the end of the **pallial oviduct**, where the eggs are fertilized. This section of the female oviduct is modified to form both an albumen (the white of an egg) gland and a large jelly gland or capsule gland. These glands enclose the fertilized eggs in jelly masses or gelatinous capsules that are then passed out of the female's body and usually attached to objects in the habitat.

Some gastropods exhibit strange and even bizarre reproductive behavior. To avoid predators, some arboreal slugs copulate in mid-air with each partner being suspended by a viscous mucal thread. In the slipper-shell snails (*Crepidula*), that are sessile, individuals may stack one atop the other with males usually on top of the stack. Males then use their long penis to inseminate the females below. Males remain males as long as they are near a female but can change into a female if isolated from a female. As if those examples were not strange enough, consider the courtship of the hermaphroditic land snail *Helix*. Prior to copulation, two individuals will intertwine and stroke each other with their cephalic tentacles. At some point in this ritual, one of the partners will drive a calcareous **copulatory dart** into the body wall of the other with such force that it is buried deep in the other's internal organs. This lancelet of love apparently acts to sexually arouse the receiving partner.

Most bivalves are dioecious. A few types, however, such as the cockles, a few oysters and scallops, and some of the freshwater clams are hermaphroditic. In fact, some clams change their gender during their lives, first developing as males and later changing into females.

Bivalves possess two gonads that encompass the intestinal loops. The sperm and eggs are released into the mantle cavity where they are swept out into the surrounding water with the exhalant current and fertilization occurs. A few types brood their eggs within the mantle cavity or on the gills. Brooded eggs are fertilized by sperm brought into the mantle cavity with the inhalant current. The females of the freshwater mussel family Unionidae brood their lava known as **glochidia** for one to ten months on their gills. When the glochidia are fully developed in the spring or fall, they are released into the water where they drift until they find a suitable fish host and attach to the fish's gills or fins. After about one to four weeks, the glochidia have matured into juvenile mussels and fall off of the host fish to burrow into the substratum and become mature mussels. This hitch-hiking adaptation allows the mussels to disperse their larva over a larger area, and greater distances than would normally be possible.

Cephalopods are mainly dioecious, and fertilization is internal. Fertilization may take place within the mantle cavity or outside, but in either case it involves copulation. A quite elaborate courtship is often a precursor to copulation.

Because the oviduct opening of females is deep within the mantle chamber, male cephalopods use one of their arms that have been modified for the purpose to transfer sperm into the female. The morphological modification of cephalopod arms for such a purpose is known as **hectocotyly**.

Precopulatory rituals in cephalopods almost always involve striking changes in coloration as a male tries to females and discourage other males in the area. Male squid often seize their female partner with their tentacles and the two swim head-to-head through the water. Eventually the male hectocotylus arm inserts a **spermatophore** (sperm packet) into the mantle chamber of his partner, near or in the oviduct opening. Mating in octopuses can be aggressive and even savage displays. In their copulatory exuberance the couple may tear at each other with their sharp beaks, or one partner may strangulate the other partner as it wraps it arms so tightly around the mantle cavity of the other that it cannot circulate water over the gills. In many octopuses, the tip of the hectocotylous arm may break off and remain in the female's mantle chamber for some time.

As the eggs pass through the oviduct, various glands cover them with membranes and coatings forming them into a capsule or grapelike cluster. Squids usually deposit their egg capsules on vegetation or the substratum whereas cuttlefish attach single eggs to seaweed or other substrata. No squid species has ever been identified in which either of the sexes cares for the fertilized eggs. This is not the case, however, with octopuses. Female octopuses often deposit their egg cases onto the ceiling of a small cave or opening in the rocks. While the eggs are incubating, the female gently caresses them and flushes them with jets of water to keep them clean and to supply a steady flow of fresh, oxygenated water.

Although we tend to associate intelligence with longevity, this is not the case with squids, cuttlefish, and octopuses. These cephalopods "live fast and die young." Quickly growing to maturity; they reproduce once and then die shortly thereafter. The pearly nautilus, however, is long-lived (perhaps to 20 years), slow-growing, and able to reproduce for many years after maturity. In the octopus at least, this self-destruction seems to be set in motion by the reproductive process. When triggered by the reproductive cycle, a pair of glands begins to secrete a hormone that speeds up the aging process and inhibits feeding in both male and female octopuses.

Development of the fertilized egg in molluscs is similar in many fundamental ways to that of the other protostomes. In gastropods and bivalves a typical trochophore larva forms (**Figure 12.17**). The trochophore is usually followed by a more highly developed, uniquely molluscan larva called a **veliger**. The veliger larva may possess a foot, shell, operculum, and other adult features. The most characteristic feature of the veliger is the swimming and feeding organ, or **velum**. It is usually during the veliger stage that gastropods undergo torsion with the shell and visceral

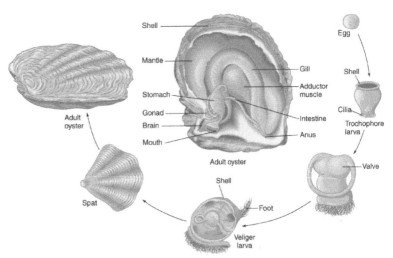

Figure 12.17 Life Cycle of Oysters. The fertilized egg develops into a trochophore larva that develops into a veliger larva. The veliger larva swims for approximately 2 weeks before settling down, attaching, and becoming a spat. Four years is required for oysters to grow to commercial size.

mass twisting 180° relative to the head and the foot. Although this phenomenon is not fully understood, it seems to have played a major role in gastropod evolution.

In cephalopods, development is always direct with the larval stages being lost entirely or passed within an egg case. The embryo grows in such a way that the mouth opens to the yolk sac, and the yolk is consumed directly by the growing and developing individual.

Molluscan Connection

Economically

Few invertebrate groups have and continue to be more important to humans than molluscs. Most important is the food connection. Molluscs of all types—snails, clams, oysters, mussels, scallops, squid, and octopus—have been important food sources for humans probably as long as humans have been a distinct species. An example from antiquity was revealed in the discovery of a Native American shell mound (or **midden** as they are known) near San Francisco Bay that contained over one million cubic feet of shells and may have taken 3,500 years of feasting to accumulate. The human-molluscs food connection continues unabated to this day. In fact, according to the National Marine Fisheries Services, consumer demand for both fish and "shellfish" (a biologically incorrect term that encompasses both molluscs and crustacean arthropods) has skyrocketed since the 1970s. The NMFS estimates commercial fresh and frozen shellfish consumption amounts to around 3.4 pounds per person yearly in the United States alone. And although statistics are lacking, the NMFS suggests that an additional 200-300 million pounds of shellfish are harvest yearly through recreational fishing. The annual world squid and octopus fishery alone exceeds 2 million metric tons per year. A 1-acre commercial mussel bed can produce as much as 10,000 pounds of mussel tissue per year. By comparison, 1 acre of pastureland produces only about 200 pounds of beef in a year.

The empty shells of molluscs have also proven to be quite useful. Humans have cut, carved, and polished them into everything from spoons and buttons to jewelry, purses, and belts, and use them as inlay on furniture and the keys of musical instruments. We even grind them up and use them to surface roadways. The inner iridescent mother of pearl layer lining the shell of some species of molluscs is especially prized. Most highly prized for their value as jewelry are those hardened mineralized lumps we call pearls.

Experts tell us that the owners of natural pearls are advised to wear them regularly so that the moisture of the wearer's skin will keep them hydrated. Otherwise, they gradually lose their luster.

Some herbivorous molluscs can be quite destructive. Introduced land snails can cause great destruction in agricultural areas, especially those with a warm, moist climate. A bivalve known as the shipworm is often described as "the termite of the sea." Its burrowing habits result in millions of dollars in damage yearly to wooden ships, docks, and pilings.

Cuttlebone, the chalky internal shell of the mollusc known as cuttlefish, is commonly used as a calcium-rich dietary supplement for caged birds and turtles. And because cuttlebone is able to withstand high temperatures and can be easily carved, it is an ideal material for making casting molds for the production of jewelry and small sculptural objects.

Mussels cling tenaciously to rocks in the face of pounding surf with natural polymer glues that humans have long envied and attempted to synthesize. The blue mussel (*Mytilus edulis*) has been the subject of intense research and has yielded both the biochemical and genetic secrets to its superstrong, waterproof glue. Early on the problem was that researchers had to use 10,000 mussels to produce just one gram of this natural super glue, at a cost of $75,000 per gram. Turning to recombinant DNA technology, researchers now have genetically engineered brewer's yeast to produce more protein in a week than could be harvested from tens of thousands of mussels. Mussel glue would have a wide variety of applications ranging from environmentally friendly wood glues for the construction industry to naval applications such as the underwater repair of vessels to the attachment of sensors and other materials to the hulls of ships and could potentially serve many uses around the home and office. Estimates place commercially available glues based on mussel attachment proteins becoming available in three to five years.

Paradoxically, research into mussel superglue has led to the potential development of "antiglues." Such chemicals might provide a natural and nontoxic barrier to molluscs, such as the infamous introduced Zebra mussel, that foul ship hulls, pilings, and the intakes of power plants. Such antiglues could be formulated in such a way as to possibly prevent the offending molluscs from ever attaching in the first place.

Ecologically

Not only are molluscs an important food source for humans, they are also a vital part of many food webs and compose a major portion of the diet of many other animals, including certain fish, birds, and mammals. In turn, some molluscs are major predators in their habitats, especially the cephalopods. Herbivorous land snails can consume a great deal of greenery and in some instances they have been known to seriously alter entire landscapes.

Medically

Being non-toxic and natural, mussel glue has a number of potential medical applications in surgery, suturing, orthodontics, and ophthalmics, and doctors in these various disciplines eagerly await its availability.

Although many people are allergic to shellfish, no molluscs are inherently toxic for consumption. Some types, however, present more of a risk than others. Being filter feeders, bivalves such as clams, oysters, and mussels, tend to accumulate chemical pollutants, toxic dinoflagellates, and various types of potentially dangerous bacteria within them depending on the season and the individual circumstances of these molluscs. One must be extremely careful about eating raw shellfish. Most foods from animals are cooked prior to consumption. However, molluscan shellfish (oysters, clams, mussels) are often consumed raw or partially cooked. Illnesses sometimes occur from eating shellfish, and although most illnesses are typically not life threatening, instances of serious illnesses and deaths have been reported. Eating raw shellfish can increase the risk of illness – cooking can reduce the risk. Shellfish that has been shucked (removed from shells) and placed in plastic or glass containers is intended to be cooked and should never be eaten raw.

An old myth specifies the best time to eat oysters is during months that contain an "R" (i.e. September through April) and to avoid eating oysters in months that do not contain an "R" (May through August). Although levels of certain naturally occurring marine bacteria, like *Vibrio*, are higher in coastal waters during warm weather months, the bacteria may still be present, although in lower levels, during cold weather months. Although most consumers are not susceptible to infection by *Vibrio vulnificus*, consumers who have certain illnesses or health conditions, such as persons with liver, stomach, or blood disorders; individuals with AIDS, diabetes, cancer, or kidney disease; and chronic alcohol abusers, should only eat molluscan shellfish that is cooked and abstain from eating it raw or partially cooked, regardless of the month. Because heat kills harmful bacteria and viruses, thoroughly cooked oysters, clams, and mussels are safe for anyone to eat all year, as long as they are legally harvested.

The golf ball-sized and beautifully-colored blue-ringed octopus (*Hapalochlaena maculosa*) of Pacific waters from Japan to Australia also possesses a potent and potentially lethal neurotoxin. Despite its small size, the blue ring contains enough venom to kill 26 humans within minutes. There is no known antidote making this octopus one of the most deadly of all sea creatures.

The subtropical and tropical cone shells or cone snails (Conidae) live under rocks, in crevices, and around coral reefs. Wielding a venomous radula modified into a harpoon, they can inject extremely toxic venom that acts very swiftly, causing acute pain, swelling, paralysis, blindness, respiratory failure, and possible death within hours. The harpoon can penetrate gloves or even wet suits. There is no antivenin, and treatment involves providing life support until the venom is metabolized by the victim.

The venom of some cone snails, however, shows much promise for providing a non-addictive pain reliever 1,000 times more powerful as, and possibly a replacement for, morphine. Many peptides produced by the cone snails show prospects for being potent pharmaceuticals effective in treating post-surgical and neuropathic pain, even accelerating recovery from nerve injury. The first painkiller derived from cone snail toxins, Ziconotide, was approved by the U.S. Food and Drug Administration in December, 2004 under the name "Prialt." Other drugs that may be used in the treatment of Alzheimer's disease, Parkinson's disease, and epilepsy are currently in clinical and pharmaceutical trials.

Culturally

Molluscs have also nourished humans culturally. In some early cultures and parts of New Guinea yet today certain shells were used as currency. Native Americans along the east coast of the United States carved white beads from the shell of the North Atlantic channeled whelk (*Busycotypus canaliculatus*) and purple beads from the shell of the Western North Atlantic hard-shelled clam (*Mercenaria mercenaria*) and wove them into intricate belts known as **wampum.** The wampum was used as a token to represent a memory and thus aided in the telling of stories, chants, and songs that is part of the oral tradition of many Native American tribes. Belts were also sometimes used as badges of office or as ceremonial devices in some cultures such as the Iroquois. When Europeans came to the Americas, they realized the importance of wampum to Native people but mistook it for money. Soon they were trading with the native peoples of New England and New York using wampum to the point that Dutch colonists began to manufacture their own wampum.

A Closing Note
The Swimming Scallop—What Zoology is Really All About

Biology textbooks by their nature are often encyclopedic compilations of concepts and facts. And I as author have erred along those same lines. Lest you get the impression that zoology is merely memorizing facts and studying smelly dead specimens pickled in jars, consider research being done on the scallop.

As we discussed earlier in this chapter, the scallop is one of only a few bivalve molluscs that can truly swim. When threatened, the scallop claps the two halves of its shell together, and thus expels a jet of water that propels it to safety. By repeatedly slamming the shell, the scallop manages to wobble unsteadily through the water.

The downward closing of the shell is accomplished by a large (and tasty) muscle, the adductor, attached to the center of each valve. This muscle can exert force only to close the shell (muscles only pull; they never push); to open the shell, the scallop relies entirely on a little rubbery pad of natural elastic called abductin just inside the hinge of the shell. This pad gets squashed when the shell closes, but as the adductor muscle relaxes, the pad rebounds and pushes the shell back open. (Culinary Hint: when shopping for live bivalves for dinner, you want the ones that are tightly closed; they are manifestly alive because they're still actively holding their shells tightly shut.)

Here is where it gets interesting, and zoology (more specifically biomechanics) springs (pardon the pun) into action. Muscles and elastic compounds do not work as well in cold water yet the Antarctic scallop, *Adamussium colbecki*, still manages (just barely) to sustain level motion. How is this possible? Mark W. Denny and Luke P. Miller, biomechanists at Stanford's Hopkins Marine Station in Pacific Grove, California traveled all the way to McMurdo Sound Antarctica to find out. They expected the shells of Antarctic scallops to be lighter than those of their warm-water kin, and they were. They also expected the adductor muscle of Antarctic scallops to be larger, but they weren't. In fact, they found that *A. colbecki* has a closing adductor only about half as big as that of a warm-water scallop of similar size. What they found was that the elastic abduction pad in these cold-water scallops was more elastic than it should have been at those cold temperatures. Somehow natural selection has fine-tuned the response of abduction to temperature. We aren't sure how just yet but the secret must lay in the arrangement of the protein polymers that store and release energy. If the details of that can be determined, a rubber that retains its bounce in extreme cold would be possible, a fact that would make materials scientists and engineers take notice. Research such as this on the scallop is but one example of what zoology is truly all about—zoologists mucking about in the field, hypothesizing, collecting and measuring, all the while encountering unexpected results and new species at every turn.

In Summary

- Because many molluscs form a hard shell, the fossil record for some mollusc groups is quite good, and this record seems to indicate that molluscs first appeared in ancient seas about 600 million years ago.
- The number of extant (living) species is estimated at anywhere from 50,000 to 110,000 depending upon the authority consulted whereas about 60,000 extinct species are known only by their fossilized remains.

- Molluscs may be grouped into seven classes:

  Class Gastropoda
  Class Bivalvia
  Class Cephalopoda
  Class Polyplacophora
  Class Scaphopoda
  Class Aplacophora
  Class Monoplacophora

- The characteristics of phylum Mollusca:

  1. They are bilaterally symmetrical coelomate protostomes with the majority possessing a definite head. The coelom is reduced to small spaces around the heart, kidney, and gonads.
  2. The ventral body wall is specialized into a large, well-defined muscular foot used for locomotion or burrowing.
  3. The dorsal body wall forms a pair of fleshy folds or lobes called the mantle that secretes calcareous spicules, shell plates, or shells.
  4. Most possess a rolled extension of the mantle known as a siphon or funnel. Usually tube-like in form, the siphon allows for the movement of water into and out of the aquatic molluscs and serves as inhalant tube in terrestrial gastropods.
  5. The viscera (internal organs) are concentrated into a visceral mass.
  6. A mantle cavity houses visceral mass. The mantle cavity opens to the outside and functions in gas exchange, excretion, elimination of digestive wastes, and the release of reproductive cells.
  7. Except for the filter feeding bivalves, they possess a file-like set of hooked teeth-called the radula.
  8. Except for the Cephalopoda, they possess an open circulatory system consisting of a heart with distinct ventricle and atria, blood vessels, and blood with respiratory pigments.
  9. They possess an excretory system consisting of one or two kidneys (metanephridia) that usually empty into the mantle cavity.
  10. Gaseous exchange is accomplished through the gills, lungs, mantle, or body surface.
  11. They possess a complete and complex digestive system with the anus usually emptying into the mantle cavity.

- Class Gastropoda—snails, land and sea slugs abalone, limpets, cowries, and conchs—are the most environmentally successful group of molluscs occupying almost every conceivable habitat on the planet. Most members of this class have a one piece shell that is typically coiled or spiraled and that usually opens on the right-hand side (as viewed with the shell apex pointing straight upward). Gastropods exhibit a wide variety of feeding strategies with some species being herbivores, carnivores, scavengers or detritivores, ciliary feeders, or parasites.

- Class Bivalvia—clams, oysters, mussels, scallops—is mainly marine being found in all ocean zone. Some types are adapted to brackish and freshwater. Unlike the gastropods, bivalves do not have a definite head and their shell is composed of two parts. Bivalves are mainly filter feeders.
- Class Cephalopoda—squid, octopus, nautilus—is exclusively marine in their distribution. All types possess tentacles and are active predators, but only the nautilus possesses an external shell.
- Class Polyplacophora—chitons—have an ovoid, flattened body covered by eight overlapping calcareous shell plates. Chitons are strictly marine and slowly creep along on their large foot eating algae, bryozoans, diatoms, and sometimes bacteria by scraping the rocky substrate with their well-developed radula
- Class Scaphopoda—the tusk or tooth shells—have the simplest shell structure and anatomy of all the molluscs. Scaphopods are relatively inactive creatures that burrow in sand at depths ranging from 6 to 200 meters (18 to 600 feet). Their food consists primarily of one-celled foraminiferans.
- Class Aplacophora is populated by exclusively benthic molluscs that have deviated from the normal molluscan form. These small creatures around 2.5 cm (1 inch) in length are worm-like without a shell, and there is no fossil evidence to suggest that any members of the class ever had one. Rather, the mantle is embedded with calcareous spicules. Furthermore, the foot is absent or vestigial at best leaving the Aplacophora with no visible means of locomotion. These out-of-the-ordinary molluscs live mostly in deep water where they parasitize hydroids and other corals.
- Class Monoplacophora was thought extinct until their discovery in April, 1952. Little is known about the monoplacophorans as only a dozen living species have been discovered so far. They have a single, flat rounded bilateral shell that is often thin and fragile and range in size from 0.5 to 3.0 cm (0.2 to 1.2 inch) in length. All the presently known living species live deep down in ocean trenches where they seem to feed on microscopic organisms in mud or bottom detritus.
- Molluscs:

1. Possess a body wall consisting of three recognizable layers: the cuticle, the epidermis, and muscles. The epidermis is composed of a protein called conchin. A great deal of the epidermis is secretory in function. Some epidermal cells secrete the cuticle, whereas others secrete mucus. Still others constitute the shell glands that secrete the calcareous spicules or shells characteristic of this phylum.
2. Wrap the visceral cavity with the dorsal body wall or mantle. The mantle cavity houses the gills (**ctenidia**), and receives the fecal matter from the anus and products of the excretory and reproductive systems. The mantle forms, repairs, and maintains the shell of a mollusc.
3. Move by means of a specialized muscle known as the foot. In cephalopods the foot is modified into tentacles.
4. Have three feeding modes: rasping herbivory, suspension feeding, and predation
5. Possess a three-chambered heart with most types having an open circulatory system. Cephalopods, however, have a closed circulatory system.
6. Exchange gases via gills in aquatic types but across thin highly vascularized areas in terrestrial species.
7. Have a nervous system patterned along the lines of the basic protostome plan of an anterior arrangement of ganglia (brain) with paired ventral nerve cords coming out of the brain. Cephalopods have the most highly developed brain and sensory system in the class.

8. Are mainly dioecious with some gastropods being monoecious.

•   Molluscs connect to humans in ways that are important economically, ecologically, medically, and culturally.

Review and Reflect

1. ***Molluscs Across Time***. Malacologists know of tens of thousands of extinct species of molluscs. Why do we know so much more of the ancient history of molluscs than we do for nearly any other type of animal?

2. ***Clam Calculations***. Some bivalves exist in tremendous numbers. At one time, it was estimated that 4.5 million clams occupied about 700 square miles of the Dogger Bank in the Atlantic Ocean east of England. If that space were equally divided among all those clams, calculate how much space each clam could call their own.

3. ***Digging In***. You are about to take a test over molluscs in your zoology class. You are sitting in lecture hall doing some last minute reviewing of the test material with fellow students. Someone says, "I don't get it. How can a clam possibly burrow into the bottom? They don't have any structures to dig with." How would you reply?

4. ***Unlikely Taxonomic Partners***. The test over molluscs is just about to begin when the person sitting next to you turns and says, "I don't get it. How can a snail, a clam, and a squid be classified in the same phylum? They seem totally different from each other, especially the squid." How would you reply?

5. ***Become One With the Mollusc***. Imagine you are a mollusc then react to each of the following:

    A. If you had to be reincarnated as a mollusc, would you rather be a land snail, freshwater clam, oyster, or octopus and why? Not only make a selection but defend your choice.

    B. You are an octopus resting on a pile of rocks minding your own business when suddenly you are threatened by a large predator. What would you do? What defensive strategies are available to you?

    C. You are a bivalve known as the pearl oyster. A tiny grain of sand has gotten into your shell and has been "rubbing you the wrong way." What are you going to do about it?

6. ***Jewel of the Sea***. What are cultured pearls and how are they different than natural pearls?

7. ***Arms and Tentacles***. A squid has eight arms and two tentacles whereas an octopus just has eight arms. In cephalopods, what is the difference between an arm and a tentacle?

8. ***Feed Me Seymour***. Your globe-trotting zoology instructor is gone again to some convention and again you have been left in charge of feeding some of the live animals in biology department's collection. This time it is the molluscs. What would you feed each of the following: land snails, marine clams, small squids?

9. ***Land Snail Love***. Terrestrial snails are hermaphroditic. Is this a beneficial reproductive adaptation? Explain.

Create and Connect

1. *In Living Color.* Most cephalopods appear to be colorblind. The Sparkling Enope Squid (*Watasenia scintillans*) or firefly squid seems to be the only exception discovered so far. Suppose a new species of octopus has been discovered, and you have been assigned the task of determining whether this new species of cephalopod has true color vision or not. Design an experiment to test the problem question: Does the newly-discovered species of octopus possess true color vision?

   Guidelines:

   A. Your design should include the following components in order:

   ➢ The *Problem Question*. State exactly what problem you will be attempting to solve.
   ➢ Your *Hypothesis*. Although this is a fictitious experiment, word your hypothesis as realistically as possible.
   ➢ *Methods and Materials*. Explain exactly what you will do in your experiment including the materials necessary to accomplish the task. Be specific, take nothing for granted, and do not expect people to read your mind as they read your work.
   ➢ *Collecting and Analyzing Data*. Explain what type(s) of data will be collected and what statistical tests might be performed on that data. It is not necessary to concoct either fictitious data or imaginary observations.

   B. Assume you have five living octopuses to work with, adequate housing for the octopuses, plenty of laboratory space, and all the equipment necessary to meet the challenge.
   C. Your instructor may provide additional details or further instructions.

2. *Molluscan Invaders.* The zebra mussel (*Dreissena polymorpha*) and the European brown garden snail (*Helix aspersa*) are species of molluscs that have been introduced onto North America and in the freshwaters of the continent. Write a report in which you detail the ecological changes and agricultural harm these introduced molluscs have caused.

   Guidelines:

   A. Format your report in the following manner:

   Title page (including your name and lab section)
   Body of the Report (include pictures, charts, tables, etc. here as appropriate). The body of the report should be a minimum of two pages long—double-spaced, 1 inch margins all around with 12 pt font.

Literature Cited A minimum of two references required. Only <u>one</u> reference may be from an online site. The *Literature Cited* page should be a separate page from the body of the report and it should be the last page of the report. Do NOT use your textbook as a reference.

B. The instructor may provide additional details and further instructions.

3. ***Head of the Class.*** The octopus is considered to be among the most intelligent invertebrates. It possess both long- and short-term memory and can learn from experience. In fact, octopuses can quickly learn how to perform a certain task merely by watching previously trained fellow octopuses.

In one experiment, an octopus was tested to see if it could distinguish between two different situations. In the first situation, a crab was placed in the same tank as an octopus. The octopus immediately attacked and ate the crab. In the second situation, another crab was placed in the tank along with a white square. Again, the octopus attacked and ate the crab. Zoologists conducted three trials of each situation a day. This pattern continued during the first two days of the experiment. On the third day, however, the octopus was given a small electric shock every time it reached for the crab in the "crab and square" situation. The experiment continued following this pattern for 13 days. The data from this experiment were plotted and appear in the graph (**Figure 12.18**).

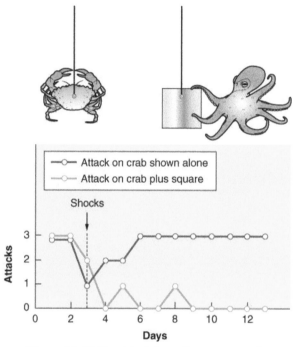

Figure 12.18 Graphical data of octopus attacks on crabs alone and on crabs plus a square.

A. Describe the response of the octopus both to the crab alone and to the crab and square during the first two days of the experiment.

B. Describe the response of the octopus to the crab alone after the second day.

C. Describe the behavior of the octopus to the crab and square after the second day.

D. What was the purpose of the trials of days 1 and 2?

E. What conclusion(s) could be drawn from this experiment?

PHYLUM ARTHROPODA: SOVEREIGNS OF THE TERRAN EMPIRE

If all mankind were to disappear, the world would regenerate back to the rich state of equilibrium that existed ten thousand years ago. If insects were to vanish, the environment would collapse into chaos.

—E. O. Wilson

Introduction

Arthropods have evolved to become the most abundant, most diverse, and most biologically successful animals on earth. Their biological prowess is demonstrated in species numbers that stagger the imagination. Of all the known and described species of animals, at least three out of every four is an arthropod. In fact, some studies indicate that as many as 80% of all known animal species are arthropods. (**Figure 13.1**) Some estimates place the total number of arthropods species at 5 to 10 million. Other studies suggest there may be as many as 6 to 9 million species of arthropods in tropical forests alone. Taxonomists can only speculate on the possible total number of extant arthropod species

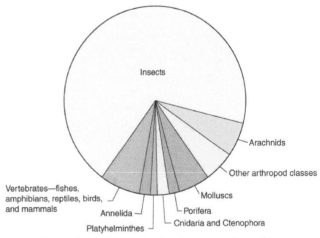

Figure 13.1 A biological nation unto themselves, the arthropods constitute far and away the largest and most diverse group of animals.

History of the Arthropods

The fossil record indicates that the first arthropods appeared in the primeval seas of the Precambrian over 600 million years ago. At some point in time and somewhere within in those ancient waters, the soft cuticle covering the segmented bodies of worm-like creatures began to harden as more proteins accumulated and the inert polysaccharide chitin was added. This developing cuticular exoskeleton offered its bearers a considerable number of benefits. A hardened cuticles offered more protection from predators; it allowed for jointed extensions on each segment to become appendages; it offered protection to joints, fostering strength in key muscle attachment areas, which, over time, became powerful levers, and it added great potential for speed of movement, including flight.

As natural selection drove the development of the exoskeleton, other changes took place in the bodies and life cycles of these protoarthropods in a process zoologists call "arthropodization." Today most zoologists judge that the modern arthropods resulting from this process represent the pinnacle of protostome development.

The phylogeny of the major extant arthropod subphyla had been an area of major interest and dispute and the validity of many of the arthropod groups suggested in earlier works is being questioned. **Figure 13.2** depicts the possible origin of the subphyla and classes comprising phylum Arthropoda as developed through a blending of traditional and cladistic interpretations.

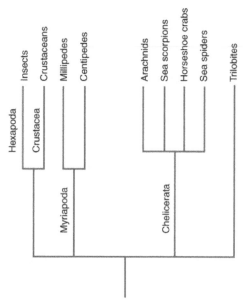

Figure 13.2 The possible origin of the subphyla and classes comprising phylum Arthropoda.

It has been long believed that the extinct Trilobitomorpha (trilobites) were ancestral to all other arthropods. Furthermore, the Crustacea have traditionally been considered as a monophyletic clade within the arthropods. The cladogram depicted in **Figure 13.3** represents this traditional view of arthropod phylogeny. Recent and ever-increasing data from molecular studies, developmental biology and new paleontological discoveries, however, have necessitated major revisions in our understanding of all things arthropod. As arthropod phylogeny is currently in a state of

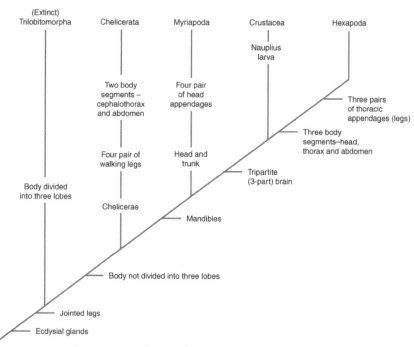

Figure 13.3 The possible phylogeny of the subphyla comprising phylum Arthropoda.

change and flux, we have opted to present a more traditional interpretation until such time as these matters become more settled.

Arthropods were some of the earliest animal forms to appear, and their lineage can be traced back to the Precambrian times some 600 million years ago. Interestingly, the earliest arthropod fossils from that time are similar to crustaceans, not trilobites as was originally hypothesized. It becomes increasingly clear that protocrustaceans and not trilobites were the ancestral stem from which other arthropods arose. If true, this means that the subphylum Crustacea is paraphyletic, not monophyletic.

Diversity and Classification

Currently, the number of identified arthropod species as listed in the literature varies from 874,000 to 1,170,000. Clearly, the total number of species of arthropods and the tremendous diversity of those species spurns our paltry attempts at classifying them. In our biological hubris we humans call the current era the age of mammals but the biological reality is that actually we live in the age of the arthropods. In a recent survey, zoologists identified over 2,000 species of insects on a *single tree* in the Amazon rain forest, many previously unknown to taxonomists. No other group of animals approaches this magnitude of species richness. It is in their individual numbers, however, that they are truly awe-inspiring. It has been suggested that if all the individual arthropods were placed in one gigantic heap and if in a single-file line all humans paraded past that pile with each human receiving their equal share of arthropods, each person would walk away with over 200 million individual arthropods. A simple calculation will yield an estimated number of total individual arthropods that defies comprehension.

In the past few decades, several specialists have suggested that arthropods are a polyphyletic group and that groups we herein consider as subphyla should each be raised to the rank of phylum. Our contention is that the evidence more strongly supports the view of Arthropoda as a monophyletic clade, and we retain its single-phylum status in this chapter.

Ignoring some minor taxa for clarity and considering only the taxa with extant members, we classify the arthropods as follows:

    Domain Eukarya
        Kingdom Animalia
            Subphylum Chelicerata
                Class Arachnida—Spiders, scorpions, ticks, mites
                Class Merostomata—Horseshoe crabs
            Subphylum Myriapoda
                Class Chilopoda—Centipedes
                Class Diplopoda—Millipedes
            Subphylum Hexapoda
                Class Insecta—Insects
                Class Entognatha—Springtails, bristletails, proturans

Subphylum Crustacea

Class Malacostraca—Crabs, shrimps, lobsters, woodlice

Class Branchiopoda—Brine shrimp, fairy shrimp

Class Maxillopoda—Barnacles, copepods, fish lice

Class Ostracoda—mussel shrimp

(A fifth subphylum—Trilobitomorpha—is represented only by several thousand species of extinct animals known as *trilobites*.)

Phylum Characteristics

Arthropods have successfully colonized virtually every habitat and exploited every imaginable life style and developmental strategy. They are adapted for life on land; in soil; in fresh, brackish, or salt water, and even in the air. Arthropods occur anywhere from heights of over 6,000 m (20,000 feet) on mountainsides to depths of over 9,000 m (30,000 feet) in the ocean and everywhere in-between.

They range in size from tiny mites and crustaceans less than 1 mm (0.04 inch) long, to the great Japanese spider crabs with leg spans exceeding 4 m (13 feet). And one species or another displays every type of symbiotic relationship and feeding mode known to biologists. Not only are they a morphologically large and diverse group, but environmentally they may be critical to the well-being of other living things around them, including humans. As the quote that opens this chapter indicates, it is quite possible and perhaps highly probable that the health of the worldwide web of life as we know it, at least terrestrially, depends on these magnificent armored animals.

The members of phylum Arthropoda exhibit the following general characteristics:

- Their segmented body is divided into **tagmata** (specialized body regions)
- The coelom is reduced to portions of the reproductive and excretory systems. Most of the body cavity consists of a **hemocoel** (sinuses or spaces within the tissues) filled with blood.
- They possess a cuticular **exoskeleton** composed of chitin, protein, and lipids. The chitinous skeleton is calcified in many groups. Growth occurs by the process of molting or **ecdysis** (Gr., *edkysis*, getting out) marked by the periodic shedding of the old exoskeleton and the formation of a new larger one.
- They possess jointed appendages from which their phylum name—arthropod (Gr. *arthros*, joint + *podos*, foot)—is derived. Ancestrally each true body segment bore a pair of jointed appendages but in modern arthropods, the number of appendages may be reduced, and they are often modified for specialized functions.
- They possess a complex muscle system attached to the exoskeleton for support and leverage. Functional cilia are absent.
- They possess a complete digestive system with mouthparts modified from ancestral appendages into structures adapted for different methods of feeding.
- The circulatory system is open with a dorsal heart, arteries, and hemocoel containing **hemolymph** (blood).

- **Coaxial glands** or **Malpighian tubules** serve as their excretory system. Coaxial glands are paired, and thin-walled spherical sacs bathed in the blood of body sinuses. Nitrogenous wastes are absorbed across the sacs, and excreted through long, convoluted tubules that empty at the base of the posterior appendages. Malpighian tubules are diverticula (pockets or pouches off the gut tract of arachnids adapted to dry environments). These tubules absorb nitrogenous wastes from the blood and then empty them into the gut tract where they are eliminated along with the digestive wastes.

- Gas exchange occurs through the body surface, gills, **tracheae** (air tubes), or **book lungs**. Tracheae are a series of branched, chitin-lined tubules that conduct gases to and from body tissues. This tubule system opens to the outside through holes called **spiracles** located along the ventrally or laterally along the abdomen. Some arachnids possess book lungs. These are paired invaginations of the ventral body wall that fold into a series of leaf-like **lamellae** (thin, flat plates or disks). Air enters the book lung through a slit-like opening and circulates between the lamellae. Respiratory gases diffuse between the blood moving among the lamellae and the air in the lung chamber.

- The nervous system is much like that of the annelids. It includes a dorsal brain made up of a ring around the gullet that attaches to a double nerve cord chain of ventral ganglia. Well-developed sense organs are present.

- The sexes are usually separate with paired reproductive organs and ducts and internal fertilization the norm; some types are capable of **parthenogenesis**. Development progresses through several stages in a process known as **metamorphosis**.

Subphylum and Class Characteristics

Each arthropod subphylum has evolved its own variations and modification on the basic theme.

Subphylum Chelicerata (Gr. *chele*, claw + keras, horn)

The following are characteristics of this subphylum:

- A body composed of two tagmata: the **prosoma** (cephalothorax) and the **opisthosoma** (abdomen). The cephalothorax represents a fusion of the head and thorax (trunk) and is often covered with a carapace-like dorsal shield. The abdomen is composed of up to 12 **somites** (sections) and a postsegmental **telson**. Antennae and wings are absent.

- **Uniramous** (unbranched) appendages attached to the cephalothorax including: **chelicerae** (anterior appendages of an arachnid often specialized as fangs.), **pedipalps** (specialized sensory appendages borne near the mouth), and four pairs (8) walking legs.

- An exoskeleton modified with projections, pores, and slits to accommodate a variety of mechanoreceptors and chemoreceptors (collectively known as **sensilla**), together with sensory and accessory cells. Vibration detectors are very important to spiders that use webs to trap prey as these detectors allow the spiders to determine both the size of the prey and its position on the web by the vibrations the prey makes while struggling to free itself. The chemical sensitivity of arachnids is comparable to taste and smell in vertebrates. Arachnids also possess two or more pairs of eyes capable of detecting movement and changes in light intensity, and some hunting spiders have eyes capable of forming images.

Class Arachnida (Gr. *arachne*, spider)—spiders, scorpions, ticks, and mites.

The members of this class are some of the most misrepresented creatures in all the animal kingdom. Their reputation as grotesque and deadly creatures is vastly exaggerated. In truth, the majority of arachnids are either harmless or very beneficial to humans.

The arachnid body is divided into two tagmata (segments): cephalothorax (prosoma) and abdomen (opisthosoma). Attached to the cephalothorax are four pairs of legs, and around the mouth are cheliceras in the form of pinchers or fangs and a pair of pedipalps. The chelicerae serve to macerate food particles or inject poison and the leg-like pedipalps are adapted for prey capture, sensory detection, and reproductive functions depending on the species (**Figure 13.4**).

Spiders (Order Araneae) are the most familiar group of Arachnids. The cephalothorax (prosoma) of spiders bears chelicerae with poison glands and fangs and the characteristic four pair of walking legs. The pedipalps are leg-like and, in males, are modified for sperm transfer. A slender, waist-like **pedicel** attaches the cephalothorax to the abdomen, a trait that allows the spider to move the abdomen in all directions. The pedicel is the last segment (somite) of the cephalothorax and is lost in most other members of Arachnida.

The abdomen has no appendages except for one to four modified pairs of conical telescoping organs called **spinnerets**, that produce **silk**. The

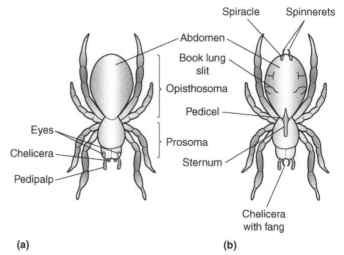

Figure 13.4 The Morphology of a Spider.
(a) Dorsal view. (b) Ventral view

protein that forms silk is stored as a liquid, but can be spun almost instantly into a solid thread without creating any clumps. The chemistry of this remarkable transformation are not completely understood but recent research in Germany and Sweden indicates that a high salt content and low acidity in the silk gland and spinning duct keeps the silk liquid, whereas reduced salt and higher acidity cause the proteins to link together rapidly during spinning.

All spiders possess spinnerets and make silk that they use for a wide variety of purposes: for safety draglines, for making durable egg cocoons, to line their homes, to trap their prey, and to immobilize their victims. In some species, males use silk to ritually immobilize the females before mating, and all male spiders make a special sperm-web or sperm-line that allows them to transfer the sperm from their genitals to their copulatory organs.

Spiders even use silk to fly. In a process of aerial dispersal known as **ballooning**, the newly hatched young (or spiderlings) of many large spiders, and the young and adults of some species can move to new areas via air currents. The spider first crawls to the top of vegetation or some other convenient launch pad, stands on "tiptoe" with abdomen pointing skyward and releases strands of silk from its spinnerets. When the breeze is judged sufficient, the spider lets go and as a wind surfer being towed by a speedboat, the spider becomes airborne. In this way, spiders may be carried for many, many miles. Spiders have been discovered over 4,500 m (14,000 feet) in the air and in the middle of the ocean 1,500 km (1,000 miles) from the nearest land. In 1839, Charles Darwin observed ballooning spiders aboard the H. M. S. Beagle when it was more than 60 miles off the coast of South America.

Flying in this manner, however, carries the risk of being eaten by insectivorous birds, such as swallows, or landing in the ocean or a lake. Nevertheless, large numbers of ballooning spiders manage to survive these risks as evidenced by their rapid colonization of fresh habitats such as freshly planted farm fields or newly formed volcanic islands.

Spider silk floating on the wind is known as **gossamer**. Gossamer can be seen on certain warm days late in the fall, when glinting in the sun, the sparkling threads wafting through the air present an almost magical scene. Gossamer has enthralled humans since ancient times. Pliny the Roman historian described a year in which it "rained wool" and over 200 years ago Charlotte Smith in her Sonnet LXIII appropriately entitled *The Gossamer* penned:

> *O'er faded heath flowers spun or thorny furze,*
> *The filmy Gossamer is lightly spread:*
> *Waving in every sighing air that stirs,*
> *As Fairy fingers had entwined the threads.*

Spiders are the largest group of arachnids, and they occur in greater numbers than most people realize. An undisturbed meadow, for example, may support as many as 2,250,000 spiders per acre (roughly the size of a football field). Spiders range in size from less than 0.5 mm (0.02 inch) in body length to a body length of 9 centimeters (3.6 inches) and leg spans as great as 25 cm (10 inches) in the case of some tarantulas.

Spiders are found all over the world, from the tropics to the poles and from the tops of mountains to underwater silken domes they supply with air. Found in ponds in Europe, northern Asia, and parts of Africa, the diving bell spider, *Argyroneta aquatica*, spends its entire life underwater. Because it must breathe air, this spider constructs a silk entrapment for a large bubble of air. The spider forms the initial bubble and replenishes the air in it periodically by rising abdomen first to the surface. There it traps a thin layer of air in the dense hairs on its abdomen and legs. It then dives back down transporting this trapped air to its larger bubble bell. This spider hunts small invertebrates, tadpoles, and frogs underwater, biting its prey to immobilize it and then swimming its prize back up and into the bell where it finishes devouring its meal.

Figure 13.5 Hidden behind a camouflaged flap, a trap door spider awaits the chance to explosively pounce on passing prey.

Most spiders feed on insects and other arthropods although a few, such as the tarantulas or "bird spiders," feed on small vertebrates as well. Spiders have an amazing array of prey catching strategies ranging from simple ambush to complex silk snares and

Figure 13.6 The golden orb spider (*Argiope*) patiently awaits for prey to blunder into its gossamer snare.

webs. Ambush hunters (or *mygalomorph spiders*) may leap from burrow entrances in the ground, in logs, or in tree trunks to capture passing prey. Some burrow dwellers lurk behind trapdoors, but others, like tarantulas and funnel-webs, will forage on the surface in the vicinity of the burrow at night (**Figure 13.5**). A few make sheet or curtain-like webs at their burrow entrances that impede both prey and predators. Others have silk or twig trip-line radiating out from the burrow entrance to alert the occupants to prey walking nearby. Others live as vagrants in leaf litter using vibration and touch to sense and ambush prey. Web-based hunters (or *araneomorph spiders*) construct webs from spider silk proteins (**Figure 13.6**).

Box 13.1
Spin Me a Tale, Weave Me a Web

Who has not reveled in the beauty of a spider web dripping with dew backlit by the early morning sun or recoiled fearfully when walking into an unexpected web? Those geometrically precise silky wonders we call spider webs rank among the most amazing constructs in the entire animal world. Spider webs are as diverse as they are beautiful:

Gum-footed webs. These webs consist of an irregular upper silk network with a closely woven, thimble-like retreat. From this upper network, vertical sticky catch threads run down to ground attachments ("gum feet"). These vertical lines are not only sticky, but they also provide strength and are under tension. When prey blunders into the array of sticky threads, its struggle breaks the lines at their weakest attachment points, and the lines contract upwards, lifting the prey off the ground. The spider then races down to deliver a quick bite and cover the struggling prey in loops of sticky capture silk.

Platform webs. These webs are constructed with a network of threads above a silk sheet (the "platform"). When insects fly into this network of threads, they are knocked down and fall through onto the silk sheet below where they are seized by the spider. In the center of the knockdown network, the spider will often place a loosely woven retreat constructed from curled leaves or some leaf fragments.

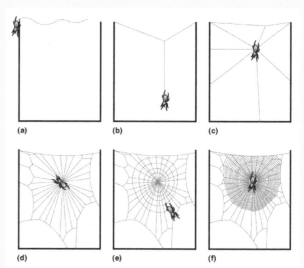

Figure 13.7 How does an orb spider spins its web?

Orb webs. These webs are the familiar large, planar (flat, two-dimensional) catching surfaces we associate with spider webs. They are sticky, elastic, and virtually invisible to flying insects, especially at night (**Figure 13.7**). The spider first releases an initial silk line extruded from the spinnerets allowing air currents to carry it across a gap in the foliage to entangle twigs or leaves on the other side (Figure 13.7a). The spider moves back and forth across this bridge line laying down more silk each time gradually strengthening the line. It then drops from the center of the bridge line to attach a vertical line to the ground (Figure 13.7b). This basic Y-shaped framework provides the scaffolding on which the supporting outer frame-lines and the radial lines (the "spokes") are laid (Figure 13.7 c-d). The spiral lines are then added (Figure 13.7 e-f). A non-sticky, temporary spiral line is laid down first, starting from the center and running outwards. This temporary spiral gives the spider a working platform from which it then lays down the more closely spaced, permanent, sticky spiral starting from the periphery toward the center or hub. The spider removes, rolls up, and eats the temporary spiral as it lays down the sticky spiral. Not every web is constructed in precisely this same manner; variations exist.

The web of a typical orb weaver spider contains 6-18 m (20-60 feet) of silk and takes anywhere from 30 minutes to several hours to build, yet it only lasts one night. Each morning, the spider eats and processes the silk from the old web and sets about spinning another.

The largest orb webs in the world are spun by spiders in the genus *Nephila*. These amazing constructs may be 2 m (6 feet) in diameter and are capable of catching small birds and bats. And not only are these webs sticky and elastic, they are incredibly strong. A spider web absorbs the proportionally equal impact of a jet Figurehter plane slamming in for a landing on an aircraft carrier every time a small insect flies full speed into it.

Horizontal line webs. A number of unrelated spiders use these simplest of webs to catch their prey. Line webs consist of a single sticky line strung either horizontally or at an angle. The spider sits in wait with a front leg holding the line. Small insects, especially flies, either hit or attempt to land on the sticky line and become stuck. The spider then pounces.

Bolas webs. At night, the bolas spiders (Araneidae) emerge to hunt male moths, which is their only prey. To do this, they sit on a horizontal line and spin a short, single vertical line with a large dollop of sticky silk on its free end (the bolas). This line hangs from a leg down into an open space among the foliage. (This is so similar to the way we hold the line over our fingers when fishing to feel for fish bites that these spiders are also commonly called angling or fishing spiders.) While the spider sits and waits, it is exuding an air-borne scent (pheromone) that mimics that used by female moths to attract males. When male moths fly in towards the spider attracted by this false scent, the spider senses the vibrations of their wing beats through long sensitive hairs. The spider then starts to swing the bolas in a circle beneath it. The moth is hit by the bolas and covered in sticky silk. Like a fisherman "playing" the moth, the spider releases coiled reserve silk from within the bolas to prevent the line from breaking. Soon the spider reels the moth in, bites it, and begins feeding upon it. Taking this practice one step further the spiders in the genus *Caelenia* in Australia do not even bother with a bolas. Instead, they simply hang from a thread (presumably emitting pheromones) legs and jaws at the ready, and the moths fly straight into them.

Net webs. *Dinopis guatemalensis*, commonly known as the ogre-eyed spider because they have huge eyes that are extremely sensitive, hunts at night. These spiders build a small platform of strands they can hang from, and then they make a small silk net. They hold the net open and hang quietly waiting for a passing insect. If an insect passes on the ground, they will see it and throw the net over it, tugging it a few times to make sure the prey is well entangled. If an insect flies past, however, the spider can detect the vibrations caused by its flight and will throw the net upwards to catch it.

Knowing this, perhaps the next time you see a spider's web you will view it with a more appreciative eye for what you are seeing is among the most amazing of nature's constructs.

Adapted from Spiders. 2002. Retrieved from http//www. austmus.gov.au>.

Other araneomorph spiders no longer build snare webs. Such spiders also have a surprising range of prey-catching strategies. Many are ambush hunters like the flower or crab spiders (Thomisidae). These spiders sit in the open, on foliage, flowers, or bark actively adjusting their body colors to conceal them from both predator and prey (**Figure 13.8**). Using well-developed sight, vibration, and tactile senses, they target flies, bees, and butterflies alighting or walking nearby. A tropical species has gone about things a little differently. Its body color and shape resemble a drop of bird dung. To enhance its disguise, the spider also secretes a chemical scent that makes it smell like dung as well. This scent attracts unsuspecting dung-feeding flies and butterflies to these unusual spiders where they are summarily dispatched. Water spiders (Pisauridae) hunt along streams and pond banks. With their legs extended into the water film, they can sense vibrations caused by fallen insects or small fish and tadpoles. The assassin spiders (Mimetidae) are spider hunters that invade the webs of other spiders and use the fangs on their elongate and slender jaws to spear their spider prey.

Figure 13.8 Small insects beware. Danger in the form of a yellow crab spider *(Misumena vatia)* lurks in the beauty of a flower.

Spiders immobilize their prey in two ways: by biting and injecting paralyzing venom, and by silk swathing and wrapping. Most hunting spiders simply grab and hold their prey in the pedipalps and front legs while biting it. Many web builders use bands of swathing silk thrown over and around the entangled prey, often before biting it (although large web builders tend to bite first). Securely wrapped and immobilized prey is sometimes stored in the web to be eaten later.

Spider venom causes paralysis in the victim and helps with the chemical breakdown of prey tissues. When feeding, the spider regurgitates enzyme-rich stomach fluids over and into its prey. This external digestion by venom and stomach chemicals, often aided by the grinding, masticating action of the fangs and toothed jaw bases and maxillae, reduces the prey's body and tissues to a chitinous soup. This liquid is sucked up through the spider's tube-like mouth, aided by the action of the pumping stomach, leaving only the hard parts of the prey behind. Spiders are also capable of digesting their own silk, and as a result, many spiders eat their used webs. When a spider drops down on a single strand of silk and then returns, it will rapidly consume the strand of silk on its way back up. In fact, many nocturnal orb spinners destroy and eat their webs as dawn approaches and then rebuild them again each night. This recycling process is very efficient as it returns the silk protein to the silk glands where it is processed into new silk.

Mating of spiders involves elaborate rituals that include chemical, tactile, and/or visual signals to allow partners to identify each other and to allow the male to approach and inseminate the female without triggering a predatory response. Females may deposit pheromones on their webs or bodies to attract males whereas males may attract a female by plucking the strands of the female's web. The pattern of plucking is species specific and helps identify the gentleman caller as a potential mate and not a possible meal.

Once a male spider has matured, he leaves his web or burrow, charges his pedipalpal mating organs with sperm and wanders off on his nomadic search for a female. Seemingly random, male wandering is usually directed by the presence of silk or air-borne pheromones put out by the female.

Having found a female, the male must first establish his identity as a mate rather than a meal. Most do this through various forms of courtship, which can involve vibrational, chemical, tactile, and visual cues. Web builders tweak the female's web in a very specific pattern; wolf spiders drum the ground; keen-sighted jumping spiders "dance," providing a visual display to the female; funnel web spiders tap the burrow entrance silk and stroke the female's legs: male flower spiders also tap and stroke but take the additional precaution of tying the female down with a few silk lines. Despite these efforts, unreceptive females will react aggressively and chase the male away, nip off the odd leg or even capturing and eating the male in some instances.

Once identities have been established, and consent given, mating occurs. Sperm transmission is accomplished by the male inserting one or both palps into the female's genital opening, known as the **epigyne**. He then transfers his seminal fluid into the female by expanding the sinuses in his palp(s). Once the sperm is inside her, the female stores it in a chamber and only uses it during the egg-laying process when the eggs come into contact with the sperm for the first time. Once fertilized, the eggs are deposited into a silk cocoon that the female then hides or carries with her.

It is a common belief that male spiders, which are usually significantly smaller than the females will be killed after or even during mating. Males are sometimes killed by females but in at least some of these cases it is likely that the males are simply mistaken as prey. The risk of this happening is greater if the female is hungry. To counter this, some male spiders offer a "bribe" to the female in the form of a prey animal, prior to mating. Even in some species of black widow, which are named exactly for this belief of inevitable post-mating doom, the male may live in the female's web for some time without being harmed.

Harvestman or daddy long legs (Order Opiliones) are often confused with spiders. Harvestmen, however, lack the narrow waist of the spider with their cephalothorax and abdomen being so broadly joined as to appear as one oval structure (**Figure 13.9**). The body typically does not exceed 7 mm (0.25 inch) even in the largest species. The signature legs of harvestmen are exceedingly long and slender given the size of their body and can exceed 160 mm (6-7 inches) in span.

Harvestmen are omnivores feeding on small insects and all kinds of plant material and fungi; some are scavengers on dead animals and fecal matter. They have neither silk glands nor poison glands. They are mostly nocturnal and dark-colored, but there are a number of diurnal species are vividly colored in patterns of yellow, green, and black with varied reddish and blackish mottling. Mating involves direct copulation rather than the deposition of a spermatophore, as is the case with spiders.

Figure 13.9 Appearance can be deceiving as the harvestman (Opiliones) known commonly as daddy longlegs is not a true spider.

Mites and ticks (Order Acarina) are typically small in stature but of all arachnids, acarines have had the greatest impact on human health and welfare. Most acarines are minute being only about 0.08 to 1.00 mm (0.0003 to 0.04 inch), but some ticks and red velvet mites may reach lengths of 10 to 20 mm (0.4 to 0.8 inch). Acarines are specialists that live in practically every conceivable habitat, including the anuses of turtles, the digestive system of sea urchins, the lungs of snakes, the trachea of bees, the shafts of feathers, the fat of pigeons, and the eyeballs of fruit bats to name only a few. Aquatic mites live mostly in lakes, ponds, and puddles, often in densities of hundreds of thousands per cubic centimeter.

Mites are so small that a dozen of them could dance on the head of a pin. They are more likely, though, to dance on your face, which they do at night before they mate, before crawling back into your follicles by day to eat.

—Rob Dunn

In mites, the cephalothorax and abdomen are fused and covered by a single carapace. An anterior projection known as the **capitulum** carries the mouthparts. Chelicerae and pedipalps are variously modified for piercing, biting, anchoring, and sucking. Most adults have four pairs of legs, like other arachnids, but some species have fewer (**Figure 13.10**).

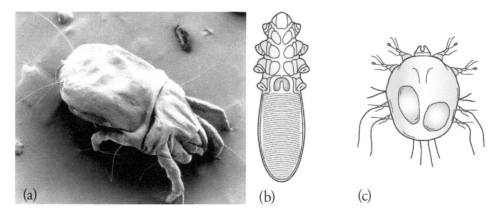

(a)   (b)   (c)

Figure 13.10 The Realm of Mites. (a) The household dust mite is a cosmopolitan inhabitant of mattresses, carpets, furniture, and bedding where they feed on flakes of shed human skin. Every gram of household dust contains nearly 200 of these tiny acarines. These mites are so ubiquitous that they have even been found on the space station. (b) Follicle mite (*Demodex*) displaying its characteristic stumpy legs. (c) A mange mite (*Sarcoptes scabiei*).

Free-living mites may be herbivores or scavengers. Scavenging mites are among the most common animals in soil and leaf litter. These mites include some pest species that feed on flour, dried fruit, hay, cheese, and animal fur. Predatory mites may be found in the soil. These monsters of the mite world have mouthparts resembling the various tools in a Swiss army knife. Some have smooth blades that snap together with tremendous force; others have jaws with sharklike teeth; still others stab and pierce with deadly sharp sabers.

Parasitic mites usually do not permanently attach to their host, but feed only a few hours or days and then drop to the ground. One mite, the notorious chigger or red bug (*Trombicula*) is a parasite during one of its larval stages on all groups of terrestrial vertebrates. The larval mite enzymatically breaks down its host's skin causing local inflammation and intense itching at the site of attack. The chigger larva drops from the host and then molts first into an immature nymph and then finally into an adult. Both nymphs and adults feed on insect eggs.

A few mites are permanent ectoparasites. The follicle mite, *Demodex folliculorum*, is a common (but harmless) inhabitant of mainly the facial hair follicles of most humans (including you, dear reader). Unlike the follicle mite, the human itch mite, *Sarcoptes scabies*, is more than a benign passenger on our bodies. Like a mole in your lawn, these mites tunnel through the epidermis of human skin releasing irritating secretions that cause intense itching. Females lay about 20 eggs each day. Such an infestation, known as **scabies**, may be acquired by contact with an infected individual.

Ticks are ectoparasites during their entire life cycle. Ticks may be up to 3 cm (1.2 inch) in length but otherwise are similar to mites (**Figure 13.11**). Hooked mouthparts are used to attach to their hosts and to feed on blood. Copulation occurs on the host, and after gorging with blood, females drop to the ground to lay eggs.

Scorpions (Order Scorpionida) are characterized by a long tail (**metasoma**) comprising six segments bearing the sting (**aculeus**) at the end. The sting has a bulbous base that contains venom-producing glands and a hollow, sharp, barbed point (**Figure 13.12**). A few species of scorpions possess venom that is highly toxic to humans. The abdomen's front half, the **mesosoma**, is also made up of six segments. The first segment contains the sexual organs; the second segment bears a pair of featherlike sensory organs known as the **pectines**; the final four segments each contain a pair of book lungs. The chelicerae are short and used for grinding food. The pedipalps are extraordinarily large, and the last segment has been modified into grasping pinchers (**chela**).

Considered to be among the most ancient terrestrial arthropods and the most primitive arachnids, all the known species of scorpions today are terrestrial predators. These largest of arachnids reach lengths of about 18 cm (7 inches) and inhabit a variety of environments, particularly, deserts and tropical rain forests where some arboreal species occur. They are notably absent from colder regions of the world.

Scorpions are nocturnal opportunistic predators of small arthropods. They use their chela (pinchers) to catch their prey initially. Depending on the toxicity of their venom and the size of their claws, they will then either crush the prey or arch the metasoma up over their back and drive the sting into the prey injecting it with venom (**Figure 13.13**).

Most arthropods are **oviparous**; females lay eggs that develop outside the body. Many scorpions and some other arthropods are **ovoviviparous**; development is internal, although large yolky eggs

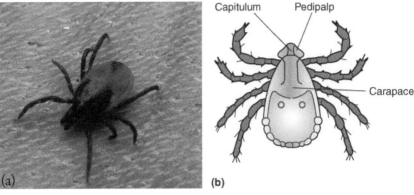

Figure 13.11 The Tick. (a) Magnified view of a tick on human skin. This tick has not yet implanted. (b) General morphology of a tick (*Dermacentor*).

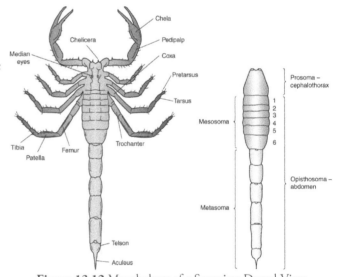

Figure 13.12 Morphology of a Scorpion-Dorsal View

Figure 13.13 Emperor Scorpion *(Pandinus imperator)*. Stinger aside, scorpions superficially resemble crustaceans. The pinchers of a scorpion, however, are elongated and highly modified mouthparts, while the pinchers of a lobster or crayfish are modified legs.

provide all the nourishment for development. Some scorpions, however, are **viviparous**, meaning that the mother provides nutrients to directly nourish the embryos. Development may take 1.5 years with 20 to 40 young being brooded inside the female's body. After being born one by one, the young are carried about on the mother's back until the young have undergone at least one molt.

Class Merostomata (Gr. *meros*, thigh + *stoma*, mouth)

This ancient class of arachnids known commonly as horseshoe crabs (Order Xiphosura) is represented by only four living species. One species, *Limulus polyphemus*, is widely distributed in the Atlantic Ocean and Gulf of Mexico (**Figure 13.14**). Scavenging sandy and muddy substrates for annelids, small molluscs, and other invertebrates, horseshoe crabs display a body form that has remained virtually unchanged for over 200 million years.

A hard, horseshoe-shaped carapace covers the prosoma of horseshoe crabs. The chelicerae, pedipalps, and first three pair of legs have pinchers and are used for walking and food handling (**Figure 13.15**). The fourth legs have leaf-like plates at their tips and are used for locomotion and digging.

Figure 13.14 Dorsal view of two horseshoe crabs in shallow water. To see horseshoe crabs in such a setting is to literally peer back in time several hundred million years.

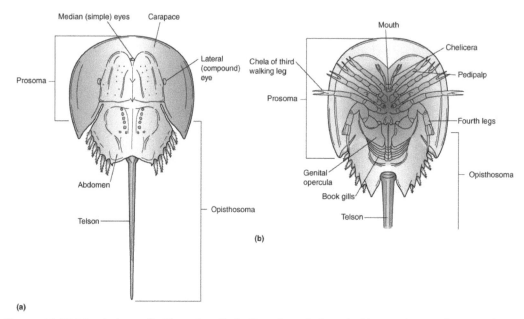

Figure 13.15 Morphology of a Horseshoe Crab (*Limulus polyphemus*). (a) Dorsal view. (b) Ventral view.

The opisthosoma of a horseshoe crab includes a long, unsegmented telson. If wave action flips a horseshoe crab over, the telson may be used to right the animal. The first pair of opisthosomal appendages called

genital opercula, covers the genital pores. The remaining five pairs of appendages are **book gills**. The name is derived from the resemblance of these plate-like gills to the pages of a closed book.

Horseshoe crabs are dioecious. During reproductive periods, males and females congregate, often in large numbers, in intertidal areas. The male mounts the female and grasps her with his pedipalps. The female excavates shallow depressions in the sand, and as she sheds eggs into the depressions, the male fertilizes them. Fertilized eggs are covered with sand and left to develop unattended.Although not detailed here, subphylum Chelicerata also includes the whip scorpions, sea spiders, pseudoscorpions, and a few others.

Subphylum Myriapoda (Gr. *myriad*, ten thousand + *podus*, foot)

This group consists of arthropods possessing:

- A body of two tagmata: head and trunk (abdomen). No carapace or wings present.
- A long and cylindrical abdomen consisting of many segments.
- Uniramous appendages.
- One or two pair of walking legs per segment.
- One pair of articulate (jointed and movable) antennae attached to the head segment.
- A few to many clustered ocelli (simple eyes) on the head segment.
- Mouthparts consisting of **mandibles** (jaws), **maxillules** (first maxillae), and **maxillae** (second maxillae). The second maxillae are fused into a single flap-like structure called the **labrum**.
- A gas exchange system composed of tracheae and spiracles.

Class Chilopoda (Gr. *cheilos*, lip + *podus*, foot)—centipedes (L. *centum*, hundred + *pede*, foot)

Centipedes are fast-moving venomous, predatory arthropods that have long bodies and many jointed legs (**Figure 13.16**). Mainly nocturnal, centipedes of the largest size and in most numbers are found in tropical climes, but their smaller kin are widely distributed in temperate zones as well.

Centipedes range in size from 10-270 mm (0.4-10.6 inches) in length. The giant redheaded centipede (*Scolopendra heros)* is the largest North American species at about 153 mm (6 inches) long but the giant of the class is the Amazonian giant centipede (*Scolopendra gigantea*) measuring in at over 30 cm (12 inches). This goliath is known to eat rodents, spiders, and even bats that it catches in midflight.

Figure 13.16 The red centipede (*Scolopendra polymorpha*) is an agile predator. Centipedes are generalist predators and large varieties have been observed eating small amphibians, reptiles, birds, and mammals.

The body of a centipede is dorsoventrally flattened and consists of 15 to 173 segments with one pair of walking legs per segment. The last pair of legs is usually modified into long sensory appendages. The first anterior trunk segment has been modified into a pair of venomous poison fangs (**maxillipeds**) that are used both for defense and for capturing and paralyzing prey.

The head has a pair of jointed antennae, jaw-like mandibles, and other characteristic mouthparts (**Figure 13.17**).

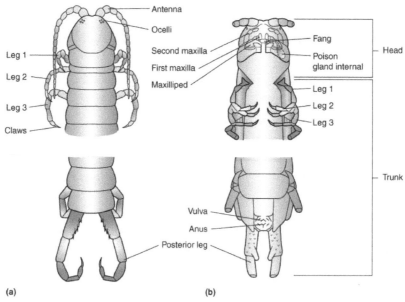

Figure 13.17 Morphology of a Centipede. (a) Dorsal view. (b) Ventral view.

Centipedes are fast and agile predators feeding on small arthropods, earthworms, and snails. The poison claws kill or immobilize prey. The poison claws, along with mouth appendages, hold the prey as mandibles chew and ingest the food. Some species are highly venomous and can produce very painful bites but only a few human deaths have ever been recorded from the bite of a centipede. The bite of a small centipede such as those found in temperate areas may be similar to a bee sting, but the bite of a large tropical species is excruciatingly painful.

Centipede reproduction may involve courtship displays in which the male lays down a silk web using glands at the posterior tip of the body. He places a spermatophore in the web, which the female picks up and introduces into her genital opening. Eggs are fertilized as they are laid. A female may brood and guard the eggs by wrapping her body around them, or they may be deposited in the soil.

Class Diplopoda (Gr. *diplos*, twofold + *podus*, foot)—Millipedes (L. *milli*, thousand + *pede*, foot).

Millipedes are relatively common leaf litter and soil animals that occur in most parts of the world (**Figure 13.18**).

Millipedes range in size from 2 mm to 300 mm (0.08-12 inches) in length. *Paeromopus paniculus*, the largest North American millipede measures in at about 160 mm (6.3 inches). Each segment of a millipede is two segments fused together resulting in two pairs of walking legs per segment. Most millipedes normally have only 100 to 300 total legs, not a thousand as their name would suggest. *Illacme plenipes*, first identified in 1926 but rediscovered recently inhabiting a tiny patch of San Benito County, California, can possess up to 750 legs or 375 pairs. Although their many legs are individually small, as a group they are powerful allowing millipedes to literally bulldoze their way through soil, leaf litter, and rotting wood.

The head of a millipede has two sections of ocelli, two antennae, two mandibles, and two maxillae (**Figure. 13.19**). The antennae are relatively short being composed of only eight segments. They have between 4 and 90 ocelli depending on the species, but some types have none and are totally blind. The millipede mouth consists of a pair of mandibles that are armed with a few blunt teeth and a lower jaw-like plate known as a **gnathochilarium**. Usually nocturnal, millipedes feed on decaying plant matter using their mandibles in a chewing or scraping fashion. A few millipedes have mouthparts adapted for sucking plant juices.

Figure 13.18 This red and black millipede (*Aphistogoniulus sp*) is a slow-moving detritovore in the Madagascar rain forest.

Millipedes roll into a ball when threatened with desiccation or when disturbed. This balling strategy alone, however, is not enough to deter the many predators seeking to dine on millipedes. To further protect themselves, millipedes have armed themselves with **repugnatorial glands** that produce defensive secretions that are foul tasting and sometimes either poisonous or sedative in nature depending on the species. Interestingly, lemurs have been observed intentionally irritating millipedes in order to rub these defensive chemicals on themselves to repel insects, and possibly produce a psychoactive effect (getting "high"?). As far as humans are concerned, this chemical brew is fairly harmless usually causing only minor effects—discoloration, pain, itching, and blisters—on the skin. Eye exposure causes general eye irritation and potentially more serious effects, such as conjunctivitis and keratitis. Many millipedes are quite docile and may safely be kept as pets and handled without risk of injury. With the proper caging and feeding, some pet millipedes can live up to seven years and grow to be as long as 38 cm (15 inches).

Male millipedes transfer sperm to the female with modified trunk appendages, called **gonopods**, or in spermatophores. Female millipedes make an underground

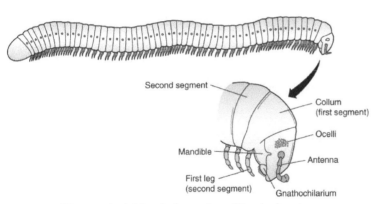

Figure 13.19 Morphology of a millipede showing an enlarged lateral view of the head.

nest into which they lay their eggs. The nest is made by excreting soil they have eaten and using anal folds to shape it as required. The eggs are fertilized as they are laid.

Although not detailed here, subphylum Myriapoda also contains class Symphyla (symphylans) and class Pauropoda (pauropodans).

Subphylum Hexapoda (Gr. *hexa,* six + *podus*, foot)

Characteristics of this subphylum are:

- A body of three tagmata: head, thorax, abdomen without a carapace.
- Uniramous appendages.
- Three pairs of walking legs attached to the thorax. Wings are present in many species.
- A head bearing a single pair of jointed antennae, mouthparts consisting of mandibles, maxilla, labium, and labrum, and compound eyes. Some groups possess ocelli as well.
- A gas exchange system consisting of tracheae and spiracles.
- An excretory system consisting of Malpighian tubules.

Class Insecta (L. *insectus*, to cut up)

One of the grandest groups of animals on the planet must surely be the insects. They exist in numbers that defy comprehension and their diversity of forms staggers the imagination.

> *We hope that when the insects take over the world, they will remember with gratitude how we took them along on all our picnics.*
>
> —Bill Vaughn

Over a million insect species have been catalogued—more than all other animal groups combined. Recent studies indicate that there may be anywhere from six to ten million additional insect species awaiting discovery. This multitude of species is organized into approximately 32 orders, far too many for us to examine each separately (**Figure 13.20**).

Figure 13.20 A few representatives of the 32 orders of insects. There are more species of insects than all other species of animals combined.

A spectacular explosion of evolutionary radiation has equipped the insects to inhabit nearly every conceivable nook and cranny of terrestrial and freshwater aquatic systems, although none is considered truly marine. They even populate the air. In fact, studies have revealed the existence of "aerial plankton" consisting of minute insects and other arthropods, extending to altitudes as high as 4,267 m (14,000 feet). Insects have become so environmentally successful and so vitally intertwined into the fabric of life in nearly all habitats that their removal might possibly result in collapse of those systems.

Insects range in size from the fairyfly (*Dicopomorpha echmepterygis*) at 0.139 mm (0.0055 inch) long and able to fly through the eye of a needle to the stick insect (*Phobaeticus serratipes*) at 55.5 cm (22 inches). The heaviest larva belongs to the Goliath beetle (*Goliathus goliatus*) at 115 gr (4.1 oz) whereas the heaviest adult insect recorded so far has been a female Little Barrier Island Weta (*Deinacrida heteracantha*) at 71 gr (2.5 oz). It is likely, however, that the adult elephant beetles (*Megasoma elephas* and *Megasoma actaeon*) that commonly exceed 50 gr (1.7 oz) could reach a larger weight.

The body of an insect is divided into three tagmata: head, thorax, and abdomen (**Figure 13.21**). The head segment bears one pair of antennae, one pair of large compound eyes and zero, two, or three ocelli as well as mouthparts. The thorax consists of three segments that from anterior to posterior are the **prothorax**, the **mesothorax**, and the **metathorax**. One pair of walking legs is attached to each thoracic segment. Wings, when present, attach dorsolaterally to the margin between the mesothorax and the metathorax. The thorax also contains two pairs of spiracles opening into the tracheal system. Most insects have 10 to 11 abdominal segments, each of which has a pair of spiracles. The abdomen also houses most of the digestive, respiratory, excretory, and reproductive organs.

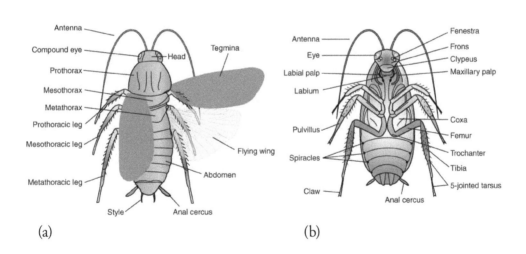

(a) (b)

Figure 13.21 Morphology of a common insect—the cockroach. (a) Dorsal view. (b) Ventral view.

Adaptation to a multitude of different habitats and lifestyles has resulted in many variations on basic arthropod mouthparts plus the challenges of competition have resulted in every conceivable kind of diet being exploited by insects. The herbivores grind and chew plants and suck their juices whereas the carnivores rip, tear, and suck the life out of other small animals. Many labor as scavengers in rotting and putrid offal whereas others have adapted to a symbiotic existence on, in, or around other creatures.

Most insects lead short lives as adults and congregate only to find suitable partners and mate or possibly perform some other survival function. The **social insects** live together in large, ordered groups known as **colonies**, however, through a phenomenon known as **eusociality**. In a eusocial society, sterile members of the species (**worker caste**) carry out specialized tasks, effectively caring for the **reproductive caste** (queens and drones). Eusociality in insects is found in the orders Hymenoperta—ants (all species), bees (few species), and wasps (few species)—and Isopetera—termites(all species)—and to a lesser extent Homoptera—aphids and thrips. The different castes within the society are often modified from each other anatomically and structurally as well as behaviorally so as to perform the survival tasks inherent with their caste (**Figure 13.22**).

> Some primal termite knocked on wood
> And tasted it, and found it good!
> And that is why your Cousin May
> Fell through the parlor floor today.
> —Ogden Nash

Some large colonies of social insects may contain as many individuals as there are people in a large human-constructed city. How are all these individuals to be controlled and coordinated? In a word—biochemistry. The queen produces pheromones that attendant workers pick up as they lick and groom the queen. As these workers then lick other workers and so on the chemical instructions issued by the queen quickly spread through the colony. It is the many chemical nuances of the pheromones produced by the queen coupled with the genetic programming of each individual that keep the colony humming along, almost as if it were one single organism.

Insect societies demonstrate a number of peculiarities:

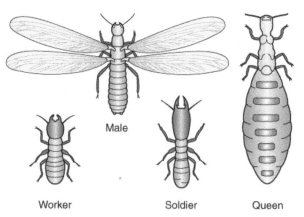

Figure 13.22 Castes of the eastern subterranean termite (*Reticulitermes flavipes*) the most common North American termite.

- *Slavery*. Slavery is wide-spread among ants. Ant slavery is biological in nature and unique because it is usually between species, unlike human slavery in which we enslave only own species doing so for cultural and/or economic gains. Invading parties of ants will seek out the colonies of other ants in an attempt to capture their larvae and pupae. Captured immatures are carried back to the invaders' nest where they acquire the nest odor eventually developing into adults that act as workers for their adopted colony. Some slaver species have become so dependent on their slave that they are no longer able to collect food or feed themselves or their immatures and would thus perish without the slave workers to perform these tasks for them.
- *Warfare*. Waging war against their own and other species, some ants engage in restless aggression, territorial conquest, and genocidal annihilation of neighboring colonies. As an example, consider the interplay between the introduced non-native Fire Ant, *Solenopsis invicta* and the native wood-

land ant, *Pheidole dentata*. The fire ants have colonies hundreds of times larger than the woodland ant and whenever they discover a woodland colony they completely destroy it. Yet woodland ant colonies are abundant around fire ants. Whenever a woodland worker discovers a fire ant scout, woodland soldiers are so rapidly deployed that the fire ant scout rarely makes it back alive to its own colony. The woodland soldiers rely on large mandibles to cut their fire ant opponent to pieces. Woodland ants, however, are no match for fire ants en masse so if their nest is discovered, the woodland soldiers fall back to form a short perimeter around the nest and sacrifice themselves in battle in an attempt to keep the fire ants at bay temporarily while the colony evacuates the nest. After the battle and the fire ants have departed, the woodland ants will return and reclaim their nest.

The evolution of social behavior involving many individuals leaving no offspring and sacrificing individuals for the perpetuation of the colony has puzzled evolutionists for many years. It may be explained by the concepts of kin selection and altruism.

- *Ranching and Farming.* Some species of ants will herd aphids like livestock. The ants will place the aphids on plants and protect the aphids as they feed on plant juices. The ants then "milk the cows" by stroking the aphids with their antennae. This stimulates the aphids to release a drop of sweet and nourishing honeydew from the tip of their abdomen. The ants will ingest the honeydew and carry it back to the colony to feed others through regurgitation.

Leaf cutter ants and some termites are gardeners. The workers collect plant material that they bring back into the nest. This plant material is not eaten directly but instead serves as compost for growing a type of fungus that is the actual food for the colony.

- *Ventilation and Temperature Regulation.* In arid tropical savannas, termites construct extremely large and complex concretions that house their colonies. Standing anywhere from 6 to 9 m (20 to 30 feet) tall, these mounds are exposed to blazing sun and low humidity, conditions that spell nearly instant doom to soft-bodied termites and the fungi gardens they depend upon for food. If one sends miniature sensors down into these mounds, however, the temperature and humidity levels within are ideal for termite survival, and they remain so often varying only ±1° C over the course of 24 hours. Opening these mounds reveals not a haphazard pile of dirt, but a complicated internal architecture: a capricious central chimney from which radiates a complex network of passages.

The mound is not the actual habitation for the millions of termites that build it. Their residence is in a nest below the mound, a spherical underground city about 2 m (6-7 feet) in diameter. These mounds represent an enormous expenditure of energy and time; a mound contains, on the average, about five cubic yards of soils and takes a mature colony about one year to build. The secret lies with the wind. By building the mound upward into stiffer breezes higher off the ground, the termites harness the wind to drive air movements in the mound's tunnels. The flow of the wind pushes air through the porous soil on the windward side and out on the leeward side, allowing the nest atmosphere to mix with fresh air from the outside. Bees also collectively regulate the hive temperature. When the weather is cold, they cluster into compact balls and shiver, warming the hive. When the weather is hot, workers fan their wings at the entrance of the hive, cooling it.

- *Communication.* Command and coordination do not always come from the queen in the form of chemical messages. Visual cues can also play a role in certain situations as exemplified by the **waggle dance** of the honeybee (Figure 5.15). First translated and understood by Austrian ethologist Karl von Frisch, the waggle dance is a peculiar Figure-eight movement performed by successful foragers through which they share with their hive mates information about the direction and distance to patches of flowers yielding nectar and/or pollen and to water sources. A waggle dance consists of one to 100 or more Figure-eight circuits, each of which consists of two phases: the waggle phase and the return phase. The direction and duration of waggle runs are closely related with the direction and distance of the patch of flowers being advertised by the dancing bee. Flowers located directly in line with the sun are represented by waggle runs in an upward direction on vertical combs, and any angle to the right or left of the sun is coded by a corresponding angle to the right or left of the upward direction of the waggle runs. The farther the target, the longer the waggle phase, with a rate of increase of about 75 milliseconds per 100 meters. Amazingly, waggle dancing bees that have been in the hive for an extended time adjust the angle of their dance to accommodate the changing direction of the sun. Therefore, bees that follow the waggle run of the dance are still correctly led to the food source even though its angle relative to the sun has changed.

One of the main reasons for the tremendous environmental success enjoyed by insects is their tremendous reproductive potential. The life of the adult insect is geared primarily to reproduction. For instance, in her lifetime, a single queen honeybee is capable of laying over 1 million eggs whereas the queens of some African termite species are reputed to lay over 30,000 eggs a day!

Insects are dioecious and most often fertilize their eggs internally. For reproduction to occur, however, the two parties must first find each other. In butterflies, the color of the female in flight can attract a male of the same species. In mayflies and certain midges, males dance in swarms to provide a visual attraction for females. In certain beetles, such as fireflies and glowworms, parts of the fat body in the female have become modified to form a luminous, glowing organ that attracts the male. Male crickets, grasshoppers, cicadas, and katydids attract females by their chirping songs, and the male mosquito is lured to the sound emitted by the female in flight. However, the most important element in mate attraction is odor. Depending on the species, the male and/or the female secrete odorous pheromones that serve as specific attractants to the opposite sex.

Males most often place their sperm into their mate's vagina during mating. In some insects, however, sperm are contained within spermatophores (packets) allowing for either direct placement at copulation or substratum placement for later collection by the female. For example, male silverfish leave a spermatophore on the ground that the female then takes into her body via her ovipositor to fertilize the eggs. It is most common for female insects to store enough extra sperm in their seminal receptacle to fertilize multiple batches of eggs. Eggs are fertilized as they leave the female and are usually laid near or on the food supply. Tiger moths, for example, will search out pigweeds on which to lay their eggs whereas the monarch butterfly prefers milkweed plants, and the sphinx moth tomato or tobacco plants. Females may use structures known as **ovipositors** to place their eggs into or on some substrate. The ichneumon wasp uses her extraordinarily long ovipositor to inject her eggs into the larva of other insects where they develop into parasites devouring the host from within. Her favorite targets are the larvae of the wood wasp or wood-boring beetles. The fact that these larvae live inside twigs and branches does not deter the female ichneumon wasp. With unerring accuracy and precision, she uses her long ovipositor to penetrate 1 to 2 cm (0.4 to 0.8 inch) of wood to find a host larva and inject her eggs into it.

Most insects experience a metamorphosis (change) at the postembryonic development stage. Insect development results in a divergence of immature and adult body forms and habits. Immature stages are a time of growth and accumulation of reserves for the transition to adulthood. The mature stage, on the other hand, is associated with reproduction and dispersal. The degree of divergence between immatures and adults can be classified into three (sometimes four) categories.

In insects that display **ametabolous metamorphosis** (Gr. *a*, without + *metabolos*, change), the primary differences between adults and larvae are body size and sexual maturity. Both adults and larvae are wingless and unlike most other insects, molting continues after sexual maturity. Silverfish (order Thysanura) exhibit ametabolous metamorphosis.

Hemimetabolous (incomplete) **metamorphosis** (Gr. *hemi*, half) is characterized by a species-specific number of molts or **instars** between egg and adult stages, during which immatures gradually take on the adult form. The external wings develop, adult body size and proportions are attained, and the sexual organs develop during this time. Immatures are called **nymphs** (**Figure 13.23**). Grasshoppers (order Orthoptera) and chinch bugs (order Hemiptera) exhibit hemimetabolous metamorphosis. When immature stages are aquatic, they often have gills (e.g. mayflies [order Ephemeroptera}and dragonflies, [order Odonata]) and are called **naiads**.

In **holometabolous** (complete) **metamorphosis** (Gr. *holos*, whole), the immatures are very different from the adults in body form, behavior, and habitat (**Figure 13.24**). The number of larval instars is species-specific, and the last larval molt forms the **pupa**. The pupa is a time of radical cellular change, during

Fertilized eggs

First instar

Third instar

Fifth instar

Adult

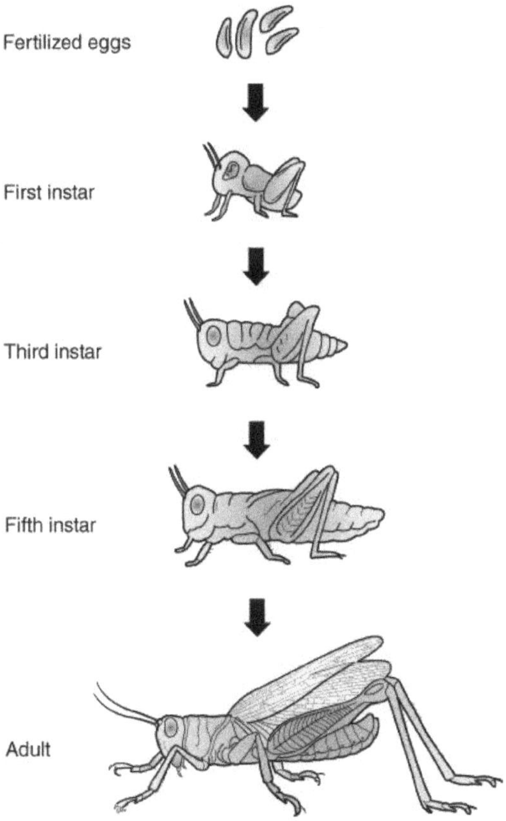

Figure 13.23 Hemimetabolous development (incomplete metamorphosis) in the grasshopper. Each development stage is referred to as an instar.

which all characteristics of the adult insect develop. A protective case may enclose the pupal stage. Some moths (order Lepidoptera) spin a silken **cocoon** around the pupa whereas the **chrysalis** of butterflies (order Lepidoptera) and the **puparium** of flies (order Diptera), are the last larval exoskeletons and are retained through the pupal stage. The final molt to the adult stage usually occurs within the cocoon, chrysalis, or puparium, and the adult, known as an **imago**, then exits.

Metamorphosis in insects is controlled and regulated by a complex interaction of hormones produced in glands and neurosecretory cells located in the prothorax and brain. **Neurosecretory cells** secrete hormones that stimulate the neurohemal **corpus cardiac** organs to release **prothoraciotropic hormone** (PTTH). Carried in the hemolymph, PTTH in turn stimulates the **prothoracic glands** to release **ecdysteroids** (molting hormones). The ecdysteroids trigger a cascade of physiological events that culminate in molting (ecdysis). All the while the neurohemal **corpora allata** organs, also regulated by neurosecretory cells, secrete **juvenile**

hormone (JH) during larval or nymphal instars, inhibiting the transition to adulthood, then reactivating once the insect is sexually mature and ready for reproduction.

When the molting hormone **ecdysone** initiates a molt in an early larval instar, the accompanying concentration of juvenile hormone is high. Such a high concentration ensures a smaller larva-to-larger larva molt. After the last larval instar is reached, the corpora allata ceases to secrete juvenile hormone. Low concentrations of juvenile hormone result in a larva-to-pupa molt. Finally, when the pupa is ready to molt, juvenile hormone is absent altogether from the hemolymph, and this deficiency leads to a final pupa-to-adult molt.

A number of insect groups, particularly those living in seasonally changing environments will undergo a process known as **parthenogenesis** (Gr. *parthenos*, virgin + *genesis*, creation or birth). Parthenogenesis is a form of asexual reproduction in which females produce eggs that can develop without fertilization. In some groups (e.g., honeybees and aphids), fertilized eggs develop into females, and unfertilized (parthenogenetic) eggs develop into males.

Subphylum Crustacea (L. *crustaceus*, hard shelled)

This subphylum is characterized by arthropods possessing:

- Sixteen to 20 segments, with some forms having 60 segments or more.
- An exoskeleton that is more pronounced and generally thicker and heavier than other taxa of arthropods.
- Two or three tagmata: cephalothorax with a shield-like carapace and abdomen or head, thorax, and abdomen.
- **Biramous** (branched) appendages.
- Head appendages consisting of two pairs of antennae, a pair of mandibles, two pairs of maxillae, and one pair of compound eyes.
- Both simple ocelli and compound eyes are often elevated on stalks.
- Gas exchange typically occurs across gills.
- Excretion by true nephridial structures.
- A developmental stage known as a **nauplius larva** that is characterized by the presence of three pair of head appendages.

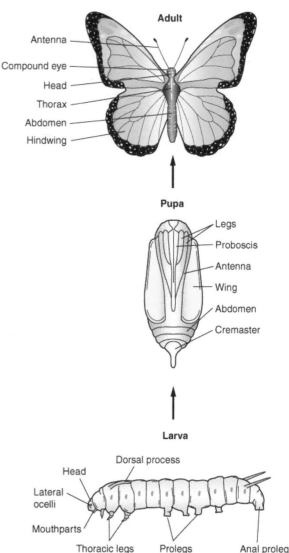

Figure 13.24 Holometabolous development (complete metamorphosis) in a butterfly. The pupa is encased in a chrysalis which is attached to a leaf or a twig by the cremaster.

Because crustaceans exist in large numbers and demonstrate a range of morphological diversity that exceeds even insects, crafting a satisfactory description of the group is difficult.

> *Creatures of such unsuspected importance and numbers stir our imagination and invite us to find out more about them. But let us beware of lightly following our curiosity in this matter. The attempt to obtain a clear-cut definition of the class Crustacea has left many a student bewildered.*

—Waldo L. Schmitt

Crustaceans are found at all depths in every marine, brackish, and freshwater environment known; they are so pervasive in marine habitats in particular that they are often referred to as "insects of the sea." Some, however, such as terrestrial crabs and isopods, have adapted to at least a semiterrestrial life, and a few taxa, collectively known as "crustacean lice," are parasitic.

Although the insects still rule in terms of numbers, the crustaceans are the most diverse in terms of form. The largest of the crustaceans include the giant Japanese spider crab (*Macrocheira kaempferi*) with its four-meter (13 foot) leg span, and the American lobster (*Homarus americanus*) at an impressive 20 kilograms (44 pounds) in weight. On the other end of this spectrum are tiny interstitial and planktonic forms that never grow larger than 0.25 mm (0.01 inch), even as adults. Although certain varieties move along the bottom, others burrow, and some, such as barnacles, are sessile, spending their life attached head down. Swimming upright or upside down, microscopic crustaceans float along in both salt and fresh water environments. The tiny, delicate members of the copepod genus *Calanus* are among the most abundant animals in the world.

Crustaceans take on many different habits and adaptations in their feeding. Those that practice suspension feeding filter plankton, bacteria, and detritus from the water whereas the predatory forms consume larvae of all types, worms, other crustaceans, snails, and fishes, and scavengers eat dead plant and animal material. Numerous types can alter the way in which they feed as their environment and availability of food change. A crayfish routinely captures and devours small invertebrates, worms, and fish, browses on water plants, or consumes decaying matter in the course of its daily survival.

Among crustacean predators, none is more heavily armed than the colorful mantis shrimp (order Stomatopoda). At about 17 cm (7 inches) long and neither true mantis nor true shrimp but superficially resembling both, these crustaceans are armed with specialized forelimbs called **raptorial claws**. These highly modified appendages are either spear-like and armed with spiny appendages topped with barbed tips for stabbing and snagging prey or club-like for bludgeoning and smashing a potential meal (**Figure 13.25**). To attack, the mantis shrimp latches the limb so it cannot move. It then contracts the muscles as much as possible storing an amazing amount of energy in a special saddle-shaped spring that due to its shape, can distribute huge loads over its surface without buckling (similar to a coiled jack-

Figure 13.25 This female mantis shrimp (*Odontodactylus scyllarus*) has emerged from her hole in shallow tropical waters. Note the club-like raptorial claws that are cocked and ready to bludgeon any prey that wanders within range.

in-the-box). When the latched limb and spring are freed, the stored energy is released with blinding speed and the claw lashes out to smash or spear the hapless prey. It has been calculated that this combination spring and muscle conFigureuration can generate 470,000 watts of power per kilogram of muscle, orders of magnitude higher than the fastest-moving muscles alone can deliver. This tremendous amount of nearly instantaneous energy can propel the claw at speeds of 23 m/s (51 mph). Furthermore, because the strike is so rapid, the claw generates cavitation bubbles between the appendage and the striking surface. The collapse of these cavitation bubbles produces measurable forces on the prey in addition to the instantaneous forces of 1,500 newtons (337 pound-force) caused by the impact of the claw. Thus, the prey is hit twice by a single strike, first by the claw and then by an even greater force from the collapsing cavitation bubbles that immediately follow. Even if the claw misses the prey, the resulting shock wave can be enough to kill or stun the prey. Captive mantis shrimp have managed to shatter double-paned aquarium glass with a single blow from this weapon.

Specialists recognize six classes of crustaceans. In this section, we will examine the four most familiar ones.

Class Malacostraca (Gr. *malakos*, soft + *ostreion*, shell)

Crabs, lobsters, shrimp, prawns, crayfish, krill, isopods, and amphipods. Malacostraca is the largest group of crustaceans and includes three subclasses, 14 orders, and many suborders. We confine our attention to a few of the most familiar orders.

Order Decapoda (Gr., *deka,* ten + *podos,* foot) As their order name implies, the crayfish, lobsters, crabs, prawns, and shrimp that comprise this order have five pairs of walking legs, the first of which is modified in many to form pinchers (**chelae**). Crayfish, lobsters, and shrimp have a body consisting of two regions: a cephalothorax and an abdomen. The cephalothorax is derived from the fusion of sensory and feeding tagmata (the head) with a locomotor tagmata (the thorax). The exoskeleton extends over and around the cephalothorax to form a shield-like carapace. A laterally compressed muscular abdomen with a "tail" extends from the cephalothorax. Decapods bristle with appendages on the head, thorax and abdomen (**Figure 13.26**).

A crab's body differs from that of other decapods in that it is dorsoventrally flattened and oval in shape. The abdomen of a crab is not apparent as it is greatly reduced and folded up under the cephalothorax (**Figure 13.27**).

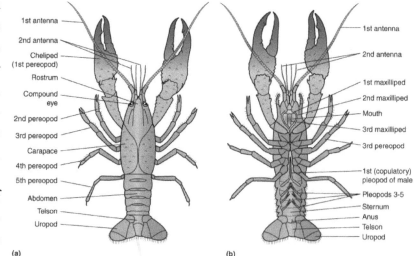

Figure 13.26 Morphology of a Male Crayfish. (a) Dorsal view. (b) Ventral view. In female crayfish the 1ˢᵗ pleopod is greatly reduced resembling pleopods 3-5.

If we live out our span of life on earth without ever knowing a crab intimately we have missed having a jolly friendship. Life is a little incomplete if we can look back and recall these small people only as supplying the course after soup and with the Chablis.

—William Beebe

Order Isopoda (Gr., *iso*, equal + podos, foot) Isopods are one of the few crustacean groups to have successfully invaded terrestrial habitats in addition to freshwater and seawater environments. Isopods are dorsoventrally flattened and lack a carapace. Common land forms are the scavenging sow bugs (*Porcellio*), or pill bugs (*Armadillidium*) that live under rocks and logs and in leaf litter (**Figure 13.28**).When threatened, pill bugs can roll into a tight ball for protection. Freshwater types are found under rocks and among aquatic plants whereas marine forms scurry about on the beach or rocky shore. Some are parasite externally or internally of fish or other crustaceans. The strangest of these isopod parasites (and perhaps the strangest invertebrate parasite of all) is *Cymothoa exigua*. This isopod parasite attaches itself at the base of the tongue of a fish with the claws, on its front three pairs of appendages, and extracts blood (Figure 2.11). As the parasite grows, less and less blood is able to reach the fish's tongue and eventually the organ atrophies from lack of blood. The parasite then replaces the fish's tongue with its own body by attaching to the muscles of the tongue stub. The fish is able to use the parasite as a functional tongue, and it appears that the parasite does not cause any other damage to the host fish. This is the only known case of a parasite functionally replacing a host organ.

Figure 13.27 Morphology of the blue crab (*Callinectes sapidus*). Dorsal view (right) shows no abdomen. Ventral view (left) reveals the abdomen folded up under and attached to the carapace.

Figure 13.28 These wood lice (*Oniscus asellus*) are residents of the damp shadowy world found beneath leaf litter and fallen branches.

Order Amphipoda (Gr., *amphis*, on both sides + *podos*, foot) Members of this order have a laterally compressed body that gives them a shrimplike appearance. Amphipods move by crawling or swimming on their sides. Some species are modified for burrowing, climbing, or hopping. Most amphipods are marine; although a small number of species are freshwater or terrestrial, and they are primarily scavengers with some predatory and a few parasitic types.

Class Branchiopoda (Gr., *branchio*, gill + *podos*, foot)

Fairy shrimp, brine shrimp, water fleas, and tadpole shrimp. Found primarily in fresh water, all branchiopods possess flattened, leaf-like appendages used in respiration, filter feeding and locomotion (**Figure 13.29**). Fairy shrimp (order Anostraca) usually live in vernal pools and temporary ponds that rains and run-off form in early spring. Embryos that have lain dormant perhaps for years revive and grow quickly to adults, racing to secure another generation of embryos before their pool once again dries to dust. Brine shrimp (order Anostraca) also form resistant embryos, but they are adapted to survive the high salinity of salt lakes and ponds such as those around the Great Salt Lake in Utah.

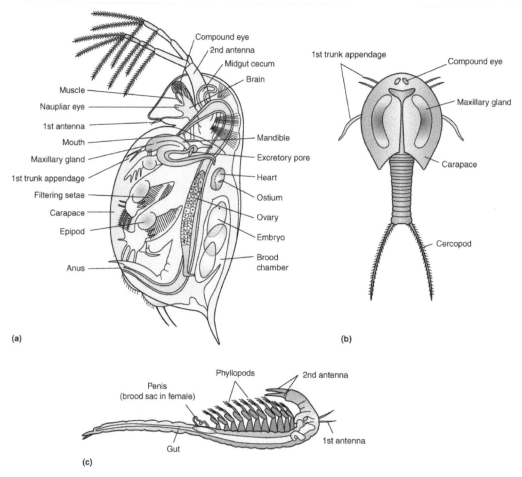

Figure 13.29 Representative Branchiopods. (a) A female water flea (*Daphnia pulex*). (b) A tadpole shrimp (*Triops*). (c) A fairy shrimp (*Branchinecta*).

Water fleas (order Cladocera) are covered by a large carapace over their bodies, and they swim by repeatedly thrusting their second antennae downward to create a jerky, upward motion. Female water fleas are seasonally parthogenetic in the spring and summer and can quickly populate a pond or lake. In response to decreasing temperatures, changing photoperiod, or decreasing food supply, females produce eggs that develop parthenogenetically into males. Sexual reproduction ensues and a generation of resistant "winter eggs" is produced to hatch in the spring.

Class Maxillopoda (L., *maxilla*, the jawbone + *podos*, foot)—copepods and barnacles

With the exception of the barnacles, members of class Maxillopoda are small and sometimes bizarre crustaceans that are recognized by their short bodies and the unique combination of five head, six thoracic, and four abdominal segments, plus a telson ("tail"). Copepods (subclass Copepoda) are so abundant in both marine and freshwater habitats that they dominate the primary consumer level of aquatic communities. Copepods have a cylindrical body with the first antennae (and the thoracic appendages in some) modified for swimming whereas the abdomen is free of appendages. Most copepods are planktonic and use their second maxillae for filter feeding. A few types live on the substrate; a few are predatory, and others are commensals or parasites of marine invertebrates, fishes, or marine mammals.

At first glance, barnacles (subclass Cirripedia) appear more molluscan than crustacean because as adults, their body is surrounded by a shell of calcareous plates and they are totally sessile as adults. One must look internally to discern their true nature. Although their head is small, and they exist without an abdomen or eyes, they do possess long, jointed thoracic **cirri** with hair-like setae. The cirri grow out of a crevice between the calcareous plates (**Figure 13.30**). Their function is to separate small, unwanted matter from desired food and discard it. All barnacles are marine, and they will attach head down by means of adhesive glands in their first antennae to almost anything, including rock outcroppings, ship bottoms and pilings, and even whales. Attachment to the substrate is direct (e.g. acorn barnacles) or by a stalk (e.g. gooseneck barnacles).

Barnacles that colonize ship bottoms reduce both speed and fuel efficiency. Much time, effort, and money have been devoted to research on keeping ships and other surface free of barnacles. In the past, coatings of toxic levels of heavy metals and chemicals were applied to ships. These paints were cheap and effective but leached easily into the surrounding environment where they caused many problems. Recently paint polymers mixed with relatively harmless pharmacological substances, such as dopamine antagonists, have shown good results in preventing the release of adhesives by barnacle larvae, thus inhibiting them from attaching.

Some barnacles have become parasites. *Sacculina* (order Rhizocephala) are highly modified to parasitize crabs in a most unusual fashion. Their first form in life is as a nauplius larva. Later, upon location of a host crab, they metamorphose into a **kentrogon**, injecting parasitic cells into the hemocoel of their host crab. Eventually, root-like absorptive structures of the parasite develop in the same location as would the crab's egg mass (if the crab had an egg mass, which it doesn't). The host crab, however, believes the parasite to be its own egg mass protecting, ventilating, and grooming it. This level of care includes well-timed spawning behavior that assists the parasite in its own reproduction. Strangely, if this parasitic barnacle infects a male crab, there is a castration effect whereby the male host crab becomes a female in its structure and behavior.

Figure 13.30 Barnacles such as these goose barnacles (order Pedunculata) encrust anything in ocean water—rocks, pilings, ship hulls, and even whales and turtles.

Class Ostracoda (Gr., *ostrakodes*, having a shell)—mussel shrimp

Wrapped in a bivalve-like chitinous or calcareous carapace ("shell") only 0.25 to 30 mm (0.01 to 0.12 inch) long, these crustaceans resemble tiny clams or mussels, thus the common name "mussel shrimp." (**Figure 13.31**) Ostracods consist of little more than a head bearing five pairs of appendages. Trunk segments have been fused and thoracic appendages number either two or zero. The head appendages are the principle force for both feeding and mobility.

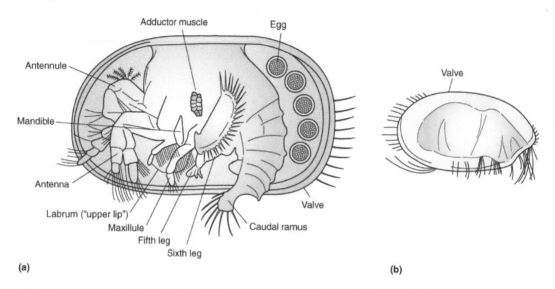

Figure 13.31 Morphology of an Ostracod. (a) The ostracod *Cytherella sp* with the left valve removed for clarity. (b) An intact ostracods showing the position of the body within the valves.

Ostracods have a worldwide distribution in the ocean where they may be planktonic or most commonly, part of the benthos, living on or in the upper layer of the sea floor. Many ostracod species are found in fresh water, and some are found in moist forest soils. They feed on varied diets; from particle, plant, and carrion feeders to predators of different types of prey. Fossil ostracods in some rock strata are often key indicators of oil deposits.

The Arthropod Body Plan

Although arthropods are the most morphologically diverse group of animals on the planet, they all share an assemblage of unifying characteristics: a tough, chitinous exoskeleton, a segmented body specialized into tagmata and jointed appendages. Ancestral arthropods may have been worm-like but an annelid encased in a rigid exoskeleton is not what arthropods came to be. In conjunction with the development of the exoskeleton, the arthropods evolved a suite of highly successful adaptations, referred to as **arthropodization**:

1. The problem of locomotion was solved by the development of flexible joints and regionalized muscles in the body and the appendages. Muscles became organized as intersegmental bands

associated with the individual body segments and appendage joints, whereas circular muscles were lost almost entirely

2. With the loss of peristaltic (squeezing) capabilities as a result of a now rigid body and the absence of circular muscles, the coelom became useless as a hydrostatic skeleton. Loss of the coelom led to the formation of an open circulatory system and the use of the body cavity as a hemocoel or blood chamber in which the internal organs are bathed directly in body fluids. Large body size, however, still demanded some sort of pumping organ for moving blood around the hemocoel, hence arthropods retained the annelid-like dorsal blood vessel but evolution modified it into a highly muscular pumping structure—a heart.

3. Instead of the open metanephridia typical of annelids, excretory organs became enclosed internally, thereby preventing the blood from being drained from the body.

4. A whole set of surface sense organs evolved with various devices for transmitting sensory impulses to the nervous system in spite of the hard exoskeleton.

5. A number of different gas exchange structures evolved to overcome the barrier of the cuticle. This makes a high metabolic rate possible which in turn allows for periods of intense activity.

6. The arthropod body itself has undergone various forms of regional specialization (tagmosis) to produce segment groups or tagmata such as the head, thorax, cephalothorax, and abdomen.

7. The process of periodic molting (ecdysis) evolved to allow the exoskeleton to be shed to make room for an increase in body size.

Such evolutionary plasticity of form and structure has been of paramount importance in establishing the diversity and dominant position of the arthropods.

Body Wall and Exoskeleton

The body wall is composed of the complex, layered **cuticle** secreted by the underlying **hypodermis**. This nonliving exoskeleton covers all body surfaces and supports a variety of functions: protection from injury and predators, preventing water loss, offering structural support, and providing a system of levers for muscle attachment.

The exoskeleton has two layers (**Figure 13.32**). Outermost is the **epicuticle**. Hard and slick because of its waxy lipoprotein composition, the epicuticle is impermeable to water and a barrier to microorganisms and natural or human-produced chemicals. Beneath the epicuticle and comprising the bulk of the exoskeleton is the **procuticle**. The procuticle is composed of chitin, a plastic-like saccharide, and several kinds of proteins. The procuticle hardens through a process known as **sclerotization** in which layers of protein are chemically cross-linked with one another resulting in hardening and darkening of the exoskeleton. (Crustaceans

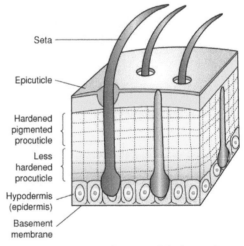

Figure 13.32 A Section of Arthropod Cuticle. The hypodermis secretes the entire skeleton. Calcium carbonate deposits and/or sclerotization harden the outer layer of the procuticle.

also incorporate calcium carbonate into their exoskeleton during sclerotization.). To retain the flexibility needed for running, jumping or flying, the innermost part of the procuticle does not fully harden.

In general, each body segment (somite) is enclosed by four exoskeletal plates, or **sclerites**: a dorsal **tergite**, two lateral **pleurites**, and a ventral **sternite** (**Figure 13.33**). Numerous variations among taxa exist, however, as the result of fusion, fragmentation, and loss of sclerites.

Molting and Growth

Although an exoskeleton endows its bearer with many advantages, it has one serious drawback. How is growth to occur? Growth by the standard gradual increase in external body size is impossible, so arthropods are forced instead to periodically shed their old exoskeleton and form a new, larger one. The process by which this shedding occurs is known as molting or ecdysis.

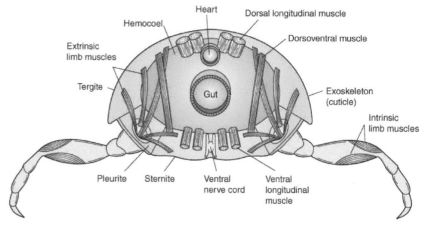

Figure 13.33 Generalized anatomy of an arthropod as seen in cross section.

Ecdysis involves shedding the old exoskeleton, forming a new exoskeleton, and then inflating the body with air before the new exoskeleton hardens. Controlled by the nervous and endocrine (hormone) systems, ecdysis proceeds through four basic stages:

1. **Preecdysis**. In preparation for ecdysis, the arthropod often hides and becomes inactive for a period of time and undergoes **apolysis**, a process in which hormones secreted by glands in the hypodermis begin to digest the old procuticle.
2. As the old exoskeleton begins to detach, the space between the hypodermis and the old exoskeleton begins to fill with a fluid known as **molting gel**.
3. The hypodermis begins secreting a new exoskeleton.
4. The arthropod takes in air or water and as its body swells; the old exoskeleton splits open along predetermined lines. The arthropod then wriggles out of the old exoskeleton.

An arthropod usually remains hidden and inactive as the new exoskeleton hardens fully by sclerotization and/or calcium carbonate deposition. In addition, color pigments are deposited in the outer layers of the procuticle.

Support and Locomotion

The arthropods have largely abandoned the hydrostatic skeleton of their coelomate ancestors. As a result, they lack discrete coelomic spaces (except for small cavities around the gonads and sometimes the excretory structures) and the associated muscle sheets that act upon them. Instead, arthropods rely on the exoskeleton for support and maintenance of body shape. Sheets of muscle are not compatible with an exoskeleton. Therefore, arthropod musculature consists of short bands of muscle that extend from one body segment to the next or across the joints of appendages and other regions or articulation (Figure 13.33).

Life inside a rigid exoskeletal box is not possible. Hence, part of the evolution of the exoskeleton included the formation of various joints where the skeleton might be articulated (moved). In contrast to most of the exoskeleton, joints between body and limb segments are bridged by very thin flexible cuticle in which the procuticular layer is much reduced and somewhat soft. Each joint is bridged by one or more pair of **antagonistic muscles**. One set of muscles, the **flexors,** bend the body or limb at the articulation point whereas the opposing set of muscles, the **extensors**, serve to straighten the body or appendage. The legs of spiders have flexor muscles, but no extensor muscles. Extension is accomplished by hydrostatic pressure from the blood. When a spider dies, blood volume decreases and the flexor muscles contract, causing the legs to bend and close in a characteristic death pose.

Some arthropod joints articulate in only a single plane as do your elbows and knees; others are constructed to allow movement in more than one plane similar to the ball-and-socket joint of your hip, and occasionally two adjacent joints will articulate at 90 degrees to one another, forming a gimbal-like arrangement that facilitates movement in two planes.

Arthropods have evolved a plethora of appendage mechanisms and contraptions for locomotion on land, through water, and in the air. Movement through the water can be accomplished in a number of different ways, smooth paddling of shrimp, jerky stroking of certain insects and small crustaceans, and the startling backward jetting propelled by the tail flexon of crayfish and lobsters. Those moving across the land walk, creep, bound, crawl, or run. Based on relative speed (body lengths per sec), the fastest arthropod is the Australian tiger beetle (*Cicindela hudsoni*) that has been clocked at 2.5 meters per second or 5.6 miles per hour. A comparable relative speed for a 6-foot human would be 1026 feet per second or 720 miles per hour, nearly the speed of sound at sea level. Moving at that rate of speed a human could cover a mile in 4.3 seconds.

Others jump and glide, whereas some, such as the flea, simply jump. To merely state that fleas jump is to dismiss the dramatic fashion through which they accomplish their jumps. Muscles in the flea's legs distort the skeleton, and a special catch mechanism locks the cocked legs in place. When the catch is released, the energy stored by distorting the skeleton explosively extends the legs catapulting the flea a distance that often exceeds 100 times its body length. A comparable relative leap for a human would be jumping the length of two football fields from a standing start!

Arthropods move in ways that are nearly as diverse as they are. The insects, however, evolved a mode of locomotion possessed by no other group of invertebrates (protostomes)—the ability to fly. In fact, insects were the first animal group to take to the air. Two different modes of flying evolved in insects—direct (synchronous) flight and indirect (asynchronous) flight. **Direct** or **synchronous flight** is accomplished when muscles acting on the bases of the wings contract to produce a downward thrust, and muscles attaching dorsally and ventrally on the exoskeleton contract to produce an upward thrust (**Figure 13.34**). Butterflies, grasshoppers, and dragonflies are examples of insects that employ a synchronous flight mechanism.

Other insects employ an **indirect** or **asynchronous flight** mechanism. In this case muscles act to change the shape of the exoskeleton on both the upward and downward wing strokes. Dorsoventral muscles pulling the dorsal exoskeleton (tergum) downward produce the upward wing thrust. Downward thrust results when longitudinal muscles contract and cause the exoskeleton to arch upwards. The flexibility and energy-storing properties of the exoskeleton enhance the power and velocity of these thrusts. Flies and wasps are examples of insects with an asynchronous flight mechanism.

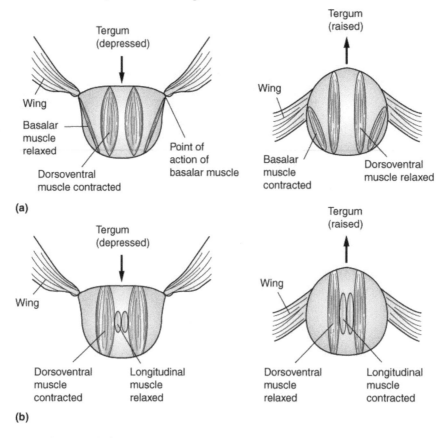

Figure 13.34 The Action of Insect Flight Muscles. (a) Muscle arrangements for the direct or synchronous flight mechanism. Note the muscles responsible for the downstroke attach at the base of the wings. (b) Muscle arrangement for the indirect or asynchronous flight mechanism. Muscles change the shape of the thorax causing the wings to move up and down.

Nerve impulses trigger muscle contractions in both types of flight. In synchronous flight, there is a one-to-one correspondence between nerve impulses and wing beats, but in asynchronous flight a single nerve impulse can result in approximately 50 cycles of the wing. Thus, rapid firing of a nerve can produce tremendous wing speeds (cycles). For instance, frequencies of 1,000 cycles per second (cps) have been recorded for some midges. High wing speeds creates the buzzing sound we associate with small insects such as mosquitoes.

Simply flapping the wings is not enough for controlled flight. The tilt of the wing must be adjusted to provide lift and forward propulsion. In most insects, such control is established through muscles that attach to sclerotized plates at the base of the wing.

Flight speeds vary tremendously. Sphinx moths and horse flies are capable of speeds up to 48 km (30 miles) per hour, while dragonflies can reach 40 km (25 miles) per hour. Other insects can partake in lengthy, nonstop flights. Migrating monarch butterflies, *Danaus plexippus*, for example, are known to fly many thousands of miles at roughly 10 km (6 miles) per hour in their annual migration—an amazing feat of endurance for such a small animal.

Appendages

Arthropod appendages are moveable outgrowths of the body wall equipped with extrinsic (connecting limb to body) and intrinsic (wholly within the limb) muscles that move the various limb segments or pieces called **podites**. Beyond this general plan, however, the variations in arthropod limbs and the myriad of terms associated them can be overwhelming to both student and experts alike.

Arthropod appendages may be categorized as **uniramous** (with only a single branch or **ramus**) and **biramous** (Y-shaped with two rami). Spiders and their kin, centipedes, millipedes, and insects possess uniramous appendages (**Figure 13.35**) whereas crustaceans bear both biramous and uniramous appendages.

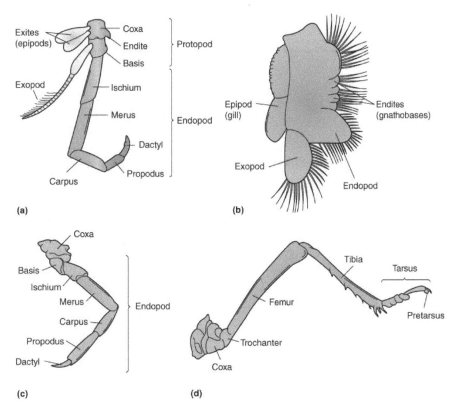

Figure 13.35 Arthropod Trunk Appendages. (a) Generalized biramous limb of a crustacean. (b) Generalized biramous phyllopodal limb of a crustacean. (c) Crustacean uniramous walking leg (stenopod). (d) Uniramous walking leg (stenopod) of a grasshopper.

The uniramous limbs of spiders and insects are typically ambulatory (walking) in function. They are long and thin, and thus are often called **stenopodia** (Gr., *steno*, narrow + *podia*, feet). Some biramous limbs of crustaceans may also be used for walking; in these limbs the inner branch is long and thin and used for walking, whereas the outer branch is greatly reduced. The swimming limbs of some crustaceans are greatly expanded and flattened and are known as **phyllopodia** (Gr., *phyllo*, leaf-shaped + *podia*, feet).

To varying degrees, arthropods have the capacity to regenerate lost appendages. In all arthropods, regeneration is associated with molting. Regenerating appendages develop within the enveloping cuticle and do not become functional until their sheath is shed at the next molt. Metamorphosis into the adult stage marks the end of molting in insects, and accordingly, adults do not regenerate amputated appendages.

Crustaceans, however, are the exception as they tend to molt and grow throughout life. Therefore, crustaceans never lose the ability to grow back a missing appendage.

Digestive System

The great diversity among arthropods is reflected in the multiplicity of their feeding strategies, mouth structures, and appendages. The only real constraint on arthropods in regards to modes of feeding is the absence of external, functional cilia that allow for filter feeding in many animal types. Many arthropods, however, have managed to overcome even this limitation and suspension feed by other means. We have already discussed specific feeding strategies in each section on specific arthropod taxa so here we generalize only about the basic structure and function of the arthropod digestive system.

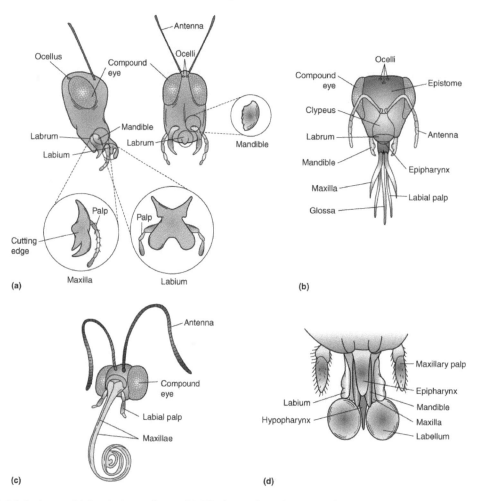

Figure 13.36 Arthropod Mouth Appendages. (a) The biting/grinding mouthparts of a grasshopper. The labrum is a sensory upper lip and the labium a sensory lower lip. The mandibles are hard and used for tearing and grinding. The maxillae have cutting edges and a sensory palp. (b) The piercing/sucking mouthparts of a honeybee. (c) The sucking mouthparts of a butterfly. (d) A close-up view of the sponging/lapping mouthparts of a housefly.

Different feeding modes require different mouthparts (**Figure 13.36**). Once food is ingested it moves into the digestive tract. The digestive tract of arthropods is complete and usually straight, extending from a ventral mouth on the head to a posterior anus. In nearly all cases there is a well-developed cuticle-lined **foregut** and **hindgut**, connected by a **midgut (Figure 13.37)**. In general, the foregut serves for ingestion, transport, storage, and mechanical digestion of food; the midgut for enzyme production, chemical digestion, and absorption; and the hindgut for water absorption and preparation of fecal material. The midgut typically bears one or more evaginations in the form of **digestive ceca** (often referred to as the **digestive gland**). The exact number of ceca and the arrangement of other gut regions varies among the different groups of arthropods,

Various terrestrial arthropods have evolved structures associated with (although not necessarily derived

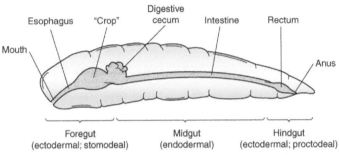

Figure 13.37 The gut regions of an arthropod.

from) the gut. For example, excretory structures called Malpighian tubules develop from the mid-or hind-guts of insects and arachnids. Many unrelated taxa have special repugnatorial glands that produce noxious substances designed to deter predators. Modified salivary glands are common silk-producing organs, but silks are also sometimes secreted by the digestive tract.

Circulation and Gas Exchange.

The body cavity is an open hemocoel with the organs being bathed directly in **hemolymph** (blood). Without a muscular, flexible body wall to augment blood movement, the elaboration of a muscular heart in arthropods became a necessity. The result is a system wherein the blood is driven from the heart chamber through short vessels and into the hemocoel, where it bathes the internal organs. The blood returns to the heart via a noncoelomic **pericardial sinus** and perforations in the heart wall called **ostia**.

The hemolymph of many kinds of small arthropods may be colorless but most larger forms contain a copper-based respiratory pigment, **hemocyanin**, that is blue when oxygenated, thus giving arthropod blood a blue color rather than the red color of vertebrate blood. Unlike the red hemoglobin in our blood, blue hemocyanin is not carried on blood cells but rather is dissolved in the plasma so that it colors not only the blood but permeates all tissues as well. If you squish an insect, you will see green hemolymph ooze out of the mangled corpse. Insect blood is green because yellow xanthophylls pigments in the leafy diet of an insect mix with the blue hemocyanin to produce a hemolymph that appears green.

As one might expect, arthropod gas exchange structures have taken one form in aquatic groups and quite another in terrestrial types. Most larger crustaceans have evolved various types of gills for exchanging gases with the water surrounding them. These gills are in the form of thin-walled cuticular evaginations. Gills are usually branched or folded, providing large surface areas.

The terrestrial insects and arachnids have evolved gas exchange structures in the form of invaginations of the cuticle. By folding the cuticle inward, the gas exchange structures are kept moist allowing oxygen to enter solution for uptake. Many arachnids possess highly folded invaginations called **book lungs** (internal gills in a sense) whereas insects possess inwardly directed branching tubules called **tracheae** that open externally through pores called **spiracles**.

Excretion and Osmoregulation

Arthropods possess nephridia, but a hemolytic circulatory system demands that they be quite different than those found in the annelids. The open nephridia of annelids would be functionally untenable in arthropods as they would drain the blood directly from the open hemocoel to the outside. Instead, arthropods are equipped with closed nephridia, and there has been a reduction in the number of nephridia as well.

In most adult crustaceans, only a single pair of nephridia persists, and these are usually associated with particular segments of the head such as the antennae (**antennal glands**) and maxillae (**maxillary glands**). In arachnids, there may be as many as four pairs of nephridial opening at the bases of the walking legs (**coxal glands**).

Another type of arthropod excretory structure exists in many terrestrial forms (e.g. arachnids and insects). These structures, known as Malpighian tubules, arise as blind tubules extending into the hemocoel from the gut wall.

The inner ends of the nephridia absorb fluid from the hemocoel that is generally similar to the blood itself, but as it passes along the plumbing system of the nephridium, a good deal of selective reabsorption occurs, particularly of salts and nutrients such as glucose. Thus, the urine exiting the nephridial pores is a concentration of nitrogenous waste products.

Malpighian tubules accomplish the same processes, but they must rely on assistance from the gut. Malpighian tubules uptake from the hemocoel is nonselective, and thus this "primary urine," emptied directly into the gut, contains nutrients, salts, water, and so on. The hindgut is mostly responsible for concentrating the urine by reabsorbing the nonwaste components.

Nervous System and Sense Organs

The arthropod brain comprises several bundles of fused ganglia. The **supraesophageal ganglion** lies in the head and is composed of the **protocerebrum**, the **deuterocerebrum**, and the **tritocerebrum**. The posterior portion of the tritocerebrum forms circumenteric (looping) connectives around the esophagus attaching the brain to the **subesophageal ganglion** that lies beneath the esophagus. From the subesophageal ganglion, a ventral nerve cord runs back to the anus. This ventral nerve cord contains a fused ganglion in each segment (**Figure 13.38**).

Possessing an exoskeleton certainly shields the bearer against the outside world, but how is an animal so clad to receive critical sensory input from the surrounding environment? Arthropods have compensated by evolving numerous cuticular processes—setae, hairs, bristles, pores, or slits—collectively known as **sensilla** that serve as mechanoreceptors and chemoreceptors.

Tactile(touch/hearing) reception. Most arthropod tactile receptors (mechanoreceptors) are movable bristles or setae. When the cuticular projections are touched, that movement is translated into a deformation of a nerve ending which in turn initiates a nerve impulse to the brain.

Sensitivity to environmental vibrations ("hearing"), pressure waves, and air currents is similar to tactile reception. Sensilla in the form of fine hairs or setae are mechanically moved by external vibrations or waves and impart that movement to underlying sensory neurons. At the base of the antennae of most insects are **Johnston's organs**, long setae that vibrate when certain frequencies of sound strike them. Vibrating setae move the antennae in its socket, stimulating sensory cells. Sound waves in the frequency range of 500 to 550 cycles per second (the range of sounds female wings produce) attract and elicit mating behavior in the male mosquito *Aedes aegypti*. **Tympanic organs** are located in the legs of crickets and katydids (order Orthoptera), in the abdomen of grasshoppers (order Orthoptera), and in the abdomen or thorax of moths (order Lepidoptera). These organs consist of a thin, cuticular membrane covering a large air sac. Sound waves resonate the air sac which in turn vibrates the membrane and stimulates the sensory nerves below. Grasshopper tympanic organs can detect sounds in the range of 1,000 to 50,000 cps. (The human ear can detect sounds between 20 and 20,000 cps.) Bilateral placement of tympanic organs allows insects to discriminate the direction and origins of a sound.

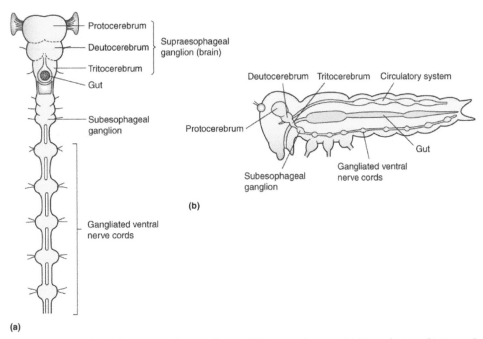

Figure 13.38 Generalized Structure of an Arthropod Nervous System. (a) Dorsal view. (b) Lateral view.

Chemical (smell/taste) reception. Chemoreceptors are usually abundant on the mouthparts, antennae, legs, and ovipositors, and take the form of hairs, pegs, pits, and plates that have one or more pores leading to internal nerve endings. Chemicals diffuse through these pores and bind to and excite sensory nerve endings. Arthropods use chemoreceptors in feeding, species identification, selection of egg laying sites, mate location, and social organization. The pheromones produced by many receptive female types must surely be among the most powerful scents in the arthropods group, and males are champions at detecting these odors. Male Chinese silkworm moths (order Lepidoptera) have been known to home in on and find perfuming females from as far away as seven miles. Odor is also a very reliable cue when it comes to arthropod feeding because it is more constant than color and shape. Some flies are so attuned to the odor of death that they can be found on the carcass of a dead animal only minutes after the animal's demise.

Photo (sight) reception. Many arthropods are capable of detecting light and use this capability for orientation, navigation, feeding, and other functions. Light is detected either through ocelli, compound eyes, or a combination of both. Ocelli consist of 500 to 1,000 receptor cells beneath a single cuticular lens. Compound eyes are much more complex and are usually well developed in most adult insects. Compound eyes consist of a few to 28,000 receptors, called **ommatidia**, that fuse into a multifaceted eye (**Figure 13.39**). The outer surface of each ommatidium is a lens. Below the lens is a crystalline cone. The lens and cone are the light-gathering structures. Certain cells of the ommatidium called **retinula cells** have a special light-collecting area known as the **rhabdom**. The rhabdom converts light energy into nerve impulses. Pigment cells surround the crystalline cone, and sometimes the rhabdom, prevent light that strikes one rhabdom from reflecting into an adjacent ommatidium.

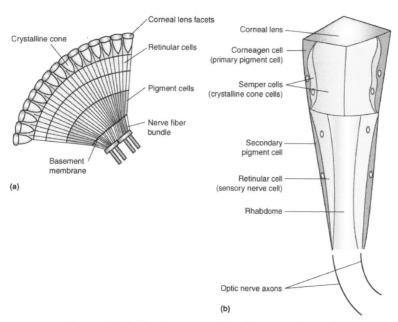

Figure 13.39 The Structure of the Compound Eye of Insects. (a)Compound eye of an insect in cross-section. (b) A single ommatidium from a compound eye.

Although compound eyes may form a fuzzy image of sorts, they are much better suited for detecting movement. In fact, movement of a point of light less than 0.1 degrees can be detected. For this reason, bees are attracted to flowers blowing in the wind, and predatory insects select moving prey. Compound eyes can detect wavelengths of light that the human eye cannot perceive, especially in the ultraviolet end of the spectrum. Some insects are also able to detect polarized light, an extremely useful adaptation for navigation and orientation. Mantis shrimp (Stomatopoda) have the most sophisticated visual system in the animal kingdom. Each eye contains 16 different types of photoreceptors. They can perceive both ultraviolet and polarized light, and discriminate up to 10,000 different colors (10 times greater than the human eye).

Arthropods are also capable of detecting changes in temperature, pressure due to increasing or decreasing depth, humidity, the position of their appendages, and even the magnetic field of the earth itself. Furthermore, research has indicated that insects are capable of some learning and have at least a limited memory.

Reproduction and Development

Arthropods are the most biologically successful group of animals on the planet, and one of the keys to their environmental dominance is their high reproductive potential (**fecundity**). Adult arthropods are reproductive machines that have the capability of reproducing their way around any obstacle—natural or human-made— they encounter. This is especially true of insects. A single Australian ghost moth (Hepialidae) female laid 29,100 eggs, and when it was dissected, 15,000 eggs still remained in the ovaries. It has been calculated that a single pair of flies beginning reproduction in April could be the progenitors of 191,010,000,000,000,000,000 flies by the end of August (assuming optimal conditions and that no flies die). Aphids may have the shortest generation time (from egg-laying adult to egg-laying adult) at only 5 days whereas the longest life cycle (from egg to reproducing adult) under normal conditions belongs to the periodical cicada (*Magicicada*). These insects require 17 years to complete nymphal development underground. The life spans of most arthropods range from a few hours to several years. However, the wood-boring beetle, *Eburia quadrigeminata*, may have its development so delayed by the poor nutritional quality of dead wood that a specimen was found emerging from a birch bookcase 40 years old.

With few exceptions, arthropods are dioecious. Sperm are commonly transferred to the female within sealed packets known as spermatophores. In this mode of delivery, the sperm are not dilute by water nor do they suffer rapid desiccation on land. Either the female is attracted to the spermatophore chemically or the deposition of the spermatophore occurs during the course of a nuptial dance, and the male afterward maneuvers the female into a position to take up the spermatophore within her genital opening. Many arthropods, such as some crustaceans, millipedes, some insects, spiders, and some mites, transfer free sperm through direct copulation and insemination.

The fertilized eggs are usually externally deposited in safe or hidden places (oviparous). The females of some arthropods, however, retain the eggs in their body but rely on the yolk within the eggs to nourish the developing young (ovoviviparous), and subsequently give birth to live young.

The eggs of many crustaceans hatch into a nauplius larvae which have fewer segments and appendages than adults (**Figure 13.40**). Additional segments and appendages then appear at regular intervals with molting. The **zoea** stage larva usually follows the nauplius stage, however, due to their accelerated development, the zoea is the first larval stage in decapods such as crabs and shrimps. There are several advantages of larval stages in the development of aquatic arthropods: less yolk is required in the eggs, and currents dispersing the larvae, allow the species to colonize new areas without planktonic larvae having to compete with benthic adults.

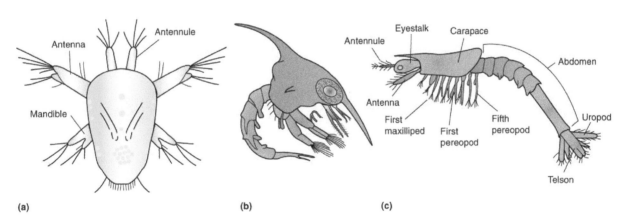

Figure 13.40 Laval Stages of Aquatic Arthropods. (a) A newly hatched copepod nauplius larva. (b) Zoea larva of a crab. (c) Zoeal stage larva of a shrimp.

The young of most arachnids are similar to the adult. The female scorpion gives birth to her young, which immediately crawl on her back. Unlike other arachnids, mites and ticks hatch as six-legged larvae, acquiring their fourth pair of legs at a later molt.

As discussed earlier, insects metamorphose from larva to adult either gradually through incomplete (hemimetabolous) metamorphosis or radically in only a few stages through complete (holometabolous) metamorphosis. Beneficially, larvae inhabit different environments and eat different foods than their parents, thus reducing competition.

Arthropod Connection

Economically

Spider silk is one of the great wonders of the natural world. That a small animal can produce an amazing substance that we humans with our technological prowess are unable to duplicate is yet another humbling reminder of how we pale in comparison to nature's works. Spider silk has a **tensile strength** (amount of force a material can withstand without breaking) greater than Kevlar (which in turn is stronger than steel). Furthermore, spider silk is highly elastic as well as waterproof. In an attempt to develop artificial spider's silk, researchers have used molecular techniques to sequence the genes coding for spider silk and have placed these genes into host organisms, including bacteria, plants, and goats. Silk proteins have been produced by these host organisms. Imagine the usefulness of a thread that is stronger than Kevlar and steel, yet elastic and biodegradable. Potential uses include body armor, parachutes, fishing nets, extremely fine sutures for microsurgery, artificial ligaments and tendons, and even personal clothing that would be nearly indestructible and never wear out. Furthermore, as anyone who wanders afield in the early morning can attest, spider webs can collect remarkable amounts of water. Artificial webs constructed in similar design could be used to harvest water from mist.

Crustaceans have been highly regarded as delectable gourmet fare since ancient times and our taste for them continues to this day as evidenced by the fact that nearly 10,700,000 tons of crabs, lobsters, crayfish, prawns, and shrimp were consumed worldwide in 2007. Over 70% by weight of all crustaceans caught for consumption are shrimp and prawns and over 80% of those are produced in Asia, with China alone producing half the world's total. Although krill have one of the greatest biomasses on the planet, these small crustaceans are not widely consumed by humans, with only about 130,000 tons caught for that purpose yearly. It takes up to seven years for a lobster to grow to 1 pound, the smallest size legal for trapping. "Big George," a lobster caught off the coast of Cape Cod in 1974, weighed in at hefty 17 kg (37.4 pounds) and was over 0.6 m (2 feet) in length. More recently, a 20 pound lobster that was caught and eventually released was estimated by experts to be over 140 years old. A quick calculation reveals that "Big George" was probably over 250 years old when he was captured.

Other arthropods provide everything from the ingredients for fabric dyes and wood preservatives to medicines. Insects, however, have the most impact on each and every one of us directly and indirectly. The economical dark side of this connection is the tremendous destruction and damage yearly to our food, clothing, and property by weevils, cockroaches, ants, moths, termites, and beetles, as well as to our pets and domestic livestock by arthropod pests and parasites. The annual lost revenue from insect damage to crops in the field and storage or insect-transmitted plant diseases in the United States alone is approximately $5 billion! As humans, we exhaust countless resources in agriculture and forestry as well as in the housing and food industries in our efforts to lessen the negative impact of insects. Termites cause around $2 billion annually in structural damage to our homes and other wooden structures, and other insects such as bark beetles, spruce budworms, and gypsy moths wreak economic havoc as well. Since 1980, the introduced gypsy moth has defoliated close to a million acres of trees a year. In 1981, a record 12.9 million acres (an area larger than Rhode Island, Massachusetts, and Connecticut combined) were defoliated.

Only about 0.5% of known insect species adversely affect human health and welfare and most contribute enormously to the richness of human life. Many others have provided valuable services and commercially valuable products, such as wax, honey, and silk, for thousands of years. Insects are responsible for the pollination of approximately 65% of all plant species, including many that we rely upon for food. In the U.S. alone, bees are annually responsible for the pollination of nearly $20 billion worth of produce, not including crops dedicated for livestock. Soil-dwelling insects play important roles in aeration, drainage, and turnover of soil, and they promote decay. It is estimated that cattle ranchers save $380 million a year because burying beetles dispose of cow dung. Another consideration is the $50 billion that would disappear from hunting, fishing, and bird-watching activities were it not for insects at the bottom of the food chain. And if it weren't for insects eating each other, we would have to spend an additional $4.5 billion in agricultural pest control.

Many new applications of arthropod chemicals are currently being investigated. For example, chitin could be sprayed onto fruit and frozen food to prevent spoilage and to preserve flavor. The natural adhesive that barnacles use to attach themselves permanently to rocks and other substrates could be useful in a number of application ranging from dentistry to underwater construction and repair. And chemicals in spider venom are being tested as potential natural pesticides.

In addition, insects are widely used in teaching and research, and have contributed to advances in genetics, population ecology, and physiology. The chemical that makes fireflies glow, for example, is used in medical tests and as a marker in genetic engineering.

Ecologically

From an ecological standpoint, the most important role arthropods play is in the composition of both aquatic and terrestrial food webs. Krill, copepods, and other planktonic crustaceans are the primary consumer foundation upon which oceanic and freshwater food chains are built whereas insects fill essentially the same role in terrestrial schemes. In short, arthropods of some sort and some size serve as food for every group of animals on the planet from protists to carnivorous plants (that actually rely on them for the mineral content of their bodies, not as actual food) and from sponges to mammals, including humans. Conversely, arthropods leave their mark on the environment as eaters. Herbivorous arthropods, especially insects in large adult swarms or as individual larva, eat tremendous amounts of plant material, including human-grown crops, fruits, vegetables, and flowers daily. The larvae of one insect—the European corn borer, *Ostrinia nubilalis*—causes $7 billion dollars in damage to one crop—corn—in the United States each year. Plagues of the desert locust, *Schistocerca gregaria*, have periodically threatened agricultural production in Africa, the Middle East, and parts of Asia for centuries. Each locust is capable of consuming its own weight (2 gr [0.07oz]) in green vegetation—leaves, flowers, bark, stem, fruits and seeds—daily. Singularly they pose no threat, but rolling swarms of these insects can contain tens of millions of individuals and are thus capable of laying waste to entire landscapes, including human agriculture and economies.

Predatory arthropods eat other animals of all types. Worms, other arthropods, and small vertebrates from fish to amphibians and even birds and mammals are eaten in large numbers by hungry arthropods. An undisturbed meadow may contain as many as 2,250,000 spiders per hectare alone. If each spider caught and consumed only one insect per day, the cumulative biomass consumed by just that one type of arthropod in just that one small area is astounding. In fact, the weight of insects eaten every year by spiders is estimated to be greater than the total weight of the entire human population.

The interactions of arthropods with other creatures are not limited to eating or being eaten, however. Arthropods form a number of different symbiotic relationships with other organisms. Some plants live more intimately with arthropods than merely being pollinated by them. Occupying hollow thorns on the bull-horn acacia tree, *Acacia cornigera*, colonies of stinging ants, *Pseudomyrmex ferruginea*, fiercely guard the tree against ravaging herbivorous insects and browsing mammals. The ants even go so far as to prune away vines and the leaves of other plants that may shade the acacia. In reward for their services, the tree provides shelter and food in the form of carbohydrate-rich nectar emanating from glands on its leaf stalk and protein-rich Beltian bodies from its leaf tips. Consider also the mutualistic relationship between cleaner shrimp and the fish who allow this bite-sized crustacean to crawl over them and even into their mouths unharmed in the search for parasites and bits of dead tissue. By allowing the cleaner shrimp to go unharmed, the fish gets a good cleaning of annoying parasites, and the shrimp gets a meal.

Many arthropods are important agents of decay and recycling and burrowing soil insects help aerate and loosen the soil and aid in soil formation.

Medically

The bites (venom delivered via mouthparts) and stings (venom delivered via a stinger usually at the tip of the abdomen) of some arthropods may be irritating, painful, or itchy but they are seldom harmful. Case in point, it would take 1, 120,000 mosquito bites to drain the average size human of blood. There are those, however, that can inject dangerous toxins to which many people are highly allergic. Bee and wasp stings, innocuous to most people, can be fatal to those highly allergic. Bee and wasp stings cause 30-120 deaths yearly in the United States. In contrast, fewer than four fatalities a year are caused by snake bite.

North Americans are fortunate in that we face few native arthropods that are individually dangerous to us. Two of the most potent are spiders, the black widow (*Latrodectus*) (**Figure 13.41**) and the brown recluse (*Laxoceles*), (**Figure 13.42**) and it is important to be able to identify them and to understand the threat they may pose to us.

The term "black widow" is almost universally recognized and feared. The name is apt for these distinctively-marked arachnids as only the female bites. The black widow's venom is a neurotoxin that blocks the nerve impulses to the victim's muscles resulting in cramps, excruciating muscle spasms, rigidity, and, in extreme cases, paralysis. These symptoms are temporary, and rarely result in death. If you are bitten, remain calm and seek medical assistance with the peace of mind that comes from knowing that there is an anti-venom for black widow

Figure 13.41 A female black widow spider displays the characteristic red or orange hourglass design on the ventral side of the abdomen. Males are gray or brown.

bite and that prior to the development of this anti-venom in 1943, only 32 of the 578 cases reported in the 200 years before that resulted in death. Widow spiders are more common in warm and dry climates but do occur throughout North America.

The brown recluse is smaller than the black widow, and they are often found in human structures such as drawers in a garage workbench or the heating ducts of your home. The distribution of the brown recluse is limited to the south central United States making them far less prevalent than most people believe. The brown recluse's venom is much nastier than that of the black widow. Recluse venom is a necrotic toxin. Instead of affecting the victim's nervous system, it acts directly on the skin and musculature to kill the tissues immediately surrounding the bite. The dead tissues heal slowly, if at all. Death is unlikely, however; since 1896, fewer than 10 of 130 or more brown recluse cases recorded in the United States have resulted in death. In severe cases, necrosis may lead to gangrene and possible amputation.

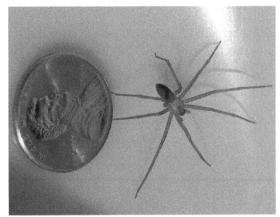

Figure 13.42 Though quite small as the comparison indicates, the brown recluse spider (*Loxosceles reclusa*) possesses a neurotoxic venom more potent than that of the black widow.

Although most species of tarantulas are not native to North America, *Aphonopelma chalcodes*, the Desert tarantula is native to southwestern deserts and can be found in Arizona, New Mexico and southern California. Tarantulas are increasingly being kept as pets. Although the bite of a tarantula may be painful, they are not considered to be venomous spiders. The greatest health risk with these spiders is the fact that when aggravated or alarmed, they can release fine hairs from their legs and abdomen. These hairs, equipped with tiny barbs, can lodge in the eyes and nasal cavities where they can be very irritating.

Only one species of scorpion in North America has venom potent enough to be dangerous to human beings. The Arizona bark scorpion (*Centruoides sculpturatus*) is found primarily in Arizona and northern Mexico. The venom of this scorpion can cause severe pain and swelling at the site of the sting, numbness, frothing at the mouth, respiratory difficulties, muscle twitching, and convulsions. Symptoms are more pronounced in the very young or very old, but death is rare.

As with some other invertebrates, researchers ply the possibility of developing drugs from the toxins of various arthropods. Scorpion venom, for example, shows potential as an immune system suppressor that may prove useful in treating autoimmune diseases or preventing the rejection of transplanted organs. On the other hand, the poison of the funnel web spider, *Hololena curta*, stimulates the immune system and could possibly prevent brain damage caused by short bouts of oxygen loss. That bee venom warms the body, reduces inflammation, and boosts the immune system may soon translate into medicines that will help relieve the pain of arthritis, reduce inflammation, and treat allergies.

Other chemicals derived from arthropods may have useful medical applications as well. An extract of horseshoe crab blood, for example, is used to test the purity of medications, and the chitin extracted from crustacean exoskeletons is used to dress wounds and produce thread for surgical stitches.

Medically, arthropods are most significant as the **vectors** (carriers) of diseases such as malaria, yellow fever, dengue fever, west Nile virus, and elephantiasis (via mosquitoes), African sleeping sickness (via tsetse flies), typhus fever (via lice), bubonic plague (via fleas), and Rocky Mountain spotted fever and Lyme disease (via ticks). The worst of these is malaria. Approximately 300 million people worldwide are infected with the malarial parasite and between 1 and 1.5 million people die from it every year.

Although fly maggots (larvae) may not be your idea of a modern medical treatment, more and more doctors believe they do have a place in modern medicine, and that place is inside open wounds. **Maggot debridement therapy** involves mixing live fly larvae into a wound's dressing and then covering the area with gauze. The maggots, which will only eat dead (**necrotic**) tissue, feed on decaying flesh and leave the underlying healthy tissue untouched. In the process, the maggots also excrete an ammonia-like anti-microbial agent that helps cleanse the wound as well. Because of the decreased circulation and **neuropathy** (degenerative changes to nerves) associated with the disease, maggot debridement therapy has proven most useful in cleansing wounds on the feet and legs of people with diabetes.

As previously discussed, bees and other pollinators are critically important in the production of many of our crops and fruits. In fact, worldwide pollination supports about 35 percent of Earth's agricultural production by weight. Less known, however, is the link between pollination and human nutrition and health. Pollination insures crops that produce essential micronutrients—vitamins, minerals, and trace elements that human health depends on.

Researchers found that in regions of Argentina, Australia, India, Iran, Mexico, Thailand, Romania, and the U.S., vitamin A availability was up to 50 percent dependent on pollination. In addition, certain hotspots in African countries, Brazil, China, and Mexico iron and folate reliance reached 15 percent. Such studies may help target hotspots for pollinator conservation efforts. Unfortunately, bees and other pollinators are dwindling due to disease, intensive land use, and wide-spread pesticide use.

A Closing Note
Backyard Battlefields and Mystery Shrimp

Beneath our feet and beyond our sight is a world of tiny things every bit as dynamic as our larger world. As a young lad, I would often take my trusty magnifying lens and crawl about the yard seeing what I could find. One day I happened upon a battle, which for its scale size, would rival anything in our larger realm. A column of black ants was marching on a nest of red ants. The red ants fought bravely but to no avail and soon the black ants were streaming into the red ants' nest. Shortly, many black ants emerge bearing the captive larva and pupa of the red ants. I followed these black kidnappers as they returned to their nest some distance away. Checking on the red ant nest the next morning, I found the entrance torn open and the ground littered with the bodies of dead red worker ants. Animals were first to develop many of the things humans now take credit for, and warfare seems to be one of them.

For many years, I taught biology in a small community surrounded by cornfields. In those days, the corn was irrigated from ditches of water called laterals running perpendicular to the rows of corn. In late spring when irrigation would commence, I would invariably have students bring in crustaceans that just mysteriously appeared overnight in the irrigation laterals (and they just as mysteriously disappeared when irrigation season was over). What they collected were the ancient-looking tadpole shrimp (*Triops*). Hatching from winter resistant eggs laid by the previous generation in the mud of the last summer's irrigation, these creatures hatched, mated, laid the next generation of mystery shrimp eggs, and then vanished as quickly as they came. Just another example of how the animal world is always with us, surrounding us and all other life forms in ways we may never fully understand, but should always appreciate.

In Summary

- The ancestors of arthropods first appeared in the appeared in the primeval seas of the Precambrian over 600 million years ago. Today most zoologists judge that the modern arthropods represent the pinnacle of protostome development.
- Arthropods have evolved to become the most abundant, most diverse, and most biologically successful animals on earth. Of all the known and described species of animals, at least three out of every four is an arthropod.
- Arthropods may be classified into four subphyla and ten classes:

> Subphylum Chelicerata
> > Class Arachnida—Spiders, scorpions, ticks, mites
> > Class Merostomata—Horseshoe crabs
> Subphylum Myriapoda
> > Class Chilopoda—Centipedes
> > Class Diplopoda—Millipedes
> Subphylum Hexapoda
> > Class Insecta—Insects
> > Class Entognatha—Spring tails, bristletails, proturans
> Subphylum Crustacea
> > Class Malacostraca—Crabs, shrimps, lobsters, woodlice
> > Class Branchiopoda—Brine shrimp, fairy shrimp
> > Class Maxillopoda—Barnacles, copepods, fish lice
> > Class Ostracoda—Mussel shrimp

- The characteristics of phylum Arthropoda:

1. Their segmented body is divided into tagmata (specialized body regions). The coelom is reduced to portions of the reproductive and excretory systems. Most of the body cavity consists of a hemocoel (sinuses or spaces within the tissues) filled with blood.
2. They possess a cuticular exoskeleton composed of chitin, protein, and lipids. The chitinous skeleton is calcified in many groups. Growth occurs by the process of ecdysis (molting) periodically shedding the old exoskeleton and forming a new larger one.
3. They possess jointed appendages. Ancestrally each true body segment bore a pair of jointed appendages but in modern arthropods, the number of appendages may be reduced and they are often modified for specialized functions.

4. They possess a complex muscle system attached to the exoskeleton for support and leverage. Functional cilia are absent.

5. They possess a complete digestive system with mouthparts modified from ancestral appendages into structures adapted for different methods of feeding.

6. Their circulatory system is open with a dorsal heart, arteries, and hemocoel containing blood.

7. Coaxial glands or Malpighian tubules serve as an excretory system. Coaxial glands are paired, thin-walled spherical sacs bathed in the blood of body sinuses. Nitrogenous wastes are absorbed across the sacs, and excreted through long, convoluted tubules that empty at the base of the posterior appendages. Malpighian tubules are diverticula (pockets or pouches off the gut tract of arachnids adapted to dry environments). These tubules absorb nitrogenous wastes from the blood and then empty them into the gut tract where they are eliminated along with the digestive wastes.

8. Gas exchange occurs through the body surface, gills, tracheae (air tubes), or book lungs. Tracheae are a series of branched, chitin-lined tubules that conduct gases to and from body tissues. This tubule system opens to the outside through holes called spiralces located along the ventrally or laterally along the abdomen. Book lungs are paired invaginations of the ventral body wall that fold into a series of leaf-like lamellae (thin, flat plates or disks). Air enters the book lung through a slit-like opening and circulates between the lamellae. Respiratory gases diffuse between the blood moving among the lamellae and the air in the lung chamber.

9. Their nervous system is similar to that of annelids with a dorsal brain connected by a ring around the gullet to a double nerve cord chain of ventral ganglia. Well-developed sense organs are present.

10. The sexes are usually separate with paired reproductive organs and ducts and internal fertilization the norm; some types are capable of parthenogenesis. Development progresses through several stages in a process known as metamorphosis.

- Subphylum Chelicerata is characterized by:

1. A body composed of two tagmata: the prosoma (cephalothorax) and the opisthosoma (abdomen). The cephalothorax represents a fusion of the head and thorax (trunk) and is often covered by a carapace-like dorsal shield. The abdomen is composed of up to 12 somites (sections) and a postsegmental telson. Antennae and wings are absent.

2. Uniramous appendages (single and unbranched) attached to the cephalothorax including: chelicerae (anterior appendages of an arachnid often specialized as fangs.), pedipalps (specialized sensory appendages borne near the mouth of an arachnid), and four pairs (8) walking legs.

3. An exoskeleton modified with projections, pores, and slits to accommodate a variety of mechanoreceptors and chemoreceptors (collectively known as sensilla), together with sensory and accessory cells.

- Subphylum Myriapoda is characterized by:

  1. A body of two tagmata: head and trunk (abdomen). No carapace or wings present.
  2. The abdomen being long and cylindrical and consisting of many segments.
  3. Uniramous appendages.
  4. One or two pair of walking legs per segment.
  5. One pair of articulate (jointed and movable) antennae attached to the head segment.
  6. A few to many clustered ocelli (simple eyes) on the head segment.
  7. Mouthparts consisting of mandibles (jaws), maxillules (first maxillae), and maxillae (second maxillae). The second maxillae are fused into a single flaplike structure called a labrum.
  8. A gas exchange system composed of tracheae and spiracles.

- Subphylum Hexapoda is characterized by:

  1. A body of three tagmata: head, thorax, abdomen without a carapace.
  2. Uniramous appendages.
  3. Three pairs of walking legs attached to the thorax. Wings are present in some species.
  4. A head bearing a single pair of jointed antennae, mouthparts consisting of mandibles, maxilla, labium, and labrum, compound eyes, and several ocelli.
  5. A gas exchange system consisting of tracheae and spiracles.
  6. An excretory system consisting of Malpighian tubules.

- Subphylum Crustacea is characterized by:

  1. Sixteen to 20 segments, with some forms having 60 segments or more.
  2. An exoskeleton that is more pronounced and generally thicker and heavier than other taxa of arthropods.
  3. Two or three tagmata: cephalothorax with a shield-like carapace and abdomen or head, thorax, and abdomen.
  4. Biramous (branching) appendages.
  5. Head appendages consisting of two pairs of antennae, a pair of mandibles, two pair of maxillae, and one pair of compound eyes.
  6. Both simple ocelli and compound eyes often elevated on stalks.
  7. Gas exchange typically across gills.
  8. Excretion by true nephridial structures.
  9. A developmental stage known as a nauplius larva that is characterized by the presence of three pairs of head appendages.

- Arthropods

1. Have a body wall composed of a complex, layered cuticle or exoskeleton secreted by the underlying epidermis.
2. Rely on the jointed exoskeleton for support and as attachments sites for the muscles that move the animal.
3. Bear many limb segments or pieces called podites. Arthropod appendages may be categorized as either uniramous (with only a single branch or ramus) or biramous (Y-shaped with two rami). Spiders and their kin, centipedes, millipedes, and insects possess uniramous appendages whereas crustaceans bear biramous appendages.
4. Possess a digestive tract that is complete and straight, extending from a ventral mouth on the head to a posterior anus. In nearly all cases, there is a well-developed cuticle-lined foregut and hindgut, connected by a midgut.
5. Have a body cavity that functions as an open hemocoel with the organs being bathed directly in hemolymph (blood). The blood is driven from the heart chamber through short vessels and into the hemocoel, where it bathes the internal organs. The blood returns to the heart via a noncoelomic pericardial sinus and perforations in the heart wall called ostia.
6. Exchange gas through gills in the crustaceans and through invaginations in the cuticle in insects and arachnids.
7. Possess nephridia for filtering wastes but a hemolytic circulatory system demands that they be quite different than those found in the annelids. In contrast to the annelids, arthropods are equipped with closed nephridia, and there has been a reduction in the number of nephridia as well.
8. Have a brain that comprises several bundles of fused ganglia. From the brain a ventral nerve cord runs back to the anus. This ventral nerve cord contains a fused ganglion in each segment.
9. Have astonishingly high reproductive rates (fecunidty).
10. With few exceptions, are dioecious. Sperm are commonly transferred to the female within sealed packets known as spermatophores. Once the eggs are fertilized the mode of development varies greatly with the subphylum.

- Arthropods connect to humans in ways that are important economically, ecologically, and medically.

Review and Reflect

1. *What's My Job?* Imagine you run an employment agency. Prospective employers have requested you find specific arthropods to fill the following job descriptions:

    - Athlete—a high jumper
    - Army—a combat soldier

- Construction—a bulldozer
- Construction—a plasterer
- Athlete—a weight lifter
- Recycling—a wood disassembler

Explain what specific arthropod you would select to fill each of the positions listed and why you selected them for that position. Your instructor may add additional jobs to be filled.

2. ***Walk a Mile in My Shoes*** If you had to be reincarnated as an arthropod, which specific arthropod would you choose to become and why?

3. ***A Giant Nightmare*** You and a group of your friends have just come out of a movie where three-story spiders battled giant ants as long as a football field. One of your friends turns to you and says, "Why don't real arthropods grow to be at least as large as elephants or whales?" How would you respond? (HINT: Fossils show that dragonflies with wingspans over 1 m (3 to 4 feet) did exist at one time but during that time it is believed that the concentration of oxygen in the atmosphere was much higher than it is today.)

4. ***The Nightmare Continues*** The movie about giant arthropods has given you bad dreams. In your dream, you are standing in a park when a giant predatory insect approaches you. What would be your best strategy—run like the wind or freeze motionless? Defend your choice.

5. ***Away We Grow*** **Figure 13.43** depicts the generalized growth pattern of a vertebrate, such as a koi fish in my pond, and an insect, such as an ant in my yard. Which line represents each type of animal? Explain how you arrived at your decision.

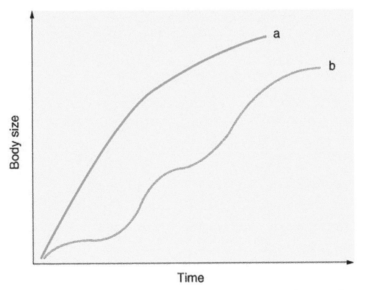

Figure 13.43 A generalized growth graph comparing the growth of an arthropod with the growth of a vertebrate.

6. ***A Sweet Set of Data*** Dr. Buzz Drosophila has been researching the relationship between the length of life in fruit flies and the amount of sugar they receive in their diet. His data is listed in **Table 13.1**. Graph the data from the table and use the graph to answer the following:

Table 13.1	
Percentage of Sugar	**Time of Survival**
0	50
0.125	75
0.500	100
0.750	150
1.000	225
2.500	200
5.000	475
10.00	480
20.00	600
30.00	580
50.00	500
80.00	400
90.00	225
95.00	195
100	100

A. Exactly what information does the graph show?

B. What percentage of sugar provides for the longest life in fruit flies?

C. Theorize why the survival longevity went down as the percentage of sugar went up.

7. ***A Sticky Situation*** Why don't spiders get stuck in the own web? Investigate and explain.

8. ***Those Pesky Bugs*** You wander over to your neighbor's garden and find him sticking grasshoppers head down into a bucket of water. Seeing the strange look on your face, he replies, "These darned grasshoppers are eating up my garden, but I don't want to spray chemical insecticides so I am going to drown the little buggers." How would you respond?

9. ***A Bug-Proof Garden?*** Several days later your gardening neighbor knocks at your door. He gathered some information from the Organic Gardening Society and he shows you the information they sent him (**Table 13.2**). Unfortunately, he lost all the information other than the table itself. As you study the table he asks:

Table 13.2	
Number of Caterpillars	**Plant Type**
105	Green bean
62	Pea
47	Tomato
4	Onion

A. What information is presented in this table?

B. Why might there be fewer caterpillars on the onion?

C. What good is this information? How can I use it? Help your neighbor draw up a garden plan for planting an organic garden that makes use of the information in the table.

10. *A Call to Arms* People who squash an annoying hornet are often unpleasantly surprised to find themselves suddenly under attack by dozens of hornets. Explain this phenomenon.

Create and Connect

1. *More Bad Dreams* Strange things happen in the laboratory. In fact, just this past week my assistant, Igor, spilled a beaker of secret formula onto a stack of zoology textbooks. This secret formula changes humans into arthropods. Unfortunately, the textbook you are now touching is from that pile, and you will change into an arthropod shortly. Science fiction writers have concocted stories about this very thing as the short piece below illustrates:

> *As Gregor Samsa awoke one morning from a troubled dream, he found himself changed in his bed to some monstrous kind of vermin (beetle). He lay on his back, which was as hard as armour-plate, and, raising his head a little, he could see the arch of his great brown belly, divided by bowed corrugations. The bed cover was slipping helplessly off the summit of the curve, and Gregor's legs, pitiably thin compared with their former size, fluttered helplessly before his eyes.*
>
> —Frank Kafka, *The Metamorphosis*

Write a short story in which you imagine what it would be like to be an arthropod.

Guidelines:

A. Specifically what type of arthropod you have changed into.

B. What you look like (diagrams and/or pictures would be appropriate).

C. How do you breathe, how and what do you eat, and how do you see, feel, and hear.

D. Set your story up in the following format:

- Appropriate Title
- Catchy Beginning. Catch and hold your reader's attention.
- Understandable Middle. Don't muddy up the middle.
- Believable Ending. Give believable (but not necessarily happy) closure.

E. The instructor may provide additional details or further instructions.

2. ***Keeping in Touch.*** Despite being encased in an exoskeleton, arthropods are remarkably in touch with the physical aspects of the environment around them. Light, sound, temperature and humidity changes, and possibly even the magnetic field of the earth itself can be sensed by arthropods. Design an experiment to test the problem question: *Do mealworm larvae respond to magnetic fields?*

Guidelines:

A. Your design should include the following components in order:

> - The *Problem Question.* State exactly what problem you will be attempting to solve.
> - Your *Hypothesis.* Although this is a fictitious experiment, word your hypothesis as realistically as possible.
> - *Methods and Materials.* Explain exactly what you will do in your experiment including the materials necessary to accomplish the task. Be specific, take nothing for granted, and do not expect people to read your mind as they read your work.
> - *Collecting and Analyzing Data.* Explain what type(s) of data will be collected and what statistical tests might be performed on that data. It is not necessary to concoct either fictitious data or imaginary observations.

B. Assume you have access to everything necessary to conduct your experiment. Within reason, money is no object.

C. Your instructor may provide additional details or further instructions.

3. ***What's in a Quote?*** In his book, *The Diversity of Life*, Edward O. Wilson states: "Humans dwell among the six-legged masses with a tenuous grip on the planet. Insects can thrive without us, but we would perish without them." Contemplate that possibility. Agree or disagree? Write a short essay in which you defend or refute Wilson's quote.

4. ***Sing the Praises*** Compose a poem or song entitled "Almighty Arthropod Armada" (or another suitable title of your choice) in which you touch on the great diversity of arthropods, their tremendous environmental success, and their importance to the rest of nature.

PHYLUM ECHINODERMATA: BIZARRE BENTHIC BEINGS

Echinoderms are a noble group especially designed to puzzle the zoologist.

—Libbie H. Hyman

Introduction

In the first 3 billion years of Earth's history animal evolution forged prokaryotes, unicellular eukaryotes, the parazoan Porifera, the radiate Cnidaria, and small, nondescript bilaterally symmetrical triploblastic aggregations of cells. Then about 535 to 530 million years ago in what has been dubbed the "Cambrian explosion," these bilateral ancestors evolved at an astonishing rate in a relatively short geologic time frame into all of the major phyla of macroscopic invertebrates; the greatest evolutionary lurch forward of animal life the planet has ever experienced. In fact, fossil evidence suggests that more phyla existed in the Paleozoic era that followed this explosion of life than exist now.

History of Echinoderms

As the grand parade of life unfolded on this planet, evolution resulted in two main lines of development: the **protostomes** that we investigated in Chapters 9 thru 13 and the **deuterostomes**, which you are about to meet in the remaining chapters of this book. The divergence between the two is based on events that occur in the earliest stages of development. Protostome (Gr. "mouth first") embryos develop via spiral cleavage of the first cells (**Figure 14.1**) and the first opening that appears (**blastopore**) eventually becomes the mouth of the animal with the anal opening developing later. Deuterostomes evolved differently. Deuterostome (Gr. "mouth second") embryos develop via radial cleavage, and the blastopore becomes the anus with the mouth developing later. This branch of the Tree of Life would eventually lead to the chordates and eventually to the vertebate chordates known as humans. If you could travel back 450-500 million years ago and dive beneath the surface of Paleozoic seas, you would encounter the first grand group of deuterostomes to develop—the echinoderms.

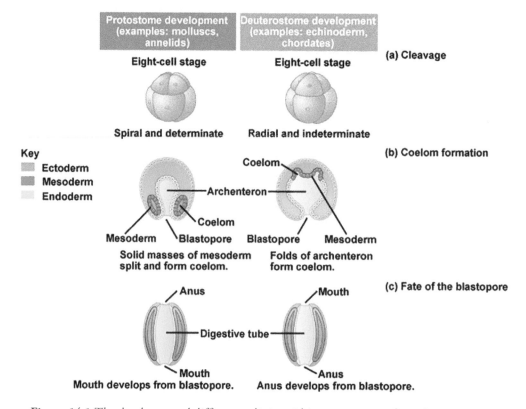

Figure 14.1 The developmental differences distinguishing protostomes from deuterostomes.

Based on an extensive fossil record, the first echinoderms appeared around 600 million years ago. Thought to have evolved from a bilaterally symmetrical ancestor, echinoderms were probably the first group of deuterostomes to appear (**Figure 14.2**).

Although only 7,000 species of echinoderms arrayed in five classes (clades) exist today, at one time there were possibly as many as 13,000 species in 20 different classes with as many as 25 anatomically distinct body forms.

Of all the modern living echinoderms, crinoids most closely resemble the oldest fossils. As crinoids do today, early echinoderms probably assumed a mouth-up position and used their water-vascular system primarily for suspension feeding rather than locomotion. The more mobile life style and mouth-down orientation of many modern echinoderms is probably secondarily derived.

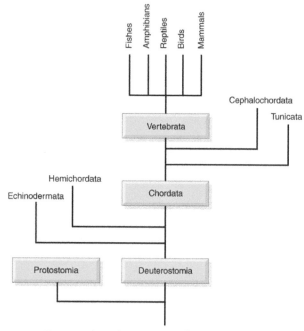

Figure 14.2 The position of Deuterostomia on the phylogenetic tree of life.

Although phylum Echinodermata is considered to be monophyletic, the definitive evolutionary relationships between living and extinct echinoderms are still a matter of research and debate. Most investigators agree that echinoids and holothuroids are closely related and form a clade, but opinions vary as to the relationship between ophiuroids and asteroids (**Figure 14.3**).

Diversity and Classification

Numbering around 7,000 extant species, the modern echinoderms are not a particularly diverse group. That does not mean, however, that living echinoderms are not major players on the stages of their various habitats. Members of at least three classes of echinoderms have flourished and often make up a major component of the benthic biota of marine ecosystems. In

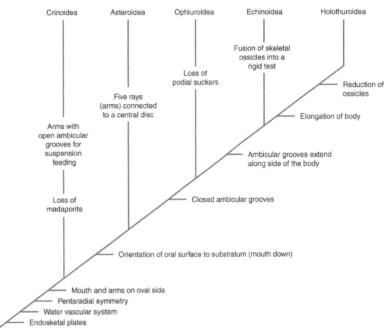

Figure 14.3 The Possible Phyolgeny of Echinoderms. Although the phylum is considered to be monophyletic, exact evolutionary relationships within the phylum are yet to be determined.

fact, they sometimes account for 90% of the benthic organisms in certain marine locales.

Limited exclusively to marine environments, echinoderms are globally distributed in almost all depths, latitudes, and environments in the ocean. They reach their highest diversity in reef environments but are also widespread on shallow shores, around the poles, and throughout the deep ocean. Lacking the ability to osmoregulate, they rarely venture into brackish water let alone freshwater.

Whereas almost all echinoderms are benthic, some sea-lilies can swim at great velocity for brief periods of time, and a few deep-sea sea cucumbers are fully floating. The larvae of many echinoderms, especially sea stars and sea urchins are pelagic, and with the aid of ocean currents can drift great distances, reinforcing the global distribution of the phylum.

The extant members of phylum Echinodermata (Gr. *echinos*, spiny + *derma*, skin + *ata*, to bear) may be classified as follows:

> Domain Eukarya
>> Kingdom Animalia
>>> Phylum Echinodermata
>>>> Class Asteroidea—Sea stars
>>>> Class Ophiuroidea—Brittle stars
>>>> Class Echinoidea—Sea urchins and sand dollars
>>>> Class Holothuroidea—Sea cucumbers
>>>> Class Crinoidea—Sea lilies and feather stars

Phylum Characteristics

Echinoderms range from tiny sea cucumbers and brittle stars less than 1 cm (2-3 inches) in size to sea stars that exceed 1 m (3-4 feet) in diameter and sea cucumbers that reach 2 m (6-7 feet) in length. Possessing a set of characteristics so distinctive that they border on the biologically bizarre, echinoderms seem like aliens on their own planet. Their unusual characteristics include:

- A calcareous endoskeleton in the form of **ossicles** or separate plates that arise from mesodermal tissue.
- **Pentaradial** adults derived from bilaterally symmetrical larvae. The body parts are arranged in fives (penta) or a multiple of five, around an oral-aboral axis.
- A water vascular system composed of water-filled canals used in locomotion, attachment, and/or feeding.
- A complete digestive tract that may be secondarily reduced.
- No excretory system or organs of any kind
- A much reduced blood-vascular (**hemal**) system derived from coelomic sinuses. Fluid is moved through the system by cilia and in some cases by muscular "pumps."
- No head and no brain. The nervous system consists of a nerve net, nerve ring, and radial nerves.
- Formidable powers of regeneration of lost parts.

This defining set of unique echinoderm characteristics seems to limit their evolutionary potential. Apparently reaching their zenith hundreds of millions of years ago, the echinoderms of today seem to constitute a phylum frozen in evolutionary time.

Class Characteristics

Echinoderms left an extensive fossil record and evolved about 25 anatomically distinct body forms which account for 20 currently recognized classes. Most of these became extinct by the end of the Paleozoic, however, and only five survive today.

Class Asteroidea (Gr. *aster*, star + *eidos*, form)—Sea stars.

Figure 14.4 Asteroids are ubiquitous in all seas at all depths. Many are drab in color whereas others are adorned in brilliant red, blue, yellows, and purple.

Known to child and zoologist alike, sea stars are the most familiar of the echinoderms. Sea star habitat ranges from the rocks, sand and mud of shallow of coastal waters to coral reefs and the deep sea floor. All sea stars are benthic creatures and no types are found living in the water column.

Many sea stars are adorned with bright colors of red, orange, blue, yellow, and purple (**Figure 14.4**). Ranging in size from a centimeter (0.4 inch) to about a meter (39 inches) across, sea stars clearly exhibit the pentaradial symmetry typical to their phylum. Sea stars are composed of a central disk from which five (penta) tapering arms (rays) typically radiate. Some species, however, have more or fewer than five, and there can even be variation in the number of arms within a species.

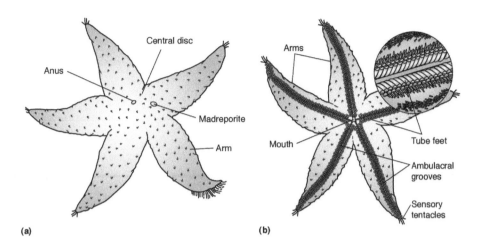

Figure 14.5 Morphology of a Typical Asteroid—the Sea Star. (a) Aboral view. (b) Oral view.

The epidermis of sea stars is ciliated and pigmented. The mouth is centered between the arms on the oral (down) surface of the animal. **Ambulacral grooves** on the underside run the length of each arm (**Figure 14.5**). The ambulacral groove houses paired rows of tube feet that protrude through the body wall on either side of the ambulacral groove (**Figure 14.6**). Movable and fixed spines project from the skeleton and give the aboral (up) surface a rough texture. Thin folds of the body wall, called **dermal gills** or **papulae**, extend between the ossicles forming the skeleton

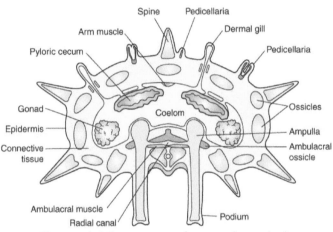

Figure 14.6 Cross-section of a sea star's arm (ray).

and function in gas exchange. In some sea stars, the aboral surface has numerous pincher-like structures called **pedicellariae**, which clean the body surface of debris as well as serving protective functions.

Class Ophiuroidea (Gr. *ophis*, snake + *oura*, tail + *eidos*, form)— Brittle stars (or serpent stars) and basket stars.

With over several thousand species, this is the most diverse group of echinoderms and likely the most abundant as well. Brittle stars exist in every benthic marine setting, often covering the sea bottom. In the velvet black of remote sea bottoms, free-roaming brittle stars roam thrive. In shallow water, they situate themselves into cracks and small holes during the day and emerge only under the cover of darkness.

Although ophiuroids possess five arms as do most sea stars, their arms are long, snake-like and distinctly set off from the central disk, giving the central disk a pentagonal shape (**Figure 14.7**). The skeletal arrangement of the arms allows for extensive lateral movement on a plane perpendicular to the oral-aboral body axis, but the arms have almost no flexibility parallel to this axis. This arrangement, coupled with the fragile nature of these animals, causes them to break easily when lifted by an appendage—hence the common name "brittle stars."

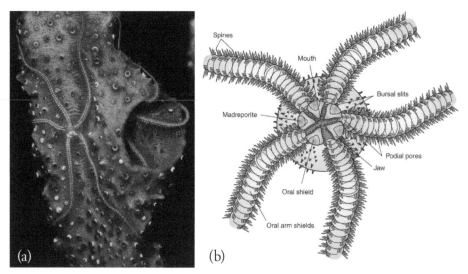

Figure 14.7 Ophiuroids. (a) With arms that writhe like snakes (serpents) and are easily detached (brittle), these brittle stars (*Ophiothrix suensonii*) are one of 1500 species of Ophiuroids living today. (b) Oral view of the morphology of a brittle star (*Ophiothrix*).

Class Echinoidea (Gr. *echinos*, sea urchin, hedgehog + *eidos*, form)—Sea urchins, heart urchins, and sand dollars.

Echinoids live in all seas, existing in both shallow and deep waters. Sea urchins have a preference for underwater floors of rock, while sand dollars and heart urchins bury themselves in more sandy settings.

Although the echinoids are radiate animals possessing five ambulacral areas, they lack the distinct arms of sea stars and brittle stars. In echinoids, the dermal ossicles have become modified into closely fitting plates forming a shell or **test**. The test of sea urchins is ball-like and slightly flattened (**Figure 14.8**), whereas the test of the sand dollar and heart urchin is highly flattened (**Figure 14.9**). An echinoid test consists of a skeleton with ten compact double rows of firmly sutured plates with movable spines. The spines of sea urchins are medium to very long whereas those of the sand dollar and heart urchins are very short. The sea urchin's elongated tube feet reach

Figure 14.8 Commonly known as the slate pencil urchin, red slate urchin, or red pencil urchin, *Heterocentrotus mammillatus* is found throughout Indo-Pacific waters. Hawaiian specimens tend to have bright red spines while specimens from the other parts of the Pacific may have yellowish or brown spines.

out through pores in each of the five pairs of ambulacral rows. Attached to the plates are small rounded

tubercles that use small muscles to work as ball-and-socket joints (**Figure 14.10**).

Several kinds of pedicellariae are present in sea urchins, most common of which are three jawed and mounted on long stalks. Pedicellariae help keep the body clean, and those of many species bear poison glands containing toxins that can paralyze small prey.

Figure 14.9 The Sand Dollar. (a) An Eccentric sand dollar *(Dendraster excentricus)* in shallow water off British Columbia. (b) Though round, the dried test of a sand dollar vividly displays it pentaradial origins.

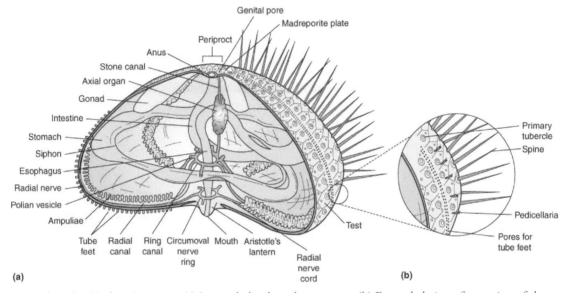

Figure 14.10 Sea Urchin Anatomy. (a) Internal glands and structures. (b) Expanded view of a portion of the test.

Class Holothuroidea (Gr. *holothourion*, sea cucumber + *eidos*, form)—Sea cucumbers.

In a phylum of unusual even bizarre creatures, the sea cucumbers may be the strangest of all. Sea cucumbers are found at all depths in all oceans, where they crawl sluggishly over hard substrates or burrow through soft sediments. In shallow waters they can form dense populations. The strawberry sea cucumber *(Squamocnus brevidentis)* of New Zealand, lives on rocky walls around the southern coast of the South Island where their numbers sometimes reach such densities (1,000 per square meter) that the whole area is simply called the "strawberry fields."

Figure 14.11 A large sea cucumber *(Bohadschia argus)* with feeding tentacles extended.

Sea cucumbers have no arms, are elongate, and lie on one side, which is usually flattened as a permanent ventral side. Large, highly modified tube feet or tentacles surround the mouth. Most adults range in length from 10-30 cm (4-12 inches), and they come in colors from black and dark brown to sky blue and orange (**Figure 14.11**).

Looking remarkably similar to their vegetable counterpart, most species of sea cucumbers have a rough, textured body wall covered with small ossicles. In rarer cases, the ossicles are large enough to act as dermal armor. The body of deep water holothurians is made of a tough gelatinous tissue with unique properties that allow the animals to control their own buoyancy, making it possible for them to both live on the ocean floor or to float over it moving to new locations with a minimum of energy. The ventral side of a sea cucumber resting on the substrate contains three rows of tube feet, which are primarily used for attachment. There may be two reduced rows of tube feet on the upper surface or absent altogether. Some types, however, have no tube feet at all anywhere, and in some groups the tube feet are scattered all over the body.

Box 14.1
Spill Your Guts to Save Your Life

Self-defense in the animal realm takes on many forms, some familiar, but some so strange as to be almost unimaginable. The North American opossum (Didelphimorphia) will snarl and hiss at an enemy, but if that fails, it will go into an involuntary trance-like state in which it remains perfectly motionless ("playing possum"). In this state, it will defecate on itself, and release a smelly green slime from its anus. The horned lizard will make rapid darting movements when threatened, swell up its spiny body to make it more difficult to swallow, and as a last resort, squirt blood up to three feet from the corners of its eyes. The Northern Fulmar (*Fulmarus glacialis*) is a sea bird that protects its nest and itself by projectile vomiting globs of acidic goo at approaching enemies, whereas the larva of the Colorado potato beetle (*Leptinotarsa decemlineata*) covers itself with its own foul-smelling feces that are foul-smelling that are most likely toxic to would-be predators. As strange-sounding as these few examples of self-defense might be, they pale in comparison to that of the sea cucumber.

To the casual observer, the sluggish sausage-like sea cucumber might appear to be defenseless. As is so often the case when it comes to animals, however, appearances can be deceiving. Many sea cucumbers produce toxins in their body wall that discourage predators. Others can evert tubules (**Cuvierian tubules**) of the respiratory tree through the anus. These tubules contain sticky secretions and toxins capable of entangling and immobilizing predators. In what is perhaps the most bizarre of any defense strategy in the whole of the animal kingdom, however, some sea cucumbers, when disturbed, forcefully expel their guts out their anus or through a rupture in the body wall in a process known as **autoevisceration**. While the would-be predator is busy with a tasty-looking pile of innards, the sea cucumber slowly creeps away on tube feet. Regeneration of lost parts soon follows, and the cucumber lives on.

For reasons unknown, sea cucumbers may also autoeviscerate during time of severe environmental stress, and there is some evidence to indicate that the process may also be used to eliminate parasites. Knowing this, and if given the choice, most people would probably elect to go with the tried-and-true human self-defense strategies of firearms and martial arts or just plain running away.

Class Crinoidea (Gr. *krinton*, lily + *eidos*, form)—Sea lilies and feather stars.

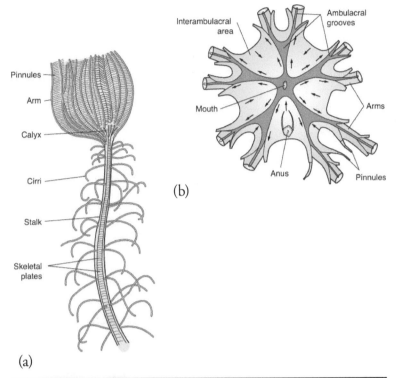

Crinoids are considered to be the most primitive of all living echinoderms, and as the fossil record reveals, they were far more numerous in ancient times than they are now. Though most crinoids prefer deep waters, feather stars are much more likely to populate shallow regions of the Indo-Pacific and West Indian Caribbean.

The flower-shaped body of most sea lilies is attached to the end of their permanent stalk. The attached end of the stalk bears a flattened disk or root-like extensions that fix the animal to the substrate. Disc-like ossicles stacked on top one another and joined by connective tissues give the stalk a jointed appearance (**Figure 14.12**).

Modern crinoid stalks rarely exceed 60 cm (24 inches) in length but some extinct types had stalks as much as 20 m (66 feet) long. The stalk usually bears projections or **cirri**, arranged in whorls. Sitting atop the stalk is a whorl of five flexible arms branching into many smaller arms. The arms are attached to the stalk by the **calyx** and bear smaller branches or **pinnules** that give them a feathery appearance. The calyx and arms together constitute the **crown**.

Feather stars are similar to sea lilies, except they lack a stalk and are swimming or crawling animals. The aboral end (opposite the mouth) of the crown bears a ring of rootlike cirri, which cling when the animal

Figure 14.12 Crinoids. (a) Sea lily with a portion of the stalk. (b) Oral view of the calyx showing the direction of ciliary food currents. Food particles touching the podia are tangled in mucus and carried toward the mouth. Debris falling on the interambuclacral areas is swept off by ciliary action. (c) A red sea lily *(Proisocrinus ruberrimus)*. Stalked crinoids are known as sea lilies whereas crinoids without stalks are referred to as feather stars.

is resting on a substrate. Feathers swim by raising and lowering their arms, and they crawl over the substrate by pulling with the tips of their arms giving them the appearance of walking on their arms.

Echinoderm Body Plan

Despite the adaptive value of bilaterality for free-moving animals and the merits of radiality for sessile animals, echinoderms are a puzzling exception of animals that are both radial and free-moving.

Body Wall and Skeleton

An epidermis covers the body of all echinoderms and overlies a dermis, which contains the interlocking skeletal elements known as ossicles. The epidermis itself consists of cells responsible for the support and maintenance of the skeleton, pigment cells, mechanoreceptor cells, which detect motion on the animal's surface, and sometimes gland cells, which secrete sticky fluids or even toxins.

Beneath the dermis and ossicles are muscle fibers or layers and the coelom. The degree of development of the skeleton and muscles varies greatly among echinoderm groups. Sea stars possess an endoskeleton of calcareous plates bound together with connective tissue, whereas in urchins and sand dollars the ossicles are fused to form a rigid test. Comparatively, sea cucumbers have only a rudimentary endoskeleton consisting of separate ossicles scattered in the leathery dermis

The endoskeleton is calcareous, mostly $CaCO_3$ in the form of calcite. The skeleton of all echinoderms begins as numerous, separate, spicule-like elements, each behaving as a single calcite crystal. From these calcite crystals an open meshwork structure (**stereom**) develops with living tissue filling the interstices. Developing from the skeleton are bumps and knobs called **tubercles**, movable and fixed spines of various lengths, and in asteroids and echinoids, unique pincher-like pedicellariae (**Figure 14.13**).

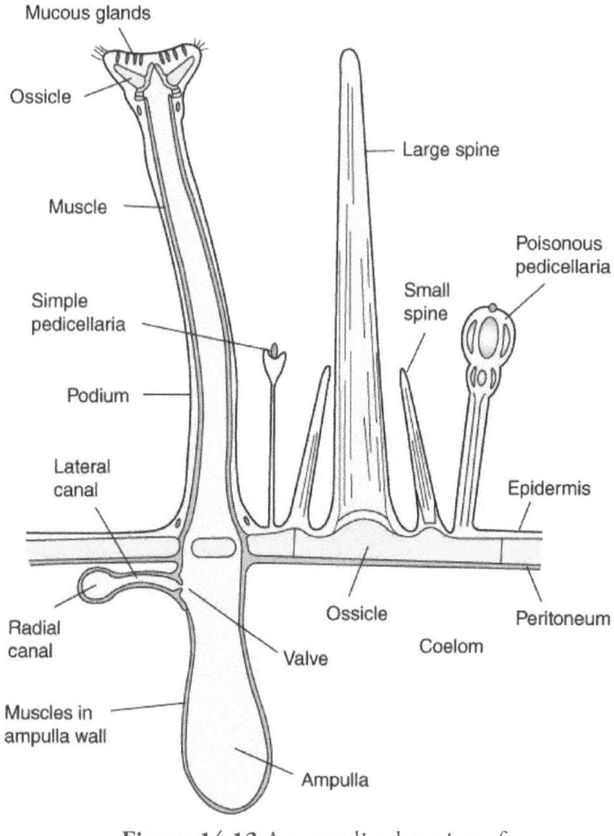

Pedicellariae differ not only in structural details, but in size and distribution on the body. Some are elevated on stalks, whereas others lie nestled directly on the body surface either singly or in clusters. Some help keep debris and settling larvae off the body, and others are used to defend against larger organisms. Some sea urchins carry toxin-producing pedicellariae with which they discourage would-be predators and in other urchins the pinchers grasp and hold objects as camouflage and protection.

Figure 14.13 A generalized section of the body wall of a sea urchin.

Water Vascular System

In no other group of animals is there a system analogous to the water vascular system of echinoderms. This hydraulic system is used by echinoderms for locomotion, food and waste transport, and respiration. The system is composed of canals connecting numerous tube feet. Echinoderms move by alternately contracting muscles that force water into the tube feet, causing them to extend and push against the substrate or an object then relax to allow the feet to retract. This whole process allows for extremely slow though powerful movement (**Figure 14.14**).

In sea stars, water enters the system through the **madreporite** or **sieve plate**, flows through the **stone canal**, and enters the circular **ring** or **circumoral canal**. The water then moves into five separate **radial canals**. The radial canals give rise to numerous **lateral canals** that terminate in a bulblike structure called an **ampulla**. The ring canal also gives rise to blind pouches called **Polian vesicles** and **Tiedemann's body**. There is some uncertainty about the function of these pouches, but it is thought that the former help regulate internal pressure within the water vascular

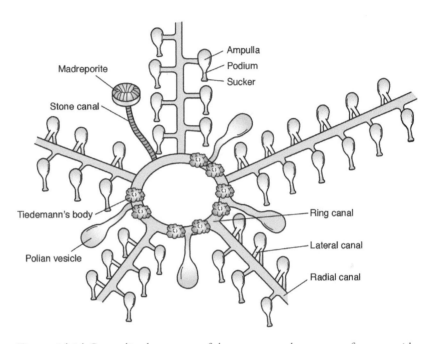

Figure 14.14 Generalized structure of the water vascular system of an asteroid.

system. The bulbous ampullae are connected to hollow, muscular suckered **podia**. Together the ampullae and podia compose a single **tube foot**.

The ability to move yet remain firmly adhered to the substrate is a tremendously useful adaptation to an animal that must sometimes withstand the surging and pounding of waves. On soft surfaces, however, such as mud or sand, suckering is not effective, and consequently the tube feet may be modified and employed to produce a stepping action. The Paxillosida, a large group of asteroids common in shallow water, have no suckered tube feet at all.

Some variations on the sea star theme exist. In ophiuroids, the ring canal gives off the usual five radial canals but also branches to a wreath of **buccal tube feet** around the mouth, the madreporite is located on the oral surface, and the podia lack suckers.

The water vascular system of the holothurians is organized to accommodate the elongation of the body. In most holothurians the madreporite is internal and opens to the coelom, the ambulacral grooves are closed and the podia of the upper surface are reduced or lost.

Crinoids have a water vascular system that operates entirely on coelomic fluid. There is no external madreporite, rather a number of stone canals arise from the ring canal and open to coelomic chambers. The number of arms in crinoids ranges from five to as many as 200, and in many cases the arms are branched.

The number of radial canals corresponds to the arm number in each species, and the radial canals branch into the pinnules of each arm. Suckerless podia occur along the pinnules, often in clusters of three. The podia are highly mobile and usually bear adhesive papillae on their surfaces; they function primarily as feeding and sensory organs.

Support and Locomotion

Echinoderms, other than the holothurians, rely on the calcareous endoskeleton for structural support and to maintain the general body shape. In the holothurians, where the skeletal plates are tiny separate ossicles, the body wall muscles form thick sheets that work on the coelomic spaces to provide a hydrostatic skeleton.

Many echinoderms possess mutable connective tissue or what is called **catch connective tissue**, a dense, white fibrous material especially prevalent in the body wall of the sea cucumber. The amazing thing about catch connective tissue isn't that it has both viscous (fluid) and elastic (solid) properties like mammalian tendons and ligaments; it is that this tissue's viscoelasticity can change dramatically. This material can go from hard and stiff to soft and flexible and back again with ease. For example, if you grasp the skin of a sea cucumber between your thumb and forefinger, at first it feels soft and pliable. Within several minutes, however, it hardens, retaining the indentations of your fingers for some time. This transformation appears to be under direct nervous control but does not involve muscular activity. The exact cellular and chemical processes involved are the subject of intense research. (See *A Closing Note* at the end of this chapter)

The asteroids exemplify locomotion using podia. The action of a single podium involves power and recovery strokes. The sea star's arms are held more-or-less stationary relative to the central disc, and movement is accomplished by the thousands of podia on the oral (bottom) surface (**Figure 14.15**). Overall, movement is generally smooth as a result of the high number of podia and the fact that at any given moment the podia are in different phases of the power and recovery strokes. Although there is some coordination of the action of the tube feet to produce movement in a particular direction, there are no waves of podial motion as seen in multi-legged animals such as millipedes. Most sea stars move slowly, but a few are relative speedsters. Some asteroids that are relatively sedentary suddenly become startlingly rapid "runners" upon encountering a potential predator (often another sea star). Sea urchins move by the use of long suckered podia and movable spines whereas sand dollars burrow and crawl largely by the action of short movable spines.

Holothurians live on the surface of various substrata or burrow into soft sediments. Crawling and burrowing in this group are accomplished by the podia or action of the body wall muscles. A few sea cucumbers are pelagic and capable of weak swimming.

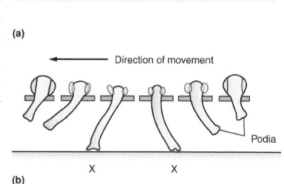

Figure 14.15 Locomotion by Tube Feet. (a) Side view of a sea star arm with tube feet in motion. (b) Changes in position of an individual podium as the animal moves in the direction of the arrow. The podium executes its power stroke while in contact with the substratum (X) and its recovery stroke while lifted from the substratum.

Feeding and Digestion

Each group of echinoderms displays unique feeding strategies. In addition, the structure of the digestive tract also differs among groups.

Most asteroids are opportunistic predators or scavengers. They feed on nearly any dead animal matter and prey on a variety of invertebrates such as snails, bivalve molluscs, crustaceans, worms, and corals. Some sea stars feed on small particles, either directly or in addition to carnivorous feeding. Plankton and other organic particles coming in contact with a sea star's surface are carried by epidermal cilia to the ambulacral grooves and then to the mouth. Sea stars will even eat each other and size doesn't seem to matter. If the attacker is smaller than the prey, the smaller attacking sea star will begin eating at the end of one of the arms of the larger prey sea star and works its way from there.

Some sea stars rely on bivalve mollusks for food and, as such, are commercially important predators on clam and oyster beds. The struggle between bivalve and sea star is one of nature's epic feeding battles albeit in slow motion. The battle of muscle power (bivalve) vs hydraulic power (sea star) begins with the sea star fully engulfing its prey while using its tube feet to attach to the shell and begin to pull it open. The constant tug of the tube feet can reach the amazing force of 12.75 newtons (1,300 pounds), and this force can be steadily exerted and maintained over a considerable period of time. Usually in less than one hour the adductor muscles (the muscles that hold shell closed) of the bivalve weaken and begin to relax. As they do so, a gap develops between the valves (shells). The sea star then everts its stomach, slipping it down into the bivalve, and begins out-of-body digestion. When digestion and absorption are complete, the stomach is withdrawn back up into the sea star. Not all sea stars evert their stomach and predigest food outside their body. Some types swallow their prey whole.

From the mouth of a sea star a very short esophagus leads to the **cardiac stomach**, which is the portion everted during feeding. Radially arranged retractor muscles serve to pull the stomach back into the body. The cardiac stomach leads to the **pyloric stomach**, from which arises a pair of **pyloric ducts** extending into each arm. These ducts lead to a paired digestive glands or **pyloric ceca** in each arm (**Figure 14.16**). A short intestine, which leads from the pyloric stomach to the anus, often bears protrusions called **rectal glands** or **rectal sacs**. The pyloric ceca and cardiac stomach are the main sites of digestive enzyme production. Chemical digestion is completed internally after the ingestion of the liquefied food.

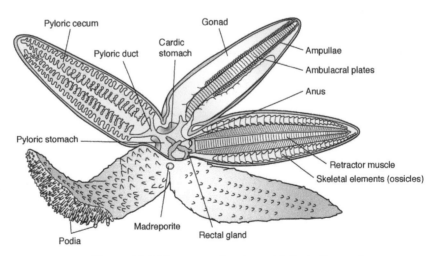

Figure 14.16 Anatomy of an Asteroid. Each dissected arm is depicted with various organs removed.

The diet of ophiuroids consists of many different types of small particles. They eat either by scavenging dead animal matter at the bottom or by suspension feeding. Some brittle stars form mucus strands between arm spines and then extend their arms into the water catching suspended particles on these mucal "nets." Trapped plankton is passed from tube foot to tube foot along the length of an arm until it reaches the mouth. Not all ophiuroids are vegetarians, and at least one type ambushes small fish. By using its arms to hold its central disk off the substrate, this fish-trapping brittle star forms an enclosure. As a prey fish swims through this enclosure the star quickly moves and twists, catching its meal with its spiny arms. The central disc then slowly settles down over the fish, and feeding begins.

Predatory suspension feeding by basket stars is mainly a nocturnal affair. At dusk, the animals emerge from their daytime hiding places and assume a feeding posture. Holding their branched arms like fans into the prevailing current, the basket stars await contact with a crustacean or worm. When contact is made with suitable prey, the arm curls to capture it and then transfer it to the mouth. The digestive tract of ophiuroids is greatly reduced in that the intestine and anus have been lost, and the remainder of the system confined entirely to the central disc.

Depending on the species, echinoids employ a number of food-getting strategies ranging from herbivory to suspension feeding and detritivory to predation. In sea urchins, feeding depends in large part on the action of a complex masticatory (chewing) apparatus that lies just inside the mouth known as **Aristotle's lantern (Figure 14.17)**. This morphological marvel consists of five protractible calcareous teeth and muscles that control protraction, retraction, and grasping movements of the five teeth. In many species, the entire apparatus can be rocked in such manner that the teeth protrude at different angles.

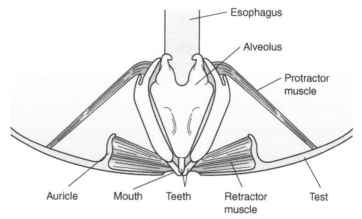

Figure 14.17 Aristotle's lantern is a complex mechanism used by sea urchins to masticate their food. Five pairs of retractor muscles draw the lantern and teeth up into the test; five pairs of protractors pull the lantern down and expose the teeth.

Most sea urchins use their teeth to scrape algae from rocks and other surfaces and to tear chunks of animal matter into smaller pieces. Heart urchins and most sand dollars burrow into soft sediments and feed on organic particles. A few species of sand dollars burrow in but leave the posterior portion of the body extended at an angle above the sediment. Diatoms and other particulate food in the water is trapped by the podia on the exposed portion of the body and then passed to the mouth.

Most holothurians are suspension or deposit feeders. Sea cucumbers extend their branched, mucus-covered tentacles into the water to trap suspended particles, including live plankton. The tentacles are then pushed into the mouth one at a time, and the food is ingested. Some sea cucumbers burrow through the soft substratum by peristaltic movements

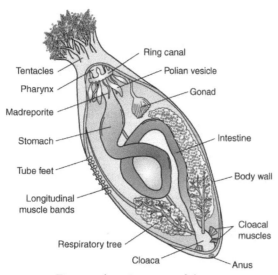

Figure 14.18 Anatomy of the sea cucumber *Sclerodactyla*.

and ingest the sediment as they move. Their worm-like movements and feeding behavior has earned the sea cucumbers the nickname of "earthworm of the sea."

The mouth of a sea cucumber is located at the anterior end of the body and is surrounded by a whorl of buccal tentacles (**Figure 14.18**). The pharynx joins an elongate intestine, the anterior end of which is enlarged into a stomach. The intestine loops forward and then extends posteriorly again to the expanded cloaca, which terminates in the anus.

Crinoids are suspension feeders that capture floating food particles from the current with the aid of mucous nets and tube feet located in open ciliated ambulacral grooves on each arm. The cilia in each groove move the captured food to the mouth for ingestion. The mouth, located at the base of the arms, leads to a U-shaped gut with the anus located near the mouth (**Figure 14.19**).

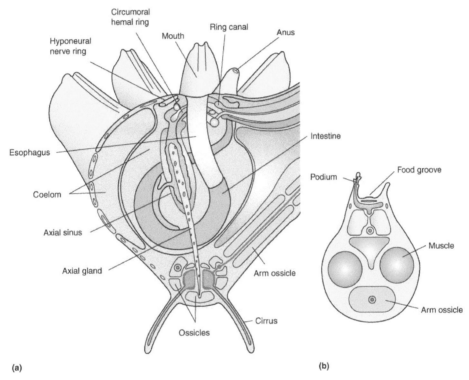

Figure 14.19 The Anatomy of Crinoids. (a) Central disc and base of one arm (vertical section). (b) An arm with open ambulacral (food) grove (cross section).

Circulation and Gas Exchange

Circulation in echinoderms is accomplished largely by the main coeloms, supplemented to various degrees by the water vascular system and the hemal system. The hemal system is complex arrangement of canals and spaces, mostly enclosed within coelomic channels called **perihemal sinuses**. The hemal system is most highly developed in holothurians (**Figure 14.20**). In this system, an elaborate set of longitudinal and connecting vessels are intimately associated with the digestive tract and the respiratory trees, and in many species includes several "hearts" or circulatory pumps. Most echinoderms rely on the fluid transport mechanisms described above to move dissolved gases between internal tissues and the body surface where thin-walled external processes serve as gas exchange surfaces.

Gas exchange in asteroids occurs across the podia and special evaginations of the epidermis and peritoneum called **papulae** or **dermal gills**. Both tissues are ciliated, and currents are produced both in the coelomic fluid by the peritoneum and in the overlying water by the epidermis. The two currents, however, move in opposite directions, thus creating a countercurrent that maintains maximum exchange gradients across the papulae surfaces.

Ophiuroids have ten invaginations of the body wall called **bursae**, which open to the outside through ciliated slits. Water is circulated though the bursae by the cilia and in some species, by muscular pumping of the internal bursal sacs. Gases are exchanged between the flowing water and body fluids. The main gas exchange structures in sea urchins are apparently thin-walled podia that operate on a countercurrent system similar to that of asteroids.

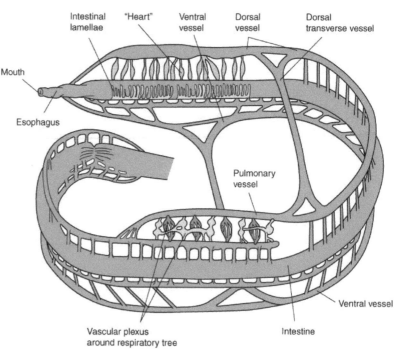

Figure 14.20 The complex hemal system of the holothurians, *Isostichopus badionotus*, and its association with the gut and respiratory tree.

Holothurians exchanges gases with the water through a pair of respiratory trees branching off the cloaca just inside the anus, so that they "breathe" by drawing water into the anus and expelling it. This apparatus is augmented by exchange across podia, which is facilitated by a countercurrent system. Several orders of holothurians, however, are defined by the absence of respiratory trees.

The pearlfish (Carapidae) have evolved a symbiotic relationship with sea cucumbers in which the pearlfish live in the sea cucumber's cloaca using it for protections, a source of food (eating the respiratory tree and gonads), and as a site for development of its eggs into adults. The pearlfish will leave the cucumber at night to feed outside as well. The pearlfish finds its way back to its host by following a chemical scent from the cucumber where it enters the anus tail first. Some worms and crabs have also specialized to use the cloacal respiratory trees for protection by living inside the sea cucumber.

Excretion and Osmoregulation

Lacking organs of excretion and osmoregulation, echinoderms are strictly confined to marine environs. In some echinoderms dissolved nitrogenous wastes (ammonia) diffuse across body surfaces to the outside. This type of excretion occurs across the podia and papulae in asteroids, and is thought to occur across the respiratory trees in holothurians. At least some excretion merely by diffusion probably takes place in most echinoderms.

Nervous System and Sense Organs

Like other radially arranged animals, the echinoderms have a noncentralized nervous system, a feature that allows them to engage the environment equally from all sides. A cerebral ganglion (brain) is lacking.

There are three main neural networks integrated with one another in the disc and arms. Chief among these is the **ectoneural (oral)** system consisting of a somewhat pentagonal **circumoral nerve ring** around the mouth (Figure 14.10) and large **radial nerves** that extend into each arm, forming a large V-shaped mass along the midline of the oral surface of the ambulacral groove. The ectoneural system is predominantly sensory in function and appears to coordinate the tube feet. The **hyponeural (deep)** system parallels the nerves of the ectoneural system and provides motor function, whereas the **entoneural (aboral)** system, absent in holothurians, consists of a nerve ring around the anus and radial nerves leading to the roof of each ray. These three systems are connected by a **nerve net** to each other and to the body wall and other structures.

Sensory receptors consist mainly of relatively simple structures derived from the epidermis. Although echinoderms have few well-developed sense organs, they are at least somewhat sensitive to touch and to changes in light intensity, temperature, and body orientation to water currents. The tube feet, spines, pedicellariae, and epidermis respond to touch, and light-sensitive organs have been found in asteroids, echinoids, and holothurians. Statocysts are known in some holothurians as well.

In spite of the rather simple nervous system with no centralized brain and the lack of specialized sense organs, many echinoderms engage in complex behaviors. Most obvious are behaviors such as coordination of the podia during locomotion and feeding, distinct righting behaviors when overturned, orientation to water currents, and reactions to light intensity, direction of light, and photoperiod. There is even evidence to support the contention that some degree of learning occurs in echinoderms; all feats that would seem nearly impossible for an animal with no organized brain.

Reproduction and Regeneration

Most echinoderms are capable of regenerating lost parts. The most striking example is surely the regeneration of internal organs in holothurians after autoevisceration as described earlier. Although certainly not as dramatic, sea stars, brittle stars, and sea lilies will easily regenerate new arms if existing ones are broken or torn off as long as a portion (about one-fifth) of the central disc remains associated with the detached arm. Sea urchins readily regenerate lost spines, pedicellariae, and small areas of the test and ophiuroids and crinoids frequently cast off arms or arm fragments when disturbed and then later regenerate the lost part.

Asexual reproduction occurs in some asteroids and ophiuroids by a process called **fissiparity**, wherein the central disc divides roughly in two with each half forming a complete animal by regeneration. Successful fission and regeneration require a body wall that can be torn and an ability to seal resultant wounds. In some asteroids fission occurs when two groups of arms pull in opposite directions, thereby tearing the animal into two pieces. In some sea cucumbers which divide transversely, considerable reorganization of tissues occurs in both regenerating parts.

Most echinoderms are dioecious, but hermaphroditic species are known among the asteroids, holothurians, and especially the ophiuroids. Asteroids and echinoids possess multiple gonads (Figure 14.10); holothurians have but a single gonad (Figure 14.18) and crinoids lack distinct gonads altogether. Echinoderms become sexually mature after 2 or three years depending on species and environmental conditions.

One life history strategy employed by echinoderms is free spawning. During spawning, eggs (potentially hundreds of millions) from females and sperm from males are shed into the water where the eggs are fertilized. Most echinoderms spawn on an annual cycle, with the spawning period normally lasting one to two months during the spring or summer; several species, however, are capable of spawning throughout the year. Spawn-inducing factors are complex and may include external influences such as temperature, photoperiod, and even salinity of the water. Many echinoderms aggregate before spawning, thus increasing the probability of fertilization of the eggs. Characteristic behaviors may be displayed during the spawning process; some asteroids and ophiuroids raise the center of the body off the sea floor; holothurians may raise the front end of the body and wave it about. These movements are presumably intended to prevent eggs and sperm from becoming entrapped in the sediment.

Another reproductive approach is brooding in which fewer, but larger eggs are produced and those eggs are protected by being held on or in the body. Brooding is especially common in boreal (cold water) and polar (very cold water) species among all groups of echinoderms and in certain deep-sea asteroids, whose environments are unfavorable for larval life.

Brooding methods vary. Most brooding asteroids hold their embryos on the body surface while crinoids cement their eggs to the epidermis from which they emerge. Many ophiuroids brood their eggs internally in the bursa sacs as do some holothurians. The majority of brooding holothurians, however, usually carry their embryos externally (**Figure 14.21**).

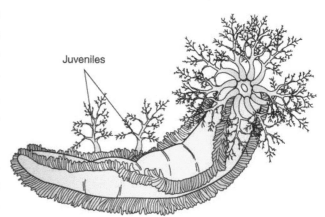

Juveniles

Figure 14.21 The sea cucumber *Cucumaria crocea* brooding its young.

The eggs produced by spawners tend to be small and develop indirectly after fertilization, whereas eggs produced by brooders tend to be large and develop directly. Brooded eggs are large and stocked with sufficient yolk to allow direct development into a juvenile echinoderm without the necessity of passing through larval stages and feeding in the water column. Spawned eggs, however, are quite small and possess little yolk. This forces a prolonged feeding period, usually as plankton, to gain sufficient energy to metamorphose through different larval stages into the adult body form. Despite their adult radial symmetry, echinoderms reveal their ancestral bisymmetry in their larval stages.

During indirect development, the fertilized egg divides many times to produce a hollow ciliated ball of cells (blastula). The blastula invaginates at one end to form a primitive gut, and the cells continue to divide to form a double-layered embryo (gastrula). Being a deuterostome, the hole through which the gut opens to the outside (blastopore) in echinoderms marks the position of the future anus; the mouth arises at the opposite end of the body from the blastopore. A pair of hollow subdivided pouches arise from the gut and develop into the coelom and water vascular system.

The gastrula develops into a basic larval type called a **dipleurula** ("little two sides") **larva**. This larva is characterized by bilateral symmetry, three coelomic sacs, and bands of cilia for locomotion and feeding.

The class specific larvae found among the groups of echinoderms are modifications of the basic dipleurula pattern (**Figure 14.22**).

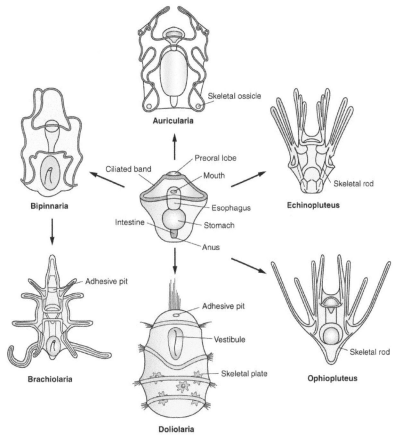

Figure 14.22 Comparison of different types of echinoderm larvae showing relationship to a hypothetical dipleurula-type ancestral larva.

After a few days to several weeks in a free-swimming planktonic form, echinoderm larvae undergo a complex metamorphosis that results in the juvenile echinoderm. During this metamorphosis, the fundamental bilateral symmetry is overshadowed by a radial symmetry dominated by formation of five water-vascular canals. Among holothurians, echinoids, and ophiuroids, the larvae metamorphose as they float, and the juveniles then sink to the seafloor. Among asteroids and crinoids, however, the larvae firmly attach to the sea floor prior to the onset of metamorphosis. The average life span of echinoderms is about four years, but some species may live as long as eight to ten years.

Echinoderm Connection

Economically

There is a constant struggle between humans and echinoderms over molluscs such as oysters, clams, and mussels that both groups consider delectable food sources. As a localized example consider that no less than

42,000 bushels of sea stars were removed from the oyster-beds of Connecticut in a single year, but not until these echinoderms had caused an estimated $631,000 in damage to the oyster crop.

Echinoderms themselves are highly regarded as food by humans in certain locales and cultures. Around 50,000 tons of sea cucumbers are harvested each year, the gonads of which are considered delicacies in Japan, Peru, and France. Often described as tasting like a mix of seafood and fruit, the quality of the gonads depends on the color, which can range from light yellow to the preferred bright orange. Sea cucumbers are also considered a delicacy in some countries of Asia and there is an increasing demand for them in these markets. Muscles of sea cucumbers are served in sushi bars, dried cucumbers are added to soup, vegetables, and meat, and the intestine is used to prepare a gourmet Japanese dish known as *konowata*.

Ecologically

Echinoderms play a key role in marine ecosystems. For example, the grazing of sea urchins reduces the rate of colonization of bare rock; the burrowing of sand dollars and sea cucumbers increase the depth to which oxygenation occurs, allowing a more complex ecological tiering to develop. Echinoderms are often vital in preventing the growth of algal mats on coral reefs, which would obstruct the filter-feeding of the coral polyps. The crucial role echinoderms play in preventing the overgrowth of algae on coral reefs is illustrated by the decline of the long-spined sea urchin, *Diadema antillarum*, in the Caribbean and along the Florida Keys. In January 1983, an epidemic of unknown cause decimated the *Diadema* population in these areas, leaving less than 5% of the original numbers. Other species of urchins were unaffected, but *Diadema* populations have not recovered. The loss of this urchin coupled with the chronic overharvesting of herbivorous fish around the island has led to algal overgrowth resulting in the destruction of a great portion of the coral reefs around Jamaica.

The crown-of-thorns sea star, *Acanthaster planci*, are notorious for eating coral polyps and they sometimes occur in aggregations or "herds" large enough to cause serious outbreaks on Pacific Ocean reefs.

Echinoderms are also the staple diet of many organisms, most notably the sea otter. Conversely, sea cucumbers provide a habitat for parasites and commensals, including crabs, worms, and snails.

As is so often the case, the human "taste" for some resource—living or nonliving—often has far-reaching environmental consequences. Thus, it is with the sea cucumber. When the demand for sea cucumbers as food resulted in the decimation of populations in the western Pacific, demand spread to the eastern Pacific. Since the 1980's sea cucumber harvesting has been a lucrative profession along the coasts of North and South America. By 1992 "pepineros" (sea cucumber harvesters) had taken between 12 and 30 million sea cucumbers from the waters around the Galapagos Islands alone. The government of Ecuador imposed a ban on cucumber fishing in a vain attempt to contain this fishing frenzy. The ban was lifted in 1993 but reestablished in 1994 when despite a three-month quota of 550,000, seven million more cucumbers were taken. The ban is nearly impossible to enforce because of the lack of man power and equipment and the expanse of ocean that needs to be patrolled.

Sea cucumbers harvesting has had devastating effects on the Galapagos Islands and threatens the existence of many of the unique species that live there. Pepineros cut large mangrove trees as fuel for drying sea cucumbers. This cutting threatens the mangrove swamps, which host hundreds of the islands' unique species. Pepineros camps also introduce nonnative species such as feral pigs, dogs, brown rats, and fire ants into the very fragile Galapagos ecosystems. Social unrest has also resulted. Conflicts between pepineros, conservationists, and the Ecuadorian government have resulted in violence and death. The islands' unique and magnificent tortoises have even been killed in protest of fishing bans. The protestors think that the government and conservationists consider the survival of the sea cucumber more important than the economic survival of some Ecuadorian people. These problems illustrate once again the complexity associated with animal conservation and the necessity of balancing multinational and multidisciplinary approaches to addressing these problems.

The calcareous tests of echinoderms are often used as a source of lime by farmers in areas where limestone is unavailable. In fact, 4,000-10,000 tons of these animals are harvested annually for this purpose.

Medically

Several kinds of toxins and secondary metabolites have been extracted from sea stars are now being subjected to research worldwide for their possible pharmacological properties and potential future use as pesticides. The strong toxins of sea cucumbers are often psychoactive, but their effects are not well studied. It does appear that some sea cucumber toxins restrain the growth rate of tumors, which has sparked interest from cancer researchers. Furthermore, some varieties of sea cucumber (known as gamat in Malaysia or trepang in Indonesia) are said to have excellent healing properties. In fact, there are pharmaceutical companies being built based on this gamat product. The effectiveness of sea cucumber extract in tissue repair has been the subject of serious study where it seems to not only help a wound heal more quickly, but also aids in reducing scarring.

The tremendous numbers of eggs produced by many echinoderms and the ease with which they can be reared in the laboratory have made these animals favorite subjects of study by embryologists. Much of our information about the biology of animal fertilization and early development comes from over a century of work focusing particularly on urchins and sea stars. In addition, the early ontogeny of some echinoderms has served as a model of deuterostome development against which many other developmental patterns are measured.

A Closing Note
Sea Cucumbers and a Quarterback's Knees

As this chapter is being written, football season has begun once again. Enthusiasm for the local high school season was seriously dampened, however, in just the second game when the quarterback of our team was lost for the rest of the season with a torn ACL. The anterior cruciate ligament is but one of several in that fragile joint we call the knee. Although ligaments and tendons—generally referred to as connective tissue—do stretch, they aren't exactly rubber bands. In fact, they have a distressing tendency to tear or break, and when they do, they are devilishly difficult to repair.

Tendon is made up mostly of collagen, a protein that spontaneously aggregates into long, thin structures known as fibrils. The fibrils interact with each other and their surroundings to form a stiff and cohesive tissue. But the process is apparently irreversible and non-renewable, and so if physical strain sunders the fibril bonds (tearing the tendon or ligament), it is impossible to reform them, at least in living tissue. The standard treatment is to tie the ruptured ends together and let scar tissue bridge the gap. Unfortunately, the scar tissue bridge is not terribly effective, and the scar tissue tends to form unwanted adhesions to surrounding tissue. As a result, the tendon or ligament never regains more than about 60 percent of its original strength.

Envision then the implications of an ointment that could cleanly break the bonds between collagen fibrils and form new ones. A surgeon could chemically undo the rest of the bonds between two disjoined fibrils in the torn ends of a tendon, add fibrils to the gap of the frayed ends, and finally stabilize the repair by reestablishing the bonds between new and old fibrils and the rest of the tissue in the matrix; no gap, no scar, no loss of strength, and a very short recovery time. In essence, this is what the catch connective tissue of sea cucumbers does on a routine basis. As mentioned earlier in the chapter, the great trick of catch connective tissue is that the tissue's viscoelasticity can change from fluid to solid and back again quite rapidly. Some cells in the dermis secrete a plasticizing protein that loosens the grip on the collagen fibrils, enabling them to slide past each other thus making the tissue soft and pliable. Other cells release a stiffening factor that causes the fibrils to "catch" and make the dermis become far stiffer. As one would expect, the exact chemistry of how this happens and how we might apply it to human injuries is the subject of intense research. Imagine how wondrous it would be if athletes and just plain folk such as myself who had to endure total rotator cuff repair in both shoulders as a consequence of arthritis or my wife who currently faces the prospect of knee surgery as the result of an accident could have their torn joints restored fully, quickly, and naturally. Potentially all this from a peculiar echinoderm that has no knees or shoulders, and one that certainly looks much more like the football than the quarterback.

In Summary

- Echinoderms, the first deuterostomes, appeared about 450 to 500 million years ago.
- Echinoderms are not a particularly diverse group. Members of three classes of echinoderms, however, have flourished and often make up a major component of the benthic biota of marine ecosystems.
- Limited strictly to marine environments, echinoderms are globally distributed in almost all depths, latitudes, and environments in the ocean.
- Echinoderms may be classified into five groups:

> Class Asteroidea
> Class Ophiuroidea
> Class Echinoidea
> Class Holothuroidea
> Class Crinoidea

- The characteristics of phylum Echinodermata:

  1. A calcareous endoskeleton in the form of ossicles or separate plates that arise from mesodermal tissue.
  2. Pentaradial adults derived from bilaterally symmetrical larvae. The body parts are arranged in fives (penta) or a multiple of five, around an oral-aboral axis.
  3. A water-vascular system composed of water-filled canals used in locomotion, attachment, and/or feeding.
  4. A complete digestive tract that may secondarily reduced.
  5. No excretory system or organs of any kind
  6. A much reduced blood-vascular (hemal) system derived from coelomic sinuses. Fluid is moved through the system by cilia and in some cases by muscular "pumps."
  7. No head and no brain. The nervous system consists of a nerve net, nerve ring, and radial nerves.
  8. Formidable powers of regeneration of lost parts.

- Class Asteroidea—sea stars—habitat ranges from the rocks, sand and mud of shallow of coastal waters to coral reefs and the deep sea floor. All sea stars are benthic creatures and no types are found living in the water column. Their bodies are composed of a central disk from which five tapering arms (rays) typically radiate. However, some species have more or fewer than five, and there can even be variation in the number of arms within a species.
- Class Ophiuroidea—brittle stars and basket stars—is the most diverse group of echinoderms and likely the most abundant as well. They abound in all types of benthic marine habitats and literally carpet the abyssal sea bottom in many areas. While ophiuroids possess five arms as do most sea stars, their arms are long, snake-like and distinctly set off from the central disk, giving the central disk a pentagonal shape.
- Class Echinoidea—sea urchins and sand dollars—has a wide distribution in all seas, from intertidal regions to deep oceans. Sea urchins prefer rocky or hard bottoms, whereas sand dollars and heart urchins like to burrow into a sandy substrate. Although the echinoids are radiate animals possessing five ambulacral areas, they lack the distinct arms of sea stars and brittle stars. In echinoids, the dermal ossicles have become modified into closely fitting plates forming a shell or test equipped with long, sharp spines.
- Class Holothuroidea—sea cucumbers—is found at all depths in all oceans, crawling sluggishly over hard substrates or burrow through soft sediments. In shallow waters, they can form dense populations. Sea cucumbers have no arms, are elongate, and lie on one side, which is usually flattened as a permanent ventral side. Large, highly modified tube feet or tentacles surround the mouth.
- Class Crinoidea—sea lilies and feather stars—is considered the most primitive of all living echinoderms. Crinoids have a flower-shaped body that is placed at the tip of a permanently attached stalk. Many crinoids are deep-water forms, but feather stars may inhabit shallow waters, especially

in Indo-Pacific and West-Indian-Caribbean regions, where the largest numbers of species are found.

- Echinoderms

    1. Are covered with an epidermis that overlies a dermis containing rock-like skeletal elements, called ossicles. The epidermis itself consists of cells responsible for the support and maintenance of the skeleton, as well as pigment cells, mechanoreceptor cells, which detect motion on the animal's surface, and sometimes gland cells which secrete sticky fluids or even toxins. Beneath the dermis and ossicles are muscle fibers or layers and the coelom.
    2. Possess a water vascular system used for locomotion, food and waste transport, and respiration. The system is composed of canals connecting numerous tube feet.
    3. Exhibit a diversity of food preferences with each group having its own unique feeding strategies. In addition, the structure of the digestive tract differs among groups.
    4. Lack respiratory organs and rely on thin-walled external processes as gas exchange surfaces.
    5. Lack excretory organs and diffuse dissolved nitrogenous wastes (ammonia) across body surfaces to the outside. At least some excretion by simple diffusion probably takes place in most echinoderms.
    6. Lack a cerebral ganglion (brain) is lacking but do possess a nervous center in the form of a somewhat pentagonal circumoral nerve ring. From each angle of the nerve ring a large radial nerve extends into each arm, forming a large V-shaped mass along the midline of the oral surface of the ambulacral groove. Although echinoderms have few well-developed sense organs, they are at least somewhat sensitive to touch and to changes in light intensity, temperature, and body orientation to water currents.
    7. Have amazing powers of regeneration, and some asteroids and ophiuroids undergo asexual reproduction by a process called fissiparity, wherein the central disc divides roughly in two and each half forma a complete animal by regeneration
    8. Are dioecious for the most part, but hermaphroditic species are known among the asteroids, holothurians, and especially the ophiuroids. Asteroids and echinoids possess multiple gonads, while holothurians have but a single gonad and crinoids lack distinct gonads altogether. Fertilized eggs develop into free-swimming planktonic larvae that undergo a complex metamorphosis that results in the juvenile echinoderm.

- Echinoderms connect to humans in ways that are important economically, ecologically, and medically

Review and Reflect

1. ***An Echinoderm Predicament.*** Imagine you are helping your instructor tidy up the biology department's collection of echinoderm specimens. You have specimens of each of the following echinoderms in separate but (oops!) unlabeled jars: a sea star, a sea urchin, a feather star, a sea

cucumber, a sand dollar, a brittle star, and a sea lily. Your instructor expects you to label each jar with the correct class and common name of the specimen therein. Explain what class characteristics you would use to accomplish this task.

2. ***What's In a Name?*** Zoologists are careful to refer to the members of class Asteroidea as sea stars and not the older and more commonly heard term starfish. Why?

3. ***Skeletons-R-Us.*** Imagine you own a company that specializes in preparing skeletons for biological supply companies and science classrooms. A private collector has come to you with the request that you prepare a cleaned and dried skeleton for the following echinoderms: a sea star, a brittle star, a sea urchin, and a sea cucumber. Could you fill this order? What would be your reply?

4. ***Naked Sea Urchins.*** You and your little sister are walking along the beach and exploring tide pools. She finds something in the sand and asks, "What is this?" You reply, "It's the dried test (skeleton) of a sea urchin." With a puzzled look she wonders, "But I thought sea urchins had long sharp spines." What would be your reply?

5. ***Sea Stars vs Bivalves.*** The two curves in **Figure 14.23** show relative changes in the populations of two mollusc species (A and B) in a coastal area over time. At the time indicated by the dotted line, a species of sea star that preys on both A and b was introduced to the area.

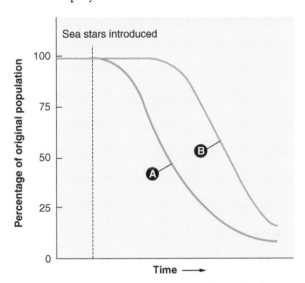

Figure 14.23 The decline of mollusc populations as the result of sea star introduction.

A. Which mollusc species is the preferred prey of the sea stars? Explain.

B. Why did the population of the less preferred mollusc also decline in size after the sea stars were introduced? Explain.

C. Questions A and B and your answers to them are based on an assumption. What is this assumption and how would you go about testing this assumption to see if it were true? (HINT: the data represented on this generalized graph was gathered under uncontrolled conditions.)

6. ***Osmoregulation Woes.*** Your study buddy says to you, "So because echinoderms have no excretory organs they are confined entirely to salt water habitats? I don't get it. If that's the case, shouldn't they be limited entirely to fresh water habitats?" How would you reply?

7. ***Oodles of Eggs.*** Why do spawning echinoderms release so many (millions) of eggs? Why do only a tiny fraction of these eggs ever develop into adults?

8. ***Away We Go.*** How does it benefit an echinoderm species to possess a planktonic larval form?

9. ***The Chordate Connection.*** What seems to be the evolutionary connection between echinoderms and chordates?

Create and Connect

1. ***From One Arm Many?*** In order to protect their oyster beds from the predation of sea stars, commercial oyster farmers used to remove the sea stars from their oyster beds, chop them into many pieces, and throw the pieces back into the water. This practice was discontinued when it came to be believed that each chopped-up piece was regenerating a whole new sea star and that such mutilations were actually increasing the population of sea stars. But is this true? If you chop a sea star into many pieces, can and will each piece regenerate a complete animal? Design an experiment to test this problem question.

   Following the tenets of a well-constructed experiment, your design should include the following components in order:

   - The *Problem Question*. State exactly what problem you will be attempting to solve.
   - Your *Hypothesis*. While this is a fictitious experiment, word your hypothesis as realistically as possible.
   - *Methods and Materials*. Explain exactly what you will do in your experiment including the materials necessary to accomplish the task. Be specific, take nothing for granted, and do not expect people to read your mind as they read your work.
   - *Collecting and Analyzing Data*. Explain what type(s) of data will be collected and what statistical tests might be performed on that data. It is not necessary to concoct either fictitious data or imaginary observations.

2. ***What's the Connection?*** Contemplate the Tree of Life and consider these questions—why are echinoderms the only invertebrates on the deuterostome branch? And what is the connection, if any, between the vertebrate deuterostomes, such as yourself, and the echinoderms? Write an essay in which you address and answer both these questions.

3. ***Sucker On.*** Among the constellation of strange characteristics echinoderms possess, the most unique may be their water vascular system. Learn more about this amazing locomotor scheme by writing a report on the water vascular system of echinoderms. Your instructor should provide additional details and specific requirements regarding length of the report, points to be covered in the report, and the number of references required.

PHYLUM HEMICHORDATA AND PHYLUM CHORDATA: THE BACKBONE ARISES

There is something about watching an animal that puts you in contact with where we came from and what we are still part of.

—John Cleese

Introduction

Compared to protostomes, the deuterostomes are far less diverse. In fact, the extant deuterostomes comprise a clade of four phyla—Echinodermata, Xenoturbellida, Hemichordata, and Chordata, and in spite of their relative lack of diversity and recent entry on the biological stage, one group of deuterostomes—vertebrate chordates—have rapidly ascended to become a group that dominates the living world around them by their size, morphological and anatomical complexity, and their great intelligence.

History of the Hemichordates

The grand force that is evolution cleaved the trunk of the Tree of Life into two main branches—Protostomia and Deuterostomia—before the start of the Cambrian some 542 million years ago.

Mitochondrial DNA analysis suggests that the deuterostomes are aligned into two clades: Olfactores (urochordates and vertebrates) and Ambulacraria (hemichordates and echinoderms). Finally, nuclear and mitochondrial data place the genus *Xenoturbella* as the sister group of the two ambulacrarian phyla. Therefore, *Xenoturbella* should be regarded as an independent phylum, Xenoturbellida, bringing the number of extant deuterostome phyla to four: Echinodermata, Hemichordata, Chordata, and Xenoturbellida (**Figure 15.1**).

PHYLUM XENOTURBELLIDA

Xenoturbella is a genus of bilateral animals consisting of two species of worm-like creatures. Since its discovery in 1949, the taxonomic placement of this enigmatic genus has been problematic. DNA analysis reveals that not only is the genus the sole member of its own phylum but also that the phylum may be basal within the deuterostomes.

These worm-like animals up to 4 cm (1.6 in) are found off the coasts of Iceland, Sweden, and Scotland. Xenoturbellidans have a very simple body plan. In fact, they have no organized organs—no brain, no digestive system, no excretory system, and no organized reproductive system (gametes occur in follicles). They do possess cilia and a diffuse nervous system.

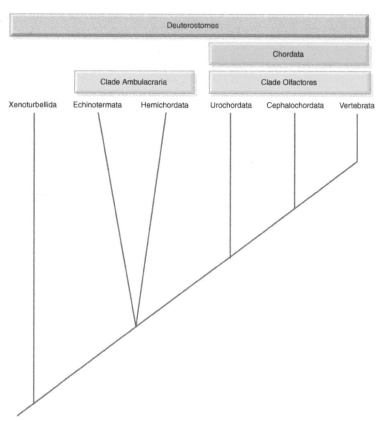

Figure 15.1 The possible phylogeny of deuterostomes.

PHYLUM HEMICHORDATA

Diversity and Classification

The phylum Hemichordata (Gr. *hemi*, half + L. *chorda*, cord) is a small group composed of only a few hundred species arranged into two classes:

> Domain Eukarya
> > Kingdom Animalia
> > > Phylum Hemichordata
> > > > Class Enteropneusta—Acorn worms
> > > > Class Pterobranchia—Pterobranchs

Hemichordates are most often found living on the sea floor in shallow waters. They are usually sedentary or sessile; colonial varieties spend their lives in secreted tubes. Although they are wide spread, they fact that they exist in such a secretive manner makes them hard to find, and their physical fragility makes collecting them a real challenge.

Phylum Characteristics

Hemichordates exhibit the following characteristics:

- Marine, deuterostome animals with a body divided into three regions: proboscis, collar, and trunk; coelom divided into three cavities
- Ciliated pharyngeal slits
- An open circulatory system
- A complete digestive tract
- A unique excretory structure ("glomerulus")
- Dorsal and ventral nerve cords with a ring connective in the collar; some species with a hollow dorsal nerve cord

Class Characteristics

Class Enteropneusta (Gr. *entero*, intestine + *pneustikos*, for breathing)—Acorn worms.

Enteropneusts are sluggish wormlike animals that live in burrows in sandy and muddy substrates between the limits of high and low tide. They range in size from 10 to 40 cm (4-16 inches) although some can be as long as 2 m (6-7 feet).

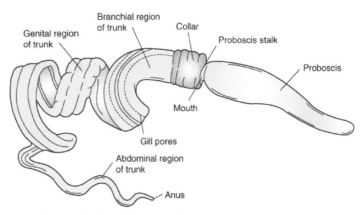

Figure 15.2 Morphology of the enteropneust *Saccoglossus*.

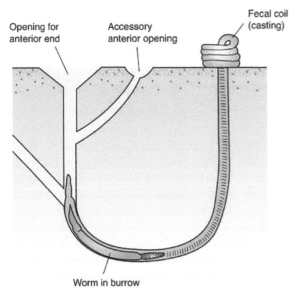

Figure 15.3 The burrow system of the enteropneust *Balanoglossus*.

Their common name—acorn worm—is derived from the appearance of the proboscis, which is a short, conical projection at the worm's anterior end. A ring-like collar is posterior to the proboscis, and an elongated trunk is the third division of the body (**Figure 15.2**).

The worm's proboscis is its most active body part as it moves around in the mud, searching for and gathering food in mucus strands. Burrow-dwelling types use a thrusting action of their proboscis to excavate their dwellings. Their self-made burrows are U-shaped and lined with mucus. The ends of the burrows are 10-30 cm (4-12 inches) apart with the bottom of the "U" reaching a depth of 50 to 75 cm (20-30 inches) below the sea floor. Out of and into opposite ends of these burrows they feed and defecate, respectively.

Feeding is accomplished with thrusts of their proboscis out of the front opening while feces accumulate in spiral mounds at the back (**Figure 15.3**).

Class Pterobranchia (Gr. *pteron*, wing or feather + *branchia*, gills)—Pterobranchs

The members of this very small class of hemichordates (only around 20 species) are found mostly in deep, oceanic waters of the Southern Hemisphere. A few live in European coastal waters, and in shallow waters near Bermuda.

Pterobranchs are small ranging in size from 0.1 to 5 mm (0.004-0.2 inch). Most types live in asexually reproduced colonies with each individual inhabiting a tube of a collagen-like material which they secrete. The colony may consist of unattached zooids (individuals) living independently in their tubes (**Figure 15.4**) or the colony may be one in which the zooids are in separate tubes, but connected to each other through a **stolon** (**Figure 15.5**).

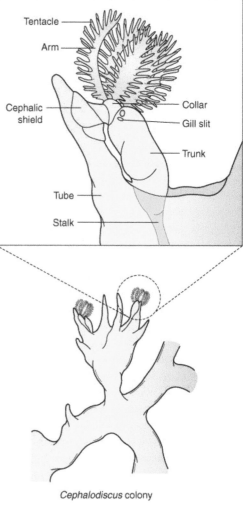

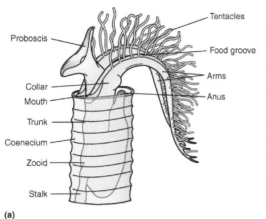

(a)

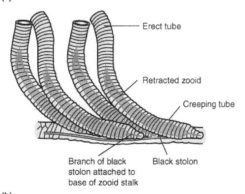

Erect tube

Retracted zooid

Creeping tube

Branch of black stolon attached to base of zooid stalk

Black stolon

(b)

Figure 15.5 The Pterobranch *Rhabdopleura*. (a) Indiviuals live attached in branching tubes and protrude ciliated tentacles for feeding. (b) A portion of a colony showing individuals attached to stolons.

Pterobranchs display the three body regions—proboscis, collar, and trunk—characteristic of hemichordates. The proboscis is expanded and shield-like and functions to secrete the tube and aid movement within the tube. The collar possesses two to nine arms with numerous ciliated tentacles.

Cephalodiscus colony

Figure 15.4 The Pterobranch *Cephalodiscus*. These tiny forms live unattached in colonial tubes in which they move about freely.

Hemichordate Body Plan

Body Wall, Support, and Locomotion

Hemichordates in general possess a ciliated epidermis richly supplied with gland cells, many of which are involved in mucus production, particularly on the proboscis and collar of enteropneusts and on the tentacles of pterobranchs. In some enteropneusts, epidermal mucal glands produce noxious compounds that may repel potential predators.

Both circular and longitudinal muscles are present in the wall of the proboscis and collar of acorn worms, but elsewhere in the body only longitudinal fibers exist. Pterobranchs possess only longitudinal muscle fibers in their body walls. The body is supported primarily by the hydrostatic nature of the body cavities, and secondarily by the structural integrity of the body wall, connective tissues, and supplemental structures such as skeletal plates in the proboscis of the acorn worm (**Figure 15.6**). The proboscis coelom is connected to a dedicated pore by a narrow canal. Located in the collar and also connected to opening pores, is a pair of coelomic cavities. The worm burrows by stiffening the proboscis and collar by taking water in through the proboscis pore then moving forward by reducing hydrostatic pressure through a contraction of muscles that force the water out through gill slits.

Hemichordates are sessile or sedentary creatures capable of only limited movement at best. Enteropneusts crawl slowly or burrow by peristaltic action of the proboscis (**Figure 15.7**). The protraction and retraction of pterobranchs within their tubular houses are accomplished by hydrostatic pressure and the contraction of longitudinal muscles. Some crawl within their tubes by using the muscular **cephalic shield** (modified proboscis).

Feeding and Digestion

Most hemichordates are ciliary-mucus feeders. Food particles are swept into the mouth by cilia on the proboscis and collar (**Figure 15.8**) of enteropneusts. Gut musculature is scant, and the food is moved along largely by cilia into the buccal cavity, pharynx, and esophagus, then into the intestine where digestion and absorption occur.

The major feeding structures of pterobranchs are the arms and tentacles. During feeding, they assume a position near an opening in their tubular cases and extend the arms and tentacles into the water. A mucus net is secreted across the latticework of tentacles to trap food particles in the water. Food trapped in the mucus is moved to the mouth by the action of cilia on the tentacles and arms. In

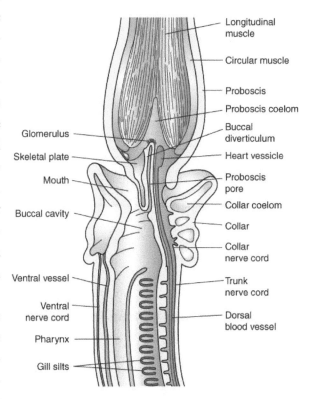

Figure 15.6 Longitudinal section through the anterior end of the acorn worm *Saccoglossus* oriented vertically.

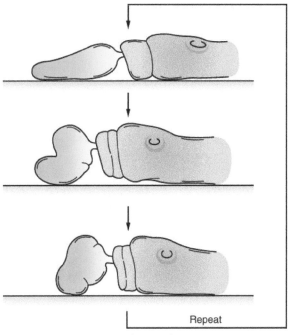

Figure 15.7 Locomotion by crawling in an acorn worm.

pterobranchs, the mouth leads to the pharynx, which connects to a pouched stomach. The intestine ascends upward from the stomach terminating in an anterior dorsal anus.

Circulation, Gas Exchange, and Excretion

Enteropneusts possess a well developed open circulatory system consisting of blood vessels, sinuses, and a contractile organ called the **heart vesicle** that is located in the proboscis. Blood move anteriorly in the dorsal vessel and posteriorly in the ventral vessel (Figure 15.6). Branches from these vessels lead to open sinuses. The blood of acorn worms is colorless, lacks cells, and carries nutrients, wastes, and gases in solution.

The circulatory system of pterobranchs is less developed than that of enteropneusts—as one might expect considering their tiny size. Although there is a central sinus and a heart vesicle, no major vessels run through the body.

Gas exchange in enteropneusts occurs between the environment and the blood mainly in the **branchial system**. Behind the collar on both sides of the trunk are rows of gill pores (Figure 15.8). Cilia drive water into the mouth through the pharynx and out the **gill slits** (Figure 15.6) and pores. Gases are exchanged across the vascular epithelium of the branchial sacs. The tiny pterobranchs have a single set of gill slits, but gas exchange occurs primarily over the high surface area of the tentacles.

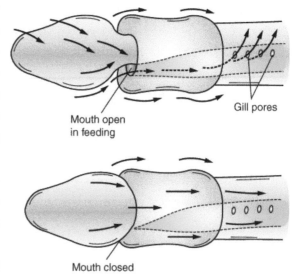

Figure 15.8 Food Currents in an Entropneust. (a) Side view of an acorn worm with mouth open, showing direction of currents created by cilia on the proboscis and collar. Food particles are directed toward the mouth. Rejected particles move toward the outside of the collar. Water leaves through gill pores. (b) When the mouth is close, all particles are rejected and passed onto the collar. Nonburrowing and some burrowing hemichordates use this feeding method.

Blood flowing anteriorly passes into the **glomerulus**, an excretory organ formed from finger-like outpockets unique to the hemichordates (Figure 15.6). The glomerulus extracts metabolic wastes and releases it into the coelom of the proboscis, and then to the outside through one or two pores in the wall of the proboscis. There is a glomerulus in pterobranchs, but it is so weakly developed that wastes are most likely eliminated primarily by diffusion across the body wall.

Nervous System and Sense Organs

The nervous system of all hemichordates consists of a netlike nerve plexus lying among the bases of the ciliated epithelial cells of the body wall. There are no major ganglia (no brain). A dorsal nerve cord (neurochord) is present in the collar of enteropneusts but is reduced to thickening of the plexus in pterobranchs. The ectoderm's invagination forms the neurochord, which, in some varieties, is hollow. It is quite similar to the nerve cord pattern we see in chordates, and it shows there is homology between the two groups for this feature.

There are few types of sensory receptors in hemichordates. Enteropneusts possess sensory cells over most of the body, probably serving as tactile receptors, and they also bear a preoral ciliary organ, presumed to be a chemoreceptor used during feeding. Little is known about the sensory apparatus of pterobranchs, but it is presumed that tactile receptors are present in the tentacles and perhaps on the cephalic shield.

Reproduction and Development

Asexual reproduction occurs in at least some enteropneusts and most pterobranchs. Small pieces of Acorn worms fragment from the trunk and each one can grow into a new individual. As is the case in most colonial invertebrates, asexual reproduction by budding is an integral part of the life history of aggregating and colonial pterobranchs.

Hemichordates are dioecious but possess no outward gender differences. Spawning in enteropneusts involves the release of mucoid egg masses by the females, followed by shedding of sperm by neighboring males. Once the eggs are fertilized the mucous coating breaks down, thereby freeing the eggs into the sea water, where all subsequent development occurs. All features of this development attest to the deuterostome nature of acorn worms. Some species produce relatively yolky eggs; others produce eggs with little yolk. Those species with yolky eggs develop directly into juvenile worms without an intervening larval stage. In those that shed nonyolky eggs, the hatching stage develops quickly into a characteristic, planktonic **tornaria larva** with ciliary bands reminiscent of echinoderm larvae. This larva soon elongates with the three body regions becoming externally apparent (**Figure 15.9**).

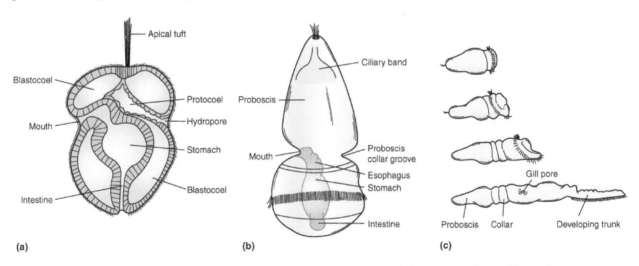

Figure 15.9 Larval Development of Enteropneusts. (a) Section of a late tornaria larva. (b) Late larva with developing proboscis. (c) Metamorphosis from late tornaria to juvenile enteropneust.

In pterobranchs, external fertilization results in the development of a planula-like larva that lives for a time in the tube of the female. This nonfeeding larva eventually leaves the female's tube, settles to the substrate, forms a cocoon, and metamorphoses into an adult.

Hemichordate Connection

Small in stature and limited in the number of total species of their kind, the hemichordates are not familiar creatures to most people, and they do not have much of an impact economically, environmentally, or medically. Instead, their importance is scientific. The study of their unique development and morphology has been of monumental importance in advancing our understanding of vertebrate evolution.

PHYLUM CHORDATA

History of Chordates

Chordate origins are shadowy as early representatives were soft-bodied and therefore left a poor fossil record. Fossils possibly attributable to all three chordate subphyla have been found in Cambrian rocks more than 505 million years old, and an extensive vertebrate fossil record seems to begin around 400 million years ago.

Though there have been several proposed lines of descent, no one is certain of the precise phylogenetic position of chordates in respect to other animals. Early theories held that the arthropod-annelid-mollusc group of the Protostomia branch gave rise to the chordates. That hypothesis, however, has largely been abandoned as molecular analysis reveals that the echinoderm-hemichordates (deuterostome branch) are the only realistic possibilities as progenitors of the chordates.

Recent phylogenetic studies of chordates and echinoderms have indicated that urochordates might be the closest invertebrate sister group of vertebrates, rather than cephalochordates as has been traditionally believed. Earlier studies seemed to suggest that cephalochordates are closer to echinoderms than to vertebrates and urochordates. The sequencing of the mitochondrial genome of the various phyla of deuterostomes, however, does not support a cephalochordate and echinoderm grouping forcing the conclusion that chordates are indeed monophyletic

Whether the ancestral chordate was more like a tunicate (sea squirt) or cephalochordate (lancelet) has been and continues to be debated. The classical theory is that the ancestor was like a cephalochordate and that one lineage became attached to hard surfaces and evolved into tunicates, whereas another remained unattached and evolved into vertebrates. An alternative theory is that the ancestor was a tunicate and that the other two subphyla arose by modification of the tadpole larva. Recent studies suggest that the tunicates probably branched off before the common ancestor of cephalochordates and vertebrates arose, for the latter resemble each other in some details of their nervous systems and biochemistry.

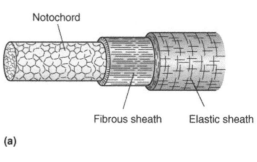

(a)

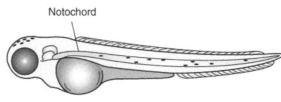

(b)

Figure 15.10 The Notochord. (a) The structure of the notochord and its surrounding sheaths. Stiffness of this primitive type of endoskeleton is the result of fluid-filled cells surrounding connective tissue sheaths, (b) The notochord of this young fish will gradually be replaced by bone vertebrae, but in lampreys and hagfish the notochord persists for life.

Diversity and Classification

Although the phylum Chordata (L. *chorda*, cord) does not have an inordinately large number of species (about 45,000), its members are conspicuous for their size, complexity of form, diversity, and familiarity.

One unique characteristic unites the chordates and gives them their phylum name—the **notochord** (Gr. *noton*, back + L. *chorda*, cord). All animal types classified as chordates possess a notochord whether it be present only at early stages or throughout the life cycle. The notochord is located within a fibrous sheath and consists of a hard, tube-shaped cluster of cells. Extending along the central nervous system, it stiffens the body in order to support the muscles (**Figure 15.10**).

We have retained the traditional view with the use of monophyletic taxa in regards to the classification of chordates. It is our belief that at the introductory level this approach maintains familiarity and provides a conceptual usefulness. Extensive change and virtual abandonment of familiar rankings may well be the outcome of ongoing cladistic analysis of chordates. We believe, however, that such ongoing changes are unsettling and confusing to students and professionals alike, and until things are sorted out, it is educationally prudent to base this and the remaining chapters on a more traditional approach.

> Domain Eukarya
>> Kingdom Animalia
>>> Phylum Chordata
>>>> Subphylum Cephalochordata—Lancelets
>>>> Subphylum Urochordata—Tunicates
>>>>> Class Ascidiacera
>>>>> Class Appendicularia (Larvacea)
>>>>> Class Thaliacea
>>>>> Class Sorberacea
>>>> Subphylum Vertebrata (Craniata)—Vertebrates
>>>>> Class Myxini—Hagfishes
>>>>> Class Cephalaaspidomorphi—Lampreys
>>>>> Class Chondrichthyes—Sharks, skates, and rays
>>>>> Class Osteichthyes—Ray-fins and lobe-fins
>>>>> Class Amphibia—Amphibians
>>>>> Class Reptilia—Reptiles
>>>>> Class Aves—Birds
>>>>> Class Mammalia--Mammals

Phylum Characteristics

The extremely diverse chordates are united by the following characteristics:

- Bilateral symmetry
- A unique combination of four features present at some stage of development: notochord, pharyngeal pouches, dorsal tubular nerve cord, and a postanal tail.

- Presence of an endostyle or thyroid gland.
- A complete digestive tract
- A ventral contractile blood vessel (heart)

Subphyla Characteristics

Subphylum Cephalochordata (Gr. *kephalo*, head + L. *chorda*, cord)—Lancelets.

The cephalochordates consist of two genera, *Branchiostoma* (formerly *Amphioxus*) and *Asymmetron*, totaling less than 50 species. These small animals are commonly known as amphioxus ("pointed on both ends") or lancelets ("tiny spears").

Lancelets are thin, flat, translucent, fishlike animals roughly 5-7 cm (2-3 inches) long. They are distributed throughout the world's oceans in shallow waters that have clean sand substrate. In spite of their streamlined shape, lancelets are relatively weak swimmers and spend most of their time in a filter feeding position—partly to mostly buried with their anterior end sticking out of the sand (**Figure 15.11**). Tampa Bay, Florida is home to a thriving population (up to 5,000 lancelets per square meter) that attracts chordate researchers from all over the world. Exceptionally high population densities are also known to exist in Lagos Lagoon (Nigeria) and Kingston Harbor (Jamaica).

The notochord of cephalochordates extends from the head (cephalus) to the tail, giving them their name. Unlike the notochord of most other chordates, the notochord of cephalochordates is mostly muscle cells. This makes the cephalochordate notochord somewhat contractile and may be an adaptation to burrowing. Muscle cells on either side of the notochord cause undulations that propel the lancelet weakly through the water. Dorsal and ventral fins help stabilize the animal as it swims and the caudal fin aids in powering the swimming motion.

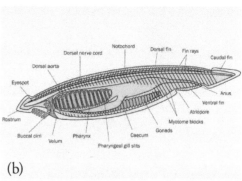

Projecting from the anterior end of the body is the **oral hood**. Ciliated, fingerlike **cirri** hang from the hood. These cirri prevent sediments and other large particles from clogging the mouth as well as playing a role in feeding. Cephalochordates are ciliary-mucus suspension feeders. During feeding, they burrow into sandy bottom sediments with their mouth pointed upward. Water is driven into the mouth by pharyngeal cilia and out through the **pharyngeal gill slits** into a surrounding **atrium**; it exits the body through a ventral **atriopore** (Figure 15.11b).

(b)

The lateral walls of the vestibule bear complex ciliary bands that collectively constitute the **wheel organ**. Cilia of the wheel organ drive food particles to the mouth and give the impression of rotation, hence the name of the structure. Pits in the roof of

Figure 15.11 The Lancelet (*Branchiostoma lanceolatum*). (a) Lancelets are normally found partially buried in coarse bottom sediments. (b) Generalized anatomy of a lancelet.

vestibule secrete mucus that flows over the wheel organ. Food particles trapped by the mucus are carried to the mouth by cilia-generated water currents.

The ventral surface of the pharynx bears the **endostyle**, an organ that binds iodine and produces strings of mucus that trap food from the water as it passes through the gill slits and into the atrium. The food-laden mucus passes into a short esophagus. The gut extends posteriorly as an elongated intestine and opens through the anus just in front of the caudal fin. Associated with the gut is the **hepatic cecum**. The cecum appears to function in lipid and glycogen storage and protein synthesis. While the endostyle in these animals is thought to have given rise to the vertebrate thyroid gland, the hepatic cecum is regarded as the precursor to the vertebrate liver and perhaps the pancreas.

This group has a surprisingly complex closed circulatory system given their relatively simple morphology. Though they are without a real heart, their blood flow resembles rather closely that of more primitive fishes, with peristaltic-like contractions of the **ventral aorta** pumping blood upward through branchial arteries and to twin **dorsal aortas**, which eventually become one. Next, microcirculation moves the blood to body tissues after which it collects in the veins on its return trip back to the ventral aorta. The blood contains no pigments or cells and is thought to function largely in nutrient distribution rather than in gas exchange and transport. Although some diffusion of oxygen and carbon dioxide may occur across the primitive gills, most of the gas exchange probably takes place across thin flaps off the body wall that lie just anterior to the atriopore.

The excretory system of lancelets consists of clusters of tubules (protonephridia) that accumulate nitrogenous wastes. These wastes are carried by a nephridioduct to a pore in the atrium.

The central nervous system of cephalochordates is quite simple. A dorsal nerve cord extends most of the length of the body (Figure 15.11b). Segmentally arranged nerves arise from the cord and innervate the muscles and other organs. The epidermis is rich in sensory nerve endings, most of which are probably tactile and important in burrowing. Some lancelets have a single simple eye spot near the anterior end of the dorsal nerve cord.

Cephalochordates are dioecious, but the genders are structurally very similar. Rows of from 25 to 38 pairs of gonads are arranged serially along the body on each side of the atrium. The volume of gonadal tissue varies seasonally, and during the reproductive period, it may occupy so much space that it interferes with feeding. Spawning usually occurs around dusk. The atrial wall ruptures and egg and sperm are released into the water. After external fertilization free-swimming bilateral larvae develop. The larvae eventually settle to the substrate before metamorphosing into adults.

This small fishlike creature peacefully filtering organic matter from sea water while partially buried on the floor of the sea gives no indication that it is one of the most recognized animals of classical zoology. During the nineteenth century, amphioxus was held in high regard because at that time it was considered by many biologists to be the direct ancestor of vertebrates. The lack of a head with special sense organs, however, leads present day zoologists to believe that even though it bears the greatest resemblance of any living animal to the chordate condition just before the origin of vertebrates, amphioxus is not a direct ancestor of the vertebrate line of chordates.

Subphylum Urochordata (Gr. *uro*, tail + L. *chorda*, cord)—Tunicates.

There are four classes of tunicates—Appendicularia, Sorberacea, Thaliacea (salps), and Ascidiacae (sea squirts). Among these four, Ascidiacea is the largest, most diverse, and best known. Often referred to as "sea squirts," some forms propel a stream of water from the excurrent siphon when frightened or otherwise bothered. Most species are sessile individual or colonial animals, attaching themselves to hard surfaces such as rocks and the undersides of boats. They are one of the most abundant animals found along numerous intertidal zones (**Figure 15.12**).

The body wall of most tunicates (L. *tunicatus*, to wear a tunic or gown) is a connective tissue-like covering called the **tunic** that appears gel-like but is, in reality, quite tough (**Figure 15.13**). Secreted by the epidermis, the tunic is composed of proteins, various salts, and cellulose, a molecular product usually associated with plants. Some mesodermal-derived tissues, including blood vessels and blood cells, are incorporated into the tunic. Rootlike extensions of the tunic called **stolons**, help anchor solitary tunicates to the substrate and may connect individuals of a colony. The **mantle** is an inner membrane that lines the tunic. The unattached end of a sea squirt contains two siphons that permit seawater to circulate through the body. One siphon is the **oral** (incurrent) **siphon**, which is the inlet for water and serves as the mouth opening, while the second siphon, the **atrial** (excurrent) **siphon**, is the opening for water being expelled by the animal.

Longitudinal and circular muscles below the body wall epithelium help to change the shape of the adult tunicate. They act against the elasticity of the tunic and the hydrostatic skeleton created by seawater confined to internal chambers.

During feeding, cells in a ventral, ciliated groove,

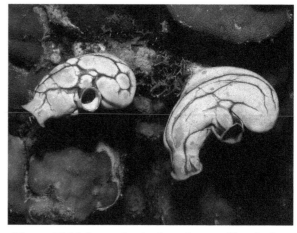

Figure 15.12 Sea squirts lead a sedentary filter-feeding existence. Despite their appearance, sea squirts are more closely related to vertebrates than they are to invertebrates such as sponges and anemones.

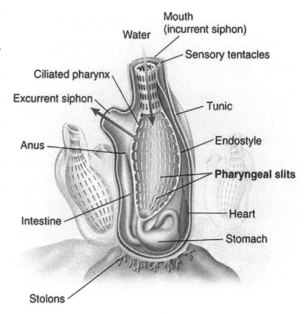

Figure 15.13 Anatomy of a tunicate (sea squirt).

called the **endostyle** form a mucus sheet. Cilia move the mucus sheet dorsally across the pharynx. Food particles in the incurrent water stream are trapped in the mucus sheet and then incorporated into a string of mucus that ciliary action moves into the stomach. Excurrent water carries digestive wastes from the anus out of the excurrent siphon.

In addition to its role in feeding, the pharynx also functions in gas exchange and excretion. Gases are exchanged, and metabolic wastes in the form of ammonia diffuse into the water that circulates over the pharynx and through the animal.

The ventral heart and one large vessel on either side of the heart make up the circulatory system. The large, twin vessels are connected to smaller vessels that circulate blood to many organs and structures, among them the pharynx, the digestive organs, and the gonads. Tunicate circulation features two oddities found in no other chordates. One is that the blood flow pauses and then reverses direction every few beats. The other peculiarity is the very high concentrations such uncommon elements as vanadium and niobium. In the sea squirt *Ciona*, vanadium can be as high as 2 million times its concentration in seawater. The reasons for this remain a mystery. Tunicate blood plasma is colorless and contains a number of different types of amoeboid cells.

A small cerebral ganglion and plexus of nerves along the dorsal side of the pharynx makes up the nervous system. This nerve plexus gives rise to a few nerves to various parts of the body, especially the muscles and siphonal areas. Sensory receptors are poorly developed in tunicates, although touch-sensitive neurons are prevalent around the siphons.

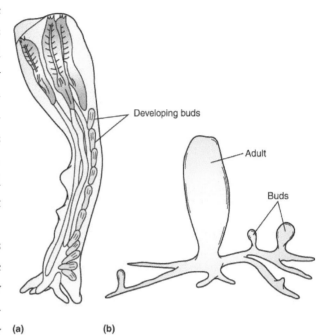

Figure 15.14 Asexual Reproduction in Ascidians. (a) Formation of buds in the colonial ascidian *Circinalium*. (b) General pattern of stoloniferous budding in an ascidian.

Although the tunicates are entirely sexual in their reproductive habits, salps and many sea squirts include asexual processes in their life histories (**Figure 15.14**). When it comes to sexual reproduction, most tunicates are monoecious (hermaphroditic), with relatively simple reproductive systems. There is a great deal of variation in the overall reproductive strategies of tunicates. Most large solitary ascidians produce high numbers of eggs with low yolk content. The eggs are shed into the sea coincidentally with the release of sperm by other individuals. External fertilization is followed by the development of a free-swimming **tadpole larva**, which eventually settles and metamorphoses into an adult (**Figure 15.15**).

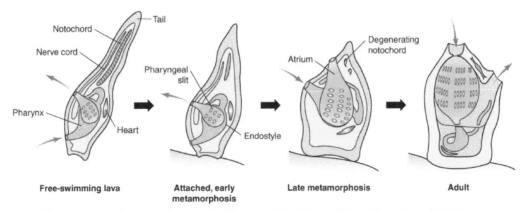

Figure 15.15 Metamorphosis of a solitary ascidian from a free-swimming tadpole larva.

In contrast, many colonial ascidians composed of tiny zooids produce relatively few eggs, but each egg has a high yolk content. These eggs are fertilized and subsequently brooded within the cloacal chamber; they are not released until the swimming tadpole larva stage is reached.

Urochordate Connection

Ecologically

Thought to have been transported in the ballast water of ships, urochordates have invaded the coastal waters of many countries, and with few natural predators, are spreading rapidly. Growing in thick mats, these animals can smother other forms of sea life.

Medically

Tunicates produce a number of different chemical compounds that have proven useful in the treatment of cancer and as antiviral and immunosuppressants.

Collectively the cephalochordates and urochordates are referred to as invertebrate chordates because they never develop true vertebrae (backbones).

Subphylum Vertebrata (L. *vertebratus*, backboned)

Although the vertebrata and the other two chordate subphyla have much in common, there are many chordate characteristics unique to vertebrata. Recent cladistic analysis suggests the name of Craniata better describes this group because although each has a cranium (a bony or cartilaginous braincase), some forms, such as the jawless fish, do not have vertebrae.

What is a vertebrate? The answer is to be found in the list of characteristics that vertebrates possess:

- A unique combination of five features—a notochord, dorsal tubular nerve cord, pharyngeal pouches or clefts, endostyle or thyroid gland, and a postanal tail—at some point in their life
- A well-developed coelom segmented into a **pericardial cavity** (cavity containing the heart) and a **pleuroperitoneal cavity** (cavity containing the lungs [pleural] and the abdominal organs [peritoneal])
- A body covering (skin) consisting of stratified epithelium derived from ectoderm with an inner skin of connective tissue made of mesoderm; among the different forms, numerous variations of outer covering exist, including scales, feathers, hair, claws, and horns
- A bone or cartilage endoskeleton with a vertebral column (except in hagfish [Myxini]) and a head skeleton (cranium and associated vertebrae)
- An appendicular skeleton with two pairs of appendages supported by limb girdles. Hagfish and lampreys are exceptions.
- A digestive tract that lies ventral to the vertebral column with a liver and pancreas

- A circulatory system made up of a ventral, multi-chambered heart attached to a confined region of arteries, veins, and capillaries; blood contains a variety of cell types suspended in plasma with hemoglobin as the main respiratory molecule
- An excretory system with a pair of tubular kidneys that feature waste-draining ducts
- An endocrine system of glands without ducts located randomly throughout the body producing various types of hormones.
- A large tripartite (three part) brain; 10-12 pairs of cranial nerves and paired sense organs.

The morphology and characteristics of each specific vertebrate class will be detailed in the chapters that follow.

Chordate Body Plan

Five distinct morphological characteristics taken together are the hallmarks of chordates. Though the forms may vary and will often disappear later, chordates experience each of the five characteristics at some point in their embryonic life. In humans, for example, the pharyngeal pouches are altered during development into the middle ear canal and the parathyroid and thymus glands.

Notochord

Found on the ventral surface of the neural tube, the notochord structure, running along the length of the body, is pliable and tube-like in chordate embryos (**Figure 15.16**). It is composed of cells from the mesoderm and serves as a key location for muscle attachment. Its pliability allows for undulatory (back and forth) movements of the body.

In some vertebrates, the notochord persists throughout life as the main axial support of the body. In most vertebrates, however, it is replaced by the vertebral column. In humans, remnants of the notochord persist as the disks between the vertebrae.

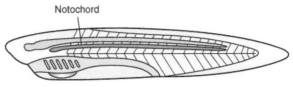

Figure 15.16

Dorsal Tubular Nerve Cord

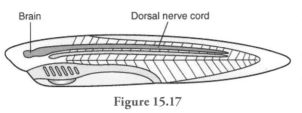

Figure 15.17

As its name suggests, the chordate nerve cord is dorsally located (back or top side) and hollow (tubular) (**Figure 15.17**). Ectodermal cells create this nerve cord as they infold above the notochord in the embryo. The hollow portion of the nerve cord is filled with fluids to nourish the nerve cells and in vertebrates, the anterior end becomes enlarged to form the brain.

Pharyngeal Pouches

Only visible during embryonic stages, the pharyngeal pouches develop into a variety of organs in chordates (**Figure 15.18**). In water-breathing animals, these pouches evolved into internal gills, while in the lung-breathing tetrapods (four-limbed) vertebrates, these pouches evolved into the Eustachian tube, middle ear cavity, tonsils, and parathyroid and thymus glands.

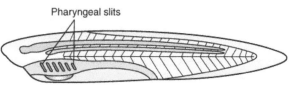

Pharyngeal slits

Figure 15.18

Endostyle or Thyroid Gland

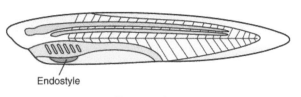

Endostyle

Figure 15.19

In hemichordates the endostyle (**Figure 15.19**) is found in the pharyngeal floor where it releases a food-gathering mucus that assists in the feeding on small particles collected in the pharyngeal cavity. Recent molecular evidence supports the traditional view that the vertebrate thyroid gland is derived from and homologous to the endostyle.

Postanal Tail

At the terminal end of the body beyond the digestive tract, the tail evolved specifically for propulsion in water (**Figure 15.20**). Its efficiency is greatly increased in fish where it has been modified into the caudal fin. In adult humans, its only trace is found in the vestigial **coccyx**, a line of smaller vertebrae at the spinal column's base. Many other animals, however, posses a well-developed and movable tail as adults.

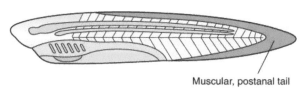

Muscular, postanal tail

Figure 15.20

A Closing Note
In the Tail of Tunicates

Is it possible that the evolution of vertebrate traits and eventually the vertebrates themselves hinged on a single genetic event? For decades, biologists studying the evolutionary origins of the backbone have bypassed the furred, feathered, and finned vertebrates and focused on larval tunicates (sea squirts) instead. The tadpole larva of these creatures is equipped with a dorsal nerve cord, a flexible support rod called a notochord, a tail, and skeletal muscles that propel them through shallow tidal flats. These are the very traits that evolutionary biologists believe were eventually modified first into chordates and then into vertebrates such as ourselves.

The larvae seem to turn back the evolutionary clock for they shed these vertebrate-like traits, reverting back to an invertebrate form as they mature.

William Jeffery and Billie J. Swalla then of the University of California located a tunicate species whose larvae naturally lacked a tail. They developed artificial hybrids by interbreeding the tailless species with a closely related tailed species. The hybrid grew tails that were shorter than normal.

These researchers then located a gene that was active in the tailed and hybrid species, but not in the tailless species. They named the gene Manx, after a breed of tailless cat. Further studies showed that Manx acts as a master control gene coding for a protein that regulates other genes. In fact, it appears to turn on most of the vertebrate traits of the tunicate larva.

This find is an exciting step, and it has great implications toward understanding the evolution of chordates and thus vertebrates. If a single gene is responsible for turning on vertebrate characteristics during larval development, then it may have been responsible for the evolution of vertebrate traits in our ancestors more than 500 million years ago. Although we'll never know for certain, it is possible that the evolution of vertebrates turned on a single, very important genetic event—the appearance of the Manx gene in the tail of the tunicate.

In Summary

- Hemichordates are worm-like marine bottom dwellers, usually living in shallow waters. Most are sedentary or sessile, and some colonial species live in secreted tubes.

- Phylum Hemichordata is a small group composed of only a few hundred species arranged into two classes:

    Class Enteropneusta
    Class Pterobranchia

- Hemichordates exhibit the following characteristics:

    1. Marine, deuterostome animals with a body divided into three regions: proboscis, collar, and trunk; coelom divided into three cavities
    2. Ciliated pharyngeal slits
    3. An open circulatory system
    4. A complete digestive tract
    5. A unique excretory structures ("glomerulus")
    6. Dorsal and ventral nerve cords with a ring connective in the collar; some species with a hollow dorsal nerve cord

- Class Enteropneusta—acorn worms—are sluggish wormlike animals that live in burrows in sandy and muddy substrates between the limits of high and low tide. Their body consists of an anterior proboscis, a collar, and an elongated trunk. The proboscis is the most active part of the animal. It probes about in the mud, examining its surroundings, and collecting food in mucus strands on its surface.

- Class Pterobranchia is a very small class of hemichordates (only around 20 species) whose members are found mostly in deep, oceanic waters of the Southern Hemisphere. A few live in European coastal waters, and in shallow waters near Bermuda. Most live in secreted collagenous tubes in asexually reproduced colonies. The colony may consist of unattached zooids (individuals) living independently in the tubes, or may be a colony in which the zooids are in separate tubes but are connected to each other through a stolon.

- Hemichordates:

1. Possess a ciliated epidermis richly supplied with gland cells, many of which are involved in mucus production, particularly on the proboscis and collar of enteropneusts and on the tentacles of pterobranchs.

2. Are sessile or sedentary and capable of only limited movement at best. Enteropneusts crawl slowly or burrow by peristaltic action of the proboscis. The protraction and retraction of pterobranchs within their tubular houses is accomplished by hydrostatic pressure and the contraction of longitudinal muscles, respectively. Some crawl within their tubes by using the muscular cephalic shield (modified proboscis).

3. Are largely ciliary-mucus feeders. Food particles are swept into the mouth by cilia on the proboscis and collar of enteropneusts whereas the major feeding structures of pterobranchs are the arms and tentacles

4. Possess a circulatory system. Enteropneusts possess an open circulatory system comprising blood vessels, sinuses, and a contractile organ called the heart vesicle that is located in the proboscis. The circulatory system of pterobranchs is less developed.

5. Exchange gases through a series of branchial chambers (enteropneusts) or across the high surface area of the tentacles (pterobranchs).

6. Excrete metabolic wastes into a glomerulus (enteropneusts) or across the body wall through diffusion (pterobranchs).

7. Have a simple nervous system consisting of a netlike nerve plexus lying among the bases of the ciliated epithelial cells of the body wall. There are no major ganglia (no brain).

8. Are capable of asexual reproduction. Acorn worms fragment small pieces from the trunk, and each one can grow into a new individual. Asexual reproduction by budding is an integral part of the life history of aggregating and colonial pterobranchs.

9. Are dioecious but possess no outward gender differences. Spawning in enteropneusts involves the release of mucoid egg masses by the females, followed by shedding of the sperm by neighboring males. Once the eggs are fertilized the mucous coating breaks down, thereby freeing the eggs into the sea water, where all subsequent development occurs. In pterobranchs, external fertilization results in the development of a planula-like larva that lives for a time in the tube of the female. This nonfeeding larva eventually leaves the female's tube, settles to the substrate. forms a cocoon, and metamorphoses into an adult.

- Although the phylum Chordata does not have an inordinately large number of species, its members are conspicuous for their size, complexity of form, and familiarity.

- Chordates may be classified into three subphyla and numerous classes:

> Subphylum Cephalochordata—Lancelets
> Subphylum Urochordata—Tunicates
>> Class Ascidiacera
>> Class Appendicularia (Larvacea)
>> Class Thaliacea
>> Class Sorberacea
> Subphylum Vertebrata—Vertebrates
>> Class Myxini—Hagfishes
>> Class Cephalaspidomorphi—Lampreys
>> Class Chondrichthyes—Sharks, skates, and rays
>> Class Osteichthyes—Ray-fins and lobe-fins
>> Class Amphibia—Amphibians
>> Class Reptilia—Reptiles
>> Class Aves—Birds
>> Class Mammalia—Mammals

- Chordates exhibit the following characteristics

1. Bilateral symmetry
2. A unique combination of four features present at some stage of development: notochord, pharyngeal pouches, dorsal tubular nerve cord, and postanal tail.
3. Presence of an endostyle or thyroid gland.
4. A complete digestive tract
5. A ventral contractile blood vessel (heart)

- Subphylum Cephalochordata—lancelets ("tiny spear")—are slender, laterally compressed translucent fishlike animals about 5-7 cm (2-3 inches) in length. They are distributed throughout the world's oceans in shallow waters that have clean sand substrate. In spite of their streamlined shape, lancelets are relatively weak swimmers and spend most of their time in a filter feeding position—partly to mostly buried with their anterior end sticking out of the sand
- Subphylum Urochordata—tunicates (sea squirts)—are small sac-like animals that move water in and out through siphons. Food particles are trapped on mucus sheets secreted by the animal. Tunicates are the source of certain chemical compounds that have proven useful in the treatment of select cancers.
- Subphylum Vertebrata—fish, amphibians, reptiles, birds, mammals—are defined by a suite of unique characteristics:

1. A unique combination of five features—notochord, dorsal tubular nerve cord, pharyngeal pouches or clefts, endostyle or thyroid gland, and postanal tail—present at some time during their live cycle

2. A well-developed coelom divided into a pericardial cavity (cavity containing the heart) and a pleuroperitoneal cavity (cavity containing the lungs [pleural] and the abdominal organs [peritoneal])

3. A body covering (skin) consisting of an outer epidermis of stratified epithelium derived from ectoderm and an inner dermis of connective tissue derived from mesoderm; many modifications of skin exist among the various classes, such as scales, feathers, hair, claws, and horns

4. A distinctive bone or cartilage endoskeleton consisting of a vertebral column (except in hagfish [Myxini]) and a head skeleton (cranium and associated vertebrae)

5. An appendicular skeleton with two pairs of appendages supported by limb girdles. Hagfish and lampreys are exceptions.

6. A complete, muscularized digestive tract that lies ventral to the vertebral column with a distinct liver and pancreas

7. A circulatory system consisting of a ventral heart of multiple chambers attached to a closed system of arteries, veins, and capillaries; blood contains a variety of cell types suspended in plasma with hemoglobin as the main respiratory molecule

8. An excretory system consisting of paired, glomerular (tubular) kidneys with ducts to drain wastes

9. An endocrine system of ductless glands scattered throughout the body producing various types of hormones.

10. A large tripartite (three part) brain; 10-12 pairs of cranial nerves and paired sense organs.

Review and Reflect

1. *Evolution Before Our Eyes?* Vertebrate embryos all bear a superficial resemblance to each other. This pattern was noted by the great 18th century biologist Ernst Haeckel and led him to conclude that "ontogeny recapitulates phylogeny." This phrase means that evolutionary history (phylogeny) is replayed (recapitulated) during embryonic development. Haeckel believed that vertebrates evolved in linear fashion. Therefore, amphibians are first fish before hatching as amphibians. By this argument, human embryos must first pass through the ancestral fish, amphibian, and reptile phases of its evolutionary history before developing into a mammal and then finally into a hominid primate. This hypothesis persists in high school biology texts, the popular press and the collective mind of the lay public. But is it regarded as a viable hypothesis by evolutionists and embryologists? Investigate and explain.

2. *What Is It?* An oceanographic research vessel has returned with many specimens that were collected on the voyage. One of the specimens is a worm-like creature that was collected from the bottom in shallow water. Zoologists think it may be a new species of hemichordate, but they aren't sure and turn to you for input. What would you say? What features must this specimen possess to be considered a hemichordate?

3. *Further Refinement*. Suppose you did tentatively identify the specimen in #2 as a hemichordate. Zoologists now want to know which class of hemichordates it belongs in. What would you say? How would you decide?

4. *What's the Connection?* What characteristics do Hemichordata share with Chordata and how do the two phyla differ? Does this evidence suggest that Hemichordata are related to chordates? Defend your answer.

5. *What Did She Say?* As you are walking out of the lecture hall after a discussion on chordates by your zoology instructor, your bleary-eyed friend says to you, "I fell asleep but kind of remember the instructor saying something about the hallmarks of chordates. What was that all about?" How would you reply?

6. *Make Another Connection*. Are you as a vertebrate biologically related to sea squirts and amphioxus? If so, explain how. Are you closely related or distantly related?

Create and Connect

1. *First You Then Me*. Sea squirts (tunicates) coordinate the release of gametes—sperm and egg—during spawning. How do they accomplish this feat given the simplistic nature of their nervous system and sense organs? Investigate hypotheses that have been advanced to explain this phenomenon. From those, develop your own hypothesis and design an experiment to determine how sea squirts coordinate the release of their gametes.

   Guidelines:

   A. Your design should include the following components in order:

   ➤ The *Problem Question*. State exactly what problem you will be attempting to solve.
   ➤ Your *Hypothesis*. While this is a fictitious experiment, word your hypothesis as realistically as possible.
   ➤ *Methods and Materials*. Explain exactly what you will do in your experiment including the materials necessary to accomplish the task. Be specific, take nothing for granted, and do not expect people to read your mind as they read your work.
   ➤ *Collecting and Analyzing Data*. Explain what type(s) of data will be collected and what statistical tests might be performed on that data. It is not necessary to concoct either fictitious data or imaginary observations.

   B. Your instructor may provide additional details or further instructions.

2. *All Hail the Chordates!* On the internet, find Phillip Pope's 1921 poem *It's a Long Way from Amphioxus*, and use it as inspiration to write a poem or song praising that pinnacle of animal evolution we call chordates. Give your work an appropriate title.

3. ***The Notochord—A Position Paper.*** It has been said, "The notochord is the single most important biological structure in the entire animal world." Agree or disagree? Write a position paper in which you defend or discount this statement.

Guidelines:

A. Compose a position paper, not an *opinion* paper. Defend your position with as many facts, Figures, quotes, and pertinent information as possible.

B. Your work will be evaluated not on the "correctness" of your position but the quality of the defense of your position.

C. Your instructor may provide additional details or further instructions.

FISHES: MONARCHS OF AN ANCIENT REALM

No human being, however great or powerful, was ever so free as a fish.

—John Ruskin

Introduction

Swimming, wriggling, crawling, and floating in the waters of this planet is an immense population of animals whose full extent we scarcely realize. Collectively known as fishes, these creatures represent the first vertebrates and the foundation lineage from which sprang all other vertebrates, including humans. From mountaintop to steamy jungle to black frigid ocean depths, there is no area on Earth that does not count fish among its inhabitants.

History of Fishes

Six hundred million years ago life was confined to the sea. Except for a few tidal algae, the land was bare and the oceans that lapped it shores were far different from the submarine world we know today. The early seas were wide, warm, and shallow, and all the life that they contained was on or near the bottom. There were no fishes as we know them, but trilobites, sponges, molluscs, cnidarians, and other invertebrate creatures teemed in the sunlit waters. Somewhere, either in the oceans or in some freshwater pond or stream of that far-off Cambrian period there arose a chordate lineage from which sprang the fishes.

The first fishes seem to have been the ostracoderms ("shell skinned"), first appearing in the Cambrian Period about 510 million years ago. The ostracoderms were jawless, small (30 cm [1 foot]) and covered with bony armor or scales, a highly evolved and complex animal for their time, but by the end of the Devonian Period about 350 million years ago, they had disappeared.

By the late Silurian, about 410 million years ago, the first fish with jaws, the acanthodians (spiny sharks), had appeared on the scene. These revolutionary fish were small and streamlined with a shark-like tail ranging from toothless filter feeders to toothed predators, but by the end of the Permian about 250 million years ago, they too had become extinct.

Another group of jawed fishes, the placoderms, appeared at the beginning of the Devonian, about 395 million years ago, and became extinct at the end of the Devonian or the beginning of the Mississippian (Carboniferous) about 345 million years ago. Placoderms were typically small, flattened bottom-dwellers with a heavily-armored body. The upper jaw was firmly fused to the skull, but there was a hinge joint between the skull and the bony plating of the trunk region.

Exactly when the two main categories of fishes, the cartilaginous *Chondrichthyes* (sharks, skates, rays, and chimaeras) and the bony *Osteichthyes* (ray-finned and lobed-finned fishes), divided and went their separate ways

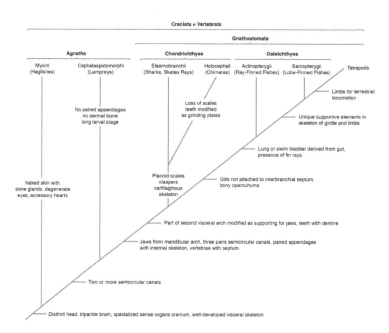

Figure 16.1 The Possible Phylogeny of Fishes and the Rise of Tetrapods. Extinct lineages have not been represented. Although Agnatha and Osteichthyes are most likely paraphyletic, they are represented here in a more traditional interpretation.

nobody knows for certain. Fossil evidence indicates this event most likely occurred, as did so many other major evolutionary milestones in fish development, during the Devonian Period or the Age of Fishes as the period has come to be known (**Figure 16.1**).

Why these lineages developed is a matter of conjecture, but it may have been simply a matter of originating in different habitats. The Chondrichthyes, which appeared about 370 million years ago in the middle Devonian, are generally believed to be descended from the armored placoderms, while the Osteichthyes, which arose during the late Silurian or early Devonian about 395 million years ago, are thought to be descended from the acanthodians.

These two groups represent different solutions to living in a watery environment, and of the two, the bony fishes have been the more evolutionarily successful in both diversity and their ecological impact on their habitats. The bony fish have been the dominant vertebrate life form in the ocean and freshwater for the last 180 million years. The cartilaginous group began to decline in diversity and impact after some 85 million years, leaving relatively few descendants represented only by the sharks, skates, rays, and deep-sea ratfish of present times.

The evolution of fishes is a history of constant adaptation to new possibilities, specialization for different modes of living, and refinement of both of these. Such useful additions to their basic form, as they did develop, were in the nature of new structures to better exploit their environment and habitat. Of these, the jaws and paired fins are prime examples.

Diversity and Classification

Fishes swarm the waters of our planet in numbers we can hardly fathom. Not even insects, with many more species, can compare to the diversity of size, shape, color, and body form of the fishes. About 70% of the Earth's surface is covered by the ocean, and about 3.5% of the land surface (1% of the Earth's total surface) is covered with fresh water. Inhabiting those waters are around 28,000 species of fishes, of which almost 27,000 species are bony fish, 970 species are cartilaginous fish (sharks, skates, rays) and 180 species are jawless fish (lamprey and hagfish). The number of species of fishes exceeds the number of all other vertebrates species combined. Ocean waters harbor 60% of all fish species while the remaining 40% are found in fresh water.

Fishes exploit every imaginable nook and niche of their watery domain, from the tops of mountains at altitudes of more than 5,000m (3 miles), as in Lake Titicaca in the Andes, to depths of 10 km (6 miles) in the ocean. Saltwater fish are found from lightless ocean abysses where pressures mount to thousands of pounds per square inch to the mid-depths where no shore, no bottom, and no surface is ever encountered to the boundless surface layers of the open sea. They are found from the high tropics to the polar regions, among wave-churned tidal rocks and coral reefs rich in life almost beyond imagining, along sandy beaches and in muddy bays and brackish estuaries. Their counterparts, freshwater fish, inhabit great rivers and tiny brooks, stagnant pools and rushing torrents, even stygian caves where blind and colorless fish are found in shallow pools as far as a thousand feet underground.

The Julimes pupfish (*Cyprinodon julimes*) lives in hot springs where the water temperature may reach 45° C (113° F), hot enough to cook most other animals, whereas the icefish (*Chaenocephalus*) lives in Antarctic waters so cold that its blood contains a natural antifreeze to keep them from freezing solid. Others can even survive in a state of partially suspended animation in mud cocoons when their seasonal pools dry and harden.

Some fishes migrate thousands of miles, whereas others spend their entire lives in the same hole. Some walk (wriggle actually) and glide as well as swim while others spend part of their existence out of the water and will drown if kept submerged.

> *No good fish goes anywhere without a porpoise.*
> —Lewis Carroll

There appear to be no boundaries in the sea and nothing to prevent fishes from ranging any part of the great ocean they desire. This is not the case, however. What limits fish distribution? Certainly regional availability of food is one factor, but the great barrier to the unrestricted spread of animals in water is temperature. Compared to the variations we know on land, the range of temperature in any body of water of any size is small. In any part of the ocean, temperature usually varies no more than about 4° C (25° F). The hottest seas in the world are the Red Sea and the Persian Gulf, where temperatures of 30° C (86° F) occur. The coldest are in the Arctic and Antarctic, where a temperature as low as 2.2° C (28° F) is not unusual. By comparison, extremes of temperature on land may range all the way from -88° C (-126° F) on the Antarctic continent to 58° C (136° F) in the North African Sahara.

Fishes are present in water of every temperature from the tropics to the poles, but they are most abundant in the temperate latitudes where temperatures range between about 6.1° C (43° F) and 20° C (68° F). However, most fish can put up with fluctuations of 12 to 15 degrees if the change is not effected too suddenly. Not surprisingly, the eggs and young are more sensitive to temperature fluctuations than are adult fishes. There are, then, broad zones of temperature, each with its own kind of fishes which breed and grow best within its temperature limits. Within these zones, much the same kind of fishes live around the globe.

Fishes range in size from the adult 7.9mm (0.31 inch) *Paedocypris progenetica* in the acidic swamps of Sumatra to the oceanic whale shark (*Rhincodon typus*) at 13.6 m (45 feet) long and weighing 22 tons (44,000 pounds). A Mekong giant catfish (*Pangasianodon gigas*) as large as a grizzly bear (2.7 m [9 feet] long and weighing 293 kilograms [646 pounds]) landed recently in Thailand now apparently holds the world's record as the largest fresh water fish. These giant fish are currently listed as critically endangered and as such, are the focus of a World Wildlife Fund and National Geographic Society project to identify and study the planet's largest freshwater fishes.

There are round fish, flat fish, and tube-shaped fish; some are brightly adorned with every known color (**Figure 16.2**), whereas others are drab. Some can even dramatically change color and others eerily glow with their own light.

As with every phylum discussed so far in this book, the classification of fishes is an area of debate and change. The reconciliation of traditional Linnaean groupings with recent and ongoing cladistic analysis of the fishes makes any scheme for classifying fish temporary and subject to change. Compounding the problem is the fact that unlike the birds and mammals, fishes are not a single monophyletic clade but rather a paraphyletic collection of taxa.

The classification system for fishes presented here represents a blending of traditional grouping with phylogenetic systematics.

Figure 16.2 As this mandarinfish (*Synchiropus splendidus*) clearly illustrates, there is no color on Nature's palette that is not found on the body of some fish.

Domain Eukarya
 Kingdom Animalia
  Phylum Chordata
   Subphylum Vertebrata (Craniata)
    Superclass Myxinomorphi (fish with no jaws)
     Class Myxini—Hagfishes
    Superclass Petromyzontomorphi
     Class Cephalaspidomorphi—Lampreys
    Superclass Gnathostomata (fish with jaws)
     Class Chondrichthyes (Cartilaginous Fishes)
      Subclass Elasmobranchii—Sharks, skates, and rays

Subclass Holocephali—Chimaeras (ratfish)

Class Osteichthyes (Bony Fishes)

Subclass Actinopterygii—Ray-finned fishes

Subclass Sarcopterygii—Lobed-finned fishes

General Characteristics

Life in water has influenced every detail of a fish's existence. There are advantages to living in water, and these advantages have played an important role in molding fishes into what they are today. Water is not subject to sudden temperature changes and is, therefore, an excellent habitat for an **ectothermic** (derive heat from external sources) animal. The changes that do occur in water temperature are slow and allow time for migration to more suitable climes or acclimatization to the existing one. The problem of supporting body weight, too, in water is far simpler than on land because cell cytoplasm has approximately the same density as water, a fish in its medium is almost weightless. This in turn means it can get along with a light and simple bone structure; and it also practically removes any limit to its size, making it possible for a fish as huge as the whale shark to move about as easily and comfortably as a goldfish.

There is one basic difficulty which fish must cope with and which more than anything else has shaped their development—water is incompressible and offers great resistance to anything moving through it. A flat and angular shape can be moved through such a medium only with difficulty (a board pushed straight down in water invariably slews off violently to one side or the other), and for this reason fishes have a basic shape that is beautifully adapted to deal with this peculiarity of water. We call such a shape streamlined or hydrodynamic: pointed at the head, bulkiest in the middle, tapering back so that the water can flow smoothly along the sides with a minimum of turbulence. There are variations on this shape, of course, but it is basic to all free-swimming fishes no matter in what specialized form they have evolved. As this basic shape was coupled with fins and movable jaws through the course of evolution, the fishes as we know them came to be.

Don't tell fish stories where the people know you; but particularly don't tell them where they know the fish.

—Mark Twain

Given the great variation in the diversity of fish and the different patterns of their evolution, a question inevitably arises—what is a fish? What characteristics must a creature possess for membership into that assemblage of animals we generally term as the fishes? In common (and older) usage, the term "fish" denotes a mixed assortment of water-dwelling animals. We speak of jellyfish, cuttlefish, starfish, crayfish, and shellfish knowing full well that when we use the word "fish" in such word associations, we are not referring to a true fish. In times past, however, even biologists did not make such a distinction. Sixteenth-century natural historians lumped whales, seals, amphibians, crocodiles, even hippopotamuses, as well as a swarm of other aquatic invertebrates, together and christened the group as fish. Later biologists, narrowing the concept of a fish, eliminated first the invertebrates and then the amphibians, reptiles, and mammals as morphological participants in the assemblage known as fishes.

With that in mind, we define a fish as an ectothermic aquatic vertebrate that extracts oxygen from water using gills, possesses appendages in the form of two sets of paired fins, and is generally covered with scales of dermal origin.

Box 16.1
Life History of a Fish As Told By….. Its Ears?

Suppose you wanted to study the growth and development of fish and other aspects of their daily life. When you attempt to do so you find that tracking and observing minute fish eggs and tiny nearly transparent larvae in the current-swept open seas or cloudy, turbid rivers presents insurmountable problems. And keeping track of individual fish in the expanse of the ocean or even a large lake is impossible. What are researchers to do?

Fortunately, there is an indirect method of reconstructing the life-history details of an individual fish. Bony fishes have three compact, mineralized structures within each inner ear, the **otoliths**, which are important in hearing, body orientation, and locomotion. By analyzing these "ear stones", ichthyologists can now tell where a fish was born, where it grew to maturity, how fast it grew, and where it migrated during its lifetime.

Otoliths grow by the accretion of mineralized layers deposited on the surface in concentric layers resembling those of an onion. The thinnest layer that can be distinguished usually reflects a day's growth. The relative width, density, and interruptions of the layers show the environmental conditions the individual encountered daily including variations in temperature and food capture. Minute quantities of elements and isotopes characteristic of the environment where the fish has spent each day are also incorporated into the bands. Thus, a day-by-day record of the growth and movement of the individual is written in its otoliths.

Analysis of otoliths may hold applications far beyond that of determining the life history of an individual fish. It is possible, for example, to "bar code" fish in order to determine years later when they are recaptured what brood they originally came from. Such a code can be imprinted on the otoliths of very young fish by subjecting them to a series of carefully controlled temperature increases and decreases.

New techniques of otolith analysis are also providing very important information on the longevity of species. Traditional aging methods often significantly underestimate fish age. Current findings are revealing that it is not uncommon for many fish to attain ages of 50 years, and some may double or even triple that Figure.

Information gathered from otoliths may also help fish biologists answer questions about spawning locations, age structure, and migration patterns, and by doing so better manage and preserve endangered stocks of fish such as the weakfish, the Atlantic bluefin tuna, and Columbia River salmon.

And there you have it; a quite amazing revelation really when you consider that most people aren't aware that a fish even has ears let alone that they have three stone flight recorders in those ears recounting every detail of that individual's life one minute layer after another.

As is nearly always the case with animals, however, to each of these there are exceptions. Body shape and the arrangement of the fins vary greatly among fishes, and some types such as pufferfish, gulpers, anglerfish, and seahorses (**Figure 16.3**) certainly do not appear to fit what we consider to be standard fish architecture. Although the surface of the skin in most fish may be covered with scales of a variety of different types, in some, such as moray eels and catfish, scales are lacking.

Swimming performance and the degree of streamlining varies greatly as well. Salmons, jacks, and tuna are highly streamlined and capable of reaching speeds of 10-20 body-lengths per second, whereas the less streamlined eels and rays clock in at a leisurely 0.5 body-lengths per second. The Atlantic bluefin tuna (*Thunnus thynnus*), for example, can attain speeds in excess of 22 m/sec (50 mph), an amazing feat given the great density of sea water.

Tuna, swordfish, and some species of sharks display some **endothermic** (heat derived from internal processes) adaptations, and consequently are able to elevate their body temperature to levels significantly higher than that of the ambient water surrounding them. The north Pacific salmon shark (*Lamna ditropis*), for example, maintains its red muscle (RM) at 20-27° C (68-86° F), much warmer than the 8.3° C (47° F) water in which it lives. Salmon sharks are lamnids, a group of sharks that also includes the mako and great white. Numerous studies have shown that lamnid sharks and tunas share many anatomical and physiological specializations that endow them with their impressive swimming power and speed. In contrast to other fish where the RM is near the skin, the RM of these sharks and tunas

Figure 16.3 Looking more like a clump of seaweed than a fish, this leafy sea dragon seahorse *(Phycodurus eques)* defies the traditional definition of fish morphology.

is near the backbone. Maintaining a higher temperature in the RM allows these muscles in lamnid sharks and tuna to produce 25-50 percent more power in their RM compared to the lower temperature RM of other fish.

Until now, no fish had ever been discovered that can warm its entire body. Enter the opah (*Lampris*), also commonly known as the moon fish or redfin ocean pan. It has been recently discovered that this fish possesses a thick web of blood vessesls near its gills called a retia mirabilia ("wonderful nets"). This network of blood vessels allow for a counter-current heat exchange between warm blood coming from the fish's body core and cold blood that has just passed through the gills. The result is a fish that can warm its entire body.

Although most freshwater fish are supremely adapted to remove oxygen from the water through gills, there are those that can extract oxygen from the atmosphere as well as from the water. The **labyrinth organ**, for example, of gouramis and betas functions as a simple lung, whereas lungfish have paired lungs similar to those of tetrapods. Mudskippers (Gobiidae) haul out onto mudflats where they feed and interact for extended periods of time. As long as they remain damp, mudskippers are able to transport necessary oxygen across their moist skin and the mucosal lining of the mouth and throat (cutaneous respiration). Mudskippers dig burrows in soft sediments for thermoregulation, to avoid predators, and for egg-laying. Even when the burrow is submerged during high tide, however, mudskippers maintain an air bubble in their burrows. This air bubble allows them to breathe when oxygen concentrations in the surrounding water fall dangerously low. As contradictory as it sounds, mudskippers will drown if held underwater.

At high tide, fish eat ants; at low tide, ants eat fish.
—Thai proverb

Class Characteristics

Class Myxini (Gr. *myxa*, slime)—Hagfishes.

The characteristics of this class are as follows:

- Strictly marine (salt water) animals
- A slender, eel-like body lacking scale and skin containing slime glands
- No paired appendages, possessing only a single fleshy caudal fin that extends anteriorly along the dorsal surface
- A fibrous and cartilaginous skeleton; notochord persists throughout life. Class Myxini is the only group of vertebrates that possess a skull but no vertebral column.
- No jaws, but two toothed plates on tongue
- Five to 16 pairs of gills with a variable number of gill openings; no operculum
- Digestive system without stomach
- Dorsal nerve cord with a differentiated brain, but no cerebellum
- Sense organs of smell, taste, and hearing, but the eyes are degenerate to the point of almost complete blindness.
- Sexes are separate (ovaries and testes in the same individual but only one is functional); external fertilization; large yolky eggs, but no larval stage.

In cold ocean waters worldwide at depths around 1220 m (4,000 feet) buried in the sand and mud in very dense groups lurks the most primitive of all fishes—the hagfish. This small group (around 65 species) has always been problematic for zoologists. There has been a long discussion in the scientific literature about whether hagfish are actually vertebrates. Recent molecular analyses, however, tend to support their traditional position as true vertebrates. Thus, hagfish are classified as vertebrates even though anatomically they lack true vertebrae. Zoologist once believed that hagfish were primitively equipped fishes that lost most of their fish characteristics as the result of leading a parasitic existence. We now know they are not strictly parasitic, and a better understanding of the fossil evidence reveals a successful body plan and lifestyle that hasn't changed for the last 300 million years.

Hagfish have elongated eel-like bodies with no scales, and a paddle-like tail consisting of a single fleshy caudal fin (**Figure 16.4**). They average about 46 cm (18 inches) in length and are generally gray in color with pinkish to blue overtones and some black or white mottling depending on the species. Hagfish have no jaws, and the mouth is typically surrounded by a ring of short sensitive **barbels** (similar to tentacles). Instead of vertically articulating jaws, they have a tongue equipped with two rows of keratinized teeth for rasping away bits of flesh from its prey (**Figure 16.5**). The skeleton of the hagfish is made up entirely of cartilage, and they retain a notochord throughout their life.

Although a small group in the number of species (around 60), hagfishes can be very abundant in individual numbers in a small area. By one estimate, the Gulf of Maine contains a population density of 500,000 hagfishes per square kilometer (0.4 square mile). Although almost completely blind, hagfishes have keenly developed senses of touch and smell that allow them to find food quickly and effectively. They eat annelid worms, molluscs, and crustaceans and will scavenge the insides of dead and dying fish. Although hagfish lack the ability and equipment to penetrate through the skin, they will enter through openings on the fish such as the mouth, gills, or anus. Once inside, the hagfish uses its rasping tongue to eat away flesh, leaving only a sack of skin and bones. They sometimes become a nuisance to fishermen by devouring the catch before it can be pulled to the surface. As with leeches, hagfish have a sluggish metabolism and can go months between feedings.

Hagfish are often called "slime eels" in reference to their legendary custom of secreting profuse amounts of sticky, gelatinous slime. When attacked or threatened, a hagfish exudes a milky fluid from special glands positioned along its body. On contact with seawater this fluid forms a thick gelatinous mass. A single hagfish can quickly produce enough slime to fill a milk jug and even their own nostril often fills with slime forcing a "sneeze" to remove it. This "sliming" may repel predators who are put off by a thick gelatinous cocoon surrounding their potential prey. Hagfish also have the ability to tie themselves in knots and slide the knot along the body (**Figure**

(a)

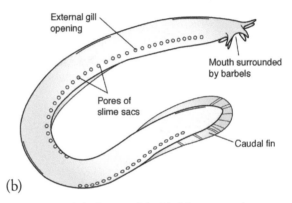

(b)

Figure 16.4 The Hagfish. (a) Of primitive lineage and possessing a skull but no vertebral column, it is debatable that the hagfish is a true fish at all. What is certain is that their feeding habits and slime-producing capabilities have earned them the reputation as "the most disgusting sea creature". (b) Morphology of the Atlantic hagfish *Myxine glutinosa*.

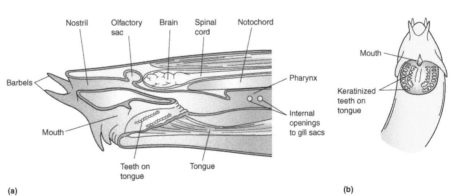

Figure 16.5 The Head Region of a Hagfish. (a) Sagittal section of the head. The tongue is shown in retracted position. (b) Ventral view of the head showing the tongue and keratinized teeth.

16.6). Such a traveling knot serves to clean slime off the body and gives extra leverage when pulling on food or escaping the jaws of predators.

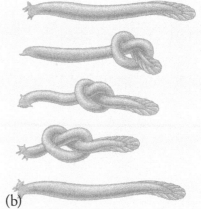

Figure 16.6 Hagfish Slime. (a) Adult hagfish secrete so much slime that they turn a 5 gallon (20 liter) bucket of water into slime in minutes. (b) Slime Removal Behavior of a Hagfish. This knotting behavior is also a defense mechanism, If a predator seizes the tail of a hagfish, the hagfish will release a thick slime all the while tying itself in an overhand knot. This knot travels to the mouth of the predator giving the hagfish leverage as it tries to pull free. This traveling knot also serves to clean the body and give leverage when pulling on food.

Class Cephalaspidomorphi (Gr. *kephale*, head + *aspidos*, shield + *morphe*, form)—Lampreys.

The following are characteristics of this class:

- A slender, eel-like body with no scales
- One or two fleshy dorsal fins, but no paired appendages
- A fibrous and cartilaginous skeleton; the notochord persists throughout life
- No jaws; a suckerlike oral disc and tongue with well-developed keratinized teeth
- Seven pairs of gills each with an external opening; no operculum
- Digestive system without a distinct stomach; spiral folds in the intestine
- Dorsal nerve cord with differentiated brain; a small cerebellum is present
- Sense organs of taste, smell, hearing; eyes well developed in adults
- Sexes separate; a single gonad without duct; external fertilization with a long larval stage

Wriggling through the waters of temperate rivers and coastal seas around the world, except in Africa, are the lampreys. The lampreys ("stone lickers") are a small group of about 40 species of elongated eel-like animals that lack scales. They range in size from 15-100 cm (6-40 inches) long and possess well-developed eyes, a fleshy dorsal and tail fin, and seven gill openings along the side of the body. Like the hagfishes, they have a cartilaginous skeleton and lack both jaws and paired fins. The mouth is round and filled with rows of horny teeth and a movable rasping tongue (**Figure 16.7**). In North America, about the half the species are the nonparasitic brook type, whereas the others are the parasitic types we most often associate with lampreys. Fastening to living fish, parasitic lampreys rasp into their flesh and feed on abraded skin, muscles, and body fluids (**Figure 16.8**).

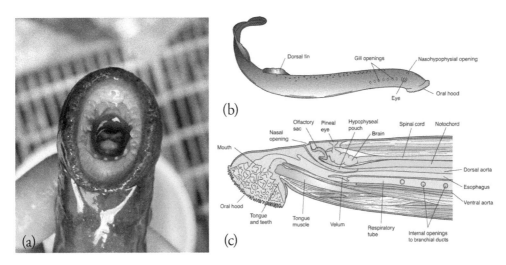

Figure 16.7 The Lamprey. (a) Though jawless, the parasitic lamprey possess formidable weaponry for attaching to and tearing through the body wall of its host. (b) Morphology of the lamprey. (c) Sagittal section of the head.

Figure 16.8 Sea Lampreys in the Great Lakes. (a) Salmon with lamprey attached. (b) Wounds caused by feeding lamprey.

Lampreys have a complete braincase and rudimentary true vertebrae. Unique among living vertebrates, lampreys also have a single "nostril" (**nasophypophysial opening**) on the dorsal side of the head—a feature they share with various fossil jawless fish, which had a similar opening.

Some species live in fresh water for their entire lives. Others are **anadromous**, meaning they are hatched and grow in freshwater, migrate to the ocean to mature, and then return to fresh water to spawn.

Because they lack mineralized tissues such as bones, lampreys are rare as fossils and their early evolutionary history is still poorly understood. The three definite lamprey species known from the fossil record are similar to living lampreys, and suggest that the group has changed very little over the course of the past 300 million years.

> *Lampreys give us a calibration point. We study lampreys because, in many respects, they're so primitive. They never had jaws, never had [true] teeth, they never had fins, they never had limbs. Lampreys provide a glimpse of conditions early in vertebrate evolutionary history.*
> —Michael Coates

Class Chondrichthyes (Gr. *chondros,* cartilage + *ichthys*, fish)

Class characteristics include:

- A relatively large **fusiform** (funnel-shaped) body (average about 2 m [7 feet]) with a ventral mouth and powerful jaws modified from the pharyngeal arch
- Paired pectoral and pelvic fins, one or two dorsal median fins; pelvic fins in male modified as claspers; fins stiff and not very movable. The tail is often heterocercal
- The skin is covered with placoid scales; naked skin in some elasmobranchs and all chimaeras; teeth are modified placoid scales
- The endoskeleton is completely cartilaginous; notochord present but reduced
- The digestive system has a J-shaped stomach (stomach absent in chimaeras); intestine equipped with a **spiral valve** that slows passage of food and increases the absorptive surface area
- five to seven pairs of gills leading to exposed gill slits (no operculum) in elasmobranchs; four pairs of gills covered by an operculum in chimaeras
- No swim bladder or lung, but an oil-filled liver aids in buoyancy
- Brain of two olfactory lobes, two cerebral hemispheres, two optic lobes, cerebellum, medulla oblongata
- Senses of smell, vibration reception (lateral line system), vision, and electroreception well developed
- Sexes separate; paired gonads with reproductive ducts opening into the cloaca; internal fertilization with direct development

The Chondrichthyes first appeared on earth almost 450 million years ago. The 850 living species that compose this class are carnivores or scavengers, and are nearly exclusively marine. They lack true bone and have a skeleton made of cartilage. (Their teeth and sometimes their vertebrae are calcified, but this calcified cartilage has a different structure than true bone.) In addition, they possess powerful jaws and muscles, paired appendages, and placoid scales. Class Chondrichthyes may be subdivided into two subclasses: Elasmobranchii and Holocephali.

Subclass Elasmobranchii (Gr. *elasmos,* plated + *branchia*, gills)—Sharks, skates, and rays. This subclass has about 820 members. The origin of sharks is obscure, but their fossil record dates to the Devonian period (408-360 million years ago). They became the dominant vertebrates of the Carboniferous period (360 to 286 million years ago) and by the Cretaceous period (144 to 66 million years ago) had developed into all present-day families.

Because they lack some of the more sophisticated features of bony fish (e.g. a swim bladder to regulate buoyancy, gill covers, several pairs of flexible fins, and a bone skeleton) the elasmobranchs were originally considered primitive in comparison. This interpretation is mistaken. The possession of certain features by one group and the lack of those features by another related group indicates only that each group took different evolutionary paths in the face of similar selection pressures.

Sharks are almost exclusively oceanic fish found in all seas, but they are most prevalent in warm waters. A few types will enter rivers and estuaries, and one species found in India and Pakistan (*Glyphis*) lives only in fresh water. Although their movements are thought to be extensive and related to reproductive or feeding activities, the actual geographic ranges of sharks are not well known. Tagging returns from large sharks on the East Coast of the United States indicate regular movement between New Jersey and Florida.

Figure 16.9 It has been said that the business of a shark is to eat anything and everything it possibly can.

With their cold black eyes and razor-sharp teeth, sharks have a sinister even evil appearance to most people, and their reputation as man-eating monsters is unmatched by any other animal (**Figure 16.9**). Contrary to popular belief, only a few sharks are dangerous to humans. Sharks are streamlined fish with a fully cartilaginous skeleton. Typically, they possess a muscular, asymmetric, upturned tail; pointed fins; and a pointed snout (**rostrum**) extending forward and over a crescent mouth set with sharp triangular teeth (**Figure 16.10**). There are paired **pectoral** and **pelvic** fins, one or two median **dorsal** fins, and one **caudal** (tail) fin. A single **anal** fin near the anal opening is present in most sharks as well. The inner parts of the male's pelvic fins have been modified into a pair of cigar-or sausage-shaped sex organs known as **claspers** that are used in copulation. Although both sharks and bony fish possess paired fins, the fins of the shark tend to be stiff and not movable to the degree of the more flexible fins of bony fish. As a result, sharks are powerful swimmers, but they are not particularly agile or graceful swimmers. Furthermore, sharks cannot stop and swim backward as can bony fish. Unlike bony fish, sharks have no gas-filled swim bladders to aid with buoyancy, but instead rely on a large liver filled with oils. Even though it constitutes 30% of

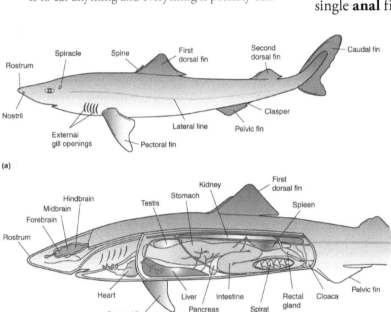

Figure 16.10 The Body Structure of a Male Spiny Dogfish Shark (*Squalus acanthias*). (a) Morphology. (b) Anatomy.

their body mass, their large, oily liver is of limited effectiveness in providing buoyancy, and so sharks must perpetually swim to keep from slowly sinking to the bottom.

Sharks range in size from the pygmy shark (*Euprotomicrus*), a deep sea species of only 22 cm (9 inches) to the whale shark (*Rhincodon*) and basking shark (*Cetorhinus*), which may reach 15 meters in length and weigh several tons. These massive sharks are harmless leviathans that subsist on plankton strained from the sea through modified **gill rakers**. All other sharks prey on bony fish, other sharks, squid, octopuses, molluscs, and mammals such as seals, and whales. Beautifully streamlined and powerful swimmers, open-ocean sharks are adept at pursuing and feeding on fast tuna, marlin, and the like. Bottom-feeding species of sharks are blunt-headed and stout in form with sluggish habits; the mollusc eaters among the bottom-dwellers among them have pavement-like, crushing teeth. The oddest shaped oceanic shark is the hammerhead (*Sphyrna*), whose heads resemble double hammers with an eye on each stalk.

Rays, including stingrays, manta rays, electric rays, sawfishes, and skates, comprise more than half of all elasmobranch species. Most rays are specialized for life on the ocean floor in shallow water where they use their blunt teeth to feed on invertebrates, mainly molluscs and crustaceans and the occasional small fish. Their most obvious modification for a bottom-dwelling existence is a lateral expansion of their pectoral fins into wing-like appendages. Gill openings are on the underside of the head with large **spiracles** (holes) on top. With the mouth and ventral side often buried in the sand, taking water in through the dorsally positioned spiracles prevents clogging the gills.

Figure 16.11 Flapping their wing-like pectoral fins, giant rays slowly glide through their watery realm.

Rays swim by slowly flapping the enlarged pectoral fins giving them the appearance of flying through the water (**Figure 16.11**). Elaborate color patterns on the dorsal surface may provide effective camouflage. Stingrays (*Dasyatis*) have a slender and whip-like tail equipped with modified scales as a venomous spine at the base of the tail. The spine is capable of inflicting wounds that are excruciatingly painful. Not only do such wounds heal slowly and often with complications, they may be fatal. Sadly, such was the case for Steve Irwin the iconic Australian conservationist and television personality known worldwide as the "The Crocodile Hunter." Irwin died in 2006 after his chest was fatally pierced by a stingray barb while filming on Australia's Great Barrier Reef.

Also included in this group are the electric rays (*Narcine* and *Torpedo*) and manta rays (*Manta*). The slow and sluggish electric rays possess large electric organs on either side of their head. Each organ is composed of numerous stacks of disclike cells connected in parallel. The electric ray is capable of discharging all the cells in the organ simultaneously, thus producing an electric current that flows out into the surrounding water. Althoughthe voltage produced is relatively low (50 volts), the amperage output may be almost 1 kilowatt—more than sufficient to stun prey and deter predators. Historically, the ancient Egyptians used electric rays as a form of electrotherapy in the treatment of miseries such as gout and arthritis.

Subclass Holocephali (Gr. *holos*, entire + *kephale*, head)—Chimaeras. Chimaeras are characterized by a large head with a small mouth surrounded by large lips. A narrow tapering tail has resulted in their common name "ratfish," although they are also known as spookfish, rabbitfish, and ghostfish. Chimeras are thought to be the remnants of a lineage that diverged from the shark line at least 360 million years ago. The fossil record reveals that chimaeras seem to have first appeared in the Devonian period, reached the pinnacle of their abundance and diversity in the Cretaceous and early Tertiary periods (120 million to 50 million years ago), and have been in decline ever since. Today there are about 50 species of extant chimaeras arrayed into six genera and three families.

Morphologically chimaeras are linked to the elasmobranchs, but they display a repertoire of unique characteristics as well. Their jaws are armed with large flat plates instead of sharp teeth. Their diet consists of algae, echinoderms, crustaceans, molluscs, and other fishes—a surprisingly eclectic diet given their lack of teeth. Also, their upper jaw is completely fused to the cranium, an unusual arrangement for a fish. They lack scales but have evolved a gill cover (**operculum**). Despite their unusual shape and appearance, chimaeras are beautifully colored and shimmer with a pearly iridescence (**Figure 16.12**).

Figure 16.12 The spotted ratfish (*Hydrolagus colliei*) gets its common name from its pointed rat-like tail. The ratfish's pectoral fins are large and triangular, and it has a venomous spine located at the leading edge of the dorsal fin.

Class Osteichthyes (Gr. *osteos*, bone + *ichthys*, fish)

With around 28,000 species, the bony fish are the major vertebrate life form in the vast expanse of earth's waters, and certainly one of the most successful from an evolutionary perspective. This lineage of fish with bony exoskeletons gave rise to a clade of vertebrates that contains not only 96% of extant fish species but all living tetrapods—amphibians, reptiles, birds, and mammals—as well.

The earliest fossils of bony fish are from the late Silurian deposits (around 405 million years ago). By the Devonian period (350 million years ago), the bony fishes had radiated extensively into two major groups: subclass Sarcopterygii and subclass Actinopterygii.

Subclass Sarcopterygii (Gr. *sarkos*, flesh + *pteryx*, fin)—Lobed-finned fishes. Characteristics of lobe-finned fishes include:

- A full bone endoskeleton; skin with embedded dermal scales with a layer of dentinelike material
- Paired and median fins present; paired fins with three large skeletal element homologous with the tetrapod humerus, radius, and ulna tipped with short dermal rays; muscles that move the paired fins are located on the limbs
- Jaws present; teeth covered with true enamel typically in the form of crushing plates restricted to the palate
- Gills supported by bony arches and covered with an operculum
- Swim bladder is used both as a lung for respiration and to regulate buoyancy in the lungfishes
- Nervous system of a small brain with a cerebrum, a cerebellum, and optic lobes

As their common name suggests, the paired fins of these fish are at the ends of leg-like lobes extending from the body. There are two groups of extant sarcopterygians: the lung fishes and the coelacanths.

The lungfishes are elongated freshwater fishes of the Amazon, western and central Africa, and Australia that possess lung-like organs as well as gills and can breathe air. The paired fins of the African and South

American lungfishes have become long, wispy sense organs while the Australian types have retained the lobed fins that characterize this subclass (**Figure 16.13**).

Only three genera of lungfishes survive today, and all live in regions where seasonal droughts are common, and the lakes and rivers they inhabit are subject to stagnation and drying. Some (*Neoceratodus*) inhabit the freshwaters of Queensland, Australia where they survive stagnation by breathing air, but they cannot withstand total drying. Others are found in freshwater rivers and lakes in tropical Africa (*Protopterus*) and tropical South America (*Lepidosiren*). They survive

Figure 16.13 The Marbled lungfish *(Protopterus aethiopicus)* is native to shallow pools and puddles of central Africa. Interestingly, this lungfish has the largest genome of any vertebrate.

when their watery habitats turn to dust by burrowing into the mud and keeping a narrow air pathway open by bubbling air to the surface. They line their retreat with copious amounts of slime mixed with mud to form a hard cocoon in which they **estivate** (a dormant state) for several months to as long as three years. When rain again fills the lake or riverbed, lungfishes emerge to again feed and reproduce.

Coelacanths were thought to be extinct for 70 million years. Thus, the ichthyologic community was stunned when people fishing in deep water off the coast of South Africa brought one up in 1938. That specimen was positively identified as a coelacanth and named *Latimeria chalumnae* after its discoverer, Miss Marjorie Courtenay-Latimer, curator in a small museum in East London. More specimens were eventually snared off the coast of the Comoro Islands just north of Madagascar. In 1997, the ichthyological community was surprised by the capture of a new species, *Latimeria menadoensis*, in northern Sulawesi, Indonesia, nearly 10,000 km (6,000 miles) from the Comoro population. Most likely these fish remained undiscovered for so long because they live at depths beyond those which humans could reach until the 20th century.

Coelacanths are stout metallic blue fish flecked with irregular patches of white or brassy yellow (**Figure 16.14**). Their paired appendages are muscular and leg-like and tipped with flexible fins. They occur in temperate waters in the "twilight zone," between 152-244 m (500-800 feet), off steep rocky slopes of volcanic islands. There they drift passively near the substrate feeding primarily on cephalopods and fishes. They are capable of moving swiftly and do so when capturing prey or avoiding danger.

Numerous characteristics are unique to the coelacanth among living fishes and other vertebrates, among them are the presence of a **rostral organ** in the snout that is part of an electrosensory system, and an intracranial joint in the skull that allows the anterior portion of the cra-

Figure 16.14 Coelacanth. The discovery of a living coelacanth thought to have been extinct for 70 million years, makes it the most well-known example of a "Lazarus taxon," a species that seemingly disappears only to be rediscovered much later.

nium to swing upwards, greatly enlarging the gape of the mouth. Other unique anatomical features include vertebrae that are incompletely formed or totally lacking bony centra, an oil-filled gas bladder, limb-like fins that are internally supported by bone, and paired appendages that move in a synchronized tetrapod-like pattern.

Modern marine coelacanths appear to be the descendants of Devonian stocks that initially inhabited fresh water lakes and rivers. The coelacanth's true evolutionary relationships are a matter of controversy and many unresolved questions remain. It seems likely that coelacanths might best be described as occupying a side branch in the basal portion of the vertebrate lineage, closely related to but distinct from the ancestor of tetrapods.

Subclass Actinopterygii (Gr. *aktis*, ray + *pteryx*, fin)—Ray-finned fishes. Characteristics of ray-finned fishes include:

- A full bone endoskeleton covered by skin with mucous glands and embedded dermal scales. The skull acts as a fulcrum, the relatively stable part of the fish. The vertebral column acts as levers against the fulcrum of the stable skull to operate the movement of the fish.
- Paired and median fins present, supported by long dermal rays or spines; muscles controlling fin movement are within the body; fins are flexible and highly movable
- Jaws present; teeth usually present with an enamel covering
- Gills supported by bony arches and covered with an operculum
- Swim bladder present, usually functioning as buoyancy control only
- Nervous system of a brain with small cerebrum, cerebellum, and optic lobes

Actinopterygians are the largest and most successful group of fishes comprising half of all living vertebrate species. Although ray-finned fishes first appear in the fossil record during the Devonian period (350-400 million years ago), it was not until the Carboniferous period (360 million years ago) that they became not only dominant in freshwater but also started to invade the ocean. At present, there are approximately 28,000 species of ray-finned fishes recognized. In addition, an estimated 5,000 to 10,000 undescribed species may still await discovery.

Traditionally three grades of actinopterygians are recognized: the *Chondrostei*—bichirs, reedfishes, sturgeons, and paddlefishes, the *Holostei*—gars and bowfins, and the *Teleostei*—all other remaining ray-finned species. The teleosts are the most familiar and largest group of fishes. Their approximately 23,600 species represent 96% of all extant fish types (or about half of all vertebrate species), and around 200 or so new species of teleosts are discovered and described each year. Although most new types are uncovered in remote or inaccessible areas such as the Amazon River of South America or deep oceanic waters, several new species are discovered each year from highly traversed and familiar areas such as the fresh waters of North America.

Ray-finned fishes are ubiquitous in all aquatic environments including extremes such as desert springs (e.g. pupfishes), subterranean caves (e.g. cavefishes), hot springs, high mountain streams, ephemeral pools, polar seas, and the inky black and crushing pressure of great ocean depths. They may live in lakes with salt concentrations three times that of seawater and swamps devoid of oxygen.

Ichthyologists have long distinguished freshwater from saltwater habitats. These chemically distinct zones, however, are often crossed by migratory **diadromous** (migrating between fresh water and salt water) species. Depending on the direction of migration, such wanderers can be **anadromous** (migrate from the ocean up rivers to spawn), as do salmon and lampreys, or **catadromous** (migrate down rivers into the ocean), as do freshwater eels. Although fresh water covers only a tiny fraction of the earth's surface (0.0093 %), approximately 41% of all fish species reside there. Most freshwater types are concentrated in tropical waters with the Amazon Basin and Southeast Asia displaying the largest most diverse assemblage of species (over 1500 different species in the Amazon region alone).

The array of body forms that exist within this subclass is truly spectacular. Consider that there are teleost fish with a body form and life style adapted to gliding, walking, or remaining immobile in addition to swimming, existing in all types of habitats except constantly dry land (though some can walk over land), feed on nearly every type of organic matter, utilize an array of impressive sensory systems, change genders, and even generate light or electricity. In addition, there is no color in nature's palette that is not found on ray-fin fishes. Their color diversity is considered essentially unlimited and ranges from uniformly dark black or red in many deep sea forms, to silvery in pelagic and water-column fishes, to the **countershading** contrasts (a dark color on top and a light color or white on the belly) of near shore or shallow water fishes to the striking and vivid colors of tropical freshwater and marine reef fish. Over the course of their evolution, several morphological adaptations have allowed teleosts to ascend to the pinnacle of fish diversity and structural development they hold today:

1. *A thin and flexible but tough body covering.* The heavy dermal armor of primitive ancient ray-finned fishes was replaced by light, thin, flexible scales. Some teleosts, such as most eels and catfishes, completely lack scales. The loss of the heavy armor scales of the ancestral bony fish increased mobility and speed and improved predator avoidance and feeding efficiency.

2. *Paired flexible fins.* Ray-fins are so named because they possess fins consisting of membranes of skin supported by bony spines (a single elongated tapering piece) or rays (bamboo-like segments) linked to the pelvic and pectoral girdles of the internal bone skeleton through proximal or basal skeletal elements (**Figure 16.15**). Such fins provide for efficient and precise movements (each ray can be controlled separately) as well as balance, braking, and even camouflage and social communication.

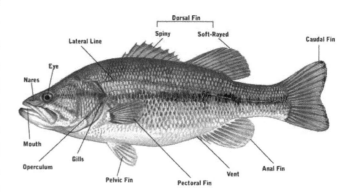

Figure 16.15 Morphology of a teleost fish.

3. *A swim bladder.* (**Figure 16.16**) In the teleost lineage, the swim bladder shifted from primarily respiratory to hydrostatic (buoyancy control) in function. Increasingly fine control of gases moving into and out of the swim bladder allowed precise regulation of buoyancy. Buoyancy control coupled with the likely coevolution of ray fins resulted in a great improvement in the efficiency of locomotion. Humans are not adapted for efficient swimming, and we expend a great deal of energy to stay afloat and move forward. If one places a floatation device around our midsection,

however, the efficiency and ease of our aquatic locomotion will be dramatically improved. Such is the benefit the swim bladder provides teleost fishes

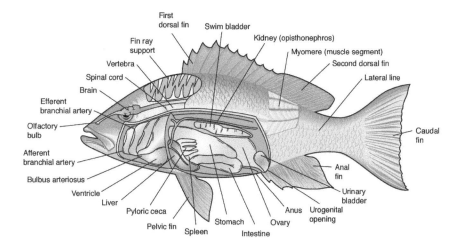

Figure 16.16 Anatomy of a female red snapper (*Lutjanus campechanus*).

4. *Improved jaws.* Adaptations in jaw structure and suspension allow for a fully movable maxilla and premaxilla. A movable upper jaw makes it possible for teleosts to protrude their jaws when opening thus creating a sophisticated suction device.

Body Plan of the Cartilaginous and Bony Fishes

Body Covering

Sharks are covered with a thick, (up to 10 cm [4 inches] in whale sharks) tough skin embedded with small placoid scales. These scales project posteriorly and give the skin a rough texture. As water passes over the streamlined body of the shark, it is channeled through the grooves on these scales reducing friction with the body and thus improving swimming efficiency.

A shark's skin is so rough that if a shark were to merely brush against you, it would most likely draw blood. In fact, shark skin was used as sandpaper by billiard ball makers working with ivory in times past. Not only is their skin rough, it is also so tough and durable that it can fashioned into leather goods such as boots, purses, shoes, and wallets. Shark teeth are actually modified scales (**Figure 16.17**). The row of teeth on the outer edge of the jaw is backed up by rows of teeth attached to a ligamentous band that covers the jaw cartilage inside the mouth. As the outer teeth wear and become useless or are broken off, newer teeth move into position from inside the jaw and replace them. This continues throughout the life of the shark. In some sharks, rows of teeth are replaced every 8-10 days, while in other species they could last several months; some sharks can lose 30,000 teeth in their lifetime. Crowns of teeth in different species may be sharp and triangular for shearing prey or flattened and stone-like for crushing the shells of molluscs.

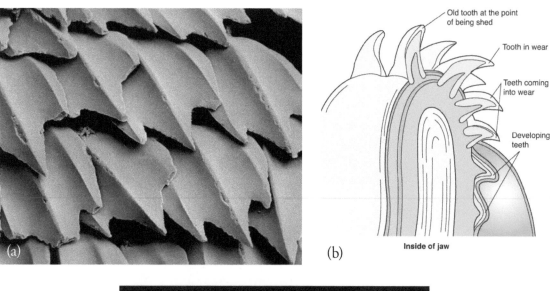

Figure 16.17 Scales and Teeth of Sharks. (a) Placoid scales (denticles) are found on sharks and rays. Unlike the scales of bony fish, placoid scales do not increase in size as the fish growns. Instead, new scales are added between the old scales (b) The teeth of sharks develop as modified placoid scales. (c) Older teeth are constantly replaced by newer teeth moving up from the inside of the jaw.

Bony fishes are typically covered with thin, two-part flexible scales embedded in the dermis. Each scale consists of a fibrous outer layer and an internal bony layer (**Figure 16.18**). Mucous glands in the epidermal covering over each scale produce a layer of slime that covers the fish even the fins. This slime layer helps protect against parasites and acts as a covering over wounds and scrapes. Some studies indicate that slime may also aid in reducing the turbulence of water moving along the body, making the fish a more efficient swimmer. Some species feed their body slime to their young, and the lungfish use it to form the cocoons that shelter them when their pools and ponds dry up.

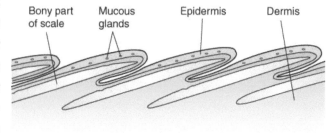

Figure 16.18 Section through the skin of a bony fish showing overlapping scales. The scales lie in the dermis and are covered by the epidermis.

The overlapping and interlocking of the scales provides a body covering for bony fish that is lightweight and flexible yet tough and durable. Teleost fish are covered by cycloid or ctenoid scales whereas other types of bony fish are outfitted with ganoid scales (**Figure 16.19**). Some bare-skinned bony fish such as catfish, eels, sturgeons, and paddlefishes have apparently taken the final step in the evolutionary reduction of the thickness and weight of the scales by abandoning scales entirely.

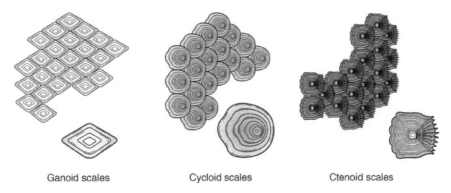

Ganoid scales   Cycloid scales   Ctenoid scales

Figure 16.19 Types of scales found on bony fish.

Fins and Locomotion

Fins are one of the hallmark traits of fishes, but the structure of the fins varies markedly between the cartilaginous fishes and the ray-finned fishes. Supported by rods of cartilage, the fins of a shark are relatively stiff and inflexible. Sharks have five different kinds of fins (Figure 16.10) performing a variety of functions:

1. Paired pectoral fins steer the shark. By rotating these fins slightly up and down, the shark can climb, dive, turn left, or turn right.
2. Paired pelvic fins stabilize the shark and prevent it from rolling as it powers forward through the water.
3. One or two dorsal fins also stabilize the shark as it swims. In some species, the dorsal fins have spines.
4. Not all sharks have an anal fin, but it provides stability for those sharks that do have one.
5. The caudal or tail fin provides the power to push the shark forward. The caudal fin of the shark is known as a **heterocercal tail** because the vertebral column turns upward when it reaches the tail, creating a larger dorsal lobe than ventral lobe in the caudal fin. This shape provides lift as the shark moves forward. (Figure 16.10) Sharks are powerful swimmers, but their relatively rigid fins prevent them from making the precise and graceful movements, including braking and swimming backward that come so easily to the bony fish.

It's not that bony fish have more fins than cartilaginous fish or that the fins of bony fish have a different function than those of cartilaginous fish. Rather, it is the structure and placement of the fins (and the presence of the swim bladder) that make bony fishes much more efficient, maneuverable, and graceful swimmers than cartilaginous fish. The fins of teleosts are membranes of skin stretched between slender rods of bone. Such fins are lighter and highly flexible compared to the thick nearly rigid fins of the shark. Furthermore, instead of having the pelvic fins located well behind the pectoral fins as is the case with sharks, the pelvic fins of bony fish are located just below or even slightly ahead of the pectoral fins. This positioning allows the pectoral fins to aid in maneuvering rather than serve simply as stabilizers.

Although the fins of bony fish are basically the same type and number and perform the same basic functions as do those of the shark, the caudal fin of bony fish is a **homocercal tail** (Figure 16.16). This type of caudal fin is symmetrical and suspended from the very tip of the vertebral column. Homocercal tails do not provide lift but do allow for uniform thrust forward and for more precise movements (each ray can be individually controlled). Fins give a bony fish control over its movements by directing thrust, supplying lift, and even acting as brakes. Any fish must control its pitch, yaw, and roll to swim effectively (**Figure 16.20**).

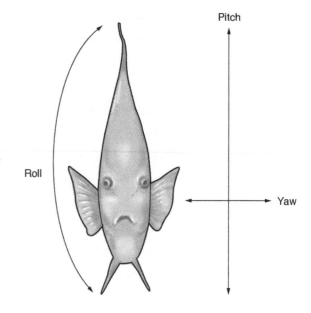

Figure 16.20 Directional control in teleost fish.

- Caudal fin provides forward thrust
- Paired pectoral fins act mostly as rudders and hydroplanes to control yaw and pitch and serve as brakes by causing drag
- Paired pelvic fins mostly control pitch
- Dorsal and anal fins control roll

The propulsive power of a fish is provided by the powerful musculature of the trunk and tail, which constitutes up to 80% of the total mass of the fish. Along the trunk, the musculature is arranged into zigzag bands, called **myomeres**. The myomeres have the external appearance of W lying on its side while internally the muscle bands are folded and nested in such a way to allow the pull of each myomere to extend over several vertebrae. (Figure 16.16) Such an arrangement allows for not only more power but also a finer control of movement since numerous myomeres are involved in bending a given segment of the body.

For a fish to swim through water, which is incompressible, it must actually shove the water aside. The fish can do this by wiggling back and forth in a serpentine motion of its head, first to the left, and then to the right, also with the curve of its body, and finally with its flexible tail (**Figure 16.21**). The water, tending to return to its original position, now flows back along the fish's narrowing sides, closing in at the tail and helping the fish forward. Three forces are at work as a fish swims:

1. **Thrust.** Thrust is a forward motion provided by the muscles bending the body and caudal fin back and forth.

Figure 16.21 The swimming motion of a fish.

2. **Lift**. Lift is an upward motion that operates in right angles to the thrust. Lift is provided by the fins and the shape of the body.

3. **Drag**. Drag is a force opposite the direction of thrust that tries to prevent the fish from moving forward. Fish must battle the pressure drag from the thickness (high resistance) of the water in front of them, the frictional drag of water moving along their body, and the vortex drag behind them caused by the movement of the tail. Drag is minimized by the streamlined shape of the fish and the slime layer covering it.

Water is 800 times denser than and 18 times more viscous (thick) than air. 10m (33 feet) of water generates the same pressure as 457,200 m (1.5 million feet) of air. Not only do fishes move through this thick medium with a seemingly effortless efficiency, some do so at impressive speeds. Most fish can swim with a top speed of about 10 body lengths per second. When these speeds are translated into kilometers per hour we find that a 30 cm (1 foot) fish can swim only about 10.4 km (6.5 miles) per hour. In general, however, the larger the fish, the faster it can swim, and some of the larger open-ocean fish can really move. On one occasion, a bluefin tuna on a line was estimated to be moving at 66 km (41 miles) per hour. Swordfish and marlin are the real speed demons of the deep being capable of incredible bursts of speed approaching or even exceeding 110 km (68 miles) per hour. Such high speeds can be sustained, however, for no more than 1 to 5 seconds.

Swimming is one of the most energy efficient modes of animal locomotion, largely because the buoyant properties of water support the animal and negate the pull of gravity. Studies have shown the metabolic cost per kilogram of body weight traveling 1 km (0.6 mile) for swimming salmon to be 0.39kcal compared with 1.45 kcal for a flying gull, and 5.43 kcal for a walking ground squirrel. Even though the medium of water is far thicker than air, fishes have evolved efficient methods of slipping through this viscous medium.

Buoyant Bladder

With body tissues denser than water, cartilaginous fish constantly battle negative buoyancy (a tendency to sink). Lacking a swim bladder, they are forced to rely on an oil-filled liver, the shape of their body, fins, and heterocercal tail, and constant swimming to generate lift. The oil in shark liver tissue has a density less than that of water, and the liver mass may contribute as much as 25 percent of the total body mass. Thus, the liver acts as a buoyant sack of oil that helps compensate for the weight of the shark. For example, a 4 m (13 feet) tiger shark (*Galeocerdo cuvier*) weighing 460 kg (1014 pounds) on land may weigh as little as 3.5 kg (8 pounds) in the sea. Not surprisingly, bottom-dwelling sharks, such as nurse sharks, have smaller livers with less oil making these sharks negatively buoyant.

Unique to bony fish, is the **swim bladder**, a thin sac nestled between the peritoneal cavity and the vertebral column that acts like a very sophisticated balloon (Figure 16.22) that may have arisen from the paired lungs of primitive Devonian bony fishes. Composed of interwoven collagen fibers and virtually impermeable to gas, the swim bladder occupies about 5 percent of the body volume of marine fishes and 7 percent of the volume of those living in freshwater. This difference in volume corresponds to the fact that salt water is denser than fresh water so a smaller swim bladder is needed.

In some extant species of bony fish (and perhaps numerous extinct types), the swim bladder functions as a lung. In most modern bony fish, however, the swim bladder functions to maintain neutral buoyancy (having the same density as water). Fish so equipped do not have to swim to maintain their vertical position in the water column. The only movement required when at rest is backpedaling of the flexible pectoral fins to counteract the forward thrust produced by water as it is ejected from the gills, and a gentle undulation of the tail fin to maintain a level posture in the water. Neutral buoyancy produced by a swim bladder works as long as a fish remains at one depth, but if a fish swims vertically up or down, the hydrostatic pressure that the surrounding water exerts on the bladder changes as does the volume of the bladder. For example, when a fish swims deeper the additional pressure of the water column above it compresses the gas in the swim bladder, making the volume smaller and reducing the buoyancy of the fish. When the fish swims toward the surface, water pressure decreases, the swim bladder expands, and the fish becomes more buoyant. To maintain neutral buoyancy, a fish must adjust the volume of gas in the swim bladder as it changes depth by either adding gas to the bladder when it swims down or removing gas when it swims up.

Two different mechanisms exist for adding or removing gas from a fish's swim bladder. Primitive teleosts—such as eels, herrings, anchovies, salmons, trout, and goldfish—retain a connection, the **pneumatic duct**, between the gut and swim bladder. These fishes are called **physostomous** (Gr. *phys*, bladder + *stom*, mouth) and because of the connection between gut and swim bladder they can gulp air at the surface to fill the bladder and can burp gas out to reduce its volume.

The pneumatic duct is absent in adult teleosts from more derived clades, a condition termed **physoclistous** (Gr. *clist* = closed). Physoclists regulate the volume of the swim bladder by secreting gas from the blood into the bladder. Both physostomes and physoclists have a **gas gland**, which is located in the anterior ventral floor of the swim bladder (**Figure 16.22**). The gas gland is the site at which gases from the blood stream enter the bladder. The gas gland secretes lactic acid, which enters the blood causing a localized high acidity in the arteries of the **rete mirabile** ("miraculous net"). This in turn forces hemoglobin to release its cargo of oxygen. The capillaries in the rete have a countercurrent arrangement so that the released oxygen diffuses from the venous capillaries to the arterial ones. As oxygen accumulates in the arterial capillaries, it eventually reaching such a high partial pressure that it diffuses into the gas bladder. If a fish needs more buoyancy, the gas gland releases more lactate. The final gas pressure attained in the swim bladder depends on the length of the rete capillaries; these capillaries are extremely short in fishes living near the surface but are extremely long in deep-sea fishes. This complex system is necessary to maintain the concentration of oxygen higher in the swim bladder than in the blood as simple diffusion would tend to pull the oxygen out of the bladder instead of pushing it in.

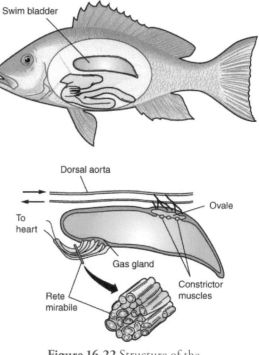

Figure 16.22 Structure of the swim bladder in teleost fish.

Physoclists have no connection between the swim bladder and the gut, so they cannot burp to release excess gas from the bladder. Instead, physoclists open a muscular valve, called the **ovale**, located in the posterior dorsal region of the bladder adjacent to a capillary bed. The high internal pressure of oxygen in the bladder causes it to diffuse out the bladder and into the blood of this capillary bed to be carried away.

Respiration and Gas Exchange

How does a fish breathe? As with any animal, a fish requires oxygen to sustain life. Oxygen is dissolved in water, and fishes extract it by taking water through their mouth, passing it through gill chambers, and expelling it through openings in the sides of the head. In cartilaginous fish the gills are covered by numerous **gill slits** and some types, especially bottom-dwellers, possess a modified slit called a spiracle located just behind the eye which assists with water uptake during respiration. In bony fish, the gills are covered with a movable flap, the **operculum**. The operculae of bony fish streamline the body, provide protection for delicate gill filaments, and make possible a pumping system to move water across the gills while the fish remains motionless, all benefits not accorded sharks by their simple gill slits.

Fish gills are composed of thin filaments, and each filament is covered with a thin epidermal membrane that is repeatedly folded into plate-like **lamellae (Figure 16.23)**. The flow of water over the gills is opposite the direction of blood flow (countercurrent flow) through the gills. This arrangement allows for extracting the greatest possible amount of oxygen from water. This counterflow system is so efficient that some bony fishes can remove as much as 85% of the dissolved oxygen from water passing over their gills.

The pumping action of the mouth and operculae (**buccal pumping**) creates a respiratory current that flows across the gills. Some filter-feeding fishes and many open-ocean fishes—such as mackerel, certain sharks, tunas, and swordfishes—have reduced or even lost the ability to pump water across the

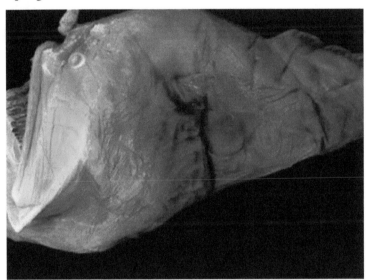

Figure 16.23 The Gill Structure of Fish. The operculum has been removed for clarity. (a) Four gill arches to each side bear numerous filaments. (b) A portion of the gill arch. The gill rakers project forward to strain out food and debris. (c) A single gill filament. Note that the direction of water flow is opposite to the direction of blood flow (countercurrent).

gills. These fish create a respiratory current by swimming with the mouth open, a method known as **ram ventilation**. Any fish that relies solely on ram ventilation must swim continuously to breathe. Many fishes rely on buccal pumping when they are at rest and switch to ram ventilation when they are swimming.

Surprisingly, a number of fishes can even live out of water for varying lengths of time by breathing air. Several methods are employed to accomplish this uncharacteristic feat. Lungfish are accurately named for possessing simple lungs, and the ability to use moist skin as a major respiratory organ allows freshwater eels to make overland treks during rainy weather.

Electric eels have degenerate gills and must supplement gill respiration by gulping air into a highly vascularized mouth cavity. The most unusual air-breathing fish may well be the Indian climbing perch (*Anabas*). Because of greatly reduced and inadequate gills, this fish actually spends most of its time on land near the water's edge, breathing air through special air chambers. As contradictory as it sounds, there are types of fish which will suffocate in water if they cannot reach the surface to breathe air.

Feeding and Digestion

More time and energy is spent in feeding or searching for food than on any other activity in a fish's daily life. According to the nature of their food, fishes have developed over the long history of their evolution feeding mechanisms and devices of ingenious and varied form; none more important and far-reaching than the evolution of jaws. Jaws freed fishes from a passive filter-feeding, mud-sucking existence, and enabled them to adopt a faster-paced predatory mode of existence. More than any other behavior, feeding shapes a fish's existence.

Although the varieties of food available to fishes are seemingly endless, most fish are carnivorous. Prey runs the gamut from tiny insect larvae to large aquatic vertebrates. Some denizens of the very deep sea are capable of eating victims nearly twice their own size and the anglerfish (order Lophiiformes) even goes fishing for other fish (**Figure 16.24**).

Because masticating (chewing) their food would block the respiratory current of water over the gills, most fish either swallow their prey whole or tear or crunch it into smaller pieces. Many fish have teeth of one sort or another to aid in feeding, and some of those teeth are fearsome and specialized tools. Piranhas, for example, have teeth with cutting edges sharp as razors. The white shark grows teeth with serrated edges like steak knives, and still others have teeth like needles that serve mainly

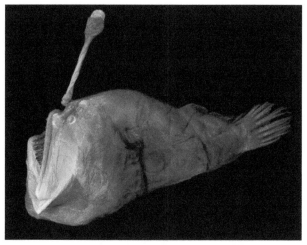

Figure 16.24 In the velvet blackness of the deep sea, the anglerfish (*Melanocetus eustales*) lures prey with a "fishing pole" tipped with a light. Equipped with a gaping toothed maw and elastic body, anglerfish can swallow prey many times larger than themselves.

to secure a firm grip on their prey. A few species of fish, like the wolf fish (*Anarhichas lupus*) of the North Atlantic, the Port Jackson shark (*Heterodontus portusjacksoni*), and the eagle ray (Myliobatidae) have developed massive crushing teeth to cope with the hard shells of their molluscan prey.

There are few herbivores in the watery realm of the fish. Those that do eat plants and macroalgae are most commonly found along coral reefs (parrotfishes, damselfishes, and surgeonfishes) and in temperate and tropical freshwater habitats (some minnows, carp, catfishes, piranha, pacu, and tetras).

The most abundant food source in the ocean on which more fishes feed than any other is plankton, the primeval fodder of the sea. The clouds of plankton that drift through the sun-lit upper levels of the ocean consists of myriads of both plants (phytoplankton) and animals (zooplankton) most of which are single-celled and microscopic in size. It has been calculated that over an entire year the net yield of plant plankton in an area such as the North Atlantic is about one ton per acre and that the total annual net production of phytoplankton amounts to nearly 500 billion tons in the entire ocean, far exceeding the biological productivity of the land. Much of this production is eaten by zooplankton, and the suspension or filter-feeding fishes in turn eat the zooplankton along with the phytoplankton. Since the plankton drifts about the oceans on or near the surface, these fishes are for the most part pelagic (open-sea) or surface dwellers. Characteristically, plankton feeders such as herring, anchovies, menhaden, capelin, pilchards, and others travel in large schools. The plankton eaters in turn are prey for larger but less abundant carnivorous fish.

To strain plankton from water, fishes need a sieve of some sort, and this they have developed in the form of structures called **gill rakers**. These are attached to the gill arches opposite the gills, and are simple or branched tooth-like processes arranged in rows as in a comb. Most plankton feeders obtain their food by swimming along with their mouths open and the gill covers expanded, straining quantities of water through them, The plankton collects on the gill rakers as the water stream goes out through the gill openings, and is diverted to the esophagus. The structure and spacing of gill rakers determine the size of food particles trapped. Fish with densely spaced, elongated. Comb-like gill rakers efficiently filter tiny prey, whereas carnivores and omnivores often have more widely spaced gill rakers.

Some types, such as hagfish, scavenge dead or dying animals while detritivores such as some suckers and minnows consume fine particulate organic matter. There are even a few parasitic sorts such as lampreys or the eye-picking cichlids which consume the flesh and/or body fluids of other animals.

Although there are several species that lack stomachs, for the most part, digestion in fishes follows the general vertebrate plan. When food is secured, it proceeds from the stomach to the tubular intestine, which may be short in carnivores but extremely long and coiled in herbivores and detritivorous types. The herbivorous grass carp (*Ctenopharyngodon idella*) has an intestine nine times the body length, an adaptation necessary to accommodate the lengthy digestion time required for tough plant carbohydrates. Digestion and absorption proceed simultaneously in the intestine. The ray-finned fishes possess numerous **pyloric ceca**, a curious adaptation found in no other vertebrate group. The primary function of these structures appears to be fat absorption, although all classes of digestive enzymes are secreted there.

To fuel their life of constant motion, the open-ocean sharks are literally eating machines. The most ferocious of these must certainly be the great white shark (*Carcharodon carcharias*). If the shark is king of the sea, then the great white must be the king of kings. With its rows of saw-edged razor-keen teeth, its speed, and its unerring sensing of prey, the great white is an instrument of death as swift and sure as a guillotine.

Great whites frequently devour their prey intact. Other sharks from 1.2-2.1 m (4-7 feet) have been found entire in the bellies of great whites, sometimes more than one at a time. A sea lion weighing 45.4 kg (100 pounds) was found intact in a great white taken off California. Even more incredibly, the remains of an entire horse in a great white taken in Australian waters were reliably reported.

Tiger sharks have a reputation for eating just about anything including conchs, skates, rays, fish, birds, sea turtles, and even human garbage. Objects such as shoes and license plates have reputedly been found in tiger sharks. Most experts agree, however, that the many stories one hears about all the strange things found in shark bellies are greatly exaggerated and most likely false.

Even though they are literally "eating machines," some sharks can to up to 6 weeks without food. The record for shark fasting was set by a Swell shark in an aquarium that went 15 months without eating.

Osmotic Regulation

An organism can be described as a leaky bag of dirty water. This is not an elegant description, but it accurately identifies the two important characteristics of a living animal—it contains organic and inorganic substances dissolved in water and this fluid is enclosed by a permeable body surface.

—F. Harvey Pough

Life in water presents distinct but different challenges for freshwater and saltwater fish when it comes to regulating the amount of water in their bodies. The salt concentration of fresh water (0.001 to 0.005 grams per mole liter [M]) is well below that of the blood of freshwater fishes (0.2 to 0.3 M). The result is that water constantly enters the body of a freshwater fish osmotically while salt is lost by diffusion outward. Freshwater fishes have several adaptations to counteract these diffusion difficulties. First, they have a kidney that forms very dilute urine allowing excess water to be pumped out of the body. Second, specialized **salt-absorbing cells** located in the gill epithelium transfer salt ions (sodium and chloride) from the water to the blood. By absorbing what little salt there is in fresh water together with salt present in food, the fish replaces the salt lost through diffusion (**Figure 16.25**).

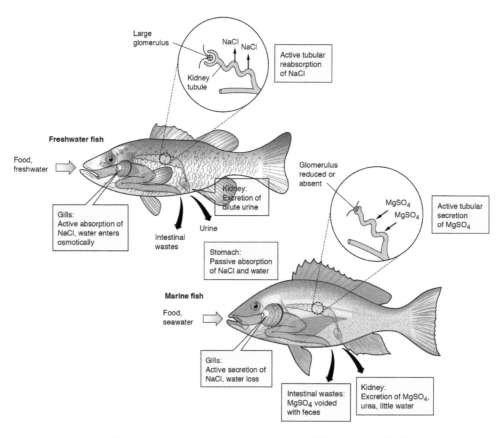

Figure 16.25 Osmotic Regulation in Bony Fish. The osmotic challenge presented to each type of fish is the opposite of the other.

Saltwater fishes encounter a completely opposite diffusion problem. They have a much lower blood salt concentration (0.3 to 0.4 M) than the seawater around them (about 1 M). As a result, they tend to lose water and gain salt. Actually, a marine fish is in danger of death by desiccation much like a desert animal deprived of water. To compensate for this water loss, a marine fish actually drinks seawater. Excess salt accompanies the seawater being drunk, however, and must be disposed of. Marine fish have two methods for accomplishing this. Some sea salt ions (sodium, chloride, and potassium) are transported by the blood to the gills where they are secreted back into the water by special **salt-secreting cells**. The remaining ions (mostly magnesium, sulfate, and calcium) are passed with the feces or excreted by the kidney.

Approximately 90% of all bony fishes are restricted to the salt conditions they have evolved to tolerate, and if placed in the opposite salt environment, they will be incapable of osmotic regulation and quickly die. Some 10%, however, can pass back and forth between both worlds with ease. These **euryhaline** fishes as they are known are of two types: those such as salmon and eels that spend part of their life cycle in fresh water and part in seawater and sculpins, killifish, and flounders that live in estuaries or intertidal zones where the concentration of salinity fluctuates greatly during the day.

Circulation and Body Temperature

The circulatory system of fishes is the simplest of all vertebrates. It is a closed system with a two-chambered heart that pumps the blood in a single loop throughout the body. It is a straightforward cycle from the heart through the gills, where it is oxygenated, to the various parts and organs of the body which consume the oxygen, and back to the heart again.

The heart of a fish consists of four parts—the sinus venosus, the atrium, the ventricle, and the bulbous arteriosus—but only two actual chambers (**Figure 16.26**). The **sinus venosus** is a thin-walled sac where blood from the fish's veins gathers before flowing into the **atrium**, a large muscular chamber. From the atrium, the blood flows into the **ventricle**, a thick-walled muscular chamber that performs the actual pumping. The blood pumped from the ventricle passes into a large tube called the **bulbus arteriosus**. The bulbus arteriosus conducts the blood to the **aorta** from which it flows into the fish's gills. The blood of fishes is thin and pale red

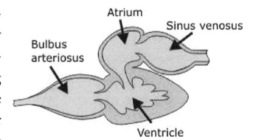

Figure 16.26 Consisting of only two actual chambers—atrium and ventricle—the heart of a fish is the simplest circulatory pump of all the vertebrates.

and moves through the body relatively slowly due to a relatively weak pumping action by the ventricle.

Fish, amphibians, and reptiles are commonly (but incorrectly) said to be "cold-blooded," whereas birds and mammals are often said to be "warm blooded" in regards to their body temperature. However, fish, amphibians, and reptiles should be more accurately designated as **ectotherms** (heat from the outside) and birds and mammals as **endotherms** (heat from the inside). As the temperature of most fish varies with the external temperature around them, they are said to be **ectothermic**. That is, as the water temperature around them rises or falls, so does the temperature of their bodies; they cannot regulate their temperature to maintain a constant "setting." Fish do, however, produce metabolic heat derived from the oxidation of food, but much of this heat is lost to the outside through the gills. Blood passing through the gills loses heat to the water quite rapidly, so that a fish's body temperature is usually within a degree or so of the actual water temperature.

Some tunas, mackerels, and sharks, however, have evolved countercurrent circulation in which the heat of the warm blood going to the gills is transferred to the cooled blood coming from the gills, and thus at least some of the metabolic heat is retained in the fish's body. This process, known as **poikilothermy**, allows fish with a variable body temperature to keep their temperature slightly higher than the surrounding water temperature. Yellowfin and skipjack tuna can keep their core temperature from about 5° to almost 12° C (9-21° F) above the water temperature. One skipjack tuna taken in warm waters registered a body temperature of 37.8° C (100° F)!

One advantage to poikilothermic fish is an increase in muscle power. Muscles contract more rapidly when warm without loss of force. With a 10° C (18° F) rise in body temperature a muscle can contract three times as fast, so three times the power is available from that muscle. More muscle power means more speed in pursuing prey, escaping enemies, and shortening the time required for long-distance migrations.

Nervous System and Sense Organs

The nervous system of a fish, as in higher vertebrates, consists of the brain and spinal cord with sensory and motor nerves branching from it. Fish have a small brain in relation to their body size. In fact, the mass of a fish brain is only 1/15 that of a bird or mammal of similar size.

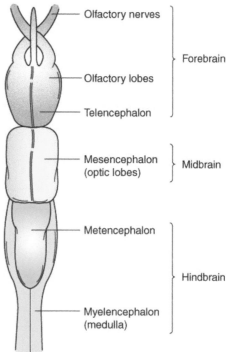

The fish brain is divided into several regions—forebrain, midbrain, and hindbrain (**Figure 16.27**). Located in front are the **olfactory lobes**, a pair of structures that receive and process signals from the nostrils via the two **olfactory nerves**. As expected, fishes that hunt primarily by smell, such as sharks, hagfish, and catfish, have large well-developed olfactory lobes. Behind the olfactory lobes is the **telencephalon**, a structure equivalent to the cerebrum in higher vertebrates. The telencephalon deals mainly with olfaction (smelling) and together with the olfactory lobes form the **forebrain.**

The **midbrain** (or **mesencephalon**) consists of two optic lobes. The optic lobes are very large and especially well-developed in species that hunt by sight, such as rainbow trout, many reef fishes, and cichlids. Lying beneath the optic lobes and connecting the forebrain to the midbrain is the **diencephalon**. The diencephalon functions in homeostasis and is associated with a number of hormones. The **pineal body** lies just above the diencephalon. This structure functions in detecting light, maintaining circadian rhythms, and controlling color changes.

The **hindbrain** (or **metencephalon**) is involved in swimming, body position, and balance. Corresponding in structure to the cerebellum of higher vertebrates, the metencephalon usually forms the largest part of the brain.

Figure 16.27 Brain of a Fish. The diencephalon and the pineal body lie out of view beneath the mesencephalon in this diagram.

The **myelencephalon** (or **medulla**) is the most posterior part of the brain. As well as controlling the functions of some of the muscles and organs, the myelencephalon, at least in bony fishes, is also concerned with respiration and osmoregulation.

Most fish possess highly developed sense organs and are very attuned to the environment around them.

Sense of smell and taste (chemoreception) Many fish have highly developed chemoreceptors that accord them extraordinary senses of smell and taste. Sharks have keen olfactory senses with some species able to detect as little as one part per million of blood in seawater. They are attracted to the chemicals in the guts of many species, and as a result, often linger around sewage outfalls. Some species, such as nurse sharks, have external barbels that greatly increase their ability to sense prey.

Bony fish have two **nares** (small openings) on either side of the head toward the tip of the snout. Nares don't lead to the throat the way nostrils do in mammals but open up into a chamber lined with sensory pads known as the **olfactory rosette**. The rosette is the organ that detects the chemicals and its size is proportional to the fish's ability to smell. Some bony fish, such as eels and salmon can detect chemical levels in water in concentrations as low as 1 part per billion! Fish use chemical cues in all sorts of ways from finding food and mates to defense. Minnows, for example, which when wounded release a chemical that incites other fish to flee. The most amazing sense of smell may belong to certain species of salmon that migrate from the ocean where they developed into adults back into the very freshwater stream where they were hatched. Their olfactory sense is so keen that they can smell the exact stream of their birth and swim up it to spawn.

Fish also possess the ability to taste. Taste buds are located not only on the tongue, but also on the lips and all over their mouths. Some fish such as the goatfish or catfish also have barbels ("whiskers") that have taste cells embedded in them.

Sense of sight (photoreception) Most fishes possess a keen sense of vision that is on par with our own vision; many can see in color, and some see extremely well in very dim light. The eye of a fish is structured somewhat differently from our own. Because water has a higher refractive index than air, the lenses of their eyes are perfectly spherical to help them better focus. They focus their eyes by moving the lens in and out instead of stretching it as humans do. Unlike humans, fishes cannot dilate or contract their pupils because the lens bulges through the iris. As is often the case with other types of animals, nocturnal (night active) fish tend to have larger eyes and pupils than diurnal (day active) fish.

Sense of hearing and touch (mechanoreception) Fishes, like other vertebrates, detect sounds as vibrations in the inner ear. The inner ears of fishes consists of two fluid-filled sacs lined with hairs. Each sac is divided into two sections, an upper section (**pars superior**) and a lower section (**utriculus**). The pars superior is divided into three semicircular canals and give the fish its sense of balance. The canals are arranged so that one gives a sense of yaw, one pitch, and the other roll. The utriculus actually gives the fish its ability to hear. It consists of several **otoliths** (ear bones) which vibrate and this in turn stimulates surrounding hair cells.

The fact that their bodies are nearly the same density as the surrounding water presents problems for fish when it comes to detecting vibrations. Because of the similarity in densities, sound waves pass through the fish's body nearly undetected. The ostariophysans, a group of teleosts that include the minnows, characins, suckers, catfishes, and other freshwater fish have developed an elegant adaptation to overcome this problem. Reception of sound begins at the swim bladder, which can easily vibrate since it is air-filled. Sound vibrations are transmitted from the swim bladder to the utriculus by a set of small bones known as the **Weberian ossicles**. The Weberian ossicles allow the ostariophysans to hear faint sounds over a much broader range of frequency than other teleosts. Other types of fish have adaptations for improved hearing as well. For example, herrings and anchovies have anterior extensions of the swim bladder that directly contacts the skull. The importance of this arrangement has been demonstrated by experiments in which the swim bladder is artificially deflated, resulting in reduced sensitivity to sounds.

Fish possess another sense of mechanoreception known as the **lateral line system** that has been described as "the sense of distant touch." The **lateral line canal** runs along the sides of the body onto the head, where it divides into three branches, two to the snout and one to the lower jaw. (Figure 16.15) Opening to the outside via **canal pores**, the canal is lined with **neuromast cells**, a cluster of hair cells which have their hairs linked together in a gelatinous glob known as a **cupula** (**Figure 16.28**). The hair cells in the lateral line are similar to the hair cells inside the human inner ear, indicating that possibly these structures share a common origin. Neuromasts may occur on the surface singly, in small groups called pit organs, or in rows within the lateral line canal.

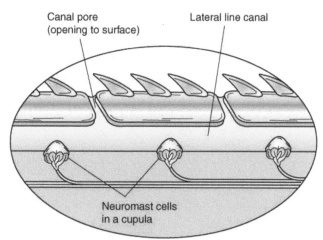

Figure 16.28 A portion of the lateral line system of a fish.

The development of the lateral line system depends on the fish's mode of life. For instance, fish that are active swimming types tend to have more neuromasts in canals than they have on their surface, and the lateral line will be further from the pectoral fins, which probably reduces the amount of "noise" generated by fin motion.

The lateral line system is sensitive to differences in water pressure due to changes in depth or in the pressure waves caused by approaching objects. As a fish approaches an object, the pressure waves around its body are distorted, and these changes are quickly detected by the lateral line system, enabling the fish to swerve or take other suitable action. Because sound waves are also pressure waves, the lateral line system is also able to detect very low-frequency sounds of 100Hz or less.

The lateral line system helps the fish to locate prey, avoid collisions, and to orient itself in relation to water currents. For instance, blind cavefish have rows of neuromast cells on their heads, which appear to be used to precisely locate food and avoid collisions in permanently pitch black conditions. Killifish can use their lateral line organ to sense the ripples made by insects struggling on the water's surface. Who cannot help but marvel at the view of a huge school of thousands of fish all moving as one in a shimmering dance of precision with each member close to but never contacting its fellows? Such is the power of the lateral line system in fish.

Electroreception A bioelectric field surrounds any living animal, and sharks and rays possess special organs for detecting this electrical potential (voltage). During the final stage of attack, sharks are guided to the bioelectric field of their prey. Electroreceptors known as the **ampullae of Lorenzini** are located primarily on the shark's head where they vary in number from a couple of hundred to thousands. These electroreceptors are modification of the lateral line system (**Figure 16.29**). With these sensors, sharks can detect prey hidden in the sand or sense the muscular contractions of struggling prey in the open. The oceanic currents moving in the Earth's magnetic field also generate electric fields that can be used by the shark for orientation and navigation.

Fish are socially intelligent creatures that do not deserve their reputation as dim-wits. Researchers have found fish to be steeped in social intelligence, pursuing strategies of manipulation, punishment, and reconciliation, exhibiting stable cultures, and cooperating to inspect predators and catch food.

Recent research has shown that fish recognized individual shoal mates, attained social prestige, and could even track relationships within the group. They have also been observed using tools, building complex nests, and exhibiting long-term memories. Such developments surely warrant a reassessment of the lowly regard in which their intelligence has been held for so long.

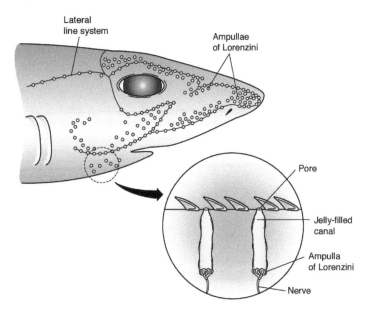

Figure 16.29 Electroreception System of a Shark. The ampullae of Lorenzini respond to weak electrical and magnetic fields and possibly to temperature, water pressure, and salinity.

Gone (or at least obsolete) is the image of fish as drudging and dim-witted pea-brains, driven largely by 'instinct,' with what little behavioral flexibility they possess hampered by an infamous three-second memory.

—Calum Brown

Reproduction and Development

Given the immense diversity of fishes, the existence of extraordinary numbers of variations on the basic theme of sexual reproduction should come as no surprise. Although the vast majority of fishes are dioecious, several different categories of hermaphroditism do occur:

- **Simultaneous hermaphrodites** possess both male and female sex organs, but they do not self-fertilize. They can, however, mate successfully with any other individual of the same species encountered. Members of the fish family Salmoniformes (salmon) and Serranidae (hamlets) are simultaneous hermaphrodites. When hamlets spawn the pair alternates the role of male and female through multiple matings, usually over the course of several nights.
- **Sequential hermaphrodites** may change their gender at some time in their life. The change may be from male to female (**protandry**) or female to male (**protogyny**) Anemonefish (clownfish) liv-

ing in symbiosis with anemones display protandry. Only the two largest fish will mate; the largest being the female and the second largest the male. When the female dies, the largest male will change gender and become a female. The remainder of the fishes are immature males. A classic example of protogyny is found in the wrasses and parrot fishes In wrasses the males form harems, with one large male sequestering and defending a group of smaller females. There are no small immature males. If the male dies, the largest female will change gender and become the male. In parrotfish, there is one large "supermale" but there may be one or more smaller males as well. These smaller males (sneaker males) cloak their presence by looking and behaving like females. While spawning the smaller males release sperm and not eggs as their appearance suggests they should, and, thus they enjoy at least a modicum of reproductive success.

Finding a mate for the purposes of sexual reproduction is usually no problem but for those fishes that live in the dark and barren waters of the deep sea, it poses immense challenges. If you find a mate in those conditions, it makes no sense, therefore, to let them go. And that is exactly what the deepsea anglerfish (superfamily Ceratiodea) do. The males are 3 to 13 times smaller than the female and unable to feed as they lack teeth and jaws. Guided to a female by her pheromones, the male attaches himself to her with hooked denticles. In some species, the tissues between the two fuse; the male becomes permanently attached and receives nourishment from the female, serving as little more than a set of parasitic testes.

Mating in fishes is usually a brief affair, but in many cases, it may be preceded by elaborate preliminaries that include color displays, "dances," sound production, and other courtship rituals. For example, the male stickleback, after a seasonal transformation in which his throat and belly become a bright orange-red, will attempt to entice females to his nest or territory by performing zigzag dancing movements and/or pricking the underside of the female's abdomen with his dorsal spines. The mating ritual of the betta (*Betta splendens*) is often violent, and the female may even be killed in the process. The male circles the female flaring his gill covers and flashing his brightly-colored and long drooping fins. A violent chase ensues in which the male may ram the female and bite chunks from her fins and body. Eventually, the male wraps himself around the exhausted female and as she releases her eggs, he fertilizes them.

Once mating has occurred, and the eggs have been fertilized, there are three basic modes of parental care toward those eggs:

- **Oviparity**. Over 97% of known fishes are oviparous. In the majority of these species, external fertilization occurs, with the male and female fish shedding their gametes into the surrounding water. In the ocean, huge schools of males and females may come together and release immense numbers of gametes into the water where they drift with the currents. Large female cod may release 4 to 6 million eggs at a single spawning. Less than one in a million of those eggs will survive to reach reproductive age.

  In most oviparous fish, the fertilized eggs are left to drift and fend for themselves. Some species, however, bury the eggs, while others attach them to vegetation or deposit them in nests. The male stickleback will build a mound-shaped nest in a hollow on the bottom of their pond using bits of plants bound together by sticky threads secreted from his kidney. After enticing as many females as possible to lay their eggs in his nest and then fertilizing them, the male aggressively stands guard over the nest and developing eggs. The nest of the Pacific salmon is little more than a depression in the gravel formed by the female using her tail fin. The most unusual nest may be that of the betta. Male bettas build a floating mucus bubble nest prior to mating. As the fertilized eggs fall from the female betta, the male quickly catches them in his mouth or retrieves them from the bottom and carries them up to the bubble nest where he deposits each egg into its own separate bubble. As in sticklebacks, the male betta then patrols beneath the nest defending it threats of all sizes with great vigor and courage.

The male yellowheaded jawfish (*Opistognathus aurifrons*) broods the fertilized eggs in its mouth until they hatch. Other mouth brooders include some cichlids and gouramis, and all arowanas, sea catfish, and cardinalfish. Depending on the group, **mouth brooding** is practiced by either the male, the female or both.

- **Ovoviviparity**. In ovoviviparous fish, the eggs are fertilized internally, and while they develop inside the mother's body, they receive no nourishment from her, depending instead on the yolk of the egg. Each embryo develops in its own egg, and the young are born alive. Familiar examples of ovoviviparous fish are the aquarists' favorite, the guppy as well as angel sharks, and coelacanths.
- **Viviparity**. In viviparous fish, the eggs are internally fertilized and the embryos receive nutrition from the mother in several possible ways. Some viviparous fish have a structure analogous to the placenta in mammals that connects the blood supply of the mother with that of the embryo. Viviparous fishes of this type include the lemon shark, surf-perches, and splitfins. The embryos of some viviparous fish exhibit **oophagy** whereby the developing embryos eat unfertilized eggs produced by the mother. This behavior, primarily attributed to sharks such as the shortfin mako and porbeagle, is also found in a few bony fish as well, such as the halfbeak (Hemiramphidae).

   Intrauterine cannibalism may be the most extreme method of vivipary. In this situation, the largest embryos in the uterus will eat their weaker and smaller siblings. This behavior is most commonly found among sharks, such as the grey nurse shark, but has also been reported for the halfbeak. After development, the young are born alive. Ovoviviparous and viviparous fishes are commonly known as livebearers to aquarists.

Most fishes reproduce continually throughout their lifetime (**iteroparity**), although some, such as Pacific salmon and lampreys, spawn but once and die shortly thereafter (**semelparity**).

Ichthyologists recognize five major developmental periods in fishes: embryonic, larval, juvenile, adult, and senescent. However, there are many gray areas within these major categories, and the developmental patterns of some species tend to defy classification into discrete categories.

Whatever the pattern of reproduction, soon after the egg is fertilized it absorbs water, and the outer layer hardens. Cleavage soon follows and the blastoderm forms soon assuming a fish-like shape. The larval stage of oviparous fish is marked by a tiny fish with an attached sac of yolk, which provides their food supply until the mouth and digestive tract develop. The live young of ovoviviparous and viviparous fishes are born fully developed without the need of a yolk sac. After a period of growth, the body shape is restructured, fin and color patterns develop, and what was once a fry becomes a juvenile bearing the unmistakable body form and markings of its species.

Growth in fish is driven by temperature. Consequently, fish living in temperate regions grow rapidly in summer when temperatures are high, and food is abundant but nearly stop growing in winter. Annual rings in the scales, otoliths, and other bony parts reflect this seasonal growth and can be used to determine a fish's age (**Figure 16.30**).

The lifespan of fishes varies greatly. In general, smaller fish have shorter lives than do larger fish. For instance, many smaller species live for only one year or less. The largemouth bass (*Micropterus salmoides*), however, can live 15 to 24 years; several species of rockfish (Moronidae) can live from 90 to 140 years, whereas some sturgeons (*Acipenser*) can survive up to 150 years. These long lifespans have become a serious issue for some fisheries because populations can be quickly decimated if most of the older reproducing fish are removed.

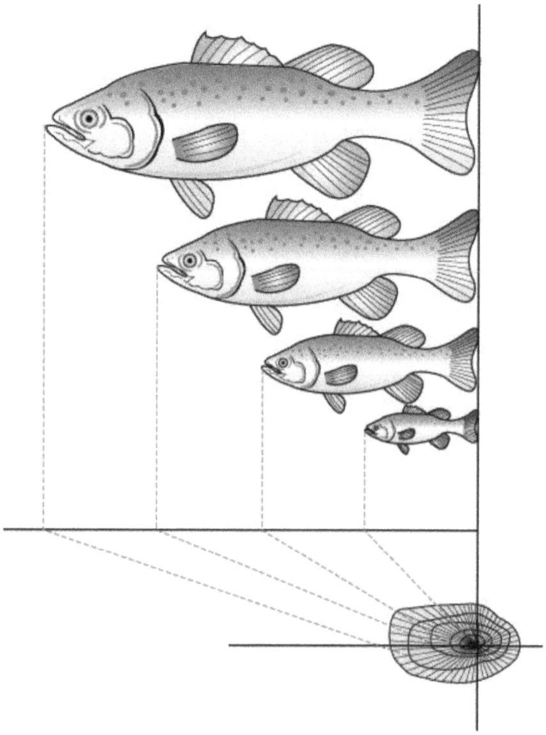

Fish Connection

Economically

Fish are an immeasurably important source of food for many types of animals, including humans. World fisheries take billions of pounds of fish worth hundreds of millions of dollars annually. According to the Food and Agriculture Organization (FAO) of the United Nations, about 1 billion people rely on fish for at least 30% of their animal protein. Some small island nations depend on fish for protein almost exclusively.

Figure 16.30 Scale Growth as a Function of Age in Bony Fish. Growth is interrupted during the winter, producing year marks (annuli). Each year's increment in scale growth is a ratio to the annual increase in body length.

Consumer demand for seafood has skyrocketed. From 1950 (18 million tons) to 1969 (56 million tons) fish consumption grew by about 5% each year. As demand continued to increase, production jumped from almost 50 million metric tons in 1976 to 95.5 metric tons in 1990, to 145 metric tons harvested worldwide annually. And the worldwide clamor for fish has continued to the point where overfishing has depleted fish populations making large scale commercial fishing, on average around the world, economically unviable without government assistance. By the 1980s, economists estimated that for every $1 earned fishing, $1.77 had to be spent in catching and marketing fish, and that imbalance has continued to grow. The fishing industry worldwide has responded by getting bigger—more vessels, larger vessels, and larger nets. A modern "factory" supertrawler can be longer than a football field and capable of catching and processing into various products up to 200 tons (40,000 pounds) of fish daily. One of the world's largest trawl nets could encircle more than a dozen Boeing 747 jetliners at its opening. Ships deploying such nets have a capture rate of about ten tons of fish per hour!

Lampreys have also long been used as food. During the Middle Ages, they were widely eaten by the upper classes throughout Europe, especially during fasting periods since their taste is much meatier than that of "true" fish. King Henry I of England is said to have died eating a "surfeit of lampreys." Lampreys are still a highly prized delicacy in Korea, the Baltic countries, and southwestern Europe. However, overfishing has greatly reduced their numbers in those locales.

Not all fish are eaten. Fish products are also economically important constituents in soap, paint, and glue and the use of fish or fish products as crop and plant fertilizer has been practiced for centuries. Furthermore, fish provide both pleasure and profit in recreation. Both marine and fresh water recreational fishing are relaxing pastimes for 34 million people in the United States alone. In addition, large sums of money are spent each year by anglers matching wits and technology against their finny adversaries. America's 34 million hunters and anglers are economic powerhouses. Their activities directly support 1.6 million jobs. They spend more than a billion dollars just on licenses, fees, and permits. They also generate $25 billion dollars annually in federal, state, and local taxes on the $76 billion dollars worth of gear and equipment they purchase each year. If these $76 billions were the Gross Domestic Product of a country, sportsmen and women as a nation would rank 57 out of 181 countries.

> *Give a man a fish, and you feed him for a day. Teach him to fish, and you feed him for a lifetime.*
>
> —Lao Tzu

Personal fish tanks have also become a global phenomenon. Estimates place the number of aquarists worldwide at around 60 million. These hobbyists spend over $4 billion dollars a year on their pastime, second only to bird watching. In the United States, there is an estimated one tank of fish in every 11 homes containing a total of at least 139 million freshwater fish and 9 million saltwater fish.

Humans have also found a way to exploit even the seemingly useless, and to some, repulsive hagfish. Their tough skin is processed into "eel leather" for use in boots, bags, wallets, purses, and other products. More importantly, Korea and Japan consume almost 5 million pounds of hagfish meat each year. Overfishing in Asia has decimated their local hagfish stocks, so the Asian hagfish fishery has turned its eyes towards North America where these "slime eels" are considered a worthless bycatch. Although it could mean a boost of millions of dollars to North American fisheries, care must be taken not to overfish these stocks as we have those of other types of fish.

Ecologically

Fish are a tremendously important part of aquatic food chains, and as predators and herbivores, fish help control the populations of the things they eat. Humans use this to their advantage by enlisting the mosquito fish (*Gambusia affinis*) to consume insects, especially mosquito larvae, and the grass carp or white amur (*Ctenopharyngodon idella*) to keep waterways clear of plant growth. Fish are ecologically threatened on a number of fronts:

Overfishing The biological diversity of the ocean and the health of many of its environs are threatened by the highly competitive race to catch enough fish to keep pace with rising international demand. **Overfishing**

occurs when fishing activities reduce fish stocks (populations) below an acceptable level. This can occur in any size body of water from a pond to the entire ocean.

Because of their low **fecundity** (potential reproductive capability) and the long time required to reach sexual maturity (up to 35 years in some species), sharks are especially susceptible to overfishing. Shark populations seem to be plummeting with most large species seeing declines of 90 percent or more since data started being collected on them. If fact, according to some marine biologists and marine ecologists sharks are the most endangered marine wildlife we have right now. In Asia, the fins of large sharks, particularly hammerheads, are highly prized. At prices in the hundreds of dollars per pound, shark fins are among the world's most expensive ingredients and are found in everything from soup to sushi and even cookies and cat food. One study estimates global trade at about half a billion dollars and 73 million sharks per year. One strategy currently being attempted is to track various species of large sharks to find where they congregate and mate, and then convince countries around those areas to establish and enforce marine protected havens.

Overcapacity The Food and Agriculture Organization of the United Nations (FAO) suggests that the maximum wild capture potential from the ocean was reached in 1996. There are too many boats on the water going after the same fish. Current estimates place the global fishing fleet at 200-300 percent larger than needed to catch what the ocean can sustainably produce. Overcapacity is a complicated issue to resolve, as removing subsidies and revoking fishing rights could have devastating effects on jobs and local or regional economies.

Bycatch As if overfishing were not bad enough, the widespread use of unselective fishing gear and indiscriminate netting practices result in one-quarter of all the fish brought on board fishing vessels each year being discarded, usually dead or dying, back into the sea. Commercial fishing vessels throw back on the average about 27 million tons of unwanted fishes annually. This bycatch amounts to about half of all the fish caught from the oceans each year that are consumed directly by humans. Along with these "trash fish," millions of other marine animals such as sea turtles, birds, and dolphins are being incidentally captured and killed in fishing operations.

A major international scientific study released in November 2006 in the journal *Science* found that about one-third of all fishing stocks worldwide have collapsed (with a collapse being defined as a decline to less than 10% of their maximum observed abundance), and if current trends continue all fish stocks worldwide will collapse within fifty years. According to marine fisheries researchers, upward of 85 percent of the world's wild fisheries either are being fished at the maximum rate that would allow for replenishment, or are already overexploited, depleted, or recovering from depletion.

Habitat degradation Aquatic ecosystems are under increasing stress from water pollution, the building of dams, removal of water for human use, and the introduction of exotic species. Freshwater fishes are particularly threatened because they often live in relatively small areas. The Devil's Hole pupfish (*Cyprinodon diabolis*), for example, occurs only in a single 3m x 6m (10 foot x 20 foot) pool within a limestone cavern in Nevada.

Overfishing can lead to the removal of so many fish of one or several species that serious negative effect to the environment results. If the depletion of a species causes an imbalance in the ecosystem, not only is it difficult for the depleted species to be brought back to sustainable levels, and other species dependent on the depleted species may become imbalanced causing further problems. Other animals in the ecosystem may be impacted by intense fishing practices as well. Long-lining results in the deaths of tens of thousands of albatross each year while driftnets indiscriminately kill millions of marine creatures as they target just one or two commercially valuable species. The use of cyanide is a popular method for capturing live reef fish for the seafood and aquarium markets, Cyanide fishers squirt cyanide poison into coral reefs where fish seek refuge. This stuns the fish making them easy to catch, but it also kills the corals of the reefs as well as other reef organisms. And less than half the fish taken in this manner survive long enough to be sold to aquarists or restaurants.

In addition, there is severe physical damage to ecosystems caused by fishing operations that use destructive gear and fishing practices, like deep-sea bottom trawling, that physically disturb marine habitats such as the ocean floor, sea grass beds, or coral reefs. In some cases, depleted fish stocks have been restored; however, this is only possible when ecosystems remain intact.

Inadequate conservation and management practices Because it is so vast and relatively unexplored, the ocean has always seemed impervious to anything humans could do to it. The hard ecological reality, however, is that the ocean's resources are finite, and depletion of these resources beyond sustainable levels is often irreversible. The management of ecosystems as opposed to managing only a few target species is the key. This entails:

- Maintaining populations of target species so they can play their natural role in ecosystems and to enable them to have sustainable reproduction rates.
- Eliminating the use of fishing gear that creates high levels of bycatch, or the incidental catch of nontarget species.
- Closing feeding, breeding, and spawning grounds to fishing to protect marine ecosystems.
- Increase production from aquaculture (aquafarming). In 2012, production of farmed sea food reached 66 million tons—3 million tons more than beef. When farmed aquatic plants and non-food products are included, world aquaculture production in 2012 was 79 million tons, worth US$125 billion. Today around 47% of all fish directly consumed by humans worldwide is produced by aquaculture. Although presented as an ecologically clean industry, farming certain species, such as shrimp and salmon, can have some serious negative environmental consequences.

The science is clear cut but the political and economic reality of the situation is that many fishing nations lack the will and/or the way to make protective management work.

Fish populations can also be negatively influenced by outside intruders. Invading fish have created environmental havoc in a number of instances. Perhaps the best example is the invasion of the Great Lakes by the sea lamprey (*Petromyzon marinus*). This species is native to the inland Finger Lakes and Lake Champlain in New York and Vermont. It is not clear whether it is native to Lake Ontario where it was first noticed in the 1830's or whether it was introduced through the Erie Canal which opened in 1825. It is thought that improvements to the Welland Canal in 1919 allowed its spread first from Lake Ontario to Lake Erie and then swiftly to the remaining lakes. Sea lampreys prey on all species of large Great Lakes fish such as lake trout, salmon, rainbow trout, whitefish, chubs, burbot, walleye, catfish, and even sturgeon. Because of its aggressive parasitic feeding behavior and lack of natural predators, the sea lamprey contributed significantly to the collapse of populations of fish species that were the mainstay of a vibrant Great Lakes fishery. For example, before sea lampreys entered the Great Lakes, Canada and the United States harvested about 7 million kgs (15 million pounds) of lake trout in Huron and Superior annually. By the early 1960s, the catch was only about 136,000 kgs (3,000,000 pounds). The fishery was devastated.

In 1955, the Great Lakes Fishery Commission was established. Working jointly with Fisheries and Oceans Canada, the U.S. Fish and Wildlife Service, and the U.S. Army Corps of Engineers, the commission undertakes sea lamprey control. This control program, known as "integrated sea lamprey management," uses several techniques to attack sea lampreys including:

- *Lampricides.* Lampricides are selective poisons that are periodically applied to lake tributaries that harbor larval sea lampreys. The lampricides kill the sea lamprey larvae in streams with little or no impact on other fish or wildlife. While about 175 streams are successfully treated at regular intervals, it is a costly control method and the commission is seeking to reduce its use by relying on alternative methods.

- *Barriers.* Sea lamprey barriers have been constructed to block the upstream migration of spawning sea lampreys; most barriers allow other fish to pass with minimal disturbance. Continuing improvements in these barriers have reduced or eliminated the necessity of lampricide treatment on many streams.

- *Traps.* Sea lamprey traps are operated at various locations throughout the Great Lakes, often in association with barriers. These traps are designed to catch lampreys as they travel upstream to spawn. Male lampreys caught in the traps are used for the sterile-male-release-technique while most females are used for research. Collectively, the traps provide the program with about 25,000 male lampreys annually.

- *Sterile male release.* If biologists are trying so hard to reduce the lamprey population in the Great Lakes, why release more lampreys into the lakes as is done in this program? Why not just destroy all sea lampreys caught in traps? Biologists believe that releasing sterilized males into the Great Lakes may actually reduce the number of sea lampreys produced in Great Lakes tributaries. A significant number of the sterilized males out-compete the normal males to mate with females and, thus, will produce nests of infertile eggs. Spawning sea lampreys (including the sterilized males) are past their parasitic phase and die after spawning.

The sea lamprey control program has brought the Great Lakes fishery back from the brink, and overall, has been tremendously successful. Although sea lamprey populations fluctuate like any other species, ongoing control efforts have resulted in a 90% reduction of sea lamprey populations in most areas of the Great Lakes. Today, the fishery is worth up to $4 billion annually to the people of Canada and the United States. Nearly 5 million people fish the Great Lakes yearly, and the fishery supports tens of thousands of jobs.

A misplaced negative attitude by humans toward certain animals can also senselessly reduce the numbers of those animals. There is no better example of this than our lowly regard for sharks. Out of hundreds of species of sharks, only four—the great white (*Carcharodon carcharias*), tiger (*Galeocerdo cuvier*), oceanic whitetip (*Carcharhinus longimanus*), and bull sharks (*Carcharhinus leucas*)—have been involved in a significant number of fatal unprovoked attacks on humans. A number of other species—shortfin mako (*Isurus oxyrinchus*), hammerhead (*Sphyrna*), blacktip reef (*Carcharhinus melanopterus*), lemon (*Negaprion brevirostris*), silky (*Carcharhinus falciformis*), and blue sharks (*Prionace glauca*)—have attacked humans without provocation, and have on extremely rare occasions even been responsible for a human death. That being said, is their reputation for ferocity against humans justified? The statistics say no. In 2000, there were 79 unprovoked shark attacks reported worldwide, 11 of them fatal. In 2005 and 2006 this number dropped to 61 and 62 respectively, while the number of fatalities dropped to only four in both years. Of these attacks, the majority occurred in the coastal waters of the United States—53 in 2000, 40 in 2005, and 38 in 2006. In 2007, the world-wide number of unprovoked attacks jumped to 71, while 2008 saw 59 unprovoked attacks. The yearly average of fatalities from the period 2001-2008 in United States coastal waters was four fatalities per year. Obviously, while one needs to exercise caution around sharks, they clearly are not the monstrous man-eaters they are often portrayed to be. To put your risk in perspective, consider that in coastal states of the U.S. during the period 1959 to 2006, there were 1,916 fatalities from lightning strikes but only 23 fatalities during that period from shark attack. Actually, sharks have more to fear from us than we do from them as overfishing coupled with the long length of time it takes them to develop sexually and to produce offspring has driven some species to the brink of extinction.

Medically

A number of chemical products derived from fish show great biomedical promise. The tetrodotoxin produced by pufferfish may prove useful in pain management, and the slime of catfish is being tested on wounds for its apparent abilities to reduce inflammation, block the growth of microbes, and accelerate tissue growth.

Research continues to confirm earlier studies that indicate the omega-3 fatty acids found in oily, dark-fleshed fishes, such as herring, mackerel, halibut, salmon, and flounder, offer genuine health benefits. Eating fish regularly appears to help ward off the serious heart rhythm disturbances associated with sudden cardiac death. It also seems to reduce the risk of heart attacks, strokes, mental decline of old age, and prostate cancer.

Fish oil pills are one way to get omega-3s but actually eating fish provides other important nutrients like selenium, antioxidants, and proteins. In fact, the American Heart Association recommends eating fish at least twice a week.

Unfortunately, there is a downside to eating this potent health food. Fish may contain environmental pollutants found in the water. Of particular concern are methyl mercury, dioxins, and polychlorinated biphenyls (PCBs). This has led to confusion among the public—do the risks of eating fish outweigh the benefits?

In a recent study, researchers from the Harvard School of Public Health (HSPH) tackled that question by undertaking the single most comprehensive analysis to date of fish and health. In the first review to combine the evidence for major health effects of omega-3 fatty acids, major health risks of mercury, PCBs, and dioxins in both adults and children/infants, the results show that the benefits of eating a modest amount of fish per week are enormous. A weekly consumption of 3 ounces of farmed salmon or 6 ounces of mackerel reduced the risk of death from coronary heart disease (CHD) by 36%. Notably, by combining results of randomized clinical trials, the researchers also demonstrated that intake of fish or fish oil reduces total mortality—deaths from any causes—by an astonishing 17%.

The investigators found no definite evidence that low-levels of mercury exposure from seafood consumption had harmful effects on health in adults. Their findings did agree with the recommendations of the Environmental Protection Agency and Food and Drug Administration that women of childbearing age, nursing mothers, and young children should eat up to two servings per week of a variety of fish (salmon, light tuna, mackerel, and up to 6 oz. per week of albacore tuna) and avoid only four species of fish—tilefish (also known as golden bass), king mackerel, shark, and swordfish. This advisory is only for women of childbearing age, nursing mothers, and young children, not the general adult population. Importantly, the evidence suggests that, for those women, it is as important for their health and for the brain development of their infants that they eat a variety of other types of fish as it is to avoid the four fish species higher in mercury.

Some studies have shown that PCBs and dioxins may be carcinogenic. The authors of the study found that the benefits of eating fish far outweighed the potential cancer risks from these chemicals. The study also points out that only 9% of the PCBs and dioxins in the U.S. food supply come from fish and seafood; more than 90% comes from other foods such as meats, vegetables, and dairy products. According to Eric Rimm, one of the researchers involved in the study, "These results from over two decades of research clearly show there is a health risk if adults don't eat fish."

The popularity of home aquariums (and those found in medical offices) might be at least partially explained by a number of studies that indicate just sitting quietly and watching fish can lower blood pressure and heart rate.

Recently it has been discovered that the mucus secreted in copious amounts by hagfish is unique in that it includes strong, threadlike fibers similar to spider silk. No other slime secretion known is reinforced with such fibers. Furthermore, an adult hagfish can secrete enough slime to turn a bucket of water into gel in just a few minutes. Research continues into potential uses for this mucus or a similar synthetic gel. Some possible applications include new biodegradable polymers, space-filling gels, and as a means of stopping blood flow in accident victims and surgery patients.

Culturally

Through the ages, fish have been featured in legend and myth, from the "great fish" that swallowed Jonah the Prophet to the half-human, half-fish mermaid. They have even been regarded as deities or gods by some cultures—Ika-Roa of the Polynesians, Dagon of various ancient Semitic peoples, and Matsya of the Dravidas of India. Fish as symbols range from the twinkling in the night sky of the constellation Pisces to the ichthys symbol used by early Christians to identify themselves to the fish as an emblem of fertility among Bengalis. Fishes have also featured prominently in art, literature, and movies.

Sharks Figure prominently in Hawaiian mythology. There are stories of shark men who have shark jaws on their backs and could change form between shark and human at any time they desired. There are also many shark gods in Hawaiian mythology. Sharks are known as Aumakua and are regarded as the guardians of the sea. In other Polynesian cultures, Dakuwanga was a shark god who was the eater of lost souls.

A Closing Note
Rule Number One: Keep All Orifices Covered!

Lurking the in muddy waters of the Amazon is a fish that locals fear even more than the legendary piranha. That fish is the candiru, also known as the toothpick fish, the vampire fish, or the penisfish.

The candiru is an eel-shaped parasitic catfish about 2.5-6 cm (1-2.5 inches) long and 3.5 mm (0.14 inch) wide that inhabits the Amazon and Orinoco Rivers of South America. Their diminutive size and nearly transparent body make them very hard to see and to find. They avoid the sun preferring to bury themselves in the mud and sand of the river bottom underneath rocks and logs.

The candiru has a voracious appetite for blood and will parasitize fish, and mammals, including humans to get it. One scientist, while holding a candiru, let it enter a small cut on his hand. It could be seen immediately wriggling under the skin towards a vein. To find a fish, the candiru first "tastes" the water trying to locate ammonia coming from the gills of a potential meal. One such an ammonia stream is detected, the candiru follows the stream to its new host and inserts itself inside the operculum. Spines around the head then pierce the host fish anchoring the candiru in place. Once attached, the candiru feeds on the blood of the host by using the long teeth on its upper jaw to rasp open a wound then employing its mouth as a slurping apparatus to gather the flowing blood.

The candiru is justifiably feared by humans bathing or swim nude in the home waters of this peculiar parasite. Natives bathe facing the current, as doing so would decrease the chance of the organism lodging itself in the rectum. Other orifices facing upstream—the vagina or penis—are either covered with clothing or tightly cupped by a hand.

If by chance the candiru should gain entrance to a human, it gorges itself on both the blood and body tissues in what is said to be a very painful experience. Because of the backwards-facing spines on the head candiru, it is impossible to remove the fish once it is inside except through surgery. If surgery is not performed in time, however, the blockage of the urinary tract will prove fatal, and the unfortunate afflicted person will die.

So if you ever find yourself, for whatever reason, naked in the Amazon or Orinoco rivers and you wish to avoid crossing paths with the only vertebrate known to parasitize humans, for heaven's sake, follow Rule One!

In Summary

- The first fishes seem to have been the ostracoderms, which appeared in the Cambrian Period about 510 million years ago and became extinct at the end of the Devonian, about 350 million years ago. These early fish were jawless, small (30 cm [1 foot]) and covered with bony armor or scales.
- The first fish with jaws, the acanthodians (spiny sharks), appeared in the late Silurian about 410 million years ago and became extinct before the end of the Permian about 250 million years ago.
- The Chondrichthyes, which appeared about 370 million years ago in the middle Devonian, are generally believed to be descended from the bony-skeleton placoderms, while the Osteichthyes, which arose during the late Silurian or early Devonian about 395 million years ago, are thought to be descended from the acanthodians.

- Fishes swarm the waters of our planet in numbers we can scarcely imagine and exploit every imaginable nook and niche of their watery domain. 60% of the fish species live in marine waters; the remaining 40% are found in fresh water.
- Fishes may be classified into two superclasses, four classes, and several subclasses:

> Superclass Myxinomorphi (fish with no jaws)
> > Class Myxini—Hagfishes
> Superclass Petromyzontomorphi
> > Class Cephalaspidomorphi—Lampreys
> Superclass Gnathostomata (fish with jaws)
> > Class Chondrichthyes (Cartilaginous Fishes)
> > > Subclass Elasmobranchii—Sharks, skates, and rays
> > > Subclass Holocephali—Chimaeras (ratfish)
> > Class Osteichthyes (Bony Fishes)
> > > Subclass Actinopterygii—Ray-finned fishes
> > > Subclass Sarcopterygii—Lobed-finned fishes

- Characteristics of Class Myxini—hagfish:

1. Strictly marine (salt water) animals
2. A slender, eel-like body lacking scale and skin containing slime glands
3. No paired appendages, possessing only a single fleshy caudal fin that extends anteriorly along the dorsal surface
4. A fibrous and cartilaginous skeleton; notochord persists throughout life
5. No jaws, but two toothed plates on tongue
6. Five to 16 pairs of gills with a variable number of gill openings; no operculum
7. Digestive system without stomach
8. Dorsal nerve cord with a differentiated brain, but no cerebellum
9. Sense organs of smell, taste, and hearing, but the eyes are degenerate
10. Sexes are separate (ovaries and testes in the same individual but only one is functional); external fertilization; large yolky eggs, but no larval stage Hagfish are found in cold ocean waters worldwide at depths around 1220 m (4,000 feet) buried in the sand and mud in very dense groups. Hagfish have elongated eel-like bodies with no scales, and a paddle-like tail consisting of a single fleshy caudal fin. Although almost completely blind, hagfishes are quickly attracted to food by keenly developed senses of touch and smell. They eat annelid worms, molluscs, and crustaceans and will scavenge the insides of dead and dying fish.

- Characteristics of Class Cephalaspidomorphi—lampreys:

1. A slender, eel-like body with no scales
2. One or two fleshy dorsal fins, but no paired appendages

3. A fibrous and cartilaginous skeleton; the notochord persists throughout life

4. No jaws; a suckerlike oral disc and tongue with well-developed keratinized teeth

5. Seven pairs of gills each with an external opening; no operculum

6. Digestive system without a distinct stomach; spiral folds in the intestine

7. Dorsal nerve cord with differentiated brain; a small cerebellum is present

8. Sense organs of taste, smell, hearing; eyes well developed in adults

9. Sexes separate; a single gonad without duct; external fertilization with a long larval stage

Lampreys are found in temperate rivers and coastal seas around the world, except in Africa. These elongated eel-like animals lack scales but possess well-developed eyes, a fleshy dorsal and tail fin, and seven gill openings along the side of the body. Like the hagfishes, they have a cartilaginous skeleton and lack both jaws and paired fins. The mouth is round and filled with rows of horny teeth and a movable rasping tongue. Although lampreys sometimes prey on small invertebrates, they are best known as predators/parasites of fish. Fastening to living fish, lampreys rasp into the flesh and feed on abraded skin, muscles, and body fluids.

- Characteristics of class Chondrichthyes—sharks, skates, rays, and chimaera:

  1. A relatively large fusiform (funnel-shaped) body (average about 2 m [7 feet]) with a ventral mouth and powerful jaws modified from pharyngeal arch

  2. Paired pectoral and pelvic fins, one or two dorsal median fins; pelvic fins in male modified as claspers; fins stiff and not very movable. The tail is often heterocercal

  3. The skin is covered with placoid scales; naked skin in some elasmobranchs and all chimaeras; teeth are modified placoid scales

  4. The endoskeleton is completely cartilaginous; notochord present but reduced

  5. The digestive system has a J-shaped stomach (stomach absent in chimaeras); intestine equipped with a spiral valve that slows passage of food and increases the absorptive surface area

  6. Five to seven pairs of gills leading to exposed gill slits (no operculum) in elasmobranchs; four pairs of gills covered by an operculum in chimaeras

  7. No swim bladder or lung but an oil-filled liver aids in buoyancy

  8. Brain of two olfactory lobes, two cerebral hemispheres, two optic lobes, cerebellum, medulla oblongata

  9. Senses of smell, vibration reception (lateral line system), vision, and electroreception well developed

  10. Sexes separate; paired gonads with reproductive ducts opening into the cloaca; internal fertilization with direct development

- Characteristics of class Osteichthyes—ray-finned fishes and lobed-finned fishes

  1. A full bone endoskeleton; skin with embedded dermal scales with a layer of dentine-like material

2. Paired and median fins present; paired fins with three large skeletal element homologous with the tetrapod humerus, radius, and ulna tipped with short dermal rays; muscles that move the paired fins are located on the limbs

3. Jaws present; teeth covered with true enamel typically in the form of crushing plates restricted to the palate.

4. Gills supported by bony arches and covered with an operculum.

5. Swim bladder is used both as a lung for respiration and to regulate buoyancy in the lungfishes

6. Nervous system of a small brain with a cerebrum, a cerebellum, and optic lobes

- Fish connect to humans in ways that are important economically, ecologically, medically, culturally, and religiously.

Review and Reflect

1. *A Candle in the Darkness* Some fish that live in the inky black depths of the ocean are bioluminescent (glow-in-the-dark). Propose as many hypotheses as possible as to why bioluminescence would be a useful adaptation for those fish that possess it.

2. *Spill Your Guts* Some deep sea fish "explode" if they are rapidly brought up to the surface from the depths at which they generally live. Review your understanding of fish anatomy and the pressures on these fish in their natural habitat to explain why this happens to these fish.

3. *Through the Senses of a Shark* Humans experience their world mainly through the visual sense. Sharks, on the other hand, have a suite of sensory equipment that includes not only vision (photoreception), but chemoreception (smell and taste) as well as electroreception. Describe a place you know as a shark might sense it.

4. *Open Water Camouflage* Have you ever sunk down to the bottom of a swimming pool and looked up? If so, you know the surface looks a bright mirror while around you and below you it is darker. Relate this observation to the fact that most open-water fish are counter-shaded.

5. *The Best of Both Worlds* What is the advantage to lungfishes of having both lungs and gills?

6. *Where Do You Stick the Thermometer?* Challenge: Can you take the temperature of a fish without physically touching the fish? Devise a way to meet the challenge and explain why your method should work based on what you know about fish anatomy.

7. *Walk a Mile in My Shoes* If you were to be reincarnated as a fish, would you rather come back as a bony fish or as a cartilaginous fish? Defend your answer.

8. *A Deadly Switch* Imagine you have two fish tanks where you live—one freshwater aquarium and one saltwater aquarium. You discover that your mischievous little brother has put one of your freshwater fish into the saltwater aquarium and one of your saltwater fish into the freshwater aquarium. Is this a serious issue to the switched fish? If so, explain why and describe what will happen to each fish if they remain switched.

9. ***In With the Good Air, Out With the Bad*** The rate at which fish consume oxygen and the corresponding effects this has on their body is illustrated in the graphs that follow. Interpret each graph and answer the questions (**Figure 16.31**).

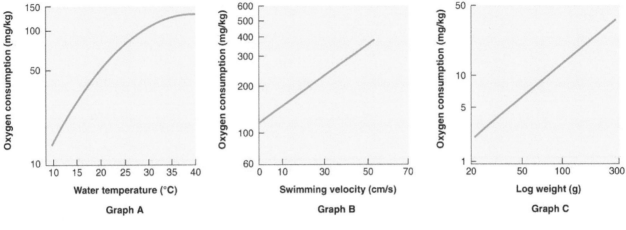

Figure 16.31

Graph A

A. What is the relationship between the temperature of the water and the amount of dissolved oxygen the water can hold?
B. What is the relationship between water temperature and the amount of oxygen consumed by fish?
C. Trout have a high oxygen demand. Would you go fishing for trout in a cold mountain stream or a shallow, warm river? Defend your choice.

Graph B

A. What is the relationship between oxygen consumption and swimming velocity?
B. How does the information displayed on the graph relate to the fact that fish that swim at extreme speeds are highly streamlined?

Graph C

A. What is the relationship between the weight of a fish and the amount of oxygen consumed?
B. Explain why this relationship exists.

10. ***More Eggs or Less Eggs?*** Cod lay eggs in the open ocean near the surface of the water and then the parents disperse. On the other hand, the male largemouth bass scoops out a nest on the sandy bottom of a lake or pond and waits for a female to deposit her eggs. The male then stands guard over the nest until the eggs hatch. Hypothesize the relative number of eggs produced by female

cod compared to the number of eggs produced by the female bass. Defend your hypothesis based on what you know about fish reproduction.

Create and Connect

1. **Fish-R-Us** Imagine you had the biological technology to create new species from scratch. Use this power to design a new species of fish specifically adapted to fit each of the following habitats and feeding situations:

   - Gulping large quantities of algae from the surface of a lake
   - Finding and getting prey from tiny cracks in a coral reef
   - Finding and getting worms buried in the bottom mud of a slow-moving river
   - Getting insects off a tree branch overhanging a pond
   - Filtering plankton from sea water

   Guidelines:

   A. Draw and completely label each fish. Add color if and where it is appropriate on each fish.
   B. Give each fish a common name.
   C. Thoroughly explain the adaptations you designed into each fish.

2. **Defend Yourself** The shark has been and continues to be one of the most misunderstood and persecuted animals on the planet. The truth is that far more people are killed by bee stings and snake bites worldwide than are killed by shark attacks. Many shark species have been pushed to the brink of extinction by a variety of factors: (1) The attitude that "The only good shark is a dead shark," (2) Commercial exploitation for fins and meat, and (3) Accidental deaths in fishing nets.

   Imagine that a deadly serious debate is raging in the great Hall of Biology. Some there are calling for the total destruction of each and every hideous and monstrous shark on the planet. Others say no. You are a great white shark and you have been called to testify to this assembly in defense of yourself and your kind. What would you say?

3. **Magna cum Fish?** As we learned in this chapter, fish have greater mental capacities than was previously imagined. While all signs point to this being true, could you prove it for yourself? Design an experiment to test the problem question: *Can fish learn?*

   Guidelines:

   A. Your design should include the following components in order:

      ➤ The *Problem Question*. State exactly what problem you will be attempting to solve.

> ➤ Your *Hypothesis*. While this is a fictitious experiment, word your hypothesis as realistically as possible.

> ➤ *Methods and Materials*. Explain exactly what you will do in your experiment including the materials necessary to accomplish the task. Be specific, take nothing for granted, and do not expect people to read your mind as they read your work.

> ➤ *Collecting and Analyzing Data*. Explain what type(s) of data will be collected and what statistical tests might be performed on that data. It is not necessary to concoct either fictitious data or imaginary observations.

B. Assume you have access to everything necessary to conduct your experiment—fish, adequate housing for the fish, plenty of laboratory space, and all the equipment necessary to meet the challenge. Within reason, money is no object.

C. Again, we have presented you with an experimental design challenge that is purposely broad and not very specific. Your greatest challenge will be to work out all the specific details leaving nothing to chance. To aid you in this endeavor your instructor may provide additional details or further instructions.

17

AMPHIBIANS: BETWEEN TWO WORLDS

The amphibians are a defeated group. They were the first vertebrates to emerge from the waters onto the lands, but they were not destined to complete the conquest, and, at first, abundant, they have shrunken into insignificance among four-footed vertebrates........ The typical amphibian is still chained to the water. In the water it is born, to the water it must periodically return. The amphibian is in many respects, little more than a peculiar type of fish which is capable of walking on land.

—Alfred Sherwood Romer

The Great Leap

Perhaps the most important milestone in the evolution of animals began some 400 million years ago when the first creatures made their way from water onto land. Known as the Devonian period, the world at that time was changing dramatically: plants had made the transition to land and complex plant ecosystems were forming there, the first woody plants appeared, and the water's edge was becoming a new kind of environment. The first animals to successfully sever their ancestral ties to water and exploit this new world were invertebrates such as snails, and later, insects.

The land held great promise for vertebrates for here was a foreign world brimming with untapped biological potential. Getting to that new realm, however, would require evolutionary modification of almost every system in the aquatic vertebrate body. No living group of animals more clearly reflects the transition from aquatic vertebrate (water) to terrestrial vertebrate (land) than do the amphibians. In fact, amphibians are the only vertebrates that show this transition in both their ontogeny (the biological development of an individual) and phylogeny (the evolutionary history of a group).

History of Amphibians

The move to land was a very gradual process, and the evolution of limbs was not a simple adaptation resulting from animals first crawling onto the shore and never looking back. In fact, the revised picture of this transformation shows that most of the changes needed for the life on dry land happened in creatures that were still living in the water. The lobed-fin fishes had already begun to evolve limbs by around 400 million years ago. They had fins that looked like fleshy paddles, and they had lungs as well as gills. Both the modern lungfishes and the coelacanths are ancient lines of fish that were common in Devonian seas. However, neither group is the likely ancestor of the amphibians; the lungfish do not have a suitable fin structure, and the coelacanths are a marine group without lungs.

(**Figure 17.1**) The most likely ancestors of the amphibians were the Rhipidistians—a type of sarcopterygian (lobed-finned) fish closely related to the modern lungfishes and coelacanth lines. Unfortunately, the fossil record of the origin of amphibians is poor. Rock deposits from the middle Devonian period contain typical rhipidistian fish, while early amphibian ancestors appear in the late Devonian, a difference of about 30 million years. No fossil species which directly link the two groups have been found to help explain what happened during those 30 million years (known as Romer's Gap, after the American paleontologist Alfred Sherwood Romer) and until more evidence is found, this important period of vertebrate evolution will remain uncertain.

The transformation from these lobe-finned fishes to the earliest **tetrapods**—four-legged animals—has long been of intense interest to zoologists. Many of the most telling fossils have been found in Greenland in the latter part of the 20th century and hopes are high that more evidence will be upturned there.

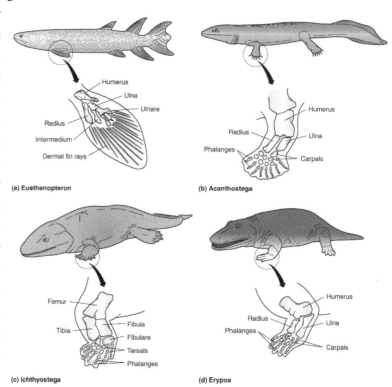

Figure 17.1 The Possible Evolution of Tetrapod Limbs. Limbs of tetrapods are thought to have evolved from the fins of Paleozoic fishes. (a) *Eusthenopteron*, a late Devonian lobed-finned fish (rhipidistian), had paired muscular fins supported by bony elements that foreshadowed the bones of tetrapod limbs. (b) *Acanthostega* was among the first vertebrates to have recognizable limbs. (c) *Ichthyostega* possessed lungs and limbs that helped it move through shallow swamps. (d) *Erypos*, had five digits on both front and hindlimbs, the basic pentadactyl model that became the tetrapod template.

The first vertebrates with limbs seem to have appeared during the upper Devonian around 360-367 million years ago. Among them was *Acanthostega*, a now-extinct genus of animals that were anatomically intermediate between sarcopterygian fishes and the first true tetrapods. With its stumpy legs and a long tail, which were probably used for propulsion in water, *Acanthostega* was basically an inhabitant of shallow water; its limbs were weakly structured and poorly positioned, and its backbone was too weak to support it on land. Despite the presence of lungs, *Acanthostega* had fishlike gills as well. Paleontologists surmise that it probably lived in shallow tropical swamps thick with plants. *Acanthostega* is seen as part of a widespread speciation event that occurred in the late Devonian period, starting with purely aquatic lobed-finned fishes, with successors showing increased air breathing capabilities, the development of land-capable limbs, and a more muscular neck that allowed freer movement of the head.

Existing around this same time period was a genus of now-extinct prototetrapods known as *Ichthyostega*. At 1.5 m (4 feet) long with massive heads, stumpy legs, and a long tail, they were big and heavy creatures for their time. Characteristics include a massive ribcage, stronger skeletal structure, and forelimbs with possibly enough power to pull the body from the water, *Ichthyostega* was anatomically closer to the true tetrapods than was *Acanthostega* In *Ichthyostega* the forelimbs seem to have been weight bearing, but the forearm was unable to extend fully. The proportions of the elephant seal appear the closest analogue among living animals. Perhaps *Ichthyostega* hauled itself up beaches, moving its forelimbs in parallel dragging its hindquarters along. Some paleontologists contend, however, that its limbs were more suited for pulling it through the swamps of its time.

From these and other finds, it now appears that the four legs common to land animals today most likely evolved for another purpose: navigating swampy wetlands, not as a means of moving directly to land. But once on land, their limbs also had a survival advantage. This is yet another example of how evolution frequently produces adaptations that come to be useful in the future for a different purpose.

The Ichthyostegoids (*Acanthostega*, *Ichthyostega* and others) were followed by temnospondyls and anthracosaurs, such as *Eryops*, a genus of extinct semi-aquatic amphibians. Fossils remains dated to about 295 million years old (lower Permian period) have been found in the Admiral Formation of Archer County, Texas, as well as in New Mexico and parts of the eastern United States. At 1.5-2.0 m (5-6 feet) long and weighing 91 kg (200 pounds), *Eryops* was one of the largest land animals of its time. The skull was large, broad, and flat reaching lengths of 60 cm (2 feet). This gave it an enormous mouth that was filled with many sharp teeth in strong jaws. Sturdy limbs supported and transported its body while out of water, and a thicker, stronger backbone prevented its body from sagging under its own weight. *Eryops* had typical amphibian posture exhibited by the upper part of the limbs extending nearly straight out from its body, while the lower part of the limbs extended downward at nearly a right angle from the upper segment. The body weight was not centered over the limbs but shifted 90 degrees outward and down through the lower limbs. Most of the animal's strength was used just to elevate its body off the ground for walking, which was probably slow and difficult.

During the Carboniferous period of 360 to 280 million years ago, amphibians became more abundant and diverse. This trend continued for another 40 million years ago leading to the Permian period (280 to 240 million years ago) being referred to as "The Age of Amphibians." The amphibian lineage was devastated by several mass extinction events during this period. The worst of these appears to have occurred at the end of the Permian in which perhaps 50% of all animal families became extinct. The amphibians were rocked again at the end of the Triassic period (around 208 million years ago) when around 35% of all animal families died out. As far as can be determined from very limited fossil evidence, the modern amphibians seem to have arisen during the Jurassic period with salamanders dating back 161 million years, while the earliest true frogs seem to have appeared around 190 million years ago. The modern orders of amphibians have clearly had separate evolutionary histories for a long time. The continued presence of such common characteristics as permeable skin, after at least 250 million years of independent evolution, suggests that some shared characteristics are central to the lives of modern amphibians. However, in regards to other characteristics, such as locomotion and reproduction, amphibians show tremendous diversity (**Figure 17.2**).

Three sets of critical evolutionary changes in amphibian lineages permitted amphibians to transition from totally aquatic to semi-terrestrial creatures. One was changes in the skeleton and muscles that allowed the development of limbs for better mobility on land. A second change involved a jaw mechanism and movable head that permitted more effective exploitation of food resources on land. A jaw-muscle arrangement that permitted rhipidistian fishes to snap, grab, and hold prey was crucial when early tetrapods began feeding on insects in terrestrial environments. Lastly, the development of lungs completed a battery of adaptations that permitted amphibians to lead the vertebrate leap onto land.

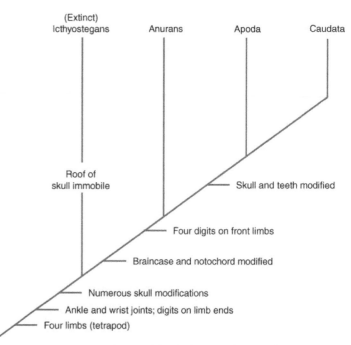

Figure 17.2 The Possible Phylogeny of Amphibians.

The emergence of paired appendages and the evolution of limbs were especially critical events in the evolution of all vertebrates. The sequence of evolutionary events leading to the appearance of tetrapod limbs has been known through the fossil record for some time, but only now are we deciphering how these events occurred at the molecular level.

Researchers have determined that fin development is associated with the presence of genes such as *HoxD*, *Fgf8*, and *Tbx18*, which are also vital in the development of human limbs. Genetic analysis of the lamprey—a living relic from the time before fish had paired fins—found the same genetic prompts in place. Later in evolution this entire genetic program was apparently reutilized in a new position to build the first paired fins and from those first paired fins tetrapod limbs eventually evolved. Strange as it may seem, human arms and legs apparently have their evolutionary roots in the dorsal, caudal, and anal fins of an ancestral line of fishes.

Diversity and Classification

The living amphibians are the smallest group of vertebrates consisting of around 5,400 species of frogs and toads, salamanders, and caecilians. Amphibians are restricted mainly to water or moist places, but because they cannot osmoregulate in salt water, none are found in the ocean. Amphibians are present on every continent except Antarctica. Their habitats include ponds, streams, and wetlands of all types, under rotten logs and rocks, in leaf litter, in trees, burrowing underground, even in pools of rainwater inside large cup-shaped leaves.

Amphibians reach their highest diversity and numbers in warm, humid climes, but some defy the rules. For example, some tree frogs are found at 3,600 m (12,000 feet) in the Sierra Nevada Mountains, and the wood frog ventures north of the Arctic Circle surviving for months in a frozen state of limbo. With its vital organs protected from damage by a natural antifreeze compound, the rest of a wood frog's body freezes solid during the winter. Some frogs and toads face an opposite challenge and live in hot, arid areas where they hide underground during the day and emerge only during the cool of the night.

The hellbender, mud puppy, and *Amphiuma* ("congo eel") are strictly aquatic. Most frogs live in or close to water, although toads usually range farther afield. Tree frogs and tree toads have taken partially to trees, with some tropical types living a completely arboreal life. Salamanders commonly hide under stones or logs, and the strange legless caecilians burrow into the moist earth of tropical areas.

In the warm and wet tropics, amphibians remain active all year round, but in the temperate zone autumnal environmental cues direct amphibians to find moist, sheltered places like muddy pond bottoms or deep leaf litter to overwinter.

Amphibians range in size from the Brazilian rainforest frog at 1 cm (0.4 inches) from snout to rump to the Chinese giant salamander (*Andrias davidianus*) at 1.83 m (6 feet) long and weighing 65 kg (143 pounds).

Investigations into the evolution and relationship of the various groups of amphibians have been hampered by a relatively sparse fossil record; key groups such as the caecilians have very few fossils. Another problem is that living anurans and caecilians are highly modified anatomically; frogs and toads are adapted for jumping and swimming while caecilians have lost their limbs entirely and are modified for a burrowing existence. Because these groups are so morphologically specialized, it is difficult to find clear-cut clues to their ancestry.

In the upheaval that is modern systematics, there has been some debate as to whether or not the traditional three orders of amphibians represent a monophyletic grouping. Shared characteristics such as the stapes/operculum complex, the importance of the skin in gas exchange, and aspects of the structure of the skull and teeth have always strongly supported a monophyletic origin, but now molecular data has effectively ended the debate and convinced most zoologists of the close relationships within in this group and the monophyletic singularity of their origin. With this in mind, we classify the amphibians as follows:

> Domain Eukarya
>> Kingdom Animalia
>>> Phylum Chordata
>>>> Subphylum Vertebrata (Craniata)
>>>>> Class Amphibia
>>>>>> Order Gymnophiona (Apoda)—Caecilians
>>>>>> Order Urodela (Caudata)—Salamanders
>>>>>> Order Anura (Salientia)—Frogs

Regarding salamanders, order Urodela is used when speaking only of living species whereas order Caudata refers to all salamander species, living and extinct. The same distinction is made with frogs. Order

Anura denotes only living species while order Salientia encompasses all frogs, living and extinct. In the popular literature, these amphibian order names are often presented as being synonymous and used interchangeably, while in the vernacular of zoology they are distinctly different.

General Characteristics of Amphibians

Amphibians are a most unusual group of vertebrates but one that retains a mere remnant of their previous glorious diversity. Although amphibians spend a great deal of time on land as adults, they have never completely broken their bond to water and become true land vertebrates. In many ways, amphibians are fish out of water leading what amounts to a double existence as their group name indicates—Gr. *amphi*, dual + *bios*, life. As adults, amphibians are encased within a thin, porous skin that performs necessary respiratory functions. The skin must remain moist at all times if it is to function properly and affect vital gas exchange. The requirements of a moist integumentary system, however, present serious environmental challenges to amphibians whose skin is exposed to air. As land animals, amphibians are in constant danger of desiccation with suffocation swiftly following. Thus, adult amphibians, with only rare exceptions, spend most of their time in or near water and are found on land only in very damp areas or close to standing water. Eggs are almost always laid and develop in water. The larval form that hatches from the egg is more like a fish than an amphibian in that they have a two-chambered heart, possess a single fleshy fin, and breathe through gills.

The living amphibians have very different body forms, but the members of the various orders share the following derived characteristics:

- A bone endoskeleton with varying numbers of vertebrae; ribs present in some, absent or fused to vertebrae in others. None possess an exoskeleton, and the notochord does not persist in adults.
- A smooth, thin, porous skin containing both mucus glands and poison glands.
- Four limbs (tetrapod) which may vary in size with the forelimbs of some being much smaller than the hindlimbs; some are legless. Limbs have varying numbers of digits and webbed feet are often present; no true nails or claws.
- The mouth is usually large with small teeth in upper or both jaws; two nostrils open into the anterior part of the mouth cavity.
- Respiration is accomplished either separately or in combination by lungs, skin, and gills; some larval types possess external gills, and these may persist throughout life.
- A three-chambered heart consisting of two atria and one ventricle.
- Body temperature regulation is ectothermic in nature.
- Separate sexes (dioecious) with internal fertilization via spermatophore in salamanders and caecilians, but external fertilization in frogs and toads. Larvae develop in water or very moist environments and undergo complete metamorphosis. Amphibians are the only vertebrates to undergo complete metamorphosis.

Order Characteristics of Class Amphibia

Order Gymnophiona (Gr. *gymnos*, naked + *ophineos*, as a snake)—Caecilians (also known as "rubber eels").

Found only in the tropical forests of South America, Africa, and Southeast Asia, the caecilians are legless, wormlike burrowers that feed on worms and other soil invertebrates (**Figure 17.3**).

Zoologists have identified and described about 160 species of these seldom-seen amphibians and arranged them into six families. Caecilians range in size from 10-150 cm (4-60 inches), and their smooth skin varies from blackish to pinkish tan. Some types have small calcite scales embedded in the skin. Caecilians appear to be externally segmented, but this is merely superficial and is caused by folds in the skin that overlie separations between muscle bundles.

Owing to their underground life, the eyes are small and covered by skin for protection giving rise to the misconception that caecilians (L. *caecus* = blind) are blind. This is not the case, but their visual sense is limited to simple light-dark perception.

Figure 17.3 Resembling snakes or large annelids worms, caecilians are actually neither. Rather, they are legless amphibians.

Caecilians possess two retractable sensory tentacles on their head. By possibly transporting chemicals from the environment to olfactory detector cells in the roof of the mouth, these tentacles are thought to function in a secondary olfactory capability in addition to the normal sense of smell based in the nose.

Except for one lung-less species, caecilians are lung breathers, but they also use the skin and the smooth skin lining the mouth for oxygen absorption. Often the left lung is much smaller than the right one, an adaptation to a cylindrical body shape that is also found in snakes.

Recent research has revealed that certain caecilians produce potent skin poisons from specialized skin glands. These poisons most likely prevent predation. In fact, the poison of the bright yellow caecilian (*Schistometopum thomense*) of West Africa kills other animals kept in the same tank within a few days. Chemical analysis has shown that these poisons are very different from those of other amphibians such as those produced by the skin of poison-dart frogs.

Order Urodela (Gr. *oura*, tail + *delos*, evident)—Salamanders.

The salamanders have the most generalized body form and locomotion of the living amphibians. Members of this order number over 500 species arrayed into ten families and are found on all continents except for Australia and Antarctica. Roughly one-third of all the known urodelans are found in North America; the highest concentration of those being found in the Appalachian Mountains. This region holds the greatest diversity of urodelans on the planet.

Salamander habitat is restricted mostly to the northern hemisphere, with the exception of a few species that live in the northernmost part of South America. Although common on the European mainland, salamanders are not native to the British Isles. Within their geographical range, salamanders and newts tend to inhabit cool, moist microhabitats in or near brooks and ponds or under logs and rocks and even in caves. Some species are aquatic throughout life (Salamandridae or newts), some take to water intermittently, and others are entirely terrestrial as adults.

In most salamanders the forelimbs and hindlimbs are of approximately the same size and set at right angles to the body. The limbs are rudimentary or absent altogether, however, in some aquatic and burrowing forms. All types possess a long and distinct post-anal tail throughout life (**Figure 17.4**). Salamanders are typically less than 15 cm (6 inches) long, but the North American hellbender (*Cryptobranchus alleganiensis)* can reach lengths of 76 cm (2.5 feet) or more. In Japan and China, the giant salamander (*Andrias japonicus*) reaches 2 m (6-7 feet) and weighs up to 30 kg (66 pounds).

Salamanders are carnivorous in all stages of their lives. Their prey consists of arthropods (mainly insects), worms, and small molluscs. Salamanders, like all amphibians, are ectothermic in nature with a low metabolic rate.

Figure 17.4 The spotted salamander (*Ambystoma maculatum*) is a common resident of the forest floor in the eastern United States and Canada.

Order Anura (Gr. *an*, without + *oura*, tail)—Frogs and toads.

> *Theories pass. The frog remains.*
> —Jean Rostand

Arrayed into 21 families, the order Anura contains over 5,200 species making it by far the largest group of extant amphibians. In fact, about 88% of all amphibian species are frogs.

The use of the common names "frog" and "toad" is more a vernacular designation than a scientific distinction. From a classification standpoint, all anurans are frogs, but only members of the family Bufonidae are considered true toads. The common usage of the term "frog" refers to species that are aquatic or semi-aquatic with smooth, moist skins, while the term "toad" generally refers to species that tend to be terrestrial with dry, warty skin (**Figure 17.5**).

Frogs are distributed from the tropics to subarctic regions worldwide, but they exist in the greatest abundance and variety in tropical rainforests. They are not found, however, on Antarctica or many oceanic islands. Frog species are often patchy in their distribution, being restricted to certain localities (specific streams or ponds) and absent or scarce in similar habitats elsewhere. Some frogs inhabit arid regions such as deserts, where water may not be easily accessible. The Australian genus *Cyclorana* and the American genus *Pternohyla* will bury themselves underground during dry periods (up to 1 m [3.3 feet]), and use its sloughed skin to create a water-proof mucal cocoon. Buried and sealed, the frog slumbers for up to five years in a process known as **estivation**. Once the rains return, they eat their skin container, dig their way out, find a temporary pool, and breed. Racing against the desert heat, the tadpoles develop at a rapid rate to become adults and burrow in before their pool dries up. Other frog types are adapted to cold, dry regions. The wood frog (*Lithobates sylvaticus*) ranges as far north as the Arctic Circle. Using both urea and glucose that it accumulates in its tissues as "cryoprotectants" to limit the amount of ice that forms and to reduce osmotic shrinkage of its cells, these frogs can survive an Arctic winter if no more than 65% of the total body water freezes.

Figure 17.5 Morphologically (a) frogs (family Ranidae) are distinguished from (b) toads (family Bufonidae) by the length of their legs—frogs, long and webbed; toads, short, stumpy, and not webbed—and their skin—frogs, smooth and damp; toads dry and bumpy.

During winter months, most frogs in temperate climates hibernate in the soft mud underlying ponds and streams. Their life processes slow to a very low ebb during this time as they live off the energy of the glycogen and fat stored in their bodies from summer feeding. Most terrestrial frogs, such as tree frogs, hibernate in the humus of the forest floor.

Compared with the other orders of amphibians, frogs are unusual because they lack tails as adults, and their hind legs are more suited to jumping and swimming than walking. Frogs range in size from the Brazilian gold frog (*Brachycephalus didactylus*) at 10 mm (0.4 inch) long to the 30 cm (1 foot) goliath frog (*Conraua goliath*) of West Africa. This giant preys on animals as big as rats and ducks. (As you contemplate these measurements, keep in mind that frogs are measured from snout to anus, not from snout to the ends of out-stretched hind legs.)

The skin texture of frogs varies from smooth to warty or folded with the skin hanging loosely on the body because of the lack of connective tissue. Frogs have three eyelid membranes: a transparent one to protect the eyes underwater, and two that vary from translucent to opaque.

The mouth of frogs is large and basketlike, and most frogs do have teeth of a sort. They have a ridge of very small cone teeth known as **maxillary teeth** around the edge of the upper jaw and several **vomerine teeth** on the roof of their mouth. Frogs lack teeth on their lower jaw and toads lack teeth altogether. Teeth are used to grasp prey and prevent its escape, not for biting or chewing. Thus, all food is swallowed whole.

Frogs are carnivores with huge appetites, and they will eat anything live they can get in their mouth. They prey mainly on insects, worms, small crustaceans, small molluscs, small fish, reptiles, birds, and even mammals as well as tadpoles and other adult amphibians. Most frogs are ambush hunters sitting camouflaged and concealed waiting for prey to approach. When prey is in range, the frog fires out its sticky tongue and pulls the prey back into the mouth. Lacking powerful throat muscles, the frog then contracts its eye muscles drawing the eyes down into the body where they exert pressure on the food and help force it down into the stomach.

Adult frogs have many enemies, such as snakes, birds, turtles, raccoons, large fish, and humans. Although they appear defenseless, many frogs have successful survival strategies they employ against potential aggressors. Some feign death, while others inflate themselves with air and push themselves up off the ground as far as possible to present as large and threatening an appearance as possible all the while emitting screeching sounds. A frog's best defense is its ability to leap, as anyone who has been startled by a frog leaping from nearly underfoot to land kerplop in the water just out of reach can attest. Some frogs are brightly colored to warn potential predators of the deadly poisonous nature of their skin, and bullfrogs in captivity do not hesitate to snap at tormentors often inflicting painful bites.

Amphibian Body Plan

Body Covering

Amphibians are covered with a thin permeable skin that varies in texture from slick and smooth in the frogs, salamanders, and caecilians to rough and warty in toads. Amphibian skin is composed of two layers: a layered outer **epidermis** and a spongy inner **dermis. (Figure 17.6)** The outer epidermal cells contain stores of **keratin**. The presence of this tough, fibrous protein provides protection against abrasion and dehydration. The rough, warty skin of terrestrial amphibians such as toads can be accounted for by the heavy deposits of keratin found

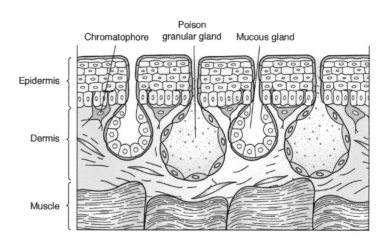

Figure 17.6 Section of a frog's skin.

within their epidermis. Amphibian keratin, however, is softer than the keratin that forms the claws, horns, hair, and feathers of other vertebrates.

Amphibian skin lacks a covering of scales, feathers, or hair, and is highly glandular. The secretions of these glands are important to the survival of the animal. Embedded in the dermis of amphibian skin are three types of exocrine glands: small **mucous glands** that secrete a protective waterproofing mucous onto the skin, **seromucous glands** that secrete enzymes and mucous, and **granular poison glands** that secrete a complex chemical mixture of biologically active compounds including alkaloids, peptides, biogenic amines, and steroids. Mucous may be the most important of the various skin secretions produced by amphibians. Death by desiccation is a constant threat to a creature whose skin is so permeable that an amphibian without skin would lose only slightly more water than an amphibian with its skin intact.

The chemicals produced by the granular poison glands perform a variety of functions. They protect the animal from bacterial and fungal infections and their neurotoxic (nerve poison) and myotoxic (muscle poison) properties discourage potential predators. The skin of the poison-dart frogs (Dendrobatidae) of Central and South America develops some of the most lethal biological toxins known. The black-legged dart frog

(*Phyllobates bicolor*) of Colombia produces skin secretions so deadly that a mere 0.00001 gram (0.0000004 ounce) is enough to kill a full-grown man. The Choco Indians poison the tips of as many as 50 blow gun darts with the secretions of one tiny frog and then use the darts to bring down large game such as jaguars, monkeys, and birds.

The skin of amphibians comes in every color of the rainbow. Some, such as the poison-dart frogs are brilliantly colored (**Figure 17.7**), while others are camouflaged with drab colorations of brown, green, and gray that blend with the natural color tones of their habitat. Amphibians owe their colorful variations to both pigment granules in the epidermis and specialized pigment-containing cells called **chromatophores** in the dermis. Most amphibians have three types of chromatophores: **xanthophores**, containing yellow, orange, and red pigments, **iridophores**, containing a silvery pigment that reflects light, and **melanophores**, containing brown or

Figure 17.7 The brightly colored poison-dart frogs display aposematic patterns that warn potential predators of the toxic nature of these tiny frogs.

black melanin. The bright, almost iridescent, colors of many tropical frogs are the result of light reflecting off the tiny mirror-like iridophores back through the xanthophores. Surprisingly, the green colors and hues so common in frogs are produced not by green pigment but by iridophores that reflect and scatter a blue light up through yellow-pigmented xanthophores. Blue light passing through the overlying yellow pigment appears green. By concentrating or dispersing various pigments by means of chromatophores, many amphibians can change their skin colors. This ability to color shift allows amphibians to adjust their body temperature (light colors reflect heat more than dark colors), and serves as a camouflage mechanism, helping them avoid visual detection by potential predators (**Figure 17.8**). With skin so equipped, amphibians display **mimicry** (a superficial resemblance of one organism to another or to a natural object), **aposematic coloration** (bright and conspicuous warning coloration), and **cryptic coloration** (matching body coloration and patterns to background colors and patterns).

The permeable skin of the amphibians may be employed to extract both oxygen and water from the environment. Although

Figure 17.8 Frogs are the masters of camouflage both in water and on land. One can look directly at them or nearly step on them without ever seeing them. This frog is using cryptic coloration to disappear into its surroundings.

most adult amphibians have lungs, their lungs are small and unable to supply the oxygen demands of the individual. Hence, they must rely on their oxygen-permeable skin to supplement their lung breathing. Several types of salamanders lack both gills and lungs and obtain all their oxygen through their skin. Water can also permeate this thin, delicate skin, and most amphibians have no need to drink water because they absorb as much as they need through their skin.

Support and Locomotion

The evolutionary transformation of amphibians from aquatic to terrestrial creatures required major adaptations and changes to their skeletal system. Paddle-like fins developed into round, stumpy limbs capable of supporting the animal's weight without the buoyant assist of water, whereas the vertebral column shortened and stiffened to accommodate the stresses and leverage required for support and movement. Also, the amphibian skull is relatively smaller and has fewer bony elements than the skulls of fishes. These adaptations lighten the skull so it can be more easily supported out of the water. Amphibians, unlike fishes, have a neck that allows for greater mobility of the head. Changes in jaw structure and musculature allow terrestrial vertebrates to grab, hold, and crush prey in the mouth.

The vertebral column of amphibians has been modified to provide both the flexibility and support required for a life on land. Similar to the arch of a suspension bridge, the weight of an amphibian's body is supported between anterior and posterior paired appendages which act as buttress columns. Supportive processes called **zygapophyses** on each vertebra prevent twisting when stress is applied (**Figure 17.9**).

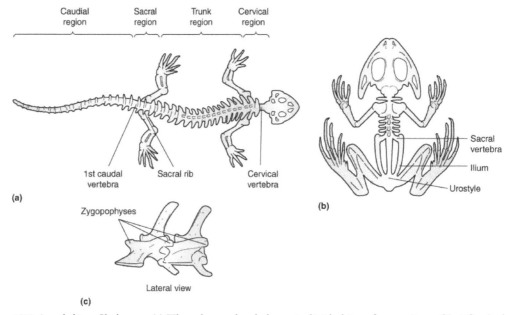

Figure 17.9 Amphibian Skeletons. (a) The salamander skeleton is divided into four regions. (b) A frog's skeleton show adaptations for jumping. The back legs are long, and firmly attached to the vertebral column through the ilium and urostyle. (c) Interlocking processes, called zygapophyses, prevent the vertebrae from twisting.

Unlike fish, amphibians depend more on appendage musculature than on the body wall musculature for locomotion. Thus, appendicular musculature predominates. To visualize this difference between the main musculature of fish and amphibians, contrast the body parts you would dine on if eating fish (body wall) as opposed to eating frog (legs).

The burrowing caecilians move by alternately folding and extending from points where the body bends making contact with the wall of the burrow. Locomotion in salamanders is reminiscent of the undulatory waves that pass along the body of a fish in which alternate movement of appendages results from muscle contractions that advance the stride of a limb by contorting the body into a curve (**Figure 17.10**). The limbs of a salamander protrude from the side of the body reducing salamanders to "belly-draggers" that must lever themselves over the ground slowly and clumsily.

Figure 17.10 The Locomotion Pattern of a Salamander. Note the fish-like bending of the body. Salamanders swim by folding the legs back against the body and propel themselves forward with back-and-forth sweeps of the tail.

The limbs of the frogs and toads are positioned more under the body than those of salamanders, and they can raise themselves more upright than salamanders. Aside from their explosive leaps, frogs and toads are slow and rather clumsy walkers on land. The long hindlimbs and the pelvic girdle of anurans are clearly modified for jumping.

Modifications of the pelvis (**ilium**) and vertebral column stiffen the posterior half of the anuran while long hindlimbs and powerful muscles form an excellent lever system for jumping. The forelimbs function as shock absorbers upon landing thanks to the elastic connective tissues and muscles that attach the pectoral girdle to the skull and vertebral column (**Figure 17.11**).

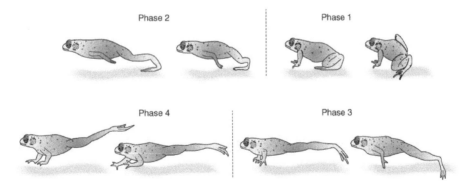

Figure 17.11 The Jumping Pattern of a Frog. Although frogs have a powerful take-off, they make a less than graceful "kerplop" landing.

Armed with these skeletal and muscular adaptations, frogs are among the greatest leapers on the planet and some are capable of jumping 20 times their own body length. That is the equivalent of a human leaping 100 feet from a standing start. By comparison, a flea can jump 150 times its own length whereas the kangaroo can manage only 4.5 times its own length. Alas, the poor elephant cannot jump at all.

In water, anurans swim with an alternate thrusting (frog kick) of webbed hind legs while salamanders swim with back and forth undulations of the tail reminiscent of the S-shaped wriggle of fishes.

Feeding and Digestion

Up high the flies are playing and frolicking and swaying.
The frog thinks: Dance! I know you'll soon end up here below.

—Willian Busch

Amphibians occupy an intermediate position in the food webs that surround them. Adults and larvae are preyed upon by all types of other creatures—large aquatic insects, fish, other amphibians, snakes, wading birds, and many types of mammals, including humans. In turn, adult amphibians are carnivores that feed on a variety of invertebrates and even the occasional small vertebrate. The main factors that determine what amphibians will eat are prey size and availability. Large bullfrogs, for example, will prey on small mammals, birds, and other anurans. The aquatic larvae of anurans (often called tadpoles) are herbivorous filter feeders that excel at straining algae and bits of organic matter from the water. Some tadpoles can extract food particles as small as one tenth of a micrometer (0.000004 inch) in diameter—an efficiency that rivals the best mechanical sieves produced by humans. In addition, some tadpoles eat so much so quickly that up to half their body weight is in their digestive system.

Many salamanders use only their jaws to capture prey, but anurans and a few salamanders use both their tongues and jaws for capturing food. Amphibians were perhaps the first vertebrates to evolve a true tongue. (The "tongue" of a fish is simply a fleshy fold anchored to the floor of the mouth and is not used for manipulating food.) The tongue of an anuran is attached in the front to the margin of the lower jaw and folds back over the floor of the mouth. When prey flits into range, the amphibian lunges while firing out its forked and sticky tongue. The tongue unfolds, turns over, and slaps around the prey. Sticky secretions on the tip of the tongue help secure the meal. The tongue and prey are then quickly flicked back inside the mouth. This capture and retrieval occurs with blinding speed in as little as 0.05 to 0.15 second!

(**Figure 17.12**) Once the prey is in the mouth, the anuran presses the prey against the vomerine teeth on the roof of the mouth while the tongue and other muscles of the mouth push the food toward the esophagus. The eyes, pulled downward during swallowing, helping force food toward the esophagus. As a result, frogs are sometimes said to be "eye swallowers" rather than "throat swallowers."

From the mouth, food passes through the esophagus into the stomach and the stomach in turn opens into the small intestine. Attached to the small intestine are the liver and gall bladder. The small intestine leads to the large intestine, which ends in a muscular cavity called the **cloaca**. Found in fish, amphibians, reptiles, and birds, the cloaca is a single posterior opening emptying the digestive, urinary, and genital tracts.

Figure 17.12 Prey Capture by a Frog. The extensible tongue is forked at the tip and attached to the front of the mouth.

Respiration and Vocalization

Breathing air is a trade off for amphibians. On the one hand, terrestrial animals expend much less energy moving air across gas-exchanges surfaces in lungs than do aquatic organisms in gills because air contains 20 times more oxygen per unit volume than water. On the other hand, exchanges of oxygen and carbon dioxide require moist surfaces, and the exposure of respiratory surfaces (skin and mouth) to air may result in rapid water loss.

Adult amphibians typically have lungs, rather than gills, for breathing oxygen. Some water-dwelling species, however, have both lungs and gills, while others lack both and obtain all the oxygen they need to survive through their permeable skin. In amphibians, gases are exchanged across three respiratory surfaces: skin (**cutaneous respiration**), mouth (**buccal respiration**), and lungs (**pulmonary respiration**). When breathing in air, carbon dioxide is lost primarily through the skin (cutaneous) while oxygen is absorbed primarily through the lungs (pulmonary). Underwater, however, gases are exchanged solely through cutaneous respiration. This ability to "skin breath" (and a greatly lowered metabolic rate) allows amphibians to overwinter at the bottom of ponds.

The lungs of salamanders are relatively simple sacs with 30% to 90% of gas exchange occurring across the skin. The lungs of anurans, however, are more subdivided, thus increasing the surface area available for gas exchange. Amphibians lack the large rib cage and powerful thoracic muscles that reptiles, birds, and mammals use to decrease the volume of their lungs (negative pressure system). Instead, they employ a **buccal pump** mechanism for ventilation (a positive pressure system) (**Figure 17.13**).

Lungs, in conjunction with vocal cords, can also be used generate sounds. Although both male and female anurans possess **vocal cords** enclosed in a **larynx**, sound production (**vocalization**) is primarily a function of males. Most salamanders and caecilians lack vocal cords but are capable of making faint squeaks.

Male frogs and toads can produce an amazing array of sounds, including clicks, whistles, chirps, trills, barks, and the typical croak. Anurans call by passing air through the larynx in the throat causing the vocal cords to vibrate. Muscles control the tension of the vocal cords and regulate the frequency of the sound. The sound is amplified by one or more **vocal sacs**, which consist of membranous pouches under the throat or on the corner of the mouth that distend and act as res-

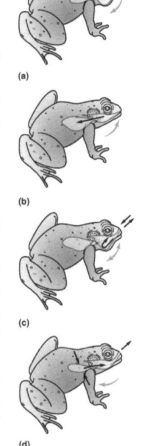

(a)

(b)

(c)

(d)

Figure 17.13 Buccal Pumping in a Frog. (a) The floor of the mouth is lowered, drawing air in through the nares. (b) With nares closed and glottis open, the frog forces air into its lungs by pulling up the floor of its mouth. (c) The mouth cavity vibrates for a time. (d) Elastic recoil of the lungs themselves and contraction of body wall muscles empty the lungs.

Figure 17.14 A vocalizing male toad. The air sacs of this male olive toad (*Amietophrynus garmani*) amplify its advertisement call.

onating structures thus amplifying the call (**Figure 17.14**). Some frog calls are so loud they can be heard more than 1 km (0.6 miles) away. The champion "sound-for-pound" amphibian caller has to be the tiny male coqui frogs (*Eleutherodactylus coqui*) of Puerto Rico and the Virgin Islands. Although they are less than 5 cm (2 inches) long and weigh only a fraction of an ounce, these tiny croakers have a call that reaches 108 decibels—louder than a low-flying jet or nearby subway train. Their call is a two-note whistle that states their name: Ko-Kee! One note (ko) of the song is thought to warn away competing males whereas the other note (kee) is thought to attract receptive females.

Box 17.1
Cost of Calling

The vocalization of male frogs is costly in two senses. The actual energy that goes into producing a call can be very large, and by calling a male increases his risk of predation. Calls sung to help female frogs locate a male can also be used as a home signal for predators as calling males are certainly easier to locate than those that are silent.

Measurement of the actual energy expended by calling males is difficult to determine because individuals must be placed in airtight metabolic chambers to measure the amount of oxygen consumed (a direct correlation of energy consumed). This procedure can frighten the frog and prevent it from calling. That difficulty was overcome when researchers took their chambers to the breeding ponds. Calling male frogs were placed in the chambers early in the evening and then left undisturbed. With the stimulus of the chorus around them, the imprisoned males would call freely. Their vocalizations were recorded with microphones attached to each chamber.

What was revealed is that the rates at which individual frogs consumed oxygen were directly proportional to their rates of vocalization. At the lowest calling rate, 150 calls per hour, oxygen consumption was barely above resting noncalling rates. At the highest calling rates, 1500 calls per hour, however, the frogs were consuming oxygen at rates even higher than during times of strenuous locomotor activity.

Researchers also found another interesting correlation. The gray tree frogs (*Hyla versicolor*) being studied gave short duration calls when they were in small choruses but lengthened their calls when part of a large chorus. Subsequent research has shown that long calls are more attractive to females than short calls. A long call, however, expends about twice as much energy as a short call. If the length of time that an isolated male can call is the main determiner of his success in attracting a female, what's a male to do? He is not physiologically capable of emitting thousands of calls per hour for hours on end every night. The best strategy seems to be giving short calls for several hours every night if the calling competition with other males is light. When faced with more intense calling from a large chorus, however, giving a longer and more attractive call is important, even if the male can call for only a short time. In other words, if competition is heavy, a male has no choice but to go all out energy-wise.

The length of call could hold additional significance other than being a locator beacon of sound. The Good Gene hypothesis predicts that certain exaggerated characteristics of males, such as length of call in gray tree frogs are a reliable indicator of inheritable genetic quality. Thus, long calls may help females identify genetically superior males. And research has shown that this may indeed be the case. In growth comparison studies tadpoles sired by long-call males grew faster than those from short-call males, especially when food was limited. Such studies suggest that a female gray tree frog can increase the fitness of her offspring by mating with a male who is giving long calls.

* Adapted from "The Energy Costs of Vocalization by Frogs." pg 241 *Vertebrate Life.* F. Pough, C. Janis and J. Heiser. Pearson Prentice Hall. 2005.

Why do frogs call? The main reason for calling is an attempt by males to attract a mate (**advertisement calls**). Males either call individually or in a group called a **chorus**. Advertisement calls attract females to breeding areas and announce to other males that a given territory is occupied. Advertisement calls are species specific, and the repertoire of calls for any one species is limited. These calls may also help induce psychological and physiological readiness to breed. **Release calls** are given by males mistaken as females by other males or unresponsive females. Advertisement calls and release calls are made with the mouth closed. A **distress call**, made with the mouth open, is emitted by both males and females in response to pain or to being seized by a predator. Distress calls may be loud enough to startle a predator into releasing the frog. The distress call of the South American smoky jungle frog (*Leptodactylus pentadactylus*) is a piercing scream similar to that of a cat in distress.

Osmoregulation and Excretion

There are three main types of osmoregulatory environments in which animals live: freshwater, marine, and terrestrial. Animals whose internal osmotic (water) concentration is the same as the surrounding environment are considered **osmoconformers**, whereas those that maintain as osmotic difference between their body fluid and the surrounding environment are **osmoregulators**. Amphibians and other vertebrates are osmoregulators.

Most adult amphibians face the challenge of living in two of the three osmoregulatory environments. In water, they are **hyperosmotic** to their environment. That is, their body fluids contain more salt and less water than there is in the external environment (water) surrounding them. This presents two problems: water is constantly diffusing into their body, and they are subject to the continual loss of body salts to the surrounding water, which has a low salt content. These problems are dealt with by producing a large amount of dilute urine. The kidney absorbs the salts that need to be retained, and the excess water is excreted out the cloaca. Amphibian skin also actively transports salt (sodium) from the surrounding water across the epithelium into the interstitial fluid and blood. Low sodium content results in the body fluids of amphibians being more dilute than that of most other vertebrates—approximately 100 milliequivalents compared with 150 milliequivalents.

In water, amphibians face the problem of excess water coming in and the loss of salts moving out. On land, however, amphibians face the potential loss of disastrous amounts of water because of evaporation across their permeable skin. Ironically, the skin is greatest source of water loss, but its permeability and high degree of vascularization make it the most important structure for rehydration. Amphibians are most abundant in moist habitats, especially temperate and tropical forests, but a surprisingly large number of species live in dry regions. Anurans have been by far the most successful of the amphibian invaders of arid habitats with all but the harshest deserts harboring some anurans.

Amphibians use two strategies to avoid excessive water loss while in air on land. The most common is to retreat to moist sites under rocks, logs, and plant litter or into holes in the ground. In extremely harsh and arid environs, the only hope is to burrow sometimes a meter or more into the bottom of temporary pools and puddles formed by the chance rainstorm, (a two-foot deep pool may only last a week in the desert) and then wait. The most extreme desert dweller is the Australian water-holding frog (*Cyclorana platycephala*) that can wait up to seven years for rain. The Aboriginal people of Australia know where to look for these frogs and in times of dire need will dig them up. By gently squeezing these small frogs, a life-saving drink of water may be obtained. Amphibians living in dry surroundings can also temporarily store water in their urinary bladder and lymph sacs equivalent to around 35% of their total body weight.

Many species of arboreal (tree) frogs have skins that are less permeable to water than the skin of terrestrial frogs. The waxy monkey tree frog (*Phyllomedusa sauvagii*) of South America loses water through the skin at a rate only one-tenth that of most frogs. It achieves this low rate of evaporative water loss by using its hind legs to spread lipid-containing secretions from dermal glands over its body surface in a complex sequence of wiping motions that appear almost comical in nature.

Some anurans also utilize behavior to control water loss. The Puerto Rican coqui lives in wet tropical forests, but has elaborate behaviors that reduce evaporative water loss during periods of activity. Male coquis emerge from their daytime retreat sites at dusk and move a short distance to calling sites on leaves in the understory vegetation. They remain at their calling sites until shortly before dawn, when they return to their daytime retreats. The behavior and activities of the frogs vary from night to night, depending on whether it rained during the afternoon. On rainy nights, the coquis extend their legs and raise themselves up off the surface while calling. In this position, they lose water by evaporation from the entire body surface. On dry nights, the behavior is quite different. The frogs call only sporadically and do so in a water-conserving posture in which the body and chin are flattened against the leaf surface, and the limbs are pressed against the body. A frog in this posture exposes only half its body surface to the air, thereby reducing its rate of evaporative water loss.

The excretory system of amphibians consists of a pair of mesonephric kidneys that lie on either side of the spine against the dorsal body wall. Urine flows from the kidneys through urinary ducts to the cloaca and then into the urinary bladder, which branches from the ventral wall of the cloaca. For many terrestrial amphibians, the urinary bladder serves as a water-storage organ.

Like the larvae of fishes, most amphibian larvae excrete the nitrogen-containing wastes as ammonia. Because ammonia is very toxic, it must be removed from the body quickly or diluted with large amounts of water. This is usually no problem for aquatic larvae, but for adult amphibians to become even semi-terrestrial, a number of physiological developments had to be made to eliminate nitrogenous wastes but also conserve water. First, adult amphibians convert ammonia into urea, a less-toxic substance that can be excreted without using as much water to dilute it. Second, in the adult amphibian kidney the rate of glomerular filtration is reduced by restriction of the blood supply, and this together with an increased release of antidiuretic hormone results in the production of a small amount of urine of the same concentration as the blood. The antidiuretic hormone (ADH or vasopressin), which increases the permeability of collecting tubules to water also increases the permeability of the bladder to water and allows the urine stored in the bladder to be reabsorbed into the body. During metamorphosis, larval amphibians change from excreting large amounts of ammonia urine to excreting small amounts of urea urine.

Circulation and Body Temperature

As in fishes, the cells of amphibians are serviced by an extensive system of capillaries that are the peripheral extensions of a closed loop system of arteries and veins. Blood is forced through this closed loop by a single pressure pump—the heart.

The amphibian circulatory system shows remarkable adaptations for a life divided between aquatic and terrestrial habitats. As lungs developed, two critical evolutionary challenges had to be met by the circulatory system as well. The first was to provide blood path to the lungs. This challenge was met by converting the sixth aortic arch into pulmonary arteries to serve the lungs and by developing a series of pulmonary veins to return oxygenated blood to the heart. The second challenge proved more difficult. Oxygenated blood from the lungs must be sent to the body while deoxygenated venous blood returning from the body must be sent to the lungs without mixing. The ideal solution to this problem required the evolution of a double circulation system consisting of separate pulmonary (lung) and systemic (body) circuits. Advanced tetrapods such as birds and mammals have completely solved this problem by evolving a heart with four chambers that act as two two-chambered hearts side-by-side but divided from each other by a muscular **septum**. The right side of the heart and attendant circulatory system sends venous blood containing carbon dioxide to the lungs while the left side receives oxygenated blood from the lungs and sends it to the body (systemic). Partitioning is incomplete in amphibians (and most reptiles) as they lack a septum.

The heart of a fish has two chambers while that of the amphibian has three—two separate atria and a single ventricle with no dividing septum (**Figure 17.15**). A **spiral valve** helps direct blood into pulmonary and systemic circuits. As we discussed earlier, gas exchange occurs across the skin of amphibians as well as in the lungs. For this reason, blood entering the right side of the heart is nearly as well oxygenated as blood entering the heart from the lungs on the left side. When an amphibian is completely submerged, all gas exchange occurs across the skin. IN this situation, blood coming into the right atrium has a higher oxygen concentration than blood returning to the left atrium from the lungs. Under these circumstances, blood vessels leading to the lungs constrict, reducing blood flow to the lungs and conserving energy. This adaptation is extremely important for those frogs and salamanders that overwinter in the mud on the bottom of a pond.

In addition to a vascular system that circulates blood, amphibians have a well-developed **lymphatic system** of vessels that filter fluids, ions, and proteins from capillary beds in tissue spaces and return these fluids to the circulatory system. The lymphatic system also transports water absorbed across the skin. Unlike

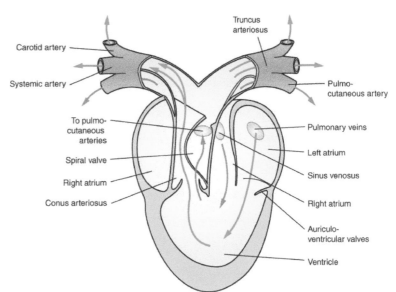

Figure 17.15 Structure of a Frog's Heart. Red arrows denote the flow of oxygenated blood; blue arrows, deoxygenated blood.

other vertebrates, the fluid in the lymphatic system is moved along by contractile vessels called **lymphatic hearts**.

Amphibians are ectothermic, and as such they depend on external heat sources to maintain body temperature. Water has powerful heat-absorbing properties, and when amphibians are in water, it is beneficial for them to take on the temperature of the water. On land, however, their body temperatures can differ from that of the air around them and fluctuates more widely.

Although some cooling results from evaporative heat loss, temperature regulation in amphibians is mainly behavioral. Many amphibians are nocturnal and remain in cooler retreats during the hottest part of the day. Basking in the sun or on warm surfaces to warm up after a cool evening may raise their body temperature 10° C above the air temperature. Basking after a meal is common and beneficial in that it increases the rate of all metabolic processes, including not only digestive functions, but also growth, and the depositing of the fat the animal will need to survive extended periods of winter dormancy.

Amphibians are found in geographical regions where the daily and seasonal temperatures often fluctuate widely. For that reason, amphibians can tolerate correspondingly wide temperature changes. Critical temperature extremes for most salamanders lie between -2° and 27° C (28°-81° F) and for most anurans between 3° and 41° C (37°-106° F).

Nervous System and Sense Organs

The amphibian nervous system is similar to that of other vertebrates. The amphibian brain (**Figure 17.16**) has three subdivisions: the forebrain (**telencephalon**) containing the olfactory (smell) centers and regions that regulate color change and visceral (gut) functions; the midbrain (**mesencephalon**), which processes sensory information; and the hindbrain (**rhombencephalon**), which functions in motor coordination and the regulation of heartbeat and respiration. Frogs have10 **cranial nerves** (nerves that pass information from the outside directly to the brain) and ten pairs of **spinal nerves** (nerves that pass information from the extremities through the spinal cord to the brain). By contrast, humans have 12 cranial nerves and 30 pairs of spinal nerves.

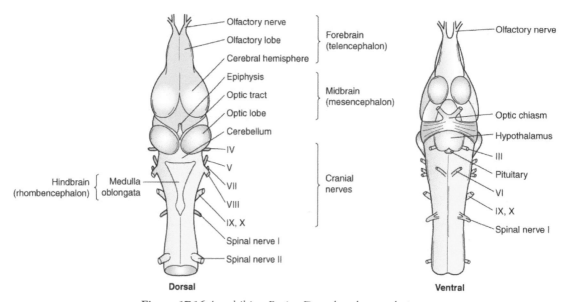

Figure 17.16 Amphibian Brain. Dorsal and ventral view.

As amphibians ventured out onto land, the sensory receptors of their fish-like ancestors were reconstituted. The lateral-line system so important and prevalent in fish is present in all aquatic larvae, aquatic adult salamanders, and some adult anurans. Evolved to detect and localize objects in water by reflected pressure waves, the lateral-line system serves no useful purpose on land.

Vision is the ascendant sense in amphibians (caecilians being an exception). As with other body systems, evolutionary modifications of ancestral aquatic eyes were required for them to function properly in air instead of water. Each eye has a closable upper and lower eyelid as well as a transparent **nictitating membrane** that provides further protection when the frog is submerged or swimming. **Lachrymal glands** (tear glands) function to keep the eyes moist (**Figure 17.17**). As in fishes, adjusting focus for near and distant objects (**accommodation**) in the amphibian eye is accomplished by moving the lens. Unlike the eyes of most fishes, however, amphibian eyes are pre-adjusted for distant objects, and the lens is moved forward to focus on nearby objects.

The retina of the amphibian eye contains photoreceptors in the form of **rods** and **cones**. Because cones are associated with color vision in some other vertebrates, their occurrence suggests that amphibians can distinguish between some wavelengths of light. The extent to which color vision is developed is unknown. The iris contains both circular and radial muscles and can rapidly expand or contract the pupil to adjust to changing light conditions. Overall, amphibians possess good vision that allows them to distinguish between moving prey, shadows that may warn of an approaching predator, and background movements, such as plants moving with the wind.

As is the case with the other body systems of the amphibians, the auditory (hearing) system of amphibians is clearly an evolutionary accommodation to life on land. The system transmits both ground vibrations and, in anurans, airborne vibrations. The ears of anurans consist of a tympanic membrane, a middle ear, and an inner ear (**Figure 17.18**). The **tympanic membrane** (eardrum) is a piece of skin stretched over a cartilaginous ring on the outside of the head. Touching the tympanic membrane is a middle-ear ossicle (bone) called the **columella**, which

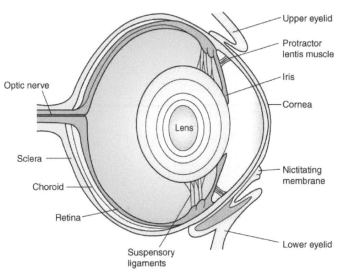

Figure 17.17 Longitudinal section of a frog's eye. The distance vision of a frog is better than its near vision. Whether frogs can see in color is debatable, but it has been shown that they respond positively to blue light.

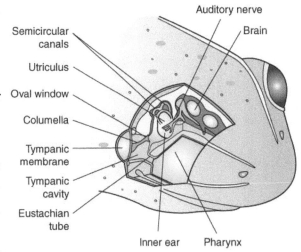

Figure 17.18 Ear Structure of a Frog. Sound vibrations are transmitted from the tympanic membrane through the columella to the inner ear. As in humans, the Eustachian tube allows pressure equalization between the tympanic membrane (outside) and the pharynx (inside).

transmits high-frequency vibrations (1,000 to 5,000 Hz) from the tympanic membrane to the inner ear. Low-frequency (100 to 1,000 Hz) vibrations from the ground are transmitted through the frontal append-ages and the pectoral girdle to the inner ear through a second ossicle, called the **operculum**. The inner ear of amphibians has fluid-filled semicircular canals. Vibrations transmitted to these canals moves the fluid within them in triggering hair cells within to generate nerve impulses to the brain. These hair cells also help detect rotational movements, whereas other sensory patches in the inner ear respond to gravity and linear acceleration/deceleration.

It has been discovered that the lungs of frogs are only slightly less sensitive than their eardrums. Frogs have an unbroken link of air from the lungs to the eardrums that researchers believe may serve two purposes: to help locate sound and to possibly protect its ear from its own raucous calls.

Suppose a male frog is calling loudly directly to the left of another frog. When the calling sound reaches the left ear of the receiving frog, it also reaches the left lung causing a pressure difference across the left eardrum thus the frog perceives the sound as coming from the left. Sound location is important to frogs. Females use it to find prospective mates and males use it to respect another male's territory.

Male frog calls can be extremely loud, even painful to the human ear. The frog's lungs protect his own ears by equalizing pressures between the inner and outer surfaces of the eardrum. Although the eardrum does vibrate in response to a frog's own call, it does so with very small amplitude.

Amphibians can screen out either high- or low-frequency sound by using muscles attached to the oper-culum and stapes to lock either or both of these ossicles. This adaptation is quite useful because anurans depend on different frequency ranges in different situations. Mating calls are high-frequency sounds that are of primary importance only during the breeding season, whereas low-frequency sounds may warn of an approaching predator.

Salamanders lack a tympanic membrane and mid-dle ear. They produce no mating calls, and the only sounds they most likely hear are low-frequency ground vibrations transmitted through the skull to the stapes and inner ear. Other sensory receptors in amphibians include tactile and chemical receptors in the skin, taste buds on the tongue and palate, and a well-developed olfactory epithelium lining the nasal cavity.

Reproduction, Development, and Metamorphosis

In temperate regions, slowly increasing spring tem-peratures seem to be the main environmental trigger that induces the physiological changes (mainly hor-monal) associated with breeding. For this reason, tem-perate breeding periods are usually seasonal, occurring in spring and summer, whereas in tropical regions, amphibian breeding correlates with rainy seasons.

(a)

(b)

Figure 17.19 Pheromone Transfer Mechanisms of Male Salamanders.. (a) A male Northern two-lined salamander (*Eurycea bislineata*) uses his enlarged teeth to scrape the top of the female's head. (b) A male smooth newt (*Triturus vulgaris*) uses his tail to waft pheromones toward the female (male is to the right and female is to the left.)

Once the drive to begin breeding has been activated, courtship begins in earnest. Courtship behavior functions to prepare individuals for mating, helps individuals identify and locate mates, ensures that eggs are fertilized and deposited in locations suitable for successful development of the hatchlings.

Salamanders rely primarily on olfactory (pheromones) and visual cues in courtship and mating. Pheromones are released primarily by males and contribute to species recognition and may stimulate endocrine activity that increases the receptivity of the female. Pheromone delivery by most salamanders that breed on land involves physical contact between a male and female, during which the male applies secretions of specialized courtship glands (**hedonic glands**) to the nostrils or body of the female (**Figure 17.19**). In anurans, male vocalizations and touch are the all-important cues. Many anuran species congregate together in small areas during times of intense breeding activity. Species-specific male vocalizations provide the initial attraction and contact between mates. After that, tactile cues become more important.

> *The solitary pool has drawn to itself the entire toad population of the surrounding country....Each toad has his own home or hermitage somewhere in that area where he spends the greatest portion of the summer season....When spring returns, he sets out on his annual pilgrimage of a mile or two....until he arrives at the sacred pool. The music and revels over, the toads vanish, each one taking his own road, long and hard to travel, to his own solitary home.*
>
> —W. H. Hudson

Most groups of salamanders employ a method of internal fertilization in which sperm is deposited in a packet called a **spermatophore**. The form of the spermatophore varies in different species, but all consist of a sperm cap on a gelatinous base. Typically, the male deposits the spermatophore on the substrate and the female picks off the cap with her cloaca. The sperm are released as the cap dissolves, and fertilization occurs in the oviducts.

Internal fertilization has been demonstrated for the Puerto Rican coqui and may be widespread among frogs that lay eggs in very moist areas on land. However, fertilization is external in most anurans. Mating begins when the male grasps the female tightly with his forelimbs so that they are both oriented in the same direction, and the male is on top the female. This mating squeeze, known as **amplexus** (**Figure 17.20**), can be maintained for several hours or even days before the female lays eggs. During amplexus, the male releases sperm as the female releases eggs.

In most cases, salamanders that breed in water lay their eggs in water. The eggs may be laid singly or in a mass of transparent gelatinous material. The eggs hatch into gilled larvae that, except in totally aquatic species, transform into terrestrial adults. The dusky salamander

Figure 17.20 When mating the male anuran mounts the female and grasps her tightly (amplexus). The male then fertilizes the eggs passing out of the female. During breeding season, male toads will mount and grasp any pliable object, often holding on for hours.

(*Desmognathus fuscus*) lays its eggs beneath a rock or log near the water, and the female remains with them until they hatch. The larvae have small gills at hatching and may either take up an aquatic existence or move directly to terrestrial life. The red-backed salamander (*Plethodon cinereus*) lays its eggs in a hollow space in a rotten log or beneath a rock. The embryos have gills, but these are reabsorbed before hatching and the hatchlings emerge as miniature versions of the adults.

Only a few salamanders give birth to live young (viviparity). The European fire salamander (*Salamandra salamandra*) produces 20 or more small larvae, each about one-twentieth the length of an adult. The larvae are released in water and have an aquatic stage that lasts about three months. The closely related alpine salamander (*S. atra*) gives birth to one or two fully developed young, each about one-third the adult body length.

Anurans lay egg masses that range from a single egg to over 25,000 eggs, depending on the species and where the eggs are laid. Typically, the fertilized eggs are laid in water and then abandoned by the parents (**Figure 17.21**).

Figure 17.21 As with other amphibians, the eggs of this European common frog (*Rana temporaria*) lack shells and must be laid in water to prevent desiccation and death of the embryos.

Some anurans, however, demonstrate remarkable examples of parental care. Male African bullfrogs (*Pyxicephalus adspersus*) guard their eggs and continue to guard the tadpoles after they hatch. The male frog moves with the school of tadpoles and will even dig a channel for them to swim from one pool in a marsh to an adjacent one. Some frogs of the families Leptodactylidae, Ranidae, and Microhylidae construct froth nests into which the eggs are deposited. These nests either float as rafts on the surface of the water or are constructed on tree branches over water allowing hatching tadpoles a short drop to safety and food.

The eggs of the Australian pouched frog (*Assa darlingtoni*) are laid in moist soil and guarded by the male. After about 11 days, the eggs hatch and about half the tadpoles manage to make it into the hip pouches of the male and emerge as baby frogs 48-69 days later. The male midwife toad (*Alytes obstetricans*) will carry a string of fertilized eggs wrapped around his legs or across his back, and when they are ready to hatch, he will carry them into water where the tadpoles are released. The fertilized eggs of the Surinam toad (*Pipa pipa*) are swept by the male onto the back of the female. There they implant and sink into her skin over a period of several days forming pockets that eventually give her back the appearance of an irregular honeycomb (**Figure 17.22**). The larvae develop through the tadpole stage inside these pockets, eventually emerging from the mother's back as fully developed toads less than 2 cm (1 inch) long.

Figure 17.22 A female Surinam toad with embedded eggs.

The most bizarre example of parental egg brooding was the Australian gastric-brooding frog (*Rheobatrachus*). Females brooded tadpoles in their stomachs, and the young emerged from the females' mouths! We will never know, however, whether the female swallowed fertilized eggs and all development occurred in her stomach or whether she swallowed tadpoles because these frogs have not been since in the wild since the 1980s and are presumed to be extinct.

In caecilians, a male organ that protrudes from the cloaca accomplishes internal fertilization. Some species lay eggs in damp protected areas on land, and the female may coil around them until they hatch, whereas other types lay eggs in water. Viviparity, however, is the order for about 75% of caecilian species. In some species, the larvae pass through stages within the oviduct where they use their fetal teeth to scrape the inner lining of the oviducts for food. The young eventually emerge from the mother's body as miniature adults being 30% to 60% of their mother's body length.

Wherever the eggs of amphibians are laid or brooded, they all undergo complete metamorphosis from egg to larvae to adult. In fact, amphibians are the only tetrapods in which this remarkable transformation occurs. Metamorphosis is under the control of hormones secreted by the hypothalamus, pituitary, and thyroid glands. A variety of environmental conditions, including temperature, crowding, and food availability, influences the time required for metamorphosis. Some desert-dwelling anurans complete the transformation in a few days as they race against the drying sun; in other species, such as bullfrogs, the larvae overwinter and totally transform the following summer (**Figure 17.23**). Metamorphic changes are the most dramatic in anurans. The large tail of the tadpole is totally absorbed, lungs replace gills, the larval "teeth" are shed, the mouth enlarges greatly, the intestine shortens, and limbs develop. Amazingly, frogs begin life essentially as a fish but change into a semiterrestrial tetrapod. Metamorphosis for salamanders is a more subtle loss of gills and acquisition of lungs as well as changes in the thickness and permeability of the skin.

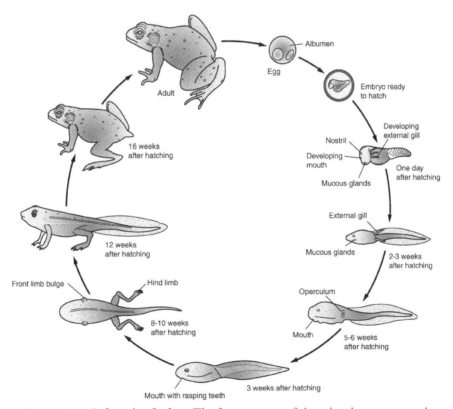

Figure 17.23 Life cycle of a frog. The frog starts as a fish and ends up a tetrapod.

Some salamanders display **paedomorphosis**, a condition in which they retain some, if not all, of the juvenile characteristics, yet become sexually mature. Some salamanders are paedomorphic because cells fail to respond to thyroid hormones, whereas others are paedomorphic because they fail to produce the hormones necessary to affect complete metamorphosis. In some salamander families, paedomorphosis is the rule, whereas in others its occurrence is variable and influenced by environmental conditions. One of several forms of paedomorphosis is **neoteny**. Neoteny delays physiological maturity but not sexual maturity. The axolotl (*Ambystoma mexicanum*) or Mexican salamander is a well-known example of a neotenic salamander (**Figure 17.24**). Prior to the growth of Mexico City, the axolotl was native to both Lake Xochimilco and Lake Chalco. The axolotl is now considered endangered in the wild, but due to its importance in scientific research, huge numbers are bred in captivity each year. Another variation of paedomorphosis, known as **progenesis**, is found in the mudpuppy (*Necturus maculosus*), a resident of pond, streams, and lakes of the eastern and midwestern United States (**Figure 17.25**). Progenesis speeds up sexual maturity, but prevents physiological maturity. In other words, the axolotl (neotenic) is an adult with some juvenile traits (gills), whereas the mudpuppy (progenetic) is a juvenile with an adult trait (sexual maturity).

Figure 17.24 The neotenic Axolotl (*Ambystoma mexicanum*). Axolotls hunt by smell, and suck their prey into the stomach with vacuum force.

Figure 17.25 The progenetic mudpuppy (*Necturus maculosus*). Mudpuppies take six years to reach sexual maturity.

Amphibian Connection

Economically

As with so many other animals, frogs are commercially important for their food value. The legs of some bullfrog species are in heavy demand in China, Europe (especially France) and parts of the United States, especially Louisiana. The worldwide harvest is an estimated 200 million bullfrogs (about 10,000 metric tons) annually. Major bullfrog suppliers include Bangladesh, China, Indonesia, and Japan, with about 80 million collected each year from rice fields in Bangladesh alone. As a result of excessive exploitation and the draining and pollution of wetlands, populations have declined drastically. With so many insect-eating frogs being removed from the ecosystem, rice production is threatened from uncontrolled, flourishing insect populations. Attempts to raise bullfrogs in farms in the United States have not been successful, mainly because bullfrogs are voracious eaters that normally require live prey, such as insects, crayfish, or other frogs, to trigger their feeding response.

Preserved frogs are often used for dissections in high school and university anatomy classes. This practice has declined in recent years with the increasing concerns about animal welfare and the decline of many frog populations.

Ecologically

If it should turn out that we have mishandled our own lives as several civilizations before us have done, it seems a pity that we should involve the violet and the tree frog in our departure.
—Loren Eiseley

The most important service amphibians provide is to serve as environmental monitors. Amphibians may amplify environmental problems and serve as excellent biological indicators of broader ecosystem health because of their intermediate position in the food webs, the sensitivity of their thin porous skin to contaminated water and air, and their typically **biphasic life** (aquatic larvae and terrestrial adults). Amphibians could be an early warning system signaling imbalances or degradation in the environment, both in local areas and worldwide. And all indications are that amphibians are indeed in peril and are frantically signaling us to that fact. Frogs and salamanders are disappearing at an alarming rate in various parts of the world whereas in other areas they are doing well, and no one knows exactly why. More than one-third of all species are believed to be threatened with extinction, and more than 120 species are suspected to be extinct since the 1980s.

Biologists suspect the declines are the result of the impact of a number of different synergistic factors that include, but are not limited to:

- Habitat destruction
- Increased levels of ultraviolet radiation due to ozone shield depletion
- Pesticides
- Parasites
- Fungal, bacterial, and viral infections. In just thirty years, the fatal and highly infectious fungal disease chytridiomycosis caused by the chytrid fungus *Batrachochytrium dendrobatidis* (Bd) has ravaged amphibian populations worldwide. This disease has led to the decline or extinction of around 200 species of frogs; the greatest disease-related loss of biodiversity ever recorded.
- Introduced predators and invasive species

In a few cases, captive breeding and release programs have been attempted to alleviate the pressure on amphibian populations; some have been successful to varying degrees and some have not. It is of the essence that governments and scientific groups worldwide and at all levels work together to halt and then reverse this alarming trend.

Some amphibian populations are in decline because of other amphibians. The voracious, aggressive African clawed frog (*Xenopus laevis*) was introduced into North America in the 1940s when it was used extensively in human pregnancy tests. When more efficient tests appeared in the 1960s, some hospitals simply dumped surplus frogs into nearby streams, where these prolific breeders have become almost indestructible pests displacing native frogs and fish from waterways.

In the 1930s, the large (often the size of a pie plate) cane toad (*Bufo marinus*) was introduced into Australia to control sugarcane pests. With no natural predators, their numbers exploded, and they themselves have become a serious pest. The resourceful Australians are Figurehting back by using large quantities of toads for educational and research purposes in schools, universities, and hospitals; and developing trade markets with China where the toads are used for medicinal and therapeutic purposes. They are even marketing wallets, purses, and vests made from toad skin leather.

Medically

Long before modern refrigeration, people in Finland and Russia placed living Russian brown frogs (*Rana temporaria*) in milk to prevent it from spoiling. We now know that this curious practice had a basis in science. Recent research shows that the skin secretions of these frogs are loaded with peptides, antimicrobial compounds as potent against Salmonella and Staphylococcus bacteria as prescription antibiotics.

Because frog toxins are extraordinarily diverse, they have captured the interest of biochemists as a "natural pharmacy." The alkaloid epibatidine, a painkiller 200 times more potent than morphine, is found in some species of poison dart frogs. Other chemicals isolated from the skin of frogs may offer resistance to HIV infection, and prove useful as anesthetics, muscle relaxants, and heart medications, and increase our understanding of strokes and seizures.

Although the skin secretions of amphibians have shown antibacterial, antifungal, and even anticancer effects and are of great interest to biologists and the pharmaceutical industry, the problem has been how to isolate and harvest these secretions. At first the amphibians were killed, and their skin harvested. Later, it was learned that electrical stimulation could induce the secretions without harming the animal. The secretions, however, were mixed together in a chemical stew that was difficult to separate. Recently researchers have found that the answer lies in molecular manipulation. Messenger RNA (mRNA) can be isolated from the glandular cells that produce the various secretions. This mRNA can then be used to produce a copy of the DNA responsible for producing the peptide toxins in the first place. Cloning this DNA could provide unlimited amounts of the various toxins in purified form for study all without harming amphibians.

Some salamanders have the ability to regenerate lost toes, limbs, eyes, spinal cords, hearts, intestines, and upper and lower jaws. The cells at the site of the injury can de-differentiate, duplicate rapidly, and differentiate again to create a new limb or organ. How salamanders can regenerate such a host of lost or damaged parts, but most other vertebrates cannot remains a mystery, a mystery that is the subject of intense study by scientists. What a near-miraculous medical advance it would be if this incredible ability of regeneration in salamanders could someday be applied to humans.

Frogs are probably the most extensively studied animals on the planet. Legions of school children and future doctors and researchers have been introduced to vertebrate anatomy by these creatures. In fact, frogs have served as important model organisms throughout the history of biology. Eighteenth-century biologist Luigi Galvani discovered the link between electricity and the nervous system through studying frogs. In 1952, Robert Briggs and Thomas J. King ushered in the age of cloning when they performed the first successful nuclear transplantation every accomplished in metazoans when they cloned a frog by somatic cell nuclear transfer. Frogs continue to be used in cloning research and other branches of embryology because they are among the closest living relatives to humans whose eggs lack shells thus facilitating observations of early development and manipulation of that development. Amphibians have added greatly to our understanding of life on earth and through the numerous occasions in which both adults and eggs have been rocketed into space to study the effects of weightlessness, they may even help us live in space.

Culturally

Numerous legends developed around the salamander over the centuries, many related to fire. This connection likely originates from the tendency of many salamanders to dwell inside rotting logs. When placed on a fire, the salamanders would attempt to escape from the log, lending to the belief that the salamander was created from the flames. The scientific name of the European salamander *Salamandra salamandra*—the fire salamander—is directly traceable to this legend. Associations of the salamander with fire appear in the Talmud and the Hadith, as well as the writings of Pliny the Elder, Conrad Lyconsthenes, Benvenuto Cellini, Paracelsus, and Leonardo da Vinci. In more modern times, when asbestos, an incombustible mineral, was discovered, it was believed to be the hair of the salamander and sometimes was referred to as "salamander's wool."

The salamander-fire connection may be more than myth and legend. Recent observations of California newts (*Taricha torosa*), which cover their bodies with slime secreted by their glands and then walk unaffected through the flame fronts of brush fires, demonstrate that these amphibians have a greater tolerance for fire than was previously understood.

Frogs and toads have featured prominently in culture, art, literature, mythology, fable and fairy tale (kiss a frog and get a prince), the occasional magical spell, and as religious deities and icons since antiquity. The connection of amphibians to earth, water, rainfall, and the underworld are reoccurring themes in many diverse cultures, and these beliefs are often depicted in indigenous art. In ancient Egypt, frogs would suddenly appear in great numbers and display a flourish of reproductive activity following the annual flooding of the Nile River Valley. Thus, frogs came to symbolize birth and resurrection. Among the hieroglyphics found on the walls of the Egyptian funerary temple of Hatshepsut (queen of Egypt during the fifteenth century B.C.) are images of the god of creation, Khnum, and his wife, the frog-headed Heqet, depicted as forming children on a potter's wheel. In fact, a number of Egyptian gods were depicted with the heads of frogs.

On the Yucatan Peninsula of Central America, the Mayan culture believed frogs and toads announced the rains with their choruses. Modern Mayans still perform ritualistic rain dances passed down from their ancient cultural past. The Maya also associated frogs with agriculture. The Madrid Codex, a fifteenth-century Mayan almanac painted on plaster-coated bark paper, shows frogs making furrows with sticks and sowing seeds. One frog, which the Maya called the *uo*, was thought to come from the sky with green corn grains in its intestines. The *uo* was probably *Rhinophrynus*, (burrowing toad) which breeds only during heavy rains. The name *uo* is onomatopoetic—the name represents the sound of the frog's call. *Uo* is also the name of the Mayan month of greatest rainfall.

In many cultures, amphibians were believed to possess supernatural and mystical powers, and their images and symbols were often used by shamans in assorted rituals. In Egypt, Greece, Turkey, and Italy frog amulets are worn for good luck or to ward off evil even to this day. The aboriginal Itelmens of the Kamchatka Peninsula of eastern Siberia considered hynobiid salamanders (*Salamandrella keyserlingii*) to be spies sent by Gaech, lord of the underground, to find and capture them for their master. In medieval Europe, extracts from the skin glands of toads were employed in witchcraft. It was believed that the toad counteracted its own poison with a special stone located in its head. This belief led shamans to use so-called toad stones—in practice, any stone the size and shape of a toad—to neutralize poisons from snakebites or bee stings.

Amphibians have also figured prominently in literature. *The Frogs*, a Greek satirical play, was first performed in Athens in 405 B.C. In this play Aristophanes used frogs to make fun of humans when the chorus repeatedly sings out to the god Dionysus, the patron of drama, as he crosses the River Styx to enter Hades and bring back the playwright Euripedes. The call, "Brekekekex, co-ax, co-ax," is thought to be the first use of phonetic imitations of animal sounds in literature. Many of Aesop's animal fables dealt with frogs and one of the most popular and well known is the traditional fairy tale of the prince turned into a frog by a wicked witch, only to be restored by the kiss of a beautiful princess. Shakespeare regularly used frog and salamander references in his plays. In *Richard III*, he derisively referred to the king as "that bottled spider, that foul hunch-back'd toad." The three witches in *Macbeth* chant, "Eye of newt, and toe of frog," as they stir those ingredients into their evil brew. In *As You Like It*, Shakespeare made yet another of his many toad metaphors: "Sweet are the uses of adversity, / Which, like the toad, ugly and venomous, / Wears yet a precious jewel in his head." The jewel, often thought to signify the toad's beautiful eye, may well refer to a toadstone. Later literary references to amphibians include those in Mark Twain's first story, *The Celebrated Jumping Frog of Calaveras County*, which featured a frog by the name of Dan'l Webster, and Karel Capek's science fiction thriller *War with the Newts* to name but several.

Frogs and toads abound in our modern secular culture as well. Portrayed as benign, ugly and clumsy, but often with hidden talents, powers, and virtues, we find them scattered through children's stories, science fiction novels, film and animation, video games, and as attractive characters in commercial advertising and as whimsical stars on television. Undoubtedly the most famous anuran of them all, Kermit the Frog, the muppet star of *Sesame Street* on public television, is loved by children around the world.

A Closing Note
A Silent Nocturnal Struggle

Along the walkway from my garage to the rear of my home, I have several low-voltage landscape lights. These lights are on timers and illuminate the walkway late into the night and for a time in the early morning. To a biologist, however, these lamps are more than a source of light. On warm summer evenings, as insects begin to flit around those lights, out come the toads. Hiding during the day, the toads come in from all directions to sit patiently under those lights snapping up every small insect that comes within range. I often position a lawn chair where I can sit and watch what to me is a far more entertaining and interesting show than what is on television.

This past summer I was fortunate enough to witness a most unusual struggle in this little microcosm. Beneath one of the lights, a garter snake had a toad in its mouth. The snake had the legs of the toad in its mouth and was struggling mightily to swallow the rest of it. The toad, however, was having none of it and was using its front limbs to alternately push against the mouth of the snake or reach out and grip a rock in an attempt to pull itself free. This silent struggle went on for over an hour. Finally, the snake opened wide, spat out the toad, and slithered off into the night. Apparently none the worse for wear, the toad nonchalantly resumed its feeding post beneath a light.

It was a happy ending for that particular toad, but on the larger global stage many amphibians are locked in a silent struggle with the forces of extinction. Tragically, the ending has not been a happy one as any number of amphibian species have apparently become extinct, and some others face an very uncertain future. Student activists have made a difference in the past and continue to make an impact on the quality of life around them. I urge you to investigate this problem in your area and if necessary, to do what you can personally to help. What better cause for a budding zoologist such as yourself than championing the unappreciated and defenseless amphibians?

In Summary

- The most likely ancestors of the amphibians were the Rhipidistians—a type of sarcopterygian (lobed-finned) fish closely related to the lungfishes and coelacanth lines. Most taxonomists agree that amphibians are monophyletic, but the exact relationships between extant and extinct taxons are controversial.
- The living amphibians are the smallest group of vertebrates numbering around 5,400 species of frogs and toads, salamanders, and caecilians. They are restricted mainly to water or moist places, but because they cannot osmoregulate in salt water, none are found in the ocean.
- Class Amphibia is composed of three orders:

  Order Gymnophiona
  Order Urodela (Caudata)
  Order Anura (Salientia)

- The living amphibians have very different body forms but the members of the class share the following derived characteristics:

  1. A bone endoskeleton with varying numbers of vertebrae; ribs present in some, absent or fused to vertebrae in others. None possess an exoskeleton and the notochord does not persist in adults.
  2. A smooth, thin, porous skin containing both mucus glands and poison glands.
  3. Four limbs (tetrapod) which may vary in size with the forelimbs of some being much smaller than the hindlimbs; some are legless. Limbs have varying numbers of digits and webbed feet are often present; no true nails or claws.
  4. The mouth is usually large with small teeth in upper or both jaws; two nostrils open into the anterior part of the mouth cavity. Larval amphibians are either herbivorous or carnivorous while adults are strictly carnivorous.

5. Respiration is accomplished either separately or in combination by lungs, skin, and gills; some larval types possess external gills, and these may persist throughout life.

6. A three-chambered heart consisting of two atria and one ventricle.

7. Body temperature regulation is ectothermic in nature.

8. Separate sexes (dioecious) with internal fertilization via spermatophore in salamanders and caecilians, but external fertilization in frogs and toads. Larvae develop in water or very moist environments and undergo complete metamorphosis. Amphibians are the only vertebrates to undergo complete metamorphosis.

- Order Gymnophiona—caecilians (also known as "rubber eels")—are legless snakelike burrowers found only in the tropical forests of South America, where they feed on worms and other invertebrates in the soil

- Order Urodela (Caudata)—salamanders—numbers over 500 species with salamanders being found on all continents except for Australia and Antarctica. Roughly one-third of all the known urodelans are found in North America; the highest concentration of these being found in the Appalachian Mountains region, which has the most diversity of salamanders worldwide.

    Most salamanders have limbs set at right angles to their body, with forelimbs and hindlimbs of approximately the same size. In some aquatic and burrowing forms, the limbs are rudimentary or absent altogether. All types possess a long and distinct post-anal tail throughout life

- Order Anura (Salientia)—frogs and toads—contains over 5,200 species making it by far the largest group of extant amphibians. In fact, about 88% of all amphibian species are frogs.

    Frogs are distributed from the tropics to subarctic regions, but they exist in the greatest abundance and variety in tropical rainforests. Some frogs inhabit arid regions such as deserts, where water may not be easily accessible. Compared with the other two groups of amphibians, frogs are unusual because they lack tails as adults and their hind legs are more suited to jumping and swimming than walking.

- Amphibians connect to humans in ways that are important economically, ecologically, medically, and culturally.

Review and Reflect

1. *Land or Water?* In what ways are amphibians terrestrial animals? In what ways are they aquatic animals? It has been said that amphibians are the animal equivalent of the mosses of the plant kingdom. Is this an accurate comparison? Explain.

2. *A Legless Amphibian?* Your study buddies are looking at a color photo of a caecilian. One says, "Obviously, this thing is a snake." Your other friend disagrees, "No, it must be a worm." What would you say to enlighten them as to the zoological truth of the matter?

3. *Affairs of the Heart* Why is the separation of oxygenated and deoxygenated blood in the heart not as important for amphibians as it is for other terrestrial vertebrates?

4. ***Thin-Skinned*** The skin of an amphibian is so permeable and such a liability that it has been suggested an amphibian would be just as well off without it. Agree or disagree? Defend your position.

5. ***In Goes the Good, Out Goes the Bad*** Your friend slept through zoology lecture again. After class he asks, "I vaguely remember the instructor saying that amphibians were mouth breathers, not chest breathers as are humans. What does that mean?" How would you reply?

6. ***Fitting Your Lifestyle*** What adaptations do caecilians have that allow them to live where they live and eat what they eat?

7. ***Strange Behavior*** Many amphibians hide from their enemies, but some advertise the fact that they are deadly poison with bright, vivid colors. When threatened, a certain harmless salamander stands on tiptoe, touches the tip of its tail to the top of its head and remains perfectly still to display a bright red underside. Explain this behavior.

8. ***A Toxic Arsenal*** Scientists have isolated over 400 toxic compounds from the skin of frogs in the family Dendrobatidae. Hypothesize the source of these toxins.

9. ***Which Way the Wind?*** Herpetologists collected data concerning migration and wind direction for a certain species of frog, as shown in **Table 17.1**. Construct a graph of this data and use the graph to answer the questions that follow:

Table 17.1 *Data Collected by Herpetologists Concerning Migration and Wind Direction for a Certain Species of Frog*	
Wind Direction	**Frogs Migrating in That Direction**
N	3
NNE	0
NE	0
ENE	0
E	4
ESE	6
SE	0
SSE	25
S	32
SSW	7
SW	21
WSW	1
W	6
WNW	0
NW	3
NNW	4

A. Specifically what did you graph?

B. What general conclusion can you draw from this data?

C. This data has been shown to be quite significant and not due to chance or coincidence. Would this conclusion hold true for other species of frogs? For the same species of frogs in different locales? Explain.

10. *A Water Connection* Why must amphibian eggs be laid either in standing water or a very moist place? Compare and contrast the eggs of amphibians with those of reptiles, birds, and mammals—the amniotes.

11. *A Race Against Time* When tadpoles undergo metamorphosis their bodies begin to produce an adult enzyme that converts ammonia into urea. The time that a tadpole takes to produce this enzyme varies among species. In the graph below, the rate of enzyme production is shown for a species that inhabits a desert environment and a species that inhabits a forest environment. Analyze the graph and answer the questions that follow (**Figure 17.26**).

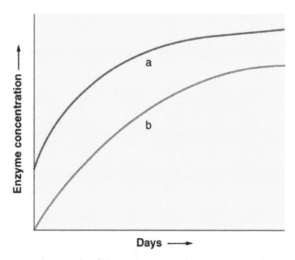

Figure 17.26 A graph of the enzyme production in two frog species.

A. Which curve represents which species of frog? Defend your answer.
B. Why is an accelerated metamorphosis necessary for desert-dwelling anurans?

Create and Connect

1. *Amphibians on Stage* From Mark Twain's celebrated jumping frog and Mr. Toad of *The Wind in the Willows* to Kermit the Frog, amphibians have found a place in literature, song, and film. Write a short story or play or compose a song or poem with a fictional frog, toad, or salamander as the main character.

2. *The Greatest Influence* Metamorphosis in amphibians is under hormonal control. However, environmental factors may play a role in the rate of metamorphosis. Design an experiment to test

the problem question: *Which environmental factor—temperature or light intensity—plays the greater role in the rate of tadpole metamorphosis?*

Guidelines:

A.   Your design should include the following components in order:

> ➤   The *Problem Question*. State exactly what problem you will be attempting to solve.
> ➤   Your *Hypothesis*. While this is a fictitious experiment, word your hypothesis as realistically as possible.
> ➤   *Methods and Materials*. Explain exactly what you will do in your experiment including the materials necessary to accomplish the task. Be specific, take nothing for granted, and do not expect people to read your mind as they read your work.
> ➤   *Collecting and Analyzing Data*. Explain what type(s) of data will be collected and what statistical tests might be performed on that data. It is not necessary to concoct either fictitious data or imaginary observations.

B.   Assume you have access to everything necessary to conduct your experiment—tadpoles, adequate housing for the tadpoles, plenty of laboratory space, and all the equipment necessary to meet the challenge. Within reason, money is no object.

C.   Again, we have presented you with an experimental design challenge that is purposely broad and not very specific. Your greatest challenge will be to work out all the specific details leaving nothing to chance. To aid you in this endeavor, your instructor may provide additional details or further instructions.

3.   ***The Current State of Affairs*** Are amphibians thriving, holding their own, or declining in your locale? Investigate this question.

Guidelines:

A.   Your instructor will detail the specific area in question. Your investigations might be statewide or perhaps just within your local county or parish.

B.   Your instructor will detail the specific amphibians to investigate. Your investigations might include all the amphibian species in your assigned locale or limited to a select number of species.

C.   Prepare a range map for the species selected for investigation.

D.   Consult local wildlife authorities as well as zoologists and ecologists on staff at your university for details and information.

E.   Prepare a short report detailing the status of the specific amphibians within the area assigned. Your instructor may provide additional details on the format and content of your report.

REPTILES: SHATTERED REMAINS

There may be a lesson for us in the rise and fall of the great reptiles. We human beings are riding high today, and many of us give little thought to the future despite the threat of global war, our fantastically burgeoning population and our profligate wastage of natural resources. Let one or more of these dangers pass beyond control, and mankind may find itself vastly reduced in numbers and struggling for survival as so many reptiles are today.

—Robert Conant

Introduction

Think of the fossil record as the muted whispers of ancient animals speaking to us down the corridors of time. Alas, the fossil record is scarce given the number of animals that roamed the land and swam the waters of the past, and thus the murmurs are faint; most often we are left with only silence. One thing that the fossils do tell us, however, is that major group of animals one after another ascended to rule the habitats of their time through their numbers and diversity, and on occasion, their size. Some, such as fishes, may still be near the pinnacle of their evolutionary radiation, whereas others, especially reptiles, are but mere shadows representing the shattered remains of their former greatness. (**Table 18.1**)

History of Reptiles

The story of reptiles begins in the late Carboniferous period some 300-320 million years ago when the world was a tropical place. The continents were rather low, and occupied by many sluggish rivers and broad swamps, with gentle rolling hills forming uplands which in some regions occupied the flanks of mountains—a world of vast green jungles of primitive plants, stretching for mile after mile across the earth. There seems to have been little of the marked zonation of climate extending on either side of the equator that we are so familiar with today. Consequently, it must have been, on the whole, a rather uniform world. However, the geological evidence indicates that the southern continents of that period experienced glacial conditions for a time during the Carboniferous and early Permian periods. Glaciation evidently did not reach into northern lands, and even in the southern hemisphere the effects of glaciation were transitory.

Table 18.1
Geologic Ages and Associated Biologic Events

Time Scale (eon)	Era	Period	Epoch	Millions of Years Before Present (approx.)	Duration in Millions of Years (approx.)	Some Major Organic events
P H A N E R Z O I C	Cenozoic	Quaternary	Recent (last 5,000 years)	0.01	1.8	Appearance of humans
			Pleistocene	1.8		
		Tertiary	Pilocene	5.3	3.5	Dominance of mammals and birds
			Miocene	23.8	18.5	Proliferation of bony fishes (teleosts)
			Oligocene	34	10.2	Rise of modern groups of mammals and invertebrates
			Eocene	55	21	Dominance of flowering plants
			Paleocene	65	10	Radiation of primitive mammals
	Mesozoic	Cretaceous		142	77	First flowering plants Extinction of dinosaurs
		Jurassic		206	64	Rise of giant dinosaurs Appearance of first birds
		Triassic		248	42	Development of conifer plants
	Paleozoioc	Permian		290	42	Proliferation of reptiles Extinction of many early forms (invertebrates)
		Carboniferous	Pennsylvanian	320	30	Appearance of early reptiles
			Mississippian	354	34	Development of amphibians and insects
		Devonian		417	63	Rise of fishes First land vertebrates
		Silurian		443	26	First land plants and land invertebrates
		Ordovician		495	52	Dominance of invertebrates First vertebrates
		Cambrian		545	40	Sharp increase in fossils of invertebrate phyla

P R E C A M B R I A N	Proterozoic	Upper		900	355	Appearance of multi-cellular organisms
		Middle		1,600	700	Appearance of eukaryotic cells
		Lower		2,500	900	Appearance of plank-tonic prokaryotes
	Archean			4,000-4,400	1,400	Appearance of sedimen-tary rocks, stromatolites and benthic prokaryotes
	Hadean			4,560	160-560	From the formation of Earth until first appearance of sed-imentary rocks; no observ-able fossil organisms

Sources: Dates derived mostly from Gradstein, F. M., et al., 2004. *A Geological Time Scale 2004.* Cambridge University Press, Cambridge. England. and from Geologic Time Scale, obtainable from http://www.stratigraphy.org, a Website maintained by the International Commission of Stratigraphy.

The freshwater environs—swamps, marshes, rivers, and lakes—and the land surrounding them were ruled by the amphibians. Although large, clumsy, and slow, they were none the less very successful because they had the place to themselves; they were the sole vertebrate pioneers on the land.

The reproductive strategy of amphibians was not a suitable means of reproduction for those tetrapods that broke with a water-based existence. Moving from shorelines to further inland during the late Carboniferous period, some types began to form eggs in which the embryos were wrapped and protected by membranes (amnion, chorion, allantois) and a shell and to lay those eggs on land, not in water as do the amphibians. These ancestral amniotes would be the stem from which would spring the modern amniotic lineages of dinosaurs, reptiles, birds, and mammals (**Figure 18.1**).

By the end of the Carboniferous, amniotes could be categorized into three groups on the basis

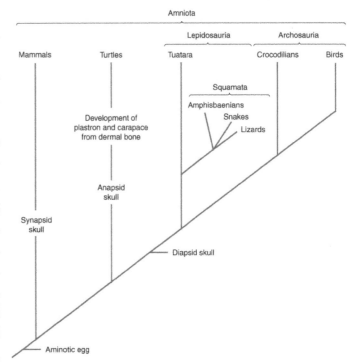

Figure 18.1 The Possible Phylogeny of Amniotes. The extinct saurischian dinosaurs would be placed just before the crocodilians and birds on this cladogram.

of skull anatomy and whether or not there were openings in the temple area behind the eye sockets. The earliest amniotes had no such openings, and they are called the **Anapsida**. Turtles are often included in this group since their skull is entirely covered with bone. The first group that diverged from the early anapsids are the **Synapsida** (one temporal opening), which ultimately evolved into mammals. The second group that separated from the basal anapsids were the **Diapsida** (two temporal openings), which evolved into dinosaurs, modern reptiles, and birds (**Figure 18.2**).

As the days of the Carboniferous period ran to their end, the balance began to change in favor of the reptiles, so that with the advent of Permian history we find reptilian tetrapods clearly dominant on the land, even though there still remained many large and powerful amphibians.

As the Permian progressed, the climate became warmer and milder, the glaciers of the Carboniferous receded, and the continental interiors became drier and perhaps even arid, with great seasonal fluctuations (wet and dry seasons). The drying climate spelled doom for many of the mighty swamp forests and along with them most of the large amphibians. Although many of the non-amniote amphibians, such as the anthracosaurs, continued into Permian, it was the synapsid amniotes that took over the role of dominant land animals.

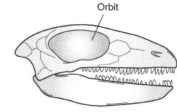

(a) Anapsid skull

Several distinct evolutionary dynasties of synapsids became prominent through the course of the Permian. The first, the pelycosaur dynasty, included the large finbacks such as *Dimetrodon* and others. These ancient reptiles attained lengths of 3 m (9-10 feet) and had a large dorsal "sail" that was most likely a thermoregulatory adaptation. Therapsids, referred to as "mammal-like reptiles," represent the second dynasty of Permian synapsids. Possessing powerful jaws and highly differentiated teeth along with legs positioned more vertically than the sprawling legs of *Dimetrodon*, therapsids were the dominant land animal by the middle of the Permian. The first therapsids were large and slow and by the end of the Permian had disappeared to be replaced by forms that were smaller and even more mammal-like. Some may even evolved fur and the ability to control their body temperatures.

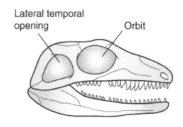

(b) Synapsid skull

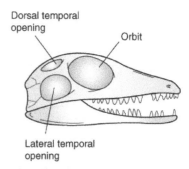

(c) Diapsid skull

The Permian proved fertile evolutionary ground for the diapsid clade. During this time two lineages appear—the Archosauramorpha which would radiate from the stem thecodonts into dinosaurs, crocodilians and birds and the Lepidosauromorpha which would radiate into several groups of aquatic dinosaurs and into the modern squamates (snakes and lizards) and the tuatara.

For whatever reasons, the anapsids did not fare as well during the Permian as did the diapsids and the only lineage to survive to modern times was that of the turtles.

Around 251.4 million years ago, as Permian life flourished, an extinction event occurred that was so massive it would form the geologic boundary between the Permian and Triassic geologic periods. Although the fossil record indicates there had been previous extinction events, the **Permian-Triassic extinction event** (or informally the **Great Dying**) was by far the most severe event. Evidence seems to indicate that the extinction was spread out over a few million years, with a very sharp peak in the last 1 million years of the Permian. Marine invertebrates suffered the greatest losses during the P-Tr extinction, with up to 96 percent of all marine species becoming extinct, while on the contrary "only" 70 percent of ter-

Figure 18.2 Skull Types. (a) the anapsid skull is characterized by no openings in the temple area behind the eye socket orbits. (b) The synapsid skull is characterized by one opening lateral to the orbit of the eye. (c) The diapsid skull is characterized by two openings— one dorsal and one lateral to the orbit. These openings give the skull extra space for the attachment of jaw muscles—the more openings, the more muscles, and the more muscles, the more powerful the jaw.

restrial species went under. Indications are that over two-thirds of amphibian and therapsid families became extinct. Too few Permian diapsid fossils have been found to support any specific conclusions about the effect of the P-Tr extinction on diapsids.

Several sets of proposed mechanisms have been advanced to explain what caused the "mother of all extinctions," including both catastrophic events such as large or multiple impact events, increased volcanism, sudden release of methane hydrates from the sea floor, and gradual processes such as sea-level change, severe anoxia (deficiency in the amount of oxygen) in the ocean, and increasing aridity. The popular notion is that mass extinction events are caused by the impact of a large comet or meteorite and that within one month or so after that pretty much everything on the planet is dead. In reality, mass extinction events take millions of years and given this time frame, scientists believe the evidence strongly suggests a sequence of a combination of both catastrophic and gradual events more accurately describes any mass extinction event. The specific catastrophes and gradual changes involved and the exact sequence of their occurrence, however, has yet to be determined.

Although many of the side branches disappeared because of the P-Tr extinction, the main stem lineages that would become the mammals, the dinosaurs, birds, and crocodiles, the modern squamates, the modern tuatara, and the modern turtle did survive. One group in particular not only survived, they thrived, and that was the dinosaurs. (The term *dinosaur*, "fearfully-great lizard," was first used as a description of giant Mesozoic reptiles in 1842 by the English anatomist Richard Owen.) The fossil record indicates that the dinosaurs flourished during the Jurassic and Cretaceous periods (mid- to late Mesozoic era), a period designated the Age of Reptiles. During this 150 million year period, the dinosaurs became the dominant terrestrial vertebrates.

Traditionally speaking, dinosaurs, a group not including birds, are paraphyletic. Crocodilians and birds, however, are also descendants of the same archosaurian lineage as are the dinosaurs. The dinosaurs may be classified into two distinct orders based primarily on the structure of their hips: the Ornithischians and the Saurischians (**Figure 18.3**). Order Saurischia ("lizard-hips") is thought to have evolved from the ornithischians in the late Triassic period. Differences in feeding habits and locomotion further divide this order into two groups: the carnivorous and bipedal theropods, and the herbivorous and quadrupedal sauropods (sauropodomorphs). With its large skull, powerful jaws with many sharp teeth and ridiculously small front limbs, *Coelophysis* was an early theropod representative of all theropods that were to follow. Large Jurassic period theropods like *Allosaurus* were replaced by even larger carnivores like *Tyrannosaurus* (6 m [20 feet] in height, 14.4 m [47 feet] in length, and 7200 kg [8 tons] in weight) during the Cretaceous. Not all theropods of the Upper Cretaceous were massive; several types such as *Velociraptor* ("speedy predator") were swift and agile.

The sauropods lived during the late Triassic. Small- to medium-sized in their early lineage, their Jurassic and Cretaceous versions grew to massive sizes, big enough to become the largest terrestrial vertebrates that ever existed before or since. Though *Branchiosaurus* was 25 m (82 feet) in length and 30,000 kg (33 tons) in weight, a still larger sauropod was *Argentinosaurus,* reaching 40 m (132 feet) long and weighing at least 80,000 kg (176,370 pounds).

Order Ornithischia ("bird-hips") includes the Hadosaurs or duck-billed dinosaurs such as *Parasaurolophus*, the Stegosaurs with their plate-like armor along the back, and the horned Ceratopsians such as *Triceratops*, all of which were herbivorous. With the Cretaceous came a steady growth in diversity for these "bird-hipped" herbivores just as the giant sauropods were experiencing their inevitable decline.

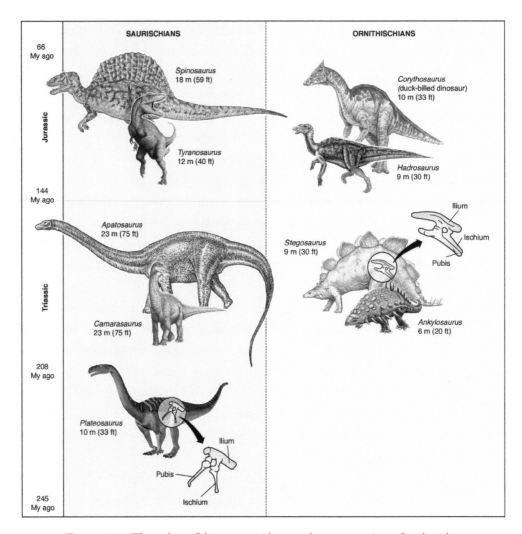

Figure 18.3 The orders of dinosaurs with several representatives of each order. Crocodilians and birds are descendants of the saurischian lineage.

The dominant animals of the Mesozoic oceans were the ichthyosaurs, some of which were as large as medium-sized whales, the long-necked plesiosaurs, and some marine crocodiles. The pterosaurs took to the air and evolved into huge spectacular flying reptiles. One of them, *Quetzalcoatlus,* was the largest flying vertebrate the planet has ever known, with a 12.1 m (40 feet) wingspan!

By the end of the Cretaceous dinosaurs had achieved a level of evolutionary success that guaranteed them a permanent place in history, and yet they were soon to disappear bringing the Age of Reptiles to a close. Around 65 million years ago, life was once again devastated by a mass extinction event. Not only did the dinosaurs disappear completely, but so did the flying pterosaurs, and the marine ichthyosaurs and plesiosaurs. In fact, between 60 and 80% of all animal species, including many marine forms, disappeared. As with other mass extinction events, the causes of the Mesozoic-Cenozoic mass extinction event were numerous catastrophic events and gradual changes that occurred sequentially over a long period of time.

One must realize that much of what is presented about dinosaurs and the world they inhabited is theory and conjecture. Approximately 700 species of dinosaurs have been named. A recent scientific review suggests, however, that only about half of these are based on fairly complete specimens that can be shown to be unique and separate species. These species are placed in about 300 valid dinosaur genera, although about 540 genera have been named. Recent estimates suggest that about 700 to 900 *more* dinosaur genera may remain to be discovered. Most dinosaur genera presently contain only one species, but some have more. Even if all of the roughly 700 published species are valid, their number is still less than one-tenth the number of currently known living bird species, and less than one-fifth the number of currently known mammal species.

Dinosaurs dominated the Mesozoic era for 165 million years, an incomprehensibly long period of time. To this day, they stir human curiosity and imagination and fill us wonder and awe. For all their might and diversity, however, they could not overcome numerous adversities and gradually faded into that long night that is extinction. In their stead arose the lineages that became the animals we know today as reptiles and birds.

It appears that sometime around 130 million years ago during the Cretaceous period, the squamates split into two major groups, Iguania (iguanas, anoles, chameleons, and agamids) and Scleroglossa (geckos, skinks, snakes and amphisbaenians). Although fossil evidence dating that split is sparse, the amount of divergence evident in the DNA of living species seems to agree with that timeline. Subsequently—about 180 million years ago—scleroglossans split into two groups, Gekkota and Autarchoglossa (**Figure 18.4**). Among the most recent of reptile lineages, snakes first appeared during the Cretaceous period some 130-135 million years ago. It was originally theorized that an ancient group of monitor-like lizards began to follow a burrowing way of life, tunneling through loose dirt and sand in search of earthworms and other prey, just as some lizards do today. Over

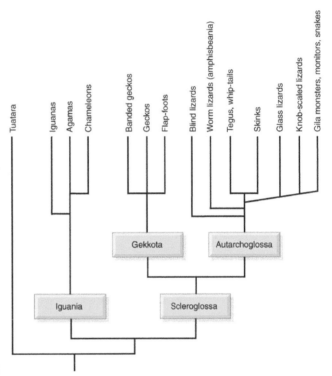

Figure 18.4 The possible phylogeny of squamates and tuataras.

a period of millions of years, these burrowing lizards lost their limbs and their external ears and replaced their eyelids with a clear **brille** (ocular scale) to protect their eyes while digging. At about the time that the dinosaurs reached their apex, one group of these burrowing lizards then gave up its subterranean lifestyle and emerged to the surface, where they rapidly diversified into the modern snakes.

The "burrowing ancestors" theory has, however, come under some attack recently. Some biologists have theorized instead that the snake's unique features are the result of a largely aquatic or semiaquatic lifestyle. In this interpretation, the lack of ears, the covered eyes, and the long limbless bodies allowed the first snakes to move efficiently through water or marsh in search of prey. Only later did snakes move from an aquatic existence to invade the dry land. Recently discovered fossils of a primitive (90million-year-old) snake from Australia support the aquatic ancestor hypothesis. In any case, the first modern terrestrial snakes to appear seem to have been relative of the living boids (boas) and pythons. These were large heavy-bodied snakes with rather primitive and heavy skull structure. The living boas and pythons have tiny clawlike toes protruding from either side of their cloaca as well as the vestiges of a pelvic girdle—remnants of the legs that their ancestors once had; an evolutionary relic tying the snakes directly to their lizard ancestors.

Diversity and Classification

Today the nearly 8,000 species of living reptiles (L. *reptus*, to creep) inhabit every continent except Antarctica. Most reptile species and individuals tend to be concentrated in the tropics and subtropics, but their numbers decline rapidly towards the poles and in high altitudes. This is reflected in the fact that there are only 340 species of reptiles in the United States and Canada but 1060 species in Mexico and Central America, and another 1560 species in South America.

The lizards and snakes are by far the most numerous and widely distributed of living reptiles. Both occur in a great variety of habitats and climatic conditions ranging from deserts to tropical rain forests. There are no fully aquatic lizards, but several groups of snakes, such as sea snakes, are highly aquatic.

Turtles and tortoises are most abundant and diversified in the tropics and extend into both the northern and southern temperate zones. Turtles are amphibious, with both fresh- and saltwater species, while the tortoises are strictly terrestrial.

The crocodilians are quite restricted both in diversity and distribution. They are limited to the tropics and a few warm temperate zones. All crocodilians are amphibious, with most being found in freshwater, although a few are tolerant of salt water.

Though the two living species of tuatara once heavily populated the two main islands of New Zealand, they are found today only on small islets of Cook Strait and off the North Island's northeast coast. There, under some government protection, they are holding their own.

The range of sizes in reptiles is much broader than it is in amphibians. The smallest reptiles are the geckos, some of which grow no longer than 3 cm (slightly more than 1 inch). Certain blind snakes (Typholpidae) are less than 10 cm (4 inches) in length when fully grown. The smallest turtles weigh less than 450 gr (1 pound) and reach a maximum length of 12.5 cm (5 inches), while the smallest crocodilians are the dwarf crocodile (*Osteolaemus tetraspis*) and the smooth-fronted caiman (*Paleosuchus*) at about 1.7 m (5.6 feet).

Truly gigantic reptiles no longer roam the earth, but some of their living representatives can grow to impressive sizes. There is much debate about the largest snake in the world. The reticulated python (*Python reticulatus*) of Southeast Asia and the East Indies can reach 8.4 m (27.6 feet) in length. The South American anacondas, however, may rival or surpass that length. For decades, persistent but unsubstantiated reports have been made of 11-45 m (33.5-148 feet) long anacondas. The average size of an anaconda is around 4.5 m (15 feet) and while some undoubtedly grow much larger, most herpetologists believe that a 15-18 m (50-60 feet) anaconda strains credibility and that a 45 m (150 feet) long specimen is an outright impossibility. It is interesting to note that since the early 20th century the Wildlife Conservation Society had offered a large cash award (US $50,000) for the live delivery of any snake 30 feet or more in length to the Bronx Zoo in New York City. The prize was never been claimed, and in 2002, the reward was cancelled to discourage people from "looking for and disturbing these animals."

What is certain is that anacondas are the largest snakes by total body mass (sheer physical bulk). With a girth of 30 cm (11.8 inches), the green anaconda (*Eunectes murinus*) can weigh 250 kg (551 pounds), considerably more than a reticulated python of the same length. One of the heaviest snakes for its length is the eastern diamondback rattlesnake (*Crotalus adamanteus*) which, though not exceeding 2.4 m (7.9 feet), can weigh as much as 15.5 kg (34 pounds). By contrast, the smallest snake is the tiny burrowing thread snake (*Leptotyphlops*) only 15 cm (6 inches) long and no bigger around than a matchstick.

Four living species of crocodilians grow larger than 6 m (20 feet) and the saltwater crocodile (*Crocodylus porosus*) and the ghavial (*Gavialis gangeticus*) may approach 9 m (30 feet). The giant among living turtles is the marine leatherback (*Dermochelys*), which reaches a total length of about 2.7 m (8.9 feet) and a weight of 680 kg (1,500 pounds). The largest of the tortoises is the Galapagos tortoise (*Geochelone nigra*) at 1.2 m (4 feet) long and 300 kg (660 pounds).

The largest lizard is the Komodo dragon (*Varanus komodoensis*). Found exclusively on small Indonesian islands, male Komodos can grow to 3 m (10 feet) and weigh up to 91 kg (200 pounds). The great size of these lizards is attributed to **island gigantism**, a biological phenomenon in which the size of animals isolated on an island increases dramatically in comparison to their mainland relatives.

With increasing reliance on cladistic methodology, important changes have been made in the traditional classification of reptiles. Although the amniotes are considered a monophyletic group, the reptiles are not. Classically, the group Reptilia excluded birds, even though birds are descendants of the reptiles' most recent common ancestor. Consequently, reptiles today are considered a paraphyletic group. Recognizing that the taxon Reptilia as historically used is not a monophyletic group, we continue to use the term and taxon in this book for the sake of familiarity and convenience. Our usage of the term "reptile" is in reference to all amniotes that have beta keratin in their epidermis that are not birds. The term, therefore, applies to the living turtles, snakes, lizards, amphisbaenians, tuataras, and crocodilians as well as such extinct groups as plesiosaurs, ichthyosaurs, pterosaurs, and dinosaurs.

There is conflict of opinion between proponents of the two major competing schools of taxonomy—cladistics and evolutionary taxonomy—as to the relationship between reptiles and birds. The cladists contend that crocodilians and birds are closely related because of their recent common ancestry. In their view and according to their rules, birds and crocodilians belong together in a monophyletic group separate from other reptiles and should be assigned to their own clade. This clade is Archosauria, which also includes the extinct dinosaurs. Because all reptiles and birds have many characteristics in common, including skull structures, and a skin that is nearly glandless with the harder "beta" keratin, cladists redefine "Reptilia" in contrast to its traditional usage and include birds in clade Archosauria.

Contrarily, evolutionary taxonomists believe birds are part of a different, novel adaptive zone and grade of organization while crocodilians belong to the reptilian adaptive zone and grade. Thus, evolutionary taxonomists place crocodilians in class Reptilia and separate birds into class Aves. These disagreements have forced zoologists to reexamine their long-held beliefs on amniote evolution and degrees of divergence. For educational purposes, "reptile," and "reptilian" as used here refer to the members of four living monophyletic orders that are combined into the paraphyletic class Reptilia.

    Domain Eukarya
        Kingdom Animalia
            Phylum Chordata
                Subphylum Vertebrata (Craniata)

Class Reptilia (traditional) or Class Sauropsida (cladistic)
    Subclass Anapsida
        Superorder Chelonia
            Order Testudines—Turtles
    Subclass Diapsida
        Superorder Lepidosauria
            Order Sphenodonta—Tuataras
            Order Squamata—Lizards, snakes, and amphisbaenians
                Suborder Amphisbaenia—amphisbaenians
                Suborder Sauria—lizards
                Suborder Serpentes—snakes
            Order Crocodilia—Crocodiles, alligators, caimans, gavials

Regardless of how we classify them, modern reptiles are the survivors of an enormous radiation of Mesozoic amniotes, including the dinosaurs, most of which became extinct at the end of the Mesozoic. And, even though their species numbers have declined drastically since the Golden Age of Reptiles, the living reptiles should not be regarded as weak or inferior to those of the distant past. In reality, modern reptiles are a biologically strong and vibrant group of successful animals.

General Characteristics of Reptiles

Unlike the amphibians, reptiles have been able to adapt to a true terrestrial existence, living and reproducing on land because they possess the following characteristics:

- A tough, dry, scaly skin that is heavily keratinized. Few cutaneous glands are present. Keratin is a resistant protein found in epidermally derived structures of amniotes. It is protective, and when chemically bonded to phospholipids, prevents water loss across body surfaces.
- Two sets of paired limbs, usually with five clawed toes. Limbs are heavily muscled and angled downward from the body allowing for more efficient and rapid locomotion than is possible with the leg structure of amphibians. Limbs are vestigial or absent in snakes, some lizards, and amphisbaenians (worm lizards).
- A strong fully ossified skeleton with a well-developed rib cage.
- Large well-developed lungs. There are no gills, and cutaneous (skin) respiration is negligible.
- A circulatory system functionally divided into pulmonary and systemic circulation pathways. The heart has a two chambered atrium and a ventricle that is more completely divided by a septum than is the amphibian heart. Crocodiles have a four-chambered heart that resembles those of birds and mammals.
- Although ectothermic, reptiles maintain a relatively high body temperature during periods of activity and are more active than amphibians.
- Paired metanephric kidneys that eliminate considerable nitrogen as uric acid, thus conserving water.

- Internal fertilization with sperm introduced directly into the female reproductive tract with a copulatory organ.
- A shelled, amniotic egg.

First distinguished in the fossil record by the mid-Carboniferous, the three evolutionary lineages of ancestral amniotes—anapsids (turtles), diapsids (tuatara, lizards, snakes, crocodilians, and birds), and synapsids (mammals)—show remarkable similarities and differences in the solutions they found to the challenges of life on land. We tend to think of mammals as the preeminent terrestrial vertebrates, but that biased opinion reflects our own position in the synapsid lineage. In reality, the living species of anapsids and diapsids greatly outnumber mammals. Furthermore, they have exploited virtually all of the terrestrial adaptive zones occupied by mammals, plus many that mammals have never penetrated. Morphologically, reptiles have also produced adaptations never seen in mammals, such as the gigantic body size achieved by some dinosaurs and the elongated body form of snakes.

Order Characteristics of Class Reptilia

Order Testudines (L. *testudo* = tortoise)—Turtles.

Turtles are the most ancient of the living reptile lines. Appearing first in the early Permian, turtles evolved a successful approach to life in the Triassic and have scarcely changed over the course of 260 million years. The 13 living turtle families numbering approximately 260 total species are characterized by a bony or leathery box-like shell, a keratinized beak rather than teeth (No turtles have teeth on their jaws), and unique among vertebrates, limbs and limb girdles located inside the ribs.

The shell, which is the key to turtle success, has also limited the group's diversity.

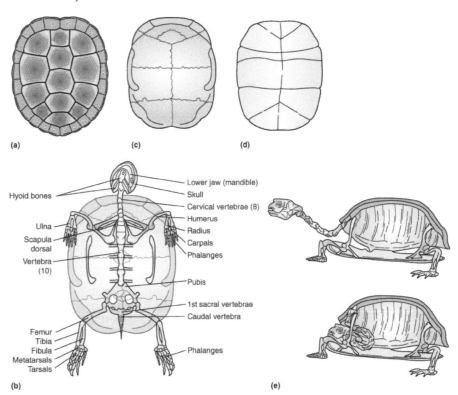

Figure 18.5 The Skeleton of a Turtle. (a) Dorsal view of the carapace. (b) Ventral view of the carapace and the appendicular skeleton. Note the expansion of the ribs outside the limb girdles as long flattened plates, which joined together form the majority of the carapace. (c) Dorsal view of the plastron. (d) Ventral view of the plastron. (e) The articulation of the neck vertebrae allow for extreme flexion.

Obviously, flying or gliding turtles have never existed, and even arboreality (life in the trees) is only slightly developed.

A turtle's shell is essentially a box of bone consisting of a domed upper shell or **carapace** and a flat lower shell or **plastron** (**Figure 18.5**). The carapace is formed from a fusion of vertebrae, expanded ribs, and bones in the dermis of the skin and consists of a number of interlocking bony plates. The plastron is formed from bones of the pectoral girdle and dermal bone. It also consists of interlocking bony plates that are larger but fewer in number than those of the carapace. The morphology of the shell reflects the ecology of the turtle species.

Based on ecology and morphology (but not taxonomy), four different categories of turtles are recognized in common North American usage:

- **Amphibious freshwater types** commonly called *turtles*. Freshwater turtles have broad paddle-like legs with webbed toes and a low carapace that offers little resistance to movement through water. Most familiar are the pond turtles that include the painted turtles (Emydidae) and the red-eared turtles (Bataguridae) often seen in pet stores and biology laboratory classes (**Figure 18.6**). The snapping turtles (Chelydridae) and the mud and musk turtles (Kinosternidae) prowl along the bottoms of ponds and slow rivers and are not particularly streamlined. They have a reduced plastron that makes them more agile than most turtles. In fact, musk turtles may climb several feet into trees.

- **Fully terrestrial types** commonly called *box turtles* or *tortoises*. Tortoises have elephant-like feet, and their toes are not webbed (**Figure 18.7**). The plastron of the box turtle (*Terrapene)* has flexible areas, or hinges, that allow the anterior and posterior sections of the plastron to be pulled up tightly against the upper shell. This allows the box turtle to pull their head, limbs, and tail in and seal them off from attack by potential predators. Smaller species of tortoises may show adaptations for burrowing. The gopher tortoises (*Gopherus*) of North America are an example—their front legs are flattened into scoops, and the dome of the carapace is reduced. The Bolson tortoise (*Gopherus flavomarginatus)* of northern Mexico constructs a burrow a

Figure 18.6 Basking Pond Turtles (Emydidae). These turtles display the flattened carapace and paddle-like legs characteristic of aquatic turtles.

(a)

(b)

Figure 18.7 Tortoises (Testudinidae) such as the (a) Galapagos tortoise (*Chelonoidis nigra*) and the (b) Desert tortoise (*Gopherus agassizi*) possess a domed carapace, hinged plastron, and elephant-like legs.

meter (6-7 feet) or more deep and several meters (12-14 feet) outward in the hard desert soil. These tortoises bask at the mouths of their burrows; when a predator appears, they throw themselves down the steep entrance tunnels of the burrows to escape, similar to an aquatic turtle diving off a log. Endemic to nine islands of the Galapagos archipelago, the Galapagos tortoise (*Geochelone nigra*) is the largest tortoise. Adults can weigh over 300 kg (660 pounds), reach lengths of 1.2 meters (4 feet) long, and live to be 200 years old.

- **Amphibious brackish water types** called *terrapins*. Terrapins have a smaller, lighter shell than do terrestrial types, and like freshwater types, their legs are paddle-like with webbed toes. They inhabit tidal marshes, estuaries, and lagoons where the water contains some salt but is not as salty as the ocean.

- **Saltwater types** called *sea turtles* or *oceanic turtles*. Sea turtles (Cheloniidae and Dermochelyidae) are totally aquatic and found in all the world's oceans except the Arctic Ocean. Their shells are generally reduced and flattened, and their limbs are flat flippers that lack toes entirely. The leatherback (*Dermochelys coriacea*) does not have a hard shell, instead it carries a mosaic of bony plates embedded in connective tissue beneath its leathery skin. The leatherback is the largest marine turtle reaching shell lengths of more than 2 m (6-7 feet) and weights in excess of 660 kg (1,323 pounds). These pelagic turtles have a wider geographical distribution than any other ectothermic amniote. Leatherbacks penetrate far into cool temperate seas and have been recorded in the Atlantic from

Figure 18.8 Sea turtles with their reduced shell, flipper legs, and tremendous lung capacity are highly adapted to an underwater existence.

Newfoundland to Argentina and the Pacific from Japan to Tasmania. They are also capable of making astoundingly deep dives. One dive that drove an attached depth recorder off the scale is estimated to have reached 1200 m (3,960 feet), which exceeds the deepest dive recorded for a sperm whale of 1140 m (3,762 feet) (**Figure 18.8**).

Turtles have adapted to a remarkable variety of environments, but the greatest number of species occurs in southeastern North America and South Asia. In both areas, most species are aquatic, living in bodies of water from small ponds, bogs and swamps to large lakes, and rivers. A few are strictly terrestrial, and others divide their time between land and water. Although turtles as a group are broadly distributed, each species has a preferred habitat and is seldom found elsewhere.

In their specific habitat, turtles are not social animals. Although members of the same species may be observed congregating along a stream or basking on a log, there is usually little interaction between individuals. Several species may inhabit the same pond or river, but each has its own foods, feeding behavior, and likely different activity periods. For example, a small lake in a costal Gulf state may be home to at least seven turtle species including snapping turtles, red-eared sliders, eastern cooters, common mud turtles, loggerhead musk turtles, common musk turtles (or stinkpots), and spiny softshell turtles.

A snapping turtle is a powerful predator that will catch frogs, snakes, small birds, and mammals, and even go fishing (**Figure 18.9**). The softshell, musk, and mud turtles, on the other hand, will pursue many of the same small aquatic animals but with different preferences; the softshell hunts mainly fish and crayfish; the musk turtle eats mainly snails, insect larvae, and carrion; and the mud turtle feeds primarily on insects, molluscs, and carrion. The slider and cooter have a mixed diet, with the cooter's more heavily vegetarian.

Tortoises are herbivores that regularly eat a variety of plants and plant parts as available. Green sea turtles eat marine grasses and algae; the leatherback dines largely on jellyfish, and the hawkbill sea turtle eats sponges.

All in all, turtles are among the most unusual vertebrates. Had they become extinct at the end of the Mesozoic, they would rival the dinosaurs in their novelty. Because they did survive, however, they are regarded as commonplace and are easily identified universally.

Figure 18.9 Settled on the bottom of any standing body of freshwater, the alligator snapping turtle (*Macrochelys temminickii*) sits mouth agape wriggling a small piece of pinkish flesh on its lower jaw. This "bait" proves irresistible to small fish who are lured to their doom in the snapper's hair-trigger jaws.

Order Sphenodontia (Gr. *sphen*, wedge + *odontos*, tooth)—Tuataras.

Today, the order Sphenodontia is represented by only one genus containing only two living species. Truly a "living fossil," the tuataras are virtually unchanged from extinct relatives that were present at mid-Permian, nearly 240 million years ago.

Tuataras have never been a diverse nor widespread group being found originally throughout the two main islands of New Zealand only. Falling victim to human encroachment and domestic animals, the two living species today are restricted to small islets of Cook Strait and off the northeast coast of North Island. Tuataras share underground burrows with ground-nesting seabirds, and venture out at dusk and dawn to feed on insects, or, occasionally, small

Figure 18. 10 Although they superficially resemble lizards, tuataras are part of a separate and distinct lineage—Sphenodontia.

vertebrates (**Figure 18.10**). At first glance tuataras appear to be lizards, but the resemblance is only superficial. Tooth arrangement and structure distinguish the tuatara from other reptiles. Two rows of teeth on the upper jaw and a single row of teeth in the lower jaw produce a shearing bite that can decapitate a small bird. They have a very low metabolic rate and grow slowly; one is recorded to have lived 77 years.

Zoologists find the tuatara intriguing because it possesses many characteristics similar to those of some Mesozoic animals from 200 million years ago. Among the common characteristics are a diapsid skull with twin, arched temporal openings, and a complete **parietal eye**. Buried beneath opaque skin and registering only changes in light intensity in modern tuataras, this "third eye" (complete with a cornea, lens, and retina) may have been an important sense organ in earlier reptiles.

Order Crocodilia (L. *crocodilus* = crocodile)—Crocodiles, alligators, caimans, gavials.

Like the tuatara, crocodilians are "living fossils" in the sense that they are the only remaining reptiles of the same archosaurian lineage from which the dinosaurs and birds descended. Remarkably, today's crocodilians (of the same lineage that started in the late Cretaceous period 170 million years ago) differ only slightly in their structure from their early Mesozoic ancestors.

All crocodilians have a massive elongated skull with powerful jaw muscles perfect for a wide opening and very quick, forceful closure. Their teeth, like those of mammals, sit in bony sockets, an arrangement called **thecodont** that was common among both archosaurs and the earliest birds. Unlike mammals, crocodilians replace their teeth throughout life though not in extreme old-age. Juvenile crocodilians replace teeth with larger ones at a rate as high as one new tooth per socket every month. As adults, tooth replacement slows to two years and even longer. As a result, a single crocodile can go through at least 3,000 teeth in its lifetime. Each tooth is hollow, and the new one is growing inside the old. In this way, a new tooth is ready once the old is lost.

The nostrils are at the tip of the snout, so the animal can stealthily breathe while mostly submerged. Air passageways of the head lead to the rear of the mouth and throat, and a flap of tissue near the back of the tongue forms a watertight seal that allows breathing without inhaling water when the mouth is filled with water or food (or both). A hard complete secondary palate of bone in the roof of the mouth separates the nasal and mouth passageways and seals them off from one another. This adaptation evolved in the archosaurs and is found today only in crocodilians and mammals. All crocodilians are also equipped with powerful legs, muscular clawed toes, and an elongated muscular tail used for swimming, defense, and prey capture.

Modern crocodilians consist of 21 species organized into three families: Gavialidae—gavials, Alligatoridae—alligators and caimans, and Crocodylidae—crocodiles.

Family Gavialidae Gavials (or gharials) are found only on the northern Indian subcontinent where most are **riverine,** being best adapted to calmer areas in deep fast-flowing rivers. In fact, they only leave the water to bask or reproduce and nest. Gavials have very elongated snouts armed with many razor-sharp teeth, an adaptation to a fish diet in adults (**Figure 18.11**). In males, the tip of the snout is enlarged. Known as

Figure 18.11 The rapier-like jaws of the gavial (*Gavialis gangeticus*) are exquisitely adapted for slashing through schools of fish in shallow water.

a *ghara* (after the Indian word for "pot"), this bulbous growth is used to generate a resonant hum during vocalization, act as a visual lure for females, and produce bubbles that have been associated with the mating rituals of the group.

Gavials lack the mechanical strength of the robust skull and jaw of other crocodilians and cannot prey upon large creatures. Adapted to eating primarily small fish, adult gavials slash the water back and forth with their rapier-like jaws and seize the fish with their needle-like teeth. Young gavials eat insects and frogs.

Family Alligatoridae True alligators are native to only two countries: the United States and China, and are now restricted to just two species, *Alligator mississippiensis* of the southern United States, and the small *Alligator sinensis* in the Yangtze River, People's Republic of China(**Figure 18.12**).

Most American alligators inhabit Florida and Louisiana, but they can also be found in southern states bordering both the Gulf of Mexico and the Atlantic. In Florida alone there are estimated to be more than one million alligators. American alligators live in ponds, swamps, marshes, and rivers. The Chinese alligator is currently found only in the Yangtze River valley and is considered extremely endangered, with only a couple of dozen believed to be left in the wild. Indeed, far more Chinese alligators live in zoos around the world than live in Yangtze River.

Figure 18.12 Alligators may be distinguished by their U-shaped snout.

The average American alligator has a length of 4 m (13 feet) and a weight of 360 kg (800 pounds). The largest American alligator ever recorded measured 5.8 m (19 feet 2 inches) and was found on Marsh Island, Louisiana. Few large alligators are ever weighed, but the larger ones could easily exceed a ton in weight. The Chinese alligator is smaller, rarely exceeding 2 m (7 feet) in length.

Alligators are opportunistic feeders, eating almost anything they can catch. When young, they eat fish, insects, snails, and crustaceans. As they grow, they progressively take larger prey, including large

Figure 18.13 Caimans (*Caiman crocodilus*) are the Central and South American representatives of the alligator family. Caimans are often sold in pet stores as "baby alligators."

fish, turtles and other reptiles as well as various mammals and birds. They will even consume carrion if they are sufficiently hungry. As humans encroach on their habitat, the odds of an attack by alligators on humans have increased. Although there were only nine fatal attacks in the U.S.A. between 1970 and 2000, thirteen people were killed by alligators in the six years between 2001 and 2007 alone. It should be noted, however, that 2007 marked the last recorded human death by alligator attack in North America. Alligators, unlike the larger crocodiles, do not immediately regard any human encountered as prey, and being wary, they most often move away from an approaching human.

In Central and South America, the alligator family is represented by five species of the genus Caiman. The Spectacled Caiman (*C. crocodilus*) has the widest distribution, from southern Mexico to the northern half of Argentina, and grows to a modest size of about 2.2 m (slightly over 7 feet) (**Figure 18.13**).

Family Crocodylidae True crocodiles live throughout the tropical regions of Asia, Africa, Australia, and the Americas. Crocodiles tend to congregate in freshwater habitats like rivers, lakes, marshes and swamps, and sometimes brackish water environments (**Figure 18.14**). The saltwater crocodile (*Crocodylus porosus*) of Australia, Southeast Asia, and the Pacific islands often lives along coastal areas and is known to venture far out to sea. The saltwater crocodile is also the largest of all the crocodilians. The record holder was shot in Australia and measured to be 8.6 m (28.2 feet) long with an estimated weight of 1134-1361 kg (2500-3000 pounds). As impressive as that is, crocodiles were evidently much larger in the past as evidenced by the fossil remains of *Sarcosuchus imperator*. First uncovered in the Sahara Desert and nicknamed "SuperCroc," this ancient beast would have been an astonishing 12.2 m (40 feet) long with a weight estimated to be around 8.75 tons (17,500 pounds), or the weight of 3-4 average cars.

Figure 18.14 Crocodiles possess V-shaped jaws that are longer and narrower than those of the alligator. Unlike alligators, when a crocodile closes its jaws, most of its teeth are visible

Like alligators, crocodiles are ambush hunters, waiting for prey to come close then grabbing the hapless victim with a lightning fast lunge and snap of the most powerful jaws in the animal world. The bite force of a large crocodile is more than 5,000 pounds per square inch (psi), compared to just 800 psi for a hyena, 690 psi for a great white shark, and a paltry 335 psi for a Rottweiler dog. Although the muscles that close the jaws of a crocodile are impressively powerful, the muscles that open the jaws are not. This allows zoologists to subdue crocodiles by taping their jaws shut or holding the jaws shut with large rubber bands. Crocodiles will eat anything they can latch their powerful jaws onto, including fish, birds, and mammals to other reptiles and even other crocodiles. Crocodiles cannot chew so their food must be swallowed whole. This presents a problem when the prey is a large animal such as a wildebeest taken at a water hole. Crocodiles overcome this limitation by clamping onto the prey and rotating their bodies violently (known as a death roll), tearing loose chunks of flesh and bone that can then be swallowed. Larger species of crocodiles can be quite dangerous to humans, especially the saltwater and Nile (*C. niloticus*) crocodiles. Accurate accounts are often difficult to obtain, but it is believed that these animals kill hundreds of people each year in Southeast Asia, Africa, and Australia.

The crocodiles are an ancient lineage dating back to the dinosaurs. Despite their prehistoric appearance, however, crocodiles are morphologically advanced compared to other reptiles for like birds and mammals; they possess a four-chambered heart, a diaphragm, and a cerebral cortex.

Members of order Crocodilia have taken everything time and nature has thrown at them for 170 million years, and although they have dwindled to only 21 living species, they have prevailed. They face an uncertain future, however, in a world dominated by the wants and greed of humans.

> *From the time I was a boy, from this house, I was out rescuing crocodiles and snakes. My mum and dad were very passionate about that and I was lucky enough to go along.*
>
> —Steve Irwin

Order Squamata (L. *squamatus* = scaly)—Lizards, snakes, and amphisbaenians (worm lizards).

Squamates make up roughly 95% of all living reptiles. Not only are they the most common of the reptiles, but they are also the most recent and diversified line of diapsid evolution. Squamates are a monophyletic group that is a sister group to the tuatara. The squamates and tuatara together are a sister group to crocodiles and birds, the extant archosaurs. Squamate fossils first appear in the early Jurassic, but mitochondrial DNA analysis suggests that they evolved in the late Permian around 260-275 million years ago at a time when Earth's landmasses were united in one giant supercontinent known as Pangea.

Although many skeletal features distinguish squamates from rhychocephalians, one of the most important is found in the lower jaw. The common ancestor had a rigid skull in which the lower jaw rotated from a pivot point on the bottom rear of the skull. As squamates evolved, the skull bone that connects to the lower jaw—**the quadrate**—lost its rigidity and became only loosely attached by ligaments to the rest of the skull. That new hinge-like configuration, known as **streptostyly**, enabled the back of the jaw to move more freely. Thus, the **kinetic skull** was born (**Figure 18.15**). In practice, a kinetic skull enables the animal to deliver a faster and more powerful bite, and to open its mouth much wider facilitating the capture and handling of prey.

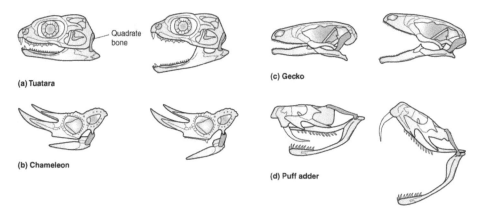

Figure 18.15 Members of the order Squamata (lizards and snakes) are distinguished from each other and from members of the closely related Rhynchocephalia (tuatara) by the joints of the jaw and skull. (a and b) The skull of a tuatara and a chameleon is relatively rigid, whereas the skull of a gecko lizard and puff adder snake (c and d) are jointed and have jaw positioning that allows for a wider gape and better manipulation of prey.

Modern squamates are aligned in three suborders: Amphisbaenia—worm lizards, Sauria—lizards, and Serpentes—snakes.

Suborder Amphisbaenia The worm lizards are peculiar elongated, legless squamates closely related to lizards and snakes. They number about 135 species (**Figure 18.16**). The tail of a worm lizard truncates in such a manner that it vaguely resembles the head (Their suborder name is derived from Amphisbaena, a mythical serpent with a head at each end.). The life of these small reptiles (less than 15 cm [6 inches] long) is poorly understood, due to their burrowing nature and general rarity. With pink skin divided into numerous rings and lacking visible eyes, the worm lizards superficially resemble earthworms. Their skin is only loosely attached to the body, and they move using an accordion-like motion, in which the skin moves and the body just seemingly drags along behind it. Uniquely, they are also able to perform this motion in reverse just as effectively. The skull of worm lizards is solidly built and specially shaped to aid in burrowing through the soil.

Figure 18.16 The peculiar legless worm lizard (Blanidae) resembles snakes or legless lizards, but has many unique features that distinguish it from other squamates.

Amphisbaenians can be found in many regions of South America as well as in the tropical regions of Africa. Only one species, *Rhineura floridana*, exists in the United States. It is found only in Florida where it is nicknamed the "graveyard snake."

Suborder Sauria Numbering approximately 4,500 species, the lizards are the largest group of reptiles. They are found almost from pole to pole in the major zoogeographical zones of every continent except Antarctica. They are also scattered over the islands of the Atlantic, Pacific, and Indian oceans as well as the island continent of Australia. Lizards are a diversified bunch; among them are terrestrial, burrowing, aquatic, arboreal, and even aerial (gliding) types.

Living lizards are aligned into 40 families. *Geckos* (Gekkonidae) (**Figure 18.17**) are among the more common and familiar. Gecko characteristics include a nocturnal nature, a relatively small size, great agility, large eyes, and toe pads that allow them to defy gravity underneath and along vertical surfaces. Unlike most lizards, geckos are capable of making chirping or clicking vocalizations. *Iguanas*

Figure 18.17 Geckos such as this Tokay gecko (*Gecko gecko*) lack eyelids and must lick their eyes to keep them clean and moist.

(Iguanidae) include the most familiar New World lizards, the marine iguana of the Galapagos Islands, and the flying dragons (*Draco*) of Southeast Asia (**Figure 18.18**). *Skinks* (Scincidae) have elongated bodies and limbs are reduced in many species. *Chameleons* (Chamaeloeonidae) are a group of arboreal lizards, mostly of Africa and Madagascar that are noted for their prehensile tail, ability to snag prey from afar with an extensible tongue, and the talent to change their body color at will (**Figure 18.19**).

A lizard's body is long and cylindrical (flattened horizontally) or compressed (flattened vertically), depending on the species. Lizards typically possess a blunt head with movable eyelids and external ear opening, four powerful legs with five clawed toes on each leg, and a long tail. A few species, such as the glass lizard (*Ophisaurus*) and the slow-worm (*Anguis fragilis*), have no legs at all. Legless lizards, however, retain the remnants of a pectoral girdle and sternum. Most lizards are colored in varying shades of greens, browns, and blacks, although some sport dazzling reds, yellows, oranges, blues, and white. Many are also adorned with ornamental crests, frills, and throat fans. Due to the smooth and shiny appearance of their scales, some lizards appear slimy or slippery. In actuality, their skin is very dry due to the absence of excretory glands.

The trademark tail of the lizard has three primary functions: grasping, balance, and storage. Reserves of fat can be stored in the tail—an important feature in those lizards that remain inactive for long periods in harsh climates. Several species, but especially the true chameleons, are noted for their **prehensile** (grasping) **tail**, which they wrap around vegetation to steady themselves as they climb. The balance function of the tail is the most important, and individuals that have lost their tails run or climb clumsily. Many species of lizards have a "break point" in the vertebrae of the tail that, if seized by a predator, separates easily from the rest of the body, a process referred to as **caudal autotomy**. Lizard tail vertebrae have a perforated fracture plane encircling

Figure 18.18 Iguanas. (a) The green iguana (*Iguana iguana*) is a large lizard native to Central and South America as well as the Caribbean. (b) The Galapagos iguana (*Amblyrhynchus cristatus*) is the world's only swimming and diving iguana. These iguanas swim out and then dive down and gnaw algae off submerged rocks.

the bone. When a tail is broken, the vertebrae break along the fracture plane, and the overlying muscle and connective tissue separates as the tail falls off. To increase the likelihood that a predator will attack the tail and not the head, many species have brightly colored tails, and they

Figure 18.19 Chameleons such as this panther chameleon (*Furcifer pardalis*) are a specialized clade of Old World lizards best known for their ability to transform their skin into an artist's palette of bright colors.

will wag them back and forth. Ideally, the attacking predator will be preoccupied with a mouthful of squirming tail while the lizard beats a hasty retreat. Tails lost in this manner will regenerate but never fully attain the size or appearance of the original missing piece. Tail regeneration is a metabolically expensive proposition and can also result in lowered social status.

Box 18.1
Your Tail for Your Life

A king snake moving slowly through short grass freezes momentarily as it senses a small lizard scurrying towards it. In the blink of an eye the snake lunges and seizes the lizard by the tail. After a brief struggle, the stubborn wriggling tail of the lizard going down the snake's gullet is the last sign of what seems to have been a rather routine predator-prey encounter—a meal for the snake and death for the lizard. A closer look in the grass of that battleground, however, would reveal the lizard beating a hasty retreat, still alive but missing most of its tail.

One unfamiliar with the nature of snakes might conclude that the snake merely bit the tail off the lizard. Not so because snakes are not structured to bite chunks off their prey; everything eaten is swallowed whole. The truth of the matter is that the lizard practiced a rather unusual defense mechanism known as autotomy or self-amputation. In the animal kingdom both invertebrates—spiders, crabs, insects, sea stars, sea cucumbers—and vertebrates—salamanders, tuatara, lizards, snakes, and some rodents are known to autotomize (Gr. *auto* = self + *tomy* = severing).

As we learned in Chapter 14, the sea cucumber practices the most extreme variety of autotomy as it literally tears out its own guts. In vertebrates, however, the tail is the only appendage known to autotomize, and in most cases, autotomy is followed by regeneration of a new tail or most of one.

The tail of the lizard that was attacked in the opening scenario is adapted for the purpose of autotomy. At points within the tail are fracture planes where the vertebrae, muscles, and blood vessels are all modified to be torn apart and bending the tail sharply to one side initiates separation. The vertebral column ruptures, and the processes of the tail muscles separate. Sphincter muscles in the arteries and valves in the veins close, preventing loss of blood. An autotomized tail may writhe about violently for several minutes, distracting a predator and allowing the lizard to scurry away to safety, wounded but still alive. When the tail of a lizard is seized, autotomy usually occurs through the plane of the vertebra immediately anterior to the point at which the tail was being held, thereby minimizing the amount of tail lost.

Leaving your wriggling tail in the mouth of a predator is certainly preferable to being eaten, but it is not without costs. Because the tail is an important status symbol among some species of lizards, one who loses their tail may fall to a lower rank in the dominance hierarchy. Losing the tail is also losing energy. Juvenile lizards that have autotomized their tail grow at a reduced rate and loss of the tail in gravid females can result in smaller clutches of eggs.

Although there are cases where humans have self-amputated trapped limbs in wilderness emergencies and survived, perhaps the best you can hope for if your are every pursued by a large predator such as a bear is to wish you were a lizard with a long tail.

All lizards possess a well-developed tongue that can be extended to varying degrees. However, none do so as dramatically or from such long range as the Old World chameleons. Using high speed video and x-ray film, zoologists have calculated that the chameleon's tongue shoots out of its mouth at more than 26 body lengths per second—21.6 kilometers per hour (13.4 miles per hour), and can snag prey one-sixth their own size located two body lengths away (**Figure 18.20**). That's the equivalent of a six-foot person snagging a full-grown turkey from 12 feet away using only their tongue.

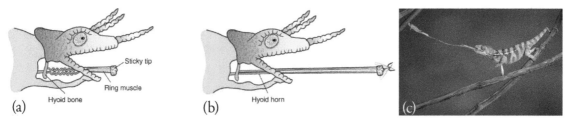

(a) (b) (c)

Figure 18.20 A Chameleon's Tongue. (a) The chameleon's long range weaponry in action. (b) The tongue, which is hollow, covers a horn-shaped piece of tapering cartilage, the hyoid horn. The horn is attached to the cent of the U-shaped hyloid bone. Ring muscles in the tongue contract against the hyoid horn and those contractions combined with the tapering of the horn create a forward thrust that (c) ejects the tongue outward for an amazing distance at speeds only time-lapse photography can resolve.

Lizards range in size from the Caribbean gecko (*Sphaerodactylus ariasae*) that at 16 mm (0.63 inch) long and can curl up on a dime to the enormous monitor lizards (*Varanus komodoensis*), often called Komodo dragons after the place in Indonesia where they were first discovered. These giants will regularly kill prey as large as pigs and small deer and have been known to bring down water buffalo. They will also attack and kill humans.

Monitor saliva is chock full of deadly species of bacteria, and most bitten prey that escape die from a fast-moving bacterial superinfection. Like sharks following a few molecules of blood to a bleeding fish, the monitor just waits until it senses a light smell of new death and follows the scent to the source.

—Jane Stevens

Most lizards are carnivores that feed mainly on insects, worms, birds, mammals, and other reptiles. About 2 % of all known species are herbivorous, especially the iguanas which consume a wide variety of plant material. The marine iguana of the Galapagos Islands (*Amblyrhynchus cristatus*) swims out and dives down to 15 m (50 feet) under water to nibble algae off rocks (Figure 18.18b). The diets of many lizards may shift with maturity and seasonal changes.

The Gila monster (*Heloderma suspectum*) of the southwestern United States desert regions and the related Mexican beaded lizard (*H. horridum*) of Mexico and Guatemala are the only venomous lizards known (**Figure 18.21**). Both types lack fangs and must introduce their venom with a chewing bite. Although often excruciatingly painful, the venom of these lizards is seldom fatal to humans.

Figure 18.21 The gila monster (*Heloderma suspectum*) is a heavy lizard with a short, swollen sausage-shaped tail. Though venomous, its sluggish nature means it presents little threat to humans.

Suborder Serpentes One might conclude that a snake is one of the least adaptable creatures in the world. It's nearly deaf and can't chew its food. Its locomotion is limited because it has no limbs, and it can't stand extreme heat or cold. Yet, surprisingly, you find the approximately 2,900 species of these legless reptiles virtually everywhere in deserts, grasslands, forests, mountains, and even oceans around the world, everywhere, in fact, except the Arctic, Antarctic, Iceland, Greenland, Ireland, New Zealand, and some small oceanic islands. It is far too cold for snakes to live in the Arctic, Antarctic, Iceland, and Greenland but New Zealand has no snakes because continental drift carved it off from Australia and Asia before snakes ever evolved there. New Zealand is now surrounded by water and snakes simply cannot get there. (As far as can be determined, no snake has ever successfully migrated across the open ocean to a new terrestrial home.) As the world's ocean levels have risen and fallen over the millennia, land bridges have come and gone between Ireland, other parts of Great Britain, and the European mainland. Any snake that may have slithered over those land bridges to Ireland, however, would have been turned to an ice cube by the glaciers that advanced and retreated across Ireland any number of times. In fact, Ireland thawed out for the last time only 15,000 years ago. Since then 12 miles of icy-cold water in the North Channel have separated Ireland from neighboring Scotland, which does harbor a few species of snakes.

An elongate and limbless requires a number of modifications to the standard vertebrate plan. Although many people find it hard to believe that a writhing snake could possess a backbone, the skeleton of a snake may, in fact, contain more than two hundred vertebrae and pairs of ribs. Joints between the vertebrae make the body very flexible. Narrowing of the body has also resulted in the reduction or loss of the left lung and the displacement of the gallbladder, the right kidney, and often the gonads into a single file arrangement like the cars of a train

Skull adaptations facilitate swallowing large prey. The upper jaws are movable on the skull and the upper and lower jaws are loosely joined so that each half of the jaw can move independently. In addition, the two halves of the lower jaw (mandibles) are joined only by muscles and skin, not rigid bone. Such flexibility allows a snake to open and spread its mouth its mouth wide, flexing its highly kinetic skull in order to swallow prey that may be considerably larger than itself (**Figure 18.22**). Reputedly, the biggest prey on record is a 59 kg (130-pound) antelope that was swallowed by an African rock python (*Python sebae*). Another way to appreciate the size of a snake's meal is to look at it in terms of relative body weights. Even the most gluttonous of us humans would have difficulty eating 10 or even 5 pounds at a sitting—and that would represent well under 10 percent of our body weight. Many snake species, however, regularly consume 25 percent of their unfed body weight. Pythons can easily eat 65 percent of their body weight and sometimes 96 percent. The record, however, appears to be held by a viper that swallowed a lizard 1.6 times its own weight. In human terms that would be equivalent to a 64 kg (140-pound) per-

Figure 18.22 With jaws held together only by highly elastic muscles and skin, snakes are able to swallow prey much bigger around than themselves.

son who normally downs a 35-pound chunk of meat occasionally swallowing a 224-pound hunk. And although no dispassionate zoologists have ever actually observed it under controlled conditions, there are numerous accounts of snakes interrupted in the act of swallowing a person, or found with a big bulge that once slit open, yielded a human being.

> *To avoid being killed by a snake, a quick rule of thumb is to stay out of the way of any snake over 11 feet long. An American man was killed recently by his hungry 15-foot pet python when he imprudently tried to handle it unassisted. No single unarmed person would stand a chance against a 30-foot snake.*
>
> —Jared Diamond

All snakes are predatory; there are no known examples of vegetarians among them. Worms, insects, spiders, fish, amphibians, birds, mammals, and other reptiles are among the usual fare, though some snakes have developed specialized tastes for exclusive diets of snails, bird eggs, and certain insects. Although smaller and weaker prey can be swallowed without great difficulty, larger and more active animals need to be subdued for the safety of the snake. Snakes use two methods to do so: constriction and venom. Some snakes constrict or hold their struggling victims by looping one or two coils of their body around the prey and maintain this grip until the process of swallowing is well advanced. The true constrictors throw many body coils around the prey and keep tightening the coils until the victim's heart stops, or it suffocates. As soon as the victim is insensible or dead it is released, reexamined with the tongue, and then slowly swallowed.

Venomous snakes have special venom glands, and specialized teeth called **fangs** for delivering the venom into their victims. In some venomous types, such as the cobras, mambas, and coral snakes, the fangs are short, fixed, and erect at all times. In rattlesnakes and others of its kind, called vipers, the fangs are longer and fold back when the mouth is closed. Most venomous snakes have their fangs in the front of the mouth, but in the back-fanged or rear-fanged snakes, the fangs are to the back of the mouth.

The clear to amber venom of venomous snakes can be grouped into a limited number of categories according to its effects: (1) substances that cause the disintegration of tissues, (2) anticoagulants that destroy the clotting capacity of the victim's blood resulting in heavy bleeding (**hemotoxins**), and (3) **neurotoxins** that act on the nervous system, especially the nerves of the respiratory system and heart. Venom is an effective offensive weapon for securing prey and can also play a defensive role in deterring enemies. The terms *venom* and *poison* are often used interchangeably and considered by many to be synonymous but they are not. Whereas poisons and venoms are both toxins, poisons can be absorbed through the skin or digestive system while venom must be injected into the tissues or blood stream by some mechanical means. One may drink snake venom with no ill effects as long as there are no lacerations inside the mouth or digestive tract. (Do not, as they say, try this at home.) There are, however, the always-present exceptions. The *Rhabdophis* snakes (keelback snakes) are rear-fanged snakes that secrete venom derived from the glands of the poisonous toads that it preys upon; similarly certain garter snakes from Oregon retain toxins in their liver from the newts they eat.

The most lethal land snake venoms are found in two Australian snakes, the inland taipan (*Oxyuranus microlepidotus*) and the king brown snake (*Pseudechis australis*). The poisons of these two species are fatal to a 20 gr (0.7 ounce) mouse at a dosage of 1 microgram (1 millionth of a gram). One bite of an inland taipan carries enough venom to kill 100 people (or 250,000 mice)! Once the prey dies or is subdued by the poison, the swallowing action of poisonous snakes is the same as for constricting snakes.

Obviously, a snake has no limbs with which to manipulate prey and push it down its throat. Instead, snakes use their jaws and muscular and flexible throat and cheeks. Once the snake bites down on the prey, backward pointing teeth prevent it from pulling back out of the snake's mouth. The snake grips the prey and begins swallowing it headfirst so that the hair, legs, spines, or feathers will be pressed back so the prey won't get stuck, and the snake won't get punctured. Snakes will even swallow porcupines with their quills, deer with their antlers, and goats with their horns, although, despite precautions, snakes do occasionally get pierced from the inside.

Once the prey is in the snake's mouth, the snake will alternately extend the opposite sides of its muscular face around it, rocking its mouth back and forth up the length of the prey. Once begun the swallowing action is automatic, but can be reversed, and the meal regurgitated if the snake is disturbed. (The appearance of regurgitated prey, mangled by the jaws and teeth and covered in saliva and mucous from the mouth, has led to the erroneous belief that constricting snakes crush their victims into pulp before swallowing them.) Eventually the prey will be totally engulfed and squeezed down the esophagus into the stomach. The process of swallowing can take up to several hours, especially if the prey is large in relation to the snake. To breathe during the long slow process that is swallowing, snakes are equipped with a tube-like tracheal opening (**glottis**) that is pushed outward between the mandibles serving as a swallowing snorkel.

Snakes eat less frequently than any other vertebrate. In the wild, feeding intervals for rattlesnakes and big constrictors range from a few weeks to a few months; snakes in zoos have refused food for over two years. The process of digestion proceeds at a leisurely pace as well. In humans, the residues of a meal typically appear in the feces within less than a day. For big snakes, on the other hand, transit times have to be measured in days or even weeks. It may take 12-14 days for rattlesnakes to process a rat; at the San Diego Zoo, a snake fed 15 times during the year defecated only eight of those meals.

Venomous snakes are classified in three taxonomic families based in part on the type of fangs:

- Elapidae—cobras, kraits, mambas, sea snakes, and coral snakes with short, permanent erect fangs in the front of the mouth.
- Viperidae—vipers, rattlesnakes, copperheads, cottonmouths, adders, and bushmasters with highly developed, movable, tubular fangs at the front of the mouth.
- Colubridae—boomslangs, tree and vine snakes, and many others, though not all colubrids are venomous. This very large family is not a natural group and has classically served as a dumping ground for snakes that don't fit neatly anywhere else. The venomous members of this group have short permanently erect fangs in the rear of the mouth.

Only four types of venomous snakes are found in the Americas—coral snakes, copperheads, water moccasin, and rattlesnakes. (**Figure 18.23**).

Three species of coral snakes are found in the United States: (1) The *eastern coral snake (Micrurus fulvius)* which is found in scattered localities in the southern Coastal Plain from North Carolina to Louisiana, including all of Florida. (2) The *Texas coral snake (Micrurus tener)* which ranges over the southern three-fourths of Texas, the western half of Louisiana, and far southern Arkansas. (There is disagreement, however, among herpetologists as to whether or not the eastern and Texas coral snakes are indeed separate species.) (3) The *Sonoran* or *Arizona coral snake (Micruroides euryxanthus)* found in northwestern Mexico and the

Southwestern United States. Although the venom of this snake is an extremely potent neurotoxin, no fatalities have ever been reported.

Figure 18.23 The Venomous Snakes of the Americas. (a) New World coral snakes (65 species). The order of bands distinguishes venomous coral snakes from nonvenomous mimics. "Red into black venom lack; red into yellow, kill a fellow." (b) Rattlesnake (32 species from Canada to Argentina). (c) Copperhead (*Agkistrodon contortrix pictigaster*). (d) Water moccasin or cottonmouth (*Agkistrodon piscivorus*).

Although the Carolinas are considered the main locale for the copperhead (*Agkistrodon contortrix*), they are also found in mixed woodlands and low-lying swampy regions in many eastern and southern states— Texas, Oklahoma, Kansas, Missouri, Arkansas, Louisiana, Mississippi, Alabama, Georgia, Florida, South Carolina, North Carolina, Tennessee, Kentucky, Virginia, West Virginia, Illinois, Indiana, Ohio, Iowa, Pennsylvania, Maryland, New Jersey, Delaware, New York, Connecticut and Massachusetts.

The water moccasin (*Agkistrodon piscivorus*) ranges along the Atlantic and Gulf coastal areas, all of Florida, and much of the Mississippi River watershed. Being the world's only semi-aquatic viper, these snakes are strong swimmers and have entered the sea to colonize islands off both the Atlantic and Gulf coast. This snake is also known as "cottonmouths" because of the fluffy white lining of the mouth.

The United States is home to 15 species of rattlesnakes while Mexico is home to 25 species. Although most abundant in the Great Plains states, rattlesnakes are found throughout the Unites States, from deserts to forests and from sea level to high in the mountains.

Reptilian Body Plan

Body Covering

> *A boa constrictor, they say,*
> *When he found his skin shedding one day,*
> *Thought, 'Oh, gosh, oh how crude*
> *To be seen in the nude,' and*
> *He blushingly slithered away.*
>
> —Gordon Krunnfusz

Reptile skin with its tough, dry scales is unique in the animal world. This characteristic skin protects the internal tissues from drying out, and in many species it also plays a role in defense and mating. The outer epidermal layer and scales are composed of a horny material called beta keratin that is unique to reptiles (**Figure 18.24**). A reptile's scales are not homologous to fish scales, which are bony dermal structures. The inner dermis layer contains many blood vessels, nerves, and connective tissue as well as the **chromatophores** (pigment cells) that give reptiles their color. By dispersing or concentrating specialized pigment cells called **melanophores**, lizards such as chameleons and anoles can alternate between drab and vibrant colors.

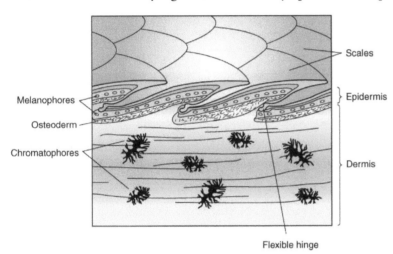

Figure 18.24 A section of reptile skin.

Most snakes are colored in drab browns, grays, or black. Some have bright red, yellow, or vivid green marking in arrangements from blotches to rings, cross bands, and stripes. However, snakes cannot change color to any great degree the way many lizards can.

In some reptiles, such as crocodilians and the box tortoise, the scales remain throughout life being replaced from underneath as they wear off. In turtles, snakes and lizards, new scales grow beneath the old, which are shed at intervals. The skin sloughed off during these regular molts may be in large flakes or a single piece leaving behind a ghostly remnant of its former occupant. The frequency of molting varies with the species and age. Young reptiles shed more than adults; snakes and lizards may shed several times a year and some turtles only once annually. The rattlesnake's rattle is a unique epidermal structure composed of interlocking horny segments; a new segment is formed with each molt, although the end segments tend to break off when the rattle gets very long. Many lizards and crocodiles also have bony plates known as **osteoderms** reinforcing the keratinized dermis scales from beneath.

577

Support and Locomotion

Most reptiles are quadrupeds with powerful limb musculature and strong, clawed toes. For the most part, reptiles move in the same pattern as do other four-legged terrestrial vertebrates.

Crocodilians can assume three main gaits or postures for locomotion on land. The low, sprawled semi-erect "belly crawl" is typically a fairly slow gait in which the crocodilian slides over a slippery substrate such as mud, using its legs to push itself along on its belly. This gait can be modified to shift that bulky body at impressively high speeds, however, normally away from a threat. "Belly run" becomes a more appropriate description of this faster gait, where the legs still operate on either side of the body rather than from underneath it.

Another gait used for rapid locomotion on land is called "galloping." With front and hind limbs moving as synchronous pairs, the crocodile bounds almost like a rabbit—hind legs moving together to push the animal forwards and into the air, and then the body bending so the front legs absorb the impact of landing as the hind legs move forward for the next bound. Galloping is used primarily as an escape response on firm ground, and enables the animal to leap over low obstacles as it heads towards the water.

Another gait is the "high walk," which is uniquely crocodilian, but resembles the erect gait of a mammal with the legs directly underneath the animal, rather than splayed out to the side. This is a slow means of getting around, but it is very useful for picking the body up to negotiate obstacles or to avoid friction with non-slippery substrates such as rocks. The high walk is normally used for short distances, but long-distance hikes using this gait are quite possible.

Most crocodiles can achieve speeds of around 12 to 14 km/hr (7.5 to 8.7 miles/hr) for short periods, which is somewhat slower than a fit human can run. Galloping Australian freshwater crocodiles, however, have been clocked at just over 17 km/hr (10.6 miles/hr) over distances of perhaps 20-30 m (66-99 feet) before they began to tire.

Bear in mind that crocodilians do not normally chase their prey preferring instead to lie in wait and surprise it with an explosive lunge at prey when it gets too close. Adult saltwater crocodiles (around 4 m [13 feet]) have been clocked at 12 m (40 feet) per second for a quarter of a second, which is long enough to capture prey standing within one body length before the prey has time to react. This isn't running, however, because the crocodile cannot maintain this acceleration for more than a very brief instant. In water, crocodilians propel themselves with slow but powerful back-and-forth sweeps of their tail using their legs as rudders and stabilizers.

Most lizards are capable of rapid acceleration and also possess great ability to change direction of motion rapidly. The aptly named racerunners (*Cnemidophorus*) can attain speeds of 24 km/hr (15 miles/hr), which in terms of their own body length, puts them in a class with fast terrestrial mammals. A tendency toward elongation of the body is found in some families of lizards and is often accompanied by reduction in limb length and even loss of limbs entirely. Such lizards propel themselves solely by lateral undulations emanating from highly complicated ventral abdominal musculature.

Many modifications of the toes are to be seen in lizards. Desert-inhabiting types often have fringes on the toes to provide increased surface area and prevent sinking into loose desert sand. Arboreal geckos and anoles have **lamellae** (fine plates) on the undersides of the toes. Each lamella is made of many tiny hairs less than 0.25 micrometer (1/100,000 inch) in diameter. These fine bristles greatly enhance the clinging ability of the lizards, allowing them to find purchase in the smallest irregularities of the substrate. Geckos and anoles can easily climb vertical panes of glass.

Several terrestrial iguanids and agamids can run for brief periods on their hind legs only. The basilisk *Basiliscus* (or "Jesus Christ Lizard" as it is sometimes called) can run across the top of still water at a speed of 1.5 m (5 feet) a second for approximately 4.5 m (15 feet) before sinking on all fours and swimming. Flaps between the toes help support the basilisk, creating a larger surface and a pocket of air that help buoy them on the surface of the water. The flying lizards (*Draco*) are even able to parachute or glide through the air and make soft landings.

How is locomotion to be accomplished by a reptile with no limbs? Snakes have evolved several adaptations to overcome this problem and these adaptations allow snakes not only to move but to do so powerfully, gracefully, and rapidly on land, in the water, and even up and across tree branches (**Figure 18.25**). The locomotion pattern most common in snakes is **lateral undulation** in which an S-shaped path is left behind when the snake propels itself by using lateral force against sticks, pebbles, vegetation, and other objects on the surface. **Concertina movement** allows a snake to navigate through narrow channels by extending forward while bracing S-shaped loops against the sides of the enclosing passage. **Side-winding** is a form of locomotion practiced by desert vipers to move with surprising speed across loose, sandy surfaces with minimum surface contact. Side winding snakes move by throwing its body forward in loops and then pulling against the loops. The body moves at an angle of about 60 degrees to its direction of travel. **Rectilinear movement** (or caterpillar crawl) allows some large snakes, such as constrictors, to advance in a straight

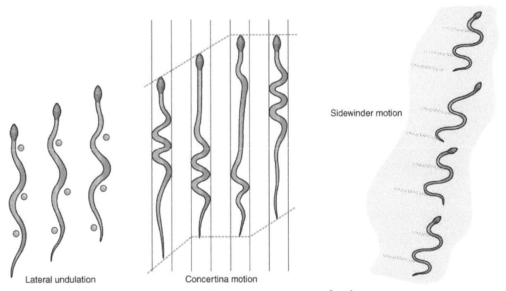

Figure 18.25 Locomotion patterns of snakes.

line. The broad belly scales (**scutes**) slide forward and catch the ground or tree branch like tire treads. Muscles attached from the scutes to the ribs then pull against the scutes. This action may alternate at several points along the snake's body moving like a wave from head to tail. Rectilinear movement allows arboreal snakes to advance along straight tree branches.

Snakes do not travel as fast as they appear to. Rattlesnakes have been clocked at 3 km/hr (2 miles/hr), and even racers never reach more than 6.5 km/hr (4 miles/hr). The record is held by an African black mamba clocked at 11.2 km/hr (7 miles/hr) while chasing a man who had been teasing it. Top speed for these snakes may be short bursts of 16-19 km/hr (10-12 miles/hr).

Feeding and Digestion

Most reptiles are carnivorous and eat all types of other animals such as worms, crustaceans, snails, spiders, insects, fish, amphibians, other reptiles, birds, mammals, and the eggs of other animals depending on the species and the size of the individual. Tortoises, some turtles, and lizards, such as the chuckwalla (*Sauromalus*), desert iguana (*Dipsosaurus*), and Galapogos iguana (*Conolophus*) are the only herbivorous representatives of class Reptilia.

The tongues of reptiles are as varied as the species and range from exceptionally protrusible (as in chameleons) to the virtually fixed tongues of freshwater slider turtles (*Trachemys*) and crocodilians, and from the forked extensible tongues of snakes and monitor lizards to the mobile fleshy tongue of geckos and tortoises. The reptilian tongue has many functions, including prey capture, the transport of food to the back of the mouth, providing the sensation of taste (in some species), and in squamates, it can deliver scent molecules to the **vomeronasal organ** in the roof of the mouth. Some skinks flash a large, brilliant blue tongue to confuse and deter predators.

Snakes, lizards, crocodilians, and tuataras all have teeth. From group to group the teeth vary in their form, their attachment, and whether or not they are shed. The teeth of herbivorous species are broadly flattened with crushing surfaces whereas those of most carnivorous species are tapered with fine points. Some snakes have modified teeth in the form of fangs that deliver venom. As the only reptile without teeth, turtles use their scissors-like beak to cut and shear prey into bite-sized chunks.

Reptile teeth may be attached in sockets (**thecodont teeth**), on the surface of the jaw (**acrodont teeth**), or on the inner side of the jaw (**pleurodont teeth**). Crocodilian teeth are thecodont, while those of the snakes and tuatara are acrodont. Most lizards have pleurodont teeth, but there are many exceptions.

As a group, carnivorous reptiles are stalkers. They make slow, deliberate movements until they are within striking distance of their prey and then suddenly snap and grab it. Powerful muscles of the head and jaws give reptiles great gripping, snapping, slicing, and tearing strength. Because reptiles cannot bite and chew, their food is swallowed whole or in large chunks torn from the prey's body. As a result, reptiles have a long digestive system with powerful digestive juices. The digestive acids of crocodiles have been known to dissolve iron spearheads and six-inch steel hooks.

Respiration

All reptiles breathe using lungs; there is no skin breathing in reptiles as there is in amphibians. Most reptiles have two lungs, but because their elongated bodies require all of their organs to be long and thin, snakes have a very small left lung or have lost the left lung entirely with only the right lung being large and functional.

Turtles are faced with a different problem. Although living within a hard shell with fused ribs is certainly protective, fused ribs do not allow turtles to expand their chests for breathing as do most reptiles. Turtles have adapted to this limitation by using abdominal, pectoral, and limb muscles to drawn air into the abdominal cavity. They then exhale by pulling the shoulder girdle back into the shell, compressing the viscera and pushing the air out. The breathing appears as bellows-like movements in the turtle's skin folds between the limbs and shell. They can also assist their breathing by the simple movements of limbs in walking. When inactive, some aquatic turtles gather enough oxygen to sustain them by using a highly vascularized hole in their mouth to intake and output water.

In squamates, the lungs are ventilated almost exclusively by the axial musculature of the trunk. As this is the same musculature that is used during locomotion, most squamates are forced to hold their breath during periods of intense body movements. Crocodilians actually have a muscular diaphragm that is analogous to the mammalian diaphragm; The difference being that the muscles for the crocodilian diaphragm pull the pubis bone (a part of the pelvis, which is movable in crocodilians) back, which brings the liver down, thus freeing space for the lungs to expand. This type of diaphragmatic setup has been referred to as the "hepatic piston."

Most reptiles lack a secondary palate and must hold their breath while swallowing. Crocodilians, however, have evolved a bony secondary palate forming the roof of the mouth that allows them to continue breathing while opening their mouth underwater to seize prey. Snakes took a different evolutionary path and extended their trachea instead. Their tracheal extension sticks out like a fleshy straw and prevents snakes from suffocating during the lengthy process of swallowing large prey whole.

Osmoregulation and Excretion

Excretion is performed by two small kidneys. Unlike the kidneys of mammals and birds, reptile kidneys are unable to produce liquid urine more concentrated than their body fluid. This is because they lack a specialized structure present in the nephrons of birds and mammals called a **Loop of Henle**, which allows a kidney to concentrate solutes in urine. To compensate, many reptiles reabsorb water in the colon, and some are also able to take up water stored in the bladder. Waste containing nitrogen is expelled as uric acid, instead of as urea or ammonia. Because uric acid has a low solubility and separates from liquid quickly, water can be conserved. In many reptile species, urine is a semi-solid waste. Marine species will also excrete excess salts through nasal (nose) and lingual (tongue) salt glands.

Circulation and Body Temperature

Reptiles have a well-developed double-loop circulatory system. Reptile hearts are either modified three-chambered with a partial septum between the ventricles or, as in crocodilians, a true four-chambered heart similar to those of birds and mammals.

Reptiles are ectotherms. Unlike the ectothermic fish and amphibians, which are more or less at the mercy of the temperature of the environment, reptiles have developed numerous behavioral strategies to actively regulate their body temperature:

1. Reptiles may be active only when the temperature is favorable. Many desert reptiles are active at night or in the early morning and avoid the scorching heat of midday.
2. Reptiles alternate between basking in hot, sunny areas to warm up and moving to shady areas to cool down.
3. Reptiles burrow into the ground or move back and forth between shallow, warm water and deeper, cool water.

4. Reptiles position their body to receive the maximum heating from the sun's rays or raise their legs and possibly their tail up off of hot rocks or sand to cool down. In what may be a cooling mechanism, alligators and crocodiles demonstrate a behavior in which they will **gape**, often resting for long periods on land with their mouths open. Because they will also gape during rain or at night, gaping may have some social function as well.

5. Some reptiles can change their color: darkening early to absorb radiant energy and become lighter later in the day to reflect radiant energy.

Control of body temperature is possible because the hypothalamus of the brain acts like a thermostat and stimulates the reptiles to warm or cool itself. Proper temperature regulation promotes growth, reproduction, and survival by controlling the rate of food processing. A snake denied access to heat may die because the food in its stomach becomes cold and rots.

We should not regard ectothermy as an "inferior" reptilian characteristic, but instead as an adaptive technique employed to cope with and exist in many challenging environments. In fact, ectothermy in reptiles does impart one advantage—slow metabolic rates (about one-tenth that of birds and mammals). This results in a slower but longer life for reptiles than for birds and mammals. In captivity, several tortoises have been documented as surviving beyond 100 years, and there are historical accounts of tortoises living to be 200 years old. Some species of lizards have lived more than 50 years in captivity. Large snakes may reach 40 years and crocodilians 100 years under such conditions.

Nervous System and Sense Organs

The reptilian nervous system is organized along the lines of the basic vertebrate plan. Despite its small size, the brain of a reptile is significantly advanced over the brains of amphibians and fish, with a slightly larger cerebrum and cerebellum.

Most reptile sense organs are well developed, but vision is the dominant sense. Reptile eyes are, depending on the species, adapted for day vision (diurnal) or night vision (nocturnal), and some lizards, crocodiles, and turtles have color vision at least as good as the human eye. The tuatara and some lizards possess a **pineal body** ("third eye") on top the head that does not form images, but may act as a light intensity meter. In addition to their eyes, some snakes such as pit vipers, pythons, and some boas have infrared-sensitive (heat) receptors in deep grooves between the nostril and eye or in pits on their upper lip which allow them to "see" the radiated heat from a warm object.

As in most higher vertebrates, the reptilian ear serves both as an organ of hearing and balance. In some reptiles, other than snakes, the tympanic membrane (eardrum) is visible on the surface of the head. In turtles, it is thin and transparent, but in tortoises it is thick and covered with skin. Some lizards and crocodilians hear through external ear holes leading to the tympanic membrane. Snakes lack a tympanic membrane but cans sense vibrations through bones in their skull.

The nasal region of reptiles is more complicated than in amphibians and allows the reptiles the potential for greater development of the sense of smell. Crocodilians, however, have little if any sense of smell and in turtles the sense is poorly developed. The sense of smell in reptiles achieves its greatest development with the lizards and snakes. In the oral cavity of many vertebrate animals, such as amphibians, some mammals and especially lizards and snakes, resides a vomeronasal organ called (**Jacobson's organ**), a region of chemically sensitive nerve endings. These nerve endings are located in a pair of pits that are continuous with the roof of the mouth. Jacobson's organ is most highly developed in snakes, in which it is used to supplement weak nasal chemical receptors in detecting airborne chemicals. Odors are sampled by flicking out the tongue then retracting it in and inserting the tips of the tongue into Jacobson's organ, where the molecules acquired by the tongue's moist surfaces are detected. The fork in the tongue gives the snake a sort of directional sense of smell. With the help of a favorable wind and their habit of swinging their head from side to side as they walk, komodo dragons may be able to use their tongue to smell carrion up 4-10 km (3-6 miles) away.

The sense of taste is relatively poor in reptiles compared to amphibians and fish, and being sealed with thick scales precludes a highly developed sense of touch.

Reproduction and Development

Sexual reproduction is the rule in reptiles, but parthenogenesis has been identified in whiptails, geckos, rock lizards, and Komodo dragons. In most species, the two sexes differ to some extent in adult size, shape, and/or color. Prior to mating, many species of reptiles engage in elaborate courtship rituals that may take hours or even days. The range of mating behavior displayed is very wide and varies enormously even within orders; male lizards may change color or erect flaps of skin around their throats; some snakes engage in complex weaving or chasing behavior, turtles and tortoises may stroke prospective mates with their forelimbs, whereas male crocodiles and alligators commonly bellow or growl to indicate their readiness for mating. In many species, the mating displays of males are intended to intimidate other males as well as to attract females.

Mating itself may be a cumbersome and potentially dangerous business, especially among large tortoises and crocodiles that are not well suited to acrobatic maneuvers on land (**Figure 18.26**). Sea turtles typically mate in the ocean where the water helps to support their heavy bodies. All sexually reproducing reptiles practice internal fertilization by introducing the sperm directly into the cloaca of the female. In the tuatara, fertilization is achieved by cloacal contact, but in all other reptiles, the males

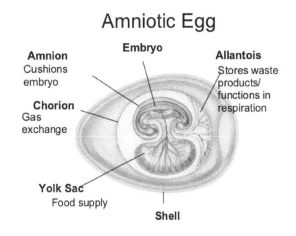

Figure 18.26 Watching large tortoises attempt to mate leads one to marvel that any fertile turtle eggs are ever laid.

Figure 18.27 The embryo develops within the amnion and is cushioned and protected by amniotic fluid. Food is provided by the yolk in the yolk sac and metabolic wastes are deposited within the allantois. Gases pass in or out of the blood vessels of the allantois and chorion through pores in the shell.

have specialized organs for insemination. After insemination, some female snakes, lizards, tortoises, and turtles can store sperm for a year or more.

Most reptiles are oviparous with the female laying eggs in which the young develop. The reptilian amniotic egg is a marvel of adaptation to a terrestrial existence (**Figure 18.27**). Though reptilian eggs still require somewhat high humidity, they can exist in much drier environments than can the eggs of amphibians. The eggs of snakes, most lizards, and turtles are covered with a tough, leathery shell; those of the tortoises, crocodilians, and many geckos are hard and calcified somewhat like birds' eggs. The number of eggs produced annually by a female reptile varies from nearly 400 in some sea turtles to a single egg in the house gecko (*Hemidactylus frenatus*). Small pond turtles lay 5 to 11 eggs, snakes and lizards about 10 to 20, and the American alligator 30 to 60. There is no larval stage, and upon hatching, the young resemble the adults except in size and become independent at once.

Many species of lizards and snakes are ovoviviparous with the embryos developing in thin-shelled eggs that are retained inside the mother's body. Nourishment for the developing young is provided solely by yolk in the egg. The eggs hatch just before birth without being laid first. Thus, ovoviviparous species often appear to bear live young. Some garter snakes have produced over 70 living young in a single brood. The line between oviparity and ovoviviparity is arbitrary. Females of some lizards and snakes retain the fertilized eggs in their bodies for a few days before laying them. Other species retain the eggs for most of the developmental period, hatching occurring shortly after laying. Also, a given species may be oviparous in parts of its range and ovoviviparous elsewhere. A few species of lizards and snakes are viviparous, truly giving birth to live young. In these species the embryonic membranes and the tissues lining the oviducts of the female come in close contact and are modified into temporary reptilian "placenta" through which food and respiratory gases are exchanged.

Parental care of the young varies among reptiles. It is the most highly developed in the crocodilians, which may build elaborate nests of rotting vegetation to hide and warm the eggs. The female lays the eggs there and then faithfully guards the nest. Peeping calls uttered by the young just before emergence signals the female to uncover the nest and assist the young to water. Female Nile crocodiles take the young hatchlings into their mouths and transport them to the water and may stay in the vicinity for several weeks after that.

Most female turtles will dig a depression in soft soil or on a sandy beach with her hind limbs, lay her eggs there, and then cover the nest again using the hind limbs as sweeps. Most snakes and lizards merely hide their eggs under preexisting cover such as rocks or logs and then crawl or slither off. Some female pythons, however, coil around their eggs and incubate them, and some female cobras guard their clutch of eggs.

Reptile Connection

Economically

Humans have exploited reptiles throughout history in a variety of ways, but mainly as food. Sea turtles, lizards, snakes, and their eggs have been fried, broiled, stewed, and eaten raw with great relish by humans worldwide for centuries.

In China, turtles large and small are used for both food and medicine. By the early 1990s, many local populations of turtles had disappeared from the country, so turtles began to be imported from around the world. Some species, such as the three-striped box turtle or golden coin turtle (*Cuora trifasciata*) are so popular for traditional Chinese celebrations that aquaculturists raise them and can sell individual turtles for more than $1,000 (U.S.), an astonishing price for a reptile less than 20 cm (8 inches) long.

Alligators and crocodiles yield high-quality leather suitable for leather goods, shoes, boots, and belts. Alligator, particularly the tail, is commonly eaten as an exotic meat in the southern U.S., especially Florida. Alligator farming is a growing industry in Florida, Texas, and Louisiana. These states produce a combined annual total of some 45,000 alligator hides with hides in the 1.8-2 m (6-7 feet) range selling for $300 each, though the price can fluctuate considerably from year to year. The market for alligator meat is also growing with approximately 400,000 kg (300,000 pounds) of meat produced annually. Lucrative crocodile farms also exist in Asia and Australia. It is estimated that 20 million crocodiles have been annihilated over the past fifty years for a variety of reasons from outright fear and loathing to use as food and leather or in folk medicine.

Although prized for food in some locales and for the bewildering array of tonics, medicines, and potions that are made from various parts of their anatomy, monitor lizards are most valued for their skins. Traditionally used for drumheads and shields, monitor skins are in great demand in the Western world to make watch straps, shoes, wallets, handbags, and other leather goods. The exquisite patterns of the skin combined with the durability of their hides make them the most popular family of lizards in the skin trade. As far as is known, not a single monitor lizard has been bred commercially and so the trade relies entirely on animals taken from the wild. Considering the vast numbers involved and the fact that only the skins of adults or subadults are suitable, it can be presumed that the trade will eventually deplete numbers to the point that they will become extinct in many areas. Recent large-scale extinctions have been suggested for several monitor species in Pakistan, India, and Bangladesh, but it is unclear whether their demise is attributable more to direct human predation or habitat destruction.

Some well-meaning but misinformed people condemn eating or using monitor lizards for any purpose. This view fails to appreciate that almost all of the countries where monitor lizards are exploited are unimaginably poorer than anywhere in North America or Europe. Because they have very little money, they experience very high infant mortality and very low life expectancy. This forces them to adopt unsustainable economic practices that result in the destruction of local ecosystems, the extinction of many animals and plants, and land fit for nothing but shanty-towns and refugee camps. Of all wild animals, monitor lizards are among the best suited to sustainable use. They breed and grow very quickly, establish large populations wherever adequate food is available, and they are not fussy about their diets. Their flesh is extremely nutritious and most importantly, their skins are valuable. (The price of a pair of monitor lizard shoes in Italy or Japan would feed a family of four for a year in most parts of the world where monitors are found.) The establishment of commercial monitor lizard farms would relieve pressure on wild populations and provide desperately poor and malnourished people with a sustainable source of food and income.

Geckos can walk up a pane of glass vertically, and biologists and engineers now know the incredible clinging power of geckos lies in the roughly 6.5 million tiny hairs on its feet. Each hair binds to surfaces because of the weak bonds that result from attractions between molecules, known as Van der Waals forces. With this information in hand, researchers are working to fashion synthetic gecko nano-hairs from silicon rubber and plastics. It appears that the artificial gecko hairs stick to various surfaces just as well as the real ones, which can hold 200 pounds per square inch of hair-covered material. Besides allowing humans to scale tall buildings and cliff facades like a certain cartoon crime fighter, these materials could be used as temporary adhesives.

Ecologically

Reptiles are important predators within the food webs of their habitats. As such, they help keep animals humans regard as pests under control. The wise farmer does not kill the snakes around his farmstead. In many parts of Asia and Africa, monitor lizards are encouraged to inhabit rice paddies, coconut groves, and other farmland to reduce crop damage by eating huge amounts of crabs, insects, and snails.

Small or immature reptiles in turn are prey to large invertebrates, fishes, amphibians and other reptiles, birds, and mammals. Eggs are the most vulnerable stage in the reptilian life cycle while the vulnerability to predators after hatching varies with the size of the hatchlings and the size and type of predator.

An estimated 11 million pet reptiles—mostly turtles, lizards, and snakes—live in U.S. households and their popularity as pets is growing, according to the American Pet Products Manufacturers Association. That Figure, while far lower than for cats and dogs, means that about one of every 25 households includes at least one reptile, and many have two or more. Nearly 100,000 ball pythons and 30,000 boa constrictors are imported annually into the United States alone as well as large numbers of lizards, especially iguanas, and turtles. The global trade in reptiles as pets contributes to depleted wild populations, damaged habitats, and the individual suffering of the animals involved.

Animals that make it to the pet store may be sold in injured or weakened condition. As many as 90 percent of wild-caught reptiles die in their first year of captivity because of physical trauma received before they were sold, or because the buyers cannot meet the animals' complex dietary and habitat needs. Captive iguanas, for instance, often suffer from malnutrition and bone disease because they don't get the diet and ultraviolet radiation they require. For humane, conservation, and public health reasons, the Humane Society of the United States recommends that reptiles not be kept as pets. Compounding the problem is the fact that many people flush or dump reptiles into habitat occupied by native reptiles where they can become an invasive species. Red-eared sliders, for example, native to the Mississippi drainage from Illinois south, are now established all over the world. In Washington, they are threatening the vanishing Pacific pond turtle, and in southern states they are compromising the genetic integrity of yellow-bellied sliders by breeding with them. The Europe Union banned the import of red-eared sliders because of the damage they are said to be doing to European pond turtles.

Since the mid-1990s, several species of non-native, large constrictor snakes, such as Burmese pythons and boa constrictors, have surfaced in localities throughout southern Florida. Probably purchased as pets then eventually released by their owners, the numbers of these snakes are increasing rapidly, and they appear to spreading northward. In doing so, they endanger native species, entire ecosystems, and even humans and their pets.

Of all the reptiles, turtles are the hardest hit by the pet trade and habitat destruction. Turtles don't spew eggs like fish and amphibians; instead, they rely on longevity, replacing themselves over decades. Some turtles, which can live for 120 years, don't become sexually mature until they are 10 to 20 years old. Some species of turtles, which are also collected for food and other uses, are in danger of disappearing in the wild altogether because of the commercial trade. The Old World turtles, especially those of China and Southeast Asia, have been decimated for a variety of reasons, and now New World turtles are being imported to meet the need. Many countries worldwide are making attempts of varying degrees to conserve turtles; such efforts are complicated by cultural practices that demand turtles for food and medicine, and in the case of the sea turtle that roams over vast areas of ocean, cooperative jurisdiction of many different nations.

Medically

Researchers have recently discovered that a turtle's organs do not gradually break down or become less efficient as they age, unlike most animals. It was found that the liver, lungs, and kidneys of a centenarian turtle are virtually indistinguishable from those of an immature turtle. This has inspired geneticists to begin to search the turtle's genome for longevity genes with an eye towards perhaps applying their findings someday somehow to humans.

Although the vast majority of snakes are not dangerous to humans, about 750 species worldwide are venomous and of those, only about 250 species have venom powerful enough to kill a human. Snakebite mortality worldwide is estimated at 30,000-40,000 people per year; the majority of these deaths (25,000-35,000) occur in Southeast Asia, owing to poor medical treatment, malnutrition of victims, and a large number of venomous species. In the United States, around 8,000 people are bitten by venomous snakes yearly, but only 9 to 15 people die—fewer deaths than are attributed to bee stings and lightning strikes. On the other hand, snake venom contains many active biological compounds that may be useful for a variety of medical purposes from preventing the growth of cancerous tumors, to serving as pain-killing drugs and ointments for cuts and burns. Today roughly a dozen diagnostic tests and drugs are derived from snake venom. ACE inhibitors, a class of drugs used to treat high blood pressure and other cardiovascular disorders, were developed from the venom of a Brazilian snake. Scientists anticipate that this is just the beginning. Venom-based drugs may replace Coumadin, a popular but troublesome blood thinner, and the venom from a pit viper snake may prove useful in lessening the damage from strokes. Even Gila monster saliva has medicinal properties. In 1992, Dr. John Eng isolated the hormone exendin-4 from a sample of Gila monster saliva he received in the mail. This hormone proved to be remarkably similar to human glucagon-like peptide-1 (GLP-1), a regulator of glucose metabolism and insulin secretion. Today Exenatide (marketed as *Byetta*), the synthetic version of the hormone Eng isolated from Gila monster saliva, is used in the treatment of diabetes mellitus type 2.

Culturally

Reptiles by their primeval appearance and behavior have crept into our fears and imagination like no other animal group has ever done. The alligator is revered in Native American and African American folklore, especially the teeth which are often worn as a charm against witchcraft and poison. Urban legend has it that full-grown alligators exist in the sewers of cities like New York City. As the story goes, people buy baby alligators as pets eventually flushing them down the toilet to get rid of them. In reality, alligators could not survive such living conditions because without UV rays from sunlight, alligators cannot properly metabolize calcium (which they would get from the sewer rats that myth states is their food source). This would result in metabolic bone disease and eventually death. Small released alligators and caimans, though, are occasionally found in northern lakes. In fact, abandoned alligators turned up in California, Nebraska, Massachusetts, Wisconsin, and Oregon in June, 2005 alone.

The ancient Egyptians worshipped many gods among them was Sobek. With a human body and a crocodile head, Sobek was an attempt to pacify crocodiles so as to reduce the danger they posed. He later came to represent the fertility of the Nile and a cult complete with temples arose in his name. As testament to his importance, mummified crocodiles have been found in the tombs of members of the Egyptian hierarchy.

Nearly every culture since prehistoric times has worshipped, revered, or loathed and feared snakes. Although Satan is depicted as a serpent in the biblical account of the Creation and snakes are regarded as symbols of paganism and evil in Judeo-Christian belief, snakes are revered by most societies. Serpent worship is one of the earliest forms of veneration, with some carving indicating such dating to 10,000 BC. Snakes were venerated in ancient Egypt, and many Egyptian gods were represented by snakes, such as the cobra goddess Neith, founder of the universe. Quetzalcoatl, the mythical "plumed serpent" was worshipped as the "Master of Life" by the ancient Aztecs of Central America. Some African cultures worshipped rock pythons and considered the killing of one to be a serious crime. In Australia, the Aborigines associated a giant rainbow serpent with the creation of life.

Other cultures have associated snakes with medicinal powers or rebirth. In India, cobras were regarded as reincarnations of important people called Nagas. The caduceus, our modern medical symbol of two snakes wrapped around a staff, comes from ancient Greek mythology. According to the Greeks, Aesculapius discovered medicine by watching as one snake used herbs to bring another snake back to life.

Owing to ignorance, a vast global compendium of superstitions and mythologies about snakes has sprung up. Many stem from the biological peculiarities of snakes; their ability to shed the skin is associated with immortality; their ever-open eyes represent omniscience; their propensity for sudden appearance and disappearance embodies procreative powers; and their ability to kill with a single bite engenders primal fears.

Reptiles, especially snakes, have been the subject of folklore and story since the dawn of mankind. From ancient cave paintings to the Bible, stories abound worldwide about reptiles that talk, jump, dance, and tempt. Recently movie producers have attempted to frighten audiences with everything from airplanes full of snakes to monstrously gigantic crocodiles in an east coast lake.

The most popular reptiles of modern times must surely be the dinosaurs of ancient times. Perhaps no other group of animals stirs our imagination, tickles our curiosity, and so fascinates young and old alike as do dinosaurs, and as paleontologists learn ever more about them and the world they lived in, this allure continues to grow. As a result, dinosaurs have been and continue to be honored and even viewed with affection in many ways from toys to books and from amusement park rides to feature films.

A Closing Note
Obey the Signs

The Savannah River forms most of the border between South Carolina and Georgia. The land surrounding the river is one of dense pine forests, small streams, ponds, and swamps and these environs are populated by a diverse array of herps—amphibians and reptiles. Nearly 60 years ago the University of Georgia established the Savannah River Ecology Laboratory (SREL) along the river. SREL has proven to be a very important biological research station and is the only site in this country in which long term and large scale ecological studies have been undertaken.

I was fortunate enough to have been able to spend an entire summer at this amazing place working with professional ecologists and herpetologists. For a biology geek such as me, this was heaven. My days were spent in the field doing everything from checking turtle traps in waist-deep mud to toe-clipping salamanders to surveying small streams for invasive molluscs (where I once nearly put my hand down on a water moccasin as I crawled over a log).

Of all the many memories I have of that experience, two stand out most clearly and both involved alligators. Many times I stood and watched alligators slowly cruising swamps shrouded in the mists of early morning. It gave me a chill then (and still does when I think about it to this day) to realize that I had the privilege, for at least a fleeting moment, to stare back through the veil of time and see a small bit of the earth as it may have been during the reign of the dinosaurs.

The other memory is less profound and professionally embarrassing. On my very first day at SREL, I was awaiting the arrival of the professor I was to be working with. Wandering around the place, I found several acres of small ponds, shelters, and large tanks containing various types of live animals. One pond was fenced with a sign that read:

!!CAUTION!!
Nesting Female Alligator
Do NOT approach the fence!

I could clearly see a pile of vegetation in the corner of the fence right behind the sign. The water of the tiny pond was green and totally covered with algae, but there was no alligator in sight either on land or in the water. So what did I do? I not only approached the fence, I put my hands on it and stared through the openings trying to see an alligator. Not having the benefit of slow-motion replay I'm not exactly sure what happened next but with a whoosh and a splash the female alligator I couldn't see launched from beneath the green water onto her nest and hit the fence hissing loudly with mouth open. I screamed and fell over backward square on my butt. I then quickly rolled over and scrambled away on all fours as fast as I could crawl. I finally stood up, dusted myself off and looked around to see if anyone had seen my embarrassing blunder. Fortunately, no one had. And while I laugh about it when I think of it today, don't repeat my mistake. If you are ever out and about in the field or somewhere where you many encounter large and potentially dangerous reptiles, know the rules and obey the signs.

In Summary

- Reptiles first appeared in the Carboniferous period (345 million years ago) and by the Permian period (260 million years ago) they had become the dominant form of terrestrial life. Several extinction events decimated the reptiles leaving only 4 living orders out of the 21 orders known to have ever existed.

- Today nearly 8,000 species of living reptiles inhabit every continent except Antarctica. Most reptile species and individuals tend to be concentrated in the tropics and subtropics; their numbers decline rapidly towards the poles and in high altitudes. The lizards and snakes are by far the most numerous and widely distributed of living reptiles.

- The paraphyletic class Reptilia is composed of four monophyletic orders:

  Subclass Anapsida
       Order Testudines—Turtles
  Subclass Diapsida
       Order Sphenodonta—Tuataras
       Order Crocodilia—Crocodiles, alligators, caimans, gavials
       Order Squamata—Lizards, snakes, and amphisbaenians

- Unlike the amphibians, reptiles have been able to adapt to a true terrestrial existence, living and reproducing on land because they possess the following characteristics:

1. A tough, dry, scaly skin that is heavily keratinized.
2. Two sets of paired limbs, usually with five clawed toes. Limbs are heavily muscled and angled downward from the body allowing for more efficient and rapid locomotion than is possible with the leg plan of amphibians. Limbs are vestigial or absent in snakes, some lizards, and amphisbaenians (worm lizards).
3. A strong well-ossified skeleton with a well-developed rib cage.
4. Large well-developed lungs. There are no gills and cutaneous (skin) respiration is negligible.
5. A circulatory system functionally divided into pulmonary and systemic circulation. The heart has a two chambered atrium and a ventricle that is more completely divided by a septum than is the amphibian heart.
6. Although ectothermic, reptiles maintain a relatively high body temperature during periods of activity and are more active than amphibians.
7. Paired metanephric kidneys that eliminate considerable nitrogen as uric acid, thus conserving water.
8. Internal fertilization with sperm introduced directly into the female reproductive tract with a copulatory organ.
9. A shelled, amniotic egg.

- Order Testudines—turtles—is characterized by a bony or leathery box-like shell, a keratinized beak rather than teeth, and unique among vertebrates, limbs and limb girdles located inside the ribs. Turtles are usually found in and around water while tortoises are more terrestrial and wider ranging.

- Order Sphenodonta—tuataras—is represented today by only two living species. Truly a "living fossil," these superficially lizard-like reptiles are virtually unchanged from extinct relatives that were present at midPermian, nearly 240 million years ago. Found originally only throughout the two main islands of New Zealand, tuataras have fallen victim to human encroachment and domestic animals to the point that the two living species today are restricted to small islets of Cook Strait and off the northeast coast of North Island. Tuataras share underground burrows with ground-nesting seabirds, and venture out at dusk and dawn to feed on insects, or, occasionally, small vertebrates

- Order Crocodilia—crocodiles, alligators, caimans, gavials—represents the only surviving reptiles of the archosaurian lineage that gave rise to the great radiation of dinosaurs and their kin and birds. All crocodilians have a massive elongated skull with powerful jaw musculature arranged to provide a wide gape and rapid, forceful closure. All crocodilians are also equipped with powerful legs and a muscular elongated tail used for swimming, defense, and prey capture. Crocodilians are found primarily in tropical or subtropical climes and in their habitats they are ferocious predators on all types of vertebrates from fish to mammals.

- Order Squamata—lizards, snakes, and amphisbaenians (worm lizards)—is composed of reptiles with a long cylindrical body either equipped with legs (lizards) or lacking legs (snakes and worm lizards). Squamates are found on every continent, and many Pacific islands, and like other reptiles are more prevalent in tropical and semi-tropical environs. Although most squamates are carnivorous animals, a few types of lizards practice herbivory.

- Reptiles connect to humans in ways that are important economically, ecologically, medically, and culturally.

Review and Reflect

1. **To Change or Not to Change** Most of the species of modern reptiles have changed little since their great period of adaptive radiation some 200 million years ago. Propose a hypothesis to explain why drastic evolutionary changes have not occurred in reptiles for hundreds of millions of years since then.

2. **Shattered Remains** The authors contend that modern reptiles represent the shattered remains of a grand and glorious period of life known as the Age of Reptiles. Do you believe this to be an accurate assessment? Explain.

3. **The Leap to Land** Prepare a table in which you compare and contrast the anatomical features and modes of reproduction of amphibians and reptiles. Use this table to explain why reptiles are true land animals but amphibians are not.

4. **Looks Can Be Deceiving** Your little cousin is also studying reptiles and she comes to you with the questions that follow. What would you answer be to each question?

   A. Why is a tuatara not a lizard?
   B. Why is a legless lizard not a snake?
   C. Aren't alligators and crocodiles the same animal?
   D. How can a crocodile open its mouth underwater without drowning?

E. Can a turtle really crawl out if its shell as they do in cartoons?

F. How can a snake swallow prey many times larger than its own head?

G. Can the flicking tongue of a snake sting or poison you?

H. Can a lizard really lose part of its tail and then grow it back? If this is true, why would they do that?

I. How can a chameleon extend its tongue out so far so fast?

J. Can geckos actually crawl up vertical pane of glass and even dash across ceilings? If this is true, how do they do it?

K. Are reptiles slimy?

5. ***Stay Away*** How can you keep from being bitten by a poisonous snake and if you are bitten, what is the proper first aid treatment for snake bite?

6. ***Pull Away*** Explain how you would go about determining which animal—a turtle or a snail—can pull the most weight per unit of body mass.

7. ***The Truth of the Matter*** Suppose you read the following imaginary headline on a tabloid newspaper—"Woman Wakes to Find Entire Watermelon Patch Eaten by Marauding Band of Snakes!" Is such a scenario biologically plausible or possible? Rewrite the headline so that would be biologically plausible or possible.

8. ***Copulation With a Purpose*** Why is internal fertilization a critical component of the reproductive cycle of reptiles?

9. ***The Amazing Amniotic Egg*** How has the amniotic egg of reptiles been modified and adapted to being laid on land and not in water as are amphibian eggs?

10. ***The Hole Thing*** When a female leatherback sea turtle crawls up on a beach to lay eggs, she digs a deep hole with her hind flippers, lays her eggs, and covers them with sand. She then crawls about 100 m (330 feet) and digs another hole. She lays no eggs in the second hole but merely covers it back up with sand. Propose a hypothesis to explain the purpose of this behavior.

11. ***How Warm Is Your Form?*** Carefully analyze and interpret the graph that follows (**Figure 18.28**). What does the data plotted on the graph suggest? What conclusions can be drawn from this graphed data?

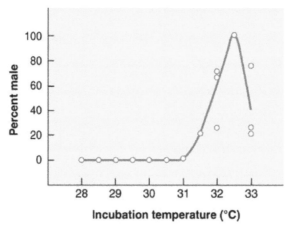

Figure 18.28 Incubation temperature vs gender selection in reptiles.

Create and Connect

1. ***What's in the Quote?***

   If you see a snake, just kill it, don't appoint a committee on snakes.

   —Ross Perot

   Although Mr. Perot's advice may prove useful in business and government, is it a sound biological suggestion? Explain.

2. ***Walk a Mile In My Shoes*** If you had to be reincarnated as a reptile, what reptile would you want to be and why?

3. ***Reptilian Rodent eRadicators*** Imagine that you are a group of snakes who wish to set up a company and sell their services as rodent exterminators.

   Guidelines:

   A. Develop a name for your company.
   B. Develop a company logo and company slogan.
   C. Write a classified ad for your company's services that could be run in a newspaper or write the script for an advertising spot that could be shown on T.V.

4. ***A Reptilian Status Report*** Are any of the local reptiles in your area endangered? Consult with local wildlife authorities and prepare a status report.

   Guidelines:

   A. If more than one endangered reptile is found in your area, the instructor may assign you to report on only one or perhaps all of them as the situation warrants. If there are no endangered reptiles in your area, the instructor may assign you to report on one or several endangered reptiles from other areas.
   B. Your report should include a detailed description of the endangered reptile including its scientific name.
   C. Prepare a range map to show the distribution of the reptile.
   D. Explain what efforts are being made to prevent the extinction of this reptile.
   E. If there currently is no organized effort to aid the endangered reptile you have selected or been assigned, suggest a course of action to remedy the situation.

BIRDS: LORDS OF THE AIR

It is easy to understand why so many of us are so fond of birds. They are lively, they are lovely; and they are everywhere. They have characteristics with which we can easily identify—cheeky and shy, gentle and viscous, faithful and faithless. Many enact the drama of their lives in full view for all to see. They are part of our world, yet, at a clap of our hands, they lift into the air and vanish into their own natural world that lies beyond our brick walls. It is hardly surprising that human beings have studied birds with a greater dedication and intensity than they have lavished on any other group of animal.

—David Attenborough

Introduction

Prehistoric cave paintings and clay Figures reveal that humans have been intrigued with birds since the foundation of our species. Clearly birds have always been valued in a utilitarian sense for their meat, feathers, and bones. Humans hunt, trap, and commercially raise many species of birds for what we regard as useful and practical purposes. Birds, however, are more than convenient, functional objects. Birds seem to strike a chord within our biological psyche. We stand in awe at the wonders of their globe-trotting migrations and we marvel at the beauty of their plumage and melodious songs all the while in envy of their effortless ability to shed the planet and soar free through the sky.

History of Birds

The evolutionary histories of birds and reptiles are clearly intertwined as revealed by the many similarities between them:

- Bird and reptile skulls are positioned against the first neck vertebrae by only one bony knob, the **occipital condyle**. In mammals there are two of these knobs.
- In birds and reptiles, the **stapes** is the lone middle ear bone. In mammals, there are three middle ear bones.

595

- Birds' and reptiles' lower jaws contain five or six bones. In mammals, the lower jaw is made up of one bone. There are many more skeletal similarities as well.
- In birds and reptiles, uric acid is excreted as nitrogenous waste. In mammals, urea is excreted.
- In all birds and most reptiles, eggs with similar yolks are laid. Inside the egg, the early embryo develops on the surface by shallow cleavage divisions.
- Physiological characteristics such the presence of nucleated red blood cells and aspects of liver and kidney function are shared by birds and reptiles.
- Behavioral characteristics related to nesting and care of the young are shared as are characteristics once thought to be the providence of birds but not reptiles, like endothermy and the presence of feathers, have been demonstrated in the dinosaurs.

The distinguished English biologist Thomas Henry Huxley was so taken by the similarities between birds and reptiles that, in 1869, he described birds as "glorified reptiles" and included them with reptiles in a single class *Sauropsida*. The first striking fossil evidence of the bird-reptile connection was unearthed in 1861 in what is now Bavaria, Germany by a quarry worker harvesting limestone. Around 147 million years before, that quarry had been the bottom of a shallow marine lagoon. During that time, an animal drowned and settled to the bottom of the lagoon where it was quickly covered with a very fine layer of loose sediment and later fossilized. When revealed, the fossil was found to be that of an animal about the size of a crow with a long, reptilian tail and clawed fingers. The complete head of the first specimen was not preserved, but what was revealed as faint wisps in that ancient mud rock was the presence of feathers; feather on a long reptilian tail and on short, rounded wings.

> *Feathers predate birds.*
> —Robert T. Bakker

This was more than a reptile; this was a bird lizard! Sixteen years later, a more complete fossil with an intact skull was discovered, revealing reptile-like teeth set in beaklike jaws. Most zoologists consider these fossilized creatures named *Archaeopteryx* (Gr. *archaios*, ancient + *pteron*, wing) to be one of the oldest bird yet discovered and very close to the main line of evolution between the reptiles and birds (**Figure 19.1**).

Birds appear to be descended from ancient archosaurs—a lineage shared by the crocodilians and dinosaurs. Although bird ancestors first appeared in the early Jurassic, the Tertiary period saw an astonishing increase in avian diversity and adaptation. For many years, it has been debated whether birds are more closely related to the crocodilians or the dinosaurs, specifically a group of dinosaurs in the Saurischian lineage of the theropods known as dromeosaurs. (This lineage also includes bipedal dinosaurs like *Tyrannosaurus* and *Velociraptor*.)

The evidence linking birds to theropods included many derived characteristics, including a **furcula** and **lunate wrist bone**. The furcula (or wishbone) is a forked bone found in theropod dinosaurs and birds that functions to strengthen the thoracic skeleton for the rigors of flight while the lunate wrist bones allow for twisting motions required for flight. Recent discoveries of fossils from late Jurassic and early Cretaceous deposits in Liaoning Province, China further support theropod ancestry. Although these fossils may not represent animals directly ancestral to birds, they do repeatedly demonstrate that the ancestral features of birds were present in diverse species within one important lineage.

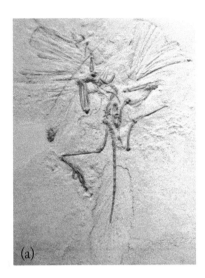

(b)

(a)

Figure 19.1 The Archaeopteryx. (a) Fossilized specimen. (b) Artist recreation.

Fossils of at least a dozen theropod dinosaurs bearing feathers have been discovered in China. The first discovered was a chicken-sized dinosaur called *Sinosauropteryx*. It had small tubular structures, similar to feathers in their early stages of development in modern birds. Another fossil, *Caudipteryx*, was a turkey-sized theropod with symmetrical feathers on forelimbs and tail. It is believed that neither of these theropods were fliers because asymmetrical feathers are required for the aerodynamics of flight. The earliest feathers may have provided insulation in temperature regulation, water repellency, courtship devices, camouflage, or balancing devices while running along the ground. Flight was apparently a secondary function of feathers. Another of the dozen feathered theropods is *Microraptor*. Living about 125 million years ago, *Microraptor* had asymmetrical feathers as well as a feathered tail. Other skeletal features suggest *Microraptor* was a climber leading some to suggest that it perhaps climbed trees and used its winds for gliding flight. Other fossils from Spain and Argentina of birds more derived than *Archaeopteryx* document the evolutionary development of more refined bird traits such as a **keeled sternum** and **alula** (a small projection on the anterior edge of a bird's wing), loss of teeth, and fusion of bone characteristic of modern birds.

A variety of fossil birds for the period between 100 million and 70 million years ago has been found. Some of these ancient birds were large, flightless birds; others were adapted for swimming and diving, and many had reptile-like teeth. Most of the lineages that these fossils represent went extinct, along with the dinosaurs, during the Cretaceous mass extinction event that ended the Mesozoic era. Studies using DNA from modern birds suggests that the Cretaceous mass extinction may not have been as devastating to

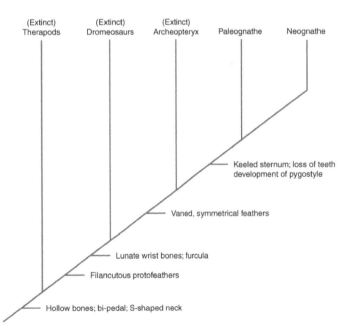

Figure 19.2 The Possible Phylogeny of Birds. The extant birds are divided into two groups: *Paleognathe*, large flightless birds, and *Neognathe*, flying birds with a keeled sternum.

birds as once thought. The results indicate that as many as 22 different avian lineages may have survived the cataclysm. The lineages that did survive into the Tertiary period were the ancestors of modern, toothless birds (**Figure 19.2**).

While the question of bird origins may not be completely settled, what is very clear is that virtually every characteristic formerly considered avian, even asymmetrical feathers, was found in the reptiles. Armed with this knowledge you should never again look at birds the same way. Dinosaurs are not completely extinct; they are with us yet today disguised as birds.

Diversity and Classification

Filling the sky with graceful flight, dazzling our eyes with brilliant plumage, and soothing our ears with melodious songs, birds are regarded by many as the most beautiful and distinctive of all the vertebrates. The many kinds of birds on Earth today have developed aerial highways to every possible habitat on earth. Birds are at home in polar regions and the tropics; in forests, deserts, and grasslands; on mountains and swimming on and under the ocean and other bodies of water. Yet, few species are truly cosmopolitan. Some shore and seabirds are worldwide in their distribution, but land birds have a seemingly haphazard distribution. Why does South America have 300 species of hummingbirds whereas Africa, with similar habitats, has not a single one? Why are finches found on even the most remote oceanic islands yet not on Australia?

Adaptive radiation has resulted in around 9,900 species of living birds divided into about 26 orders within the class Aves (L. *avis*, bird) (The number of orders varies, depending on the authority consulted and the classification system used.). But just how many birds are there? The total number of birds on the planet is very difficult to estimate because their populations fluctuate seasonally, but ornithologists have suggested that there may be between 100 billion and 200 billion adult or near adult birds in the world at any one time, roughly 20 billion of them in North America alone. The world's most abundant birds are oceanic. A single flock of shearwaters off Australia once was computed to number more than 150 million birds. Penguin colonies often number hundreds of thousands, and an estimated 5 million Adelie penguins have been recorded on a single group of small islands. The Atlantic puffin has been estimated to number 15 million, and the guano birds on islands off Peru gather in aggregations of more than 10 million.

The most common bird in the world is the red junglefowl (*Gallus gallus*) most regularly seen as the common domestic chicken (**Figure 19.3**). The most widespread commonly seen wild bird in the world is probably the European house sparrow (*Passer domesticus*) which has been transported all over the world by European settlers and can now be found on at least 2/3 of the land masses of the world including New Zealand, Australia, North America, India, and, of course, Europe.

Figure 19.3 First domesticated at least five thousand years ago in Asia, the red junglefowl is thought to be ancestral to the domestic chicken.

The rarest bird in the world is much harder to estimate because though a large number of birds are rare, in most cases the exact number of birds living for any given species is impossible to ascertain. Some species have been rare (extinct?) for some time. The orange-necked partridge (*Arborophila davidi*), for example, was last seen in 1927, and the Sudanese Red Sea cliff swallow (*Hirundo perdita*) is known from only a single dead specimen found in 1984 at Sanganeb lighthouse, north-east of Port Sudan, Sudan. The magnificent but secretive ivory-billed woodpecker (*Campephilus principalis*) was thought wiped out in 1944 with the logging of its last old-growth habitat. A confirmed sighting (a videotape of a single large male) in April of 2004 (and the last to date), however, galvanized ornithological circles, raising hopes that somehow a small number of the birds had managed to survive in remote pockets of pristine habitat. Not all experts agree on the identification of that single specimen leading to what is currently one of the most rancorous debates in the history of American ornithology.

The smallest bird in the world is generally agreed to be the Bee hummingbird (*Mellisuga helenae*) from Cuba that weighs a mere 1.6 gr (0.056 oz). The smallest flightless bird is the inaccessible island rail (*Atlantisia rogersi*). Reaching a mere 12.5 cm (5 inches) in length and weighing 35 gr (1.45 oz) this little beauty can only be found in the southern Atlantic Tristan de Cunha Islands.

When it comes to determining the largest bird, there are three possible ways of measuring: heaviest, tallest, or longest wingspan. Whatever method you choose, however, the records are all held by extinct species. The heaviest bird ever was probably the extinct *Dromornis stirtoni* from Australia. This flightless giant lived between 1 and 15 million years ago and probably stood nearly 3 m (10 ft) tall and weighed in at a massive 500 kg (1100 lbs). The tallest bird, as far as we know so far, was *Dinornis maximus*, a giant moa from New Zealand. This monstrosity, weighing only half as much as the Australian *Dromornis*, stood an incredible 3.7 m (12.1 ft) tall. From South America and South Carolina come the record for largest flying bird and longest wingspan, both extinct. The giant teratron, *Argentavis magnificens*, ("magnificent Argentine bird") had a wingspan of at least 6 m (19.5 ft) whereas the 25-million-year-old fossil *Pelagornis sandersi* had a wingspan estimated to be up to 7.4 m (24 ft). Gliding over ancient waves, this soaring seabird had the wingspan of a small airplane.

As with insects, spiders, amphibians, and reptiles, the physical size of extant species of birds tends to be smaller than that of their extinct ancestors. Even so, the dimensions of modern birds are still impressive. The largest living bird is without doubt the ostrich (*Struthio camelus*). This popular bird stands an amazing 2.74 m (9 ft) high and can weigh as much as 160 kg (353 lb) (**Figure 19.4**). The ostrich is also the fastest bird and can run 70 km (42 miles) per hour, and claims of speeds of 96 km (60 miles) per hour have been made.

The heaviest flying bird is the kori bustard (*Ardeotis kori*) of Africa with a weight of 19 kg (42 lb). Close runner-ups are the Great bustard (*Otis tarda*) and the mute swan (*Cygnus olor*) both of which have been recorded at 18 kg (40 lb).

Figure 19.4 This male ostrich (*Struthio camelus*) displays both the regal nature and massive size characteristic of this magnificent bird.

The title of 'bird with the longest wings' among land birds also has several close contenders with the Andean condor (*Vultur gryphus*) with a well recorded wingspan of 3 m (10 ft) and the maribou stork (*Leptoptilos crumeniferus*) at 2.87m (9.5 ft) but with an unconfirmed report of a specimen with a 4.1 m (13.3 ft) span. However, the real record holders for wingspan are the oceanic soaring birds. A male wandering albatross (*Diomedea exulans*) holds the crown for having the longest officially recorded wings in a living bird. A specimen caught and measured by the Antarctic research ship *USNS Eltanin* in the Tasman Sea in 1965 had a wingspan of 3.63 m (11.9 ft). As with all records, however, there are unconfirmed reports of even larger specimens.

The classification of birds is a contentious issue with frequent debate and constant revision the norm. The system presented here are the taxonomic orders in the subclass Neornithes, or modern birds. Most evidence seems to suggest that the taxonomic assignment of orders here is correct, but ornithologists disagree about the relationships between the orders themselves. Evidence from modern bird anatomy, fossils, and molecular analysis have all been brought to bear on the problem, but no strong consensus has emerged.

Domain Eukarya
  Kingdom Animalia
    Phylum Chordata
      Subphylum Vertebrata (Craniata)
        Class Aves
          Subclass Neornithes
            Superorder Palaeognathae
              Order Struthioniformes—Ostrich, rheas, cassowaries, emus, kiwis
              Order Tinamiformes—tinamous
            Superorder Neognathae
              Order Passeriformes—Perching songbirds
              Order Sphenisciformes—Penguins
              Order Gaviiformes—Loons
              Order Podicipediformes—Grebes
              Order Procellariiformes—Albatrosses, shearwaters, petrels
              Order Pelecaniiformes—Pelicans, boobies, cormorants, anhingas, frigate-birds
              Order Ciconiiformes—Herons, egrets, storks, wood ibises, flamingos
              Order Anseriformes—Ducks, geese, swans
              Order Falconiformes—Vultures, hawks, eagles, ospreys, falcons
              Order Galliformes—Grouse, quail, pheasants, turkeys, domestic fowl
              Order Gruiformes—Cranes, limpkins, rails, coots
              Order Charadriiformes—Shorebirds, gulls, terns, auks
              Order Columbiformes—Pigeons, doves, sandgrouse
              Order Psittaciformes—Parrots, lories, macaws
              Order Cuculiformes—Cuckoos, roadrunners
              Order Strigiformes—Owls
              Order Caprimulgiformes—Whippoorwills, goatsuckers, nighthawks

Order Apodiformes—Swifts, hummingbirds
Order Coraciiformes—Kingfishers, bee-eaters, rollers
Order Piciformes—Woodpeckers, toucans, honeyguides, barbets
Order Musophagiformes—Turacos
Order Coliiformes—Mousebirds
Order Trogoniformes—Trogons
Order Coraciiformes—Kingfishers, hornbills, and others

Regardless of the details and disagreements about bird taxonomy, one can only marvel at the explosive radiation of species that occurred once birds made their initial appearance on evolution's stage. It is not practical in a textbook of this nature to examine each of the many orders of birds thus a general discussion of birds follows.

General Characteristics of Birds

Despite approximately 150 million years of evolution, during which they proliferated and adapted to specialized ways of life, we have little difficulty recognizing a living bird as a bird. There is great uniformity of structure among birds with little variation from the basic theme:

- They are covered with feathers.
- They have paired limbs with the forelimbs being in the form of wings and the hind limbs as scaled legs with clawed toes.
- They usually have a spindle-shaped body with four divisions: head, relatively long neck, trunk, and tail.
- They are endothermic and possess a four-chambered heart.
- Their bones are lightweight and usually hollow; no teeth and each jaw is covered with a keratinized sheath forming a beak
- Their lungs are relatively small with nine supplemental air sacs nestled in among the internal organs and skeleton; a syrinx (voice box) is located near the junction of the trachea and bronchi.
- Their excretory system of a bird consists of metanephric kidneys. However, birds lack a bladder and possess a cloaca similar to reptiles.
- All are oviparous and lay amniotic eggs with much yolk and hard calcareous shells.

The reason for this sameness of structural and functional uniformity among birds is that they evolved into flying machines. The ability to fly greatly restricted the evolution of morphological diversity compared to the mammals, a group that includes such diverse and dissimilar forms as whales, bats, porcupines, and humans.

I never for a day gave up listening to the song of our birds, or watching their peculiar habits, or delineating them in the best way I could.

—John James Audubon

Body Plan of Birds

The ability to fly has molded all aspects of bird morphology from a body covering of feathers to adaptations of their skeletal system, musculature, respiratory system, and circulatory system.

Body Covering

The single most iconic feature that distinguishes birds from other living animals is their feathers. Simply put, it if has feathers, it is a bird; if it lacks feathers, it is not. No bird is without them, and no other living vertebrate group possesses anything similar to them. Feathers are horny, keratinous outgrowths from specialized areas of the skin called **papillae**. Except for the penguin, feathers do not arise from all parts of the bird's skin but grow in specific tracts or **pterylae**. Early development of a feather is similar to that of a reptile's scale. Feathers continue to grow outward, however, eventually losing their blood supply becoming a dead structure that receives nothing from the body except physical support. There can be enormous variation among species in the number of feathers. For example, a tundra swan (*Cygnus columbianus*) has around 25,000 contour feathers, whereas a ruby-throated hummingbird (*Archilochus colubris*) has only around 950.

There is no color on nature's palette that is not found on some bird. The vivid almost breath-taking colors of bird feathers are the result of two different processes: pigmentation and structure. Red, orange, and yellow feathers are colored by pigments, called **lipochromes**, deposited in feather barbules as they are formed. Black, brown, gray, and red-brown colors are due to a pigment called **melanin**. The spectacular red feathers of certain parrots owe their vibrancy to a rare set of pigments found nowhere else in nature (**Figure 19.5**). Pigments are not limited to the plumage but may also be found on the beak, legs, feet, and sometimes neck and head.

Figure 19.5 Red, orange, and yellow feather colors in most birds are the result of carotenoids pigments, but in parrots a novel class of lipochrome pigments called psittacofulvins account for these bright colors.

Certain blue and green colors and metallic sheens are due to a structural texture effect in microscopic portions of the feather itself known as the **Dyck texture**. This texturing scatters the shorter wavelengths of light. Structural colors are either iridescent or noniridescent. Iridescent colors are caused by the refraction of incident light by the microscopic structure of the feathers giving the feather a metallic appearance. Iridescent colors change according to the angle at which light strikes the feather surface (**Figure 19.6**). Noniridescent colors are produced by the scattering of light through minute air-filled cavities in the feathers, resulting in a specific, noniridescent color. Blue colors, such as those of bluebirds, blue jays, indigo buntings and Stellar's jays, are almost always produced in this manner. Dyck texture can also enhance pigment colors. The under-neath proximity of wavelength-absorbing melanin gives an added intensity to the blue feathers. If a blue jay' feather is back-lit, the light transmitted directly through the feather will cause it to look brown. The blues are lost because the light is no longer being reflected back, and the brown shows up because of the melanin in the feather. Greens are usually a blend of yellow pigment and structural texturing that scatters blue light. The feather structure of many species also reflect light in the ultraviolet range, and since some birds can see into the UV range, they may appear quite different to each other than they do to us. Diet may also play a role. The pink color characteristic of adult flamingos is due to the beta-carotene found in the tiny crustaceans they feed on.

Figure 19.6 The iridescent metallic sheen of this hummingbird's plumage is the result of the unique structure of its feathers.

Plumage is the term given to the arrangement and appearance of all the feathers on a bird's body. Within species, plumage can vary with age, gender and breeding season, and social status. As many birds mature, successive molts bring different plumage colors and patterns. Once adulthood is reached, plumage is often determined by the breeding cycle. Male birds especially tend to have bright-colored and often large feathers during breeding and then molt to drabber colors and smaller feathers after breeding.

In time, as feathers become worn, they loosen and drop out and are replaced through a process called **molting**. New feathers develop through the same follicle from which the old ones formed. There is much variation in the pattern of molting. For example, penguins shed their old feathers nearly simultaneously so that they come off in sheets, and many water birds, such as ducks and geese, lose all their primary wing feathers at once rendering them flightless for several weeks after molting. Most birds lose and replace a few feathers at a time in regular sequence. Molt is annual in most species, but some species may have two molts a year, and large birds of prey may only molt once in two or three years.

The feathers of birds perform five essential functions:

1. They insulate and hold heat in the body by their own weight or thickness and by trapping air beneath and between them. In a process known as "ruffling the feathers," muscles attached to each feather allow

Figure 19.7 The lack of flight feathers and a heavy covering of glistening oil gives the mistaken impression that these stately emperor penguins (*Aptenodytes forsteri*) have no feathers at all.

birds to pull their feathers up trapping even more air during cold weather. Penguins, which live in extremely cold climes, have very thick layers of puffy and highly insulating down feathers, which they coat with a thick layer of oil to further seal, waterproof, and insulate them. The oil gives them the deceptive appearance of having a slick covering that resembles skin more than feathers (**Figure 19.7**).

2. Feathers create wing and tail surfaces essential for lift and control during flight.
3. Feathers waterproof the body. This is accomplished in most birds by taking oil on their beak from the **uropygial gland** (preen gland) at the base of the tail and applying it to the feathers.

Strangely, a few water birds, such as the cormorants (Phalacrocacidae) and the anhinga (*Anhinga anhinga*), lack waterproofing and must spread their soaked wings to dry after swimming.

4. Feathers are colored to camouflage birds by blending them with the backgrounds of their habitats or against the sky as they fly.
5. Feathers are important in species recognition. The plumes, crests, and brilliant colors on male birds shown during courtship displays enable the females to select a mate of the same species and the best possible mate from those assembled (**Figure 19.8**).

Figure 19.8 A male peacock in shimmering, full breeding plumage.

To serve all the functions required of them, feathers have been specialized into several categories: Contour (flight) feathers, down feathers, semiplume feathers, and filoplume feathers (**Figure 19.9**).

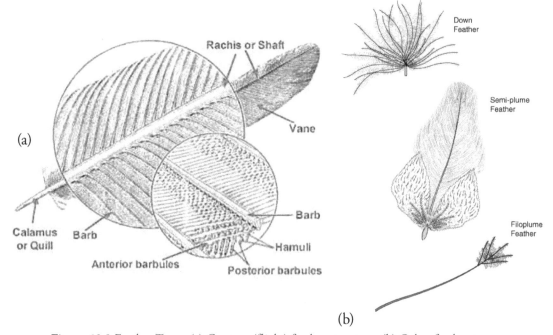

Figure 19.9 Feather Types. (a) Contour (flight) feather structure. (b) Other feather types.

Contour feathers are composed of a stiff, hollow central **calamus** (or quill) and a rachis (or shaft) which is a continuation of the calamus. The shaft bears numerous barbs arranged in a closely parallel manner. The barbs spread diagonally outward from both sides of the rachis to form a flat surface. Each barb is composed of many parallel filaments (**barbules**) that stick out along each side perpendicularly. Barbules can number up to 1200 per barb and can exceed 1 million per single feather. Barbules of one barb overlap barbules of another and are held together with tiny hooks known as **barbicels**. If adjoining barbs are ever separated, the bird can quickly zip them together by running the feather through the bill in a process called **preening**.

Contour feathers are the large feathers that cover a bird's body, wings, and tail. The longer contour feathers of the wing and tail are called **flight feathers**, while those covering the rest of the body are smaller in size and comprise the general body feathers.

Down feathers grow beneath and between the contour feathers. They are short and fluffy and function to trap air and thus insulate the bird. Down feathers lack the hooks (barbicels) for cross-attachment found in contour feathers, allowing them to float free of each other and trap much more air than would be possible than if they were linked together. Young birds are covered by down feathers before they develop adult plumage. Powder down is a light dusting of powder on some birds thought to derive from disintegrating down feathers. This powder, especially in herons and bitterns, may help waterproof the bird and remove fish slime from the feathers.

Semiplume feathers are a cross between contour feathers and down feathers. They have a shaft like contour feathers but lack barbicels as do down feathers. Semiplume feathers grow beneath the contour feathers and provide physical protection and insulation.

Filoplume feathers are incredibly small. They have a tuft of barbs at the end of a bare shaft. Unlike contour feathers, which are attached to a muscle for movement, filoplume feathers are attached to nerve endings. These feathers send signals to the brain about the placement of feathers for flight, preening, and insulation.

Feathers are high maintenance structures and birds spend about 10% of their daily time preening and grooming their feathers. Preening and grooming involve using the bill to hook breaks in contour feathers back together, to brush away foreign matter on the feathers and to apply oily secretions from the preen gland to the feathers. Birds can find it difficult to preen hard-to-reach places, like the back of the head and neck. Most birds solve the problem by using their feet. Hummingbirds, for example, spend almost half their grooming time using their feet. Secretions from the preen gland enhance feather flexibility, make the feather waterproof, and act as anti-microbial agents inhibiting the growth of feather-degrading bacteria. Bathing in water or dust also helps keep the feathers clean and may help remove parasites. One of the strangest forms of feather care is called anting, and it is practiced by over 250 species of birds. These birds find a suitable ant nest and spread themselves out over it encouraging the ants to crawl into their feathers. They also often pick up individual ants and stroke them across their feathers. It is believed that formic acid produced by the ants helps kill parasitic fleas and feather lice on the bird. The behavior of the bird during this process seems to indicate that it may derive an almost ecstatic sense of pleasure from the experience.

Feathers most likely originated as filamentous insulation or waterproofing structures, or possibly as markers for mating, with their use in flight emerging later as a secondary purpose. It has long been thought that feathers evolved from the scales of reptiles, but recent research suggests that although there is a definite relationship between these structures, their evolutionary connection (if any) is unclear. Exactly when and how feathers originated and evolved is a hotly debated topic among ornithologists and evolutionary biologists and remains one of the great mysteries of avian biology.

Support and Locomotion

Compared with reptile or mammal skeletons, there are fewer bones in a bird's skeleton (**Figure 19.10**). Flying animals need a light but rigid skeleton, and this has been accomplished in birds by the fusing of some small bones and the elimination or reduction of other bones. For example, birds lack the heavy tail, heavy jaws, and heavy teeth of their reptilian kin. Many birds have bones containing air spaces or hollow bones (**pneumatized bone**) stiffened by internal struts (**Figure 19.11**). As a general rule, large flying birds have proportionally greater pneumaticity in the skeleton than small ones. The bones of modern birds are phenomenally light yet strong. The skeleton of a frigate-bird of 1.4 to 1.8 kg (3 to 4 pounds) with a 2.1 m (7 feet) wingspan would weigh only 112 gr (4 ounces)—around 6% but less than the weight of all the feathers. By comparison, the human skeleton weighs about 20% of the total body mass.

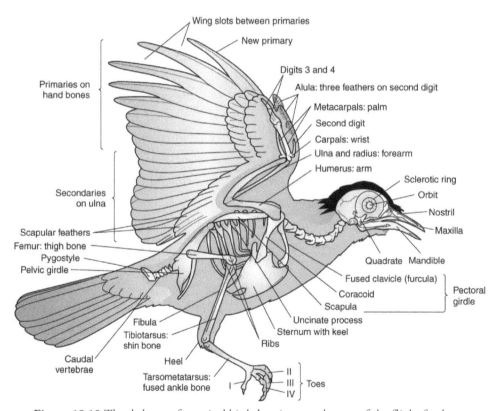

Figure 19.10 The skeleton of a typical bird showing attachment of the flight feathers.

The shoulder and hip girdles of a bird's skeleton are strong to provide support and to withstand the rigors of flight and, except in flightless birds, the sternum is modified into a flat keel to provide a greater surface area for the attachment of flight muscles. Fused clavicles form an elastic **furcula** that stores energy as it flexes during wing beats.

The vertebral column is rigid in birds with most vertebrae except the **cervicals** (neck vertebrae) fused together. Interestingly, mammals have seven cervical vertebrae no matter how long their neck, but most birds have from 13 to 25 cervical vertebrate. Thus as contradictory as it may sound, a small bird has more neck bones than a giraffe. The many unfused cervical vertebrae give birds extraordinarily flexible necks. The 14 cervical vertebrae of an owl's neck allow it to turn its head through a range of 270 degrees—not, however, in a full circle as myth would have it. Each of the 5 to 8 thoracic (chest) vertebrae normally bears a pair of complete ribs. Each rib in turn bears a flat, backwards-pointing spur, the **ulcinate process**, characteristic of birds. The sternum, ribs, and their articulations form the structural basis for a bellows action by which air is moved through the lungs.

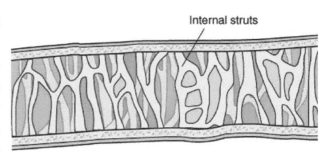

Figure 19.11 The hollow wing bone of a songbird displaying the stiffening struts and air spaces that replace bone marrow. Such bones are remarkably light but amazingly strong.

The skull of a bird is lightly built and mostly fused into one piece. It consists of a large cranium (braincase), huge eye sockets (**orbits**), and a lightweight toothless beak consisting of hard keratin molded over bony jaws (**Figure 19.12**). Birds differ from mammals in being able to move the upper jaw rather than the lower relative to the cranium. When the mouth is opened, both the lower and upper jaws move; the lower jaw (**mandible**) by a simple hingelike articulation with the quadrate bone at the base of the jaw; the upper jaw (**maxilla**) through the flexibility provided by a hinge between the frontal and nasal bones.

Figure 19.12 The Skull of a Bird. Birds have the largest eyes relative to the size of their skull in the animal kingdom.

The forelimbs of a bird are adapted as wings for flight. Although modified in form, a bird's wing is clearly a rearrangement of the basic tetrapod limb with all the parts—arm, forearm, wrist, and fingers—represented. Major modifications include restrictions of the motion of the elbow and wrist joints to one plane, reduction in the number of digits, loss of functional claws, fusion of the metacarpals and most of the carpals (the "hand"), and modification of the elements toward the tip of the limb for the attachment of feathers. The humerus, radius, and ulna are well developed. The secondary flight feathers are attached to the ulna, which thus directly transmits force from the flight muscles to these feathers. The primary flight feathers are attached to the digits. All modern birds have **allulas**, tufts of feathers on the first digit of each wing. Allulas allow birds to make smooth and accurate landings. Without them, a bird would drop quickly when braking.

Most birds have about 175 different muscles controlling the movements of its wings, legs, feet, tongue, eyes, neck, lungs, sound-producing organs, and body wall. Bird muscles come in two main types: one designed to provide explosive bursts of speed for short distances and others structured for long endurance. Largest of all the muscles are the breast muscles or **pectorals**. These muscles form the bulk of the fleshy mass in the breast and constitute about 15 to 20 % of the bird's total weight. The pectorals provide the powerful downstroke of the wing, and bear most of the burden of supporting a bird in flight (**Figure 19.13**). The **supracoracoideus**—the muscles that raise the wing—acts as the adversary to the pectorals. Though one might expect to find this muscle on the backbone, it is actually positioned on the breast, underneath the pectorals. A tendon attaches the supracoracoideus to the top of the humerus of the wing giving the bird the ability to pull it from below. Both the pectorals and the supracoracoideus muscles are attached to the keel and together constitute about 25 to 35 percent of a bird's total weight. It benefits the aerodynamics efficiency of the bird to locate this large muscle mass so low in the body.

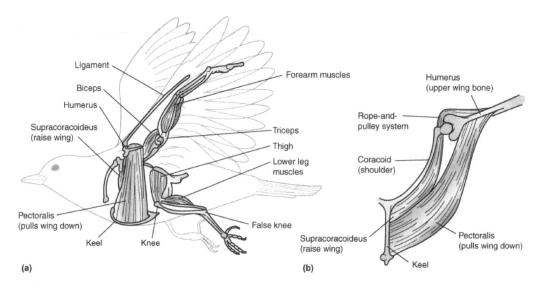

Figure 19.13 The Major Muscles of the Avian Flight System. (a) Side view of the left half of the system. (b) Close up of the two major muscle groups responsible for flapping flight.

The size of the leg muscles depends on how a bird uses its legs and feet. Active swimmers, runners, and climbers have larger, stronger leg muscles than species that do few or none of those things. The legs of the highly aerial whippoorwills and swallows are so weak and tiny that these birds find any walking difficult; such activity for hummingbirds is almost impossible. Ground birds such as the quail, turkey, and domestic chicken have flight muscles that tire easily (this muscle is light colored when cooked i.e. white meat) but powerful leg muscles for running and scratching (this muscle is dark colored when cooked i.e. dark meat). In birds that are strong fliers, somewhat the reverse is true.

Birds no longer carry a long reptilian tail; all that remains is the **pygostyle** (Figure 19.10). Formed over the few bones of the pygostyle is a soft area of muscle to which the tail feathers attach. This muscle mass is composed of up to 1000 very small muscles that are used in movement and control of the tail feathers.

Certainly the most remarkable movement birds are capable of is the ability to fly. Mammalian bats do it, and the extinct pterosaurs did it, but none have done it more gracefully, powerfully, or precisely than the birds. The simplest form of flight is **gliding**. Gliding flight—several strong wing stokes and then a glide—saves energy, but gravity and air conditions determine how far a bird can skim before flapping again. Pheasants and quail beat their wings rapidly to takeoff and gain altitude, tire quickly, and then glide on stiff wings. Many other birds use the glide when they have gained enough height and want to save energy on descent. Some birds have elevated gliding to a specialized skill called **soaring** (**Figure 19.14**). Hawks, eagles, and vultures with broad wings and tails search for food from high in the air and must be able t spend hours aloft. These birds travel by rising on thermal currents over land to great heights, and then glide down to be carried up by another column of rising warm air known as a **thermal**.

The long narrow wings of the albatross, shearwaters, and frigatebirds and the lack of thermal currents over water have forced these soaring seabirds to develop a different kind of soaring known as **dynamic soaring (Figure 19.15)**. Their flight requires continuous winds that do not carry them high but give them enough lift over waves to make a long wind-pushed glide until they descend to wave level, where they turn into the wind and are lifted like a kite for the next glide.

Figure 19.14 Soaring flight is a special kind of glide in which the bird uses rising currents of warm air to increase its height relative to the ground without flapping.

Sustained or **flapping flight (Figure 19.16)** involves an up-and-down motion of the wings to keep the bird aloft and a nearly imperceptible forward-and-backward motion to make it move forward. During flapping flight, different parts of a wing have different functions. The proximal part of the wing (half closest to the body) moves less and provides most of the lift, whereas the distal part of the wing moves through a wide arch and generates the thrust that propels a bird forward. Power is applied on the downstroke. The large primary feathers on the wingtips are bent upward and twisted in the air; as they are brought down, they pull the wing and the entire bird with them. On the upstroke, the primaries separate to permit easy passage of air between them, and the wingtips move upward and backward, still providing slight propulsion. Then the cycle begins again. Most species of birds do not flap their wings continuously

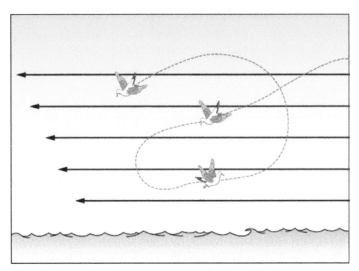

Figure 19.15 Dynamic soaring flight in which energy can be extracted from horizontally moving air and transferred to the bird so that an energy gain is achieved which enables it to fly continuously without flapping.

during flight. Rather they exhibit on of two intermittent flight patterns: **flap-gliding** in which a bird flaps the wings several times and then glides for a period slowly losing altitude. **Flap-bounding** occurs when a bird rises when flapping then falls when not flapping. A bird that is flap-bounding bobs up and down as it flies. From an energy efficiency standpoint, flap-gliding works best for large birds, whereas flap-bounding works best for small to medium-sized birds. Some small birds such as Budgerigars and European Starlings, do exhibit both types of intermittent flight, with flap-gliding being used at lower speeds, and flap-bounding at higher speeds.

Figure 19.16 Flapping Flight. As the bird is flapping along, the wing twists to make sure each part of the wing maintains the proper angle of attack into the air. As the wing twists, the outer part of the wing moves downward and the lift force on the outer part of the wing is angled forward. This allows the bird to generate a large amount of forward propulsive force or thrust, without any loss of altitude.

The most remarkable aerial acrobats are the hummingbirds. Their tremendous wing speeds of 50 to 200 beats per second (so fast the wings produce the humming sound for which they are named) and the swiveling and sculling actions of their wings allow these "feathered helicopters" to fly straight up, hover, and even fly backwards (**Figure 19.17**).

Figure 19.17 The Wing Action of a Hummingbird. The secret to a hummingbird's ability to hover, change direction, and even fly backwards lies with a wing that is nearly rigid but hinged at the shoulder by a swivel joint and powered by an unusually large supracoracoideus muscle.

In level flight, most perching birds fly at 22 to 40 km (15 to 25 miles) per hour whereas geese and ducks travel between 65 to 70 km (40 to 60 miles) per hour, and some shorebirds have been clocked from airplanes at 176 km (110 miles) per hour. In a dive, however, the peregrine falcon (*Falco peregrinus*) is in a class alone. This feathered missile climbs high above its prey, folds its wings, and drops for a thousand feet or more at speeds of about 320 km (200 miles) per hour before slashing into its victim.

The highest-flying birds ever recorded were a flock of geese over Dehra Dun, India at an altitude of 9,000 m (29,700 feet). Even small birds such as warblers and vireos have been known to fly as high as 6,364 m (21,000 feet) during migration—a remarkable physical feat for such small birds considering the cold temperatures and reduced oxygen levels at such altitudes. Going the other direction, emperor penguins (*Aptenodytes forsteri*) dive (fly) as deep as 265 m (875 feet) into Antarctic waters to feed.

The longest sustained flights may be made by sooty terns (*Onychoprion fuscatus*). Like some other seabirds, the terns spend years at sea and do not come to land until maturity. However, unlike most other seabirds, terns quickly become waterlogged and cannot come to rest on the water. Thus, they spend virtually all of their first six to eight years in ceaseless flight.

No aspect of bird flight is as awe-inspiring as those mass movements of birds we call **migration**. What is the destination or the origin of that whirling cloud of dark, winged bodies against a gray autumn sky? What primal forces drive those silhouettes flashing across the face of a full moon on a bright October night? Of all known bird species, more than half regularly migrate. Other types of animals migrate, but only birds do it in such epic proportions (the total number of birds migrating each season is estimated to be in the billions). Migration can be as casual as a short trek up and down a single mountainside or a grueling inter-continental journey. Some types of birds migrate in flocks, whereas others make solitary journeys.

Why do birds migrate? Food, water, and a suitable breeding place are vital to a bird's survival. Changing seasons can transform a safe and comfortable habitat into a hostile and forbidding one. When this happens, birds have three choices: go dormant and hibernate, ride it out, or migrate. Some types are adaptable enough to stay put and ride it out, but only the common poor-will (*Phalaenoptilus nuttallii*) hibernates, usually concealed in piles of rocks. Many birds opt for migration and the stimulus to begin the process seems to be changes in the length of the day. As days grow shorter, a bird's internal clock triggers hormonal changes that cause not only frenzied feeding to accumulate maximum fat stores, but also cause grouping into larger and larger flocks. At some magic moment discernible only to the birds, the time seems right, and away they go.

Some migrations schedules do not always closely follow seasonal changes, and for some nomadic species like the crossbills and redpolls, fluctuations in food supply may force migration in some winters but not in others.

Each migratory species chooses a well-defined route of travel between it nesting site and winter range. These migration routes are generally quite broad. Waterfowl tend to be confined to narrower migration corridors by the availability of suitable habitat. Because most birds breed in the Northern Hemisphere, most birds migrate south in the northern winter and north again to nest in the northern spring and summer. Some follow the same route while others use a different route in the fall than they do in the spring (**Figure 19.18**). The trip may be completed by some in a very short time while others enjoy a more leisurely journey. Some warblers take up to 60 days in their migration from the winter grounds of Central America to the summer breeding grounds of Canada. Some smaller species use the protection of nighttime skies to migrate and the daytime to feed and rest; others migrate mainly during the daytime stopping to feed and rest in late afternoon.

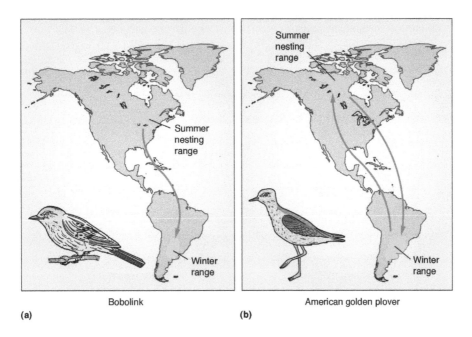

Bobolink (a)   American golden plover (b)

Figure 19.18 The bobolink (*Dolichonyx oryzivorus*) (a) migrates the same route between winter and summer ranges. Even the colonies in western areas of the summer breeding range take no shortcuts but adhere to the ancestral seaboard routes. American golden plovers (*Pluvialis dominica*) (b) fly a loop migration pattern. Flying north in the spring by way of Central America and the Mississippi valley presents more favorable route at that time.

The highest migrators are bar-headed geese (*Anser indicus*), which fly over Mount Everest at altitudes exceeding 9,000 m (30,000 feet), a dramatic testament to the remarkable efficiency of a bird's respiratory system. Most migration, however, occurs much lower, with exact altitude determined by wind direction, wind speed, and cloud cover. The longest annual migration is undertaken by the sooty shearwaters (*Puffinus griseus*). These birds nest in New Zealand and Chile but spend the northern summer feeding in the North Pacific off Japan, Alaska, and California, an annual round trip of 64,000 km (39,800 miles). The most arduous migration journey has to be that of bar-tailed godwits (*Limosa lapponica*). Several of these small long-billed wading birds were tagged in New Zealand and tracked by satellite as they migrated as a flock all the way to the Yellow Sea in China, a nine-day nonstop flight of 11,026 km (6, 851 miles). One specific female of the flock, nicknamed "E7," flew onward from China to Alaska. On August 19, 2007 she departed Alaska, on a nonstop flight back to New Zealand, setting a new flight record of 11,570 km (7,189 miles), a Herculean feat for a small ball of feathers and muscle that would fit in your hands.

Box 19.1

Migration Mysteries: How Do Birds Find Their Way?

How do birds know where to go and how do they know when they have arrived, especially young single juvenile birds making their first journey? Ornithological research and observations indicate the answer lies both inside and outside the birds. Bird navigation has a strong genetic/evolutionary component. To a certain degree, birds seem to have an "instinct map" imprinted on their genetic code. Case in point—the European or common cuckoo (*Cuculus canorus*), as cuckoos are famous for, deposit their eggs in the nests of other birds that unwittingly serve as foster parents. This in and of itself is astonishing evolutionary behavior, but even more amazing is that once raised, the young cuckoo makes its own solitary way to ancestral wintering grounds in the tropics before returning single-handedly to northern Europe to seek a mate.

The cuckoo and many other such examples clearly demonstrate that innate (genetic) forces are strong in migrating birds, but genetics alone are apparently not enough. Migrating birds seem to depend heavily on environmental cues as well. Sun compasses for day migrators and star maps for night migrators seem to play a role. Most small birds migrate at night, and some seem to calibrate their internal compass for a night flight from the setting sun, as well as the plane of the polarized light emitted as the sun sets. Birds employ a number of other navigation "instruments" as well, including the ability to not only detect but possibly even see magnetic fields, keenly remember visual landmarks, and perhaps even detect olfactory (smell) cues.

Bird migration clearly involves more than internal compass headings because true navigation also requires an awareness of position and time, especially when blown off course and lost. Case in point: a Welsh manx shearwater (*Puffinus puffinus*) transported to North America and released was back in its original burrow on Skokholm Island off the Pembrokeshire coast of Wales one day before a letter to the press announcing its release.

Migrating birds are also quite adept at accurate weather forecasting. All migrations present considerable risk and part of the trick to surviving the ordeal is setting off at the right time to utilize favorable winds. Birds often react to weather changes before there is any visible sign of them, and in tests, some birds have been shown the ability to detect the minute difference in barometric pressure between the floor and ceiling of a room. Lapwings (subfamily Vanillinae) are ground feeders that flee the Netherlands for warmer climes in France and Spain just before the onset of a cold snap (freezing ground = no food) and return to Holland just ahead of a thaw. Their departure and arrival are precisely linked to pressure changes that foreshadow a change in the weather.

As ornithologists work diligently to peel back the many layers of mysteries that surround migration, our understanding of bird migration has improved. Some mysteries, however, have deepened even further. Fortunately, we do not have to be able to fully understand all the details and fine print of bird migration to appreciate and marvel at the globetrotting feats of migrating birds.

Even though their ancestors were most likely flying birds, some 40 species of extant birds have forsaken flight, relying instead on their ability to run or swim. The best known of the flightless birds are the ostrich, emu, cassowary, rhea, kiwi, and the penguins (**Figure 19.19**). Flightless birds have smaller wing bones than their flying kin and the keel on their breastbone is greatly reduced or absent.

Figure 19.19 Representative Flightless Birds. (a) Kiwi (*Apteryx*). Kiwis are endemic to New Zealand where they use their long thin bill unique among birds with nostrils at the tip to probe for invertebrates beneath leaf litter. (b) Cassowary *(Casuarius)*. These large birds are native to the tropical forests of New Guinea, nearby islands, and northeastern Australia. These fruit eating birds are quite shy, but when provoked, they can inflict serious injuries.

Limbs and Feet

The hind limbs of all birds are heavily muscled at the thigh. The lower portions of the limbs are composed of very light yet very strong tendons that run through sheaths all the way to the toes. The lower leg and toes lack muscle and are composed mostly of bone and tendon wrapped with tough, reptile-like scaly skin. Each toe has a claw at the tip. Most birds have four toes, some have three, and a few, such as the ostrich, have only two.

To compensate for the lack of lower leg musculature, the legs are controlled by a series of tendons from the muscles at the top of the legs. When a bird stands, these tendons are relaxed, and the toes spread out. When a passerine (perching bird) perches by squatting and bending its legs, however, tension is applied to the tendons and the toes lock in a grasp so firm and tightly coiled that birds so equipped can sleep on the perch without falling. This tendon feature is also useful to owls and hawks in their predatory strikes as they sink their talons into prey while bending their legs for the impact.

When an eagle grips in earnest, one's hand becomes numb, and it is quite impossible to tear it free, or to loosen the grip of the eagle's toes with the other hand. One just has to wait until the bird relents, and while waiting one has ample time to realize that an animal such as a rabbit would be quickly paralyzed, unable to draw breath, and perhaps pierced through and through by the talons in such a clutch.

—Winwood Reade

Nothing so clearly reveals the habitat a bird occupies as the form and structure of its feet and legs. The length of the toe and claw, the position of the toes, the number of toes, the thickness of the toes, and the presence of webbing are specific adaptations related to habitat, feeding, and locomotion (**Figure 19.20**).

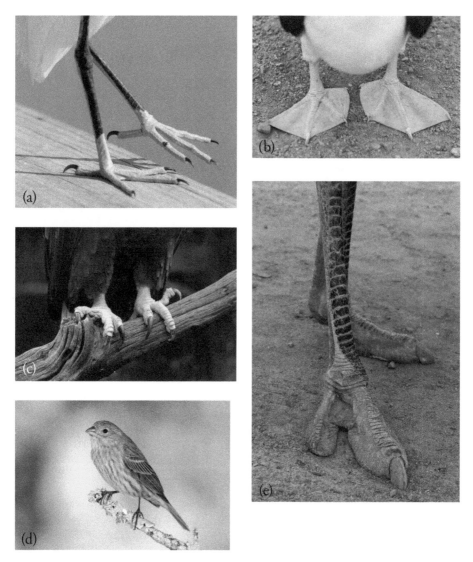

Figure 19.20 Foot Types in Birds. (a) waders have long slender toes for support on mud and vegetation. (b) Swimmers have either lobes (grebes) or webs (duck) on their toes to increase the surface area for a more powerful stroke. (c) Raptors have powerful toes and large hooked talons. (d) Perching birds have long slender toes—three in front, one in back—suited for wrapping around twigs and branches. Woodpeckers have two toes in front and two in back for bracing. (e) Ground birds have short muscular toes adapted for running and walking.

Feeding and Digestion

Birds are **endothermic** (heat from the inside) and, as a result, have a high metabolic rate, which means they burn food quickly and must feed often to refuel. In general, the smaller the bird, the faster they use energy. Land birds weighing 100 to 1,000g (0.25-2.5 pounds) eat approximately 5% to 9% of their body weight each day. Birds weighing 10 to 90g (0.34 to 3.2 ounces) may have to eat 10% to 30% of their body weight each day. Imagine a 150 pound human having to eat 45 pounds of food a day just to survive.

Among birds, there are herbivores, carnivores, omnivores, and scavengers. The digestive system of a particular bird is structured to the types of food eaten and to the lack of teeth. Herbivorous birds feed on the parts of plants that yield the highest energy return—seeds, fruits, buds, and nuts. Because their diet consists of tough, hard-to-digest plant material, herbivorous birds have a more complex digestive system than do carnivorous birds. Herbivores have a storage sac, called the **crop**, leading to a two-part stomach where food is mixed with digestive juices in the first part (**proventriculus**) and then ground to a pulp in the second part, the thick-walled muscular **gizzard** (**Figure 19.21**). Studies on turkeys have shown that they can grind up 24 walnuts, shell and all, in just four hours and can even grind steel needles to pieces. To aid the gizzard in this grinding action, herbivorous birds often swallow small stones. Carnivorous birds feeding on more easily digested flesh—insects, fish, amphibians, reptiles, mammals, and even other birds—usually do not have a crop (or only a small one) and their gizzard functions as a normal stomach.

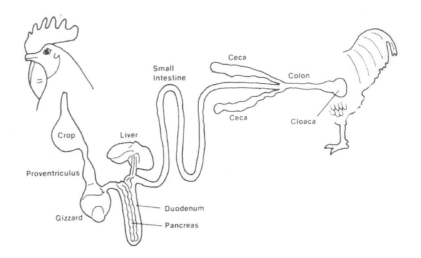

Figure 19.21 Lacking teeth, herbivorous birds must "chew" their food internally using the muscular ventriculus (gizzard). Tiny pieces of swallowed grit in the gizzard act as a bird's "teeth."

The powerful and swift action of a bird's digestive juices can be seen in the speed with which foods are digested. Shrikes are known to digest an entire mouse in just three hours. A thrush fed elderberry fruits will defecate the seeds 30 minutes later, and the seeds of berries eaten by waxwings can appear in their feces in as little as 16 minutes. Owls do not attempt to digest the bones and fur, feathers, or skin of their prey; they regurgitate this material in the form of a compact pellet instead.

Some birds have very specialized diets. Hummingbirds live primarily on nectar from flowers (although some larger types feed extensively on insects as well). Tree sap is the specialty of the aptly named sapsuckers, and the wax from beehives is a major part of the diet of honeyguides. Special bacteria in their digestive system make them the only vertebrate able to digest wax.

The need for constant food is so critical that birds must be perfectly structured and adapted to getting it, and no other body structure of a bird demonstrates this so clearly as does the beak (**Figure 19.22**). The beaks of birds are adapted to many feeding tactics:

- Straining small animals from the water
- Catching fish by grasping, scooping, or spearing
- Crushing seeds and nuts
- Tearing flesh
- Sipping nectar
- Sweeping for insects
- Chiseling for insects
- Probing soil, leaf litter or mud for prey

Figure 19.22 Types of Bird Beaks. (a) Straining. (b) Spearing/grasping. (c) Probing. (d) Tearing. (e) Chiseling. (f) Scooping. (g) Sipping.(h) Crushing.

Many birds could be said to have nondescript, multipurpose beaks adapted to picking and manipulating a wide variety of potential food items. Some birds such as the woodpecker finch (*Camarhynchus pallidus*) of the Galapagos Islands and the crow (*Corvus*), can manipulate tools with their beak. Using a twig or thorn, the finch or crow pries insects out of cracks and holes, then drops the tool and catches the insect.

Although there are no true parasitic birds, there is a poisonous one—the hooded pitohui (*Pitohui dichrous*) (**Figure 19.23**). Researchers attempting to capture the bird of paradise in New Guinea also trapped an exotic orange and black jungle songbird known as the "rubbish bird" by the locals. The sharp beak and claws of these birds cut and scratched the researchers as they removed them from their nets. One researcher licked his wounds, and felt his tongue and mouth immediately go numb. Curious, other researchers put pitohui feathers on their tongues and suffered immediate reactions of numbness, burning, and sneezing on contact with the feathers.

Figure 19.23 Lurking in the shade and shadow of New Guinea forests is the brightly colored hooded pitohui, the only known poisonous bird.

Feather samples analyzed by the National Institutes of Health and the Smithsonian Institution were found to contain a poison called batrachotoxin (BTX). The only other creature known to possess this poison is the Columbian poison-dart frog, although the concentration of the poison in the frogs is about 1,000 times higher. Extracts of the skin, feathers, and muscles of the pitohut injected into mice caused the mice to convulse and die within 2 minutes. The birds probably do not produce the toxin themselves, but most likely acquire it from certain beetles that are a regular part of their diet. The BTX is produced in the skin of the pitohui and spread as a dander throughout the feathers. This toxin is thought to repel both predators of and parasites on the pitohui.

Respiration

The high metabolic rate required to maintain an endothermic body and the rigors of flight require a rapid and efficient uptake of large amounts of oxygen. Birds have one of the most complex respiratory systems of all animal groups and this system shows several adaptations that allow it to not only meet the demands placed on it but yet fit into a small, streamlined body.

The lungs, although not particularly large, are divided into many air passages, each exposed to blood vessels. This arrangement presents a large surface area for gas exchange while occupying a small space. The lungs are supplemented by a series of nine interconnecting air sacs that allow the bird to intake more air than the lungs can hold at one time (**Figure 19.24**). When a bird inhales, about three-fourths of the fresh air bypasses the lungs entirely and flows directly into posterior air sacs. These sacs extend from the lungs and connect with air spaces in the bones filling them with air. The other one-fourth of the air goes directly into the lungs. When a bird exhales the depleted air flows out of the lungs, and the stored fresh air from the posterior air sac is simultaneously forced into the lungs. Thus, a bird's lungs receive a constant supply of fresh oxygenated air during both inhalation and exhalation.

Breathing rate varies among species. Most small birds breathe between 100 and 200 times per minute, whereas very large birds breathe 6 to 12 times per minute at rest but faster during strenuous activity.

Sound production is achieved in birds using the **syrinx**, a bony structure at the bottom of the trachea (unlike mammals, which have a larynx at the top of the trachea) **(Figure 19.24)**. The syrinx and sometimes a surrounding air sac resonate to vibrations that are made by when the bird forces air past membranes in the syrinx. The bird controls the pitch by changing the tension on the membranes and controls both the pitch and volume by changing the force of exhalation. The two sides of the trachea can be controlled independently allowing some species to produce two notes at once.

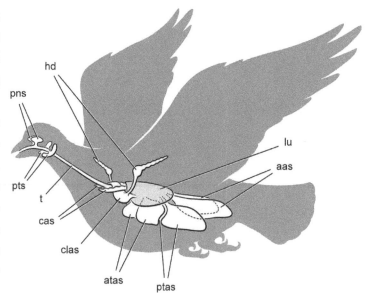

Figure 19.24 Lung and Air Sac System in Birds. All pneumatic spaces are paired except the clavicular air sac, and the lungs are shaded. Abbreviations: *aas*, abdominal air sac; *atas*, anterior thoracic air sac; *cas*, cervical air sac; *clas*, clavicular air sac; *hd*, humeral diverticulum of the clavicular air sac; *lu*, lung; *pns*, paranasal sinus; *ptas*, posterior thoracic air sac; *pts*, paratympanic sinus; *t*, trachea.

A bird doesn't sing because it has an answer, it sings because it has a song.

—Maya Angelou

Ornithologists categorize bird vocalization into songs and calls (although the distinction is somewhat arbitrary). Songs are considered to be longer and more complex than calls and are associated with courtship and mating, whereas shorter calls tend to serve such functions as alarms, keeping members of a flock in contact, or stimulating parents to feed begging chicks. Bird song is best developed in the order Passeriformes (and for that reason they are commonly called "songbirds"), with most song being emitted by male rather than female birds. Early in the breeding season, songs are employed by many birds to attract a mate, establish territory, and warn off potential rivals. Individual variation in songs of many species is well known, and it is believed that some birds can recognize their mates and neighbors by this variation. It is hypothesized that bird song has evolved through sexual selection, and experiments suggest that the quality of bird song may be a reliable indicator of fitness and good health.

Birds can hear in a range below 50 Hz (infrasound) to above 20 kHz (ultrasound). The range of frequencies at which birds call in an environment varies with the quality of the habitat and the ambient sounds. In urban areas with a great deal of low-frequency noise, it has been noted that birds sing louder as well as at a higher pitch.

Some birds can mimic not only the songs and calls of other birds but other sounds as well. Analysis of a recording of an African gray parrot (*Psittacus erithacus*) in Zaire revealed the unmistakable reproduction of sounds from nine different bird species and one species of bat, and captive African grays also demonstrate amazing abilities to mimic human speech. The champion bird mimic, however, may well be the appropriately named mockingbird (*Mimus polyglottos*), which has been known to imitate the songs of 20 or more bird species within 10 minutes. Mockingbirds have also been known to perfectly mimic many sounds around them as well, ranging from the barking of dogs to automobile horns and even the sirens of fire engines.

Nonvocal sounds (**sonation**) are not uncommon among birds. Some snipe, hummingbirds, and the American woodcock (*Scolopax minor*) have narrow tail feathers that produce loud sounds when the birds are in flight. The elaborate courtship of grouse includes vocalizations as well as stamping of the feet and noises made with the wings. Bill clapping is a common element in the courtship rituals of storks, and bill snapping is a common threat of owls.

Excretion

Birds have a pair of relatively large metanephric kidneys. Like the reptiles, their kidneys extract nitrogenous wastes from their bloodstream and excrete it as uric acid rather than urea. Urine flows to the cloaca by way of **ureters**. Birds have no urinary bladder. In the cloaca, the uric acid is combined with fecal matter, concentrated as water is reabsorbed from it, and then eliminated as a pasty semisolid waste (the familiar bird droppings on your clean windshield).

Birds have less efficient kidneys than do mammals. Although some mammals concentrate urine to 25 times that of blood, birds do not have the same ability to concentrate and remove sodium, potassium, and chloride. Most birds, in fact, are unable to concentrate their urine to even six times that of blood. To make up for this, some ocean birds that take in a heavy load of salt with their food and water supplement kidney action by removing excess salt through special glands located over each eye. The excreted salt solution runs out the nostrils.

Circulation and Body Temperature

Birds have a well-developed and powerful four-chambered heart that moves red blood through two separate circulatory loops. As one might expect, powerful flying birds have larger hearts than nonfliers, weak fliers, or soaring birds.

The heartbeat is extremely rapid in birds, especially the smaller species. The hummingbirds have a rate that ranges from 500 beats per minute at rest to 1,200 beats per minute during extreme activity. Their blood hurls through their body at pressures and speeds that would destroy the human circulatory system. On the other hand, a very large bird, such as the ostrich, may have a range of 38 to 176 heartbeats per minute.

Birds are endothermic and maintain a steady temperature independent of the external environment (although their body temperature can vary as much as 10° under differing conditions). The normal temperature range for most birds is between 38° and 40° C (100.4° to 104° F), depending on the species. Body temperature is highest in the passerines (perching songbirds). When one handles a live wild bird, you are immediately struck with how light they are and how warm they feel.

Young birds born without a covering of down have almost no ability to regulate their body temperature and thus require brooding (warming) by a parent. Some birds, such as swifts and swallows, can lower their body temperature, heartbeat, and breathing rate during short cold periods and enter a state of torpidity in which they consume practically no energy. Hummingbirds use this same ability at night to avoid starving to death before morning light presents the next opportunity for feeding.

In hot weather, birds employ a variety of cooling strategies. They may become less active, move into the shade, depress their feathers to reduce their insulating powers, and/or pant. Birds also lose heat through their unfeathered legs and feet. When exposed to long periods of extremely hot temperatures, storks, and the New World vultures defecate on their legs. As moisture in the feces evaporates, it cools the blood in the birds' legs.

Nervous System and Sense Organs

The central nervous system (brain and spinal cord) of birds is structurally and functionally similar to that of mammals. The brain of a bird weighs about 10 times as much as the brain of a reptile of the same size, but slightly less than that of a mammal of similar size (**Figure 19.25**). There is, however, considerable variation between birds of similar size. Birds have traditionally been considered inferior in intelligence to mammals, and derogatory terms such as "bird brain" have often been used colloquially. Such perceptions are no longer considered scientifically valid.

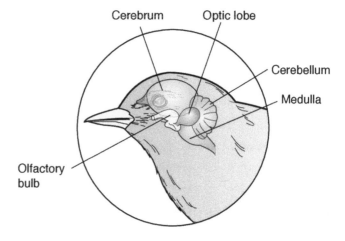

Figure 19.25 The Brain of a Bird. A bird has a very large brain for the size of its head. The visual and auditory senses are particularly well-developed in most species.

Bulging laterally from the midbrain, the large **optic lobe** gives birds the acute vision necessary for flight and food-getting. The **cerebrum**, consisting of two cerebral hemispheres, is large and well-developed, as in mammals, and controls behavior patterns, eating, singing, navigation, mating, and nest building. The **cerebellum** interprets visual cues and coordinates muscles and equilibrium for movement and balance; it may be thought of as the flight control center.

Vision Birds are highly visual animals; their daily existence, especially flying, demands acute visual perception. Birds have the largest eyes relative to body mass of all vertebrates—an indicator of their importance. Some hawks, owls, eagles, and buzzards often have eyes as large or even larger than human eyes. In fact, the eyes of a bird may weigh more than its brain.

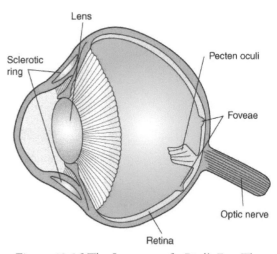

Figure 19.26 The Structure of a Bird's Eye. The pecten oculi is a structure peculiar to birds' eyes and is thought to provide nourishment to the retina.

The eye of a bird is similar to the eye of other vertebrates in anatomy (**Figure 19.26**). However, bird eyes are flatter and nearly immobile in the skull. Birds cannot turn their eyes to see to the side. Instead, they turn their entire head with reflex movements of the neck replacing movement of the eyes. The entire retina of birds is thicker than that of mammals and the rods (for dim light vision) and cones (for acuity and color vision) are more abundant. The density and proportion of rods to cones vary with the species and lifestyle. Diurnal birds have retinas packed with cones while nocturnal birds have mostly rods in their retinas. The retina of most nocturnal predatory birds is packed with rods. Owls have almost a million rods per square millimeter (humans have only around 200,000 rods per square millimeter). As a result, owls can see an object at 2 m (6-7 feet) with an illumination of 7.3 x 10^{-7} (0.00000073) foot candles, the equivalent of a human seeing an object a mile (1.6 km) away illuminated only by the light of a single match.

Color vision varies among birds. Diurnal birds tend to have keen color perception, whereas nocturnal birds are probably colorblind. Many birds are **tetrachromatic**, possessing ultraviolet light-sensitive cones in their retinas that allow them to perceive light in the UV range. The plumage patterns of many birds are discernible in ultraviolet and are thus invisible to the human eye. Male Eurasian blue tits (*Cyanistes caeruleus*) display an ultraviolet reflective crown patch during courtship by posturing and raising of the nape feathers. UV detection is also used in foraging. Kestrels have been shown to search for prey by detecting the UV reflecting from urine marks left on the ground by rodents.

Any area of the retina with a denser concentration of receptor cells perceives images more sharply. One such area of densely packed receptor cells is called a **fovea.** Not only are cells denser in the fovea but they are also arranged in a pit shape that serves to reflect light, allowing a larger image to be formed. Humans have a single fovea so when you focus directly on an object, the image falls on the fovea resulting in poor peripheral vision. Most birds have a one-fovea eye that functions similar to ours, but birds that need to be good judges of speed and distance, such as hawks, terns, parrots, swifts, and hummingbirds, are bifoveal (Figure 19.26). A few terns and swallows even have a third fovea, whereas owls and sandpipers have only one poorly developed fovea and must bob their heads to gain perspective. There has been much debate about the acuteness of avian vision. It appears to be better than human vision but perhaps not significantly so in some cases. A vulture sees about as sharply as humans, but a chicken appears to see about only one-twenty-fifth as well as humans, and hawks about eight times better. Such acute vision enables a hawk to clearly see a scurrying mouse more than a mile away.

The position of a bird's eyes reflects its lifestyle (**Figure 19.27**). The eyes of most birds are on the sides of their heads. Known as **monocular vision**, this side placement allows them to better see things on each side at the same time as well as in front of them. Monocular vision gives birds a wide area of vision (up to 300 degrees in some birds) altering them to danger as quickly as possible without turning their head. Some birds, like woodcocks (*Scolopax*), have their

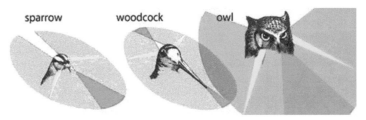

Figure 19.27 Visual Fields of Different Birds. The orange areas indicate where the fields overlap to give binocular vision.

eyes placed far back on the sides of their heads, allowing them to even see danger behind them. Some birds, such as most raptors (birds of prey), have their eyes set far to the front giving them an overlapping field of vision similar to that of humans called **binocular vision**. These birds have much sharper vision to the front than do their monocular cousins.

The eyelids of birds are not used for blinking and are usually closed only for sleeping. The eye of a bird is lubricated by a third eyelid, the **nictitating membrane** that sweeps across the eye horizontally. In diving birds, this membrane serves as a contact lens improving underwater vision. In the black-billed magpie (*Pica hudsonia*), this third eyelid has an orange spot that is displayed during courtship or aggressive displays.

Hearing Birds, just as mammals, have a three-part ear: the **external ear** with a sound conducting canal that runs to the eardrum, the **middle ear** with rod-like, vibration-transmitting columella, and the **inner ear** where the "hearing organ" or **cochlea** is located. Birds lack an external **pinna** (auricle) such as the funnel-shaped flaps of skin and cartilage found on the human head. The most extensive variations of the ear are found among the owls. In several species of owls, the ears are bilaterally asymmetrical (unique among vertebrates) to help localize sound. Owls can also change the position of their **auricular feathers** (feathers found around the external ear opening), and even the heart-shaped face of the owl helps collect sound and directs it to the ears. As a result, owls can detect prey in total darkness with an error of only 1 degree horizontal and vertical.

The frequency range any bird species can perceive is narrower than that of mammals, and although birds are less sensitive to the high and low ends of their range than mammals, they are similar in the middle frequencies. Birds, however, are about 10 times more sensitive to rapid fluctuations in pitch and intensity than humans. Birds can distinguish between frequencies that differ by 1% or less, and they can distinguish between sounds separated in time by only 0.6 to 0.25 milliseconds! Some cave-dwelling species, including the oilbird (*Collocalia*) and swiftlets (*Aerodramus*) emit sound at a frequency between 2 and 5kHz to echolocate like bats to navigate (but not find food) in the darkness of caverns, and recent evidence indicates that penguins may use echolocation to locate prey.

Olfaction (Smell) In general, the olfactory senses of birds are considered to be poor, but there are exceptions, and we are perhaps being misled by the lack of evidence and experimentation. Both Darwin and Audubon demonstrated that vultures did not find carcasses if they were covered, and more recent studies reveal that migrating vultures have been disoriented by odors released into the air.

Birds have been tested for their ability to smell by training them to discriminate between air that contains odor and air that does not. In one experiment pigeons were trained to peck at a disk when they smelled an odor.

Albatrosses, skuas, and petrels in the Antarctic seem to be able to detect flesh, fat, or blood spread on the sea's surface by smelling it. Lecah's petrel (*Oceanodroma leucorhoa*) navigates at least partially by olfactory clues. They can find their nests at night by flying from downwind into the breeze coming from the colony and find their own nest even if it has been moved. Homing pigeons may also navigate using olfactory clues. Pigeons with their nostrils purposefully plugged took significantly longer to find their way home. The flightless kiwi (*Apteryx*) of New Zealand is one of the few birds that seems to be specifically adapted to a keen sense of smell. With nostrils located at the tip of their long probe-like bill, these nocturnal feeding birds sniff the ground while foraging for earthworms and other concealed food (Figure 19.22c).

Olfaction (Taste) The taste buds of birds are concentrated on the posterior part of the tongue and pharynx floor, and although they are similar in structure to the taste buds of mammals, there are far fewer of them. Most birds swallow their food quickly, giving little indication of taste discrimination. Hummingbirds, however, can distinguish different kinds of sugars and differing concentrations of sugar, and some birds quickly learn to associate bitter taste with certain plants or animals and avoid them.

Tactile (Touch) A bird's skin possesses nerve endings that detect heat, cold, pressure, and pain. However, birds are so completely covered by feathers that few places on their body are exposed enough to give them a direct sense of touch. As a result, birds rely on the filoplume feathers to keep them indirectly in touch with their surroundings. The tip of the bill of ducks and geese, the tongues of woodpeckers, and the bill types of many baby altricial (born bare with eyes closed) birds have concentrated nerve endings in the tips of their bills. Many shorebirds have pits with sensory cells along the bill that are very sensitive to pressure. Wood storks (*Mycteria americana*), for instance, can fish in turbid water and close their beak on live fish they touch but cannot see in 0.019 seconds after initial contact. By comparison, it takes humans 0.040 seconds to blink their eyes. Tactile sensory cells are also found at the base of flight feathers, and it is believed that they play an important sensory role in flight.

One of the most amazing sense birds possess is the ability to navigate and find their way during migration. Various techniques of observing bird migration—banding of individual birds, direct observation of day and night movements, radar observations, and controlled experiments—have shown that migratory birds use a variety of navigation techniques. Day migrators navigate chiefly by sight. In their migration they follow remembered and sought-out landmarks in the topography and may use the sun as a visual compass to determine direction. Night migrators orient themselves to certain star patterns and can become disoriented on cloudy nights. Some birds also seem to possess a magnetic compass. Laboratory experiments with birds in orientation cages show that birds do respond to magnetic direction and that their orientation can be shifted by changing magnetic directions. Olfactory clues may also play a role. There is a strong genetic component to migration in terms of timing and route, but this may be modified by environmental influences. Although birds are naturally able to navigate, they also benefit from any number of landmarks and cues. Learning also seems to be an important part of navigation as a bird's ability to navigate correctly usually improves with experience. For example, satellite tracking data of day migrating raptors have shown that older individual are better at correcting for wind drift. The ability to successfully perform long-distance migration can probably only be fully explained with an accounting for the ability of birds to recognize landmarks and form mental maps.

Reproduction and Development

There are almost as many different avian reproductive strategies as there are types of birds. All birds, however, share certain reproductive features: the sexes are separate (dioecious), male and female physically mate with internal fertilization resulting, the fertilized eggs are covered with a thin, brittle calcareous shell and always pass outside the mother's body (oviparity), the eggs are laid in a safe place where they will be warmed, and the young receive varying degrees of parental care until they become independent.

The breeding cycle of birds is timed to take place at a season when the maximum amount of food is available to the female forming eggs inside her or when the maximum amount of food is available for feeding the young. Sometimes the breeding season represents a compromise between these two periods.

Most North American birds do not winter where they breed. With the coming of spring migrants return to the region where they were born or to the territories they previously occupied. As males begin to come into the spring breeding area, they quickly establish and begin to defend some territory. This territory may be no larger than the area within the reach of their beak (colonial nesting birds), or it may be as large as several square miles (raptors). Territorial ownership is most frequently conveyed and contested by songs and feather display but, if necessary, fights and chases may develop. Establishing territories serves several functions: (1) it spaces out the pairs of birds and provides each with a place to mate, (2) it spaces out the nests and makes them harder for predators to find, and (3) it spaces out the adults over the available habitat and thus reduces competition for food for the young.

Ducks and geese that nest in the Arctic arrive already paired and lose none of the little nesting time available in preliminaries like courtship. However, with most birds, once territories have been established, and females begin to arrive, the competition of courtship begins. Courtship is a necessary prelude that helps establish a bond between the male and female that results in fertilization of the eggs. For most birds, the bond between male and female is established after courtship rituals in the male's territory, whether nesting takes place there or not. Usually, it is the female that selects a mate from a number of male suitors on the basis of display and/or suitability of territory.

Courtship rituals may be one-sided, with the male displaying or dancing. Collectively, the male prairie chicken (*Tympanuchus cupido*) strut, dance, and make booming sounds with brightly-colored air sacs on their necks on a display ground known as the **lek**, while the females calmly circle them (**Figure 19.28**). The male birds of paradise (Paradisaeidae) are without equal in the brilliance and oddity of their displays. The display often involves the erection of brilliant iridescent plumage accompanied by bizarre cracking, whirring, or buzzing calls, and dancing. In one species, the display culminates with the male hanging upside down on its perch (**Figure 19.29**). Male displays may involve more than brightly-colored feathers, strange calls, and odd dances. Male bowerbirds (Ptilonorhynchidae) exhibit an extraordinarily complex set of behaviors in which they build a **bower** to attract mates. Depending on the species, the bower ranges from a circle of cleared earth with a small pile of twigs in the center to a complex and highly decorated structure of sticks and leaves—usually shaped like a walkway, a small hut, or a maypole—into and

Figure 19.28 A male prairie chicken displaying on a lek.

Figure 19.29 Count Raggi's bird-of-paradise is widely distributed in New Guinea. These birds are polygamous and during breeding season males such as this one congregate on leks.

around which the male places a variety of objects he has collected. These objects—usually strikingly colored—may include hundreds of shells, leaves, flowers, feathers, stones, berries, and even discarded plastic items or pieces of glass. The male bowerbirds spends hours carefully sorting and arranging his collection, with each object in a specific place; if an object is moved while the bowerbird is away, he will carefully put it back in its place. Some go so far as to strip away the leaves of surrounding vegetation allowing more light to stream in and illuminate their creation. At mating time, the female will go from bower to bower inspecting the quality of the bower, watching as the male owner conducts an often elaborate mating ritual. Numerous females often end up selecting the same male, and many underperforming males are left with an empty bower and no mate.

In some birds, courtship involves rituals in which both sexes participate. Western grebes (*Aechmophorus occidentalis*) perform several elaborate courtship dances. In one dance, the two birds begin rapidly "running" across the water side by side and culminate with both diving down into the water (**Figure 19.30**). In another variation, the two birds raise their breasts out of the water and caress each other with vegetation held in their bills. Cranes bow to each other, jump into the air with outstretched wings, and bow again. Storks face each other, throw back their head and long necks, and make loud clapping noises with their bills. Many birds of prey swoop and dive at each other high

Figure 19.30 The synchronized ballet that is part of the courtship ritual of Western grebes (*Aechmophorus occidentalis*).

in the air. Once attracted to a potential partner, the bald eagle (*Haliaeetus leucocephalus*) may begin an elaborate flying courtship ritual known as **cartwheeling**. In this magnificent display, the potential pair soars to great heights, lock talons, and begin a breathtaking death-defying plunge towards the ground. Just moments before striking the earth, the eagles disengage and once again soar into the heavens. If the timing is not perfect, certain death awaits the falling cartwheelers.

The vast majority (95%) of bird species are socially **monogamous** with the pair remaining together for at least one breeding season. Some may nest together in following years, but only because they both return to the same territory. In such a case, the loyalty is to the territory, not the mate. Among larger birds that live longer, such as geese, cranes, swans, raptors, and many seabirds, pairs often bond for as long as both partners live. Crane researcher George Archibald tells the story of a wild female sandhill crane that was hit and killed by a car in Wisconsin. After her untimely death, her mate spent hours every day all summer and fall standing in solitary loneliness by the roadside where she had been killed.

In a few monogamous birds that raise more than one brood per mating season, the females may switch mates between broods. The briefest kind of monogamous pairing is found in birds that come together to mate but leave the raising of the young to the female. In many passerines such as blackbirds, bobolinks, meadowlarks, wrens, and some warblers, the male bonds and mates with more than one female (**polygyny**). The opposite situation in which the female mates with several males (**polyandry**) and leaves the male to incubate the eggs and raise the young also occurs. In the Arctic nesting red-necked phalaropes (*Phalaropus lobatus*), the female is larger and more brightly colored than the male. The females actively court and pursue the males, compete with other females for territory, and aggressively defend their nests and chosen mates (at least for a short time). Once the females lay their eggs, they begin their southward migration, leaving the males to incubate the eggs and raise the young.

The advantage of monogamy for birds is that both parents share the burden of care. In most types of animals, any significant parental care on the part of the male is rare, but in birds it is quite common; in fact, it is more extensive in birds than in any other class of vertebrates. In birds, male care can be seen as essential to female fitness for in some species the females are unable to successfully raise broods without the help of the male. There is sometimes a division of labor in monogamous species, with the roles of nest building, incubation of eggs, nest site defense, and chick feeding being either shared or undertaken by one mate.

Throughout much of the year, male birds' testes are very small and bean-shaped. The increasing day length of early spring triggers the production of hormones by the pituitary gland and the testes begin to grow, reaching up to 300 times their normal size. After breeding, the testes shrink rapidly back to tiny bodies.

In most female birds only the left ovary and oviduct develop. On the right side, these organs shrink to almost nothing (**Figure 19.31**). The **infundibulum** (the expanded end of the oviduct) picks up the discharged eggs and, while passing them down the oviduct, secretes **albumin** (egg white), shell membranes, shell, and shell pigments around the egg.

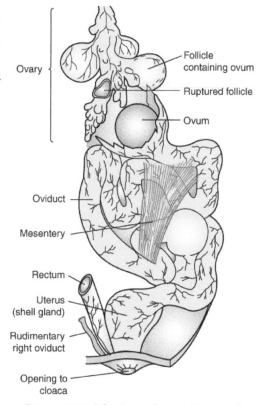

Figure 19.31 The Reproductive System of a Female Bird. In most birds only the left ovary and reproductive tract are functional.

Except for ostriches, cassowaries, kiwis, swans, geese, and ducks males of most bird species lack a retractable penis. As a consequence, sperm must be transferred by contact of the male's cloaca with the cloaca of the female, an act known as the "cloacal kiss." For this to be achieved, the male must mount the female, and she must move her tail to one side. The positioning is delicate, and the male is dependent on the full cooperation of the female. Many attempts at copulation appear to be unsuccessful because the male fails to get in the proper position. Once proper positioning is achieved, sperm transferal can occur in less than a second. Many pairs of birds mate numerous times over the course of several days.

The egg is fertilized in the upper oviduct many hours before the additional layers are added. After mating, the female's oviduct may hold live sperm for several days. Domestic hen chickens are still fertile 5 or 6 days after mating, with fertility dropping off rapidly after that. On occasion, hens may lay fertile eggs 30 days after copulation.

Once the eggs have been fertilized, the next order of business is nest building. Birds seek to protect their nests by concealing them or by building them in places inaccessible to predators. Nests design and construct can be extremely varied. Most passerines build a familiar nest of twigs and plant fibers in an open-topped cup design (**Figure 19.32**).

Figure 19.32 Bird nests are some of the most amazing constructs found in the animal world. (a) A robin's nest with eggs. (b) A male southern masked weaver (*Ploceus velatus*) constructing a woven nest, the most elaborate of nest types.

The swifts and swallows use mud and saliva to form their nests and cement them in place, whereas the weaver birds and orioles weave elaborate and complex chambers from strands of grass or other plant material, with the nest of the plover being nothing more a slight depression in the sand. Some birds build no nests at all. The fairy tern (*Sterna nereis*) lays a single egg on a horizontal tree branch, from which it never seems to roll. After hatching, tern chicks cling tenaciously to the branch with large clawed feet. The male emperor penguin incubates a single egg on the top of its feet and hatches out a youngster in the dead of the Antarctic winter.

Some birds forgo parenting and nest building entirely. In an act known as **brood parasitism**, the female lays her eggs in other bird's nests and then departs, leaving the foster parents to incubate her eggs and raise her young. The foreign eggs are often accepted and raised by the host species, often at the cost of the loss of their own brood. There are two kinds of brood parasites: **obligate brood parasites**, which are incapable of raising their own young and must lay eggs in the nest of other species; and **non-obligate brood parasites**, which are capable of raising their own young but lay eggs in the nest of **conspecies** (birds of the same species) in order to increase their reproductive output. Some colonial nesting birds, such as bank swallows and African weavers, often practice non-obligate brood parasitism. The most famous (or infamous) obligate brood parasites are the cuckoos (Cuculidae) and cowbirds (*Molothrus*) (about 1% of all known bird species are obligate brood parasites). The

Figure 19.33 the cowbird chick on the left will grow to be many times larger than its host parent on the right. In fact, the cowbird may grow so fast that the host parent(s) may die of exhaustion trying to keep up with its demands for food.

eggs of some obligate brood parasites are adapted at hatching more quickly than their host's eggs and the foreign chick then pushes the host's eggs or chicks out of the nest ensuring that all the food brought to the nest is fed to them (**Figure 19.33**).

The smallest nest is built by the hummingbird; it is a scant 2.5 cm (1 inch) across. Raptors build huge nests of limbs and sticks and add to them year after year. A bald eagle nest in Ohio measured 2.5 m (8.5 feet) across and 3.5 m (11.6 feet) deep and was estimated to weigh 3600 kg (2 tons).

Once nest building is complete, the female lays eggs in the nest. The size, color, and markings of bird eggs are consistent within each species but vary considerably from one species to the next. The color of eggs is controlled by a number of factors. Those of hole and burrowing nesting species tend to be white or pale, whereas those of open nesters tend to have camouflage coloration patterns. All birds are oviparous (lay amniotic eggs), and all bird eggs are covered with a thin, brittle, porous calcareous shell. The tiniest eggs are those of the hummingbirds. They are pea-sized, always white, and always number two. The largest are those of the ostrich at 13 to 18 cm (5 to 7 inches) across and weighing 1.5 kg (3.3 pounds). Females can lay fertilized eggs or unfertilized eggs, but only those that are fertilized will develop into young. The word "egg" has a confusing connotation when applied to

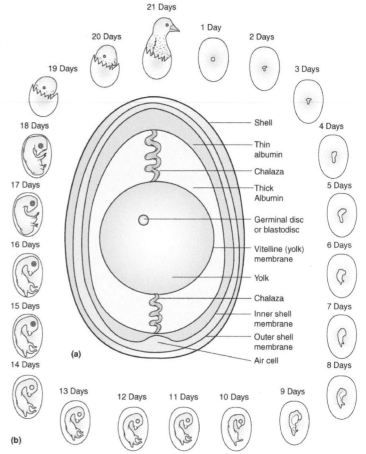

Figure 19.34 The Amniotic Egg of a Bird. (a) Structure of a chicken's egg. The germinal disc is the cells that are beginning to arise from the fertilization of a single ovum. (b) The development of a chicken from fertilized egg to hatching.

birds. Prior to fertilization, the single-celled **ovum** (the actual "egg") rests on top of the yolk. Once the ovum is fertilized, it begins to grow into an embryo feeding off the yolk as it does so. The other structures—yolk, albumin, shell membranes, and shell—are merely noncellular nutritive, supportive and protective accessory layers wrapped around the actual egg (ovum) or developing embryo (**Figure 19.34**).

Some birds are **determinate layers**. That is, they always lay a certain number of eggs and then stop, even if some the eggs are taken. Birds that are **indeterminate layers** continue laying eggs if their eggs are taken. A flicker whose eggs were removed each day laid 71 eggs in 73 days! Domestic fowl fall into this category. The record for a chicken is 361 eggs in 365 days and for a domestic duck the record is an astonishing 363 eggs in 365 days.

Fertile bird eggs must be kept warm to develop properly, so one or both parents maintain proper temperature through incubation, usually by covering the eggs with their body and feathers. Some types of incubating birds often develop **brood patches**, or places where the feathers fall out on the underbody of the parent to expose warm, bare skin that can be pressed directly against the eggs. Eggs need not be constantly covered. In most perching birds where the female alone is responsible for incubation, 6 to 10 minutes may be spent off the eggs feeding for every 15 to 30 minutes spent on the eggs in incubation.

The embryo inside the egg is very sensitive to high temperatures. This requires that in some situations eggs must be protected from the direct rays of the sun. Ducks with open nests, for example, will pull camouflaged downy feathers (originally plucked to form their brood patches) over the nest to cover the eggs when they leave providing shade if it is hot and helping to retard heat loss when it is cold. Other species may stand over the nest and shade the eggs when temperatures rise. Some shorebirds soak the feathers of their bellies and use them to wet the eggs before shading, thus helping to cool the developing embryo by evaporative heat loss.

Embryos are less sensitive to cold than to heat, particularly before incubation has started. Mallard duck eggs have been known to crack by freezing and still hatch successfully. Egg temperature is regulated in response to changes in the temperature of the environment by varying the length of time that a parent bird sits on them and the tightness of the "sit." For instance, female house wrens (*Troglodytes aedon*), which incubate without help from the males, sat on the eggs for periods averaging 14 minutes when the temperature was 15° C (59° F), but an average of only 7.5 minutes when the temperature rose to 30° C (86° F). Eggs are also turned periodically from about every eight minutes in some species to once an hour or more in others. The turning presumably helps to warm the eggs more evenly and to prevent embryonic membranes from sticking to the shell.

A unique form of incubation is found in the turkey-like megapodes (Megapodiidae) of Australia. They heat their eggs by depositing them in a large mound of decaying vegetation, which the birds have scratched together. By opening and closing the mound as needed, the attending males carefully regulate the heat of decomposition, which takes the place of the parental body heat used in normal incubation. Incubation to hatching may take as little as 11 days in passerines, woodpeckers, and cuckoos to as long as 80 days in albatrosses and kiwis.

Newly hatched young fall into several categories. Those that are helpless, blind, naked, and feeble upon hatching are called **altrical**. Those that are fully feathered with down, bright eyed, and able to run and peck at food almost immediately after hatching are called **precocial** (**Figure 19.35**).

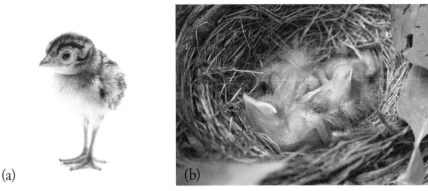

(a) (b)

Figure 19.35 Precocial young vs altrical young. In general, ground nesting birds produce precocial young, whereas birds that nest up off the ground produce altrical young. (a) A precocial chick and (b) Altrical chicks in the nest.

- **Precocial**: Eyes open covered with feathers or down, leave nest after one or two days

  o   Independent of parents: megapodes
  o   Follow parents but find own food: ducks, shorebirds
  o   Follow parents and are shown food: quail, chickens
  o   Follow parents and are fed by them: grebes, rails

- **Semiprecocial**: Eyes open, covered with down, able to walk but remain at nest and are fed by parents: gulls, terns
- **Semialtrical**: Covered with down, unable to leave the nest, fed by parents

  o   Eyes open: herons, hawks
  o   Eyes closed: owls

- **Altrical**: Eyes closed, little or no down, unable to leave nest, fed by parents: passerines

Young, growing birds place extreme demands on their parent(s) to supply enough food. One pair of phoebes was reported making 845 trips to their nest in one day. A larger bird like the eagle may make only two or three trips daily, but the prey it brings to its hungry young consists of relatively large animals. The length, amount, and nature of parental care varies widely across the different orders and species of birds. At one extreme are the megapodes whose parental care ends at hatching. The newly hatched megapodes chick digs itself out of the nest mound and without parental assistance can fend for itself immediately. At the other extreme are many seabirds that have extended periods of parental care. The chicks of the great frigatebird (*Fregata minor*) take up to six months to **fledge** (become mature enough to fly) and are fed by the parents for up to another 14 months.

In some species the care of the young is shared between both parents; in others it is the responsibility of just one parent. Crows, the Australian magpie (*Gymnorhina tibicen*), superb fairy-wrens (*Malurus cyaneus*), the rifleman (*Acanthisitta chloris*), and red kite (*Milvus milvus*) exhibit **alloparenting** in which members of the same species will help the breeding pair in raising the young, These helpers are usually close relatives such as the chicks of the breeding pair from previous breeding seasons.

> *Some tribes of birds will relieve and rear up the young and helpless of their own and other tribes when abandoned.*
>
> —William Bertram

The time required for chicks to fledge varies dramatically. The chicks of the ancient murrelet (*Synthliboramphus antiquus*) leave the nest the night after they hatch, following their parents calls out to sea, where they are raised away from the threat of terrestrial predators. Some ground breeding species, especially the ducks, move their chicks out of the nest at an early age. In bird species that nest up off the ground, the chicks leave the nest soon after, or just before, they can fly. Parental care after fledging varies. Chicks may leave the nest alone and receive no further help, whereas in some cases parent(s) continue some supplementary feeding after fledging. Chicks may also follow their parents during their first migration. Once young birds are mature enough to fend for themselves and have been abandoned by their parents, a crucial point in their survival has been reached, as mortality is the higher in the first year of any bird's life than at any other time.

Bird Connection

Economically

Birds and their eggs have been and continue to be major sources of food for humans in most societies. In fact, bird meat and eggs provide around 15 percent of the total animal protein consumed by humans worldwide. As urban human populations have grown, so has the need for mass production of foods. Factory farming of poultry now provides considerable protein to populations around the world. During the last two decades of the 20th century the production of chicken meat increased by an average of 6% per year and by the year 2000, factory farming of chickens was producing more than 20 billion broiler chickens per year.

With the development of agrarian human cultures, several species of wild chickens, ducks, geese, and pigeons have been exclusively bred into many commercial varieties. Guinea fowl (*Numida meleagris*) from Africa have also been widely exported and bred not only for food but also because they are noisy when alarmed, thus warning of the approach of intruders (avian watchdogs). Ostriches are reared as livestock not only for their plumes and meat but also for their skin, which can be processed into high-quality leather.

Besides meat and eggs, birds provide feathers for clothing, bedding, and decoration. Down feathers are the best insulation against cold ever developed. Humans have never improved on it and use down feathers in pillows, coats, vests, sleeping bags, and comforters. Feathers have long been used on arrows and fishing lures.

With the rise of agricultural, humans' relationship with birds became more complex. Guano (the excrement of fish-eating birds) has been recognized as the best organic fertilizer since the days of the Incas in Peru. Vast quantities were mined from island breeding colonies and in most parts of the world the birds that produce it have been severely depleted. In regions where grain and fruit are grown, birds may cause serious damage to those crops. In North America various species of blackbirds are serious pests in grain fields; in Africa a grain-eating finch, the red-billed quelea (*Quelea quelea*) occurs in flocks like locusts, in plague proportions so numerous that alighting flocks may break the branches of trees.

From pigeon racing to falconry and sport hunting, birds have had a strong role in many traditional recreational pursuits. According to a survey by the Department of the Interior, watching, feeding, and attracting birds rank second only to gardening as America's favorite pastime. This interest is due mainly to the fact that birds are the most visible and accessible form of wild animal life on the planet. It has been estimated that some 55 million people in the United States alone are involved in feeding birds at some level, and as many as 70 million people in the United States regard themselves as bird watchers. A multibillion dollar industry has developed to supply the travel, equipment and feeders, and bird seed demanded by bird watchers and bird feeders.

Many birds are also kept as pets and companions. Small finches and parrots are especially popular and easy to keep. The most popular are the canary (*Serinus canaria*), and the budgerigar (*Melopsittacus undu-*

latus) of Australia (often called a parakeet). These birds have been bred for a variety of color types. Zoos in many cities import strange and exotic birds from many lands and are a source of recreation and enjoyment for millions of people each year. One recent estimate suggests there may be as many as 31 million pet birds in the United States alone.

Ecologically

By virtue of their visibility and the high esteem with which we regard them, birds aid all animal life, including humans, by serving as monitors of global environmental quality. For example, in the 1960s falcons and eagles warned of DDT pesticide contamination. In present times, one of the best indicators we have of the level of oil pollution in the ocean is the grisly toll of blackened bird bodies washed up on beaches. Birds appear to be under siege. A recent World Conservation Union report states that 12% of all birds on the planet are threatened with extinction, and some ecologists fear that up to 14 % worldwide could be extinct within a century. By then a quarter of all birds may be as good as gone, with populations becoming so small that they pass the genetic point of no return. The factors causing the decline of many birds are varied, complex and interrelated:

1. **Habitat destruction**. To cite one of many examples, the tropical rainforests have been reduced by around 45 % of their original area, and this destruction continues with up to 20.4 million hectares (78,700 square miles) lost each year.

2. **Introduced species**. Introduced predators such as cats, rats, snakes, fish, and mongooses have been the major cause of extinction of many island birds. From a study of radio-collared farm cats in Wisconsin, researchers estimated that in that state alone, cats may kill 19 million songbirds in a single year. Introduced herbivores such as rabbits and goats can be as damaging as predators as their eating destroys native vegetation the birds depend on for food, cover, and nesting.

3. **Human predation**. A well-documented example is that of the passenger pigeon (*Ectopistes migratorius*). Estimated to number 5 billion birds when Europeans first landed in America, migrating flocks of these birds took days to pass an observer. On September 1, 1914, "Martha" the last known living passenger pigeon died in a zoo in Cincinnati, Ohio. Over the course of the nineteenth century, the species went from being one of the most abundant birds in the world to extinction, blasted into oblivion for their meat and feathers by commercial hunters.

4. **International trade**. Prized types of wild birds tempt trappers and dealers with rich rewards. Parrots are the most popular, and the trade has had devastating effects on them. Spix's macaw has been illegally trapped to the last bird in the wild, and the only hope for the species lies with 68 individuals held in captivity. Of those, only nine are found in the breeding programs of zoos. The others are privately owned with the majority being in a breeding program in Doha, Qatar.

5. **Pesticides and pollution**. The DDT alarm of the 1960s has resulted in restricted use of chlorinated hydrocarbons, but new classes of deadly chemicals have emerged to take their place. In addition, chemical herbicides, oil, acid rain, industrial chemicals, and nuclear contamination threaten birds and their habitats worldwide.

The potential decline or loss of species after species of birds could be ecologically devastating. In their many habitats, birds are important pollinators and seed dispersers (some seeds will not germinate unless they pass through the digestive tract of a bird); their absence would have wide ecological ramifications. Others are important two-way links in the food web consuming immense quantities of insects, rodents, and weed seed while also serving as prey for other animals. Still others birds serve as scavengers that clear away carcasses or keep pests in check. In India, for example, a rapid decline of vultures in the 1990s led to a rise in feral dogs and a subsequent outbreak of rabies.

For the most part, humans are to blame. Ecologists and conservation groups worldwide rank the loss of native habitat and the introduction of invasive species as the most crucial problems, but unchecked activities like fishing, hunting, and logging play a role as does human-induced climate change. Worst off are the specialists among birds—birds that eat only one type of food or live in only in very specific habitats. Marine birds are also particularly at risk because they are long-lived, slow-breeding, and prone to accidental death by certain commercial fishing practices such as long-line fishing.

Since the 1600s at least 120 species of birds are known to have gone extinct, mostly as a result of human interference of one sort or another, and as of 2011, 1,221 species were listed a threatened by Birdlife International and the International Union for Conservation of Nature IUCN). However, there are bright spots amid the doom and gloom. Teetering on the brink of extinction in the early 1970s, the whooping crane (*Grus americana*) now numbers around 340 wild birds and another 145 living in captivity. The Californian condor (*Gymnogyps californianus*) was down to 27 individuals all in captivity after the last male was caught in 1987. Today their numbers have increased to approximately 200, and recently, for the first time in 22 years a California condor born in the wild fledged from a nest near Hopper Mountain National Wildlife Refuge. Even more extreme is the plight of the Mauritius Kestrel (*Falco punctatus*). Once down to 4 wild individuals, it now numbers more than 300 individuals. These are but several of many examples of successful conservation and restoration efforts being made on the behalf of birds worldwide.

> *God loved the birds and invented trees. Man loved the birds and invented cages.*
>
> —Jacques Deval

Medically and Scientifically

Research on hair cell regeneration in birds' ears may one day aid humans who are hearing impaired and lead to a better understanding of the workings of our own brain. Birds are also used as subjects for research relating to drugs, genetic engineering, diet and nutrition, sclerosis and fibrosis, muscular dystrophy, visual impairment, organ development and deformity, pain, aging, trans-species brain tissue implants, and toxicology.

Although birds are subject to a wide range of diseases and parasites, only a few of these are known to be capable of infecting humans (zoonotic). A notable exception is ornithosis psittacosis, or parrot fever, a serious and sometimes fatal disease resembling viral pneumonia. The microorganism of the disease is transmitted directly to humans via the excrement of pigeons, parrots, and a variety of other birds. Wild birds may also act as reservoirs for diseases such as encephalitis and West Nile virus that are transmitted from birds to humans via biting arthropod vectors, mainly mosquitoes. Thought to have jumped directly from birds to humans, the virus subtype H_1N_1 first appeared in the United States in 1918 and raced around the world, even to the

Arctic. The resulting worldwide pandemic, known commonly as Spanish flu, resulted in the deaths of 50 to 100 million people, possibly more than died during the plague of Black Death in Asia and Europe in the mid1300s.

More recently, avian influenza (virus subtype H_5N_1) is an emerging avian influence virus that is causing global concern as a potential pandemic threat. The common result of this illness among wild birds is minor but domesticated poultry (chickens and turkeys) are quite susceptible to a strain of avian flu virus adapted to waterfowl. In domestic poultry, this virus appears to mutate rapidly into a deadly form that can kill over 90% of a flock in days and rapidly spread to other flocks with similar devastating results. It can only be stopped by killing every domestic bird in the area of outbreak. Across a growing number of countries in Asia, Europe, and Africa, millions of sick poultry have died from the disease while millions more healthy birds have been put to death at the hands of authorities attempting to halt an epidemic.

Since the first outbreak of H_5N_1 in 1997, there have been an increasing number of bird-to-human transmissions of the virus resulting in severe and even fatal infections. According to data from the World Health Organization, as of November, 2007, 206 people in twelve countries have died from the virus. Although some experts on avian flu believe "the world is teetering on the edge of a pandemic that could kill a large fraction of the human population," currently there is no evidence that human-to-human transmission is occurring.

The study of birds has contributed much to both the theoretical and practical aspects of many areas of biology. Darwin's study of the Galapagos finches was an important component in his formulation of the idea of the origin of species by natural selection. Collections of birds in research museums still provide the basis for important studies of geographic variation, speciation, and zoogeography. The study of animal behavior (ethology) has been based to a large extent on studies of birds by Konrad Lorenz, Nikolaas Tinbergen, and their successors, and work on the domestic fowl added to the development of both genetics and embryology.

Culturally

Drawings of birds on the wall of caves in southern France and Spain, bird images of ancient Egyptian and American cultures, and even Biblical writings are evidence that humans have marveled at birds and bird flight for millennia. From the Greek myth of Icarus with his feather and wax wings to Leonardo da Vinci's early drawing of a flying machine (1490) to Orville Wright's first successful powered flight at Kitty Hawk, North Carolina December 17, 1903, humans have longed to break the bonds of earth and experience the soaring life of a bird.

> *The desire to fly is an idea handed down to us by our ancestors who looked enviously on the birds soaring freely through space on the infinite highway of air.*
>
> —Wilbur Wright

Birds feature prominently in folklore, religion and popular culture. In religion they may serve as messengers as in the case of Hugin and Munin, two common ravens who were reputed to fly around the world and then whisper their finding into the ears of the Norse god Oden, or as leaders for a deity, such as the cult of Make-make where the Tangata manu (birdmen) of Easter Island serve as chiefs. They may also serve as religious symbols and icons as in Christianity where Jonah is symbolized as a dove. Birds themselves have been deified as with the Dravidians of India who perceive the peacock as Mother Earth. Eagles have been adopted as national symbols in the United States, Russia, Mexico, and Germany. The vulture was an important icon in ancient Egypt as was the mythical phoenix in Greece.

Birds and their feathers have been and continue to be an important component of modern culture through art, poetry and literature, song, and film. Perceptions of individual bird species vary from culture to culture. In Africa, owls are associated with bad luck, witchcraft, and death, whereas across much of Europe they are regarded as wise with the ability to see the future. Hoopoes (*Upupa epops*) were considered sacred in ancient Egypt, symbols of virtue in Persia, thieves across much of Europe, and harbingers of war in Scandinavia.

The decorative properties of feathers have always attracted human attention. Very few of the world's cultures are without some tradition of feather decoration for fashion and finery. But birds have paid a heavy price for our fascination with feathers. Around 1800, 80,000 Hawaiian mamo birds (*Drepanis pacifica*), extinct since 1899, were sacrificed to make the royal cloak worn by King Kamehameha I. From the simple and practical to the ridiculously flamboyant, hats of all kinds have adorned heads around the world since time immemorial. The natural beauty and longevity of bird feathers make them an ideal trimming for almost any style of hat. From the single-feathered fedora to the extravagantly plumed Shakespearean hats, feathers were always in demand by the world's milliners. Birds of all kinds were used for their feathers. Ostrich, heron, peacock, and bird of paradise were enormously popular, but common garden fowl, such as pigeon, turkey, and goose were also used. There remained, however, a certain status associated with sporting real egret, osprey, and heron feathers. The feathers of these species fetched good prices among "plumassiers," the merchants who prepared feathers for the fashion industry. And some dealers would stop at nothing to obtain just the right feathers of just the right birds.

Herons and egrets in particular suffered at the hands of plume Shunters. Their feathers were the most attractive during the breeding season, when hatchlings are abundant and helpless in their nests. Thus, the carnage was twofold. Thousands of young starved to death as their parents lay dead and skinned nearby. Entire colonies were eradicated quickly and easily in this manner, and it was not long before these birds were virtually gone. The demand for feathers reached epidemic proportions. In 1892, a single order of feathers to a London dealer included 6,000 bird of paradise feathers, 40,000 hummingbird feathers, and 360,000 feathers from other East Indian birds.

The cruelty of the fad did not go completely unnoticed by some very influential people. In 1906, Queen Alexandria announced that she would not longer wear wild bird feathers in her hats, and others followed suit. In 1915, the importation of plumage other than ostrich and garden fowl was banned from Canada. However, legislation, individual fashion statements, and even the establishment of bird conservation groups such as the Nation Audubon Society were not enough to stem the trade in bird pelts for fashion purposes. As absurd as it may sound, it was a trendy new hairstyle that would ultimately save the birds. In 1913, Irene Castle introduced the bob and other short hairstyles which would not support large extravagant hats, and when many women changed their hair style, most plume-hunters were forced to abandon their grisly trade.

Fortunately for all animals, today's fashions tend to be characterized by a more global awareness of conservation issues and most of the sought after birds of times past have returned from the brink of extinction.

A Closing Note
Out My Window

As I write this, it is early winter, and a light snow is falling. I have just come indoors after having one of the most amazing biological "wow!" moments of my life. While putting food in one of my several bird feeders, a nuthatch flitted in and landed on my gloved hand. It cocked its head in that funny manner that nuthatches have and looked me directly in the eye. It then proceeded to hop into the scoop of bird seed I was holding rock steady in my hand. After selecting a seed, it looked me in the eye once again, and then flitted off as quickly as it came. For a person like me who tries to connect with nature as often as possible, you can't get much more connected than to hold a wild bird in your hand.

I have several bird feeders situated where I can see them from the window of my second story home office. This gives me the happy opportunity to participate in two of the most popular pastimes in this country—feeding and watching birds. And I do spend a great deal of time just sitting and staring out the window, often through binoculars, at the amazing avian world that gathers every morning around my feeders. (Don't tell my editors. They think I am busy writing.)

I am a casual bird watcher (note how I said, "sitting and watching" in the previous paragraph) as opposed to being a birder. Birders take bird watching to the next level. These are people, who, ruggedly dressed, tromp through forest, field, and swamp equipped with binoculars, notepads, and often tape recorders. Some birders are professional ornithologists, but most are just common folk who wish to interact with birds in the bird's habitat and on the bird's terms.

If you've never paid much attention to birds, I urge you to start doing so. As a busy student that probably doesn't have much money, keep it casual. Start with just be more observant of the birds around you in the course of your daily life. If you do this, you will soon begin to wonder exactly what birds you are observing. If so, your first purchase should be a good bird identification book. Such books may be purchased at most book stores or online at a reasonable price. Next you will find you probably need a good pair of binoculars. Buy the best binoculars you can afford, and although they won't be cheap, they will be worth it.

Once you start identifying birds, experts suggest you start keeping a life list. A life list is a record of the bird species you have seen and identified over time. Typically a life list is kept in a journal with each entry detailing the bird species, time, location, and any other pertinent information you chose to add. You may not think you need to keep such a list at first, but if you don't, you will most assuredly regret not doing so in the future. Be warned—you may turn into a "twitcher"—someone who goes to great lengths and journeys great distances to add birds to their life list.

Whether bird watching for you is only occasional and casual or grows into an extremely personal passion, connecting on any level to those feathered miracles we call birds can be as rewarding and satisfying as anything a person can do.

Watching birds has become part of my daily meditation affirming my connection to the earth body.

—Carol P. Christ

In Summary

- Birds are closest phylogenetically to theropods, a group of Mesozoic dinosaurs with several bird-like characteristics.
- Birds are at home in polar regions and the tropics; in forests, deserts, and grasslands; on mountains and swimming on and under the ocean and other bodies of water. However, few species are truly cosmopolitan. Some shore and seabirds are worldwide in their distribution, but land birds have a seemingly haphazard distribution.
- Adaptive radiation has resulted in around 9,900 species of living birds divided into about 26 orders within the class Aves. The number of orders varies, depending on the authority consulted, and the classification system—traditional or cladistic—used.
- There is great uniformity of structure among birds with little variation from the basic theme:

  1. They are covered with feathers.
  2. They have paired limbs with the forelimbs being in the form of wings and the hind limbs as scaled legs with clawed toes.
  3. Their body is usually spindle-shaped with four divisions: head, disproportionately long neck, trunk, and tail.
  4. They are endothermic and possess a four-chambered heart.
  5. Their bones are lightweight and usually hollow; no teeth and each jaw is covered with a keratinized sheath forming a beak
  6. Their lungs are relatively small with nine supplemental air sacs nestled in among the internal organs and skeleton; a **syrinx** (voce box) is located near the junction of the trachea and bronchi.
  7. Their excretory system consists of metanephric kidneys but they lack a bladder and possess a cloaca similar to reptiles.
  8. All are oviparous and lay amniotic eggs with much yolk and hard calcareous shells.

- Birds connect to humans in ways that are important economically, ecologically, medically, scientifically, and culturally.

Review and Reflect

1. *A Blast from the Past* Bird fossils are rarer than reptile or mammal fossils. Why? Propose an explanation to explain this discrepancy. The ancestors of birds had teeth, but modern birds lack them. Form a hypothesis to explain why modern birds have no teeth.
2. *Surviving a Catastrophe* DNA evidence seems to indicate that the Cretaceous-Tertiary (K-T) mass extinction event was not as devastating to birds as it was to other types of animals. Why not? Propose an explanation.

3. *Skeletal Differences* Even though both penguins and ostriches are flightless birds, penguins have a large keeled sternum like birds that do fly, but ostriches do not. Provide an explanation for this difference. Your explanation should include a discussion of the structure and function of the keeled sternum.

4. *Going Down in Flames?* Birds have exquisitely crafted and fine-tuned by evolution into flying machines. List and discuss each adaptation—structural and physiological—for flight that birds possess. Use your list to react to this scenario: Your friend comes to you with a scheme he has concocted to fly by flapping (not gliding and no engine). He has been to the gym and developed tremendously large chest and arm muscles. He has also built a pair of wings out of ultralight materials and covered them with feathers. He tells you he plans to jump off a tall building and use his big muscles and light wings to fly. What advice would you give him?

5. *Eats Like a Bird?* A person that doesn't seem to eat much food is often described as "eating like a bird." Is this comparison biologically accurate?

6. *Mixed Up Meals* Imagine you are the assistant curator in a large aviary. Your boss has left on vacation and told you to feed the birds while she is gone. She gave you several diets but forgot to tell you which diet goes with which bird. The birds to be fed are hawks, hummingbirds, chickens, and robins. Match the diets with the correct type of bird.

   Diet A: raw ground meat, oyster shell flour, powdered skim milk, iodized salt, mineral mixture, and vitamin A-D feeding oil.
   Diet B: honey, sweetened condensed milk, vitamins, and almond oil.
   Diet C: minced cooked meat, ground carrots, and ground hard-boiled eggs with shell.
   Diet D: ground milo, ground corn, ground whole wheat, ground barley, ground oats, soybean meal, alfalfa meal, and vitamins.

7. *Stand Up and Be Counted* Several techniques exist for censusing birds, but none is completely satisfactory. Census by direct count is possible only in certain situations; for example, with colonial birds or birds of very restricted habitat. The simplest procedure is to use what is called a **strip census**, which is taken by counting and identifying the birds in a strip of measured width through an entire area. Extrapolating the number of birds in the strip to a larger area gives an *approximation* of the total number of birds in a larger area. Imagine you are working in a wooded area that is 600 m (1,980 feet) wide by 300 m (990 feet) long. You are counting the woodpeckers in a strip of this larger wooded area that is 10 m (33 feet) wide by 330 m (990 feet) long.

   A. How large an area does the entire wooded area occupy?
   B. How large an area does your census strip occupy?
   C. In your census, you count seven woodpeckers. Approximately how many woodpeckers are there in the entire wooded area?
   D. How might such a census be made more accurate?

8. ***Nest Building: Learned or Instinctive?*** Consider the following information and then answer the question that follows:

- Birds of the same species build nests that look exactly the same generation after generation.
- Four generations of weaver birds were reared completely isolated from nesting materials. The fifth generation had access to nesting materials and wove perfect examples of their species' very elaborate nests.
- Introducing eggs and young into unfinished nests does not interrupt the normal course of nest building activity.
- When surplus moss was placed in front of the nest burrows of Wilson's petrels (*Oceanitis oceanicus*), they carried so much inside that they left hardly any room for themselves.

Based on this information, does it seem that birds must learn to build their nests or is it an instinctive behavior? Defend your answer.

9. ***So Many birds, So Little Space*** Your friend attended a lecture by the famous ornithologist Dr. Bob White. After the lecture, he approached Dr. White with a question that has puzzled him for some time—"How can so many different types of birds be packed into small areas yet all find sufficient food." Being in a hurry, Dr. White only had time to hand your friend the illustration found in **Figure 19.36**. Your friend has brought the illustration to you and asked you to explain the connection between the illustration and the question he asked Dr. White. What would you say?

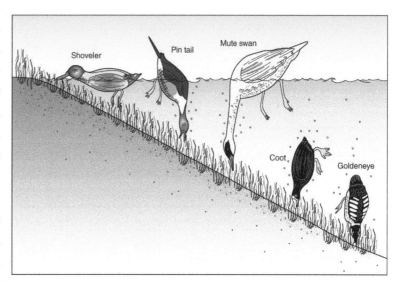

Figure 19.36 The relative feeding depths of several species of water birds.

10. ***Star Light, Star Bright*** Do nocturnal bird migrants have an innate (inborn) sense of direction or do they learn directions as nestlings? To investigate this intriguing question, Stephen Emlen designed a set of elegant experiments using indigo buntings (*Passerina cyanea*) raised in a planetarium in which star patterns could be modified. One group of nestlings (A) was allowed to see stars in a normal night sky rotating around the North Star. A second group (B) of nestlings saw a night sky pattern that was rotating around Betelgeuse, a bright star in the constellation Orion, as if Betelgeuse were the North Star. A third group (C) was raised seeing only points of light at night that did not rotate.

When the birds grew to an age for migration, they were placed in cages under a normal night sky that allowed recording of the direction in which they tried to orient or migrate. Group C that had seen only points of light during their development, with no rotation of the sky, showed no ability to detect direction and moved randomly. Group A that had developed seeing the normal sky rotated around the North Star oriented correctly for migration; and Group B that developed seeing the sky rotated about Betelgeuse showed consistent orientation as if Betelgeuse were the North Star, even though now exposed to a normal sky with stars rotating around the North Star. What conclusions can be drawn from these experiments? In other words, what is the answer to the problem question presented at the beginning of this passage?

11. ***Distant Damage Hits Close to Home*** How can the destruction of tropical rainforests affect populations of birds in the continental United States?

Create and Connect

1. ***A World Without Birds?*** Imagine that a mysterious virus has spread rapidly worldwide, wiping out every single bird on the planet but not harming any other form of animal life. Describe what the world would be like without birds.

2. ***In Living Color*** Some birds wear plumage of bright colors, especially during breeding season, whereas others are drab their entire lives. We assume birds can see those colors, otherwise why go to the expense of growing them and the risk of wearing them? But is this true for plainly colored birds as well as brightly colored birds? Design an experiment to test the problem question: *Do plainly colored birds see colors?*

   Guidelines:

   A. Your design should include the following components in order:

      ➢ The *Problem Question*. State exactly what problem you will be attempting to solve. You should test one specific species, and if you test others, you should test them one species at a time.
      ➢ Your *Hypothesis*. While this is a fictitious experiment, word your hypothesis as realistically as possible.
      ➢ *Methods and Materials*. Explain exactly what you will do in your experiment including the materials necessary to accomplish the task. Be specific, take nothing for granted, and do not expect people to read your mind as they read your work.
      ➢ *Collecting and Analyzing Data*. Explain what type(s) of data will be collected and what statistical tests might be performed on that data. It is not necessary to concoct either fictitious data or imaginary observations.

B. Assume you have access to everything necessary to conduct your experiment. Within reason, money is no object.

C. As usual, we have presented you with an experimental design challenge that is purposely broad and not very specific. Your greatest challenge will be to work out all the specific details leaving nothing to chance. To aid you in this endeavor, your instructor may provide additional details or further instructions.

3. ***All Hail Birds*** Birds have been revered in song and as symbols throughout human history. How would you symbolize these beloved biological icons?

A. Develop a family coat of arms that features a bird symbol. Explain your choice of bird(s) and its meaning as a symbol on your coat of arms.

B. Imagine you have been given the task of developing the national seal for a newly-formed country. A bird (or birds) should figure prominently into your design. (The United States has a bald eagle as the centerpiece of our national seal.) Explain your design and the reason for your choice of bird(s) used in your design.

C. Compose an ode to birds. This may be in the form of a song, poem, or limerick. As inspiration consider the following penned by Helga M. Seter:

> *A dove is an interesting lot*
> *For its coo is more soothing than*
> *Not but when walking outside*
> *Keep an eye on the sky or your*
> *Plain dress will be polka-dot!*

MAMMALS: THE MAGNIFICENT HAIRY ONES

With mammals, the male appears to win the female much more through the law of battle than through the display of his charms.

—Charles Darwin

Introduction

Humans have developed no closer relationships with or bonds of dependence on any other animal group than the mammals, and those bonds reach no higher pinnacle than that between a person and their mammalian pets. We even share the physicality of our bodies with them for in form and structure we too are mammals. However, in spite of the many ways our lives are intertwined with and affected by mammals, we do not seem to hold them in very high esteem. Mammals are judged for their usefulness and utility; valued more for what they can do and provide for humans than for the marvelous creatures they actually are.

History of Mammals

The evolutionary history of mammals is perhaps the most fully documented progression of any amniote lineage because the primary structures used to classify mammals—skulls, teeth, and ear bones—fossilize well. When we look at the broad picture of mammalian evolution, we find it stretches back over 150 million years to hairless, ectothermic ancestors.

Paleontologists distinguish mammals by a feature found in all living mammals (including monotremes) that was not present in any of the early Triassic synapsids: mammals use two bones for hearing that were originally used for eating by their ancestors. The earliest synapsids had a jaw joint composed of the **articular** (a small bone at the back of the lower jaw) and the **quadrate** (a small bone at the back of the upper jaw). Most reptiles, birds, dinosaurs, and extinct therapsids and dinosaurs have and had such a jaw conFigureuration. In mammals, the articular and quadrate bones have become the **incus** and **malleus** bones in the middle ear, part of the hearing apparatus. Mammals also have a double **occipital condyle**. They have two knobs at the base of the skull that fit into the topmost neck vertebrae, whereas other vertebrates have a single occipital condyle. With these few but distinct bone structures, paleontologists have been able to determine the evolutionary path of mammals with a great degree of certainty.

The first fully terrestrial vertebrates were amniotes that possessed eggs wrapped in a shell and internal membranes that allowed the developing embryo to breathe, but kept water in. The first amniotes arose in the late Carboniferous. Within a few million years, two important lineages became distinct: the synapsids, from which mammals are descended and the diapsids, from which dinosaurs, reptiles, and birds descended.

The first synapsids evolved into a diverse paraphyletic group of herbivorous and carnivorous forms known collectively as **pelycosaurs**. Outwardly resembling lizards, these early synapsids were the most common land vertebrates of the early Permian and included the largest land animals of the time. About 260 million years ago at the start of the Permian, a branch of pelycosaurs, the **therapsids,** first appear and eventually replace the pelycosaurs as the dominant land vertebrates. The therapsids went through a series of stages, beginning with animals that were very much like their pelycosaur ancestors and ending with the Triassic **cynodonts**, some of which could easily be mistaken for mammals.

The Permian-Triassic extinction event (P-Tr event) ended the dominance of the therapsids, and in the early Triassic all of the medium to large animal niches were taken over by archosaurs, the ancestors of crocodilians, pterosaurs, dinosaurs, and birds. During the Triassic, the dinosaurs became diverse and abundant, but all non-mammalian synapsids passed into extinction. The cynodonts and their descendants survived through the Triassic as small, shrew-like, mainly nocturnal insectivores. Although the early mammals of the mid-Triassic had developed all the elements of modern mammals, they would have to spend another 150 million years literally living at the feet of the dinosaurs before they could achieve the mantle of dominance.

Around 65 million years ago the Mesozoic era came crashing down with the Cretaceous-Tertiary mass extinction event (K-T event) in which about 70% of all species, including all the dinosaurs and pterosaurs, disappeared in a relatively short time geologically. This disaster left huge niche vacancies in all habitats, and it was the mammals alone of the surviving groups of animals that began an explosive

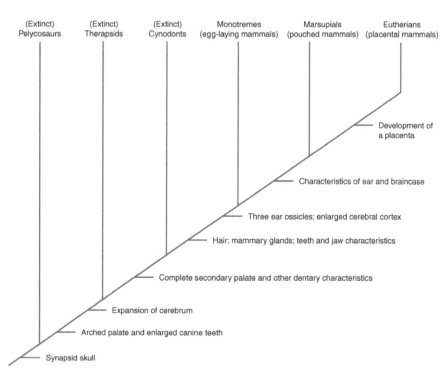

Figure 20.1 The Possible Phylogeny of Mammals. The single lines leading to the extinct pelycosaurs, therapsids, and cynodonts represent various groups of these animals that arose and then eventually became extinct.

radiation of types and forms to fill those vacant niches. By the end of the Cretaceous, 15 mammal families that we know about were in existence. Over the next 15 million years the remaining ten families (five became extinct along with the dinosaurs during the K-T event) expanded to become 78 families by the early Eocene period of the Cenozoic era. The number of genera increased from about 40 to over 200 during the same time period. By the middle of the Eocene epoch some 45 million years ago, all major groups of mammals alive today had come into existence. Thus, the Tertiary has been and continues to be the Age of Mammals (**Figure 20.1**).

> *Most of our ancestors were not perfect ladies and gentlemen. The majority of them weren't even mammals.*
>
> —Robert Anton Wilson

Diversity and Classification

Class Mammalia (L. *mammas*, breast) with nearly 5000 species arrayed into 26 orders, is not a large group compared to arthropods, fish, and birds, but they are very diverse lot. Mammals have adapted their basic body plan to an astonishing array of sizes, shapes, and appearances. The smallest mammal in the world is the recently discovered Kitti's hog-nosed bat (*Craseonycteris thonglongyai*) of Thailand, which weighs just 1.5 gr (0.05 ounces). The blue whale (*Balaenoptera musculus*), weighing 117 metric tons (130 tons or260,000 pounds) and reaching 30 meters (100 feet) in length is not only the largest mammal, it is the largest animal to have ever existed on the planet.

Mammals have a presence in nearly every conceivable environment on this planet that supports life. They are found from polar ice to steamy tropical jungle on every continent except Antarctica. They also swim and dive on and under the waters of ocean, lake, and pond. Some burrow ceaselessly into the soil and others scurry or gallop across deserts and grasslands and through forests. Some sail through the sky and others swing from tree to tree. Not only are mammals found practically everywhere, but by diversity of form, efficiency of endothermy, plasticity of their genetics, and the power of their intelligence, they tend to dominate those environments in which they are found.

The classification of mammals, as with other groups of animals discussed previously in this book, is presently debated between two camps: the morphological and the molecular. Molecular studies based on DNA analysis have revealed over the last few years what appear to be new relationships among mammal families. The most recent classification systems based on molecular studies establishes four lineages of placental mammals that diverged from common ancestors in the Cretaceous. It should be noted, however, that these molecular results are still controversial mainly because they are not reflected by morphological data and thus not accepted by many taxonomists. Again, as we have stated in every chapter of this book from Chapter 9 on, until the two camps come up with a common (or at least a generally agreed upon) classification system for mammals, we feel it is educationally prudent to take the more traditional approach to mammal classification. As was the case with birds, the large number of mammalian orders precludes detailed discussion and description of each individual order.

Domain Eukarya
  Kingdom Animalia
    Phylum Chordata
    Subphylum Vertebrata (Craniata)
      Class Mammalia
        Subclass Prototheria—Egg-laying mammals
          Order Monotremata—Platypuses, echidnas
        Subclass Theria—Marsupial mammals
          Order Didelphimorphia—American opossums
          Order Paucituberculata—Shrew opossums
          Order Microbiotheria—Monito del Monte
          Order Dasyuromorphia—carnivorous marsupials
          Order Peramelemorphia—Bandicoots, bilbies
          Order Notoryctemorpha—Marsupial moles
          Order Diprotodontia—Koalas, wombats, possums, wallabies, kangaroos
        Infraclass Eutheria—Placental mammals
          Order Xenarthra—Anteaters, armadillos, sloths
          Order Insectivora—Shrews, hedgehogs, tenrecs, moles
          Order Scandentia—Tree shrews
          Order Dermoptera—Flying lemurs
          Order Chiroptera—Bats
          Order Primates—Prosimians, monkeys, apes, humans
          Order Carnivora—Dogs, wolves, cats, bears, weasels, seals, sea lions, walruses
          Order Cetacea—Whales, dolphins, porpoises
          Order Sirenia—Sea cows, manatees
          Order Proboscidea--Elephants
          Order Perissodactyla—Horses, zebras, asses, tapirs, rhinoceroses
          Order Tubulidentata—Aardvark
          Order Artiodactyla—Swine, camels, deer, hippos, antelope, cattle, sheep, goats
          Order Hyracoidea—Hyraxes
          Order Pholidota—Pangolins
          Order Rodentia—Squirrels, rats, mice, woodchucks, beavers, porcupines, voles
          Order Lagomorpha—Rabbits, hares, pikas
          Order Macroscelidea—Elephant shrews

Based on reproductive differences, mammals fall into three main morphological (not taxonomic) groups: *monotremes* (egg-layers), *marsupials* (pouched), and *placentals*. The monotremes are by far the smallest group consisting of one species of duck-billed platypus and four species of echidna (spiny anteaters). Like reptiles and birds, monotremes possess a cloaca and lay eggs, characteristics unique among the mammals.

Marsupials give birth to live young, but the young are born in a very undeveloped state. After birth, the young complete their development inside a special pouch (**marsupium**) in the mother's abdomen, feeding on milk supplied by her nipples.

Unlike a young marsupial, a developing placental spends a relatively long time (**gestation period**) inside the mother's uterus. The developing young are nourished by a spongy mass embedded in the wall of the mother's uterus called a **placenta**. By the time a young placental mammal is born, it is fully formed, although it may not yet have hair or functioning eyes or teeth.

The most numerous placental mammals in both species and individuals are the rodents (Rodentia). In contrast, order Tubulidentata is represented by a single living species, the aardvark (*Orycteropus afer*). The elephants (Proboscidea) and the horses, rhinoceroses, and their allies (Perissodactyla) are examples of orders in which far greater diversity occurred in the middle and late Tertiary period than today.

General Characteristics of Mammals

A mammal is any amniotic vertebrate that possesses or demonstrates the following characteristics:

- The body covered partially or wholly with hair
- Skin containing various types of glands—sweat, scent, sebaceous (oil), and mammary glands
- A synapsid skull with two occipital condyles, a secondary **palate**, and **turbinate bones** in the nasal cavity. Their lower jaw is a single enlarged bone (**dentary**). Their middle ear consists of three **ossicles** (malleus, incus, stapes), the skeleton has seven cervical vertebrae (in most types), and the pelvic bones are fused
- Teeth that are **diphyodon**t (milk or deciduous teeth replaced by a set that lasts the rest of the lifetime) and **heterodont** (teeth varying in size and shaped based on function)
- The eyelids are movable and possess fleshy external flaps (**pinnae**) around the ear openings
- The body is endothermic, and a four-chambered heart powers the circulatory system.
- The excretory system consists of metanephric kidneys that contain ureters emptying into a bladder. A cloaca is found only in monotremes.
- The brain is highly developed, especially the cerebral cortex, the site of intelligence
- They are viviparous except for monotremes which are oviparous. The developing young are nourished by milk from the mother's mammary glands.

Birds vary little from their basic body form. Mammals, on the other hand, have deviated and diversified greatly from theirs. In body form mammals run the gamut from the skittering shrew to the prowling lion and from the long-snouted anteater to the sleek porpoise, and from the hooting howler monkey to the industrious beaver.

The life span of mammals varies as enormously as their size. Most shrews survive less than a year, exhausting themselves in a life of frenzied activity. By contrast horses can live 20 years, chimpanzees can live to be over 50, and elephants can survive into their 60s. Humans are thought to have the longest live span of any mammals, with a few individuals living over 110 years.

Mammalian Body Plan

Body Covering

Hair is to mammals what feathers are to birds and scales are to reptiles. Hair grows from a hair follicle sunk into the epidermis of the skin. A hair never stops growing, fueled by a constant production of cells in the follicle. As the hair shaft is being pushed up and out, the cells in it eventually die. The hair eventually consists mainly of dense fibrous keratin protein, the same material that constitutes nails, claws, and hooves in mammals and feathers in birds. Hair, however, is more than just a strand of keratin. It consists of three layers: the outer **cuticle**, middle **cortex**, and the inner **medulla**. Differences in these three layers allow mammalian hair to exhibit a many different structures. Deer hair tends to be brittle, for example, because of deficiencies in the cortex of the hair, whereas deficiencies in the medulla result in the hollow, air-filled hairs of wolverines. Furthermore, rabbits and other mammals have scaled hairs that interlock when pressed together. The curly hair of sheep is the result of growth from curved follicles.

The hair of mammals usually performs one of two basic functions. They may be either sensory (whiskers) or protective. Most hair serves to insulate, keeping heat in and, in many cases water out, to conceal, to signal, and to protect. However, some mammals are hairless or nearly so. The elephant, rhinoceros, and hippopotamus live in warm areas and have a thick skin and a large body that retains heat so an insulating coat of hair would actually be a detriment. Other nearly hairless mammals that live in cold areas or the water, such as the walrus and whale, are insulated by thick layers of fat (blubber) beneath the skin. We humans are much hairier than we appear. In reality, a human has about the same number of hair follicles as a gorilla, but most human body hair does not grow as long and coarse or as dark as those of a gorilla.

The hair of mammals varies in length, density, texture, and color in different species. It is heaviest on arctic mammals but often thin and short on tropical species. Two kinds of hair form the **pelage** (coat) of most mammals (**Figure 20.2**): (1) dense, soft, highly insulative **underhair** and (2) coarse and longer **guard hair** that protect against wear and to provide coloration. The underhair traps a layer of insulating air next to the body. In many species that spend a great deal of time in the water, such as otters, beavers, and fur seals, it is so dense that it almost impossible for water to penetrate it.

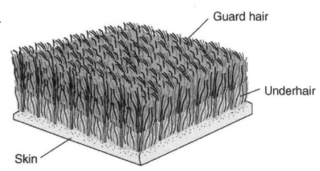

Figure 20.2 A section of mammalian skin with hair.

The coat of hair is shed periodically (molted), and new hairs grow in. Like birds, mammals vary in their pattern of molting. In most mammals, there are periodic molts of the entire coat. In some, such as foxes and seals, the coat is shed once every summer. Most mammals, though, shed twice a year (usually in the spring and fall). The spring and summer bring thinner, lighter coats while cooler winters call for much thicker, fuller coats. In some mammals, there is even a color differentiation between the seasonal coats. For example, weasels and the Arctic fox (*Alopex lagopus*) have a dark summer coat but replace it with a white one for winter. This white winter coat serves to camouflage many arctic animals against their snowy surroundings. The snowshoe rabbit or varying hare (*Lepus americanus*) benefits from three annual coats: a brownish summer coat, a gray autumn coat, and a white winter coat (**Figure 20.3**).

Unlike birds, mammals tend to be cloaked in rather drab colors that serve to hide rather than advertise their owner. Appearances to the contrary, no mammal has hair that is naturally blue or green in color. Some cetaceans (whales, dolphins, porpoises), along with the mandrill (*Mandrillus sphinx*) (**Figure 20.4**), have shades of blue skin whereas other cases of blue hair or fur have always been found on closer examination to be a shade of gray. Sloths and polar bears occasionally appear to have green patches of fur that in reality are growths of symbiotic cyanobacteria or algae (**Figure 20.5**). Many species blend into their environments with clever color patterns and markings such as stripes on a tiger or spots on a leopard. A few, such as the skunk, warn of their presence with conspicuous coloration.

Many mammalian species enjoy benefits of hair beyond its power of insulation. Many mammals have tactile sensory "whiskers" (**vibrissae**) around their snouts. Vibrissae are especially long in nocturnal and burrowing animals that need to find their way under low to no light conditions. A few, such as the porcupines, hedgehogs, and echidnas, have guard hairs that are modified into a spiny amour. The North American porcupine (*Erethizon dorsatum*) may have as many as 30,000 barbed hairs or **quills** (**Figure 20.6**) with as many as 100 to 140 growing from each square inch of skin. When attacked, the animal quickly turns around and thrashes its barbed tail. (Contrary to popular belief, the porcupine does not throw its quills.) On contact, the quills that sink into the victim's skin break off at the bases from which they came. Equipped with reverse, locking hooks (like a fishing hook), the quills work their way deep into tissues. Because of these backward facing hooks, quills cannot be easily pulled out once they are embedded. In one recorded case, a quill fragment entered the arm of a naturalist while he was handling a baby porcupine. Some 45 hours later the point emerged

Figure 20.3 Seasonal variation in the coat color of hares. (a) summer, and (b) winter

Figure 20.4 Bright red-and-blue markings identify this mandrill as a mature male. The area around the male anus (rump) is similarly colored.

Figure 20.5 The greenish tinge to the hair of this three-toed sloth (Bradypodidae) is caused by an algae that lives on the sloth. This algae lives nowhere else, and is passed directly from mother to offspring.

from the man's skin at a spot 42 mm (1.7 inch) from the initial entry point. It traveled through the arm muscles at the rate of about an inch a day. As long as the quill passes through muscle no permanent damage usually results. Animals can be blinded and even killed, however, if the quill penetrates the eye or a vital organ.

Mammalian skin is unique not only for hair but also for the numerous glands it contains that are not found in other vertebrates. There are four main categories of skin glands in mammals: mammary, sweat, sebaceous, and scent. All glands are derivatives of the epidermis. The most important of these glands are the **mammary glands**, from which this group derives its name. Mammary glands are modified sweat glands that occur on all female mammals and in a rudimentary form in all male mammals. Mammary glands grow as the female matures, getting increasingly larger and functional during pregnancy. The mammary glands provide nutritious secretions called **milk** for the initial period of development by the young. In all

Figure 20.6 Porcupine quills or spines come in varying lengths and colors depending on the animal's age and species.

mammals except monotremes, the openings to the mammary glands are projecting nipples or teats, from which the milk is sucked by the young. Monotremes lack nipples. Instead, the milk oozes out, and the young suck it from tufts of hair on the belly of the female.

Sweat glands, found only in mammals, are of two types: eccrine and apocrine. **Eccrine glands** release a clear, watery fluid onto the surface of the skin. As this fluid evaporates it cools the skin by drawing heat away. In most mammals, eccrine glands are restricted to certain areas. Dogs, for example, have eccrine glands only on their paws and their nose. In horses and primates, they are scattered over the body but in rodents, rabbits, and whales they are very small or nonexistent. **Apocrine glands**, larger than eccrine glands, open either into a hair follicle or into a space where there once was a hair follicle. In humans, these glands develop with the onset of puberty and are found primarily in the armpits, pubic region, breasts, and scrotum. The milky fluids secreted by these glands form a film where they dry on the skin. Their purposes are purely associated with the reproductive cycle, and they play no role in temperature regulation.

Secretions from the sebaceous (oil) glands keep skin and hair soft and oily and provide insulation and waterproofing. Sebaceous glands are usually associated with the hair follicle though they can open directly onto the surface of the skin. They become swollen with a fatty substance that secretes in a greasy liquid known as **sebum**. Sebaceous glands in most mammals are found across the whole body. In humans, however, they are localized on the scalp and facial areas.

Scent and musk glands are used by mammals for attracting mates (pheromones), marking territories, identification and communication during social interactions, and protecting themselves. There are many known locations on the body for these glands. They may be behind the eyes and on the cheek (pica and woodchuck), at the base of the tail (wolves and foxes), on the back of the head (camels), and even on the penis (muskrats, beavers, and many canines). Perhaps the best-known scent organs are found in the anal regions of the skunks. The secretions of these glands, certainly the most odoriferous of all glands, can be forcefully discharged out the anus for 2 to 3 m (7-10 feet). Pity the poor creature in the line of fire of that choking, sickening blast.

Where it is subjected to heavy wear, mammal skin forms a dense keratinized area such as the calluses on human palms or soles or the foot pad of bears, dogs, and mice. The claws, nails, and hoofs of mammals are hard keratinized outgrowths of the skin.

Support and Movement

The active life-style of mammals requires a skeleton that is well braced for the attachment of many muscles. To accomplish this, the bones of a mammal are nearly completely ossified (solid), and considerable fusion of bones occurs, as in the pelvic girdle. The skeletal system of mammals (and other vertebrates) is broadly divisible functionally into axial and appendicular portions. The **axial skeleton** consists of the **cranium** (braincase) and the backbone and ribs and it serves primarily to protect the central nervous system. The limbs and limb girdles (attachment sites of the limbs) constitute the **appendicular skeleton**. In addition, there are skeletal elements derived from the gill arches of primitive vertebrates including the jaws, the hyoid apparatus supporting the tongue, and the auditory ossicles of the middle ear. The vast majority of mammals have seven cervical vertebrae; sloths with six or nine and sirenians (manatees and dugong) with six are exceptions.

The mammalian skull consists functionally of the cranium and the jaws. In general, it is the head of the mammal that meets the environment. The skull houses and protects the brain, the eyes, and other sensory apparatus, the teeth and tongue, and contains the entrance to the pharynx. Thus, the head functions in sensory reception and interpretation, food acquisition, defense, respiration, and in higher groups, communication.

The skull of mammals differs markedly from that of reptiles in several respects. The cranium of mammals is exceptionally large. In fact, the sphenoid bones that compose the entire reptilian cranium form only the floor of the cranium in mammals. The nasal cavities of mammals are separated from the oral cavity by a **secondary palate (Figure 20.7)**. This allows continuous breathing while chewing or suckling. Other specializations of the mammalian skull include two paired articulating surfaces at the neck (occipital condyles) and an expanded nasal chamber with complex folded turbinal bones, providing a larger surface area for detection of odors. In mammals, the lower jaw is a single bone, the dentary.

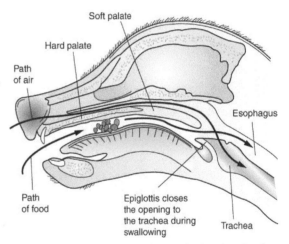

Figure 20.7 Consisting of both the hard and soft palate, the secondary palate of a mammal almost completely separates the nasal and oral cavities. This alignment requires only a momentary cessation of breathing when the mammal swallows.

The bones of the mammalian middle ear are a diagnostic feature of this class. The three **auditory ossicles**—malleus, incus, stapes—form a series of levers that serve mechanically to increase the amplitude of sound waves reaching the tympanic membrane (eardrum), produced by disturbances of the air. The **malleus** ("hammer") rests against the eardrum and articulates with the **incus** ("anvil"). The incus in turn articulates with the **stapes** ("stirrup"), which rests against the oval window of the inner ear. The mechanical efficiency of the middle ear has been increased by the incorporation of two bones of the reptilian jaw assemblage.

Growing from the skull of some mammals are horns and antlers. Although these terms are often used interchangeably, they are, in reality, quite different structures. **Antlers** are a pair of bony branched structures that protrude from the frontals of the skull and are shed annually; **horns** are also paired and protrude from the frontals, but they are permanent, unbranched, and composed of a bony core overlain with a keratinized sheath.

Antlers are one of the most easily recognized characteristics of the deer family Cervidae (**Figure 20.8**). They are present only in males (except for caribou) and are capable of growing astonishingly large. The growth cycle is regulated by testicular and pituitary hormones. Secretions from the pituitary initiate growth in April or May. In the north-

Figure 20.8 Mammals with Antlers. (a) Deer (Cervidae), and (b) Moose (Cervidae).

ern hemisphere increasing day length also plays a role. As they grow, antlers are covered with skin and soft hair called **velvet**, which carries blood vessels and nerves. As antlers near the end of the growing process, they harden. The velvet dies and comes off in tatters, removed in part by the animal rubbing and thrashing their antlers against vegetation. The antlers are stained during this action, giving them a brown, polished wooden appearance. Over the years, each new set of antlers becomes larger and more intricate than the pair it replaces.

Full-grown antlers are used during the breeding season in competitions for females. As winter approaches and day length shortens, antler-growth hormone secretions lessen. As a result, the connection between the antler and their bony attachment point, the **pedicel,** weakens, and eventually the antlers are shed. For a few months in late winter, males are without antlers until the cycle begins again. Because an older moose or elk accumulates upwards of 50 pounds of calcium salts to form the bone needed to fashion new antlers, their yearly growth is a constant strain on the metabolism of these animals.

Horns occur in males of all species of family Bovidae (buffalo, bison, wild and domestic cattle, antelopes, gazelles,

Figure 20.9 Mammals with Horns. (a) Bighorn sheep (*Ovis Canadensis*), and (b) Water buffalo (*Bubalus bubalis*).

sheep, and goats), and females often bear them as well. Horns vary from species to species in shape and size (**Figure 20.9**). The growth of horns is completely different from that of antlers. In many species, the horns never stop growing, and neither the sheath nor the core are ever shed. Possessing their own centers of ossification, horns are not attached to the skull but fuse with the skulls bones secondarily.

As with antlers, horns are used by males in fights and displays during the breeding season. In some species, members of both sexes have horns. In those groups, some degree of sexual dimorphism is usually the rule. Horns on males are thicker at the base and able to withstand the forces of the head-on battering ram crashes with other males. On females, they are straighter and thinner, which may make them more suited for stabbing (defense).

Giraffe horns are paired, short, unbranched, permanent bony processes that are covered with skin and hair (**Figure 20.10**). They differ from other horns in that they do not project from the frontal bones, but lie over the sutures between the frontal and parietal bones of the skull. Giraffe horns begin as cartilaginous structures in the fetus and may not fuse to the skull until the animal is four years old. Horns are present at birth and occur on both male and female giraffes.

Figure 20.10 Female giraffes display tufts of hair on top of the horns, whereas male horns are larger and tend to be bald on top.

The horns of the rhinoceros differ from true horns in that they have no core or sheath. Rhino horns are composed of hair like keratinized filaments forming a clustered, cemented grouping of **dermal papillae**. The horn is situated over the nasal bones. In species with two horns, the second horn lies over the frontal bones but is not attached to them (**Figure 20.11**).

Mammals and birds have abandoned the pattern of skeletal growth typical of reptiles. Reptiles display **indeterminate growth** in that the skeleton continues to grow throughout most of their lifetime. Reptile growth is also seasonal, which is reflected in the typical pattern of growth rings formed in their bones. Mammals and birds, on the other hand, exhibit **determinate growth** and stop growing in their adult stage. Growth in mammals and birds is less influenced by seasonality than is growth in reptiles, and so their bones seldom show pronounced growth rings.

Figure 20.11 The horn of a rhino is not simply a clump of modified hair as is widely believed. Rather, it is entirely keratin with dense patches of calcium and melanin in the middle.

The muscular system of mammals is comparable to that of reptiles, and although the proportions and specific function of muscular elements have been altered in mammals, the relationships of these muscles remain essentially the same. Exceptions to this generalization are the muscles of the skin and the number and placement of jaw muscles. A sheath of dermal (skin) muscle developed in many mammals allows the movement of the skin independent of the movement of deeper muscle masses. These skin movements function in such mundane activities as twitching of the skin to thwart insect pests or more important tasks such as shivering to generate heat in some species. The dermal musculature of the facial region is particularly well developed in primates and carnivores but occurs in other groups as well. Refined facial musculature allows for many expressions that may be of importance in communication and social structure.

Limbs, Locomotion, and Migration

Several terms describe how and where an animal moves. **Natotorial** animals swim; **volant** animals fly. **Cursorial** animals walk or run for long distances. **Scansorial** animals are climbers; in the extreme, they are **arboreal**, spending most of their lives in the trees. Hoppers are said to **saltatorial**. If they use their hind limbs only in a fast succession of hops, they are said to **ricochetal**. **Fossorial** types are diggers, usually living in burrows.

The limbs of mammals range from the slender, tapering legs of the agile deer and antelopes to the round, stumpy legs of the elephant; the paddle-like limbs of the seal and porpoises; and the long, delicate forelimbs and fingers covered with skin of the bats.

Habitat specialization in cusorial mammals has been accompanied by adaptations in the structure of their foot. Some mammals, such as humans, raccoons, opossums, bears, rabbits, weasels, mice, pandas, rats, skunks, and hedgehogs, walk and run with the sole of the foot touching the ground (**plantigrade locomotion**) (**Figure 20.12**). Wolves, foxes, dogs, coyotes, cats, and lions walk/run on their digits (toes) with the heel raised (**digitigrade locomotion**), whereas deer, horses, goats, antelopes, and sheep walk/run on a single toe capped by a hoof and with the heel raised (**unguligrade locomotion**).

A leaping style of movement (**saltatory locomotion**) has arisen in several unrelated mammal groups—some marsupials, rabbits and hares, and several independent lineages of rodents, such as the kangaroo rats (Heteromyidae), jerboas (Dipodidae), and hopping mice (Muridae) (**Figure 20.13**). The jumpers and hoppers have long lower limb segments, rigidly hinged joints, and in most, a long, heavy tail counterbalances the weight of the forelimbs and head.

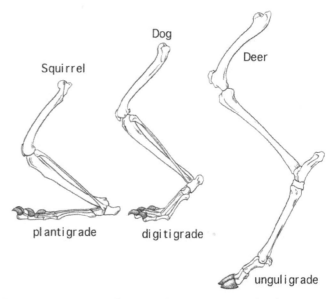

Figure 20.12 Mammalian Foot Structure. (a) In the plantigrade foot of the squirrel, the sole contacts the ground when walking/running. (b) In the digitigrade foot of the dog, only the toes touch when walking/running. (c) In the unguligrade foot of the deer, only a single toe touches the ground.

Figure 20.13 Jerboa (*Jaculus jaculus*) are hopping desert rodents found throughout Africa and Asis. They hop in a manner similar to a kangaroo, although they are not related.

Mammals of several orders have attained great size. Convergent evolution in elephants, hippopotamuses, and rhinoceroses resulted in a ponderous mode of locomotion referred to as **graviportal**. These huge mammals have short lower limb segments in which the diameter of the limb bones is disproportionately large, with the digits in a circle around the foot for maximum support, like the pedestal of a column (**Figure 20.14**).

Scansorial locomotion (climbers) requires animals possess mobile limbs and the ability to grasp with hands and feet. Squirrels and monkeys are familiar scansorial mammals that lead an arboreal lifestyle (**Figure 20.15**). Some climbers, such as sloths or gibbons, suspend themselves from tree branches with elongated limbs; others, such as tree squirrels, scamper on limbs and climb by clinging to projections and irregularities, and yet others, such as tarsiers, spring from place to place in the trees. Some scansors spend their entire lives in trees whereas some climb only occasionally.

Fossorians, such as moles, badgers, and naked mole rates have digging adaptations for use in obtaining food, and in the construction of burrows or tunneling. Moles (Talpidae) have a pointed snout, thick velvety fur, rudimentary eyes, and broad paddle-like front feet with long, powerful claws (**Figure 20.16**).

Mammals adapted to life in water are referred to as being **natatorial**. The most extreme examples are cetaceans (whales, dolphins, and porpoises) and sirenians (dugongs and manatees). Both are totally aquatic and helpless out of water (**Figure 20.17**). Buoyed by their liquid environment, whales have evolved into the largest mammal and indeed the largest animal to have ever lived. Pinnipeds (seals and walruses) are strong swimmers and deep divers, but they do come out onto land to mate, give birth, and nurse their young (**Figure 20.18**). The forelimbs of natatorians are elongate and paddle-like while the hind limbs are flippers (seals and walruses) or a tail (whales).

Bats (order Chiroptera) are the only true flying mammals. In bats, the finger bones of the forelimbs are exceptionally long, extremely flexible, and covered with skin to form the only true wing in class Mammalia, thus making them the only mammal capable of flapping flight (**Figure 20.19**). The flying phalangers or wrist-winged gliders (*Petarus*) are arboreal mar-

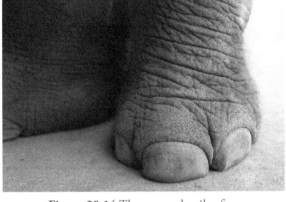

Figure 20.14 The toes and nails of an elephant. Columnar legs are necessary to support the great weight of these animals.

Figure 20.15 All monkeys have dexterous grasping hands and feet, but only New World monkeys have a tail that is prehensile (grasping).

Figure 20.16 The mole (Talpidae) can tunnel at the rate of 18 feet per hour and construct 150 feet of tunnel in a single day.

supials that have a fold of skin (**patagium**) running from the wrist to the ankles which they spread and use to glide from tree to tree.

Perhaps the most unusual mammalian locomotory specialization is non-saltatory bipedalism. **Bipedal locomotion** means habitually walking upright on the hind legs. Although some primates and bears may occasionally rear up on their hind legs, humans are considered to be the only true bipedal mammals. Our skeleton, legs, and feet are adapted to this unusual stance in several ways. Our foot is plantigrade, and our

Figure 20.17 Representative Natatorial Mammals. (a) Manatee (*Trichechus*) and (b) humpback whale with calf (*Megaptera novaeangliae*).

pelvis is short, vertically oriented and flared; our femur is extremely long; our heel is down-turned, and we possess a large, elongated big toe.

Many mammal species are extremely restricted in range and make extraordinary efforts to return to their birth area if removed from it. Studies on marked brown rats showed that most after recapture had moved no more than 40 feet from the place where they were originally captured. In larger species, the range is proportionately increased. The grizzly bear may wander across an area of 19 km (12 miles), whereas the movements of wolves often encompass as much as 50 kilometers (30 miles) in any one direction.

Figure 20.18 Representative Pinnipeds. (a) Weddell seal (*Leptonychotes weddellii*), and (b) walrus (*Odobenus rosmarus*).

True migration does occur among many types of mammals. On land, the large hoofed mammals show the most striking movements. For example, large herds of North America caribou (*Rangifer tarandus*) travel between 650 to 800 km (400 to 500 miles) on their seasonal journey, pressing onward despite all obstacles.

Sea and air provide fewer barriers to migration than land. As a result, migrations in these environments are particularly common. The movements of gray whales (*Eschrichtius robustus*) from the Arctic Ocean to sheltered lagoons

Figure 20.19 Bare membranous wings allow bats to do what no other mammal can do—fly.

along Baja California are well documented. The great blue whales make spectacular journeys as well. One tagged blue whale traveled 490 km (300 miles) in 32 days; another traveled 800 km (500 miles) in 88 days. Fur seals also travel great distances to return to breeding grounds.

Bats are also known for remarkable migrations considering their size and the distance traveled. For example, the tiny European pipistrelle (*Pipistrellus pipistrellus*) travels 1,000 to 3,000 km (600 to 800 miles) between southeastern Europe and central Russia.

The strangest migration is that practiced by the lemmings. Lemmings are small, largely nocturnal rodents (Cricetidae) that inhabit the plateaus and mountain slopes primarily of the Scandinavian Peninsula. Periodically, and with favorable conditions, their numbers increase dramatically for several years. Every three or four years, the carrying capacity of the habitat is exceeded, and a mass migration outward from population centers begins. The lemmings travel one by one at first, but natural barriers and topography funnel them into groups of ever-increasing size. Many perish crossing rivers or in the jaws and beaks of predators. Some march all the way to the sea and attempt to swim across what seems to be just another water obstacle. Many become exhausted and drown. These mass movements, apparently triggered by overcrowding, cause thousands to perish. However, not all lemmings are seized by this wanderlust, and the cycle begins anew. Contrary to widely held popular belief, lemmings do not commit mass suicide.

The urge to migrate in mammals is based on physiological stimuli. Glandular secretions and other physical transformations operating in a seasonal rhythm trigger and dictate the pattern of mammal migration.

Feeding and Digestion

In the broadest terms, mammals can be classified based on their food intake into four groups: insectivores, eating insects and small invertebrates; strictly herbivores, eating only plant material; strictly carnivores, eating only flesh; and omnivores, eating both plant material and flesh. More specifically, however, the feeding habits and exact diets of mammals are an interplay between age, season, health, and environmental conditions and circumstances. Hunger brings necessity, and few mammals are so specialized for one diet that they cannot make do in an emergency with food they normally would not accept.

The size of a mammal is also a factor in determining diet type. Since small mammals have a high ratio of heat losing surface area to heat generating volume and lose heat at a faster rate than do large mammals, they tend to have a high metabolic rate and great energy demands. Mammals that weigh less than about 500 gr (18 ounces) are mostly insectivores because they cannot tolerate the slow, complex digestive process of an herbivore. Driven by the fire of an unusually high metabolic rate, the voracious predatory shrews must eat 80-90 percent of their body weight a day to survive. If shrews were the size of cows, they would be the most ferocious predator on the planet.

Larger animals, on the other hand, generate more heat, but less of this heat is lost. Therefore, they can tolerate either a longer period between feeding (large carnivores) or a slower digestive process (herbivores). An African elephant weighing 6 tons must consume 135 to 150 kg (300 to 400 pounds) of rough plant fodder each day to obtain enough nourishment to sustain its life. Insectivorous mammals that weigh more than 500 gr (18 ounces) usually cannot collect enough insects during their waking hours to sustain themselves. The only large insectivorous mammals are the anteaters that feed on huge colonies of insects—ants or termites. The anteaters make do on an insect diet by having a comparatively low metabolic rate and are inactive much of the time.

Mammalian evolution during the Mesozoic resulted in major changes in teeth and jaws. The uniform homodont teeth of the first synapsids evolved into the differentiated heterodont teeth of modern mammals capable of a wide range of actions, including cutting, seizing, grinding, gnawing, tearing, and chewing. The heterodont teeth of mammals are differentiated into four types: **incisors** with sharp edges for snipping and biting; **canines** for piercing; **premolars** and **molars** for shearing, slicing, crushing, and grinding (**Figure 20.20**). A mammal's jaws and teeth,

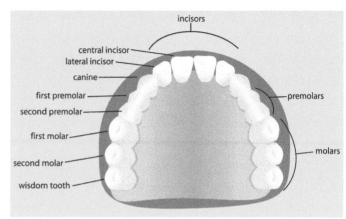

Figure 20.20 Human teeth demonstrate the highly specialized dentition characteristic of mammals.

tongue, and digestive system are all designed for and adapted to its individual feeding habits. In general, the more difficult the food is to digest, the more involved the digestive system, especially in length (**Figure 20.21**).

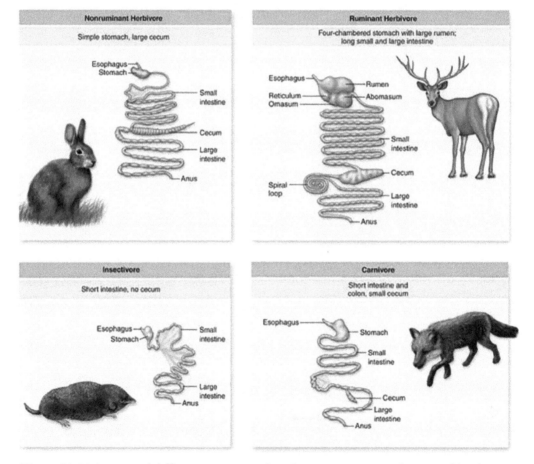

Figure 20.21 Anatomical differences in mammalian digestive systems are determined by the type of food eaten. Plants are difficult to digest, whereas flesh is relatively easy to digest. Therefore, the digestive systems of plant-eaters are longer and more complicated than those of flesh-eaters.

Insectivorous mammals, such as shrews, moles, anteaters, and most bats, tend to be small animals (Anteaters tend to be an exception as they grow to be relatively large animals) that eat insects, worms, grubs, and other small invertebrates. Shrews are one of the few types of venomous mammals. The toxins in their saliva can paralyze a small animal in seconds. Other than anteaters that have no teeth, the teeth of insectivorous mammals tend to be pointed, allowing them to pierce the exoskeleton of prey, and their digestive tracts are usually short in comparison to other mammals.

Most mammals are **herbivorous**. The plants they eat vary according to the adaptations of their anatomy, especially their teeth, and the availability of plants. Grass, shoots, and leaves are commonly utilized, but some mammals dine on nectar, fruit, bark, fungi, nuts, and seeds.

In a green world such as ours, a vegetarian existence might seem to be an easy lifestyle. However, plant-eating mammals must contend with the cellulose armor encasing plant cells, and battle a devilish array of spikes, thorns, and poisons that plants use to protect themselves. To counter these defenses, herbivorous mammals usually have long intestines and even pouched stomachs to provide extended exposure of ingested plant material to the rich flora of microorganisms in the gut necessary to fully digest such a chemically tough meal. Digesting plant material is so difficult that the **ruminants**, such as cattle, goats, sheep, camels, alpacas, llamas, giraffes, bison, yaks, water buffalo, wildebeest, antelope, and deer grind their food with broad, flat teeth, swallow it, and then later bring the food (known as **cud**) back up from the pouched stomach for even further chewing. Rabbits chew and eat some of their own fecal pellets passing them through the body for a second round of digestion. The largest group of herbivorous mammals is the rodents, followed by the **ungulates** or the hoofed mammals.

Carnivorous mammals eat mainly herbivores, but they must work harder for food than herbivores—green plants at least are stationary. Great demands are made on the wits, stealth, and precision of carnivores if they are to survive. In the ever-escalating evolutionary arms race between carnivore and herbivore, the herbivores match the intelligence and guile of the carnivores with keen senses of detection, speed, and agility. For some herbivores, their size alone allows them to survive against predatory carnivores (rhinos and elephants), for others, group defense mechanisms keep them relatively safe (muskoxen).

Most carnivores have biting and piercing teeth set in powerful jaws and their muscular limbs terminate in toes with sharp, deadly claws. Because the protein found in flesh is more easily digested than cellulose, the digestive tract of carnivores is relatively short and absent of pouches (Figure 20.21). Unlike herbivores that feed nearly continuously, carnivores feed at irregular intervals and often eat as much as possible when they do feed. Lions, for example, gorge themselves with their kill, then sleep and rest for periods as long as one week before eating again.

> *There was a young lady from Niger*
> *Who smiled as she rode on a tiger.*
> *They returned from the ride*
> *With the lady inside,*
> *And the smile on the face of the tiger.*
>
> —Anonymous

The largest groups of carnivores are the cats (Felidae or felines), such as the cat, lion, tiger, cougars, lynx, cheetah, and jaguar and the dogs (Canidae or canines), such as dogs, wolves, coyotes, and foxes. Carnivores hunt singly, in pairs, or in packs and depend on speed, cunning, and strength to bring down prey and sharp teeth and powerful jaws to tear it into chunks for swallowing (**Figure 20.22**).

Omnivorous mammals are the most generalized and opportunistic feeders of the group eating both plant material and other animals for food. Pigs, raccoons, bears, foxes, many rodents, badgers, skunks, coyotes, and most primates, including humans are accomplished omnivores. In the world of omnivores, however, the rat reigns supreme. Be it organic or inorganic, rats will eat or attempt to eat practically anything, including their own kind. They will gnaw paper, cloth, wood, plastics, water pipes, electric cables, and building materials. Rodents gnaw because their incisors grow continuously and must constantly be worn down. This incessant gnawing of inorganic materials causes immense damage to humankind in the form of power outages and fires.

Figure 20.22 The lion (*Panthera leo*) has a social hierarchy that extends to its feeding behavior and hunting practices. In open area, lionesses hunt in coordinated packs while males watch. In wooded areas, both genders hunt equally, but often keep their kills separate.

Respiration

In mammals, the lungs are large and well developed but lack the accessory air sacs found in birds. Mammalian lungs have a spongy texture because they are filled with innumerable air pockets called **alveoli**. In humans, the lungs contain about 300 million alveoli, which provide a total respiratory surface of about 70 square meters (6,000 feet), or roughly the area of a regulation tennis court.

Mammalian lungs, like those of reptiles and birds, inflate using a negative-pressure mechanism. Unlike reptiles and birds, however, mammals possess a muscular **diaphragm** that separates the thoracic (chest) cavity from the abdominal cavity. Contractions of the diaphragm and rib muscles increase the size of the lung but reduce the pressure allowing air to be pushed in from the outside (inhalation). Relaxation of the diaphragm and rib muscles decreases the size of the lung but increases the pressure forcing air to be pushed out (exhalation). The mammalian lung is known as a **bellows lung** as it resembles a blacksmith's bellows.

The respiratory rate varies according to the size of the animal, its age, amount of work being done, and other factors. Respiratory rate, expressed in cubic centimeters of air per kilogram of weight per hour, is 200 for a resting human, 460-850 for a rabbit, and over 5,000 for one species of shrew. The rate for a hibernating dormouse is 15.

Sitting on top of the trachea (windpipe) is the voice box or **larynx**. It is made up of several separate cartilages with membranous folds—the **vocal cords**—lying within it. Air from the lungs passes over the vocal cords and causes them to vibrate. The volume of air expelled from the lungs, the tautness of the cords, the shape of the pharynx, and the position of the mouth and lips modulate the voice. Mammals' voices range from the eerie howl of the wolf to the soft squeak of mice and from the braying of the donkey to the soft lullaby sung by a human mother to her child. In dogs the voice box engages with the nasal passages; to bark, a dog must raise its head to disengage the voice box. Hence, if a dog is placed in a kennel with a low roof so that it cannot fully raise its head, it will be prevented from barking. Whale vocalizations ("songs") travel the greatest distance of any mammalian sound. The low-frequency song of male humpback whales can be heard 16,093 km (10,000 miles) away! The spoken languages and songs of humans are probably the most complex sounds made by mammals. The sounds and calls of mammals serve to (1) warn of danger, (2) intimidate or frighten enemies, (3) assemble members of gregarious species, (4) attract mates, and (5) locate and coordinate parents and young.

Excretion

Mammals, like all amniotes, have a metanephric kidney. Unlike reptiles and birds that excrete mainly uric acid in a semisolid form, mammals excrete urea in a liquid form. Urea is less toxic than ammonia and does not require large quantities of water in its excretion. Unlike uric acid, however, urea is highly water soluble and cannot be excreted in a semisolid form.

The primary adaptation of the mammalian nephron is a portion of the tubule system called the **loop of the nephron** (**Figure 20.23**). This long loop and the remainder of the tubule system allow mammals to produce urine that is anywhere from twice as concentrated as blood (beavers) to 22 times more concentrated (Australian hopping mice). Highly concentrated urine accomplishes the same function as salt glands do in reptiles and birds.

The amount of water lost depends on activity level, physiological state, age, and environmental temperature. Mammals in very dry environments have metabolic and behavioral adaptations to reduce water loss. The Egyptian gerbil (*Gerbillus pyramidum*) and the kangaroo rat (*Dipodomys*) of the southwestern

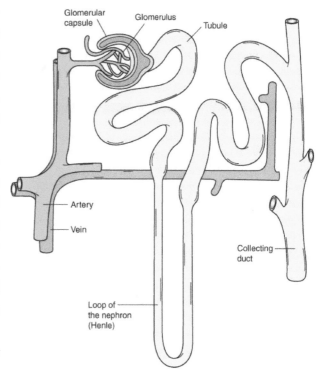

Figure 20.23 The long loop of the nephron (Henle's loop) of a desert animal conserves water and prevents dehydration. The excretory system of some desert rodents is so efficient that they live their entire life without ever having taken a single drop of liquid water.

United States can subsist on seeds not only as a source of food but also as a source of water. Drinking very little if any water, these animals prosper on dry seeds with less than 10 percent water content. The nearly dry seeds are a rich source of carbohydrates and fats, and the metabolic oxidation of these carbohydrates produces water as a by-product. Their feces of these mammals are a dry pellet, and their nocturnal habits reduce evaporative water loss furthering conserving water.

Circulation and Body Temperature

Mammals, like birds, have a four-chambered heart and a double-loop circulatory system that keeps the pulmonary (lung) and systemic (body) circulations separate and independent. Although both the hearts of birds and mammals evolved from the hearts of ancient reptiles, the mammalian heart evolved from the synapsid reptilian lineage whereas the avian heart evolved from the archosaur lineage (**Figure 20.24**).

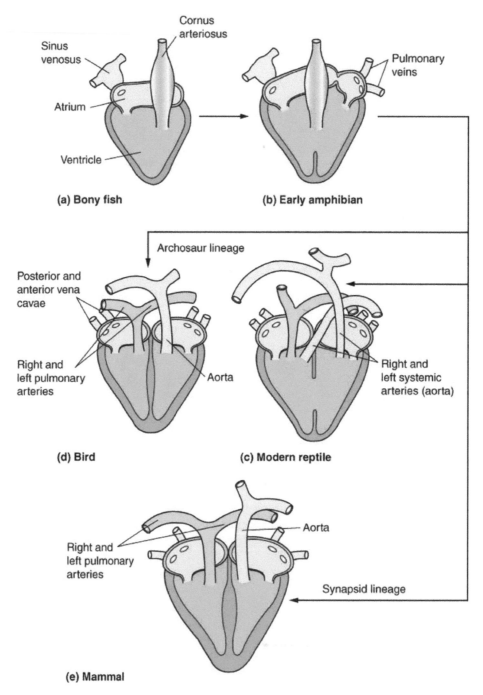

Figure 20.24 Possible Sequence in the Evolution of the Vertebrate Heart. The four-chambered heart of birds and mammals most likely arose independently.

Heart rates vary greatly in mammals but are a function of size with small mammals having a higher resting heart rate than larger mammals. Shrews have a heart rate of 800 beats per minute; a mink 272 bpm; humans 70 bpm; horses 50 bpm; elephants 35 bpm, and whales 7 bpm. Diving mammals change their heart rate whilst diving; seals have a surface rate of 120 bpm but reduce this on the way down to about 10 bpm while diving. Hibernating mammals can have greatly reduced heart rates. Hedgehogs have a normal heart rate of between 200 and 280 bpm; this decreases to 147 bpm during sleep but drops to around 5 bpm during hibernation. Bats also reduce their heart rates when in torpor during winter in temperate climates. The heart rate of a medium-sized bat can go from 700 bpm at rest to 1000 bpm while flying to only 25 bpm while overwintering.

Like birds, mammals are endothermic animals. The body temperature of mammals varies from 30° C (86° F) in the platypus to 37° C (98.6° F) in humans to 39° C (102.2° F) in rabbits and cats. Endothermy is a costly metabolic proposition, and some mammals minimize the cost by not maintaining a constant temperature. Many bats, for example, allow their temperatures to fall when at rest. They get so cold that when they awaken, they must move and jerk their wings to raise their temperature enough to fly.

Being widely distributed over the earth, mammals face some of the harshest environmental temperature conditions of any vertebrates. Nearly all face temperatures that require them to dissipate excess heat at some times and to conserve and generate heat at other times. Heat is generated (**thermogenesis**) in mammals in two ways: (1) metabolically through the cellular metabolism of special fat deposits called **brown fat** and (2) through the muscular activity of shivering. Once heat is generated, it is held in the body by the insulative properties of the pelage and/or fat deposits under the skin. A deer is so well insulated that it can spend the night sleeping on snow, and when the deer arises in the morning, none of the snow beneath it will have melted. Mammals without a pelage can conserve heat by allowing the temperature of surface tissues to drop. A walrus in cold, arctic waters has a surface skin temperature of near 0° C (32° F); however, a few centimeters below the skin surface, body temperatures are about 35° C (95° F). Upon emerging from the icy water, the walrus quickly warms its skin by increasing the blood flow to peripheral vessels.

Mammals adapted to cold climes also employ countercurrent heat exchange systems in their limbs. In such a system arteries passing through the core of an appendage are surrounded by veins carrying blood back toward the body from the appendage. Heat transfers from the arterial blood on the way out to the venous blood headed back in resulting in heat being returned to the core of the body instead of lost to the environment. During winter the lower part of a reindeer's leg may be at 10° C (50° F), while the core body temperature is about 40° C (104° F) due to its countercurrent heat exchange system.

Getting rid of excess heat presents few problems in cool, moist environments as heat can be radiated into the air from the evaporative cooling of sweat off the skin or respiratory surfaces by panting. Hot, dry environments present a greater challenge because evaporative cooling may upset water balance. Elephants and jackrabbits use their large ears to radiate heat without water loss. Small mammals may be active only during the cooler temperatures at night, and others seek shade or water holes during the heat of day.

Box 20.1
How Warm is Your Form? Thermoregulation in Endotherms

Endothermy is a derived characteristic of mammals and birds, and even though the two lineages evolved endothermy independently, the costs and benefits are essentially the same for both groups. Endothermy is a superb method of becoming relatively independent of many of the challenges of the physical environment, and it allows animals possessing the characteristic to sustain activity at high levels for prolonged periods of time. It is, however, extremely expensive metabolically and demands a reliable supply of sufficient food.

Mammals and birds maintain a constant high body temperature by adjusting heat production to equal heat loss from their bodies under ever-changing and often harsh even brutal environmental conditions. If asked to picture a mammal facing severe environmental conditions, most people would likely visualize a furry creature battling subzero temperatures in the face of howling winds and driving snow. Such a scenario, however, would be only part of the endothermal temperature regulation saga for deserts present challenges to mammalian physiology and homeostasis that are just as great, if not greater, than those posed by extreme cold.

In the coldest places on earth, the polar regions, it is always cold all of the time. It is just slightly less cold during the day than night and in some seasons than others. In desert areas, however, as exposed sand and rock heat quickly it can become bake-oven hot during the day but near freezing at night as the extreme heat of the day is quickly lost without soil and plants to hold it in. Desert air temperatures can climb to 40 or 50° C (104 to 122° F) during summer, and the ground temperature may exceed 60 to 70° C (140 to 158° F). An endotherm in these conditions is continually absorbing heat and that heat plus metabolic heat must somehow be dissipated to maintain the body temperature in the normal range. In actuality, maintaining a body temperature 10° C below the ambient temperature can be a greater challenge for an endotherm than maintaining it 100° C above ambient.

What's an endotherm to do? When faced with extreme cold mammals can either migrate to more favorable conditions or fall into a state of hibernation or torpor, which lowers their body temperature. Some desert endotherms employ torpor during times of water or food shortages, others relax the limits of homeostasis in order to tolerate greater-than-normal ranges in body temperature or body-water content, and some employ behavioral strategies such as going underground during the day and becoming active only in the cool of the night. During summer, daytime ground temperatures in the Sonoran Desert approach 70° C (158° F), and even a few minutes of exposure would be deadly. Kangaroo rats (*Dipodomys*) spend the day in burrows 1 to 1.5 m (3.3 to 5 feet) underground, where air temperatures do not exceed 35° C (95° F). In the evening, when the external air temperatures have fallen to about 35° F the kangaroo rats emerge to forage.

Not all desert rodents, however, are nocturnal. The antelope ground squirrel (*Ammospermophilus leucurus*) with its tail held back over its body like a parasol can be seen frantically running across the desert surface even in the middle of the day. High temperatures limit the time that squirrels can be active to no more than 9 to 13 minutes, which they spend in almost frenetic activity. Once their bodies heat to a certain point, they must pop into a burrow deeper than 60 cm (2 feet) where the soil temperature is 30 to 32° C (86 to 90° F). Once passive cooling (the squirrels neither sweat nor pant) brings their body temperature down sufficiently, off they go again. Consider that the next time you turn down the air conditioning thermostat on a hot, humid day.

Hibernation is a period of inactivity usually occurring in the winter in which the metabolic, heart, and respiratory rates slow. True hibernators include the echidnas, the duck-billed platypus, moles, shrews, chipmunks, ground squirrels, woodchucks, and bats. The heartbeat of the ground squirrel, normally about 150 bpm, slows to 5 bpm. The animal's respiration also drops from about 200 breaths per minute to 4 or 5 breaths per minute, and its body temperature may plummet as much as 30 degrees. In fact, the temperature of some hibernating mammals may drop to within several degrees of freezing. Excess body fat is accumulated prior to hibernation and serves as bodily fuel through the period of hibernation, resulting in some mammals losing as much as a third to a half of their body weight during this period. Bears and raccoons do not truly hibernate but instead undergo a sort of stupor called **winter sleep**. Their body temperatures and metabolic rates decrease somewhat, but they do not necessarily remain inactive all winter.

Nervous System and Sense Organs

The basic vertebrate nervous system reaches its greatest refinement in the mammals, and nowhere is this more evident than in the mammalian brain. Mammals have the largest brain relative to body size of any vertebrates. This large brain and accompanying well-developed sense organs give the mammals an intelligence, curiosity, and vitality unmatched by any other animal.

> *Cats are smarter than dogs. You can't get eight cats to pull a sled through snow.*
>
> —Jeff Valdez

The mammalian brain consists of three parts: cerebrum, cerebellum, and medulla. The large size and complexity of the **cerebrum** distinguishes the brain of a mammal from that of other vertebrates. The massive, wrinkled cerebrum is the site of thinking, learning, memory, and interpretation of senses. The smaller **cerebellum** coordinates muscular movements and the **medulla** regulates autonomic body functions such as heart rate and respiratory rate. The size of the brain relative to total body weight is not a reliable guide to mammalian intelligence. The degree of folding and convolution on the surface of the cerebrum is perhaps a better indicator.

Mammals are equipped with an impressive array of senses related in degree of development to their necessity for survival in specific situations and habitats. For example, the carnivores, rodents, and hoofed mammals have an acute sense of smell; this sense is reduced in the whales and primates and absent in the porpoises and dolphins.

No other vertebrate seems to depend so heavily on the sense of hearing as do mammals. Mammals alone have an external funnel-like **pinna** that serves to locate and direct sound waves. The pinna is especially large and well-developed in bats (**Figure 20.25**), which depend on echolocation for flight and food-getting. Bats are capable of generating and hearing ultrasonic frequencies. Dogs, cats, and small rodents can hear high frequencies beyond those the human ear can detect, and elephants and whales can hear sounds of very low frequencies, often over great distances.

Figure 20.25 The over-sized ears of a bat are the biological equivalent of a radar dish.

The eyes of most mammals are large, well-developed, and quite sensitive to movement. Many species preyed upon by others react to motion alarms by freezing, but the opposite applies to predators; the slightest movement in their visual field attracts their attention. The retina of most mammals contains few of no cone cells for detecting color, and consequently most are color-blind. Color-blindness may be correlated with the fact that most mammals are drab in color. Only the squirrels, the primates, and a few other mammals enjoy color vision.

All mammals have two eyes, but their development in the skull varies widely. In hoofed mammals, which are always on the lookout for predators, the eyes lie on the side of the head so that a very wide visual field is produced. These animals can see from straight ahead to far behind without turning the head. In carnivores, the eyes lie more on the front of the face, giving a more extensive stereoscopic view, and great sense of depth and distance. The primates, with flat faces and eyes set relatively close together, have developed stereoscopic vision to its highest level.

The sense of smell is highly developed in some mammals. The hoofed mammals can detect the odor of an approaching predator over long distances forcing carnivores to work to stay constantly downwind of their intended victims. Dogs are also noted for their great sense of smell. This remarkable canine ability has been put to use by modern law enforcement to detect illegal contraband of all types.

> 'Smelling isn't everything,' said the elephant. 'Why?' said the bulldog. 'If a fellow can't trust his nose, what is he to trust?' 'Well, his brains perhaps,' she replied mildly.
>
> —C. S. Lewis

In mammals, the sense of touch is well developed. Receptors are associated with the bases of hair follicles and are stimulated when a hair is displaced.

The most striking manifestation of the advanced brain of mammals is the wide and complex variety of behaviors they exhibit. No other creatures have such varied and complicated behaviors of food-getting, defense, reproduction, and social interaction. No other animals band together in such intimate and intricate social hierarchies as do mammals.

Reproduction and Development

For most mammals there are well established seasons in which to mate, often during the spring or in wintertime. These mating seasons are almost always aligned with the time of year it is best to rear young after delivery. At these times internal and behavioral changes may occur in both sexes. The primary stimulus to mating in most mammals is the condition of the female, who at certain times enters a state known as **estrus**, or heat. In most mammals, estrus occurs once only during certain breeding seasons (**monoestrus**). In others, such as humans and other primates, bats, cats, whales and squirrels, **estrous cycles** occur at regular intervals throughout the year (**polyestrous**).

Like birds, many mammals possess secondary sex characteristics (sexual dimorphism) that are often of great importance in mating behavior. The mane of the male lion, the shoulder cape of some male baboons, the antlers of male deer, and the inflatable nose pouches of the male elephant seal are all important traits distinguishing the sexes. Colored faces, coats, and rumps, as well as size, also play a role. A bull fur seal may weigh 270 kg (600 pounds) and its mate only 34 kg (75 pounds). Male weasels and minks may weigh twice as much as their mates.

Courtship displays in mammals generally are less spectacular than in birds but may be quite elaborate. Various rodents engage in rough-and-tumble fights, the male short-tailed shrew emits excited clicks as he approaches his potential mate, some sea lions engage in long caressing sessions on shore, and platypus pairs swim in a tight circle beak to tail.

Humans imagine prolonged monogamous relationships between mammals, but such pairings are extremely rare. Mating in many species, especially rodents, and bats, is promiscuous and indiscriminate. Other types, such as seals and deer, practice polygamy, with male gathering and serving harems of females. In some carnivores, the male remains with the female and helps gather food until the young are **weaned**. Foxes and wolves tend to remain faithful to one mate, and a high degree of fidelity is shown by humans and other primates and the American beaver.

The climax of courtship is copulation. This is accomplished in different ways by different species. Sometimes the period of union is very long. Ferrets may remain locked in a single union for over an hour and the rhinoceros for several hours. Alternatively, many small rodents will copulate scores of times a day, each act taking only a few seconds. Shaw's jird (*Meriones shawi*), a small North African desert rodent, has been observed copulating 224 times in two hours, nearly twice a minute. Based on where the fertilized egg develops, the mammals may be divided into three groups developmentally:

Monotremes. In the platypus and echidnas (**Figure 20.26**), a thin, leathery shell is deposited around the eggs, which pass out of the female's body (oviparity). The platypus (*Ornithorhynchus anatinus*) lays two or three eggs, each about 1.25 cm (0.5 inch) in diameter, in an underground nest about two weeks after mating. Curling about the eggs, the female incubates them for about 10 to 12 days until they hatch, still somewhat under-developed, blind, and hairless. Once hatched, the young lap milk that oozes from the mammary glands onto the fur on the female's underside (monotreme females have no nipples). The young are suckled for three or four months with the female leaving the burrow only for short periods to forage. At around four months, the young leave the burrow.

The female echidna (four species in the family Tachyglossidae) lays a single, soft-shelled leathery egg 22 days after mating and deposits it directly into a pouch on her body. Hatching takes 10 days; the young echidna, called a **puggle**, then sucks milk from the pores of two milk patches and remains in the pouch for 45 to 55 days at which time the spines begin to develop. The mother digs a nursery burrow and deposits the

Figure 20.26 Representative Monotremes. (a) Platypus (*Ornithorhynchus anatinus*) and (b) Echidna (Tachyglossidae).

puggle, returning every five days to suckle it until it is weaned at seven months.

Marsupials. The marsupials—opossum, kangaroo, wallaby, koala, bandicoot, and wombat—are pouched viviparous mammals (**Figure 20.27**). The fertilized egg develops into an embryo (**blastocyst**) inside the reproductive tract of the mother lying in shallow grooves in the uterine wall and absorbing nutrients from the uterine mucosa by way of a vascularized yolk sac. **Gestation** (the period of intrauterine development) is brief in marsupials ranging from 12 to 14 days in the American opossum to 38 to 40 days in the largest kangaroos. The undeveloped young (**neonatus**) are born (viviparity) onto the fur of the mother's underside while still embryos, both anatomically and physiologically. They then begin a long arduous and dangerous journey from the area of the vaginal opening (vent) up and into the mother's pouch. The mother may lick the fur to mat it down in front of the advancing young, but there is little else she can do to help. Once the safety of the pouch is reached, the embryo searches for a nipple, swimming with stubby forelimbs and head through a tangle of fur. If and when the embryo finds a nipple, it seizes it, bites down and hangs on (**Figure 20.28**). If a nipple cannot be located, the embryo dies. The embryonic young marsupial remains attached to the nipple, completing its development in the pouch until it becomes sufficiently large and independent enough to leave the pouch. In kangaroos, the young will return to the pouch for some time after their "second birth" for protection and will continue to nurse from outside the pouch for a period of time.

Marsupials are among mammals in seven different orders that employ **embryonic diapause**, a strategy in which the embryo (blastocyst) does not immediately implant in the uterine wall but is maintained in a state of dormancy. Kangaroos have a complicated reproductive pattern in which the mother may have three young in different stages of development depending on her at the same time (**Figure 20.29**).

Placentals. Monotremes and marsupials account for only 5 % of the mammals. In the other 95 percent, the placentals, the developing embryo attaches to and develops a link with the mother. Nutrients, oxygen, carbon dioxide, and water are exchanged between the embryo as it develops and the mother through a network of intertwined blood vessels called the **placenta**. The placenta allows the embryo to develop for a much longer time inside the mother and frees the mother to move about and feed while still protecting the embryo. Gestation (develop-

Figure 20.27 Representative Marsupials. (a) Kangaroo (Macropodidae) and (b) Opossum (Didelphimorphia).

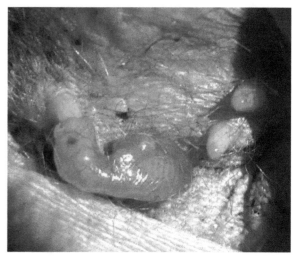

Figure 20.28 Once a nipple is found, the marsupial embryo clamps down and holds on literally for dear life.

ment) is prolonged in placentals whereas in marsupials it is lactation (feeding) that is prolonged. The gestation period for placentals ranges from 21 days in mice, to 30 to 36 days in rabbits and hares, to 60 days in cats and dogs, to 280 days in humans to 22 months (665 days) in elephants, the longest gestation period.

Some mammals, such as most rodents, carnivores, and primates, give birth to young that are naked and helpless. Others, such as the hoofed mammals, give birth to active, fully furred young able to feed, run, and hide shortly after birth. Bear cubs, for example, born blind and helpless, weigh only 280 grams (10 ounces). A baby giraffe, on the other hand, may weigh 39 kg (85 pounds), stand 1.8 m (6 feet) tall, and be able to stand alone in few minutes after birth and run effectively several days after birth. Rodents give birth to blind, naked, helpless young but compensate for this initial disadvantage by speedy maturation. Young field mice may be weaned before they are three weeks old.

The potential reproductive capacity (fecundity) of mammals varies greatly. Usually the larger the animal, the fewer young produced. Large mammals, such as elephants

Figure 20.29 Kangaroos have a complicated reproductive pattern in which the mother may have three young in different stages of development. She may have (1) an embryo in diapause in her uterus; (2) a Joey on teat in her pouch arresting development of the embryo in the uterus; and (3) a juvenile kangaroo returning to drink from her pouch.

and horses, give birth to a single young with each pregnancy. An elephant can be expected to produce, on average, four calves over her 50 years of fertility. Small rodents, the food for many a predator, are known to parent at least one litter of numerous offspring every season. For example, meadow voles (*Microtus pennsylvanicus*) give birth to up to 17 litters of four to nine baby meadow voles every year. The amazing fecundity of small rodents and our ignorance at removing the predators that hold them in check is expressed in this excerpt from Thornton Burgess's "Portrait of a Meadow Mouse."

He's fecund to the nth degree
In fact this really seems to be
His one and only honest claim
To anything approaching fame.
In just twelve months, should all survive,
A million mice would be alive—
His progeny. And this, 'tis clear,
Is quite a record for a year.
Quite unsuspected, night and day
They eat the grass that would be hay.
On any meadow, in a year,
The loss is several tons, I fear,
Yet man, with prejudice for guide
The checks that nature does provide
Destroys. The meadow mouse survives
And on stupidity he thrives.

Once the young of any mammal are born, they are nourished by milk from the mother's mammary glands until they can feed and hunt for themselves. The exact components of mother's milk vary from species to species but all contain significant amounts of fat, protein, and calcium, as well as vitamin C. The milk of seals may be the richest containing 50 percent fat.

Mammals are noted for the length and quality of parental care devoted to their young. Quality lavished on a few offspring has replaced the quantity of offspring as the key to survival in mammals. Carnivores often spend months teaching their young the intricacies of hunting and social order. Young elephants may remain with their mother for several years after they are born and mother and child continue to recognize each other for the rest of their lives. Humans are the most caring of mammalian parents (or the young take longer to train). Even in the most primal societies, young humans do not become truly independent until at least 12 years old, and in more advanced societies with protracted periods of education, youngsters may not be emancipated until the age of 21 or older. Even after that and for the remainder of their lives, there is a bond of caring, concern, and support between human parent and child.

Just as mammals vary greatly about most aspects of their lives, so they vary when it comes to the length of their lives. As far as we can determine, the records (in years) appear to be:

- shrew: 1
- mice: 4
- dogs: 29
- cats: 36
- horses: 62
- chimpanzees: 75
- elephants: 78
- humans: 122.4 (set by Jeanne Clament, a French woman who was born in 1875 and died in 1997)
- Galapagos tortoise: 190
- Bowhead whale: 211

The prevailing wisdom has always been that, in general, small mammals "live fast and die young" whereas large mammals live longer, but it depends on which clock you use. This generalization holds true when using solar time (the way we normally mark minutes, days, years) but perhaps not if we use metabolic time. Elephants live over 70 years but mice, at best, last only four years. Yet the elephant lives a slightly shorter metabolic life than does the mouse. Indeed the elephant gets slightly less than 1 billion heartbeats per lifetime and the mouse slightly more.

Metabolism is the key. Apparently, a body has only so many heartbeats and so many breaths. A larger animal with a lower metabolism, breaths more slowly and its heart beats slower. It reaches sexual maturity later and lives longer—by solar time. However, when researchers measured the total amount of energy animals of different sizes require and then calculated the amount of energy each gram of tissue in those animals burn in their lifetime, they found a startling result. Tissues in smaller animals use more energy than tissues in larger animals. So, equating "living" with total energy used, a gram of mouse tissue lives more (during its short solar-timed lifespan) than a gram of elephant tissue. (Be that as it may, I think it best to continue to set our watches by the sun and hope metabolically for the best.)

Mammal Connection

Economically

No other group of animals is so much a part of human history or as intertwined with our daily lives and needs as are our fellow mammals. Humans have hunted, trapped, domesticated, and in general exploited mammals for their leather and fur for clothing, and their blubber, milk, and flesh for food the entirety of human existence. Dogs, donkeys, horses, oxen, camels, and elephants have carried our burdens and plowed our fields and continue to do so in many parts of the world today. Others have been trained to race, hunt, stand guard, and guide the sight impaired, and yet others provide many useful commercial products such as leather, wool, and lanolin.

Trapping and hunting of wild mammals is deeply intertwined in human history so that today trapping and hunting for sport and profit (meat, fur, and leather) is a multibillion-dollar enterprise. In the United States alone, for example, it is estimated that more than 2 million deer are harvested annually by licensed hunters. On the other hand, ranch-raised mammals such as the mink, fox, and chinchilla, are also important to the fur industry, which directly or indirectly accounts for many millions of dollars in revenue each year in North America alone.

Some mammals are directly detrimental to the economic well-being of humans. Rats and mice of Old World origin now occur virtually throughout the world, and each year cause substantial damage and economic loss mainly to agricultural products, especially cereal grains. Herbivorous mammals may eat or trample crops and compete with livestock for food, and native carnivores sometimes prey on domestic herds. The reintroduction of wolves in Yellowstone National Park in the winter of 1995-96 was highly controversial and stiffly challenged by ranchers around the park who worried that wolves from the park would prey on their cattle and sheep. Although some predation on livestock does happen, the National Fish and Wildlife Service estimates wolves kill, on the average, about four or five head a year of cattle and/or sheep. This is a minute portion of the tens of thousands of free-ranging livestock that die of natural causes each year. Ranchers are allowed to kill only wolves that attack livestock. The attitude of "the only good predator (wolf, bear, coyote, etc.) is a dead predator" that prevailed throughout the history of exploration and settlement of this continent still exists. Large sums of money are spent annually to control populations of "undesirable" wild mammals, a practice long deplored by ecologists and conservationists because of its negative impact on habitats and the fact that trapping, poisoning, and hunting select animals as a means of wildlife control or management simply does not work, never has, and never will.

Ecologically

> *We have enslaved the rest of animal creation and have treated our distant cousins in fur and feathers so badly that beyond doubt, if they were to formulate a religion, they would depict the Devil in human form.*
>
> —W. R. Ingles

Because of their cleverness, large size, and large numbers, mammals affect the environment and ecology of their habitats like no other animal group. Burrowing moles and ground squirrels alter soil structure and improve soil aeration. Beavers dam streams and create ponds that serve as habitat for many other creatures. Grazing mammals disperse seeds far and wide, and fruit-eating bats help pollinate many tropical flowers. Whales and walruses stir up the ocean floor in their search for food, adding productive material to sea water. However, no mammal has made the widespread, monumental, and permanent changes to the planet that the bare, upright human mammal has made.

In many habitats, mammals are important links in the food web as predator or prey. In fact, large apex mammals may be the key to maintaining the balance of an ecosystem. Case in point—when wolves were reintroduced into Yellowstone National Park data began to accumulate showing a clear and remarkable linkage between the presence of wolves and the health of an entire stream side ecosystem. With the removal of wolves from the park, nearly 70 years of grazing by elk had devastated stream-side cottonwoods, willows, and berry-laden shrubs. This in turn began to play havoc with an entire streamside ecosystem (the cottonwoods in some areas were nearly extinct) and associated wildlife, including birds, insects, fish, and others. Food webs had broken down, and erosion was becoming a problem in some areas. The successful reintroduction of wolves into the park led to fewer elk that in turn seems to have begun the process of returning balance to at least some areas of stream-side ecosystem. Furthermore, wolves have boosted biodiversity in and around Yellowstone in that there are fewer elk and coyotes but more eagles, pronghorns, foxes, and wolverines

Many domestic and some wild mammals thrive in association with humankind. Most, however, have not fared well under the unrelenting human pressure to dominate the planet. Many large mammals have been exterminated entirely while others exist today only in parks and zoos; others are in danger of extinction. One of the most noteworthy cases is the Stellar sea cow (*Hydrodamalis gigas*). These inoffensive marine mammals up to 10 m (33 feet) long evidently lived only along the coasts and shallow bays of the Komandor Islands in the Bering Sea. Discovered in 1741, the slow-moving and easily captured sea cow were easily killed by sealers and traders for food. The last known individual was taken in 1768, only 27 years after their first discovery by Europeans.

Endangered mammals come in all sizes from shrews to whales, and nearly all orders of the class have some members that are listed as endangered species. The World Conservation Union estimates that at least 25 % of all mammal species face a high risk of extinction. In nearly every case, the mammal in question has become endangered primarily as the result of loss of habitat. As human population swells, our farms, roads, factories, homes and cities push ever deeper into the wild leaving less and less for our mammalian kin. Our fellow primates seem to be particularly hard hit. Most nonhuman primates are forest dwellers. Tragically, it has been estimated that the tropical rainforest is being destroyed at a rate of 1 hectare (2.5 acres) per second—an area the size of the state of Colorado every year at least. As a consequence, the populations of forest primates are dwindling fast, and many species soon will become extinct or found only in isolated preserves. Reports over the last several years by world conservation and primatological groups suggest that 25% of the 625 species and subspecies of primates are at risk of extinction. The golden-headed langur of Vietnam (*Trachypithecus poliocephalus*)and China's Hainan gibbon (*Nomascus nasutus hainanus*) number only in the dozens. The Horton Plains slender loris of Sri Lanka (*Loris tardigradus nycticeboides*) has been sighted just four times since 1937. Perrier's sifaka of Madagascar (*Propithecus diadema perrieri*) and the Tana River red colobus of Kenya (*Procolobus rufomitratus*) are now restricted to tiny patches of tropical forest.

Miss Waldron's red colobus of Ghana and the Ivory Coast (*Piliocolobus badius waldronae*) and Bouvier's red colobus of the Republic of Congo (*Piliocolobus pennantii bouvieri*) are most likely already extinct in that it has been decades since a live sighting of either species has occurred.

Primates are under assault from all sides. Hunters kill and butcher primates to sell the meat (bushmeat), and for organs and body parts to be sold as components of folk medicines. Trappers capture them for live sale, mainly as pets, and loggers, farmers, and ranchers cut down and burn off their habitat.

Medically and Scientifically

Our relationship with mammals medically speaking is a two-edged sword that cuts both for good and bad as mammals contribute to both human suffering and in improving human health. Harboring the fleas that carried the bacteria of bubonic plague, rats contributed to the devastating plague known as Black Death that killed as many as half the people in medieval Europe. Some mammals continue to serve as important reservoirs or agents of transmission for a variety of disease that infect humans, such as plague, tularemia, yellow fever, rabies, leptospirosis, Lyme disease, hemorrhagic fevers such as Ebola, and Rocky Mountain spotted fever. The annual economic cost resulting from mammal-borne diseases that affect humans and domestic animals is incalculable. Countless, monkeys, chimps, dogs, rats, and rabbits, on the other hand, have lost their lives in the development of medicines, vaccines, and surgical techniques that have relieved human suffering and saved untold lives.

Today, domesticated strains of the house mouse, European rabbit, guinea pig, hamster, gerbil, and other species provide much-needed subjects for the study of human-related physiology, psychology, and a variety of diseases from dental caries (cavities) to cancer. The study of nonhuman primates (monkeys and apes) has opened broad new areas of research relevant to human understanding and welfare. The quills of porcupines may even provide hypodermic needles that are easier to insert and timed to degrade to avoid tearing tissues on removal, useful in applications such as immunizing livestock animals and performing battlefield medicine.

Many people keep mammalian companions ranging from dogs and cats to horses and ferrets. Such relationships impart a sense of purpose and feelings of affection, joy, and companionship for all who touch the lives of those wonderful animals we call pets, especially the elderly. Pets and companion animals can actually improve the medical condition of the ill in some cases. A Veteran's Administration hospital in the authors' locale has a live-in dog on the floor where permanent residents are housed. The nurses there would be quick to tell you that that dog greatly improves the mental health and outlook of those around him and improves the quality of daily life for all those that live and work in such an institutionalized setting.

Culturally and Historically

The quest for marine mammals was responsible for the charting of areas in both the Arctic and Antarctic regions, and the presence of furbearers, particularly beavers and the carnivorous martens, and fishers, was one of the principal motivations for the opening of the American west, Alaska, and the Siberian taiga.

Big cats are not only feared, but they have also been revered and symbolized throughout the course of history. For millennia, jaguars have served as potent cultural and religious icons for many indigenous

American people from the Mayans and Aztecs to the Guarani Indians of the Gran Chaco. The Maya believed the jaguar's skin symbolized the night sky, and the Aztecs fed the hearts of sacrificial victims to the big cats. Among Amazonian societies, the jaguar with its large reflective eyes was thought to connect to the spirit world. The lion has come to symbolize both religious and stately power and leadership. Many ancient cultures associated the lion with various deities and his likeness decorated Solomon's throne as well as the thrones of the King of France and medieval bishops.

The elephant was valued for not only its military utility but for its symbolic value as well. These great behemoths graced religious processions for the same reason that they accompanied armies, their majestic and awesome presence. Towering over the field of combat equipped and girded for war, they must have been an awesome sight and a foot soldier's worst nightmare.

The elephant has been a cultural icon of Thailand since ancient days. Their great strength was harnessed in many ways and they quickly became man's ally in labor. But the elephant came to symbolize much more to the Thai people than a mere beast of burden. It has become a symbol of fortune, and the superstitious will pay to walk beneath the animal's body in hopes of receiving some of the good fortune that it carries. Rare white elephants, the exclusive property of the reigning monarch, were considered so sacred that they were placed on the flag and coins of Siam (formerly Thailand) and their images can still be found in the compounds of many older temples.

I like pigs. Dogs look up to us. Cats look down on us, but pigs treat us as equals.
—Sir Winston Churchill

Mammals permeate popular culture and our society. We see and interact with them every day in some form or another. Serving as everything from school mascots to advertising symbols and from the stars of documentary television and full-length feature films to the theme of works of literature and song, our mammalian kin surround us constantly.

A Closing Note
Badger On a Stick

I grew up poor but happy in a family that hunted. We weren't sport hunters; only small game (pheasants, rabbits, etc.) were our targets and the family ate what was shot. One wintery day while on a hunting mission, my father, younger brother and I spied a badger going down a hole. (In case you do not know anything about badgers, let me assure you that they are large evil-tempered mammals with powerful legs, long claws, and sharp teeth.) Seeing that badger heading down that hole gave my father an idea he would quickly come to regret. You see, my father had heard that because badgers have such loose skin (true), you could pull them out of a hole by winding a tree branch into them (false). As fate would have it, there was a large tree near the badger hole and my father used a hand ax to quickly fashion a long pole from one of the limbs.

Leaving my brother and I (and his rifle) watching from the safety of our beat up old pickup truck, my father stuck that pole down that hole and began to twist. Did it work? Yes to the degree that it did get the badger out of its hole, but not to extent that the badger was pulled out wound up on the end of the branch as my father had anticipated. No, that badger came boiling out of that hole clearly very angry and very agitated about being poked and rammed with a sharp stick.

As the badger approached him growling, my father managed to pin the animal to the ground momentarily with the branch. Turning his head back towards us as he held the growling writhing badger at bay, you could see several emotions and questions flit across my father's face. He seemed frightened, yet he was laughing as he calculated the distance to the safety of the truck, clearly wondering how fast a badger can run. With a mighty thrust against the branch, my father pushed off, spun around, and sprinted for the truck, screaming for us to open the door. As Dad leaped into the truck, the badger shuffled off in the opposite direction still growling. We laughed about it so hard on the way home that tears streamed down our faces. Whenever the family gathers to this day, that story invariably comes up, and we laugh about it still.

Mammalian myths, legends, and folklore abound and if you chose to perhaps test some of the old beliefs, I strongly suggest you think twice about trying anything that requires you to lure a big, angry mammal out of its hole by poking it with a stick.

In Summary

- The evolutionary history of mammals is perhaps the most fully documented transition in vertebrate history because the primary structures used to classify mammals—skulls, teeth, and ear bones—fossilize well.
- By the middle of the Eocene epoch some 45 million years ago, all major groups of mammals alive today had come into existence.
- Mammals have a presence in nearly every conceivable environment on this planet that supports life. They are found from polar ice to steamy tropical jungle on every continent except Antarctica. They also swim and dive on and under the waters of ocean, lake, and pond.
- Class Mammalia with nearly 5000 species arrayed into 26 orders is not a large group compared to arthropods, fish, and birds, but they are very diverse lot.
- Based on reproductive differences, mammals fall into three main morphological groups: monotremes (egg-layers), marsupials (pouched), and placentals.
- Mammals exhibit a unique suite of characteristics:

  1. The body is covered partially or wholly with hair
  2. Skin containing various types of glands—sweat, scent, sebaceous (oil), and mammary glands
  3. A synapsid skull with two occipital condyles, a secondary palate, and turbinate bones in the nasal cavity. Their lower jaw is a single enlarged bone (dentary). Their middle ear consists of three ossicles (malleus, incus, stapes), the skeleton has seven cervical vertebrae (in most types) and the pelvic bones are fused
  4. The teeth are diphyodont (milk or deciduous teeth replaced by a permanent set that lasts the rest of the lifetime) and heterodont (teeth varying in size and shaped based on function)
  5. The eyelids are movable, and they possess fleshy external flaps (pinnae) around the ear openings.

6. The body is endothermic, and a four-chambered heart powers the circulatory system

7. The excretory system consists of metanephric kidneys with ureters that usually open into a bladder. No cloaca except in monotremes.

8. The brain is highly developed, especially the cerebral cortex, the site of intelligence

9. They are viviparous except for monotremes which are oviparous. The developing young are nourished by milk from the mother's mammary glands.

- Mammals connect to humans in ways that are important economically, ecologically, medically, historically, and culturally.

Review and Reflect

1. ***Skulls on Parade*** As part of a practical exam in zoology laboratory you have been given the four skulls you see depicted (**Figure 20.30**). Examine the skulls carefully. Which skull A or B is that of the mammal? Justify your answer. Which skull C or D is that of an herbivorous mammal and which is that of a carnivorous mammal? Justify your answer.

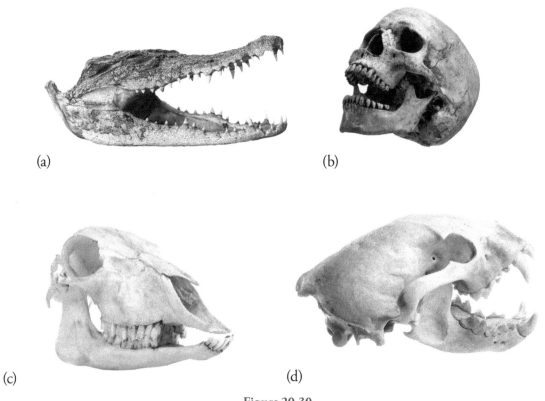

(a)

(b)

(c)

(d)

Figure 20.30

2. ***Which Order?*** Below are the brief descriptions of several mammals. Use the description and your knowledge of mammals to classify each animal into its proper order.

Mammal *A* is a flying mammal that has sharp teeth and large ears.

Mammal *B* stands upright, has a long, heavy tail for counterbalance and gestates its young in a pouch.

Mammal *C* is small with gnawing front teeth and eats plant material. It has a short gestation period and produces many offspring.

Mammal *D* has webbed feet, a broad flat tail, and a soft bird-like beak. It lays eggs.

Mammal *E* has a thick layer of subcutaneous fat, lacks pinna and hind legs, and mates and bears its young underwater.

3. ***Does Hair Help?*** The metabolic rates of two groups of sheep were measured as the amount of oxygen consumed per hour. One group was sheared before the experiment and the other was not. The results are shown in **Figure 20.31**. Do these results support the theory that hair is an evolutionary advantage for mammals? Describe a situation where having a thick pelage would be a disadvantage.

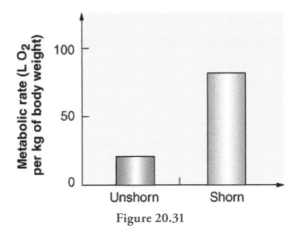

Figure 20.31

4. ***Does Size Matter?*** The relationship shown in **Figure 20.32** is often called the "mouse-to-elephant curve." Analyze the graph and explain the relationship(s) shown. What is the connection, if any, between a mammal's place on the curve and its longevity?

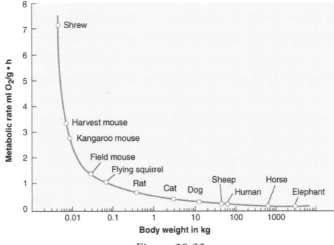

Figure 20.32

5.   **Tough to Digest** The digestive system of many mammals contain an accessory pouch known as a **cecum**. The cecum of an herbivorous mammal is often quite large compared to that of a carnivore. Why? Do humans have a cecum that functions in the same manner as those of other mammals? Explain.

6.   **Our Curious Thumb** Humans have a structure unique to the primate mammals—the opposable thumb. Our thumb sticks out at nearly a right angle to the rest of our digits (fingers) and gives us and most of our fellow primates the ability to easily grasp and precisely manipulate objects. We usually just take this amazing structure for granted so let's do an activity that will make you aware of just how much you depend on your curious thumbs.

Guidelines:

A.   Time how long it takes you to perform the following tasks:

> Untie and remove shoes.
> Put shoe back on and retie.
> Shell and eat a peanut.
> Open a door.
> Unscrew a bottle cap.
> Put on and button up or zip up a jacket or coat.
> Write your name and address

B.   Now tape both your thumbs to your palms. (You may need some assistance with this.). Perform and time the same set of tasks as before.

Analyze:

A.   Did your time to complete the same set of tasks differ when you taped your thumbs? Explain

B.   React to this statement: "The monumental works of man and his great and ever-advancing technology rests on the evolution of the opposable thumb."

7.   **A Hard Learned Lesson** In 1905 the Kaibab deer found on the Kaibab Plateau of the Grand Canyon in Arizona were estimated to number 4,000 deer in 727,000 acres of range. The average carrying capacity of the range was estimated to be 30,000 deer. A campaign was begun in 1906 to protect the deer herd by increasing their population. The first step in the campaign was to establish a hunting ban in the area. Then, in 1907, the Forest Service stepped up the campaign to "help" the deer by exterminating the natural predators of the deer.

Table A shows the number and kinds of predators killed from 1907 to 1939. Table B shows the population of Kaibab deer from 1910 to 1939. (**Table 20.1 and Table 20.2**).

Table 20.1	
Predators Removed	
1907-1917:	600 mountain lions killed
1907-1923:	11 wolves killed
1907-1923:	3,000 coyotes killed
1918-1923:	74 mountain lions killed
1924-1939:	142 mountain lions killed
1923-1939:	4,388 coyotes killed

Table 20.2			
Deer Population			
1910:	9,000	1928:	35,000
1915:	25,000	1929:	10,000
1920:	65,000	1930:	25,000
1924:	100,000	1931:	20,000
1925:	60,000	1935:	18,000
1926:	40,000	1939:	10,000

Guidelines:

A. Use the information in Table B to construct a line graph of the deer population from 1905 to 1939.

B. Draw a horizontal line at the 30,000 point on your graph. This marks the original estimated carrying capacity of the range.

C. Use Table A and your graph to analyze this situation.

Analyze:

A. What two methods did humans use in an attempt to protect the Kaibab deer by increasing the size of their population?

B. Were these methods successful? Did the deer population increase as they had hoped it would? Did they "help" the deer? Explain using data from your graph.

C. Was the original estimated carrying capacity of 30,000 deer an accurate Figure? Explain using data from your graph.

D. Why did the population of deer begin to decline in 1925 although the elimination of many predators continued to occur?

E. Signs that the deer population was out of control began to appear as early as 1920. The Kaibab Deer Investigation Committee recommended reducing and removing livestock owned by local residents from the range and by reducing the number of deer in the herd by 50 percent (culling) as quickly as possible. Deer hunting was reopened in 1924.

Suggest what YOU would have done in the years 1915 and 1923 to manage the deer herd.

F. If the Forest Service had not interfered, what do you think would have happened to the deer population?

G. Is predator removal an effective population management tool? Explain.

H. What future management plans would you suggest for the Kaibab deer herd?

8. ***Slowing Things Down*** A curious phenomenon that lengthens the gestation period of many mammals is delayed implantation. The fertilized egg (blastocyst) remains dormant while its implantation into the uterine wall is postponed for periods of a few weeks to several months. Explain as many situations as you can think of where delayed implantation would improve the survival chances of the developing young.

Create and Connect

1. ***Defend Yourself*** When the first dried skin of a platypus arrived in Britain from the Australian colonies around 1789, it was deemed a fake and thought to be the beak of a bird and various mammal parts sewn together. Imagine that you are the attorney for a duck-billed platypus that has been arrested for impersonating a mammal. Write a defense for monotremes as mammals.

2. ***My Kin, My Servant?*** What should be our relationship with our furry kin? Write a position paper on humans and their relationships with mammals.

Guidelines:

A. Compose a position paper, not an <u>opinion</u> paper. Defend your position with as many facts, Figures, quotes, and pertinent information as possible.

B. Your work will be evaluated not on the "correctness" of your position but the quality of the defense of your position.

C. Your paper might cover the entire spectrum of human-mammal interactions or it might focus more specifically on one aspect of the issue.

D. Your instructor may provide additional details or further instructions.

3. ***Close Quarters*** Rodents develop and live their lives in close contact with each other and objects in their environment. Do such contacts play a role in the development of rodents? Design an experiment to test the problem question: *What effect, if any, does touching and handling by humans have on rodent growth?*

Guidelines:

A. Your design should include the following components in order:

> ➤ The *Problem Question*. State exactly what problem you will be attempting to solve. You should test one specific species, and if you test others, you should test them one species at a time.

> ➤ Your *Hypothesis*. While this is a fictitious experiment, word your hypothesis as realistically as possible.

> ➤ *Methods and Materials*. Explain exactly what you will do in your experiment including the materials necessary to accomplish the task. Be specific, take nothing for granted, and do not expect people to read your mind as they read your work.

> ➤ *Collecting and Analyzing Data*. Explain what type(s) of data will be collected and what statistical tests might be performed on that data. It is not necessary to concoct either fictitious data or imaginary observations.

B. Assume you have access to everything necessary to conduct your experiment. Within reason, money is no object.

C. As usual, we have presented you with an experimental design challenge that is purposely broad and not very specific. Your greatest challenge will be to work out all the specific details leaving nothing to chance. To aid you in this endeavor, your instructor may provide additional details or further instructions.

THE HUMAN CONDITION: RISE OF THE CULTURAL APE

We are just an advanced breed of monkey on a minor planet of a very average star. But we can understand the Universe. That makes us something very special.

—Stephen Hawking

Introduction

Humans may well be the most unusual species we have, thus far, encountered on our journey through the animal realm. Humans are creatures with an almost alien-sized brain that can write poetry, raise giant cities, travel in space, and even contemplate and explore their own mind; an advanced technological being wrapped in the body of a primate, a naked primate at that. This duality of mind and body distinguishes humanity for we what we are biologically—a speaking cultural ape.

As a species that is capable of contemplating itself, we have always sought, and **anthropology** continues to seek answers to certain questions about ourselves: What and where are our origins? How are we anatomically different than our closest primate kin? What does it mean to be "human"? The field of anthropology unites many other disciplines as it searches both the past and present to understand the human condition (**Figure 21.1**).

Rise of Primates

To comprehend our biological nature, we need to understand our place in the hierarchy of animals. As previous chapters have revealed, humans are clearly vertebrate in morphology and placental mammals by our reproductive patterns. On one branch of the placental mammal portion of the tree of life, we find the primates, our closest mammalian kin. Primates, especially humans, are remarkably recent animals on the evolutionary stage. Primate-like mammals (or proto-primates) first arose in the early Paleocene epoch about 65 million years ago. At that time, the world was a very different place than it is today. The continents were in other locations, and they had somewhat different shapes. North America was still connected to Europe, but not yet to South America. India was not yet part of the Asia, but headed towards it, and Australia lay close to Antarctica. Most land masses had warm tropical or subtropical climates. The flora (plants) and fauna (animals) of the time would be unrecognizable to us since most of the plants and animals that are familiar to us had not yet evolved. From fragmentary fossil evidence (mostly from North Africa) it appears there lived a group of creatures at that time resembling modern squirrels and tree shrews that ranged from chipmunk-size to rat-size that were primarily insectivores. These were the proto-primate mammals (Plesiadapiformes). Their numbers declined into the Eocene, and they were extinct by the end of the epoch, possibly as a result of competition with rodents that first appear and radiate in the late Paleocene.

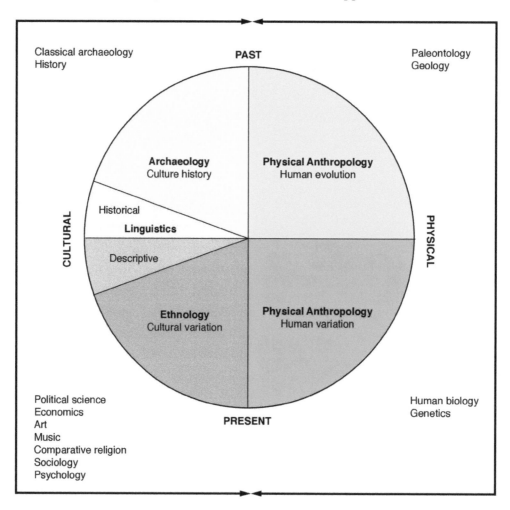

Figure 21.1 Anthropology and related disciplines in the study of humankind.

The first true primates (Euprimates) appear in the fossil record during the early Eocene epoch some 55 million years ago in North America, Eurasia, and Northern Africa (**Figure 21.2**). These creatures were somewhat squirrel-like in size and appearance, but apparently they had grasping hands and feet that were efficient in manipulating objects and climbing trees. It is likely that they were developing effective stereoscopic vision as well. From this ancestral type developed the **Strepsirrhini** (or traditionally **prosimians**— lemurs, aye-aye, bush babies, potto, and lorises. The Eocene epoch was a period of maximum prosimian adaptive radiation with at least 60 genera developing. This is nearly four times greater prosimian diversity than today. These primates lived in what would become North America, Europe, Africa, and Asia, and it was during this time that they reached the island of Madagascar. The Eocene prosimians flourished as a result of a combination of lack of competition from monkeys and apes since these more highly developed primates had not yet evolved, and because of a favorable climate. With the transition of the Eocene epoch into the Oligocene epoch some 35 million years ago, temperatures cooled, and monkeys appeared for the first time resulting in many prosimian species becoming extinct. Today most modern prosimians either live in locations where monkeys and apes are absent, or they are normally active only at night when most of the anthropoids are sleeping

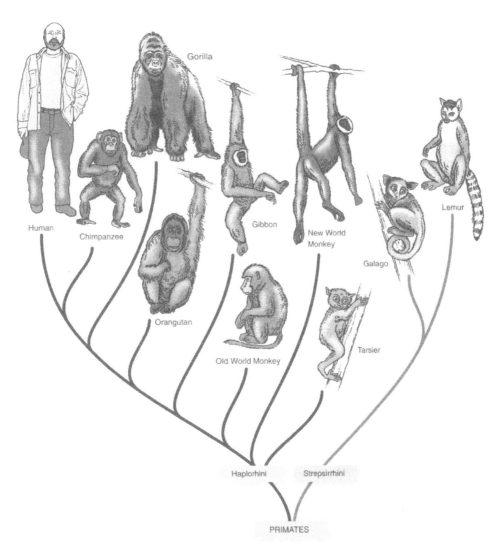

Figure 21.2 Primate Family Portrait. This tree depicts the possible phylogeny of primates. The green box at the base of the tree represents ancestral protoprimates.

From the prosimians arose the **Haplorhini** (or **anthropoids)**—tarsiers, tamarins, baboons, monkeys, apes, and humans. Monkeys were first of the anthropoid lineage. Appearing during the Oligocene or slightly earlier, the ancestral monkeys ranged from the size of a large squirrel to a large domestic cat. They most likely were fruit and seed-eating forest tree-dwellers. Compared to the prosimians, early monkeys had fewer teeth, a less fox-like snout, larger brains, and more forward-looking eyes.

The Oligocene was an epoch of major geological changes with resulting regional climate shifts. By the beginning of the Oligocene, North America and Europe had drifted apart becoming distinct continents. The Great Rift Valley system of East Africa also was formed during the Oligocene along a 1200 mile volcanically active fault zone. The collision of India with Asia that had begun in the Eocene continued into the Oligocene resulting in the progressive growth of the immense barrier we call the Himalayan Mountains and the forcing upward of the Tibetan Plateau beyond. These and other major geological events during the Oligocene are thought to have triggered global climatic changes. The cooling and drying trend that had begun in the late Eocene accelerated, especially in the northern hemisphere. This in turn resulted in the general disappearance of primates from these northern areas.

Geologic change continued apace into the Miocene epoch as continental drift ground slowly on, creating new mountain chains that altered local weather patterns. In addition, the progressive global cooling and drying trend continued. Growing polar ice caps reduced sea levels exposing previously submerged coastal lands. As a result of this and tectonic movement, a land connection was reestablished between Africa and Asia providing a migration route for primates and other animals between these continents. Much of the continuous East African and South Asian tropical forests were broken into a mosaic of forests and vast dry grasslands because of the climate changes. This resulted in new selection pressure being applied to the primates of the time.

Early in the Miocene epoch about 25 million years ago the first apes appear apparently having evolved from monkeys. Under pressure to adapt to less forest and more grassland, African apes became largely terrestrial, exploiting grains, tubers, and dead animals for food. Under these conditions, an upright posture became a distinct advantage as it provided a better view of predators while leaving the hands free for gathering food, caring for the young and using tools. This was the beginning of the Hominidae lineage that would lead eventually to orangutans, gorillas, chimpanzees, and a number of different species of *Homo*.

Diversity and Classification of Primates

The order Primate (L. *primas,* = excellent or noble) embraces 230-270 species aligned into two suborders and contains animals as primitive as the insect-eating aye-aye to the highly complex human. The development of sharp vision and depth perception, grasping hands, and a tremendous brain capacity have given primates a unique combination of specialized talents.

Figure 21.3 Members of the Suborder Strepsirrhini. (a) Ring-tailed lemur (*Lemur catta*) and (b) Loris (Lorisidae).

Geographically, primates are almost totally confined to the tropical latitudes, although the Barbary macaque (*Macaca sylvanus*) is found in northern Africa and the Japanese macaque (*Macaca fuscata*) occurs on both main islands of Japan. Humans, with cleverness of invention and aptitude for technology, have managed to dwell or at least visit everywhere on earth, including the greatest depths of the ocean, and have even ventured off the planet into space and walked on the moon.

Primatologists have traditionally classified primates into two suborders, prosimians—lemurs, lorises, bush babies, and tarsiers—and anthropoids—monkeys, baboons, gibbons, orangutans, gorillas, chimpanzees, and humans. Recent DNA and chromosomal analysis strongly support dividing living primates into one lineage comprising lemurs and lorises and another lineage composed of anthropoids (monkeys and apes) and tarsiers. Lemurs and lorises comprise the suborder Strepsirrhini—moist dog-like muzzle between nose and lip—whereas anthropoids and tarsiers comprise the suborder Haplorhini—dry skin or fur between nose and lip.

Strepsirrhines retain a greater number of earlier mammalian features (*e.g.* claws, long snout, lateral-facing eyes) than do anthropoid primates. With the exception of large-bodied species in Madagascar, an island that separated from Africa during the late Cretaceous before anthropoids evolved, strepsirrhines are small bodied and nocturnal (**Figure 21.3**).

Haplorhines are mostly larger than strepsirrhines and are generally diurnal rather than nocturnal. Haplorhines possess a shortened face, forward-directed eyes, and a larger, more complex brain (**Figure 21.4**).

Figure 21.4 Members of the Suborder Haplorhini. (a) Gorilla (*Gorilla*), (b) Orangutan (*Pongo*), and (c) Chimpanzee (*Pan*).

The classification of primates:

    Domain Eukarya
      Kingdom Animalia
        Phylum Chordata
          Subphylum Vertebrata (Craniata)
            Class Mammalia
              Order Primates
                Suborder Strepsirrhini
                    Family Cheirogaleidae—dwarf and mouse lemurs
                    Family Lemuridae—Lemurs
                    Family Lepilemuridae—Sportive lemurs
                    Family Indriidae—Wooly lemurs and allies

Family Daubentoniidae—Aye-aye
Family Lorisidae—Lorises, pottos, and allies
Family Galagidae—Galagos
Suborder Haplorhini
Family Tarsiidae—Tarsiers
Family Callitrichidae—Marmosets and tamarins
Family Cebidae—Capuchins and squirrel monkeys
Family Cecopithecidae—Mandrils, baboons, and macaques
Family Hylobatidae—Gibbons
Family Hominidae
Subfamily Ponginae—Orangutan
Subfamily Homininae
Tribe Gorillini—Gorillas
Tribe Hominini—Chimpanzees and humans

Until recently humans were classified in a family separate from the gorilla, chimpanzees, and the orangutan. Molecular evidence, however, reveals that chimpanzees and humans are as similar to each other as many sister species. (That is, they are as similar as different species within the same genus.) The primate classification system used here keeps the chimps and humans in different genera, but in the same subfamily (Homininae). The term "hominin" is now used to refer to the chimpanzees, humans, and extinct members of direct human lineage, whereas the term "hominid" is used in reference to the entire family Hominidae—orangutans, gorillas, chimps, and humans.

General Characteristics of Primates

The traits and tendencies found in primate groups are:

- Shortening of the face accompanied by reduction of the snout
- The general retention of five functional digits on the fore and hind limbs. Enhanced mobility of the digits, especially the thumb and big toes, which are opposable to the other digits.
- A semi-erect posture that enables hand manipulation and provides an optimal position preparatory to leaping.
- Claws modified into flattened and compressed nails to support the sensitive digital pads on the last phalanx of each finger and toe.
- A shoulder joint allowing a high degree of limb movement in all directions and an elbow joint permitting a rotation of the forearm.
- A reduced snout and olfactory apparatus.
- A reduction in the number of teeth compared to primitive mammals.
- Teeth and digestive tract that are adapted to an omnivorous diet

- A complex visual apparatus with high acuity and a trend toward development of forward-directed binocular eyes and color perception. Well-developed hand-eye motor coordination.
- A large brain relative to body size, in which the cerebral cortex is particularly enlarged.
- Only two mammary glands (some exceptions).
- Typically, only one young per pregnancy associated with prolonged infancy and pre-adulthood.

Primates are eclectic in their food choices. Most species eat a wide array of foods ranging from insects and other small animals to fruits, flowers, and foliage. However, different species occupying the same habitat differ considerably in the time of feeding, the levels of the forest from which they feed, the type of food eaten, and how far they range to find food.

The shape and structure of the primate body are adapted and suited to its environment. Some walk on arms (forelimbs) and legs (hind limbs) nearly the same length; others that **brachiate** (swing) through the trees with extra-long arms and reduced legs. In humans, the arms are shorter, and the legs elongated to accommodate upright bipedal (two legs) locomotion.

The contrast in locomotion is reflected in the shape of the hands and feet. Some brachiating monkeys lack a thumb, but in apes and humans the thumb is well developed, agile, and opposable to provide a precise but powerful grip. The feet of lemurs are long and narrow, and those of the apes are broader and adept at grasping. Humans have a foot with reduced toes and an arch designed to carry the weight of the body on the heel and balls of the feet.

The structure and length of the tail vary considerably. Tails are retained in some strepsirrhinians, such as lemurs, but lost in the slow-moving potto and lorises. Tails are present in most monkeys. Some Central and South American monkeys (New World monkeys) have a **prehensile** (grasping) tail, but none of the African and Asian monkeys (Old World monkeys) demonstrate such ability with their tail. Apes and humans have no tail at all.

What an ugly beast the ape, and how like us.
—Marcus Tullius Cicero

Whether we like it or not, there can be no doubt that modern humans are ape-like primates. However, we are certainly much more. How did we become *Homo sapiens sapiens*? What does it mean to truly be human? Let us begin to attempt to answer those questions by examining the saga of human evolution. As we do so, it will become apparent that the human ape has gone through a series of changes and advances since the human lineage diverged from the chimpanzee lineage. Over the course of his evolutionary history, the human ape evolved first into a bipedal speaking tribal ape then into a hunting social ape, an intellectual tool-making ape, an agricultural ape, and finally a cultural ape. Some of these stages developed simultaneously, and all blended together to form the human species we have become today.

Rise of the Speaking Bipedal Tribal Ape

Charles Darwin, in his book, *The Descent of Man and Selection in Relation to Sex* (1871), proposed the notion that humans and apes shared a common ancestry. Such an idea was not necessarily original to Darwin con-sidering that Linnaeus recognized that humans are primates well before their evolutionary connection well before Darwin. Darwin was, however, the first to subject such an idea to scientific scru-tiny. During Darwin's time virtually no fossil evidence linking humans with apes existed. This forced Darwin to formulate his hypotheses mostly on anatomical comparisons between humans and apes.

The idea of a common ancestry between humans and apes was met with indignation by the Christian Victorian world of that time. Since then the fires of controversy surrounding this idea have not dimmed but have, if anything, grown brighter. Today in the whole of biology there is nothing more controversial or hotly debated than is the origin of humans. And this debate has spilled over into many different venues ranging from religious pul-pits to science classrooms, and even into legislative chambers and courtrooms.

Today the concept that humans and apes have a common ancestral past is supported by two main bodies of evidence. On one hand, we have the data of DNA analysis and comparative biochemical and cytological studies of modern humans and apes that can be used to construct a reverse calendar of evolutionary events. On the other hand, we have fossil evidence that has been accumulating since the publication of Darwin's book. The early search for hominin fossils was driven by a desire to find a con-nection in lineage between humans and apes, a "missing link." First there was the discovery of two Neanderthal skeletons in the 1880s. Then, in 1891, Eugene Dubois uncovered Java man (*Homo Erectus*). Between 1967 and 1977 (labeled the "golden decade" by paleoanthropologist Donald C. Johnson) many spectacular dis-coveries were made in Africa. Fortunately, important Hominin fossils continue to be discovered, but unfortunately, most homi-nin fossils are far from complete and may consist of only shards of bone, a single tooth, or piece of skull (**Figure 21.5**).

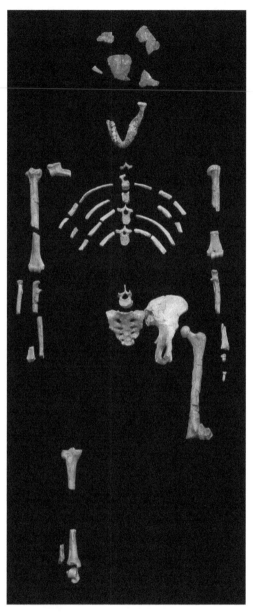

Figure 21.5 Pieces of bone of specimen AL 288-1 (commonly known as Lucy). These bones represent about 40% of the skeleton of an individual female *Australopithecus afarensis*

Based on best current evidence, one possible evolutionary sequence in human evolution is illustrated in **Figure 21.6**. Some pathways are debated, and not all fossil species are shown, especially some of the different species currently identified as *Australopithecus*. There is also considerable dispute concerning many overlap-ping species, especially the possible overlap between *Homo habilis* and *Homo erectus*. It could well be that the two are continuing examples of the same species. The same dispute exists with *Homo erectus*, *Homo sapiens* (archaic), and *Homo sapiens sapiens*.

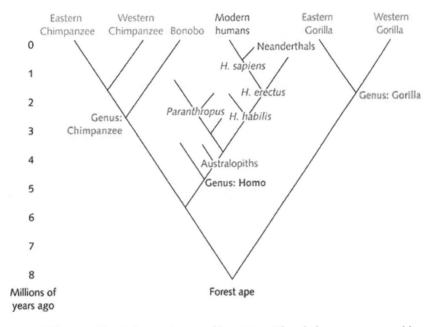

Figure 21.6 The possible phylogenetic tree of hominins. The phylogenies presented here are just a few of the working hypotheses from different authorities. Phylogenies and cladograms are constantly being revised as more fossil and genetic evidence becomes available.

Humans are more closely related to the African apes than to orangutans and even more closely related to chimpanzees than to gorillas. According to recent molecular dating, the divergence between the lineages leading to humans and chimpanzees occurred about 6.3 million years ago, and so the earliest species in the tribe Hominini, to which humans belong, would have appeared shortly thereafter (**Table 21.1**).

Table 21.1			
Evolutionary Details of the Main Groups of Hominins			
Lineage[a]	**Approximate Time (Years ago)**	**Adaptations, Behavior and Habitats**	**Fossil and Archaeological Evidence**
Hominin ancestors	8 to 5 My	Relatively large-bodied apes distributed in Central and Eastern Africa across forest-woodland mosaics	No fossil evidence yet, but when found, expected to be a group or groups ancestral to humans and chimpanzees
Australopithecines	4 to 2 My	Bipedal on the ground, occasionally arboreal Open savanna, and mosaic grasslands and woodland habitats Fibrous plant diet that may also have included meat[b]	Extensive fossils in Eastern and Southern Africa Large teeth and jaws

Homo habilis	Pliocene-Pleistocene 2 to 1.5 My	Improved bipedalism Tools to procure and process food. Habitats in drier areas indicating larger home ranges Scavenging and active animal hunting	Skeletal changes and increase in brain size Early stone tools
Homo erectus	Early-Mid Pleistocene 1.5 to 0.5 My	Entry into new habitats and geographical zones Definite preconception of tool form Manipulation of fire Increased level of activity and skeletal stress	Fossils found in formerly unoccupied areas of Africa, and outside Africa Development of a stone tool industry Archaeological hearths Increased cranial and postcranial development
"Archaic" *Homo sapiens*	Mid Pleistocene 500 to 150 thousand years	Geographical divergence and ecological adaptations More complex tools	Old World distribution with some distinct regional morphologies Bifacial axes: Acheulean-Mousterian stone tool industries
H. sapiens neanderthanlensis	Late Pleistocene 150-35 thousand years	Large and robust individuals More social complexity and development of ritual Increasingly sophisticated tools	Massive cranial and postcranial development Intentional Burial of the dead Increase number of stone-tool types
H. sapiens sapiens	Late Pleistocene to Present	Decreased levels of activity and skeletal stress Expansion of technology Development of complex cultures Increase in population size	Appearance of "anatomically modern" humans From Upper Paleolithic (Aurignacian) stone tools to satellite communication Beginnings and expansion of agriculture

[a]   "Lineage" designates the name commonly given to a major group found in the specified period. As the text indicates, other names have been used for fossil groups in these periods (for example, *H. ergaster, H. rhodesiensis, H. heidelbergensis*). Various groups also overlapped.

[b]   New findings of 2.5-My-old hominins in Ethiopia suggest that behavioral changes associated with lithic (stone tool) technology and enhanced carnivory (butchered antelopes, horses, and other animals) may have been coincident with the emergence of the *Homo* clade that arose from *Australopithecus afarensis* in East Africa (de Heinzelin et al.,, 1999).

Modified from tables in Foley R. 1996. The adaptive legacy if human evolutions: A search for the environment of evolutionary adaptedness. Evol. Anthropol, 4, 194-203; and Potts, R., 1992. The hominid way of life. In *The Cambridge Encyclopedia of Human Evolution*, S Jones, R. Martin, and D. Phibeam (eds) Cambridge University Press, Cambridge, England, pp. 325-334.

Point(s) of Origin?

Where did modern humans originate? Roughly 100,000 years ago, the world was occupied by a morphologically diverse group of hominins. In Africa and the Middle East there was *Homo sapiens*; in Asia *Homo erectus*, and in Europe, *Homo neanderthalensis*. However, by 30,000 years ago this diversity vanished, and humans everywhere had evolved into the anatomically and behaviorally modern form—*Homo sapiens sapiens*. There is great deliberation between two schools of thought regarding the nature of this transformation: one stresses multiregional continuity, and the other suggest a single origin for modern humans. The **Multiregional Continuity Model** contends that after *Homo erectus* left Africa and dispersed into other parts of the Old World, regional populations slowly evolved into modern humans. The emergence of Homo sapiens was not restricted to any one area but occurred throughout the entire geographic range where *Homo erectus* dispersed. Since their original dispersal, natural selection in regional populations is responsible for the regional variants (races) we see today. In contrast, the **Out of Africa Model** asserts that *Homo sapiens* evolved relatively recently in Africa and the Middle East and migrated into Eurasia replacing all other hominin species, including *Homo erectus*.

At this time, the question of the origin(s) of modern humans remains unanswered. There is molecular evidence supporting both models, but a lack of fossil specimens muddies the waters. For the future, both strategic fossil discoveries and refinement of genetic information will hopefully decide the matter. Some investigators are becoming more open to a somewhat intermediate position in which there was an African origin with some mixing of populations.

Earliest Hominins

Three African fossil species—*Sahelanthropus tchadensis, Orrorin tugenensis,* and *Ardipithecus kadabba*—compete for the designation of earliest hominin. These three forms, which were added to the known hominin fossil record within the past decade, date from around the period of the human (hominin) and chimpanzee (panin) divergence and show a mix of hominin features including evidence that suggests bipedal locomotion. Perhaps the most surprising aspect of these finds is the diversity of forms that appeared around the divergence. Clearly, the traditional view of a single lineage leading progressively toward the human form is an oversimplification.

Figure 21.7 An artist's reconstruction of *Sahelanthropus tchadensis*

Sahelanthropus tchadensis (**Figure 21.7**), discovered in 2001 in Central Africa, has been dated between 7 and 6 millions old, suggesting that it may have lived slightly before the time when the chimpanzee and human lineages diverged. The nearly complete cranium and fragments of lower jaw and teeth reveal a protohominin mixture of ape and hominin features.

Orroin tugenensis (**Figure 21.8**) is represented so far by lower jaw fragments and lower limb fragments including three femora that show clear indications that it was a habitual biped when on the ground. The upper limb fragments also show arboreal traits shared with chimpanzees, indicating that this species may have spent a fair amount of time in trees.

Figure 21.8 An artist's reconstruction of *Orrorin tugenensis*

Ardipithecus kadabba (**Figure 21.9**) at 5.6 million to 5.8 million years ago is slightly younger than *Orrorin*. This third contender for the title of basal hominin is relatively well represented by limb remains but less well by cranial remains. It was clearly a bipedal hominin and shares cranial and dental features with later Australopithecines along with more primitive features such as fairly large canine teeth.

Australopithecines

Ardipithecus ramidus, dated to 4.4 million years old, may well be the earliest true hominin. The remains are incomplete with most *A. ramidus* fossils being teeth. Because limb fossils have not been found, questions about size and type of locomotion remain unanswered. Other fossilized animals found with the ramidus fossils would suggest that ramidus was a forest dweller.

Australopithecus anamensis was named in 1995 after its discovery in Allia Bay in Kenya. Living between 4.2 and 3.9 million years ago, its body showed advanced bipedal features, but the skull closely resembles the ancient apes.

Figure 21.9 An artist's reconstruction of *Ardipithecus kadabba*

Australopithecus afarensis lived between 3.9 and 3.0 million years ago. It retained the apelike face with a sloping forehead, a distinct brow ridge over the eyes, flat nose, and a chinless lower jaw. It stood between 107 cm and 152.4 cm (3 feet 6 inches and 5 feet tall and was fully bipedal. Its build (ratio of weight to height) was about the same as the modern human, but its face and head were proportionately much larger, and the thickness of its bones showed that it was quite strong. The discovery of a nearly complete *A. afarensis* dubbed "Lucy" is one of the most famous hominin discoveries of all times.

Australopithecus africanus was similar to *A. afarensis* and lived between 3 and 2 million years ago. It was also bipedal but was slightly larger in body and brain size, although the brain was not advanced enough for speech. This hominin was an herbivore and ate tough, hard to chew, plant materials. The shape of the jaw was like modern humans.

Australopithecus robustus lived between 2 and 1.5 million years ago. It had a body similar to *A. africanus*, but a larger and more massive skull and teeth. Its huge face was flat with large brow ridges but no forehead. Brain size most likely could not support speech capabilities. *Australopithecus boisei* existed at about the same time as *A. robustus* with an even more massive face but about the same size brain. Many authorities believe that *robustus* and *boisei* are variants of the same species. We should regard the Australopithecines as a group in which considerable evolutionary change was occurring, exemplifying rapid radiation of bipedal tropical apes (**Figure 21.10**).

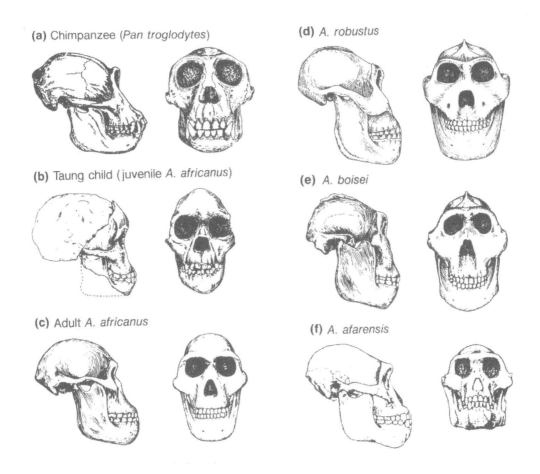

(a) Chimpanzee (*Pan troglodytes*)

(b) Taung child (juvenile *A. africanus*)

(c) Adult *A. africanus*

(d) *A. robustus*

(e) *A. boisei*

(f) *A. afarensis*

Figure 21.10 A comparison of the skulls of a chimpanzee (a) and several australopithecines (b-f).

Origin of Homo

Homo habilis is often referred to as the "handy man" because tools were found with its fossil remains. These associated artifacts indicate that these new hominins were engaged in making regularly patterned stone tools known as **Oldowan tools**. Australopithecine species had probably earlier embarked on using simple stone tools, bones, and sticks. This is likely because chimpanzees engage in simple tool use and modification.

> *Man is a tool-using animal. Without tools, he is nothing, with tools he is all.*
> —Thomas Carlyle

Habilis existed between 2.4 and 1.5 million years ago (**Figure 21.11**). The brain size in earlier fossil specimens was about 500 cc but rose to 800 cc toward the end of this species existence. The increased brain size shows evidence that some speech may have developed. *Habilis* was about 152.4 cm (5 feet) tall and weighed about 45.4 kg (100 pounds). Some paleoanthropologists believe that *habilis* is not a separate species and should be considered as either a later *Australopithecus* or an early *Homo erectus*. It is possible that the smaller early examples are in one species and the later larger examples in the other.

Figure 21.11 An artist's reconstruction of *Homo habilis*.

Homo erectus lived between 1.8 million and 300,000 years ago and was a successful species for a million and a half years (**Figure 21.12**). Early examples had a brain size of 900 cc on the average, but the brain grew steadily during its reign reaching almost the same size as modern human, at 1200 cc. This species definitely had speech, developed tools, and weapons, captured and kept fire, and learned to cook food. Though proportionately the same, *erectus* was sturdier in build and much stronger than modern humans. Only the head and face differed much from our own. Traveling out of Africa into China and Southeast Asia, *erectus* turned to hunting for food and developed clothing suitable for northern climates. These hominins became geographically widespread and existed for a very long period of time. There seem to be a number of variants in this species, and if we knew all the distinctions among different *H. erectus* groups, it might be possible and prudent to mark off new species. However, these distinctions encompass anatomical and behavioral traits that we cannot always discern solely through fossils.

Figure 21.12 An artist's reconstruction of *Homo erectus*.

Archaic *Homo sapiens* (also *Homo sapiens heidelbergensis*) were types that may have provided a bridge between *erectus* and *Homo sapiens sapiens* during the period 500,000 to 200,000 years ago. The archaic *sapiens* represents a diverse group of skulls found with features intermediate between the two. Brain size averaged about 1200 cc and skulls are more rounded and with smaller features. The skeleton shows a stronger build than modern humans but was well proportioned.

Figure 21.13 An artist's reconstruction of *Homo neanderthalis*.

Homo sapiens neanderthalensis (also *Homo neanderthalensis*) lived in Europe and the Mideast between 150,000 and 35,000 years ago. Neanderthals coexisted with the various archaic species and early *Homo sapiens sapiens*. Their brain size was slightly larger than that of modern humans at about 1450 cc, and the braincase was longer and lower than that of modern humans, with a marked bulge on the back of the skull. Like *erectus*, they had a protruding jaw and receding forehead with a weak chin. Neanderthals lived mostly in cold climates, and their body proportions were similar to those of modern cold-adapted peoples: short in height and solid of build (**Figure 21.13**). Males averaged about 168 cm (5 feet 6 inches) in height. Their bones were thick and heavy and show signs of powerful muscle attachments indicating that Neanderthals would have been extraordinarily strong by modern standards. Their skeletons also reveal that they endured brutally hard lives. A large number of tools and weapons have been found; most more advanced that those of *erectus*. Neanderthals appear to have been formidable hunters and are the first people known to have buried their dead, complete with flowers and artifacts. They also apparently used herbal medicines. Plaque on Neanderthal teeth yields traces of chamomile and yarrow, two anti-inflammatory drugs.

Neanderthals disappeared abruptly around 25,000 years ago with Gibraltar seemingly their last refuge. Although they are no longer with us as a species, traces of Neanderthals are with us still if nothing more than as segments of DNA within the human genome. Ancient liaisons between humans and Neanderthals as recently as 37,000 years ago in Europe reverberates today in the genomes of billions of modern Asian and European humans In fact, recent research indicates that the DNA of living Asians and Europeans is, on the average, 2.5 percent Neanderthal.

Homo sapiens sapiens first appears about 195,000 years ago and for the last 24,000 years or so has been the sole survivor of genus *Homo* to walk the planet. Modern humans are characterized by a brain size of about 1,350 cc, a flat forehead, small or absent eyebrow ridges, a prominent chin, and a **gracile skeleton** (slender bones).

The long-term trend toward smaller molars and decreased robustness of the skeleton can be discerned even within the last 100,000 years. The face, jaw, and teeth of Mesolithic humans 10,000 years ago were about 10% more robust than our own. Upper Paleolithic humans 30,000 years ago were about 20% to 30% more robust than their modern counterparts in Europe and Asia. Interestingly, some modern humans, such as aboriginal Australians, have tooth sizes more typical of archaic *sapiens*. The smallest tooth sizes are found in those areas where food-processing techniques have been used for the longest time, a probable example of natural selection that has occurred within the last 10,000 years.

Box 21.1
Are Humans Still Evolving or Are We Ancient History?

For quite some time, the consensus among not only the general public but the world's preeminent biologists has been that human evolution is over. The late Stephen Jay Gould, eminent Harvard paleontologist and evolutionary theorist, once stated, "Natural selection has almost become irrelevant to us."….There have been no biological changes. Everything we've called culture and civilization we've built with the same body and brain." This view has practically become doctrine, and the doctrine states that modern *Homo sapiens* emerged around 50,000 years ago and that our bodies and brains were mostly sculpted during the long period when we were hunters and gatherers.

In other words, the modern human skull houses a Stone Age brain, and to suggest otherwise was nothing short of blasphemy. At the risk of being branded heretics, however, some researchers are challenging this long-held view and contend that not only is evidence beginning to mount that humans are still evolving, but that human evolution may possibly be occurring at a faster and faster pace.

The main evidence for continuing human evolution is etched in the human genome in the form of millions of rare gene variants. Furthermore, the data indicates that over the past 10,000 years human evolution has occurred a hundred times more quickly than in any other period in our species history. In fact, human evolution may be in overdrive as the result of the rapid increase in rare gene variations and DNA

---

methylation. Humans have five times as many rare gene variants as would be expected (About 61 million instead of 12 million). On average, every duplication of the human genome includes 100 new errors. Thus as our population ballooned from as estimated 5 million individuals just 10,000 years ago to more than 7 billion today, our DNA was given many opportunities to accumulate mutations and variations. Unfortunately, evolution hasn't had time to weed out the harmful mutations so that an estimated 80 percent are probably deleterious to us.

These mutations relate to virtually every aspect of what it means to be a functioning human—our skin and eye color (apparently no one on earth had blue eyes just 10,000 years ago), our brain, our life span, our digestive system, our skeleton and bones, our immunity to pathogens, and even sperm production.

Many of these DNA variants seem to be unique to a continent of origin, a finding that has provocative implications. University of Utah anthropologist Henry Harpending states, "It is unlikely that human races are evolving away from each other. We are getting less alike, not merging into a single mixed humanity." Harpending continues, "We aren't the same as people even a thousand or two thousand years ago. Almost every trait you look at is under strong genetic influence." Chiming in is John Hawks, University of Wisconsin at Madison paleoanthropologists and bone specialist, "You don't have to look hard to see that teeth are getting smaller, skull size is shrinking, and overall stature is getting smaller." Some changes are similar in many parts of the world, but other changes, especially over the past 10,000years, are distinct to specific ethnic groups. In Europeans, the cheekbones slant backward, the eye sockets are shaped like aviator glasses, and the nose bridge is high. Asians have cheekbones facing more forward, very round eye orbits, and a very low nose bridge. Australians have thicker skulls and the biggest teeth, on average, of any human population today. Hawks wonders, "It beats me how leading biologists could look at the fossil record and conclude that human evolution came to a standstill 50,000 years ago."

Although populations of humans may be evolving apart in modern times, DNA studies conducted by genetic anthropologists show we all share a common female ancestor (mitochondrial Eve) who lived in Africa about 140,000 years ago. In addition, all living men share a common male ancestor (genetic Adam) who lived in Africa about 60,000 years ago. Around 50,000 years ago humans began pouring forth from Africa eventually spreading around the world.

Such information is divined by genetic anthropologists through the preparation of *haplotype maps*. Haplotype mapping makes use of the fact that genetic variants in the male Y chromosome and maternal mitochondria are often inherited together in segments called haplotypes, with each haplotype representing a unique mutation away from the original ancestral male and female.

As our ancestors spread across the face of the planet and came to occupy niches as diverse as the frigid Arctic Circle, to dry deserts and steamy rainforests, the cultural and demographic shifts necessary to survive those inhospitable conditions sparked a transformation in both the body and brain of our species that still continues. Over the past few centuries, and accelerating ever more quickly in the past 50 years, a steady stream of human innovations has begun to drastically speed up processes that were, until very recently, the sole providence of nature. It appears that our technology has created ways of accelerating change (genetic engineering) and new habitats (modern cities and space stations), essentially fracturing our species and transforming our future as a species.

The next time you look in the mirror consider that the image you see is not so much one of a Stone Age relic but rather that of a work in progress; a creature some experts believe is no more than a generation or two away from the emergence of *Homo evolutus*: a hominin that controls its own evolution.

Mark of Humankind

Based on the evolutionary evidence such as it is, a modern human seems to be defined anatomically by two main characteristics: bipedalism and a large brain with developed sites supporting speech and intelligence.

Bipedalism

Although paleoanthropologists have established bipedalism as a long-standing feature in hominin lineages, and although the questions of why and how bipedalism originated have been discussed for over a century, the origin of bipedalism remains a matter of controversy. What we do know is that bipedalism evolved very early in hominin evolution and that it represents a key innovation leading to a number of adaptations we associate with modern humans. Three different arenas may have exerted selection pressures that influenced bipedalism, each of which may have been influenced by the others.

Improved Food Acquisition Beginning with the Late Miocene and extending through Pliocene-Pleistocene times, there were periodic decreases in global temperatures, marked by the onset of ice-sheet formation in Antarctica and the northern hemisphere. As ice locked up water, various terrestrial areas became relatively dry, and open environments such as woodlands and grasslands replaced rain forests in many tropical regions.

Early hominins lived in a patchy environment of mixed woodland and **savanna** (relatively dry grassland and bushland with occasional trees) that provided seasonal food supplies. Their habitat dictated an omnivorous diet, demanding relatively more time devoted to searching for food over longer distances. An upright stance and bipedal striking would have enhanced long-distance foraging by enabling them to carry food gathered in different places.

Bipedalism may have arisen as a byproduct of adaptations that reduced forelimb involvement in quadrupedal support and movement. As hands became increasingly specialized for grasping, manipulating and carrying food, tools, weapons, and offspring, selection occurred for an upright stance and for transferring locomotion to hind limbs.

Improved Predator Avoidance Bipedalism enhances height, and so improves a hominin's ability to see over tall grass and obstructions and to wade in deeper water to pursue game or seek protection from predators. Bipedalism fosters the use of manual weapons such as stick wielding and stone throwing, which extends the reach of hominins beyond the teeth, claws, and other defenses of animal competitors, predators, and prey. Bipedalism also allows hominins to carry tools and weapons from place to place as well as to move offspring from one camp to another or from one food resource to another.

Improved Reproductive Success It has been proposed that bipedalism enabled adult males to carry food manually to their females and offspring who could remain sequestered in a single locality, the **home base**. This mode of provisioning reduced the need for females to be continuously mobile in foraging for themselves and their clinging offspring and accorded three important advantages:

- A relatively stable home base that provided for more constant social relationships and perhaps closer mother-infant relationships that improved infant survival;
- Reduced infant injuries because infants were no longer bring dragged around by their continuously mobile mother.
- Reduction in the spacing between births by allowing parents to successfully care for more offspring.

Speech

The most advanced of primate vocalizations is human speech. Compared with animal vocalizations that are often limited to a sequence of sounds in single tones, human speech provides a rapid means of communication. For example, we can interpret a sequence of dots and dashes, as in Morse code, at a rate of much less than 50 words per minute, whereas we can often easily understand a sequence of spoken syllables delivered at 150 words per minute.

As in other mammalian vocalizations, the larynx, in the upper part of the tracheal tube, provides the basis for speech. In producing sound the larynx acts like a woodwind reed, controlling vocal pitch by opening and closing rapidly so that expired air from the lungs is interrupted to form puffs; the greater the frequency of puff formation, the higher the pitch. To produce the vowels of human speech, laryngeal puffs must pass through a tube-like airway (the pharynx) whose lengths and shape determine the eventual frequency patterns emitted and thus the quality of the different vowel sounds (**Figure 21.14**).

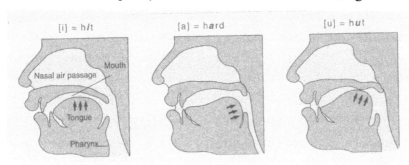

Figure 21.14 A diagrammatic view of how adult humans produce vowel sounds by positioning the tongue (arrows) in different parts of the oral cavity.

Figure 21.15 diagrams the upper respiratory and vocalization systems of an adult human, a chimpanzee, an infant human, and a reconstruction of the presumed upper respiratory and vocalization system of an adult australopithecine. The pharynx is much longer in human adults than in chimpanzees because the larynx is displaced downward in the neck, and the bulging tongue formed by shortening of the mandible now forms the anterior wall of the pharynx. As a result, humans can enunciate vowels and syllables more clearly by positioning the tongue in both mouth and pharynx. Newborn human infants show the same overlap between epiglottis and soft palate as nonhuman primates, but the pharynx lengthens considerably during infancy and childhood, transforming humans from obligate nose breathers as infants, to the adult condition of voluntary mouth breathers.

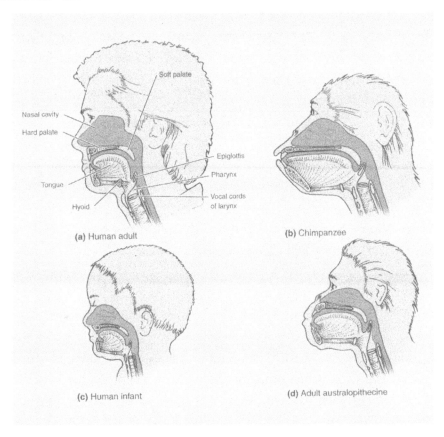

Figure 21.15 The upper respiratory systems of (a) an adult human, (b) a chimpanzee, (c) a human infant, and (d) an adult australopithecine illustrating important structures associated with vocalization.

Because speaking depends so much on soft tissues that do not readily fossilize, researchers find it difficult to uncover the phylogenetic history of speech. We can, however, make important correlations between skull structure and the positioning of the larynx, size of the tongue, and length of the pharynx. According to such studies, the vocal tract of australopithecines was no different from that of nonhuman primates and was probably also true for most, if not all, of the lineage classified as *H. erectus*.

Although an intact **hyoid bone** (used for laryngeal muscle control) has been found with a 60,000 year old Neandertal skeleton, we still do not know the actual position of the larynx and other soft tissue in this fossil or earlier ones and thus remain uncertain as to the extent of Neanderthal vocalization.

Rise of the Social Hunting Ape

Humans have been tribal (group) animals since they first walked erect, more than four million years ago. We were forced to gather in groups or tribes out of sheer necessity. Being bipedal, we usually could not out-climb nor out-run either our predators or our prey. Only through tribal cooperation could we hold predators at bay and secure food. For the first two million years of their existence, early hominins were tribal herbivores, but for the last 2 million years they have been social hunters (and in many aspects still are).

Gathering together in groups for the common good is not unique to humans because from ants to birds and orcas to gorillas, other types of animals form tribes and cluster together for survival. With humans, however, tribal (safety in numbers) grouping gave rise to complex social systems; human collectives gradually became ordered and more structured. Some anthropologists believe that one contributing factor in the move from tribal humans to social humans was when males began to assume more of a parental rather than strictly reproductive role in the collective.

Early human societies would have displayed several basic characteristics:

- *Roles*. Each segment of the group—mature men, young men, mature women, young women, small children, and the very old—would have certain duties (roles) they had to perform for the common good.
- *Rules*. For the group to function successfully as a coordinated whole, a code of conduct (rules and regulations) of some type would have to have been established.
- *Rulers*. To enforce the rules, rulers (chiefs or tribal elders) would need to be selected.

The first human societies were most likely very similar to hunter-gatherer societies that continue today in places such as Central Africa (*Mbuti Pygmies*), South Africa (*Kalahari Bushmen*), and Australia (*Aborigines*). These groups consist of social communes or bands where males are usually the hunters and females the plant gatherers. Because their omnivorous diet depends on highly variable and often seasonal plant and animal food sources, each band moves several times a year over fairly wide ranges to different home bases or settlements.

> *Problem solving is hunting. It is a savage pleasure, and we are born to it.*
> —Thomas Harris

Although we do not know when hunting began in human history, it was a significant enterprise for a long enough period—in many societies, up to the agricultural (Neolithic) revolution about 15,000 to 10,000 years ago—to have seriously influenced human behavior and societal development. First, successful medium and large game hunting requires active cooperation among hunters. Hominin hunters, empowered with simple weapons such as wooden spears, clubs, and hand axes, used techniques such as tracking, stalking, and chasing game into swamps, over cliffs, into ambushes, or by continuing the chase until the animal tired. With cooperative hunting, humans could bring down larger animals and feed more people than could single hunters alone.

Second, cooperative hunting and killing of large animals emphasized increased social cohesion during both hunting and the food sharing that followed. Transfer of information in successful hunts became vital, performing a necessary function in many later social interactions of the entire group. Improved communication became especially advantageous as individuals took more complex leadership roles in planning, hunting, helping, food gathering, food sharing, infant care, child training, and other vital activities.

Third, successful hunting emphasized perceiving and retaining information on migratory pathways, watering sites, and home base settlements, whose geographical positions extended over ranges greater than those occupied by most other primates or carnivores. Hominin hunters had to mentally dissect their experiences and observations into component geographical and ecological features, prey behaviors, weather effects, and seasonal changes, and store and synthesize this information into communicable mental maps that enabled prediction, planning, and modification.

Fourth, hominin hunting involved stresses that fostered increased locomotory adaptations such as persistence in the chase (humans can continue jogging for distances that are generally longer than many large animals can continue running), maneuverability in the kill, and long-distance traveling to or between home bases while carrying heavy burdens. Recent anatomical and physiological evidence indicates that adaptations for long-distance endurance running characterized the genus *Homo* from *Homo erectus* onward. Others suggest that the need for increased diffusion of metabolic heat during these pursuits would have selected for the loss of body hair and increased number of sweat glands, features that among primate are restricted to humans.

Finally, hunting placed further social emphasis on the home base to enable food exchange among foraging subgroups, particularly when the food supply was irregular, as it often is in hunting. A home base has value for nursing and pregnant females who could not always or easily cover the long distances necessary for large-scale hunting. The home base would have become a center for food sharing, shelter, hunting preparations, sexual bonding, childcare, and other social interactions in which communication skills tied all members together.

Rise of the Intellectual Tool-Making Ape

Possessing **intelligence** (the capacity to reason, plan, solve problems, and learn) is not unique to humans. In fact, a number of different vertebrate species are regarded as being highly intelligent. To be human is not merely a matter of possessing intelligence, but also a matter of possessing a pliable multifaceted **intellect**. Human intellect is unique in that it developed as a control over our animalistic instincts, allowing us to voluntarily adapt what would otherwise be unchangeable instinctive behavior. Thus, when humans are presented with challenges and situations beyond the scope of our instincts, our refined, multifaceted intellect allow us to adjust, adapt and overcome. To fully understand what it means to be truly and uniquely human from the standpoint of intellect, one only need contrast humans to our closest vertebrate and primate kin—the chimpanzee. A chimpanzee is an ape with raw intelligence driven mainly and usually by instincts, whereas a human is an ape with instincts driven mainly and usually by a refined intellect. (Nothing about animal behavior is absolute. The use of the disclaimers "mainly" and "usually" is meant to account for those times when nonhuman intelligent animals break free of their instincts and, if only for a moment, operate on raw intelligence. However, the phrase was inserted more in consideration of those times when humans lock out their intellect and behave as an instinctive animal.)

> *Intelligence is what you use when you don't know what to do.*
>
> —Jean Piaget

703

Brain size (specifically the cerebral cortex) and thus the intelligence and intellectual capabilities of humans developed slowly at first and then more rapidly over the 4-million year span of our development. As time passed, selection pressures and fortunate opportunities allowed humans to develop their basic intelligence to a level achieved by no other animal on the planet—a pliable, multifaceted intellect. Intelligence is but one facet of intellect. Other facets would include the ability to think abstractly and comprehend ideas; the ability to accumulate knowledge or wisdom; the ability to use language; the development of an individual personality and character; and introspection.

We see the first hints of an emerging intellect in the social skills associated with cooperative hunting and other social activities and in the stone implements and tools that date back at least to *Homo habilis*. *Homo erectus* continued making tools, and there is clear evidence of progressive improvement in tool making over time. By 400,000 years ago, *H. erectus* sites commonly had tens of thousands of discarded tools.

With the appearance of the Cro-Magnon culture about 40,000 years ago, tools started becoming markedly more sophisticated. Tool-making skills involve not only manual dexterity, hand-eye coordination, and considerable concentration and focus, but also the ability to plan and visualize an object that is not apparent in the raw material from which it is created. The artisan must conceptualize the final form of a tool in its three dimensions and implement such concepts by mastering a series of techniques. These included finding and recognizing appropriate, workable stones in outcrops that were often widely dispersed, carrying these stones back to a base, and then shaping them into tools by a sequence of precise strokes. The toolmakers also had to supplement the considerable mental abilities they used in tool making with social and communication abilities, in order to transmit such skills to other individuals who could continue the industry.

A wider variety of raw materials such as bone and antler were used, and new implements for making clothing, engraving, and sculpting were developed. Fine artwork, in the form of decorated tools, beads, ivory carving of humans and animals, clay figurines, musical instruments, and spectacular cave paintings appeared over the next 20,000 years.

Rise of the Agricultural Ape

The late arachaic *Homo sapiens* populations were most likely tribal wanderers who exploited a wide range of food sources. This most likely required that they range seasonally over vast territories. Exposed to the elements, hounded by predators, and with reliable and accessible food always in short supply, life for these early hominins must have been a brutal proposition, especially for infants, young children, and the old. The evolution to hunter-gatherer societies improved the lot of *Homo* but life was still hard. The average Neanderthal male, for example, could look forward to a life span of about 30 to 35 years.

The next major milestone in the rise of modern humans happened when hunter-gatherers became agriculturists and began to grow rather than seek out food. Greek legend has it that in a burst of goodwill, Demeter, goddess of crops, bestowed wheat seeds on a trusted priest, who then crisscrossed Earth in a dragon-drawn chariot, sowing the dual blessings of agriculture and civilization.

For decades, paleontologists and archaeologists also regarded the birth of agriculture to be such a dramatic and important milestone that they christened it the **Neolithic Revolution**. The contention was that farming was born after the end of the last Ice Age, around 10,000 years ago, when hunter-gatherers settled in small communities in the Fertile Crescent, a narrow band of land arcing across the Near East. Here they quickly learned to grow their own food by sowing cereal grains and breeding better plants. Societies then raised more children to adulthood, enjoyed food surpluses, clustered in villages and developed culture, and headed down the road to civilization. This novel agricultural way of life then diffused across the Old World.

As is the case with many a hypothesis, this scenario did not stand the test of time and has fallen beneath an onslaught of new data. By employing sensitive new techniques from sifting through pollen cones to measuring minute shape changes in ancient cereal grains, researchers are building a new picture of agricultural origins. They are pushing back the dates of both plant domestication and animal husbandry around the world, and many now view the switch to an agrarian lifestyle as a long, complex transition rather than a dramatic revolution. The latest evidence suggests, for example, that hunter-gatherers in the Near East first cultivated rye fields as early as 13,000 years ago. But for centuries thereafter they continued to hunt wild game and gather an ever-increasing range of wild plants, only becoming full-blown farmers living in populous villages around 8500 B.C. And in some cases, villages appear long before intensive agriculture.

> *The transition from hunters and gatherers to agriculturists is not a brief sort of thing. It's a long developmental process and one that did not necessarily go hand in hand with the emergence of settlements.*
>
> —Bruce Smith

A new picture of the origins of agriculture has begun to emerge as evidence continues to be gathered worldwide. In the Near East, not only were some villages born before agriculture, but may have even spurred the development of agriculture in some cases. In China, North America, and Mesoamerica, plants appear to have been cultivated and domesticated by nomadic hunter-gathers during the dramatic environmental shifts that accompanied the final phase of the last Ice Age. To many it no longer makes sense to suppose a strong casual link between farming and settled village life. Indeed, in many regions, settled villages of agriculturists emerged only centuries or millennia after cultivation, if at all. Many ancient peoples simply straddled the middle ground between foraging and farming, creating societies and economies that blended both together. These societies in the middle ground certainly are not failures, but rather societies that found an excellent long-term solution to their environmental challenges.

Rise of the Cultural Ape

Eventually, for reasons still unclear, many of the early domesticators became true agriculturalists by 10,500 years ago in the Near East, 7,000 years ago in China, and later in the Americas and Africa. And during this transition, human populations did indeed soar as hamlets became villages and villages became cities. Archeological sites in the intensively studied Fertile Crescent, for example, show that fields increased more than 10-fold in size from 0.2 hectares to 2.0 to 3.0 hectares during this time of transition. As food supplies became more reliable, the lifestyle became more stable and easier to bear. In turn, a safer, more stable, and longer life permitted humans more opportunities to develop their intellectual side, and as they did so, human societies were transformed into human culture and civilization.

Consider New York City as an example. The physical structure of the city—buildings, roads, electrical and plumbing systems, transportation systems, and so on—are the monuments of present-day **civilization**. The pyramids of Egypt are the remains of a past civilization, whereas the space station may be a preview of future civilizations.

The institutions and systems that build and maintain the physicality of the city as well as those that protect, feed, clothe, house and entertain the human residents of that city-sized chunk of civilization are the **culture**. Even rural areas have civilization and culture, just not on such a grand scale. However, there is no one standard human culture homogeneously layered around the world. Humans in different geographical areas put their own twist on human culture, and so regional differences crop up often within the same country or even within a large city.

Our Cultural Hallmarks

What are the cultural components and systems that make modern civilization possible?

Communication systems

As the necessary centers in the human brain evolved, simple vocalization gave way to speech. As our intellect evolved, speech gave way to language with the result that today *H. sapiens sapiens* is the only animal with written language.

The ability to communicate swiftly and precisely is so critical to modern culture and civilization that all other cultural systems are totally dependent on it. In fact, humans love to communicate and seem almost addicted to doing so. The evolution of our intellect has produced communication systems that are truly amazing. We now live in age of electronic communication in which it is possible to sit at a computer and converse with someone on the other side of the planet as it they were in the next room or use a cell phone in Boston to wish your cousin in Seattle happy birthday. And each day billions of people worldwide watch human culture and civilization fly out of their television sets.

One benefit of communication-based culture is the easy access to knowledge and information. The dark side, however, is information overload as so succinctly described by E.O. Wilson when he said: "We are drowning in information, while starving for wisdom."

Our compulsion to communicate has driven us to even attempt to communications with other sentient beings on other planets somewhere in the void beyond Earth. The Pioneer 10 and 11 spacecraft launched in 1972 and 1973, were designed for a close flyby of two giant outer planets—Pioneer 10 past Jupiter and Pioneer 11 past Saturn. Bolted to the main frame of each spacecraft was a 6 x 9 inch gold anodized plaque displaying graphic representations of who sent the craft and from what planet they had originated. These plaques were added in the outside chance that sometime in the far distant future these tiny spacecraft might be picked up by a space faring alien race. Launched in 1977, the twin satellites Voyager I and II were designed to take advantage of a rare alignment of the outer planets by flying close to these planets one after the other collecting data and taking pictures. Inside each of these spacecraft is a golden record containing 115 images and variety of natural sounds as well as "Greetings from Earth" in 55 languages as well as music from around the world. Again, these records were added to the spacecraft in hopes that someday somewhere out among the starts where they might be found and understood. Pioneer 10 and 11 have long since left the solar system and are no longer in contact with us. Voyager I has become the most distant of all human-made objects at a distance of 16 billion km (10 billion miles) it flies another 1.6 million km (1 million miles) farther from the sun each day. Voyager II is about 80 percent as far away as its twin. The power supply of each satellite should last for another 20 years or so. Many thousands of years from now these spacecraft will be closer to other stars than to our sun. We have tossed a spacecraft bottle into the cosmic ocean in hopes that someone or something may eventually open it and read the message inside.

Currently, a number of organized efforts collectively known as SETI (**S**earch for **E**xtra-**T**errestrial **I**ntelligence), survey the sky to detect the existence of transmissions from civilizations on distant planets. Although SETI searches for communication from aliens, METI (**M**essages to **E**xtra-**T**errestrial **I**ntelligence) projects create electronic messages and beam them out into space in hopes they will be received by an alien intelligence which in turn might communicate with us.

Legal systems

Humans are the only animals with morals (sense of right and wrong) and a value system of justice (fair play). Ancient human societies must surely have realized the necessity of establishing codes of conduct and rules to account for moral and justice issues. Those ancient rules and codes evolved into the legal systems of modern times. Legal systems vary from culture to culture around the world, but most have three basic components: (1) those that make the regulation or laws (legislators), (2) those who interpret the regulation or laws (judges), and (3) those who enforce the regulation or laws (governments, police, and armies). The modern legal system functions as did the ancient system of rules to prevent the culture from slipping into chaos and anarchy. This fact has been repeated time and again throughout modern history as countries are caught in a civil war or the government of a country is forcibly disrupted and displaced either by forces from outside or within its borders.

Protective systems

Although not exclusively a trait of the human intellect, **altruism** (concern for others not related to us) is perhaps more powerful in humans than in other animals for no other animals cares as much for the welfare of others of its own kind as does the human ape. As such, a number of different protective cultural systems designed to ensure the safety and protection, health, and general well-being of other humans have been developed over the course of our evolution. Protective systems may be public (governmental) or private in nature and funding.

Public Safety and Protection is ensured by police and fire departments at the local level and armies at the state and national level. Some governmental agencies attempt to ensure the safety of everything from the food we eat to the drugs we take while others set and enforce regulations to improve the safety of working conditions for employees in all kinds of industries and jobs.

Public Health is protected by our medical system that ranges from governmental agencies such as the Surgeon General and the Center for Disease Control to your personal physician. Ongoing research into the prevention, improved treatment, and even cure of the host of microbial and degenerative diseases that plague humankind is also an important part of the modern human health cultural system.

Public well-being refers to the medical, domestic, and financial health of an individual. Medical well-being is a function of the health system of our culture. Domestic well-being (being properly fed, clothed, and housed) and financial well-being are the providence of a host of different agencies and groups, both private and governmental. From feeding the hungry in soup kitchens to building housing for the homeless to job training and financial assistance, our modern culture attempts to aid those less fortunate.

Economic systems

An economic system may be defined as a particular subset of cultural institutions that deal with the production, distribution, and consumption of goods and services. Economic systems vary from place to place and from country to country, but they all have money at their root. Be it in the form of paper, coin, or polished pieces of shell, money functions as a medium of exchange for goods and services rendered, a unit of account (the price or value of goods and services), and a store of value (financial capital that can be reliably saved, stored, and retrieved. Simply put, the economic system generates money and the money funds the other cultural systems.

Aesthetic systems

The rise of human intellect allowed humans to develop an aesthetic sense. That is, humans view and value certain aspects of the world around them not for their practicality but for the pleasure they provide. From Niagara Falls to the Grand Canyon to a fiery sunset, humans appreciate the beauty of natural things and are filled with a sense of awe by the majesty these things.

Not content to appreciate only natural beauty, humans have developed their own aesthetically pleasing schemes such as art, music, dance, acting, and literature. What we regard as aesthetically pleasing is one of the most intensely personal parts of our individual being.

Innovation systems

The rise of intellect fostered a literal explosion in the creativity and inventiveness (science and engineering) of humans. Innovation driven by evolving intellect has gone through a number of phases or stages: stone tools, metal tools, machines, electric machines, and technology. The first stage appears to have tool-making. At first tools and weapons were crafted from natural objects such as rocks, bones, and antlers. With the discovery of techniques for processing metals, tools and weapons were forged from copper and bronze at first and later from iron and steel. The next historical leap in innovation saw the construction of machines of all types from vacuum cleaners to steam engines. Modern times have seen the application of electricity to machines to the extent that our modern culture is literally driven by electricity. Often taken for granted, our deep dependence on electricity was driven home to my family personally several years ago when a huge ice storm devastated the area where my family lives. Those without generators were forced to live for days as if the calendar had been turned back to the early 1900s.

The latter part of the 20th century found humankind entering yet another and still current phase of innovation we call the technology stage. Science and engineering taken to the technological level have given us everything from space travel and computers to artificial hearts and genetically altered plants and animals and cellular phones and flat-screen televisions to artificial intelligence and nanobots.

Civilization and all the components and systems of culture that accompany it are the result of the ever increasing refinement of our multifaceted intellect. Our intellect created and drives human culture and civilization, but this culture and civilization in turn challenges our intellect to maintain and improve it. Presumably, as our intellect continues to evolve so will our culture and civilization in unknown ways that **futurists** take great pleasure in trying to predict.

A Final Note
The Beast Within

Each human and even the whole of the human species is two separate but intertwined beings. One side of us—the intellectual being—is basically good, gentle, and compassionate. Our other side—the instinctive being—is animalistic and can be violent, selfish, or cruel as needed to ensure its survival or advance its own agenda. Part of the power of the human intellect is to keep the instinctive beast chained away as much as possible. However, there is not a single person who at some time or another in their life will not experience a brief loss of control and, for a moment, his or her dark, animalistic side will emerge. Rage may cause you to lash out in anger or paralyzing fear may cause you to lose control of your bladder. In whatever form you encounter it, the human dark side is not a pleasant or healthy place to spend much time.

Unfortunately, the dark side of human nature is plundering the planet and putting many of the various groups of animals we have discussed in this book in peril. It is also plundering our species. What a grand and glorious culture and civilization we might have if not for hate, racism, religious prejudice, greed for money, moral corruption by position of power, and all the other evils of the dark side. We really are a strange creature: a being with the intellect to reach for the stars yet one who still battles and kills both his human and animal kin in the mud. Charles Darwin may have described us best when he said: "Man with all his noble qualities…still bears in his bodily frame the indelible stamp of his lowly origin."

The fate of humanity and our animal kin rests with each of us and with all of us. I have written this book with two fervent hopes in mind. One is that by reading this book you will have gained both a respect for our wonderful world and the amazing animals we share it with and a better understanding of our animal kin and yourself. The second is that you will use this respect and understanding as inspiration to do whatever you can to protect and safeguard your planet, your fellow animals, and your own species. Goodbye and best wishes for continued success in your future biological endeavors.

In Summary

- Humans are creatures with an almost alien-sized brain that can write poetry, raise giant cities, travel in space, and even contemplate and explore their own mind; an advanced technological being wrapped in the body of a primate, a naked primate at that.

- Primates, especially humans, are remarkably recent animals on the evolutionary stage. Primate-like mammals (or proto-primates) first arose in the early Paleocene epoch about 65 million years ago.

- The order Primate embraces 230-270 species aligned into two suborders and 13 families:

>   Suborder Strepsirrhini
>   Family Cheirogaleidae—dwarf and mouse lemurs
>   Family Lemuridae--Lemurs
>   Family Lepilemuridae—Sportive lemurs
>   Family Indriidae—Wooly lemurs and allies
>   Family Daubentoniidae—Aye-aye
>   Family Lorisidae—Lorises, pottos, and allies
>   Family Galagidae—Galagos
>   Suborder Haplorhini
>   Family Tarsiidae—Tarsiers
>   Family Callitrichidae—Marmosets and tamarins
>   Family Cebidae—Capuchins and squirrel monkeys
>   Family Cecopithecidae—Mandrils, baboons, and macaques
>   Family Hylobatidae—Gibbons
>   Family Hominidae
>   Subfamily Ponginae—Orangutan
>   Subfamily Homininae
>   Tribe Gorillini—Gorillas
>   Tribe Hominini

- The characteristics of primates:

1. Shortening of the face accompanied by reduction of the snout
2. The general retention of five functional digits on the fore and hind limbs. Enhanced mobility of the digits, especially the thumb and big toes, which are opposable to the other digits.
3. A semi-erect posture that enables hand manipulation and provides an optimal position preparatory to leaping.
4. Claws modified into flattened and compressed nails to support the sensitive digital pads on the last phalanx of each finger and toe.
5. A shoulder joint allowing a high degree of limb movement in all directions and an elbow joint permitting a rotation of the forearm.

6. A reduced snout and olfactory apparatus
7. A reduction in the number of teeth compared to primitive mammals.
8. Teeth and digestive tract adapted to an omnivorous diet
9. A complex visual apparatus with high acuity and a trend toward development of forward-directed binocular eyes and color perception. Well-developed hand-eye motor coordination.
10. A large brain relative to body size, in which the cerebral cortex is particularly enlarged.
11. Only two mammary glands (some exceptions).
12. Typically, only one young per pregnancy associated with prolonged infancy and pre-adulthood.

- Modern humans seem to be defined anatomically by two main characteristics: bipedalism and a large brain with developed sites supporting speech and intelligence.
- Modern humans have gone through a number of different stages and types of evolution: anatomical, agricultural, social, and technological.

Review and Reflect

1. *My Father or My Cousin?* Some people believe that human evolution means we evolved directly from apes. Is this a scientifically accurate interpretation of human evolution?

2. *Is it Just Me or is it Getting Colder?* Explain what role climate change may have played in the evolution of humans.

3. *Compare and Contrast* In what ways are humans similar to apes—chimps, gorillas, and orangutans? How are we different than apes?

4. *A Critical Difference* A lecture on human evolution you attended has just ended. Your friend seated next to you says. "The professor made an issue out of the difference between intelligence and intellect. Aren't they the same thing really?" What would you say?

5. *The Cultural Ape.* We have used the term "cultural ape" to describe modern humans at the peak of their evolution (if what we are today actually is the "peak" of our evolution). Do you agree or disagree with our usage of that term to describe modern humans? Defend your answer.

6. *Look into the Crystal Ball.* Given what you now know about where we may have come from (evolution) and present day human culture, civilization, and behavior, play the role of a futurist and predict what human culture and civilization will be like 500 years from now.

7. *What's in a Quote?* Arthur Koestler said: "The evolution of the brain not only overshot the needs of prehistoric man, it is the only example of evolution providing a species with an organ which it does not know how to use." What do you think he meant by that remark? Interpret it in your own words.

Create and Connect

1. ***Humans Arise*** Write a position paper on the origin of humans and how they developed into modern *Homo sapiens sapiens*.

   Guidelines:

   A. Compose a position paper, not an <u>opinion</u> paper. Defend your position with as many facts, Figures, quotes, and pertinent information as possible.

   B. Your work will be evaluated not on the "correctness" of your position but the quality of the defense of your position.

   C. Your instructor may provide additional details or further instructions.

2. ***A Day in the Life*** Describe one day in the life of an australopithecine and one day in the life of *Homo erectus*.

3. ***The Human Condition*** You are the science editor of a large metropolitan newspaper. Your editor has given you the following assignment: Write a column entitled, *The Human Condition—What Does It Mean to be Human?* What would you write?

APPENDIX A
SCIENTIFIC WRITING

For the scientific enterprise to be successful, scientists must clearly communicate their ideas and work. Scientific findings are never kept secret. Instead, scientists share their ideas and results with other scientists, encouraging critical review and alternative interpretations from colleagues and the entire scientific community. Communication, both verbal and written, occurs at every step as the science cycle turns.

How well we communicate is determined not by how well we say things but by how well we are understood.

—Andrew Grone

One of the objectives for each chapter in this book is to develop your communications skills through a variety of written review questions and creative challenges. In these assignments, you will be asked to write everything from formal scientific reports to essays to position papers to short stories. The exact format and details will be given with each assignment but before you write and as you write consider these guidelines, hints, and suggestions:

- ☑ *We can't read your mind.* Your instructors can only evaluate what you have actually written. Do not make the mistake of using the old excuse that, "They will know what I mean" to rationalize incomplete work.
- ☑ *Mean what you say and then write what you mean.* Refer to a dictionary and thesaurus to ensure clarity and proper word usage. For example, you allude to a book but you elude a pursuer. Also, use correct composition and grammar.
- ☑ *Be clear and concise.* Delete unnecessary words such as adjectives and adverbs that have limited use in describing your work. Write clearly in short and logical but not choppy sentences. For example, "The biota exhibited 100% mortality response." is a wordy and pretentious way of saying, "All the organisms died." Keep it simple and straightforward. Do not use vague or ambiguous words or phrases. For example, "An adequate supply of planaria will be needed" reveals no useful information. Instead employ a specific phrase, "100 planaria will be the sample size."
- ☑ *Keep related words together.* The following sentence is taken from an actual scientific publication. "Lying on top of the intestine, you perhaps make out a small transparent thread." Do

we really have to personally lie on top of the intestine to see the thread? The author should have written, "A small transparent thread lies atop the intestine."

☑ ***Use active voice whenever possible***. Using "I" instead of "me" makes the paper easier to read and understand. However, in the Methods and Materials section of a formal scientific paper, passive voice should be used so that the focus of the writing is on the methodology, rather than the investigator.

☑ ***Think metric***. Use metric units for all measurements and numerals when reporting measurements, percentages, decimals, and magnification. When beginning a sentence, write the number as a word. Numbers of ten or less that are not measurements are written out. Numbers greater than ten are given as numerals. Decimal numbers less than one should have a zero in the one position (e.g., 0.153; not .153).

☑ ***Proofread and edit.*** Carefully proofread your work even though your word processor has checked for grammatical and spelling errors. These programs cannot yet distinguish between "your" and "you're," for example. Also, edit then revise and then do it again.

☑ ***Do your own work and write your own words***. Plagiarism is a serious offense, and it can result in serious consequences.

☑ ***Regarding references:***

A. No footnotes are used, but each reference should be cited in the body of thepaper. For example, Bicak (1999) reported that red-tailed hawks nest in large trees.

B. How to cite:

Journal articles	Watson, J.D. and F. H. Crick. "Molecular Structure of Nucleic Acids: A Structure for Deoxyribose Nucleic Acid. " *Nature,* 1953, vol 171, pp. 737-738.
Books	Darwin, C. R. *On the Origin of Species.* London: John Murray, 1859.
Government Publications	Office of Technology Assessment. *Harmful Non-indigenous Species in the United States.*Publication no. OTA-F-565. Washington, D. C. U.S. Government Printing Office, 1993.
Encyclopedia-	Bergman, P. G. editor. (2002). "Annelids." In the *Encyclopedia Britannica.* (Vol 1, pp. 223-225). Chicago: Encyclopedia Britannica
Internet Sources-	Council of Biology Editors. 1999. Oct. 5. home page. <http://www. councilscienceeditors.org> Accessed Oct 7, 2004.

NOTE: Personal or popular websites are normally not legitimate references. Check with your instructor if in doubt about citing an on-line reference.

Interview- An interview with Dr. Joseph Springer regarding Sandhill cranes. Oct 12, 2004.

NOTE: Direct quotations must be cited. For example:

"Sandhill cranes have been staging along the Platte River for thousands of years." (Springer, 2004)

GLOSSARY

Abiotic: Any nonliving factor of the physical environment

Abyssal Plains: The seafloor beyond the continental slope

Abyssopelagic: The open sea beyond the continental shelf below 4,000 meters

Acanthor Larva: Shelled larva of an Acanthocephalan

Accommodation: Adjusting for near and distant objects

Acetabulum: A ventral, hookerless sucker of digenetic flukes

Acoelomate: An animal lacking a true coelom

Acontium or Acontial Filament: A free thread attached to the aboral end of the mesentery

Acrodont Teeth: Teeth attached on the surface of the jaw

Acrorhagi: Inflatable tentacles loaded with nematocysts found in sea anemones

Actinopterygii: Ray-finned fishes

Actinula Larvae: One possible larval stage of a hydrozoan that forms directly from the planula larva

Active Transport: Movement of molecules across a cell membrane from an area of low concentration to an area of high concentration

Aculeus: The stinger at the end of the metastoma in scorpions

Adhesive Tubes: Tubes that contain cement glands that secrete adhesive substance used to temporarily attach a gastrotrich to objects in the environment

Advertisement Calls: Calls produced to attract a mate

Aeroplankton or Aerial Plankton: Small invertebrates (insects and spiders) that float in the air.

Age Structure: The distribution of individuals across different cohorts (ages) in a population

Agroecosystem: An artificial system of agriculture created by humans that has replaced prairies

Albumin: The white of an egg

Alloparenting: Members of the same species help the breeding pair in raising the young

Allopatric: Existing in different geographical areas

Altricial: Birds born bare with eyes closed

Altruism: Concerns for others not related to us

Altruistic Behavior: A behavior in which an individual appears to act in a way that benefits others at cost to itself

Alula: A small projection on the anterior edge of the wing; it functions as a freely moving first digit and bears three or four feathers; also called a bastard wing or a spurious wing

Alveoli: Air pockets that fill the mammalian lung

Ambulacral Grooves: Grooves located down the length of each arm of sea stars where tube feet protrude

Ametabolous Metamorphosis: The process of development in which the larval body structure is a smaller, immature version of the adult (incomplete metamorphosis)

Amictic: Females produced via parthenogenesis (asexual)

Amino Acids: Small molecules that contain an amine group and a carboxylic acid and make up proteins

Amphids: Chemoreceptors found on the head of nematodes

Amplexus: Mating squeeze by a male frog on a female frog

Ampullae of Lorenzini: Electroreceptors that form a modified lateral line in sharks

Anadromous: Hatch and grow in freshwater, migrate to the ocean to mature, and then return to freshwater to spawn

Anal Canals: Tubular structures in cnidarians that collect undigestible wastes that are expelled out of the body

Anal Fin: A single fin near the anal opening of fishes

Anal Pores: Small holes that connect the anal glands to the outside

Analogous: Similarity in structure that results from independent evolutionary events

Anaphase: The stage of mitosis when chromosomes are split and the sister chromatids move to opposite poles of the cell

Anapsida: The earliest amniotes that lack openings in the skull

Ancestral Characteristics: Characters found in the ancestors of a group: basal traits

Annulations: A form of sexual reproduction by which an organism regularly breaks along preformed lines

Annuli: Rings found on the body segments of annelids

ANOVA: A statistical test that is used to determine if the means of a trait or characteristics are significantly different among more than two groups

Antagonistic Muscles: Opposing muscles

Antennal Glands: Excretory glands that contain a single pair of nephridia that are found on the head in association with the antennae

Anterior: Pertaining to the head end

Anthropoids: Halplorhini: group consisting of tarsiers, tamarins, baboons, monkeys, apes, and humans

Anthropology: Study of humans

Anthropomorphism: The application of human characteristics to nonhuman animal behavior

Antlers: A pair of bony, branched structures that protrude from the frontals of the skull

Aorta: The largest artery in the body that transports blood from the heart to the body

Apocrine Glands: A type of sweat gland that opens either into a hair follicle or into a space where there once was a hair follicle

Apolysis: Digestion of the old procuticle by hormones secreted from glands in the hypodermis

Aposematic or Warning Coloration: A bright, conspicuous color pattern used to advertise toxicity to predators

Appendicular Skeleton: The limbs and girdles of the skeleton

Aquiferous System: The cells and tissues that carry water through the canals of the body of a sponge

Aboreal: Pertaining to tree-dwelling

Arboreal: The bottom end of polyps that attaches to the substratum

Archaeocytes: Large motile ameboid cells that can change into any other type of sponge cell and, thus, assist with building the skeleton, nourishing eggs, distributing food, and mediating chemical and physiological responses

Aristotle's Lantern: A feeding structure that consists of five protractible calcareous teeth and muscles that control protraction, retraction, and grasping movements of the five teeth

Anthropodization: A suite of highly successful adaptations that evolved in the arthropods

Articular: A small bone at the back of the lower jaw

Ascites: Accumulation of excess fluid in the abdomen

Asconoid Sponges: Simple, small, vase-like sponges that do not contain body folding

Associative Behavior: Behavior that is learned

Associative Learning: Learning that occurs by association with certain events

Asymmetrical: Lacking any definite geometric design in body parts

Atmosphere: Layers of air surrounding the planet

Atoke: the anterior portion of some tube-dwelling and burrowing worms that does not differentiate into specialized structures during the breeding season

Atoms: Matter composed of a central nucleus with positively charged protons and neutral neutrons and an outer cloud of negatively charged electrons

Atrial (excurrent) Siphon: An outlet for water and waste expelled by tunicates

Atriopore: A ventral pore or opening in the atrium

Atrium: A large, muscular chamber of the heart

Atrium: In the flatworm, a chamber that contains the penis and often the terminal portion of the female tract

Atrium: The interior cavity of a sponge

Auricles: Spike-like projections on the side of the head of turbellarians that contain chemoreceptors

Auricular Feathers: Feathers found around the external ear opening in birds

Autotomize: To detach one's own limbs (occurring in some molluscs, echinoderms, and arthropods)

Autozooids: the conspicuous secondary polyps of sea pansies, and sea pens

Axial Skeleton: the cranium, backbone, and ribs of the skeleton

Axial Skeleton: A stiff internal axial rod embedded in the body tissue of many colonial anthozoans

Ballooning: A means of dispersal of small spiders in which they release a loop of parachute silk that is caught by the wind and blows them into the air

Barbels: Short sensitive tentacle-like structures that surround the mouth of hagfish

Barbicels: Tiny hooks that hold together the barbules of feathers

Barrier: Any physical obstacle that prevents a species from occupying an otherwise biologically and geographically acceptable area

Batesian Mimicry: The process of palatable species copying the coloration, structure, and behavior of a poisonous or unpalatable species

Bathypelagic: The open sea beyond the continental shelf from 1,000 meters to 4,000 meters

Behavior: The organized and integrated pattern of activity by which an organism responds to its environment

Bellows Lung: Mammalian lung

Benthic Zone: The seafloor

Bilateral Symmetry: A body design in which one side of the body is a mirror image of the other

Binocular Vision: Vision that produces a sharp image in front of the head due to the eyes being set far to the front of the head

Binomial Nomenclature: A Latin naming system that provides each organism with a unique genus and species combination

Biodiversity: The number of different species of plants and animals

Bioerosion: the process by which invertebrates bore into the shells of molluscs and corals and kill them

Biogenesis: The notion that any organism arises from another organism(s) of the same species

Biogeochemicals: Chemical elements that cycle between biota and the physical environment

Biological Community: All of the populations of different species living together in the same place at the same time

Biological Magnification: The process by which toxic chemicals become increasingly concentrated as they pass up the food chain

Bioluminescent: Producing light by biological mechanisms

Biomass: The largest geographical biotic unit that consists of various ecosystems and their attendant communities with similar environment requirements

Biome: The largest geographical biotic unit and consists of various ecosystems and their attendant communities with similar requirements of environmental conditions.

Biomineralization: The process of removing minerals from the blood and crystallizing them for another purpose

Bioprospect: Search for organisms with medicinal value

Biosphere: A layer 100 feet or so above and within the planet where life occurs

Biotechnology: The manipulation of living organisms for practical purposes and/or to produce useful products

Biotic: Describes anything living (living organisms)

Bipedal Locomotion: Habitually walking upright on the hind legs

Biphasic Life: Aquatic larvae and terrestrial adults

Biradial Symmetry: A combination of radial and bilateral symmetry

Biramous: Branching appendages

Bladder: An organ that stores excess liquid

Blastocyst: An embryo

Blastopore: The first opening that develops in an embryo

Blastula: The fluid-filled ball of cells that forms a developing embryo after multiple cleavage stages

Bond: To join together (as in a chemical bond or union)

Book Gills: The posterior of five pairs of appendages on the opisthosoma

Bower: A structure built by male bower birds to attract mates where they display a complex set of behaviors

Brachial Heart: A heart that is located at the base of the gills in cephalopods and collects unoxygenated blood, which is then pumped to the gills

Brachiate: Move by swinging through the trees

Branchial System: Gas exchange system of enteropneusts including the gill pores, branchial sacs, and gill slits

Brille: Ocular scale

Brood Parasitism: The process by which an organism lays its eggs along with a host's eggs causing the host to provide for the intruder's young often at the expense and death of the host's young

Brood Patches: Places where the feathers fall out on the underbody of the parent to expose the warm, bare skin that can be pressed directly against the eggs

Brown Fat: Special fat deposits that produce extra heat when metabolized

Buccal Cavity: The external opening/cavity into the digestive system

Buccal Pump: A positive pressure system for ventilating the lungs

Buccal Pumping: Pumping action of the mouth and opecula to create a respiratory current across the gills

Buccal Respiration: Gases exchanged across moist membranes of the mouth

Buccal Tube Feet: A wreath of tube feet that extend from the ring canal (in addition to the five radial canals) around the mouth

Budding: The process by which bumps or protrusions that form on the adult body and break off from the adult body eventually form a new individual (asexual reproduction)

Buds: Bumps or protrusions that form on the adult body and break off from the adult body to eventually form a new individual

Bulbus Arteriosus: A large tube that conducts blood into the aorta

Calamus: Quill

Calcareous Skeleton: The skeleton of the stony coral, which is made of secreted calcium carbonate

Calciferous Glands: Calcium secreting glands that function to remove excess calcium taken in with food, to remove excess CO_2, and to regulate the acid-base balance of body fluids.

Calyx: A cup-like body form

Calyx: Attaches the arms of a crinoid to the stalk

Cameral Fluid: Liquid that a *Nautilus* uses to fill the chambers of its shell to provide buoyancy

Canal Pores: Openings to the outside from the lateral line system

Candling: Holding fish fillets up to bright light to check for the presence of worms

Canines: Teeth used for piercing

Capitulum: Anterior portion of a mite that houses the mouthparts

Carapace: Upper, domed shell of a turtle

Carbohydrates: Any of a large group of organic compounds occurring in foods and living tissues and including sugars, starch, and cellulose. They contain hydrogen and oxygen in the same ratio as water (2:1) and typically can be broken down to release energy in the animal body.

Cardiac Stomach: A portion of an echinoderm's stomach that is inverted during feeding

Carnivores: Pertaining to animals that eat other animals

Carotenoids: Fat-soluble yellow, orange, and red pigments

Carrying Capacity: The maximum number of individuals of a population that can be supported by the available resources of an area for an indefinite period of time

Cartwheeling: A mating behaivor in birds in which a potential mating pair soars to great heights and then plunge to the ground and goes back up again just before reaching the ground

Catadromous: Migrating from fresh water to saltwater to breed

Catch Connective Tissue: A dense, white fibrous and mutable connective tissue found in the body of sea cucumbers

Caudal Autonomy: Easy separation of a tail from the body

Caudal Fin: Tail fin

Cecum: An accessory pouch of the mammalian digestive system

Cells: The smallest unit of life; living building blocks

Cellulose: A biologically important polysaccharide that composes plant cell walls

Centriole: A cylindrical cell structure composed mainly of a protein called tubulin that is found in most eukaryotic cells

Centrosome: An organelle that serves as the main microtubule organizing center of the animal cell as well as a regulator of cell-cycle progression

Cephalic Shield: A muscular, modified proboscis

Cephalic (head) Tentacles: Tentacles that extend from the head and may bear eyes as well as tactile and chemoreceptor cells

Cephalization: Development of a defined head end

Cerebellum: A section of the mammalian brain that interprets visual cues and coordinates muscles for movement and equilibrium

Cerebral Ganglion: A rudimentary brain

Cerebrum: A section of the mammalian brain responsible for behaviors such as eating, singing, navigation, mating, and nest building in birds

Cervicals: Vertebrae of the neck

Chaetae or Setae: Bristle-like structures that extend from the body

Chaparral: A shrubland ecosystem found primarily in California

Character Displacement: A type of reservoir partitioning in which over time natural selection acts on the plasticity of animal forms and selects different forms and shapes of an organism and changes the morphology of populations (or species) that shared fundamental niches

Chelae: Pinchers

Chelicerae: Anterior appendages of an arachnid often specialized as fangs

Chemoreception: Detection of chemical substances such as food chemicals and hormones

Chemoreceptors: Receptors that bind specific chemicals in the environment that lead to physiological changes in an organism

Chi Square Test: A statistical test that is used to determine if two or more samples are significantly different from one another (if they are drawn from different populations

Chloragogen Cells: Specialized nitrogen-accumulation cells that line the coelom and extract nitrogenous wastes from the blood

Choanocytes: Flagellated sponge cells that help move water currents through the chambers and canals of a sponge

Choanoderm: The inner lining of the atrium of a sponge

Chorus: Calls produced by a group

Chromosomes: Coiled and thickened chromatin

Chromatin: Mass of genetic material composed of DNA and proteins that condense to form chromosomes during eukaryotic cell division

Chromatophores: Specialized pigment-containing cells in the dermis

Chrysalis: The last larval exoskeleton of butterflies that surrounds the pupa during metamorphosis into the adult

Circumoral Nerve Ring: A pentagonal ring of nerves around the mouth of an echinoderm

Cirri: Ciliated, finger-like projections extending from the hood of cephalochordates and function to trap and prevent particles from clogging the hood

Cirri: Long, jointed appendages in barnacles used for feeding

Cirri: Projections that extend from the stalk of a crinoid and are arranged in whorls

Cirri: Sensory tentacles

Civilization: Complex human societies that have developed social structure and technologies

Clade: A group that includes a common ancestor and all the descendants of that ancestor

Cladogram: A branching line diagram that reveals the phylogenetic relationships between select groups

Claspers: Male sex organs used during copulation

Classical Approach: the philosophical approach to science (founded by the ancient Greeks) in which the natural world is explained by reason and logic rather than by experimentation

Classical Conditioning or **Pavlovian Conditioning**: A learning process by which a response that was initially triggered by stimulus is transferred to another stimulus that had no effect

Cleavage: The process by which the zygote divides into increasingly smaller cells

Climax Stage or **Climax Community**: The period of time marked by the presence of the final successors, typically *k*-selected species in an area

Clitellum: An external mucus producing band found on oligochaetes

Cloaca: A single posterior opening that empties the digestive, urinary, and genital tracts

Closed System: A system in which matter (chemicals, energy, etc) is recycled within the system

Cnidae: Stinging or adhesive structures found in cnidarians

Cnidoblasts: Cells that form from interstitial cells in the epidermis of cnidarians and that produce cnidocytes

Cnidocil: The trigger –hair on the cnidocyte once stimulated by mechanoreception causes the cnidocyte to discharge

Cnidocyte: A fully formed cnidocyte that contains the cnidae

Coaxial Glands or **Malphigian Tubules**: Paired, thin-walled spherical sacs that absorb nitrogenous wastes from the hemolymph

Coccyx: A line of small vertebrae at the base of the spinal column of chordates

Cochlea: Part of the inner ear concerned with hearing

Cocoon: A sealed capsule that forms around the fertilized eggs in the mucus ring of some worm species

Coelom: A fluid-filled body cavity that forms within the mesoderm

Coelomic Sinuses: Openings in the coelom

Cognition: Conscious intelligent acts

Cohort: A group of individuals of the same age within a population

Collar: Neck-like structure

Collencytes: Ameboid cells found in the mesohyl of sponges that secrete fibrous collagen

Colloblasts: Special adhesive cells of ctenophores

Colonies: Large, ordered groups of the same species

Comb Plates: Rows of ciliated cells on ctenophores used for locomotion

Community: An ecological unit consisting of populations of different species that coexist

Competitive Exclusion: The removal of a species from an area due to negative interactions with another species sharing that same habitat and niche

Concertina Movement: A type of locomotion in snakes in which they extend forward while bracing S-shaped loops against the sides of an enclosing passage

Conchin: A protein that composes the cuticle of molluscs

Conditioned Stimulus: A stimulus that initially fails to cause a response without trial and error

Cones: Photoreceptors that receive information about color vision

Conispiral: Coiled in the shape of a cone

Conspecies or **Intraspecies Competition**: Interactions between individuals of the same species

Conspecies: Animals of the same species

Consumers or **Heterotrophs**: Organisms that consume other organisms as an energy source

Control Group: The group in an experiment that does not receive the variable under examination and is used to compare the results with the test group

Convergent Evolution: Evolution of similar traits in species that are not closely related due to adaptation to a similar environment or niche

Conversion Efficiency: In ecology the ratio of net production from one trophic level to the next highest level

Coprolites: Fossilized fecal droppings

Copulatory Dart: A sharp, calcareous structure that is driven into the body wall of a partner as part of a mating ritual

Copulatory Spicules: External reproductive structures of male nematodes

Coral Bleaching: The process by which corals lose their color because of the loss of their photosynthetic symbionts, zooanthellae, due to an environmental stressor

Corona: The anterior, ciliated portion of a rotifer

Coronal Muscles: Circular sheets of epidermal muscle fibers around the bell margin and over the subumbrellar surface of medusae

Corpora Allata: Neurosensory cells that store and release PTTH and produce juvenile hormone

Cortex: The middle layer of a hair

Countershading: Dark colors on top (dorsal) and light colors on the belly (ventral)

Courtship Rituals: Elaborate and complex behaviors associated with mating

Coxal Glands: Excretory glands that contain as many as four pairs of nephridia as the bases of walking legs

Cranial Nerves: Nerves that pass information from the outside directly to the brain

Cranium: Braincase

Crop: A large thin-walled chamber that stores food coming down the pharynx and esophagus

Crown: Composed of the calyx and arms

Cryptic Coloration or **Camouflage**: A defensive coloration strategy that allows organisms to effectively disappear against certain backgrounds

Ctenidal Vessels: The site of gas exchange of hemolymph before it is drawn back into the heart and pumped to the hemocoel

Ctenidia: Gills of bivalves

Cud: Partially chewed food

Culture: The institution and systems that build and maintain the physicality of an area as well as those that protect, feed, clothe, house, and entertain the human residents of that area

Cupula: A gelatinous structure that surrounds the neuromast cells

Cursorial: Walking

Cutaneous Respiration: Gases exchanged across the skin

Cuticle: The outer layer of a hair

Cuticle: A nonliving thing and transparent outer body layer

Cyclomorphosis: Morphological variations in body form due to seasonal or nutritional changes

Cydippid Larvae: Planktonic larval stages of ctenophores

Cynodonts: A type of therapsids that superficially resembles mammals

Cysticercus: A larval stage of a tapeworm that hatches from the oncosphere larva and develops in the muscle of the host where it breaks out of a cyst and develops into the adult worm

Cytokinesis: The process in which the cytoplasm of a single eukaryotic cell is divided to form two daughter cells

Cytology: The study of the structure and function of cells

Cytoplasm: The part of the cell enclosed by the plasma membrane

Cytoskeleton: A scaffolding of protein filaments and tubules found within the cytosol

Cytosol: A translucent gel in which all internal cell components are suspended

Dactylozooids: The defensive polyps of a hydrozoan

Decay: Break down

Decomposers: Organisms that feed on dead organisms

Definitive Host: Animals that harbor the adult form of a parasite

Dehydration Synthesis: A process that bonds molecules together through the removal of water molecules

Density: The number of individuals living in a particular area

Density Dependent: Abiotic factors whose effects on a population changes with the density of the population

Density Independent: Abiotic factors whose effect on a population is not affected by the density of the population

Dentary: The bone that forms each half of the mandible in mammals

Denticulated: Possessing teeth

Deoxyribonucleic Acid (DNA): The double-stranded genetic code, composed of nucleic acids, for the formation of proteins

Dermal Gills or **Papulae:** Thin folds in the body wall that function in gas exchange

Dermis: Spongy, inner layer of skin

Desiccation: Drying out

Determinate Growth: A pattern in which growth stops in the adult

Detritus: Waste or remains of living organisms

Deuterocereburum: A section of the supraesophageal ganglion

Deuterstomes: Organisms with embryos that develop via radial cleavage of the first cells and from an anus first at the blastopore

Diadromous: Migrating between fresh water and saltwater

Diaphragm: A muscle that separates the thoracic cavity from the abdominal cavity

Diapsida: Amniotes that possess two temporal openings

Diencephalon: Connects the forebrain and midbrain and functions in homeostasis

Diffusion: The process by which atoms and/or molecules move from an area of high concentration to an area of low concentration

Digestive Ceca: The digestive gland formed as an evaginations off the midgut

Digitgrade Locomotion: Walking and running with raised heels

Dioecious: Possessing separate sexes

Diphyodont: Milk or deciduous teeth replaced by a set that lasts the rest of the lifetime

Dipleurula Larva: A basic larval type of an echinoderm with bilateral symmetry, three coelomic sacs, and bands of cilia

Diploblastic: Possessing two cell layers

Direct or **Synchronous Flight**: A mode of flying in which the muscles acting on the bases of the wings contract to produce a downward thrust and other muscles attach dorsally and ventrally on the exoskeleton to produce upward thrust

Distal: Farther down the body

Distress Call: A call produced in response to pain or being seized by a predator

Domain: The largest/highest taxonomic category above kingdom

Dorsal: Pertaining to the top or back side

Dorsal Aortas: Arteries that receive blood from the ventral aorta in cephalochordates

Dorsal Fins: A fin located on the back (dorsal) region

Dorsal Vessel: A blood vessel located over the digestive tract on the back side of an organism

Dyck Texture: Structural texture effect in microscopic portions of the feather that lead to certain blue and green colors and metallic sheens

Dynamic Soaring: Flight that requires continuous winds in which the bird is carried high; a wind-pushed glide

Eccrine Glands: A type of sweat gland that releases a clear, watery fluid onto the surface of the skin

Ecdysis: The process of molting (shedding the outer skeleton layer and forming a larger one

Ecological Niche: All of the resources and physical conditions that a species requires for survival

Ecological Pyramids: Diagrams that represent the relationship between energy and trophic levels within an ecosystem

Ecological Succession: The process by which communities gradually change from more simple to more complex assemblages

Ecology: The branch of biology that studies the interactions between live and the physical environment within the biosphere

Ecosystem: A complex of communities and their interactions with the physical environment in a particular area

Ectoderm: The outer cellular layers of the developing gastrula

Ectoneural (oral): A sensory neural network that coordinates the tube feet and consists of a pentagonal circumoral nerve ring and radial nerves

Ectoparasites: An organism ((parasite) that feeds on the outside of its host

Ectothermic: Derived heat from external soruces

Elements: Types of atoms

Embryology: The study of the development of embryos starting with the fertilization of the zygote

Embryonic Diapause: A strategy in which the embryo (blastocyst) does not immediately implant in the uterine wall but is maintained in a state of dormancy

Emigration: Process of individuals leaving an area or population

Empiricism: The approach to science (founded in Western culture during the Renaissance) in which the natural world is explained through accurate observation, measurement, and experimentation

Encystation: The process by which a foreign object that enters a bivalve is covered in many layers of nacre

Endocytosis: A process by which cells engulf large extracellular particles

Endoderm: The inner cellular layers of the developing gastrula

Endoparasites: An organism that feeds on the inside of its host

Endoplasmic Reticulum: A network of tubules and flattened sacs that perform a number of functions within eukaryotic cells

Endostyle: An organ in cephalochordates that binds iodine and produces strings of mucus that trap food from the water as it passes through the gill slits and into the atrium

Endosymbiotic Theory: An explanation for the rise of eukaryotic cells

Endosymbiont: An organism living inside another organism benefitting both organisms

Endothermic: Heat derived from internal processes

Enterocoely: The process of the blastopore developing into the anus

Entoneural (aboral): A neural system that consists of a nerve ring around the anus and radial nerves in echinoderms

Enzymes: Biological catalysts responsible for thousands of metabolic reactions that sustain life

Enzymatic–gland Cells: Wedge-shaped ciliated cells of the gastrodermis of cnidarians that produce the enzymes for extracellular digestion

Epicuticle: The outermost layer of the exoskeleton of arthropods that functions as a barrier to water and microorganisms

Epidermis: The outer most layer of cells in animals with a tissue grade of organization

Epidermis: Outer layer of skin

Epifaunal: Bottom dwelling but not a burrower

Epigenetics: A field of genetics that examines the reversible and heritable changes that can occur without a change in DNA sequence

Epigyne: Female genital opening of spiders

Epiparasitism: The process of one parasite feeding on another parasite

Epipelagic: The open sea beyond the continental shelf from the surface down to 200 meters. The sunlit layers

Epitoke: The posterior portion of some tube-dwelling and burrowing worms that develop gonads during the breeding season

Equilibrium Model: A biological model depicting stable and unchanging communities

Esophagus: A tube that extends from the buccal cavity (or pharynx) to the stomach and is part of the digestive system

Estivation: A type of hibernation that occurs under stressful conditions in which the metabolic rate approaches zero until conditions are favorable again

Estrous Cycle: Estrus cycle that occurs at regular intervals throughout the year

Estrus: Reproductive receptivity in mammals, heat

Estuary Ecosystems: An ecosystem that is located at the junction of fresh and salt water and contains a mixture of the two (brackish)

Etching Cells: Specialized cells in boring invertebrates that chemically remove fragments or chips of calcareous material

Episodic Behavior: A behavior that is adjusted based on recalled social content of a previous event

Ethology: The study of animal behavior

Eukaryotes: Organisms that have cells with membrane-bound organelles

Eukaryotic Cell: Cells that contain membrane-bound organelles

Euprimates: True primates

Euryhaline Fishes: Fishes that can live in both salt and freshwater

Eusociality: Members group of the same species living together in a social group in which there is a division of labor and reproductive and nonreproductive individual

Eutely: Phenomenon in which adults of a cell line always have the same number of cells

Eutrophic Lakes: Lakes with an abundant supply of organic matter and minerals and high biological activity

Eutrophication: The process by which oxygen is depleted from the water due to the rapid/explosive growth of algae, which in turn causes the suffocation of fish and other aquatic fauna

Evolutionary Theory of Aging: The theory that ageing is a result of natural selection acting on a species' life span

Exchange Pools: Places where chemical are stored for a short period of time (typically within biota)

Exocytosis: A process by which a cell directs the contents of secretory vesicles out of the cell membrane and into the extracellular space

Exoskeleton: A chitinous skeleton found primarily in arthropods

Experimental Theory: A theory that is derived from an experiment and has a median level of truth probability

Exponential Growth: Rapid growth of a population

External Ectoderm: The outer germ layers of an embryo that will become the skin, skin glands, hair, and nails

External Ear: Anatomical structure from the sounds-conducting canal to the ear drum

Extinction: Cessation of a behavior

Extinction: No longer existing or living; an extinct species

Extracellular: Outside the body

Extrinsic: Relating to something outside of itself

Exumbrella: The convex upper surface of the bell of a cnidarian

Eyespots or **ocelli**: Structures that contain photoreceptors and can detect light but are not able to form images

Fangs: Specialized teeth in snakes used to deliver venom into victims

Fat: A biologically important hydrophobic organic molecule

Fauna: Animals

Fecundity: Potential reproductive capacity

Filial Imprinting: Bonding between parent and offspring

Fitness: Chance of reproductive success

Flagellated Canals: Evaginations of sponge body into finger-like projections that are lined by cells possessing flagella and can beat to move water through the canal

Flame Cell: A cell with a tuft of beating cilia that functions to bring water and wastes into the canals of the protonephridia for elimination

Flapping Flight: Flight produced by up and down motion of the wings

Fledge: Become mature enough to fly

Flexors: Muscles that bend the body or limb at the articulation point

Flight Feathers: The longer contour feathers of the wings and tail

Flooded Forests: Forests covered completely or at least occasionally in water.

Flora: Plants

Flying Phalangers or **Wrist-winged Gliders**: Animals with a fold of skin running from the wrists to the ankles and used in gliding

Food Chain: A linear map of who eats who (or what) in an ecosystem

Food Web: Overlapping and interlocking food chains

Foot: A muscular structure used for locomotion and burrowing in molluscs, which develops from the ventral body wall

Foot: The attachment organ of a rotifer that possesses pedal glands

Foraging: The process of finding food

Forebrain: Composed of the olfactory lobes and the telencephalon

Foregut: The anterior most section of the arthropod digestive track that serves to ingest, transport, store, and mechanically digest food

Fossil: Any remains, impressions, or traces of living organisms from past geological ages

Fossorial: Pertaining to digging and burrowing

Fovea: A portion of the eye with dense concentrations of receptor cells that perceives images more sharply

Frontal Plane: A plane through the body that cuts the body into dorsal and ventral sections

Fundamental Niche: The total range of environmental conditions suitable for survival of a species that discounts the effects of interspecific competition and predation

Furcula or **Wishbone**: A forked bone found in theropod dinosaurs and birds that functions to strengthen the thoracic skeleton for the rigors of flight

Fusiform: Funnel-shaped

Futurists: Those who try and predict the future

Gametes: The sex cells; cells that will combine with another cell to form a new organism; sperm and egg in animals and pollen and egg in plants

Gape: Resting for long periods of time on land with the mouth open

Gas Gland: A gland that is the site where gases from the bloodstream enter the swim bladder

Gastric Filaments: A series of filaments in scyphozoans and cubozoans that possess cnidocytes

Gastric Pouches: Extensions of the stomach of a scyphozoan and cubozoans that aid in the circulation of water

Gastrocoel: The inner body cavity that forms due to the invagination of the endoderm of the gastrula

Gastrodermis: The inner most layers of cells of diploblastic organisms

Gastrovascular Cavity or **Coelenteron**: A sac-like or branched cavity in a cnidarian that contains a single opening that serves as both mouth and anus

Gastrozooid: The feeding polyp of a hydrozoan

Gastrula: The embryonic strucrture that forms after the invagination of the outer cells to form an inner body cavity called the gastrocoel

Geckos: Lizards that are nocturnal, small in size, agile, lack eyelids, and possess large eyes and toe pads

Gemmules: Small spherical structures produced by sponges at the onset of winter or during drought that are resistant of freezing or drying

Gene Pool: The genetic makeup of a population

Genetic Mass: The amount of DNA in an organism

Genetically Modified Organism (GMO): An organism that has had it DNA manipulated in some way; also called transgenic

Genital Opercula: The first pair of appendages on the opisthosoma that covers the genital pores

Genophagy: A theory that dominant genes push recessive genes out of existence

Geological Uniformitarianism: The proposal that the Earth is shaped by slow-acting natural forces over very long periods of time

Germ Layers: Cellular layers that arise during embryonic development

Germinative Zone: The area just behind the scolex in a tapeworm where new proglottids arise

Germline Mutation: Mutations that occur in the DNA of reproductive cells and can be passed to offspring

Gestation Period: Period of time spent developing and growing inside the mother's uterus

Gestation: The period of intrauterine development

Gill Rakers: A fan-like projection that filters plankton and particles for feeding

Gill Slits: Opening along the pharynx for gas exchange in enteropneusts

Gill Slits: Slits in the body of an animal that cover the gills and allow water to flow out

Girdle: An extension of the body that encircles the outer edge of a chiton shell and can be used to clamp down on rocks and trap prey items

Gizzard: A muscular chamber that leads from the crop and is where food particles are ground up

Gladius or **Pen**: A small, thin internal shell of some molluscs

Gliding: The simplest form of flying in which a bird produces several strong wing strokes and then glides

Glochidia: The larvae of certain species of freshwater mussels which are released into the water and attach to the gills or fins of a fish and develop onto juvenile mussels

Glomerulus: An excretory organ of hemichordates with finger-like outpockets that extract metabolic wastes and release them to the outside

Glottis: Tube-like tracheal opening that allows snakes to breathe while swallowing

Glucose: A biologically significant simple sugar

Glycogen: A biologically important polysaccharide used to store energy by animals

Gnathochilarium: The lower plate-like jaw of a millipede

Golgi Complex: An organelle consisting of layers of flattened sacs that take up and processes secretory products

Gonopods: Modified trunk appendages used to transfer sperm to females in some arthropods

Gossamer: Spider silk that floats on the wind

Graviportal: Locomotion produced from animals possessing short hind limbs in which the diameter of the bones is disproportionately large

Guard Hair: Coarse, long protective hairs that produce coloration

Gut: A tube that runs down the length of the body from the mouth to the anus

Habitat: The place where an organism normally lives or where the individuals of a population live within their range

Habituation: A form of nonassociative learning in which the responder stops responding to a stimulus over time

Half-life: The amount of time it takes for an isotope to break down by fifty percent

Haplorhini: Anthropoids; group consisting of tarsiers, tamarins, baboons, monkeys, apes, and humans

Hardy-Weinburg Equilibrium Model: A probability model that shows that gene frequencies are inherently stable, but that evolution is expected in all populations all of the time

Heart Vesicle: A contractile circulatory organ located within the proboscis of enteropneusts

Hectocotyly: The modification of a cephalopod arm for transferring sperm

Hedonic Glands: Specialized courtship glands in salamanders

Hemal: A blood-vascular system derived from coelomic sinuses

Hemimetabolous: Having larval stages that undergo gradual changes to become adults

Hemocoel: A body cavity that contains blood or hemolymph

Hemocoel: The body cavity in some invertebrates that is bathed in hemolymph

Hemocyanin: A copper-containing protein found in the hemolymph of molluscs that transports oxygen throughout the circulatory system.

Hemolymph: The watery substance that makes up the blood of invertebrates

Hemotoxins: Anticoagulants that destroy the clotting ability of blood resulting in heavy bleeding

Hepatic Cecum: A structure associated with the gut that functions in lipid and glycogen storage and protein synthesis in cephalochordates

Herbivores: Animals that consume plants

Hermaphroditic: Possessing both male and female reproductive capabilities

Heterocercal Tail: Caudal fin of sharks in which the vertebral column turns upward at the tail

Heterodont: Teeth varying in size and shape based on function

Hibernation: A period of inactivity in which metabolic and respiratory rates slow and body temperature drops

Hindbrain or **Metencephalon:** Posterior part of the brain that controls respiration and osmoregulation as well as controlling some muscles and organs

Hindgut: The posterior most section of the arthropod digestive tract that serves to absorb water and prepare fecal matter

Hirudotherapy: The use of leeches to restore blood circulation to grafted or severely damaged tissue

Holistic Concept: The notion that as the climate and other factors change species respond to these changes in body structure

Holometabolous: Having immature that are very different in body form to the adult and require a drastic change in body structure to become an adult

Home Base: A single location where individuals reside

Homeostasis: The proposition that all organisms maintain stable internal conditions that do not change as conditions in the external environment do change

Homocercal Tail: Caudal fin of bony fish that is suspended from the tip of the vertebral column

Homologous: Similarity of structure that results from descent from a common ancestor

Home Range: The area an individual covers or patrols in its normal routine

Horns: Paired, permanent structures that protrude from the frontals

Host: An organism fed upon by a parasitic organism

Hydrosphere: Layers of water found on and in the planet

Hydrostatic Skeleton: A support structure that consists of a water-filled interior constrained by the muscular walls of the body

Hydrothermal Vents: Fissures in the deep sea where magma rises close to the seafloor and heats water that then rises out of these fissures

Hyperosmotic: body fluids contain more salt and less water than in the external environment surrounding them

Hypertrophy: Excessive enlargement

Hypodermic Impregnation: A form of reproduction whereby the copulating partners stab each other with the penis and inject sperm through the body wall of the other

Hypodermis: The body wall layer that secretes the cuticle in arthropods

Hyponeural: A deep system of parallel nerves that controls motor function

Hypothesis: A tentative explanation or answer to a problem or question

Iguanas: The most familiar New World lizards including marine iguanas and flying dragons

Ilium: Bone of the pelvis

Imago: The adult that emerges from the cocoon, chrysalis, or puparium

Immigration: The process of individuals entering an area

Immunocompetence: The ability to distinguish between self and nonself

Imprinting: Species recognition and bonding

Incisors: Teeth with sharp edges for snipping and biting

Inclusions: Small organisms or pieces of organisms that become trapped and are preserved in amber

Incurrent Canals: Canals/passageways in a sponge formed by finger-like invaginations of a sponge body

Incurrent Pores: Opening in a sponge's body through which water flows into the sponge

Incus: A bone of the middle ear in mammals

Independent (control) Variables: Factors that are kept constant among different experimental groups over the course of an experiment

Indeterminate Growth: A pattern of growth in which the skeleton continues to grow throughout most of the lifetime

Indirect or **Asynchronous Flight**: A mode of flying in which muscles act to change the shape of the exoskeleton to produce upward and downward strokes

Individualistic Concept: The notion that as the climate and other factors change species respond to these changes independently

Infaunal: Sediment dwelling burrower

Infundibulum: The expanded end of the oviduct

Innate: Behaviors based on preset neural pathways

Inner Ear: Location of the hearing organ

Inquilinism: The use of a second organism for housing

Insectivorous: Animals that eat insects, worms, grubs, and other small invertebrates

Instars: Molt stages in between the egg and adult

Inorganic: Does not contain carbon atoms

Instinctive: Something that is not learned but possessed at birth

Integrational Theory: A theory that is supported by a large body of evidence and has a high level of truth probability

Intellect, Intelligence: Ability to think, reason, and learn

Interglacial Periods: Periods of warming between glacial periods

Interspecies Competition: Between individuals of different species

Interspecific: Describes interactions between members of different species

Interstitial Cells: Totipotent cells with a large nucleus that give rise to sperm and egg cells and are located between the epidermal layer and epitheliomuscular cells in cnidarians

Interstitial Fauna: Animals that inhabit the water-filled spaces between grains of sand

Intertidal or Littoral Region: A zone in the marine ecosystem where the ocean meets the land

Intestine: A tube that leads from the stomach to the posterior end of the digestive system

Intracellular: Within the body

Intraspecies Competition: Between individuals of the same species

Intraspecific: Describes interactions between members of different species

Intrauterine Cannibalism: Eating of siblings by the largest embryos in the uterus

Intrinsic Rate of Increase: The exponential growth rate of a population

Introduced Species: Species that are new residents to a particular area where they are not normally found

Iridophores: A type of pigment cell, or chromatophore, containing a silvery pigment that reflects light

Island Gigantism: The observation that some animals on islands grow to be considerably larger than their continental counterparts

Isotopes: Elements that possess the same number of protons but different numbers of neutrons

Iteroparity: The ability to reproduce continually throughout the lifetime

Jacobsen's Organ: An organ responsible for the sense of smell in reptiles that contains a region of chemically sensitive nerve endings

Johnston's Organs: Long setae located at the base of antennae in most insects that vibrate when hit with certain frequencies

Jump Dispersal: A means of dispersal (way to spread out in space) in which an actively reproducing population becomes established beyond the normal range of the population

Juvenile Hormone: A hormone produced in the corpora alta and secreted during molting of the larval stages

***k*-selected**: Species or populations with lower reproductive rates that survive best when the population is close to its carrying capacity

Keeled: possessing a sternum

Kentrogens: A developmental stage of a parasitic barnacle that injects parasitic cells into the hemocoel of their crab host

Keratin: Tough, fibrous protein that provides protection against abrasion and dehydration in outer epidermal cells

Keystone Species: An influential species that has an unusually strong direct effect on the composition of a community

Kin Selection: Changes in gene frequency across generations that are driven by interactions between related individuals

Kinetic Skull: Adapted to enable animals to deliver a faster and more powerful bite and open its mouth wider to facilitate the capture and handling of prey

Kleptocnidae: The process of incorporating cnidae (stingers) from another animal by digesting them

Kleptoparasitism: The process of a parasite stealing food that belongs to the host

Labrum: The flap-like mouth structure formed by a fusion of the second maxillae

Labyrinth Organ: A simple lung-like structure found in some freshwater fish that takes in oxygen from the air

Lachrymal Glands: Tear glands

Lacunae: Thin walled spaces that sit among blood vessels in a closed circulatory system

Lacunar System: Complex circulatory channels found in the epidermis

Lamellae: Plate-like folding of gills

Lamellae: Thin, flat plates or discs

Larynx: Structure that encloses the vocal cords; the voice box

Lateral Canals: Many smaller canals that are part of the water vascular system of a sea star, which extend from the radial canals and end in a bulbous ampullae

Lateral Diverticula: Side pouches of an organ

Lateral Line Canal: A variation of the lateral line system of fish in which receptor cells (neuromasts) lie beneath the skin rather than be directly exposed to the environment

Lateral Line System: Mechanoreception that occurs along the sides of a fish

Lateral Undulation: Locomotion patterns in snakes in which an S-shaped path is left behind the moving snake

Law of Superposition: A geologic principle that states that the lower a layer of rocks is in the Earth, the older the rocks

Learned or **Associative Behavior**: A behavior that develops from modification of a behavior through experience

Lek: An area where males display (dance and make noises) to attract females

Leuconoid Sponges: Sponges that typically lack an atrium but possess many small flagellated chambers

Leucophores: Pigment cells that reflect light

Life Span: The average length of time from birth to death of an organism

Limnetic Zone: Open freshwater from shore to shore and down as far as light penetrates

Lipids: Any of a class of organic compounds that are fatty acids or their derivatives and are insoluble in water but soluble in organic solvents.

Lipochromes: Red, orange, and yellow pigments found only in the feathers of parrots

Lithosphere: The rocky outer part of the Earth consisting of the crust and uppermost mantel

Littoral Zone: The shallow edge around the shore where freshwater meets land

Loop of Henle: A specialized structure in the nephron that allow kidneys to concentrate urine

Lophophore: A fringe of hollow tentacles used for feeding

Lorica: The plate-like encasement of the body of a Loriciferan

Loricate Larva: The larval form of a priapulid that lives in mud and morphologically resembles a loriciferan

Lunate Wrist Bone: A bone found in theropod dinosaurs and birds that allows for twisting motions required for flight

Lung: The respiratory organ found in most terrestrial organisms

Lymphatic Hearts: Contractile vessels that move around the lymphatic fluid

Lymphatic System: A network of vessels that filter fluids, ions, and proteins from capillary beds in tissue spaces and returns them to the circulatory system

Lysosome: An organelle in the cytoplasm of eukaryotic cells containing degradative enzymes enclosed in a membrane

Macroecosystems: A large area containing living biota and abiotic media in which life exists, such as forests, prairies, deserts, lakes, ponds, streams, and rivers

Macroevolution: Evolution on the large scale

Macrofossils: Large fossils

Macromolecules: Large and complex assemblages of atoms joined together

Macropores: Pores or openings in sediment produced by burrowing animals

Madreporite or **Sieve Plate**: The opening to the water vascular system of a sea star

Maggot Debridement Therapy: The placement of maggots into a wound dressing allowing maggots to eat the dead, decaying tissue leaving the wound clean

Male Pore: Openings on the body where sperm is released during mating

Malleus: A bone of the inner ear of mammals

Malpighian Tubules: Paired, thin-walled spherical sacs that absorb nitrogenous wastes from the hemolymph

Mammary Glands: Modified sweat glands that secrete milk for the initial development period of young mammals

Mandible: The lower jaw

Mandibles: Invertebrate jaws

Mantle: Fleshy lobes that form from the dorsal body wall and secrete calcareous spicules, shell plates, or shell in molluscs

Mantle: The inner membrane that lines the tunic of tunicates

Mantle Cavity: A body cavity that houses the visceral mass and opens to the outside to allow the exchange of gases, food, wastes, and reproductive cells with the outside

Manubrium: A fold of the body wall that hangs from the center of the subumbrella and surrounds the mouth of a hydromedusa

Mark and Recapture: A method of calculating the density of a species in which a small portion of the population is captured, tagged, released, and ideally recaptured

Marsupium: A special pouch where the young of marsupials complete their development

Mastax: The pharynx of a rotifer

Maxilla: The upper jaw

Maxillae: Either of a pair of irregularly shaped bones of the skull fused in the midline supporting the upper teeth and forming part of the eye sockets, hard palate, and nasal cavity

Maxillary Glands: Excretory glands that contain a single pair of nephridia that are found on the head in association with the maxillae

Maxillary Teeth: Small cone teeth around the edge of the upper jaw in frogs

Maxillipedes: Paired venomous poison claws in centipedes

Maxillules: The first maxillae in crustaceans

Means of Dispersal: A mechanism for species to spread out in space and time

Medulla: The inner layer of a hair

Meiosis: A type of cell division that results in four daughter cells each with half the number of chromosomes of the parent cell, as in the production of gametes and plant spores

Melanin: Brown, black, gray, and red-brown pigments

Melanophores: A type of chromatophore containing brown or black melanin

Mesencephalon: Midbrain

Mesenchyme or **Mesoglea**: A layer of partially cellular gelatinous material derived from the ectoderm and located between the epidermis and gastrodermis of cnidarians

Mesenteries: Membranous sheets

Mesoderm: The middle germ layer that develops in some animals from the migration of endodermal cells in the developing gastrula

Mesoecosystem: A medium-sized ecosystem within a macroecosystem

Mesohyl: A middle layer of a sponge between the pinacoderm and choaoderm that contains the ameboid cells, collenocytes, spongocytes, sclerocytes, and archaeocytes

Mesopelagic: The open sea beyond the continental shelf from 200 meters to around 1,000 meters down

Mesosoma: The first six segments of the abdomen in scorpions

Mesothorax: The middle-most segment of an insect body

Metabiosis: A mode of living in which one organism is indirectly dependent on another for the preparation of the environment in which it can live

Metaphase: A stage in mitosis in which the chromosomes align in the equator of the cell before being equally separated into each of the two daughter cells

Metameric: Segmented body

Metamorphosis: The process of development in which the body of the larvae changes drastically to become an adult

Metanephridia: A type of nephridia that has an open ciliated funnel on one end and opens to the outside on the other end and functions as a kidney in invertebrates

Metasoma: A long tail composed of six segments that bears the stinger in scorpions

Metathorax: The posterior-most segment of an insect body

Metazoans: Complex, multicellular animals

Microecosystem: A small-sized ecosystem within a mesoecosystems

Microevolution: Evolution on a small-scale

Microfilaments: Contractile fibers found within the cytosol

Microfossils: Tiny fossils

Microhabitat: An extremely localized small-scale environment

Microtriches: Tiny folds within the tegument of flukes and tapeworms that function to increase surface area

Mictic: Amictic females that produce ova by meiosis

Microtubules: A component of the microskeleton

Midbrain or **Mesencephalon**: A portion of the brain associated with vision, hearing, motor control, sleep/wake, alertness, and temperature regulation

Midden: An ancient mound of shells that were discarded by early humans

Middle Ear: The portion of the ear internal to the eardrum and external to the oval window of the cochlea

Midgut: The middle section of the arthropod digestive tract that serves to produce enzymes, and to digest and absorb nutrients

Migration: Mass movement of animals

Milk: Nutritious secretion of the mammary glands

Mimicry: The process of copying the color patterns, behavior, or structural adaptations of another species

Mitochondria: An organelle found in large numbers in most cells, in which the biochemical processes of respiration and energy production occur

Mitosis: The process, in the cell cycle, by which the chromosomes in the cell nucleus are separated into two identical sets of chromosomes, each in its own nucleus

Mitotic Spindle: The subcellular structure that segregates chromosomes between daughter cells during cell division

Moist Deciduous and Semi-evergreen Forests: Forests in areas with a cool dry winter during which time many of the trees drop some or all their leaves

Molars: Teeth for crushing and grinding

Molecular Clock: The average rate at which a particular kind of gene or protein evolves

Molecule: Atoms bonded together

Molting: The replacement of old feathers as they become worn and drop out

Molting Gel: A fluid that enters the space between the hypodermis and the old exoskeleton during molting

Monocular Vision: Vision in which an animal can see things on each side of their head at the same time as well as in front because the eyes are on the sides of the head

Monoecious: Possessing both male and female gonads in the same body

Monoesterous: Possessing one estrus period per breeding cycle

Monogamous: Mated pair remains together for at least one breeding season

Monogamy: A reproductive strategy in which one male and one female mate exclusively

Monophyletic: A group of species whose members are related to one another through a unique history of common descent

Monosaccharides: Simple sugars such as glucose, galactose, and fructose

Montane Rain Forests: Forests that experience at least 40 inches of rain and are found in mountainous areas with cooler climates

Morphology: The exterior form and structure of an organism

Mortality: Death

Motility: Possessing the ability to move

Mouth Brooding: Fertilized eggs are brooded in the mouth until hatching

Mullerian Mimicry: The process of several unrelated species resembling one another

Mutagens: Something that induces mutations in DNA

Medulla or **Myelencephalon**: Posterior part of the brain that controls muscles and organs

Myomeres: Zigzag band arrangement of musculature

Myoneme: Contractile myofibril in epitheliomuscular cells in cnidarians

Nacre: A smooth iridescent, aragonite-based secretion deposited on the inner surface of some bivalve shells

Naiads: The aquatic, immature Hemimetabolous stages of terrestrial insects

Nares: External openings of the nasal cavity

Nasohypophyseal Opening: An opening on the dorsal side of the head of a lamprey

Natality: Birth

Natatorial: Swimming

Nauplius Larva: An arthropod larval stage characterized by the presence of three head appendages

Neck: The region located immediately behind the scolex in a tapeworm

Necrosis: Death of tissue

Necrotic: Dead

Nematocysts: Harpoon-like tubes that are connected to toxin sacs and are used to impale and paralyze prey

Neo-Darwinism or **Modern Evolutionary Synthesis**: The combination of Darwinian natural selection and Mendelian inheritance that explains both micro- and macroevolutionary changes

Neolithic Revolution: The birth of agriculture

Neonates: Undeveloped young

Neoteny: A form of paedomorphosis in which an organism displays physiological maturity but not sexual maturity

Nephridia: Tube or funnel-like structures that filter wastes from the body

Nephridioducts: Excretory pores that lead from the protonephridia to the external environment

Nephridiopore: An opening to the outside (or to the mantle cavity) that connects to the nephridia

Nephrostome: An open ciliated funnel of a metanephridia

Neritic Zone: A zone in the marine ecosystem from the low-tide line to the edge of the continental shelf

Nerve Net: Connection between the between the hyponeural, ectoneural, and entoneural systems of echinoderms

Nerve Net: A diffuse network of nerve cells forming a primitive nervous system in cnidarians, ctenophores, and certain other organisms

Nerve Rings: Bundles of nerves concentrated around the bell margin of a hydromedusae

Nested Hierarchy: Clades nested within one another

Neural Crest: An embryonic ectodermal structure that becomes facial cartilage, skin pigment cells, ganglia of the autonomic nervous system, dentin of teeth, spiral septum of the heart, and ciliary body of the eye

Neural Tube: An embryonic ectodermal structure that becomes the brain, spinal cord, retina, and posterior pituitary of the adult

Neuromast Cells: A cluster of hair cells that are joined together in a gelatinous cupula

Neuropathy: Degenerative changes to nerves

Neurotoxins: Toxins that act on the nervous system

Niche Differentiation: A type of resource partitioning in which populations (or species) utilize different parts of a shared resource

Nictitating Membrane: A transparent or translucent third eyelid in some animals that can be drawn across the eye for protection

Nitrogenous Base: A molecule that acts as a base and contains nitrogen

Nonassociative Learning: The simplest form of learning in which a behavioral change occurs due to a repeated presentation of a stimulus

Nonequilibrium Models: Biological models depicting unstable and changing communities

Nonobligate Brood Parasites: Birds that are capable of raising their own young but lay eggs in the nest of conspecies to increase reproductive output

Notochord: A hard tube-shaped cluster of cells that are wrapped in a fibrous sheath and extends along the central nervous system and provides a stiff structure of the attachment of muscles

Nuchal Organs: Sensory pits or grooves that function as chemoreceptors located in the head region

Nucleus: A dense organelle present in most eukaryotic cells, typically a single rounded structure bounded by a double membrane, containing the genetic material

Nucleic Acid: A complex organic substance present in living cells, especially DNA or RNA, whose molecules consist of many nucleotides linked in a long chain

Nucleolus: An organelle found within the nucleus of eukaryotic cells

Nucleotides: Molecules that contain a sugar, nitrogenous base, and a phosphate group

Nutritive-muscle Cells: Large, elongated cells of the gastrodermis of cnidarians that engulf small food particles by phagocytosis

Nymphs: Immature hemimetabolous individuals

Obligate Brood Parasites: Birds that are incapable of raising their own young and must lay eggs in the nest of other species

Obligate Mutualism: The process of two species being so tightly associated that they cannot live separately

Obligate: Necessary for survival

Occipital Condyle: Bony protrusions from the base of the skull that seat the skull on top the first vertebrae

Ocelli: Discs or pits of photoreceptor cells that allow the animal to respond to light

Ocelli: Small, simple eyes of some invertebrates

Oldowan Tools: Regularly patterned stone tools associated with Homo habilis

Olfactory Nerves: Nerves that transport sensory information from the olfactory lobes to the brain

Olfactory Rosette: Sensory pads that detect chemicals and line the chamber that extends from the nares

Oligotrophic Lakes: Lakes with a small amount of organic matters and minerals and low biological activity

Ommatidia: The fused receptor units of a compound eye

Omnivores: Organisms that eat both plants and other animals

Oncosphere Larva: A larval stage produced by some tapeworm species that hatches and lives within an intermediate host

Oophagy: Eating of unfertilized eggs by developing embryos

Ooze: Thick blanket of mud and decaying matter that has accumulated over millions of years on the deep sea floor

Open System: A system in which matter or energy flows through and can be lost from the system

Operant Conditioning: Learning through trial and error

Operculum: A middle-ear bone that transmits vibrations from the pectoral girdle to the inner ear in amphibians

Operculum: Gill cover in fishes

Opisthaptor: The posterior attachment organ of monogenetic flukes

Opisthosoma: The posterior tagmata of arthropods that consists of the abdomen

Optic Lobe: A region of the brain attached to the midbrain and is responsible for vision, flight, and food acquisition in birds

Optimal Foraging Theory: The most successful animals tend to feed on prey that maximizes their net energy intake per amount of foraging time

Oral (incurrent) Siphon: An inlet for water and the mouth opening of tunicates

Oral Arms: Extensions of the manubrium of cnidarians that contain cnidocytes and aid in the capture and ingestion of prey

Oral Disc: The flared oral end of a sea anemone that possesses eight to several hundred hollow tentacles

Oral End: The top end of polyps that contains the mouth and tentacles

Oral Hood: An anteriorly projecting structure that surrounds the mouth on the body of a cephalochordate

Oral Sucker: An anterior, hookerless sucker that surrounds the mouth of diagenetic flukes

Orbits: Eye sockets

Organelles: Membrane bound structures found in eukaryotic cells

Organic: Containing carbon atoms

Orthogenesis: The hypothesis that life has an innate tendency to progress from simple forms to higher, complex forms

Osculum: An opening found at the top of some sponge

Osmoconformers: Animals whose internal osmotic concentrations is the same as the surrounding environment

Osmoregulators: Animals that maintain the osmotic difference between their body fluid and the surrounding environment

Osmosis: The diffusion of water molecules from an area of high concentration to an area of low concentration

Osphradia: Chemoreceptors patches on the gill or mantle wall

Ossicles: Inner-ear bones consisting of the malleus, incus, and stapes

Ossicles: Separate plates that arise from mesodermal tissue and form a calcareous skeleton

Osteoderms: Bony plates located beneath the keratinized dermis scales and reinforces them in lizards and crocodiles

Ostia: Pores that perforate the body of a sponge

Ostia: Perforations in the heart wall that allow blood to flow into the heart in some animals

Otoliths: Earbones

Outgroup: An organism that is not contained within a group of related organisms

Ovale: A muscular valve in the posterior dorsal region of the bladder that opens to release the high internal pressure in the bladder

Ovaries: Female reproductive organs where eggs are released

Overfishing: Reduction of fish stocks below an acceptable level

Overturn: A seasonal mixing of the zones in large, temperate lakes

Oviduct Funnels: Tubes that connect the ovaries to the oviducts and allow for the passage of eggs

Oviducts: Openings on the body where eggs are released

Oviparity: Egg-laying in which eggs are externally fertilized

Oviparous: A developmental strategy in which egg development occurs within the female's body until they are ready to hatch

Ovipositors: Female structures used to place eggs on or in a substrate

Ovoviparous: A developmental strategy in which egg development occurs within the female's body until they are ready to hatch

Ovum: The single-celled egg

Ozone: A form of atmospheric oxygen that blocks ultraviolet radiation from the sun; the Earth's "sunscreen"

Paedomorphosis: Adults retain some but not all juvenile characteristics and are sexually mature

Palate: Roof of the mammalian mouth

Pallial Oviduct: A tubular structure of the female reproductive system of molluscs in which eggs are fertilized and sperm is temporarily stored

Pallial Tentacles: Sensory tentacles located in the margin of the mantle of a bivalve and contain tactile and chemoreception cells

Papillae: Horny, keratinous outgrowths from specialized areas of the skin

Paraphyletic: As assemblage in which member species are all descendants of a common ancestor but which does not contain all the species descended from that ancestor

Parapodia: Fleshy lobes that extended from the body

Parasitoids: Parasites that lay eggs of living hosts that then feed on the host's tissues until the host dies

Parenchyma: Tightly packed cells that are derived from the mesodermal layers

Parietal Eye: An eye buried beneath opaque skin that detects changes in light intensity

Pars Superior: Upper section of the inner ear of a fish that determines balance

Parthenogenesis: A mode of reproduction involving the development of an unfertilized egg into a function individual

Patagium: The fold of skin that runs from the wrist to the ankles in flying phalangers

Pearl: Formed by layers of nacre that were laid down by a bivalve in response to the presence of a foreign object between the shells

Pectines: A feathery response organ in scorpions

Pectoral Fins: Anterior pair of fins

Pectorals: Breast muscles

Pedal Cords: Ventral nerve cords that innervate the muscle of the foot in molluscs

Pedal Disc: The flattened aboral end of a sea anemone used for attaching to the substrata

Pedal Gland: An organization of cells that produces large amounts of mucus to lubricate the sole of a snail's foot during locomotion

Pedal Glands: Located on the foot of a rotifer and produce an adhesive secretion

Pedal Laceration: A method of asexual reproduction in sea anemones in which parts of the pedal disc are left behind while moving; these remains form new adults

Pedicel: Connection between the antler and their bony attachment

Pedicellariae: Pincher-like structures that protrude from the aboral surface of some sea stars and function to clean the body of debris and protect the body

Pedipalps: Specialized sensory appendages located near the mouth of an arachnid

Peer Review: A process in which scientific work is evaluated and reviewed by experts in the field

Pelagic Zone: The open sea beyond the continental shelf

Pelvic Fins: Posterior pair of fins

Pelycosaurs: The collective term for the first herbivorous and carnivorous synapsids

Penis Fencing: A reproductive behavior in hermaphroditic flatworms in which the first male to successfully pierce the skin of another flatworm becomes the male of the mating pair and delivers its sperm to its mate (the loser in the fight assumes the female role)

Pentaradial: A body symmetry in which the body parts are arranged in fives or a multiple of five around an oral-aboral axis

Peptide Bonds: The bonds that join amino acids together

Pericardial Cavity or **Pericardial Coelom**: The cavity containing the heart

Pericardial Sinus: A circulatory structure in some invertebrates that collects the blood as it returns to the heart

Perihemal Sinuses: Coelomic channels that enclose the hemal system of echinoderms

Periostracum: A thin, organic surface coat on the inside surface of a molluscan shell

Peristomium: The second anterior segment that surrounds the mouth and bears sensory tentacles

Peritoneum: A slick epithelial lining of the coelom

Permafrost: Permanently frozen ground

Permian-Triassic Extinction Event: A mass extinction event informally known as the "Great Dying" spread out over a few million years when 96% of marine species and 70% of terrestrial species went extinct

Phagocytosis: Engulfment of a particle or organism by a cell

Phagosome or **Food Vacuole**: A vesicle formed around a particle (usually food) absorbed by phagocytosis

Pharyngeal Gill Slits: Slits in the body of an animal that cover the gills and allow water to flow out of the body of chordates

Pharynx: A tube that leads from the mouth into the gastrovascular cavity of an anthozoans

Pharynx: A muscular structure in the anterior portion of the digestive system superior to the esophagus

Phasmids: Unicellular sensilla in the lateral tail region of certain species of nematodes

Pheromone: A body secretion that functions as a mode of communication

Phoresy: The use of a second organism for transportation

Phosphate Group: A molecule that contains phosphate

Phospholipids: A class of charged lipid that contains a phosphate group

Photoreceptors: Receptors that detect the presence of light

Photosynthesis: A process carried out by plants and photosynthetic bacteria that uses solar energy to convert carbon, hydrogen, hydrogen, and water molecules into simple sugars and carbon dioxide

Phototaxis: Ability in animals to respond to light

Phyllopidia: The expanded and flattened swimming limbs of some crustaceans

Phylogenetic Trees: Depiction of evolutionary relationships among more than two groups of taxa

Phylogenetics: A biological classification scheme that groups organisms based on their evolutionary relationships

Phylogeny: The history of organismal lineages over time

Physoclistous: Lacking a pneumatic duct

Physostomous: Possessing a pneumatic duct

Phytoplankton: The microscopic primary producers in aquatic ecosystems

Pinacocytes: A layer of flattened cells on the outer surface of a sponge

Pinacoderm: The outer surface of a sponge

Pineal Body: Located above the diencephalon and functions to detect light, maintain circadian rhythms, and control color changes

Pinna (pl.Pinnae): Funnel-shaped flap of skin at the external ear opening

Pinnules: Smaller branches with a feathery appearance

Pioneer Species: The first species that inhabit newly disturbed land, typically *r*-select species

Pioneer Stage: The period of time in an area marked by the presence of the pioneer species

Placenta: A spongy mass embedded in the wall of the mother's uterus, which nourishes the embryo

Planospiral: A coiled, spiral shaped that is flattened

Plantigrade Locomotion: Whaling and running on the soles of the feet

Planula Larva: A free-swimming cnidarian larva that turns into either a polyp or medusa

Plasma Membrane: A bilayered membrane that surrounds and contains a cell

Plasmids: A circular DNA molecule in bacteria that is separate from chromosomal DNA and can replicate independently

Plastron: The flat lower shell of a turtle

Plesiadapiformes: Proto-primate mammals

Pleurites: The lateral sclerite (exoskeletal plate)

Pleurodont Teeth: Teeth attached to the inner side of the jaw

Pleuroperitoneal Cavity: The cavity containing the lungs

Plumage: The arrangement and appearance of all feathers on a bird's body

Pneumatic Duct: A connection between the gut and swim bladder in basal teleost fish

Pneumatized Bone: Hollow bones with air spaces in them and stiffened by internal struts

Podia: Hollow, muscular sucker that connects to the bulbous ampullae as part of the water vascular system of an echinoderm

Podites: Limb segments or pieces of arthropods

Point Mutations: A small-scale mutation in the DNA

Polian Vesicles: Blind pouches that arise from the ring canal, which are thought to help regulate internal pressure of the water vascular system of a sea star

Polyandry: A reproductive strategy in which females mate with more than one male

Polyestrous: Estrus cycle that occurs at regular intervals though out the year

Polygamy: A reproductive strategy in which males or females have more than one mate

Polymers: Large macromolecules composed of repeating units

Polynomial: Possessing several names

Polyphyletic: A group comprising species that arose from two or more different ancestors

Polysaccharides: Complex sugars; many simple sugars units bonded together (starch and cellulose)

Polyunsaturated: A fat that contains more than one double bond in the carbon chain

Population: An ecological unit consisting of a group or groups of individuals of the same species living in the same place at the same time

Population Bottleneck: A decrease in the genetic variation of populations due to genetic drift

Population Equilibrium: The state of a population when emigration and immigration are more or less in balance

Population Explosions: Times of rapid, exponential growth of a population

Porocyte: A ring-shaped cell of a sponge that forms the incurrent pore

Posterior: Pertaining to the rear or tail end

Precocial: Birds born with eyes open, down feathers, and the ability to move and collect food

Predator: An organism that kills and consumes another organism

Preening: The process of zipping together barbs of feathers that have become separated by running the feathers through the beak

Prehensile Tail: A tail adapted for grasping

Premolars: Teeth for shearing, slicing, crushing, and grinding

Prey: An organism that is consumed by another organism

Primary Consumers (Herbivores): Organisms that feed directly on the primary producers (plants)

Primary Producers (Autotrophs): Organisms that produce their own food through photosynthesis (plants)

Primary Productivity: The amount of photosynthetic activity

Primary Succession: The first stage in ecological succession in which new colonizers inhabit an open space

Primers: Something that elicits slow, long lasting behavioral responses

Proboscis: A piercing anterior structure that extends from the mouth in some insects and nemerteans

Proboscis: The flexible conspicuously long snout of some mammals (tapirs, shrews); especially the trunk of an elephant

Procuticle: The hard inner layer of an insect exoskeleton

Profundal Zone: The area below the limits of light penetration in freshwater

Progenesis: A form of paedomorphosis in which sexual maturity is sped up in juveniles

Proglottids: Individual sections of tapeworms

Prohaptor: A posterior adhesive organ of monogenetic flukes

Prokaryotes: Organisms that have cells lacking membrane bound organelles

Prokaryotic Cell: A type of cell that lacks membrane bound organelles

Prophase: A stage of mitosis in which the chromatin condenses into double rod-shaped structures called chromosomes in which the chromatin becomes visible

Prosimians: Strepsirrhini; group consisting of tarsiers, lemurs, aye-aye, bush babies, potto, and lorises

Prosoma (Cephalothorax): The anterior tagmata (section) of arthropods that is a fusion of the head and thorax

Prostaglandins: A group of 20 lipids that act as local chemical messengers in vertebrate tissues

Prostomium: The anterior most segment of annelids that contain antennae in sea worms

Protandric: Gonads first produce sperm and then later produce eggs

Protandry: Changing sex from male to female

Proteins: Any of a class of nitrogenous organic compounds that consist of large molecules composed of one or more long chains of amino acids and are an essential part of all living organisms

Prothoracic Gland: A hormone producing organ in the head of insects that releases ecdysone

Prothoracicotropic Hormone (PTTH): A hormone produced by cells in the brain and ganglia that stimulate the release of ecdysone

Prothorax: The anterior-most segment of an insect body

Protocerebrum: A section of the supraesophageal ganglion

Protogyny: Changing sex from female to male

Protonephridium: A tube-like nephridia with a closed bulb at one end and an opening to outside at the other end

Protostomes: Organisms with embryos that develop via spiral cleavage of the first cells and form a mouth first at the blastopore

Proventriculus: First section of a bird's stomach where food is mixed with digestive juices

Proximal: Close to the body core

Proximate Causation: The immediate causes of an animal's behavior

Pseudocoel: A "false" body cavity that develops between the mesoderm and endoderm

Psuedogenes: Gene remnants that no longer function but continue to be part of an organism's genome

Pterylae: Tracts on a bird's skin from which feathers grow

Ptychocysts: Specialized cnidocytes that produce the tubes of burrowing anthozoans

Puggle: A young echidna

Pulmonary Respiration: Gases exchanged across lungs

Pupa: The last larval molt of a homometabolous species

Puparium: The last larval exoskeleton of flies that surrounds the pup during metamorphosis into the adult

Pygidium: The terminal segment of a worm

Pygostyle: The remains of the reptilian tail in birds

Pyloric Ceca: Structures in ray-finned fish that function to absorb fat

Pyloric Ducts: Extensions of the pyloric stomach that stretch into each arm of a starfish

Pyloric Stomach: A portion of an echinoderm's stomach that is not inverted for feeding

Pyramid of Biomass: A diagram that represents the biomass present at each trophic level within an ecosystem

Pyramid of Energy: A diagram that represents the number of calories per trophic level

Pyramid of Numbers: A diagram that represents the relative numbers of organisms at each trophic level

Quadrate: A small bone at the back of the upper jaw

Quadrate: The skull bone that connects to the lower jaw

Qualitative Data: Any data that contains numbers; "hard data"

Quantitative Data: Any data that are observational or anecdotal; "soft data"

r-selected: Species or populations with high reproductive rates that survive best when the population is far below the carrying capacity

Radial Canals: Part of the water vascular system of a sea star that extends from the circumoral canal into each of the five arms of a sea star

Radial Nerves: Large sensory nerves that extend into each arm of an echinoderm

Radial Symmetry: A body design in which the body parts radiate outward from a central axis

Radula: A file-like set of hooked teeth used for rasping and tearing

Ram Ventilation: Swimming with an open mouth to create a temporary current across the gills

Ramus: A single, branching appendage

Range: A geographical area that provides the necessary requirements for a population to exist

Range of Tolerance: The environmental and physical requirements for a population to exist

Raptorial Claws: Specialized forelimbs modified for piercing or smashing prey in mantis shrimp

Rate-of-living Theory of Aging: The theory that ageing (and deterioration) is an inevitable outcome of the harmful act of living and cell metabolism

Reaggregation: the process of separated cells fusing together again

Realized Niche: The actual habitat that an organism or population occupies after accounting for competition and predation

Reciprocal Altruism: A behavior in which one individual benefits at the immediate cost to another; however, this individual is guaranteed future benefit

Reciprocity: Mutual exchange between partners

Recombinant DNA: A form of biotechnology in which DNA is isolated from one organism, manipulated in some way, and then reintroduced into the cells of another organism

Rectal Glands or Rectal Sacs: Protrusions from the short intestine of an echinoderm

Rectilinear Movement: A type of caterpillar crawl locomotion in some snakes

Reciprocity: A hypothesis suggesting that individuals may form "partnerships" in which mutual exchanges occur because they benefit both participants

Red Tide: A bloom of toxic red algae

Regeneration: The process of producing viable adults from small fragments

Release Calls: Produced by either male amphibians that are mistaken as females by other males or by unresponsive female amphibians

Releaser: A trigger that initiates a behavior

Renopericardial Canals: Blood vessels that extend from the heart to the kidneys

Reproductive Caste: The queens and drones in a eusocial species

Repugnatorial Glands: Cells that produce defensive secretions that are foul tasting and can be poisonous or sedative in nature

Reservoir: An area of the environment that serves as a storage place for chemicals over a long period of time

Residence Time: The amount of time an element or chemical remains in a reservoir or exchange pool

Resin: A syrupy material excreted from species of evergreen trees

Resource Partitioning: A subdivision of a shared resource by populations or species with overlapping fundamental niches

Rete Mirabile: A mesh of arteries that is a site of high acidity from the release of lactic acid by the gas gland

Retia Mirabilis. A mesh of blood vessels that allows for counter-current heat exchange.

Retinula Cells: Specialized cells of the ommatidium that possess special light-collecting areas called rhabdom, which convert light energy into nerve impulses

Rhabdite: A type of rhabdoid that secretes mucus to prevent desiccation in a turbellarian

Rhabdom: Special light-collecting areas of retinula cells that convert light energy into nerve impulses

Rheoreceptors: Receptors that are stimulated by water currents and allow an organism to orient its body to various water movements

Rhombencephalon: Hindbrain

Rhopalia: Specialized projections between the scalloped margins of a hydromedusan's bell that contain statocysts and ocelli

Rhynchocoel: The hollow "nasal cavity" of nemerteans, which houses an eversible proboscis

Ribonucleic Acid (RNA): The single-stranded genetic code that serves as messenger and organized in the construction of those proteins

Ribosomes: A minute particle consisting of RNA and associated proteins, found in large numbers in the cytoplasm of living cells

Ricochetal: Using just the hind limbs to hop in fast succession

Ring or Circumoral Canal: Part of the water vascular system of a sea star that forms a ring in the body of the sea star and has five separate radial canals branching off it

Ritualized Threat Displays: Threatening displays that rarely result in injury or death because they have been used time and again to send information to opponents

Riverine: Best adapted to calmer areas in deep fast-flowing rivers

Rods: Photoreceptors that function in less intense light than other kinds of photoreceptors

Rostral Organ: A large gel-filled cavity in the snout of a coelacanth

Rostrum: Snout of a fish

Ruminants: Mammals that grind their food with broad, flat teeth, swallow it, and later bring it back up for further chewing

Salt-absorbing Cells: Cells in the gill epithelium that transfer salt ions from the water to the blood

Salt-secreting Cells: Special cells that excrete salt from the blood to the surrounding water

Saltatorial: Hopping

Saltatory Locomotion: Leaping movement

Saprotrophs: Organisms that feed on nonliving organic matter

Saturated Fats: Fats that contain only single bonds between carbon atoms

Savanna: Relatively dry grasslands and bush lands with occasional trees

Scabies: An infestation of itch mites that tunnel through the epidermis of the skin and excrete irritating chemicals

Scalids: Thin pointed structures; stylets

Scansorial: Climbing

Scavenger (or Detritivores): Organisms that feed on dead and decaying matter

Schistosomiasis or Bilharzia: A debilitating disease in humans caused by the fluke *Schistosoma* living within the circulatory system

Schizocoel: The embryonic gut of an annelid

Schizocoely: The process of the blastopore developing into the mouth

Scientific Law: A theory with a considerable amount of evidence that has proven to be correct and true over a considerable amount of time; the highest probability level of truth in science

Sclerites: Four exoskeletal plates of arthropods

Sclerocytes: Amoeboid cells found in the mesohyl of sponges that produce calcareous and siliceous spicules

Sclerosepta: Thin, radiating calcareous septa within the skeletal cup of a stony coral

Sclerotization: The process of chemically cross-linking the layers of proteins in the protocuticle to harden and darken the exoskeleton

Scolex: The anterior head region of a tapeworm that contains suckers and hooks

Scutes: Broad belly scales in snakes

Scyphistoma: A polyploidy larval stage of scyphozoans and cubozoans that lives attached to hard surfaces

Sebum: The fatty, greasy secretion from sebaceous (oil) glands, which provides insulation and waterproofing

Secondary Consumers or Carnivores: Organisms that feed on the primary consumers

Secondary Palate: The partitioning that separates the oral and nasal cavities

Secondary Productivity: The activity (energy processing) or amount of herbivorous animals

Secondary Succession: The second stage in ecological succession in which the first colonizers are replaced by different plants and animals

Semelparity: Reproduce once and then die shortly thereafter

Seminal Vesicles: Part of the male reproductive tract where sperm cells complete differentiation

Senescence: The process of aging

Sensilla: A collective term for several mechanoreceptors and chemoreceptors

Sensory Cells: Cells within the epidermis that detect information from the surroundings

Sensory Palps: A sensory appendage located in the prostomium

Septa: Vertical structures between the gastric pouches of scyphozoans and cubozoans that possess an opening to help circulate water

Septal Filament: A glandular, ciliated band located on the free edge of the mesentery of anthozoans

Septum: A muscular structure that divides the heart into chambers

Sequential Hermaphrodites: Possess male and female sex organs but not at the same time; they change sex at some point during their life

Seral Stage: The period of time in an area that is between the pioneer and climax stages and marked by the presence of both pioneer and climax species

Serendipity: Something good or beneficial that happens by chance

Serous Glands: Large glandular cells embedded in the dermis of amphibian skin and produce a complex chemical mixture of biologically active compounds

Sessile: Not able to move freely

Sexual Imprinting: The process by which a young animal learns the characteristics of a desirable mate

Sexual Selection: The process by which interactions between members of the same gender and interactions between males and females (mate choice) causes changes in a population over time

Shell Glands: Cells that secrete calcareous spicules or shells in molluscs

Side-winding: A form of locomotion in snakes in which they move side-to-side quickly minimizing surface contact

Sign Stimulus: A releaser that an animal responds to

Silk: A proteinaceous liquid that hardens as it is drawn out of the spinneret of an arachnid

Simultaneous Hermaphrodites: Possess both male and female sex organs at the same time

Sinus Venosus: A thin-walled sac where blood from the fish's veins collect prior to entering the atrium

Sinuses: Open spaces around the organs of the body

Siphon or Funnel: A rolled extension of the mantle that functions as an inhalant tube in terrestrial gastropods and allows water in and out of the body of aquatic mollusc's visceral mass

Siphonoglyph: A ciliated groove in the center of the oral disc responsible for pulling water into the gastrovascular cavity of a sea anemone

Siphonozooids: Highly modified polyps used to pump water into the interconnected gastrovascular cavities of corals

Siphuncle: A small tube that is used to control buoyancy in the *Nautilus*

Skinks: A group of lizards with elongated bodies and reduced limbs

Soaring: Flight in which birds rise on thermal currents and then glide down until carried up by another thermal current

Social Insects: Insects that live together in large, ordered groups

Sociobiology: A field of scientific study which is based on the assumption that social behavior has resulted from evolution

Sole: The flat ventral foot of a gastropod

Somatic Cell Mutations: Mutations that occur in the DNA of body cells that cannot be passed to offspring

Somatic Tube: The outer portion of a eucoelomate that contains the sense organs and muscles

Somites: Sections

Somitogenesis: The formation of somites or body divisions

Sonation: Nonvocal sounds

Spatial Memory: A type of memory in which an animal can recall the location of an object based on its comparative position to other objects in the environment

Speciation Events: Rise of a new species

Species: The smallest cluster of organisms that possess at least one diagnostic character and that are reproductively isolated from all other species

Species Diversity: A measurement of the different number of species in a community

Sperm Funnel: A ciliated structure part of the male reproductive tract that connects to the vas deferens in annelids

Spermatophore: A packet of sperm

Spherical Symmetry: A body design that is round with its parts concentrically arranged around a central point

Spicules: Either calcareous or siliceous skeletal elements of a sponge

Spinal Nerves: Nerves that pass information from the extremities through the spinal cord to the brain

Spinnerets: Conical telescoping organs that produce silk in arachnids

Spiracles: Holes that lead from the tracheae to the outside

Spiral Valve: A valve in the intestine of chondrichthyes that slows the passage of food and increases absorptive area

Spiral Valve: A valve that directs blood into the pulmonary and systemic circuits in some amphibians and lungfish

Spirocysts: Sticky tubes or loops that wrap around and stick to prey

Sponges: The simplest multicellular invertebrate animals that lack symmetry, have a cellular grade of organization, and are sessile

Spongin: Modified collagen fibers found in a sponge

Spongocoel: The internal cavity of a sponge

Spongocytes: Ameboid cells found in the mesohyl of sponges that produce supportive collagen called spongin

Stapes or Columella: Middle ear bone that transmits high-frequency vibrations from the tympanic membrane to the inner ear

Starch: A biologically important polysaccharide used to store energy by plants

Statocysts: Pits or closed vesicles containing tiny calcareous statoliths that when stimulated inhibit muscular contraction on that side

Statocysts: Sensory structures that allow an organism to orient to gravity

Statoliths: Calcareous bits found in statocysts that aid in the detection of body orientation

Stenopodia: The long, thin uniramous walking appendages of arthropods

Stereogastrula: A gastrula with no cavity

Stereom: An open meshwork structure with living tissue that fills the interstices that develop from the calcite crystals of an echinoderm exoskeleton

Sternite: The ventral sclerite plate of an exoskeleton

Steroids: A class of lipids that can be found in animal cell membranes or that function as a hormone

Stolon: Connections between the zooids of pterobranchs

Stomach: The digestive organ that processes food

Stone Canal: Part of the water vascular system of a sea star that connects the madreporite to the circular ring

Strata: Layers; pertaining to rock

Strepsirrhini: Prosimians, lemurs, aye-aye, bush babies, potto, and lorises

Sterptoneury: The twisted internal anatomy of some molluscs in which the nervous system takes the shape of a figure eight due to torsion

Streptostyly: Hinge-like configuration of the skull that enables the back of the jaw to move freely

Strip Census: A method to determine the population size of birds by identifying and counting the birds in a strip of measured width throughout an entire area

Strobila: The region of a tapeworm just behind the scolex that contains reproductive structures called proglottids

Stylets: Thin pointed structures

Subesophageal Ganglion: The posterior portion of the tritocerebrum that loops around the esophagus and connects to the ventral nerve cord

Subumbrella: Undersurface of the bell of a medusa

Sugar: A type of carbohydrate

Summer Eggs: A type of turbellarian egg that possesses a thin shell and hatches rapidly

Supracoracoideus: The muscles that raise the wing of birds

Supraesophageal Ganglion: A bundle of fused ganglia in the head of arthropods

Survivorship: The percentage of the population that survives to a given age

Survivorship Curve: A graph of the percentage of the population that survives to a given age

Swim Bladder: A thin sac located between the peritoneal cavity and the ventral column that functions to maintain buoyancy in fish

Symbiosis: Interactions between different species or populations (interspecies)

Sympatric: Species existing in the same geographical area

Sympatric Speciation: The process through which new species evolve from a single ancestral species while inhabiting the same geographical area

Symplesiomorphies: Common ancestral characteristics

Synapomorphies: Unique derived characteristics

Synapsida: Amniotes that possess one temporal opening

Synconoid Sponges: Sponges with evaginated flagellated canals that create one layer of folding

Syncytial: Multinucleated mass of fused cells

Syrinx: The voice box of birds located near the junction of the trachea and bronchi

_t_-Test: A statistical test that is used to determine if the means of a trait or characteristic are significantly different between two groups

Tadpole Larva: The free-swimming larval forms of a tunicate which metamorphs into an adult

Tagmata: Specialized body regions

Tagmatization: The modification of segments for specialized functions

Taxa: Taxonomic groups

Taxonomic Categories: Hierarchical rankings of organisms into domain, kingdom, phyla, classes, orders, families, genera, and species

Taxonomy: The scientific grouping (classification) and naming (nomenclature) of living things

Tegument: The nonciliated external covering of flukes and tapeworms, which functions to protect the body

Telencephalon: Forebrain equivalent to the cerebrum and functions in olfaction

Telomeres: A region of repetitive nucleotide sequences at each end of a chromatid, which protects the end of the chromosome from deterioration or from fusion with neighboring chromosomes

Telophase: The final stage of mitosis in which the nucleus reforms and the cell membrane or cell wall separates the two daughter cells

Telson: The tail section of certain crustaceans

Tensile Strength: The amount of force a material can withstand without breaking

Tergite: The dorsal sclerite plate of the exoskeleton of an arthropod

Terpenes: A class of lipids that comprise many biologically important pigments

Territory: Any space an animal defends against intruders

Tertiary Productivity: The activity (energy processed) or amount of carnivores

Test: The shell of echinoderms formed by closely packed dermal ossicles

Testis: Male reproductive organs where sperm cells form and mature

Tetrachromatic: Possessing four different types of color-sensing cone cells in the eye

Tetrapods: Four-limbed animals; all vertebrates above fish

Thecodont Teeth: Teeth set in sockets in the jawbone

Therapsids: A branch of the pelycosaurs that arose in the Permian

Thermogenesis: Generation of heat

Tiedermann's Body: Blind pouches that arise from the ring canal as part of the water vascular system of a sea star

Toes: Protrusions from the foot of rotifer

Tool: Anything not part of the body that is used to accomplish a given task

Tornaria Larva: A planktonic larval stage of hemichordates with nonyolk eggs that possess ciliary bands

Torsion: Twisting

Totipotent: Possessing cellular plasticity in which a cell is capable of changing form and function

Tracheae: A series of branched, tubes that conduct gases to and from body tissues in insects

Tracheae: A thin-walled cartilaginous tube descending from the larynx to the bronchi and carrying air to the lungs

Transverse Fission: A form of asexual reproduction in which an organism splits in half separating head from tail

Transverse Plane: A plant through the body that cuts the body into anterior and posterior sections

Triglyceride: A lipid composed of three fatty acid chains each attached to a glycerol molecule

Trinomial: Possessing three names

Triploblastic: Possessing three tissue layers

Tritocerebrum: A section of the supraesophageal ganglion

Trochophore Larva: The unique larval stage of the Trochozoa

Trochophore Larvae: Larval stage of a polychaete

Trophi: Hard jaws of some rotifers that are located on the mastax and used to grind food

Trophic Cascades: The indirect interactions between species at different levels of the trophic web

Trophic Levels: A measure of how far an organism is removed from its original source of energy

Trophic Structures: The feeding relationships within communities and ecosystems

Tropical Rain Forests: Forests with diverse biota that are constantly warm between 68° and 77° F and receive at least 40 inches of rain per year

Tube Foot: The ampullae and podia of the water vascular system of an echinoderm that is used for locomotion

Tunic: The body wall of most tunicates, which is composed of a tough connective tissue secreted by the epidermis

Turbinate Bones: Bones of the nasal cavity

Tympanic Membrane: Eardrum, a piece of skin stretched over a cartilaginous ring at the end of the ear canal

Tympanic Organs: Organs composed of a thin, cuticular membrane that covers a large air sac, which, when vibrated by sound waves, leads to the stimulation of sensory nerves in insects

Typhlosole: A ridge or fold that increases the internal surface area of the intestine

Typolite: A fossil that forms as a mold of the external features of an organism

Ultimate Causation: The origin of a behavior

Unconditioned Response: A response that initially occurs due to a stimulus without any conditioning

Unconditioned Stimulus: A stimulus that initially initiates a response

Underhair: Dense, soft and highly insulative hair

Understory: Layers of plant life beneath the taller trees/plants

Ungulates: Hoofed mammals

Unguligrade Locomotion: Walking and running on a single toe capped by a hoof and with the heel raised

Unicellular Glands: Mucus producing glands embedded in the epidermis that connect to the outside through pores

Uniramous Appendages: Single, branched appendages

Unsaturated Fats: Fats that contain at least one double bond between the carbons

Ureters: Tubes that carry urine to the cloaca or urinary bladder

Uropygial Gland: An oil gland located at the base of the tail in birds

Utriculus: Lower section of the inner ear of a fish

Vas Deferens: Part of the male reproductive tract that connects to the sperm funnel and to the male pore in annelid worms

Vas Deferens: The coiled duct that connects the testes with the urethra

Vectors: Carriers

Veliger: A larval stage unique to molluscs that possess a foot, shell, operculum and some other adult feature

Velum: A shelf-like inward projecting fold of the bell of a hydro medusa that helps increase the force of the water jet for locomotion

Velum: The swimming and feeding structure of the veliger larva

Velvet: Skin and soft hair that cover antlers as they grow

Ventilation: The process of actively moving air or water over the respiratory surfaces for gas exchange

Ventral: Pertaining to the bottom or belly side

Ventral Aorta: An artery that pumps blood to the dorsal arteries and then to body tissues in cephalochordates

Ventral Vessel: A blood vessel that is located under the digestive tract on the ventral side of an organism

Ventricle: A thick-walled muscular chamber of the heart that pumps blood from the heart

Vermiculture: A farming/gardening practice that uses earthworms to process soil

Vermiform: Resembling a worm in form or movement

Visceral Tube: the inner portion of a eucoelomate that contains the gut

Viviparous: A developmental strategy in which the female provides nutrients directly to the embryos

Vocal Cords: Cartilaginous membranous folds located in the larynx, which vibrate and produce sound when air passes over them

Vocal Sacs: Membranous pouches under the throat or on the corner of the mouth that distend and act as resonating structures that amplify the call of a frog or toad

Vocalization: Sound production

Volant: flying

Vomerine Teeth: A ridge of small cone teeth on the roof of the mouth in frogs

Vomeronasal Organ or Jacobson's Organ: An organ in the roof of the mouth of many animals that detects scent molecules

Waggle Dance: A coordinated dance conducted by honey bees in which a bee informs its hive mates of the direction and distance to nectar and pollen by the direction, angle, intensity, and rate of the dance

Wampum: An intricate Native American belt woven out of white and colored beads made from the shell of the channeled whelk and hard-shelled clam

Weaned: Period when young no longer feed off the milk of their mother

Wheel Organ: The lateral walls of the vestibule bearing complex ciliary bands that drives food particles into the mouth of cephalochordates

Winter or Dormant Eggs: A type of turbellarian egg that possesses thick, resilient shells, which are capable of withstanding cold and desiccation

Worker Caste: Sterile members of a eusocial species

Worm Casts: The excrement that passes out of worms

Xanthophores: A type of chromatophore containing yellow. Orange, and red pigment

Zoea: A larval stage that typically follows the nauplius stages in crustaceans

Zonites: Divided segments of kinorhynchs that bear spines but no cilia

Zooanthellae: Photosynthetic yellow-brown unicellular algae

Zoogeography: A field of biology that studies the patterns of animal distribution in time and space

Zooids: Asexual buds that differentiate along the length of an adult turbellarian's body

Zoology: The study of the animal world

Zoonotic: Ability of a parasite to infect more than one species

Zygapophyses: Supportive processes on each vertebrae of an amphibian skeleton that prevent twisting when stress is applied

Zygote: The single, large cell that develops immediately after fertilization

Zygotes: Fertilized eggs

INDEX

Note: *f* indicates a figure, *t* indicates a table

PHOTO CREDITS

Chapter 1

1.1 ©Marcio Jose Bastos Silva/Shutterstock; **1.2** ©Midosemsem/Shutterstock; **1.4** ©Everett Historical/ Shutterstock; **1.5a** Wellcome Images

Chapter 2

2.1 ©Barnaby Chambers/Shutterstock; **2.2** ©Tomsickova Tatyana/Shutterstock; **2.8** ©steve estvank/ Shutterstock; **2.9a** ©Paul Reeves Photography/Shutterstock; **2.9b** ©Isaac Mok/Shutterstock; **2.10** ©Henrik Larsson/Shutterstock; **2.11** ©Marco Vinci

Chapter 3

3.11 ©Geoff Ruth

Chapter 4

4.8 ©La Gorda/Shutterstock; **4.9** ©DnBr/Shutterstock; **4.11** ©Blamb/Shutterstock **4.12** OpenStax College; **4.16** OpenStax College; **4.18** OpenStax College **4.24** ©LSkywalker/Shutterstock **4.31a** and **b** ©Alila Medical Media/Shutterstock

Chapter 5

5.1 Max Planck Gesellschaft; **5.4** Public Library of Science Computational Biology, July, 2009; **5.9** ©Sorin Colac/Shutterstock; **5.10** Whatiguana; **5.11** ©Steve Schlaeger/Shutterstock; **5.14** ©Ondrej Prosickey/Shutterstock; **5.15** ©P. Kirk Visscher; **5.16** ©Megan McCarty **5.19** ©Eric Isselee/Shutterstock

Chapter 6

6.2a ©Gilles San Martin; **6.2b** ©Martin Fowler/Shutterstock; **6.3** ©Professor marginalia

Chapter 7

7.3 ©Philippe Bourjon; **7.5** ©Nick Hobgood; **7.11a** ©Lamiot (Own work)

Chapter 8

8.4 © Jubal Harshaw/Shutterstock, **8.6** ©Frank Fox, **8.7b** ©Sierra Blakely; **8.9b** ©Alexander Vasenin; **8.10** ©R. Gino Santa Maria/Shutterstok; **8.12** ©Andrea Issotti/Shutterstock; **8.14** ©Louis Wray; **8.16** (WT-shared) Pbsouthwood at wts wikivoyage; 8.17 ©George Berninger Jr.; **8.18** ©Nhobgood Nick Hobgood; **8.21** ©Designua/Shutterstock

Chapter 9

9.3 ©Eduard Sola; **9.4a** ©Jubal Harshaw/Shutterstock; **9.5a** ©D. Licjarsol K. Kucharska/Shutterstock; **9.18** ©Courtesy of Theodore E. Nash, M.D.;**9.20** ©Marlin Harms

Chapter 10

10.11a Courtesy of Center for Disease Control; **10.11b** CSRIO; **10.11c** Sustainable Sanitation Alliance; **10.11d** ©Carolina K. Smith/Shutterstock

Chapter 11

11.5a ©Nick Hobgood; **11.6a** ©Michael Linnenbach; **11.8c** ©Alex Hyde; **11.10** ©John Hallmen

Chapter 12

12.2b ©Stephano Buttafuco/Shutterstock; **12.3** ©Nick Hobgood; **12.6** ©Zikamoi; **12.7a** ©kikujung-boy/Shutterstock; **12.8a** ©Nhobgood Nick Hobgood; **12.8b** ©Hans Hillewaert; **12.9** ©Boris Pamikov; **12.10a** ©NatureDiver/Shutterstock; **12.12** ©zcw/Shutterstock; **12.13** ©aluxum/Shutterstock; **12.16** ©BlueOrange Studio/Shutterstock

Chapter 13

13.5 ©Graham Manning; **13.6** ©Vaikoovery; **13.8** ©Jannes Pockele; **13.10a** CSRIO; **13.11a** ©Alf1; **13.13** ©Audrey Snider-Bell/Shutterstock; **13.14** ©Wasu Watcharadachaphong; **13.16** ©photospirit/Shutterstock; **13.18** ©Ryan M. Bolton/Shutterstock; **13.20** ©Protosov AN/Shutterstock; **13.28** ©Hugh Lansdown/Shutterstock; **13.30** ©Manuel Alvarez/Shutterstock

Chapter 14

14.7a ©Laszlo Ilyes; **14.8** ©Scott Roy Atwood; **14.9a** ©D. Gordon E. Robertson; **14.9b** ©Wildcat Dunny; **14.11** ©Ethan Daniels/Shutterstock; **14.12c** Courtesy NOAA

Chapter 16

16.2 ©Luc Viatour/www.Lucnix.be; **16.3** ©Jim G; **16.4** Courtesy NOAA; **16.8** Courtesy Great Lakes Fishery Commission; **16.9** University of Aberdeen; **16.13** ©Open Cage; **16.11** ©Kiersti Joergensen/Shutterstock; **16.12** ©Greg Amptman/Shutterstock; **16.14** ©Alessandro Zoc/Shutterstock; **16.17a** Sue Lindsay/Australian Museum; **16.17c** ©MP cz/Shutterstock; **16.23** ©Maysaki Miya

Chapter 17

17.3 Wilkinson M, Sherratt E, Starace F, Gower DJ Public Library of Science (2013); **17.4** ©Matt Jeppson/Shutterstock; **17.5a** ©Tom Reichnter/Shutterstock; **17.5b** ©Paul Reeves Photography/Shutterstock; **17.7** ©Brandon Alms/Shutterstock; **17.8** ©alslutsky/Shutterstock; **17.14** ©EcoPrint/Shutterstock; **17.20** ©Ryan M. Bolton/Shutterstock; **17.21** ©Birute Vijeikiene; **17.22** © By Dein Freund der Baum; **17.24** ©Maslov Dmitry/Shutterstock; **17.25** Carnivora© ER Degginger

Chapter 18

18.6 ©Orhan Cam/Shutterstock; **18.7a** ©Fotos593/Shutterstock; **18.7b** ©Matt Jeppson/Shutterstock; **18.8** ©orlandin/Shutterstock; **18.9** ©Ryan M. Bolton/Shutterstock; **18.10** ©CreativeNature R.Zwerver/

Shutterstock; **18.11** ©Nazzu/Shutterstock; **18.12** ©Orhan Cam/Shutterstock; **18.13** ©Erni/Shutterstock; **18.14** ©defpicture/Shutterstock; **18.16** ©Henrik Larsson/Shutterstock; **18.17** ©Audrey Snider-Bell/Shutterstock; **18.18a** ©granat/Shutterstock; **18.18b** ©Natursports/Shutterstock; **18.19** ©David Havel/Shutterstock; **18.20c** ©Cathy Keifer/Shutterstock; **18.21** ©fivespots/Shutterstock; **18.22** ©tsnebula23/Shutterstock; **18.23a** ©Patrick K. Campbell/Shutterstock; **18.23b** ©Tom Reichner/Shutterstock; **18.23c** ©Dennis W. Donohue; **18.23d** ©Leighton Photography/Shutterstock; **18.26** ©HUANG Zheng

Chapter 19

19.1a ©Natursports/Shutterstock; **19.1b** ©Linda Bucklin/Shutterstock; **19.3** ©Vasin Lee/Shutterstock; **19.4** ©Volodymyr Burdiak/Shutterstock; **19.5** ©SantiPhotoSS/Shutterstock; **19.6** ©Humming Bird Art/Shutterstock; **19.7** ©Volt Collection/Shutterstock; **19.8** ©percom/Shutterstock; **19.12** ©Gavran333/Shutterstock; **19.19a** ©Jason Stitt/Shutterstock; **19.19b** ©Worakit Sirijinda/Shutterstock; **19.20a** ©Nagel Photography/Shutterstock; **1920b** ©Steffen Foerster/Shutterstock; **19.20c** ©Photoman29/Shutterstock; **19.20d** ©raulbaenacasado/Shutterstock; 19.20e ©nadi555/Shutterstock; **19.22a** ©Nadezda Boltaca/Shutterstock; **19.22b** ©Don Mammoser/Shutterstock; **19.22c** ©LeonP/Shutterstock; **19.22d** ©Michael Wick/Shutterstock; **19.22e** ©Menno Schaefer/Shutterstock; **19.22f** ©Nancy Bauer/Shutterstock; **19.22g** ©birdiegal/Shutterstock; **19.22h** ©Brian lansbey/Shutterstock; **19.21** ©ErikBeyersdorf; **19.23** ©Markus Lilje; **19.24** ©C. Abraczinskas; **19.28** © GregTheBusker; **19.29** ©markaharper1; **19.30** ©Pictureguy/Shutterstock; **19.32a** ©D and D Photo Sudbury/Shutterstock; **19.32b** ©Villiers Steyn/Shutterstock; **19.33** © Dario Sanches; **19.35a** ©pelena/Shutterstock; **19.35b** ©Daleen Loest/Shutterstock

Chapter 20

20.3a ©Martin Hesko/Shutterstock; **20.3b** ©Howard Sandler/Shutterstock; **20.4** ©Alessandro Caroli/Shutterstock; **20.5** ©BMJ/Shutterstock; **20.6** ©Jerocflores/Shutterstock; **20.8a** ©Tom Reichner/Shutterstock; **20.8b** ©Paul Tessier/Shutterstock; **20.9a** ©Tom Reichner/Shutterstock; **20.9b** ©J Reineke/Shutterstock; **20.10** ©andamanec/Shutterstock; **20.11** ©AndChisPhoto/Shutterstock; **20.13** ©reptiles4all/Shutterstock; **20.14** ©Suppakrit Boonsat/Shutterstock; **20.15** ©worldswildlifewonders/Shutterstock; **20.16** ©Grimplet/Shutterstock; **20.17a** ©Greg Amptman/Shutterstock; **20.17b** ©seb2583/Shutterstock; **20.18a** ©Dmytro Pylypenko/Shutterstock; **20.18b** ©Vladimir Melnik/Shutterstock; **20.19** ©Hugh Lansdown/Shutterstock; **20.20** ©Athanasia Nomikou/Shutterstock; **20.22** ©creativex/Shutterstock; **20.25** ©Eric Isselee/Shutterstock; **20.26a** ©worldswildlifewonders/Shutterstock; **20.26b** ©mark higgins/Shutterstock; **20.27a** ©Milosz_M/Shutterstock; **20.27b** ©Quadxeon/Shutterstock; **20.29** ©Andy Lim; **20.30a** ©MongPro/Shutterstock; **20.30b** ©Baimieng/Shutterstock; **20.30c** ©Wallenrock/Shutterstock; **20.30d** ©Chris Kolaczan/Shutterstock

Chapter 21

21.4a ©LeonP/Shutterstock; **21.4b** ©Praiseng/Shutterstock; **21.5a** ©Dennis W. Donohue/Shutterstock; **21.5b** ©Matej Hudovernik/Shutterstock; **21.5c** ©Kjersti Joergensen/Shutterstock; **21.6** ©120 **21.8** ©Didier Descouens; **21.12** ©Linda Bucklin/Shutterstock; **21.13** ©Linda Bucklin/Shutterstock; **21.16** ©Pichugin Dmitry

CPSIA information can be obtained
at www.ICGtesting.com
Printed in the USA
JSHW021047161119
2479JS00002B/6

9 781457 542121